Fractional Vibrations with Applications to Euler-Bernoulli Beams

This book examines vibration phenomena with an emphasis on fractional vibrations using the functional form of linear vibrations with frequency-dependent mass, damping, or stiffness, covering the theoretical analysis potentially applicable to structures and, in particular, ship hulls.

Covering the six classes of fractional vibrators and seven classes of fractionally damped Euler-Bernoulli beams that play a major role in ship hull vibrations, this book presents analytical formulas of all results with concise expressions and elementary functions that set it apart from other recondite studies. The results show that equivalent mass or damping can be negative, depending on fractional orders. Other key highlights of the book include a concise mathematical explanation of the Rayleigh damping assumption, a novel description of the nonlinearity of fractional vibrations, and a new concept of fractional motion, offering exciting additions to the field of fractional vibrations.

This title will be a must-read for students, mathematicians and physicists, and engineers interested in vibration phenomena and novel vibration performances, especially fractional vibrations.

Ming Li is a professor at Ocean College, Zhejiang University, China, and an emeritus professor at East China Normal University. He has been a contributor for many years to the fields of mathematics, statistics, mechanics, and computer science. His publications with CRC Press also include *Multi-Fractal Traffic and Anomaly Detection in Computer Communications* and *Fractal Teletraffic Modeling and Delay Bounds in Computer Communications.*

Fractional Vibrations with Applications to Euler-Bernoulli Beams

Ming Li

CRC Press
Taylor & Francis Group
Boca Raton London New York

CRC Press is an imprint of the
Taylor & Francis Group, an **informa** business

First edition published 2024
by CRC Press
2385 NW Executive Center Drive, Suite 320, Boca Raton FL 33431

and by CRC Press
4 Park Square, Milton Park, Abingdon, Oxon, OX14 4RN

CRC Press is an imprint of Taylor & Francis Group, LLC

ISBN: 978-1-032-60360-5 (hbk)
ISBN: 978-1-032-60891-4 (pbk)
ISBN: 978-1-003-46094-7 (ebk)

DOI: 10.1201/9781003460947

Typeset in Minion
by codeMantra

To my wife Yonglan Zhang, my daughter Joanna Jiayue Li, and my parents Xidong Li and Fanggui Yin — for making it both possible and worthwhile.

Contents

Preface

VIBRATION IS AN IMPORTANT subject in many fields, ranging from mechanical engineering to electronic engineering. In education, it is an obligatory course for undergraduates majoring in shipbuilding and ocean engineering in China. In fact, the vibration performance of a ship is now one of the obligatory items required by most ship bureaus over the world.

This book consists of four parts. The first part is for conventional linear vibrations in Chapters 1–6 based on my lectures on the course of ship hull vibrations for undergraduates and postgraduates in Ocean College, Zhejiang University, China. That may be an obligatory part for university teaching in the major of shipbuilding and ocean engineering for 32 teaching hours. It is the fundamental part of the book from the point of view of both classroom teaching in Chapters 1–6 and academic research in Chapters 7–33. Based on Chapters 1–6, one may see that why we restrict the absolute value of damping ratio to be smaller than one, why we often pay attention to the vibration phenomena with negative damping in Chapters 8–14, 16–19, and 17–27, and why we are so interested in explaining the Rayleigh damping assumption. The second part, Chapters 7–16, contains my research in the theory of fractional vibrations with ordinary differential equations. The third part, Chapters 17–28, is my research in vibrations of fractional Euler-Bernoulli beams. It is actually applications of Chapters 8–14 to Euler-Bernoulli beams. My preference is given to Euler-Bernoulli beams because they play a role in ship hull vibrations. The fourth part is about my research in test techniques relating to ship science and ocean engineering.

The main highlights in this book are as follows.

1. Methods discussed in Section 4.4 are not on the track of modal analysis, but they may be useful for the analysis of a decoupled vibration system with multiple degrees-of-freedom.

2. In Chapter 5, we clarify that the function form of the special solution to a vibration system with multiple degrees-of-freedom has to be harmonic-type instead of the assumed harmonic one as conventionally described.

3. The author introduces the convolution representation of a fractional derivative in Theorems 7.1–7.3 and the classification of fractional vibration systems in Chapter 7.

4. The author proposes equivalent motion equations of six classes of fractional vibrators in Chapters 8–13.

5. We put forward a set of expressions of quasiequivalent mass, damping, stiffness, damping ratio, natural frequencies, logarithmic decrements, and Q factors for six classes of fractional vibrators in Chapters 8–13.
6. The author brings forward the analytical formulas of response functions (free, impulse, unit step, and frequency transfer) to six classes of fractional vibrators using elementary functions in Chapters 8–13.
7. We establish a mathematical explanation of the Rayleigh damping assumption in Chapter 14.
8. We note that the range of quasiequivalent mass may be $(-\infty, \infty)$ in Chapter 15.
9. In Chapter 16, we bring forward the results of six classes of fractional vibration systems with variable fractional orders, including that orders vary randomly and the mathematical explanations of the Rayleigh damping assumption based on variable fractional orders.
10. We present the analytic expression of free response to the longitudinal vibration of a beam with fractional coordinates in Chapter 17.
11. The author proposes the analytic expression of free and forced responses to an Euler-Bernoulli beam with fractional coordinates in Chapters 18 and 19, respectively.
12. The concept of seven classes of fractionally damped Euler-Bernoulli beams is introduced in Chapter 20.
13. We establish the analytic expressions of forced responses to seven classes of fractionally damped Euler-Bernoulli beams in Chapters 21–27, respectively.
14. In Chapter 28, we introduce the concepts of super-displacement, super-velocity and sub-velocity, super-acceleration and sub-acceleration in fractional vibrations addressed in this book. We also explain the non-Newtonian forces encountered in fractional vibrations.
15. The usefulness of anti-aliasing filtering used in sampling in vibrations is discussed in Chapter 29.
16. The author sets up a method for the record length requirement of ocean surface waves in Chapter 30.
17. The Hilbert–Huang transform used in the time-frequency distribution of encountered waves is described in Chapter 31.
18. We propose an optimal controller of irregular wave maker in Chapter 32.
19. In Chapter 33, we present the fractal dimension of the von Kármán spectrum that is commonly used in ocean engineering.

This monograph may be a reference for engineers, mathematicians, and physicists. It may be a textbook for undergraduates or postgraduates in universities with a curriculum of 32 teaching hours for Chapters 1–6 and 32 hours for the rest.

Preferred preliminaries are calculus, university physics, theoretic mechanics, material mechanics, and linear systems for Chapters 1–28, and random processes for Chapters 29–33.

This monograph is dedicated to humanity, world peace, and the memory of my graduate supervisor, the academician of the Chinese Academy of Engineering, the Chairman of the 12th International Ship and Offshore Structure Congress, ex-Director of Science Committee of China Ship Scientific Research Center, ex-vice Director of China Ship Scientific Research Center, Professor, Dr. Bing-Han Xu.

Ming Li, Ph.D., Professor, Grand Secretary Li XVI (Rizhao)
Ocean College, Zhejiang University, PR. China

Acknowledgments

THANKS GO TO PROF. Jianxing Leng and Ocean College, Zhejiang University, China, for providing me with the professor position to lecture on the course of ship hull vibrations for undergraduates and postgraduates. Without my teaching in Zhejiang University, it would have been impossible for me to arrange my time to write this monograph. Particular thanks go to Wei-Hui Wang, emeritus Professor with the National Taiwan Ocean University, for his valuable comments on the book. Prof. Xuekang Gu, the vice technical director of China Ship Scientific Research Center (CSSRC), is appreciated for discussions and data regarding Chapter 31. Prof. Maoxiang Gu (Ex-Director of the CSSRC, the academician of the Chinese Academy of Engineering) is appreciated for his care about my research of wave maker control (Chapter 32). Prof. Jingjian Chen (CSSRC) is acknowledged for his encouragement about my derivation of the optimal controller in Chapter 32. Prof. Yousheng Wu (Honorary-Director of CSSRC, the academician of the Chinese Academy of Engineering) is sincerely appreciated.

I would like to take the opportunity to acknowledge a group of professors in Tsinghua University (Beijing) for their education in hard times. They are Shiliang Xu, Qiji Yang, Desheng Wang, Deyun Lin, Dingyue Kou, Baoqin Liu, Jingzhao She, Daimao Lin, Jingxian Zou, Xiaoqing Ding, Xiguang Ma, Xueli Qiao, Xuexia Zhang, Shichang Hou, Zhenming Feng, Jiaguang Fang, Jiaqing Li, Bo Lv, Siming Luo, Guoxiang Zhao, Mengtao Wang, and Xiyuan Yan.

Prof. Yang Quan Chen (University of California, Merced) and Prof. Jun-Sheng Duan (Shanghai Institute of Technology) are sincerely appreciated for their discussions about fractional calculus. Prof. Swee Cheng Lim (Multimedia University) is appreciated for his instructions and discussions in fractional processes. I am grateful to my students, Miss Ying Dong, for her assistance on some of drawings in Chapters 1–6 and Dr. Peiwei Shan for some drawings in Chapter 31. Mr. Yong Chen is acknowledged for improving the resolution of some drawings.

The views and conclusions contained in this book are those of the author and should not be interpreted as representing the official policies, either expressed or implied, of the Chinese government.

I

Fundamentals

CHAPTER 1

Harmonic Vibrations

When a structural element suffers from a loading within the range from point O to B in Figure 1.1, the relationship between the stress σ and strain ε on the element is expressed by Hooke's law in the form

$$\sigma = E\varepsilon \text{ for } 0 \le \sigma \le \sigma_p, \tag{1.1}$$

where E is the Young's modulus.

Statically, a structure may not be failure for $0 \le \sigma \le \sigma_p$. However, if a loading is dynamic and when resonance or negative damping occurs, the structure may fail even if $0 \le \sigma \le \sigma_p$.

1.1 THREE ELEMENTARY PARTS OF VIBRATION SYSTEM

1.1.1 Mass and Inertia

Consider the rigid mass of a body, which moves at the acceleration $a = x''$. The unit of a is m/s^2. The body moves at a because the force F_1 exerts on it, see Figure 1.2. The unit of

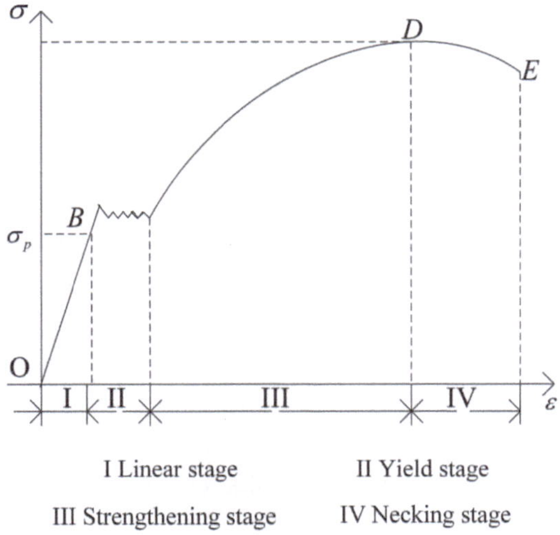

FIGURE 1.1 Stress-strain curve (σ-ε curve).

DOI: 10.1201/9781003460947-2

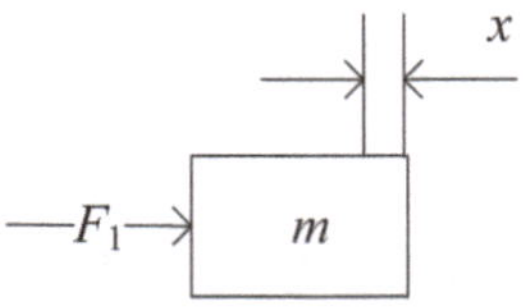

FIGURE 1.2 Rigid mass of body.

F_1 is N. According to Newton's second law, F_1 is proportional to a with the parameter m in the form

$$F_1 = \frac{d}{dt}\left(m\frac{dx}{dt}\right).$$

When m is a constant in terms of t, F_1 is in the form

$$F_1 = ma. \tag{1.2}$$

The quantity m is called the mass of that body. Its unit is kg. It is a measure of the inertia with which the body moves. Often, we use m as a symbol for the body of interest for simplicity.

1.1.2 Stiffness and Restoration

Suppose that the body connects with a spring as indicated in Figure 1.3. When the body m moves at a, the spring gives a force F_2 on m in the opposite direction. Thus, F_2 is inversely proportional to the displacement x in the form

$$F_2 = -kx. \tag{1.3}$$

The quantity k is called the stiffness of the spring. Its unit is N/s. It measures the restoration of a spring to its equilibrium state.

1.1.3 Damping and Resistance

When a damper is supposed to be viscous, as shown in Figure 1.4, a vibration system produces a force F_3. It is proportional to the velocity x' in the form

$$F_3 = cx'. \tag{1.4}$$

The quantity c is called the damping coefficient. Its unit is N·s/m. It is a measure of resisting the movement of a rigid body m. A damper dissipates energy.

1.2 UNDAMPED FREE VIBRATIONS

If there is no driven force in a vibration system, we say that the system is in free vibration. Its response is called free response. When there is no damping in a system like that in Figure 1.3, we say that the vibration system is undamped or damping free. In general,

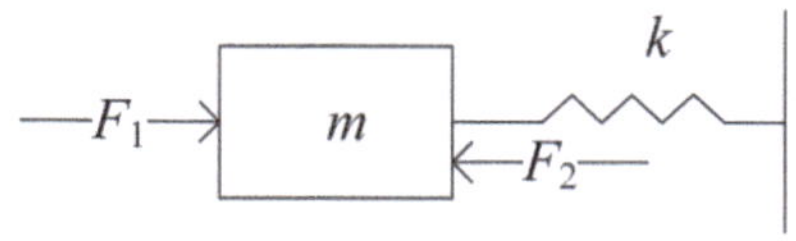

FIGURE 1.3 Spring connected to rigid body.

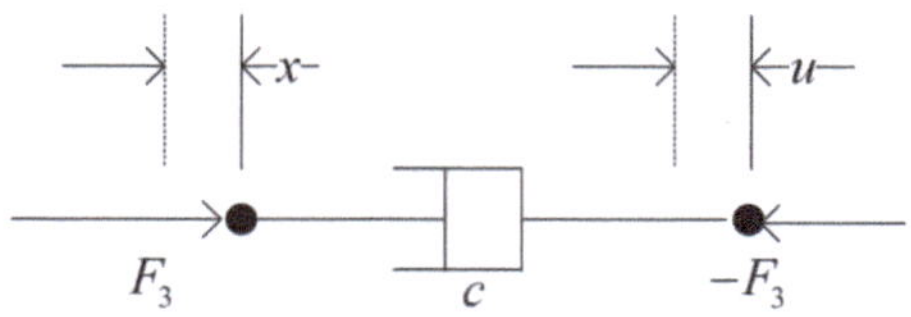

FIGURE 1.4 Viscous damper.

a system that consists of two elements, namely, inertia and restoration, may vibrate. For example, denoting initial displacement and initial velocity by x_0 and v_0, respectively. Figure 1.3 implies a vibration equation given by

$$\begin{cases} mx'' + kx = 0, \\ x(0) = x_0,\ x'(0) = v_0. \end{cases} \tag{1.5}$$

The solution to Eq. (1.5) is expressed by

$$x(t) = x_0 \cos\omega_n t + \frac{v_0}{\omega_n}\sin\omega_n t, \tag{1.6}$$

where

$$\omega_n = \sqrt{\frac{k}{m}}, \tag{1.7}$$

is with the unit of rad/s. Since it is intrinsically determined by a vibration system with m and k, it is called the natural angular frequency without damping. The natural frequency of a damping free system is denoted by $f_n = \frac{\omega_n}{2\pi}$. The unit of f_n is Hertz. Denote by T_n the period of vibration. Then, $T_n = \frac{1}{f_n}$. Its unit is s.

Equation (1.6) can be rewritten by

$$x(t) = A\sin(\omega_n t + \alpha), \tag{1.8}$$

where A is given by

$$A = \sqrt{x_0^2 + \left(\frac{v_0}{\omega_n}\right)^2}, \tag{1.9}$$

and α is expressed by

$$\alpha = \arctan \frac{x_0 \omega_n}{v_0}. \tag{1.10}$$

The amplitude A represents the maximum displacement that a vibrator deviates from its equilibrium position. It reflects the amount of energy that a vibration system contains. It is determined by the initial displacement x_0 and the initial velocity v_0. In fact, x_0 produces the initial potential energy E_{p0} in the form

$$E_{p0} = \frac{1}{2} kx^2(0) = \frac{1}{2} kA^2 \sin^2(\alpha). \tag{1.11}$$

Denote by $v(t) = x'(t)$. Then, one sees that v_0 results in the initial kinetic energy E_{k0} given by

$$E_{k0} = \frac{1}{2} mv^2(0) = \frac{1}{2} m\omega_n^2 A^2 \cos^2(\alpha) = \frac{1}{2} kA^2 \cos^2(\alpha). \tag{1.12}$$

According to the principle of work and energy, one has

$$E = E_{p0} + E_{k0} = \frac{1}{2} kA^2. \tag{1.13}$$

The above implies that the total mechanical energy of a vibration system stays constant without energy dissipation due to being damping free.

The quantity $\omega_n t + \alpha$ is called phase. It represents the position of a vibrator at time t. The initial phase α is determined by x_0, v_0, and ω_n. Figure 1.5 indicates free responses of harmonic vibrations without damping.

1.3 DAMPED FREE VIBRATIONS

When considering three parameters m, k, and c, as indicated in Figure 1.6, we have $F_1 + F_2 + F_3 = 0$. Thus,

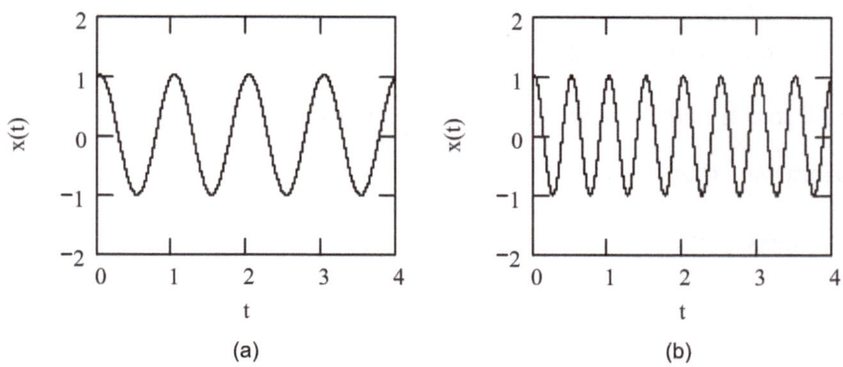

FIGURE 1.5 Free responses to harmonic free vibration without damping. (a) $x(t)$ for $f_n = 1$ (Hz), $x_0 = 1$ (m), $v_0 = 1$ (m/s). (b) $x(t)$ for $f_n = 2$ (Hz), $x_0 = 1$ (m), $v_0 = 1$ (m/s).

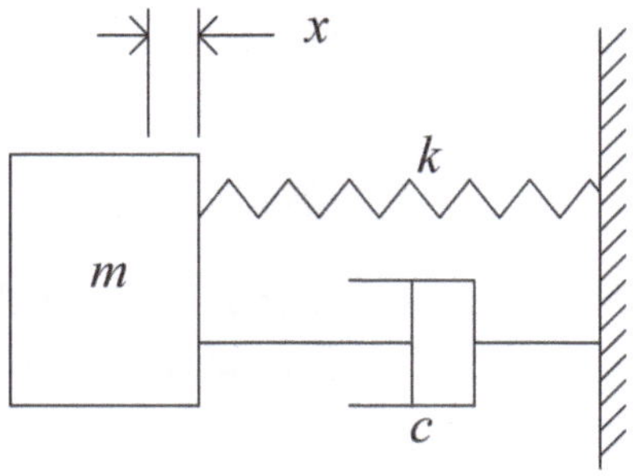

FIGURE 1.6 Single degree-of-freedom vibration system with viscous damper.

$$mx'' + cx' + kx = 0. \tag{1.14}$$

In physics, it is often expressed by

$$x'' + 2nx' + \omega_n^2 x = 0, \tag{1.15}$$

where

$$2n = \frac{c}{m}. \tag{1.16}$$

The quantity n is called damping parameter.

As the characteristic equation of (1.15) is given by

$$s^2 + 2ns + \omega_n^2 = 0, \tag{1.17}$$

the characteristic roots are

$$s_{1,2} = -n \pm \sqrt{n^2 - \omega_n^2}. \tag{1.18}$$

The general solution to Eq. (1.15) is expressed by

$$x(t) = A_1 e^{s_1 t} + A_2 e^{s_2 t}, \tag{1.19}$$

where A_1 and A_2 are determined by initial conditions.

Denote by the critical damping coefficient c_c. By critical damping coefficient, one means that $c = c_c$ such that $s_{1,2} = -n$. Consequently, one has

$$n = \frac{c_c}{2m}, \qquad c_c = 2m\omega_n = 2\sqrt{mk}. \tag{1.20}$$

Further, define a dimensionless parameter ζ by

$$\varsigma=\frac{c}{c_c}=\frac{c}{2\sqrt{mk}}. \tag{1.21}$$

It is called damping ratio. It contains all three quantities, namely, m, c, and k, in a vibration system. In engineering, a vibration equation is usually written by

$$\begin{cases} x''+2\omega_n\varsigma x'+\omega_n^2 x=0, \\ x(0)=x_0, x'(0)=v_0. \end{cases} \tag{1.22}$$

The characteristic roots of the above are given by

$$s_{1,2}=-\varsigma\omega_n\pm\omega_n\sqrt{\varsigma^2-1}. \tag{1.23}$$

The above facilitates the discussion of four types of vibrations from the point of view of damping.

1.3.1 Vibration with Excessive Damping

When $c>c_c$ or $\zeta>1$, a vibration system is said to be overdamped or excessively damped. In this case, $s_{1,2}<0$. Note that

$$\begin{aligned} s_1&=-\varsigma\omega_n+\omega_n\sqrt{\varsigma^2-1}, \\ s_2&=-\varsigma\omega_n-\omega_n\sqrt{\varsigma^2-1}. \end{aligned} \tag{1.24}$$

Since

$$x(t)=A_1e^{s_1t}+A_2e^{s_2t}, \tag{1.25}$$

where A_1 and A_2 are determined by initial conditions x_0 and v_0. They are

$$\begin{cases} A_1=\dfrac{x_0s_2-v_0}{s_2-s_1}=\dfrac{x_0\omega_n\left(\varsigma+\sqrt{\varsigma^2-1}\right)+v_0}{2\omega_n\sqrt{\varsigma^2-1}}, \\ A_2=\dfrac{v_0-x_0s_1}{s_2-s_1}=\dfrac{v_0-x_0\omega_n\left(-\varsigma+\sqrt{\varsigma^2-1}\right)}{-2\omega_n\sqrt{\varsigma^2-1}}. \end{cases} \tag{1.26}$$

The complete solution to a vibration system with excessive damping is given by

$$x(t)=\frac{x_0\omega_n\left(\varsigma+\sqrt{\varsigma^2-1}\right)+v_0}{2\omega_n\sqrt{\varsigma^2-1}}e^{-\left(\varsigma-\sqrt{\varsigma^2-1}\right)\omega_n t}-\frac{v_0-x_0\omega_n\left(-\varsigma+\sqrt{\varsigma^2-1}\right)}{2\omega_n\sqrt{\varsigma^2-1}}e^{-\left(\varsigma+\sqrt{\varsigma^2-1}\right)\omega_n t}. \tag{1.27}$$

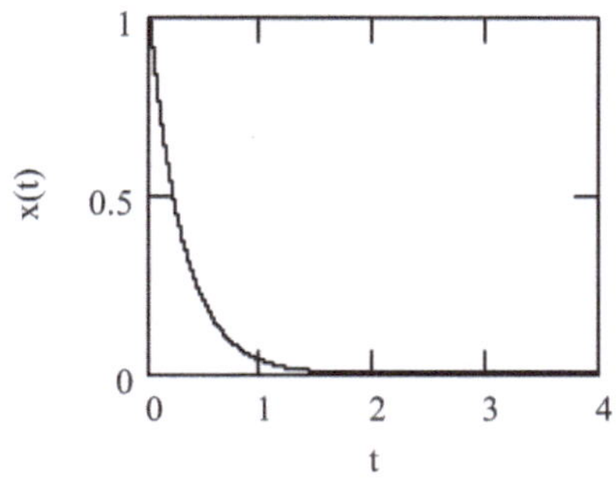

FIGURE 1.7 Overdamped vibration response $x(t)$ for $f_n=2$ (Hz), $x_0=1$ (m), $v_0=1$ (m/s), and $\zeta=2$.

Since $x(t)$ is monotonically decreasing without vibration, we say that it is just a trivial vibration case. From a view of energy, we say that the initial energy provided by x_0 and v_0 is quickly dissipated when $\zeta>1$. Figure 1.7 illustrates a case of overdamped vibration.

1.3.2 Vibration with Small Damping

When $0<\zeta<1$ or $0<c<c_c$, we say that a vibration system is with small damping. In this case, we write

$$s_{1,2}=-\varsigma\omega_n \pm \omega_n\sqrt{\varsigma^2-1}=-\varsigma\omega_n \pm i\omega_n\sqrt{1-\varsigma^2}, \tag{1.28}$$

where $i=\sqrt{-1}$. Denote by

$$\omega_d=\omega_n\sqrt{1-\varsigma^2}. \tag{1.29}$$

It is called damped natural angular frequency. Let T_d be the period of a damped vibration. Then,

$$T_d=\frac{2\pi}{\omega_d}=\frac{2\pi}{\omega_n\sqrt{1-\varsigma^2}}. \tag{1.30}$$

Theoretically,

$$\omega_d<\omega_n, \quad T_n<T_d. \tag{1.31}$$

However, since $\zeta<1$ and $\varsigma^2 \ll 1$ so that $\varsigma^2 \approx 0$, we have $\sqrt{1-\varsigma^2}\approx 1$. Besides, there are uncertainties and errors in measurement, structure manufacture, computations, and so on in engineering, it is reasonable to consider

$$\omega_d \approx \omega_n, \quad T_n \approx T_d. \tag{1.32}$$

The complete solution to a small damped vibration system is given by

$$x(t)=e^{-\varsigma\omega_n t}\left(x_0\cos\omega_d t+\frac{v_0+\varsigma\omega_n x_0}{\omega_d}\sin\omega_d t\right), \quad t\geq 0. \tag{1.33}$$

The above can be rewritten by

$$x(t) = Ae^{-\varsigma\omega_n t}\cos(\omega_d t - \theta), \quad t \geq 0, \tag{1.34}$$

where A is in the form

$$A = \sqrt{x_0^2 + \left(\frac{v_0 + \varsigma\omega_n x_0}{\omega_d}\right)^2}, \tag{1.35}$$

and θ is given by

$$\theta = \arctan\frac{v_0 + \varsigma\omega_n x_0}{\omega_d x_0}. \tag{1.36}$$

Figure 1.8 indicates a curve of the response to a small damped vibration system.

From the point of view of energy, we see that the initial energy provided by x_0 and v_0 is slowly dissipated by a damper with $0<\zeta<1$.

1.3.3 Vibration with Critical Damping

If $c=c_c$ or $\zeta=1$, one says that a vibration system is critically damped. For $\zeta=1$, the complete solution to Eq. (1.22) is given by

$$x(t) = \left[x_0 + (v_0 + x_0\omega_n)t\right]e^{-\varsigma\omega_n t}, \quad t \geq 0. \tag{1.37}$$

Note that a vibration system with critical damping does not oscillate as indicated by Figure 1.9.

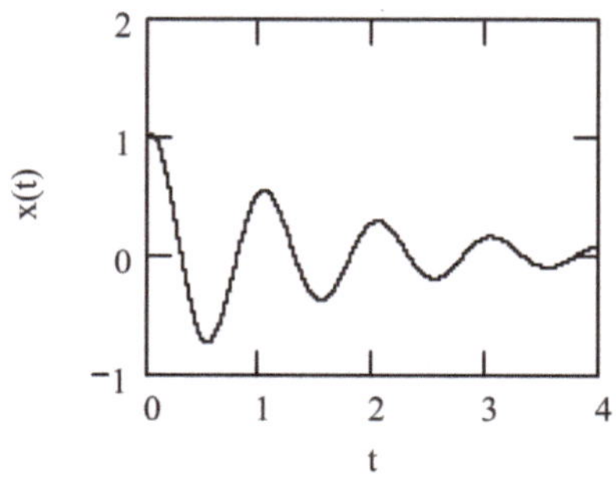

FIGURE 1.8 Small damping response $x(t)$ for $f_n=2$ (Hz), $x_0=1$ (m), $v_0=1$ (m/s), and $\zeta=0.1$.

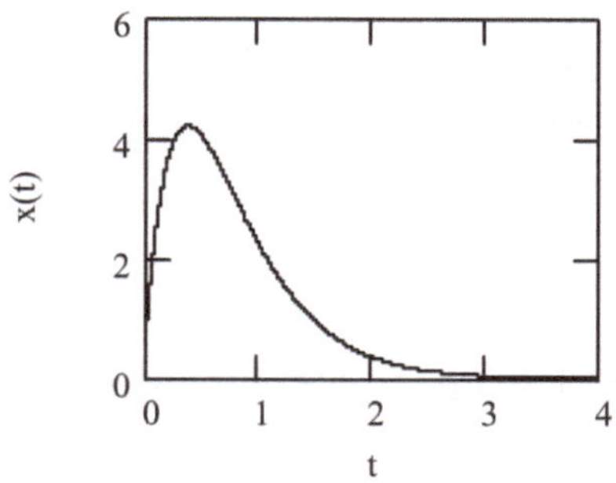

FIGURE 1.9 Critical response $x(t)$ for $f_n=4$ (Hz), $x_0=1$ (m), $v_0=1$ (m/s), and $\zeta=1$.

1.3.4 Vibration with Negative Damping

If $c<0$ or $\zeta<0$, we say that a vibration system is negatively damped. In this case, the complete response is in the form

$$x(t)=e^{|\varsigma|\omega_n t}\left(x_0\cos\omega_d t+\frac{v_0-|\varsigma|\omega_n x_0}{\omega_d}\sin\omega_d t\right),\quad t\geq 0. \tag{1.38}$$

It approaches infinity for $t\to\infty$ until a structure is collapsed, or out of control, or out of order. Figure 1.10 gives a plot of a vibration response with negative damping.

Investigating negative damping is crucially meaningful in either academic research or practice, referring Den Hartog [1], Nakagawa and Ringo [2] for some examples of structures that may cause vibrations with negative damping.

1.4 COMPUTATIONS OF NATURAL FREQUENCY AND DAMPING RATIO

From the previous discussions, one sees that two essential quantities are f_n and ζ in a vibration system. With the fast development of measurement instrumentations, measuring f_n and ζ is nowadays quite mature. For the sake of the completeness of the book, however, traditional contents in computations of f_n and ζ are briefed in this section.

1.4.1 Computations of Natural Frequency

1.4.1.1 Static Deflection Method

Consider a mass-spring system with stiffness k in Figure 1.11,

Denote by δ_{st} the static deflection of the spring. Then, according to the theory in theoretic mechanics, one has

$$\delta_{st}=\frac{mg}{k}, \tag{1.39}$$

where g is the gravity acceleration. Since $\omega_n=\sqrt{\dfrac{k}{m}}$, one has

$$f_n=\frac{1}{2\pi}\sqrt{\frac{g}{\delta_{st}}}. \tag{1.40}$$

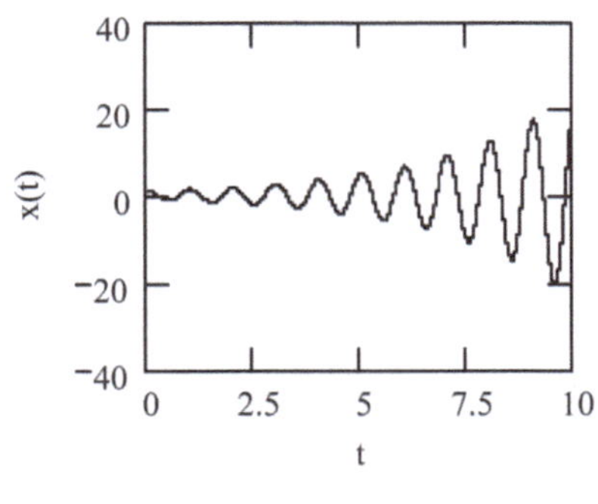

FIGURE 1.10 Response to a negatively damped vibration system for $f_n=5$ (Hz), $x_0=1$ (m), $v_0=1$ (m/s), and $\zeta=-0.01$.

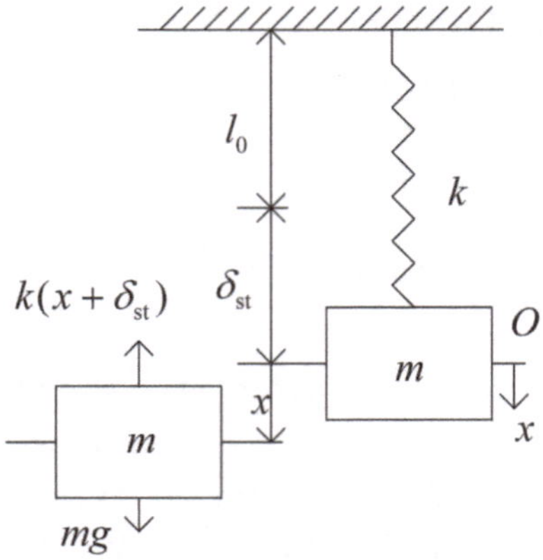

FIGURE 1.11 Vertical mass-spring system.

Therefore, f_n can be immediately obtained once δ_{st} is got [3].

1.4.1.2 Energy Method

A free harmonic mass-spring system follows the principle of work and energy. Thus,

$$E = E_p + E_k = E_{p0} + E_{k0} = \text{const.} \tag{1.41}$$

Therefore, one has

$$\frac{dE}{dt} = \frac{d}{dt}\left(E_p + E_k\right) = 0. \tag{1.42}$$

Hence,

$$E_{p,\max} = E_{k,\max}. \tag{1.43}$$

For a mass-spring system in Figure 1.11, one has

$$E_p = \frac{1}{2}kx^2, \quad E_k = \frac{1}{2}m\left(\frac{dx}{dt}\right)^2. \tag{1.44}$$

When $x(t) = A\sin\omega_n t$, the above becomes

$$E_p = \frac{1}{2}kA\sin^2\omega_n t, \quad E_k = \frac{1}{2}mA\omega_n^2\cos^2\omega_n t. \tag{1.45}$$

Thus,

$$E_{p,\max} = \frac{1}{2}kA, \quad E_{k,\max} = \frac{1}{2}mA\omega_n^2. \tag{1.46}$$

According to Eq. (1.43), one has $\omega_n = \sqrt{\frac{k}{m}}$.

Natural frequencies of many structural elements are summarized in the literature; see e.g., Peirce et al. [4], Palley et al. [3].

1.4.2 Computation of Damping Ratio

Let t_n and t_{n+1} be two time points where $x(t)$ reaches its successive peak values, respectively. Denote by ψ the attenuation rate of $x(t_n)$ to $x(t_{n+1})$. Then,

$$\psi = \frac{Ae^{-\varsigma\omega_n t}}{Ae^{-\varsigma\omega_n (t+T_d)}} = e^{\varsigma\omega_n T_d} = e^{\frac{2\pi\varsigma}{\sqrt{1-\varsigma^2}}}. \tag{1.47}$$

Let Δ be the logarithmic decrement of $x(t)$. It is defined by

$$\Delta = \ln\psi = \frac{2\pi\varsigma}{\sqrt{1-\varsigma^2}}. \tag{1.48}$$

By measuring Δ from the response $x(t)$ to a vibration system, one can obtain ζ with the simple computation of Eq. (1.48).

1.5 FORCED VIBRATIONS

Consider a mass-spring system driven by an external force $e(t)$. Then, one has a vibration motion equation expressed by

$$\begin{cases} mx'' + cx' + kx = e(t), \\ x(0) = x_0, x'(0) = v_0. \end{cases} \tag{1.49}$$

When

$$e(t) = F\sin(\omega t + \beta), \tag{1.50}$$

$x(t)$ is called the harmonically forced vibration response or forced response in short.

The topic of harmonically forced vibrations appears simple, but it is important in practice. Since a mass-spring system is linear, it satisfies the superposition. By superposition, we mean that if an excitation $e_j(t)$ results in a response $x_j(t)$ and if the excitation is given by

$$e(t) = \sum_j e_j(t), \tag{1.51}$$

the solution to the vibration system is

$$x(t) = \sum_j x_j(t). \tag{1.52}$$

When $e(t)$ is a periodic signal with period T,

$$e(t) = \frac{a_0}{2} + \sum_{n=1}^{\infty} a_n \cos n\omega t + b_n \sin n\omega t,$$

where

$$\omega = \frac{2\pi}{T},$$

a_0, a_n, and b_n are Fourier coefficients of a Fourier series. Thus, if we obtain harmonic responses for $\cos(n\omega t)$ and $\sin(n\omega t)$, we have the response in terms of a periodic driven force based on superposition and the Fourier series. We mention the Fourier series at the moment just for enhancing the importance of harmonically driven vibrations. Details for the Fourier series will be given later.

Usually, a vibration system driven by a harmonic force is expressed by

$$\begin{cases} x'' + 2\omega_n \varsigma x' + \omega_n^2 x = \dfrac{F}{m}\sin(\omega t + \beta), \\ x(0) = x_0,\ x'(0) = v_0. \end{cases} \tag{1.53}$$

Unless otherwise stated, we only consider vibrations with $0 < \zeta < 1$.

1.5.1 Stationary Vibration

Denote by $x_p(t)$ the stationary solution to Eq. (1.54). In mathematics, $x_p(t)$ is called the particular solution to Eq. (1.54)

$$x'' + 2\omega_n \varsigma x' + \omega_n^2 x = \frac{F}{m}\sin(\omega t + \beta). \tag{1.54}$$

Suppose that

$$x_p(t) = A_1 \sin(\omega t + \beta) + A_2 \cos(\omega t + \beta). \tag{1.55}$$

Substituting the right on Eq. (1.55) to the left on Eq. (1.54) produces

$$\left[\left(\omega_n^2 - \omega^2\right)A_1 - 2\omega_n \omega \varsigma A_2\right]\sin(\omega t + \beta) + \left[\left(\omega_n^2 - \omega^2\right)A_2 - 2\omega_n \omega \varsigma A_1\right]\cos(\omega t + \beta) = \frac{F}{m}\sin(\omega t + \beta).$$

The above implies

$$\begin{aligned} \left(\omega_n^2 - \omega^2\right)A_1 - 2\omega_n \omega \varsigma A_2 &= \frac{F}{m}, \\ \left(\omega_n^2 - \omega^2\right)A_2 + 2\omega_n \omega \varsigma A_1 &= 0. \end{aligned} \tag{1.56}$$

Let γ be the frequency ratio. It is defined by

$$\gamma = \frac{\omega}{\omega_n} = \frac{f}{f_n}. \tag{1.57}$$

Then, Eq. (1.56) is rewritten by

$$\begin{aligned} \left(1-\gamma^2\right)A_1 - 2\gamma\varsigma A_2 &= \frac{F}{k}, \\ \left(1-\gamma^2\right)A_2 + 2\gamma\varsigma A_1 &= 0. \end{aligned} \tag{1.58}$$

Two coefficients are expressed by

$$\begin{aligned} A_1 &= \frac{F}{k}\frac{1-\gamma^2}{\left(1-\gamma^2\right)^2+(2\varsigma\gamma)^2}, \\ A_2 &= \frac{F}{k}\frac{-2\varsigma\gamma}{\left(1-\gamma^2\right)^2+(2\varsigma\gamma)^2}. \end{aligned} \tag{1.59}$$

Therefore,

$$x_p(t) = \frac{F}{k}\left[\frac{1-\gamma^2}{\left(1-\gamma^2\right)^2+(2\varsigma\gamma)^2}\sin(\omega t+\beta) - \frac{2\varsigma\gamma}{\left(1-\gamma^2\right)^2+(2\varsigma\gamma)^2}\cos(\omega t+\beta)\right]. \tag{1.60}$$

The above may be rewritten by

$$x_p(t) = A\sin(\omega t+\beta-\theta), \tag{1.61}$$

where the amplitude A is

$$A = \frac{F}{k}\frac{1}{\sqrt{\left(1-\gamma^2\right)^2+(2\varsigma\gamma)^2}}. \tag{1.62}$$

The phase θ is

$$\theta = \arctan\frac{2\varsigma\gamma}{1-\gamma^2}. \tag{1.63}$$

The function $x_p(t)$ in Eq. (1.61) implies that the stationary response to the harmonically driven mass-spring system is also a harmonic function. In other words, if an excitation of a vibrator is a sinusoidal signal $F\sin(\omega t+\beta)$, the response is also a sinusoidal signal $A\sin(\omega t+\beta-\theta)$. It has the three properties described as follows:

- The response $x_p(t)$ has the same frequency ω as that of the excitation.
- The vibration system weighs the amplitude F of the excitation to the response amplitude A, see (1.62).
- In comparison with the excitation $F\sin(\omega t+\beta)$, the response $x_p(t)$ is delayed by the vibration system. The delay is reflected by the phase θ.

1.5.2 Analysis of Stationary Response

For stationary response $x_p(t)$, its amplitude A and phase θ are with two arguments, namely, ω and ζ. One may write ω with γ. There are following properties in $A=A(\gamma, \zeta)$.

1. If $\omega=0$, $\gamma=0$ and $A(0, \zeta)=\text{constant}$. In fact, $\omega=0$ corresponds to static loading of a structure. Hence, it is trivial in vibrations.
2. When $\zeta=0$ and $\gamma=1$,

$$A(\gamma,\varsigma)\big|_{\varsigma=0,\gamma=1}=\infty. \tag{1.64}$$

   The above says that a resonance occurs when $\gamma=1$ or $f=f_n$. To be precise, a mass-spring system resonates if the excitation frequency f encounters the natural frequency f_n.

3. When f is far away from f_n, the system response becomes small. In the extreme,

$$\lim_{\gamma\to\infty} A(\gamma,\varsigma)=0. \tag{1.65}$$

4. The normalized dynamic coefficient A_0 is defined by

$$A_0=\frac{A}{F/k}=\frac{2\varsigma}{\sqrt{\left(1-\gamma^2\right)^2+(2\varsigma\gamma)^2}}. \tag{1.66}$$

   It is also called amplitude-frequency response function. Denote by $A_{0,\max}$ the maximum of A_0. Then,

$$A_0(1,\varsigma)=A_{0,\max}=1. \tag{1.67}$$

   Note that $\gamma=1$ means $f=f_n$.

   As a matter of fact, when $\zeta\neq 0$, $A_{0,\max}$ exactly occurs at $\gamma=\sqrt{1-2\varsigma^2}$. Thus,

$$A_{0,\max}=\frac{1}{\sqrt{1-\varsigma^2}}. \tag{1.68}$$

In engineering, it is reasonable to suppose $\varsigma^2 \approx 0$ for small ζ. Thus, the resonance takes place at

$$\gamma = \sqrt{1-2\varsigma^2} \approx 1. \tag{1.69}$$

Accordingly, the resonant amplitude is approximately given by Eq. (1.67).

5. If $\varsigma \geq \frac{1}{\sqrt{2}} \approx 0.707$, there is no maximum in the amplitude-frequency response function A_0 when $\gamma \neq 0$, see Figure 1.12. $\sqrt{2}/2$ is a useful constant in vibrations. It may be regarded as the critical value of large damping from a view of vibration engineering.

   Resonance is a phenomenon that plays a role in vibration engineering. When resonance takes place, the driven force amplitude F is amplified to $\frac{F/k}{2\varsigma}$. Figure 1.12 gives a few illustrations of A_0.

The phase-frequency response function θ has the following properties.

1. When $\gamma \to 0$, $\theta = 0°$.
2. For $\gamma = 1$, $\theta = 90°$.
3. If $\gamma \to \infty$, $\theta = 180°$.
4. At the neighborhood of $\gamma = 1$, θ is a linear function in terms of γ.
5. If $\zeta = 0$, $\theta = 0°$.
6. For a practical vibrator, $\theta \geq 0°$. That may be interpreted as the causality that a physical vibration system must obey.

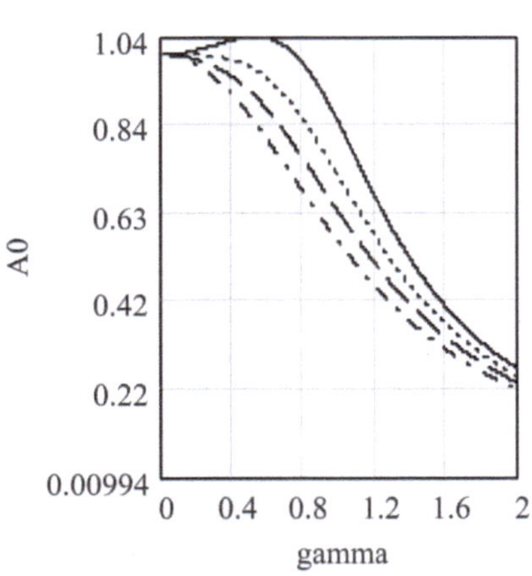

FIGURE 1.12 Amplitude-frequency response function A_0 for $\zeta = 0.6$ (solid), 0.707 (dot), 0.8 (dash), 0.9 (dadot).

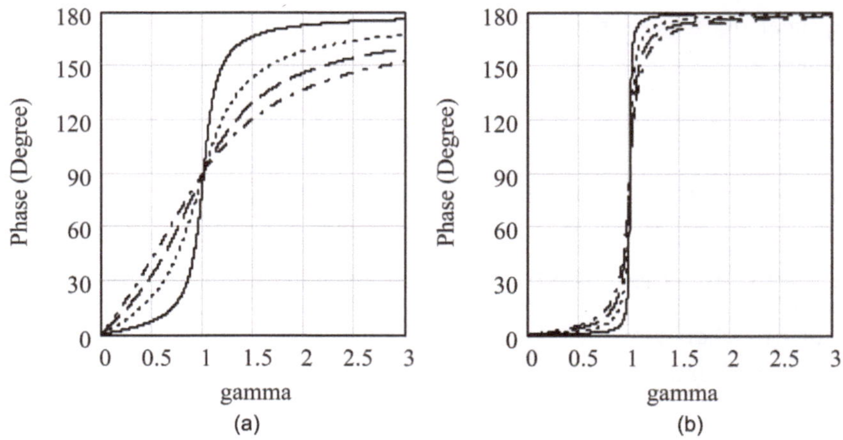

FIGURE 1.13 Phase-frequency response function $\theta(\gamma, \zeta)$. (a) $\theta(\gamma, \zeta)$ for $\zeta=0.1$ (solid), 0.3 (dot), 0.5 (dash), $\zeta=0.7$ (dadot). (b) $\theta(\gamma, \zeta)$ for $\zeta=0.01$ (solid), 0.03 (dot), 0.05 (dash), 0.07 (dadot).

Note that there is a singularity at $\gamma=1$ in Eq. (1.63). We use the following equivalent formula of $\theta\,(\gamma, \zeta)$ to draw its curves, see Nakagawa and Ringo [2, p. 50]

$$\theta(\gamma,\varsigma)=\arccos\frac{1-\gamma^2}{\sqrt{\left(1-\gamma^2\right)^2+(2\varsigma\gamma)^2}}. \tag{1.70}$$

Figure 1.13 shows several curves of θ.

1.5.3 Resonance Bandwidth

As previously mentioned, a resonance happens when $f=f_n$. In engineering, due to uncertainties and errors, taking $f=f_n$ as a resonance condition does not make much sense. In practice, we say that a resonance may occur when $f\in(f_n+\varepsilon, f_n-\varepsilon)$ for $\varepsilon>0$. The question is how to quantify ε.

To answer the aforementioned question, we first describe a concept of half-power points of an amplitude-frequency response function. For the stationary response $x_p(t)=A\sin(\omega t+\beta-\theta)$, its power is $P_x=A^2$. Since A is a function of ω and ζ, we write

$$P_x=P_x\left(f,\varsigma\right)=A^2\left(\frac{f}{f_n},\varsigma\right)=\left(\frac{F}{k}\right)^2\frac{1}{\left[1-\left(\frac{f}{f_n}\right)^2\right]^2+\left(2\varsigma\frac{f}{f_n}\right)^2}. \tag{1.71}$$

It is clear that

$$P_{x,\max}=P_x\left(f_n,\varsigma\right)=A_{\max}^2=A^2\left(\frac{f_n}{f_n},\varsigma\right). \tag{1.72}$$

Approximately, $A(\gamma, \zeta)$ is of symmetry around $\gamma=1$. Let f_1 and f_2 be such that

$$P_x(f_1,\varsigma)=P_x(f_2,\varsigma)=\frac{1}{2}P_{x,\max}(f_n,\varsigma). \tag{1.73}$$

Then, f_1 and f_2 are called half-power points. Let $f_2>f_1$. Define

$$\mathrm{B}_{0.707}=f_2-f_1. \tag{1.74}$$

$\mathrm{B}_{0.707}$ is called the resonance frequency bandwidth (resonance bandwidth for short) of a mass-spring system. When considering the half-power points, we have

$$A\left(\frac{f_1}{f_n},\varsigma\right)=A\left(\frac{f_2}{f_n},\varsigma\right)=\frac{1}{\sqrt{2}}A\left(\frac{f_n}{f_n},\varsigma\right)=\frac{1}{\sqrt{2}}A_{\max}\approx 0.707A_{\max}. \tag{1.75}$$

That is the reason we write the bandwidth with the subscript 0.707. Figure 1.14 indicates the bandwidth with half-power points.

In the sense of $\mathrm{B}_{0.707}$, a vibration system resonates if $f\in(f_1, f_2)$. It does not resonate if $f\notin(f_1, f_2)$. When $f\notin\mathrm{B}_{0.707}$, the energy in the vibration system is too weak to produce a resonance. From the above, one has

$$\frac{f_n}{\mathrm{B}_{0.707}}=\frac{1}{2\varsigma}. \tag{1.76}$$

In engineering, we often need to compare the sharpness of the resonance curves $A(\gamma, \zeta)$s of several mass-spring systems. In this case, one usually uses a normalized amplitude-frequency response function. Define

$$A_3=\frac{A_0}{A_{0,\max}}=\frac{1}{A_{0,\max}}\frac{1}{\sqrt{(1-\gamma^2)^2+(2\varsigma\gamma)^2}}. \tag{1.77}$$

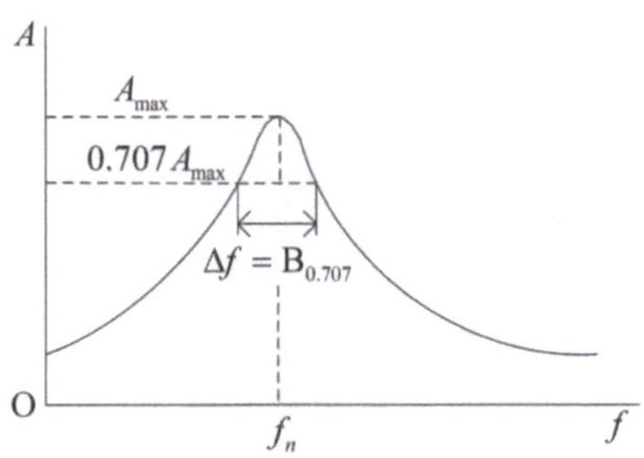

FIGURE 1.14 Resonance bandwidth $\mathrm{B}_{0.707}$.

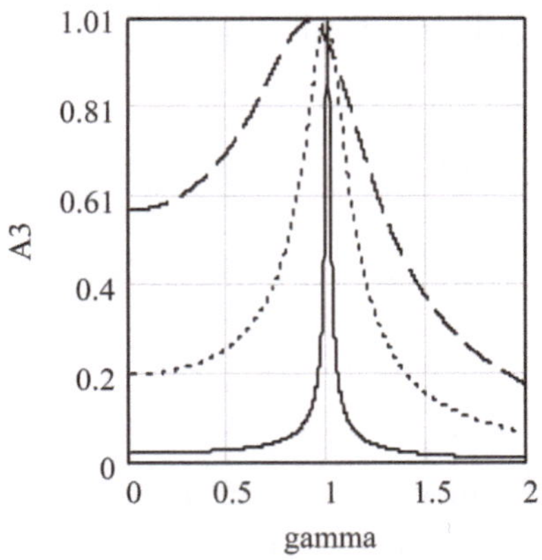

FIGURE 1.15 Normalized amplitude-frequency response function A_3 when $\zeta=0.01$ (solid), 0.1 (dot), 0.3 (dash).

Then, the maximum of A_3 is 1. Taking into account (1.68), we have

$$A_3 = \frac{A_0}{A_{0,\max}} \approx \frac{2\varsigma\sqrt{1-\varsigma^2}}{\sqrt{\left(1-\gamma^2\right)^2+(2\varsigma\gamma)^2}}. \tag{1.78}$$

Figure 1.15 gives several curves of the normalized amplitude-frequency response function.

1.5.4 *Q* Factor

As can be seen from the previous discussions, a small ζ corresponds to a sharp curve of $A(\gamma, \zeta)$. In other words, $A(\gamma, \zeta)$ attenuates faster in terms of γ if ζ is smaller. In fact, ζ is a quantity to measure the energy dissipation of a mass-spring system.

In the field of vibrations, there is another parameter, denoted by Q, is conventionally used to describe the sharpness of a curve of $A(\gamma, \zeta)$. It is called Q-value or Q factor in short (Nakagawa and Ringo [2, Section 3.4]). Its full name is quality factor (Harris [5, p. 2.18]). The Q factor of a mass-spring system is defined by

$$Q = \frac{\omega_n m}{c} = \frac{k}{\omega_n c} = \frac{\sqrt{mk}}{c} = \frac{1}{2\varsigma}. \tag{1.79}$$

Therefore, the larger the Q the sharper the curve of $A(\gamma, \zeta)$.

According to Eq. (1.76), we may use Q to express $B_{0.707}$ in the form

$$B_{0.707} = \frac{f_n}{Q}. \tag{1.80}$$

Alternatively,

$$B_{0.707} = 2\varsigma f_n. \tag{1.81}$$

1.5.5 Logarithm Resonance Curves

A resonance curve $A(\gamma, \zeta)$ is important in engineering. However, $A(\gamma, \zeta)$ may be narrow in shape if it is indicated with ordinary coordinates for small ζ in particular. For instance, in Figure 1.15, the curve with $\zeta=0.01$ may be too narrow to provide a satisfactory view regarding $B_{0.707}$. Thus, we need other coordinates for viewing $A(\gamma, \zeta)$.

Three types of coordinates are commonly used in engineering.

1. Use $\log_{10} A(\gamma, \zeta)$ to extend the resonance curve. The unit of $\log_{10} A(\gamma, \zeta)$ is B (Bel). Thus, the symbol $A(\gamma, \zeta)$ (log) stands for $\log_{10} A(\gamma, \zeta)$ or $\log_{10} A(\gamma, \zeta)$ (B). Figure 1.16 indicates $\log_{10} A_3(\gamma, \zeta)$ in Figure 1.15.

2. When the extension of A_3 using $\log_{10} A(\gamma, \zeta)$ is not enough, $20\log_{10} A(\gamma, \zeta)$ may be utilized. The unit of $20\log_{10} A(\gamma, \zeta)$ is dB (Decibel). Figure 1.17 gives $20\log_{10} A_3$ for A_3 in Figure 1.15.

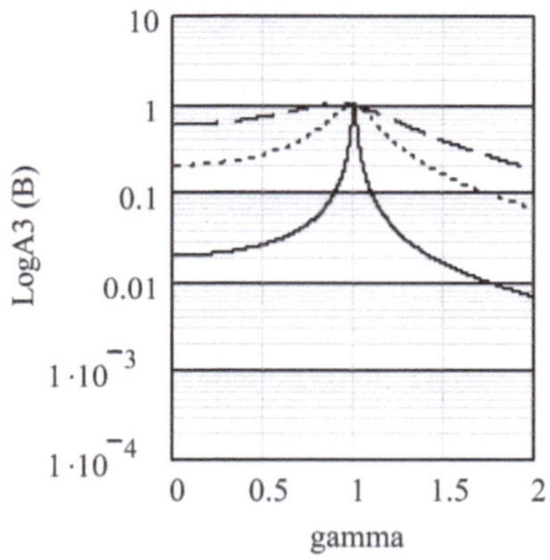

FIGURE 1.16 Normalized amplitude-frequency response function $\log_{10} A_3$ (B) for $\zeta=0.01$ (solid), 0.1 (dot), $\zeta=0.3$ (dash).

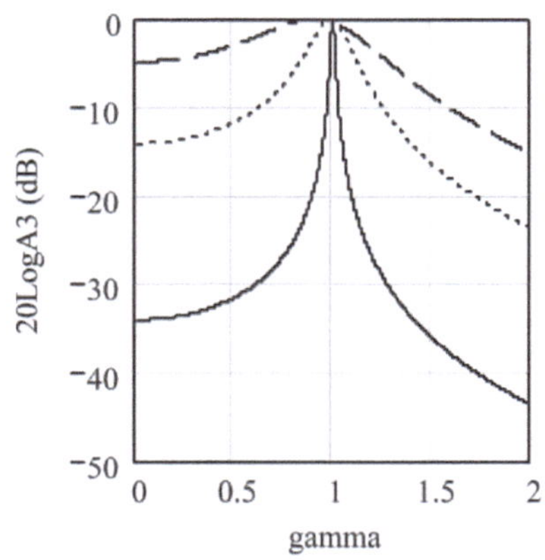

FIGURE 1.17 Normalized amplitude-frequency response function $20\log_{10} A_3$ (dB) when $\zeta=0.01$ (solid), 0.1 (dot), 0.3 (dash).

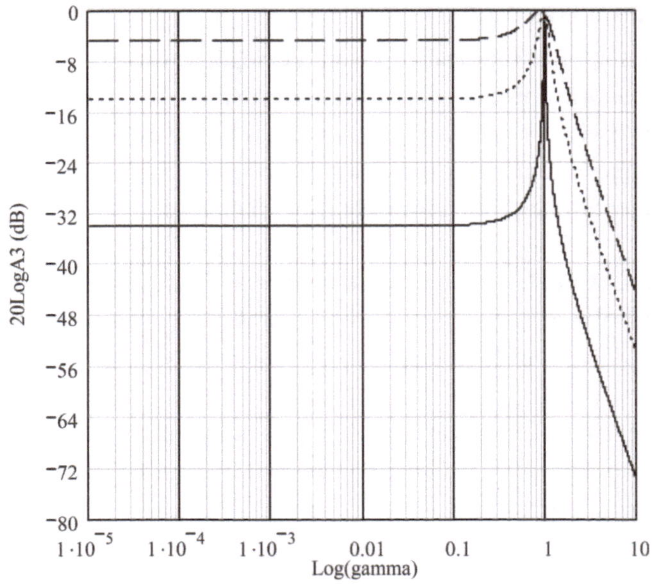

FIGURE 1.18 Bode diagram of A_3 in Figure 1.15 when ζ=0.01 (solid), 0.1 (dot), 0.3 (dash).

3. The Bode diagram means that the vertical axis uses $20\log_{10}A$ (dB) while the abscissa axis takes $\log_{10}\gamma$. The Bode diagram of A_3 is indicated in Figure 1.18.

To describe the frequency points f_1 and f_2 in (f_1, f_2) for $B_{0.707}$, we have used the terms half-power points owing to Eq. (1.71) and 0.707 points due to Eq. (1.75). Since

$$20\log\frac{1}{\sqrt{2}}\approx -3. \tag{1.82}$$

The term of minus 3 dB points or −3 dB points in short is used in the field. Three terms, namely, half-power points, 0.707 points, and −3 dB points are used to describe the same thing, i.e., $B_{0.707}$.

Before the days that computers were popularized, any skill in computation saving was desired. Then, the logarithm in computation was emphasized. For instance, suppose a mass-spring system consists of two subsystems in series, as shown in Figure 1.19. Let B_1 and B_2 be the amplitude-frequency responses of two subsystems in series, respectively. Denote by B the total amplitude-frequency response of the system. Then,

$$B = B_1 B_2. \tag{1.83}$$

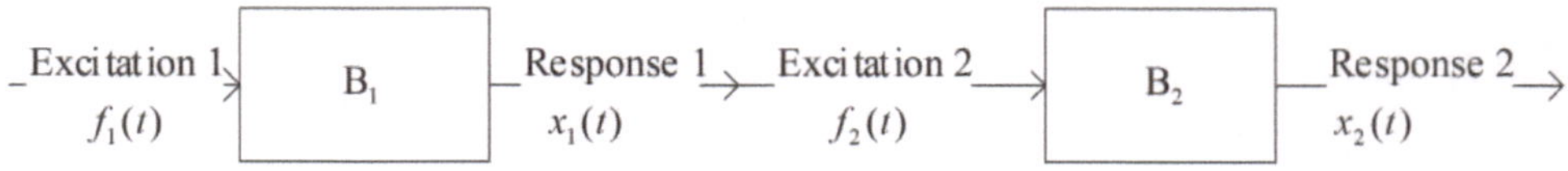

FIGURE 1.19 Two mass-spring systems in series.

If using logarithm operation, one has

$$20\log \mathrm{B} = 20\log \mathrm{B}_1 + 20\log \mathrm{B}_2. \tag{1.84}$$

Therefore, the multiplication on the right side of Eq. (1.83) changes to the addition on the right side of Eq. (1.84) by using logarithm operation. Such a computation saving may unnecessarily be emphasized today because computers are now popular. Therefore, the logarithm operation used today is mainly for curve viewing.

1.6 COMPLETE RESPONSE

Taking into account the initial conditions in Eq. (1.53), the complete solution is given by

$$\begin{aligned} x(t) = {} & e^{-\varsigma\omega_n t}\left(x(0)\cos\omega_d t + \frac{x'(0)+\varsigma\omega_n x(0)}{\omega_d}\sin\omega_d t \right) \\ & - e^{-\varsigma\omega_n t} A\left[\sin(\beta-\theta)\cos\omega_d t + \frac{\omega\cos(\beta-\theta)+\varsigma\omega_n\sin(\beta-\theta)}{\omega_d}\sin\omega_d t \right] \\ & + A\sin(\omega t+\beta-\theta). \end{aligned} \tag{1.85}$$

For the purpose of describing the physical meaning of $x(t)$, one often utilizes two classifications.

1.6.1 Classification Method 1

The first item on the right side of Eq. (1.85) is the transient oscillation part in $x(t)$. It is excited by the initial conditions x_0 and v_0. Accordingly, it is a free response. It is called zero input response and denoted by $x_{zi}(t)$. The second and third items are excited by driven force irrelevant of x_0 and v_0. One calls it a zero-state response and denotes it by $x_{zs}(t)$. Thus, one has

$$x(t) = x_{zi}(t) + x_{zs}(t), \tag{1.86}$$

where

$$x_{zi}(t) = e^{-\varsigma\omega_n t}\left(x(0)\cos\omega_d t + \frac{x'(0)+\varsigma\omega_n x(0)}{\omega_d}\sin\omega_d t \right), \tag{1.87}$$

$$\begin{aligned} x_{zs}(t) = {} & -e^{-\varsigma\omega_n t} A\left[\sin(\beta-\theta)\cos\omega_d t + \frac{\omega\cos(\beta-\theta)+\varsigma\omega_n\sin(\beta-\theta)}{\omega_d}\sin\omega_d t \right] \\ & + A\sin(\omega t+\beta-\theta). \end{aligned} \tag{1.88}$$

With that classification, we see that initial conditions produce $x_{zi}(t)$ while externally driven force results in $x_{zs}(t)$. Denoting by $x_{\mathrm{free}}(t)$ and $x_{\mathrm{forced}}(t)$ the free response and forced one, respectively, we have

$$x_{zi}(t) = x_{\mathrm{free}}(t), \; x_{zs}(t) = x_{\mathrm{forced}}(t). \tag{1.89}$$

Note that the first item on the right side of Eq. (1.88) is transient while the second is stationary. Denote by $x_{zs,tr}(t)$ and $x_{zs,sta}(t)$ transient part and stationary one of $x_{zs}(t)$, respectively. Then, $x_{zs}(t)$ can be

$$x_{zs}(t) = x_{zs,tr}(t) + x_{zs,sta}(t), \tag{1.90}$$

where

$$x_{zs,tr}(t) = -e^{-\varsigma\omega_n t} A\left[\sin(\beta-\theta)\cos\omega_d t + \frac{\omega\cos(\beta-\theta)+\varsigma\omega_n\sin(\beta-\theta)}{\omega_d}\sin\omega_d t\right], \tag{1.91}$$

and

$$x_{zs,sta}(t) = x_p(t) = A\sin(\omega t + \beta - \theta). \tag{1.92}$$

Example 1.1

Let $x(0)=1$, $v(0)=1$, $\zeta=0.1$, $\omega_n=3.14$, $\omega=3.14$, $F/k=1$, $\beta=0$. Then, we have $x_{zi}(t)=x_{free}(t)$ shown in Figure 1.20a. $x_{zs,\ tr}(t)$ is indicated in Figure 1.20b. The plot of stationary response $x_p(t)$ is given in Figure 1.20c. The curve of $x_{zs}(t)=x_{forced}(t)$ is shown in Figure 1.2d. Figure 1.20e shows the plot of complete solution $x(t)$.

The stationary response $x_p(t)$ is certainly important. The zero input response can be avoided by releasing initial conditions to zero before a structure begins moving. However, the transient response part $x_{zs,\ tr}(t)$ is excited by external force and cannot be avoided in any case. It has obvious effects on the zero-state response and total response, as can be seen from Figure 1.20d and e. In some applications, for instance, loading in a fatigue test machine, $x_{zs,\ tr}(t)$ might be a cause to make a test machine work anomalously, see Swanson [6], Li [7]. The issue of transient response to a vibration system will be mentioned again late from the point of view of impulse response and step one.

1.6.2 Classification Method 2

The first two items in Eq. (1.85) approach zero if t is sufficiently large. They are transient responses. The last item is the stationary one. Denote by $x_{tr}(t)$ the transient part in $x(t)$. Then, $x(t)$ may be expressed by

$$x(t) = x_{tr}(t) + x_p(t), \tag{1.93}$$

where

$$\begin{aligned} x_{tr}(t) &= e^{-\varsigma\omega_n t}\left(x(0)\cos\omega_d t + \frac{x'(0)+\varsigma\omega_n x(0)}{\omega_d}\sin\omega_d t\right) \\ &\quad - e^{-\varsigma\omega_n t} A\left[\sin(\beta-\theta)\cos\omega_d t + \frac{\omega\cos(\beta-\theta)+\varsigma\omega_n\sin(\beta-\theta)}{\omega_d}\sin\omega_d t\right]. \end{aligned} \tag{1.94}$$

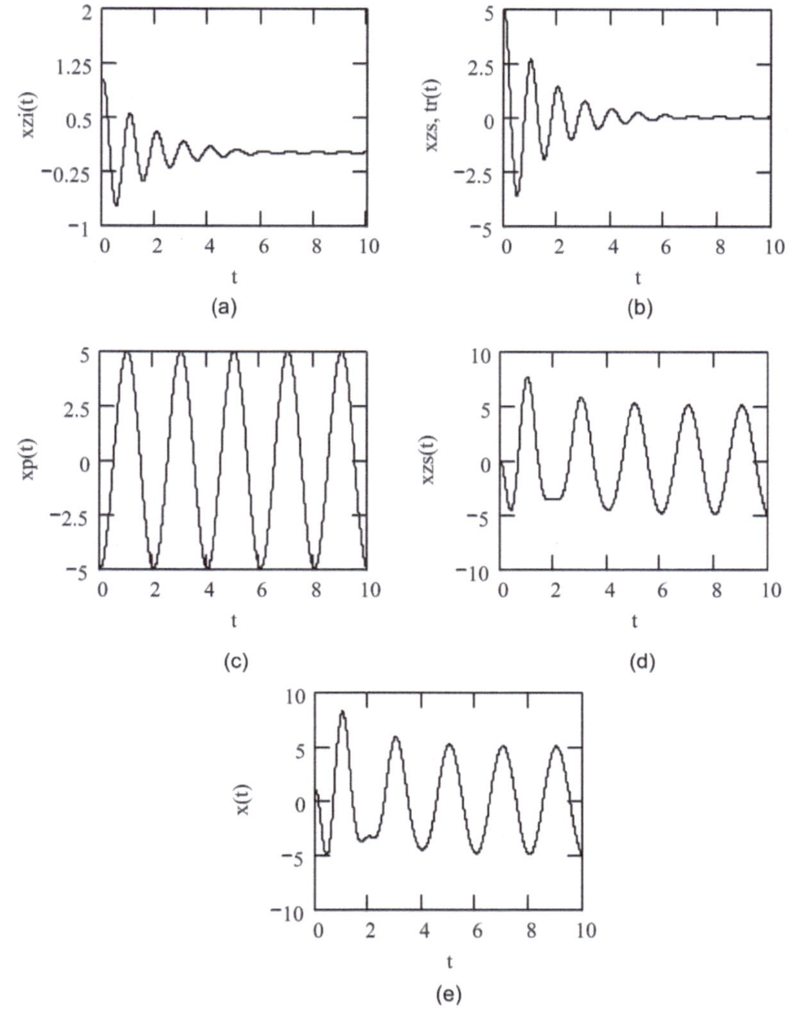

FIGURE 1.20 Plots of response components with $x_0=1$, $v_0=1$, $\zeta=0.1$, $\omega_n=3.14$, $\omega=3.14$, $F/k=1$, $\beta=0$. (a) Zero input response $x_{zi}(t)=x_{free}(t)$. (b) Transient response part excited by external force $x_{zs,\,tr}(t)$. (c) Stationary response $x_p(t)$. (d) Zero-state response $x_{zs}(t)=x_{forced}(t)$. (e) Complete response.

The above says that $x_{tr}(t)$ consists of two items. One is $x_{zi}(t)=x_{free}(t)$ and the other $x_{zs,\,tr}(t)$. Thus,

$$x_{tr}(t)=x_{zi}(t)+x_{zs,tr}(t). \tag{1.95}$$

1.7 PHASOR AND STATIONARY HARMONIC RESPONSE

Stationary response to a system driven by sinusoidal force plays a key role in vibrations. Its computation is necessary for engineers. Phasor is a tool in this regard (Harris [5]).

1.7.1 Phasor

The theory of phasor and its applications to handling stationary solution to a differential equation of order 2 with a sinusoidal driven signal were introduced by Steinmetz [8,9]. Its applications to mechanical vibrations were not mentioned in the early famous literature by Timoshenko [10,11]. Den Hartog's work [1] might be an earlier piece of literature mentioning phasor in vibrations.

Consider Euler's formula in the form

$$e^{it} = \cos t + i \sin t, \quad i = \sqrt{-1}. \tag{1.96}$$

Since

$$e^{it \pm i2n\pi} = e^{it}, \quad n = \pm 1, 2, \ldots \tag{1.97}$$

one says e^{it} is the principal sheet of $e^{it \pm i2n\pi}$. Unless otherwise stated, we only consider the principal sheet of a complex function in what follows.

From Eq. (1.96), one sees that

$$e^{it} = \mathrm{Re}\left(e^{it}\right) + i\,\mathrm{Im}\left(e^{it}\right). \tag{1.98}$$

When considering a sinusoidal function, one has

$$Fe^{i(\omega t+\beta)} = F\cos(\omega t+\beta) + iF\sin(\omega t+\beta). \tag{1.99}$$

Thus,

$$\begin{aligned} F\cos(\omega t+\beta) &= \frac{1}{2}\left[Fe^{i(\omega t+\beta)} + Fe^{-i(\omega t+\beta)}\right], \\ F\sin(\omega t+\beta) &= \frac{1}{2i}\left[Fe^{i(\omega t+\beta)} - Fe^{-i(\omega t+\beta)}\right]. \end{aligned} \tag{1.100}$$

Alternatively,

$$\begin{aligned} F\cos(\omega t+\beta) &= \mathrm{Re}\left[Fe^{i(\omega t+\beta)}\right] = \mathrm{Re}\left(e^{i\omega t}Fe^{i\beta}\right), \\ F\sin(\omega t+\beta) &= \mathrm{Im}\left[Fe^{i(\omega t+\beta)}\right] = \mathrm{Im}\left(e^{i\omega t}Fe^{i\beta}\right). \end{aligned} \tag{1.101}$$

A phasor of a sinusoidal function $F\sin(\omega t+\beta)$ or $F\cos(\omega t+\beta)$ is defined by

$$\overline{F} = Fe^{i\beta}. \tag{1.102}$$

It is complex irrelevant of frequency ω. A phasor $\overline{F}$ corresponds to $F\sin(\omega t+\beta)$ or $F\cos(\omega t+\beta)$. We write a mapping for either

$$\overline{F} \leftrightarrow F\sin(\omega t+\beta), \tag{1.103}$$

or

$$\overline{F} \leftrightarrow F\cos(\omega t+\beta). \tag{1.104}$$

To reconstruct a sinusoidal function *f*(*t*) from its phasor $\overline{F}$, one has to multiply $\overline{F}$ by $e^{i\omega t}$ and get the imaginary part from $\overline{F}e^{i\omega t}$ by

$$F\sin(\omega t+\beta)=\mathrm{Im}\left(\overline{F}e^{i\omega t}\right). \tag{1.105}$$

Similarly, multiplying $\overline{F}$ by $e^{i\omega t}$ and getting the real part from $\overline{F}e^{i\omega t}$ yields

$$F\cos(\omega t+\beta)=\mathrm{Re}\left(\overline{F}e^{i\omega t}\right). \tag{1.106}$$

Either $F\sin(\omega t+\beta)$ or $F\cos(\omega t+\beta)$ is real while $\overline{F}e^{i\omega t}$ is complex. Thus, according to Eq. (1.101), we see that

$$\begin{aligned} F\cos(\omega t+\beta) &\neq \overline{F}e^{i\omega t}, \\ F\sin(\omega t+\beta) &\neq \overline{F}e^{i\omega t}. \end{aligned} \tag{1.107}$$

Based on Euler's formula, we have the correspondence between the derivatives of a sinusoidal function *f*(*t*) and the derivatives of complex exponential function $Fe^{i(\omega t+\beta)}=e^{i\omega t}Fe^{i\beta}=e^{i\omega t}\overline{F}$ in the form

$$\begin{aligned} & f'(t)=\omega F\cos(\omega t+\beta)\leftrightarrow i\omega\overline{F}e^{i\omega t}, \\ & \ldots \\ & f^{(n)}(t)\leftrightarrow(i\omega)^n\,\overline{F}e^{i\omega t}. \end{aligned} \tag{1.108}$$

With the notion of phasor, we ignore the frequency factor $e^{i\omega t}$. The relationships between the derivatives of a sinusoidal function *f*(*t*) and the phasor $\overline{F}$ are given by

$$\begin{aligned} & f'(t)\leftrightarrow i\omega\overline{F}, \\ & \ldots \\ & f^{(n)}(t)\leftrightarrow(i\omega)^n\,\overline{F}. \end{aligned} \tag{1.109}$$

1.7.2 Application of Phasor to Harmonically Stationary Response

By using a phasor, one may transfer a differential vibration equation with sinusoidal driven force to a phasor one,

$$\begin{gathered} mx''+cx'+kx=F\sin(\omega t+\beta) \\ \Downarrow \\ (i\omega)^2 m\overline{X}+i\omega c\overline{X}+k\overline{X}=\overline{F}, \end{gathered} \tag{1.110}$$

where $\overline{X}$ is the phasor of *x*(*t*).

A vibration equation in phasor is algebraic. With the phasor equation in Eq. (1.110), we have the response in phasor in the form

$$\bar{X} = \frac{\bar{F}}{(i\omega)^2 m + i\omega c + k}. \tag{1.111}$$

The response $x(t)$ is given by

$$x(t) = \mathrm{Im}\left(\bar{X}e^{i\omega t}\right). \tag{1.112}$$

If the excitation is $F\cos(\omega t + \beta)$, we have

$$x(t) = \mathrm{Re}\left(\bar{X}e^{i\omega t}\right). \tag{1.113}$$

By using phasor, we see that the solution $x(t)$ is obtained by solving an algebraic phasor equation instead of differential one. Therefore, phasor method is computation saving in solving stationary response.

When rewriting Eq. (1.111), we have

$$\begin{aligned}\bar{X} &= \frac{\bar{F}}{\left(k - \omega^2 m\right) + i\omega c} = \frac{\bar{F}}{m} \frac{1}{\left(\frac{k}{m} - \omega^2\right) + i\frac{\omega c}{m}} \\ &= \frac{\bar{F}}{m} \frac{1}{\left(\omega_n^2 - \omega^2\right) + \omega_n^2 i\frac{\omega c}{\omega_n^2 m}} = \frac{\bar{F}}{k} \frac{1}{\left(1 - \frac{\omega^2}{\omega_n^2}\right) + i\frac{\omega c}{\omega_n^2 m}} = \frac{\bar{F}}{k} \frac{1}{\left(1 - \gamma^2\right) + i2\varsigma\gamma}.\end{aligned} \tag{1.114}$$

In polar coordinates, we have

$$\bar{X} = \left|\bar{X}\right| e^{-i\theta}, \tag{1.115}$$

where

$$\left|\bar{X}\right| = \frac{F}{k} \frac{1}{\sqrt{\left(1-\gamma^2\right)^2 + (2\varsigma\gamma)^2}}, \theta = \arctan\frac{2\varsigma\gamma}{1-\gamma^2}. \tag{1.116}$$

Thus,

$$x(t) = \mathrm{Im}\left(\bar{X}e^{i\omega t}\right) = \frac{F}{k} \frac{1}{\sqrt{\left(1-\gamma^2\right)^2 + (2\varsigma\gamma)^2}} \sin(\omega t + \beta - \theta). \tag{1.117}$$

In the field, the ratio of the response phasor $\bar{X}$ to the driven one $\bar{F}$ is called frequency transfer function. Denote it by $H(\omega)$. Then,

$$H(\omega) = \frac{\bar{X}}{\bar{F}} = \frac{1}{k} \frac{1}{\left(1-\gamma^2\right) + i2\varsigma\gamma} = \frac{1}{k} \frac{1}{\sqrt{\left(1-\gamma^2\right)^2 + (2\varsigma\gamma)^2}} e^{-i\theta}. \tag{1.118}$$

The function $H(\omega)$ describes the intrinsic characteristics of frequency response of a mass-spring system.

1.8 RESPONSE MEASUREMENT BY SCANNING

In engineering, measurement is an essential way to identify a mass-spring system. One way to measure $H(f)$ of a system is scanning, whereby one exerts driven force $F\sin[(2\pi(f_0+n\Delta f)t)$ at each frequency point f_0, $f_0+\Delta f$, $f_0+2\Delta f$, ..., $f_0+n\Delta f$, ... ($n=0, 1, 2, ..., N$) to record the corresponding responses $A\sin[(2\pi(f_0+n\Delta f)t-\theta)$ (Li [12]). Figure 1.21 indicates a measurement diagram.

As long as Δf is small enough, we can encounter the maximum or an appropriate approximation of maximum of $H(f)$ at f_n or sufficiently near f_n [7,12]. In addition, we can measure the bandwidth $B_{0.707}$. Therefore, we may obtain f_n from $H(f_n)$ and ζ based on $\varsigma = \frac{B_{0.707}}{2f_n}$. Thus, after mass m is measured, stiffness k is attained according to $k = m\omega_n^2$. Besides, damping c is obtained based on $c = 2\varsigma\sqrt{mk}$.

1.9 NOTES ON NONLINEAR VIBRATIONS

The radical requirement in the classroom teaching for undergraduates is linear vibrations as discussed above. In practice, however, structures may suffer from nonlinear vibrations. There are two categories in nonlinear vibrations. One is conventionally nonlinear vibrations and the other is fractional ones.

1.9.1 Conventionally Nonlinear Vibrations

By conventionally nonlinear vibrations, we mean that a motion equation is of integer order but nonlinear. A well known one is described by the Duffing's equation given by

$$mx'' + k\left(x + \beta x^2\right) = 0, \tag{1.119}$$

where β is a dimensionless coefficient. When $\beta > 0$, we call k is the stiffness of a hard spring. If $\beta < 0$, k is the stiffness of a soft spring. Since the unit of kx is N but $k\beta x^2$ is not, the restoration force in Duffing's equation is non-Newtonian when $\beta \neq 0$.

Another well-known nonlinear vibration system is characterized by the Van de Pol equation. It is in the form

$$mx'' + c|x'|x' + kx = 0. \tag{1.120}$$

Because the unit of cx' is N but $c|x'|x'$ is non-Newtonian, one has a damping force in the form $c|x'|x'$ in the Van de Pol equation.

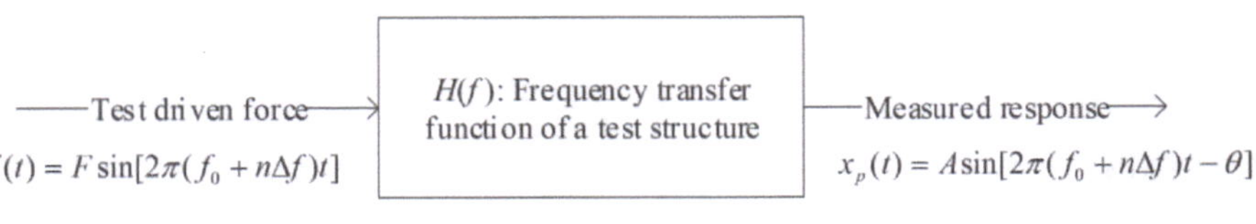

FIGURE 1.21 Diagram of measuring mass-spring system.

1.9.2 Fractional Vibrations

Fractional vibrations attract the interests of researchers, see e.g., Uchaikin [13,14], Rossikhin and Shitikova [15–18], Rossikhin [19], Li [20,21], Duan et al. [22–25]. According to Chapter 7 of this book, a general form of fractional vibrations is expressed by

$$m\frac{d^{\alpha}y(t)}{dt^{\alpha}}+c\frac{d^{\beta}y(t)}{dt^{\beta}}+k\frac{d^{\lambda}y(t)}{dt^{\lambda}}=f(t),\quad 1<\alpha<3,\ 0<\beta<2,\ 0\leq\lambda<1. \tag{1.121}$$

Note that Eq. (1.121) follows superposition. To be precise, when excitation $f_j(t)$ $(j=1, 2,\ldots)$ produces the response $y_j(t)$, the excitation $\Sigma f_j(t)$ yields the response $\Sigma y_j(t)$. An interesting thing about fractional vibrations is that Eq. (1.121) is nonlinear in mechanics. As a matter of fact, the forces of inertia, damping, and restoration in Eq. (1.121) are non-Newtonian when $\alpha\neq 2$, $\beta\neq 1$, and $\lambda\neq 0$. Precisely, the unit of the fractional inertia force $my^{(\alpha)}(t)$ is not Newtonian for $\alpha\neq 2$. Neither are the fractional damping force $cy^{(\beta)}(t)$ if $\beta\neq 1$ and the fractional restoration one $ky^{(\lambda)}(t)$ when $\lambda\neq 0$. Thus, Eq. (1.121) belongs to nonlinear vibrations.

1.10 RECTANGULAR COEFFICIENT

Let f_3 and f_4 be such that

$$A\left(\frac{f_3}{f_n},\varsigma\right)=A\left(\frac{f_4}{f_n},\varsigma\right)=\frac{1}{10}A\left(\frac{f_n}{f_n},\varsigma\right)=0.1A_{\max}. \tag{1.122}$$

Suppose that $f_4>f_3$. We call f_4-f_3 the 0.1 bandwidth of a mass-spring system. Denote it by $B_{0.1}$. Then,

$$B_{0.1}=f_4-f_3. \tag{1.123}$$

Denote the rectangular coefficient by $K_{0.1}$. It is defined by

$$K_{0.1}=\frac{B_{0.1}}{B_{0.707}}. \tag{1.124}$$

Therefore, in addition to the Q value previously discussed, we may use another quantity, say $K_{0.1}$, to measure the sharpness of a resonance curve. Figure 1.22 indicates an ideal rectangular resonance curve.

Different from Q, $K_{0.1}$ takes a rectangle as a reference to characterize the sharpness as can be seen from Figure 1.23.

Obviously, if a resonance curve is rectangular, vibrations disappear completely outside the scope of the resonance bandwidth. Unfortunately, for a simple mass-spring system, that is impossible because it is not difficult to show that

$$K_{0.1}\approx 10. \tag{1.125}$$

The above implies that

$$B_{0.1}\approx 10B_{0.707}. \tag{1.126}$$

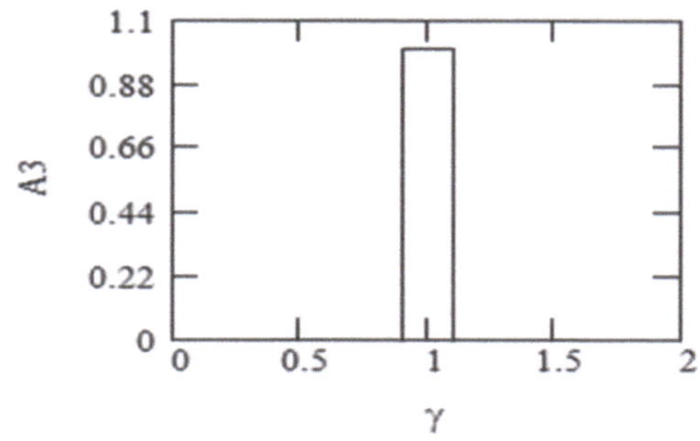

FIGURE 1.22 Plot of an ideal rectangular resonance curve.

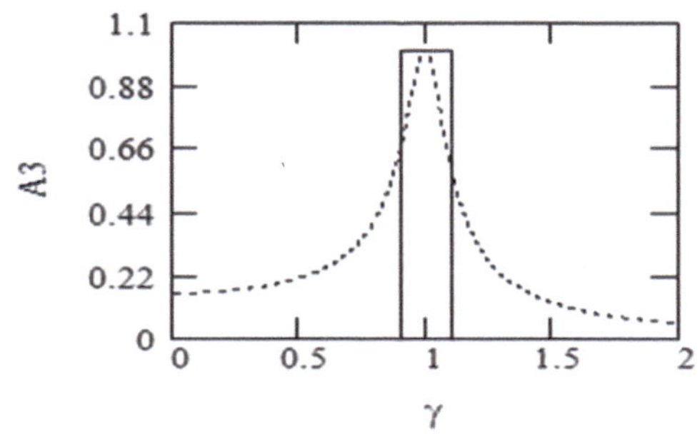

FIGURE 1.23 Illustrations of resonance curves. Rectangle (solid). Conventional curve (dot).

In general, a vibration system with multiple degrees-of-freedom may have a smaller value of $K_{0.1}$.

Example 1.2

Suppose there are two simple mass-spring systems. Their frequency transfer functions are, respectively, given by

$$H_1(\omega)=\frac{1}{m_1}\frac{1}{\left(\omega_{n1}^2-\omega^2\right)+i2\varsigma_1\omega_{n1}\omega},\quad H_2(\omega)=\frac{1}{m_2}\frac{1}{\left(\omega_{n2}^2-\omega^2\right)+i2\varsigma_2\omega_{n2}\omega},\tag{1.127}$$

where $\omega_{n1}=1$, $\omega_{n2}=1.28$, $\zeta_1=0.13$, $\zeta_2=0.08$, $m_1=0.85$, and $m_2=0.2$. Consider the frequency transfer function of a system in the form

$$H(\omega)=H_1(\omega)H_2(\omega)$$
$$=\frac{1}{m_1m_2}\left[\frac{1}{\left(\omega_{n1}^2-\omega^2\right)+i2\varsigma_1\omega_{n1}\omega}\right]\left[\frac{1}{\left(\omega_{n2}^2-\omega^2\right)+i2\varsigma_2\omega_{n2}\omega}\right].\tag{1.128}$$

Now we estimate its $K_{0.1}$. The resonance curve of $|H(\omega)|$ is shown in Figure 1.24, from which we estimate that $K_{0.1}\approx 1.29$. Finally, we note that there are two resonance

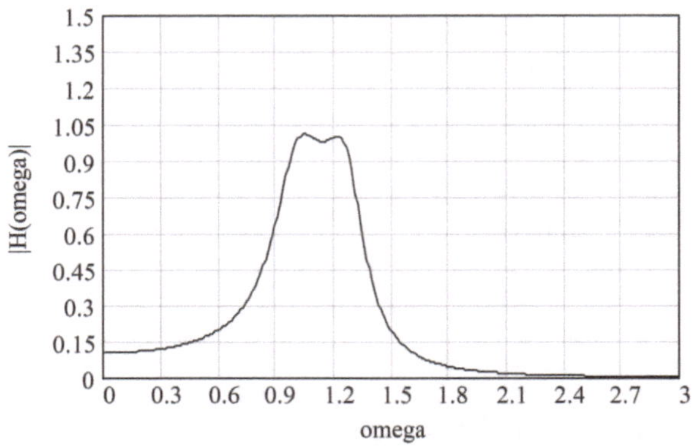

FIGURE 1.24 Illustration of $|H(\omega)| = |H_1(\omega)H_2(\omega)|$ when $\omega_{n1}=1$, $\omega_{n2}=1.28$, $\zeta_1=0.13$, $\zeta_2=0.08$, $m_1=0.85$, and $m_2=0.2$.

frequencies in $H(\omega)$, implying $H(\omega)$ is a frequency transfer function of a system with two degrees-of-freedom, referring Chapter 5 for the meaning of multiple degrees-of-freedom.

1.11 EXERCISES

1.1. Denote the frequency transfer function of a simple mass-spring system by $H(f)$ as that in Eq. (1.118). Let f_3 and f_4 be such that $|H(f_3)| = 0.1|H(f)|_{\max}$ and $|H(f_4)| = 0.1|H(f)|_{\max}$, where $f_3 < f_4$. Let $B_{0.1} = f_4 - f_3$. We call $B_{0.1}$ the 0.1 bandwidth of the system. Prove that $B_{0.1} \approx 10\ B_{0.707}$.

1.2. Following Eq. (1.118), we define normalized frequency transfer function by $H_1(\omega) = \dfrac{H(\omega)}{H(\omega_n)}$. Prove that $|H_1(\omega)| \le 1$ if $\zeta > \dfrac{\sqrt{2}}{2}$.

1.3. Prove that $|H_1(\omega)| \le 1$ if $Q < \dfrac{\sqrt{2}}{2}$.

1.4. Suppose that mass of a mass-spring system is frequency varying. It is denoted by $m(\omega)$. Prove that the inertia force of this system is given by $m(\omega)\dfrac{d^2x}{dt^2}$, where x is displacement.

1.5. If a mass-spring system is with frequency varying mass $m(\omega)$ but with viscous damping c, please express its damping ratio ζ. (Hint: Use Eq. (1.21))

1.6. For a mass-spring system with frequency varying mass $m(\omega)$ but with viscous damping c, please write its expression of damping free natural angular frequency ω_n. (Hint: Use Eq. (1.7))

1.7. Let ω_d be damped natural angular frequency of a mass-spring system with frequency varying mass $m(\omega)$ but with viscous damping c. Give its representation for $0 < \zeta < 1$. (Hint: Use Eq. (1.29))

1.8. Denote by γ the frequency ratio of a mass-spring system with frequency varying mass $m(\omega)$. Give its expression. (Hint: Use Eq. (1.57))

1.9. Let Q be Q factor of a mass-spring system with frequency varying mass $m(\omega)$ but with viscous damping c. Write its expression. (Hint: Use Eq. (1.79))

1.10. Let Δ be logarithmic decrement of response to a mass-spring system with frequency varying mass $m(\omega)$ but with viscous damping c. Give its representation. (Hint: Use Eq. (1.48))

1.11. Based on Eq. (1.118), please derive its generalized expression of frequency transfer function of a mass-spring system with frequency varying mass $m(\omega)$ but with viscous damping c.

1.12. Denote by $x_p(t)$ the stationary solution to a mass-spring system with frequency varying mass $m(\omega)$ but with viscous damping c. When the excitation is $F\sin(\omega t+\beta)$, give its expression for $0<\zeta<1$. (Hint: Use Eq. (1.60))

1.13. Let $m(\omega)$ be frequency varying mass of a mass-spring system with damping free. The initial conditions of system are expressed by $x(0)$ and $x'(0)$. Write the free response of this system. (Hint: Use Eq. (1.33))

REFERENCES

1. J. P. Den Hartog, *Mechanical Vibrations*, McGraw-Hill, New York, 1956.
2. K. Nakagawa and M. Ringo, *Engineering Vibrations*, Shanghai Science and Technology Publishing House, Shanghai, China, 1981. In Chinese. Translated from Japanese by S. R. Xia.
3. O. M. Palley, Г. B. Bahizov, and E. Я. Voroneysk, *Handbook of Ship Structural Mechanics*, National Defense Industry Publishing House, Beijing, China, 2002. In Chinese. Translated from Russian by B. H. Xu, X. Xu, and M. Q. Xu.
4. C. A. Peirce, W. B. Carver, and C. E. O'rourke, *Handbook of Formulars and Tables for Engineers*, 3rd Ed., McGraw-Hill, New York, 1929.
5. C. M. Harris, *Shock and Vibration Handbook*, 5th Ed., McGraw-Hill, New York, 2002.
6. S. R. Swanson, Ed., *Handbook of Fatigue Testing*, ASTM Special Technical Publication 566, Philadelphia, PA, 1974.
7. M. Li, Experimental stability analysis of test system for doing fatigue test under random loading, *Journal of Testing and Evaluation*, 34(4), 2006, 364–367.
8. C. P. Steinmetz, *Theory and Calculation of Alternating Current Phenomena*, McGraw-Hill, New York, 1897.
9. C. P. Steinmetz, *Engineering Mathematics: A Series of Lectures Delivered at Union College*, McGraw-Hill, New York, 1917.
10. S. P. Timoshenko, *Vibration Problems in Engineering*, 2nd Ed., Fifth Printing, D. Van Nostrand Company, Inc., New York, 1955.
11. S. P. Timoshenko, *Mechanical Vibrations*, D. Van Nostrand Company, Inc., New York, 1955.
12. M. Li, An iteration method to adjusting random loading for a laboratory fatigue test, *International Journal of Fatigue*, 27(7), 2005, 783–789.
13. V. V. Uchaikin, *Fractional Derivatives for Physicists and Engineers*, vol. I, Springer, Berlin, 2013.
14. V. V. Uchaikin, Relaxation processes and fractional differential equations, *International Journal of Theoretical Physics* 42(1), 2003, 121–34.

15. Yu. A. Rossikhin and M. V. Shitikova, Analysis of the viscoelastic rod dynamics via models involving fractional derivatives or operators of two different orders, *The Shock and Vibration Digest*, 36(1), 2004, 3–26.
16. Yu. A. Rossikhin and M. V. Shitikova, Application of fractional calculus for dynamic problems of solid mechanics: Novel trends and recent results, *Applied Mechanics Reviews*, 63(1), 2010, 010801.
17. Yu. A. Rossikhin and M. V. Shitikova, Application of fractional operators to the analysis of damped vibrations of viscoelastic single-mass systems, *Journal of Sound and Vibration*, 199(4), 1997, 567–586.
18. Yu. A. Rossikhin and M. V. Shitikova, Classical beams and plates in a fractional derivative medium, Impact response, in: H. Altenbach and A. Öchsner (eds.), *Encyclopedia of Continuum Mechanics*, vol. 1, Springer, Berlin, 2020, pp. 294–305.
19. Yu. A. Rossikhin, Reflections on two parallel ways in progress of fractional calculus in mechanics of solids, *Applied Mechanics Reviews*, 63(1), 2010, 010701.
20. M. Li, *Theory of Fractional Engineering Vibrations*, Walter de Gruyter, Berlin/Boston, MA, 2021.
21. M. Li, Three classes of fractional oscillators, *Symmetry-Basel*, 10(2), 2018, 91 pp.
22. J.-S. Duan, Y.-J. Lan, and M. Li, A comparative study of responses of fractional oscillator to sinusoidal excitation in the Weyl and Caputo senses, *Fractal and Fractional*, 6(12), 2022, 692.
23. J.-S. Duan, M. Li, Y. Wang, and Y.-L. An, Approximate solution of fractional differential equation by quadratic splines, *Fractal and Fractional*, 6(7), 2022, 369.
24. J.-S. Duan, L. Jing, and M. Li, The mixed boundary value problems and Chebyshev collocation method for Caputo-type fractional ordinary differential equations, *Fractal and Fractional*, 6(3), 2022, 148.
25. J.-S. Duan, D.-C. Hu, and M. Li, Comparison of two different analytical forms of response for fractional oscillation equation, *Fractal and Fractional*, 5(4), 2021, 188.

CHAPTER 2

Vibrations Excited by Periodic Forces

We discuss responses to a mass-spring system under the excitation of deterministic forces. Deterministically driven forces are classified into two classes. One contains periodic forces, and the other contains non-periodic ones. We address the former in this chapter and the latter in Chapter 4.

When a signal $f(t)=f(t+T_1)$, one says that it is a periodic signal with the period T_1. Figure 2.1 indicates the response to a mass-spring system under a periodic excitation force.

For facilitating discussion, preliminaries about the Fourier series are first briefed. After that, we explain vibration responses.

2.1 FOURIER SERIES

A periodic function $f(t)=f(t+T_1)$ can be expressed by a trigonometric series in the form

$$f(t)=\frac{a_0}{2}+\sum_{n=1}^{\infty}(a_n\cos n\omega_1 t+b_n\sin n\omega_1 t), \tag{2.1}$$

where $\omega_1=2\pi/T_1$ is the base frequency of $f(t)$, a_n, b_n, and $a_0/2$ are all called the Fourier coefficients [1–3]. They are given by

$$a_0=\frac{2}{T_1}\int_{-T_1/2}^{T_1/2} f(t)dt, \tag{2.2}$$

$$a_n=\frac{2}{T_1}\int_{-T_1/2}^{T_1/2} f(t)\cos n\omega_1 t dt, \tag{2.3}$$

DOI: 10.1201/9781003460947-3

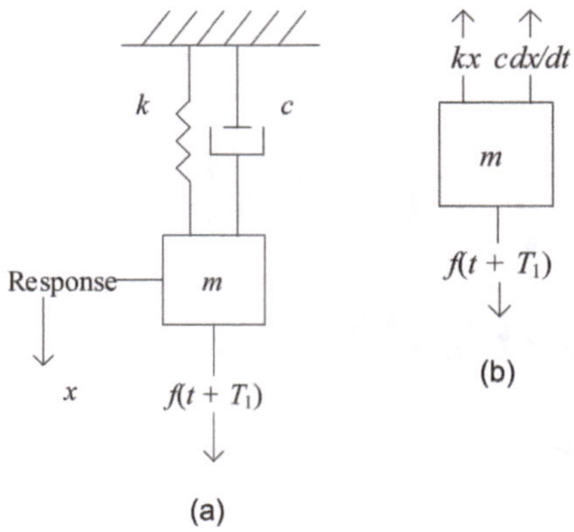

FIGURE 2.1 Mass-spring system under the excitation of periodic force. (a) Mass-spring system with viscous damping. (b) Force diagram.

$$b_n = \frac{2}{T_1} \int_{-T_1/2}^{T_1/2} f(t) \sin n\omega_1 t dt. \tag{2.4}$$

The quantity $a_0/2$ is the average of $f(t)$. In mechanics, it represents a static loading. One may also take it as a sinusoidal function with 0 Hz frequency. The coefficients a_n and b_n are the amplitude of $\cos n\omega_1 t$ and $\sin n\omega_1 t$, respectively.

Fourier used the expression (2.1), see Fourier [4, p. 204], [5, p. 204], [6, p. 162]. He further utilized its equivalent representations. By using the sum of cosine functions [4–6], Eq. (2.1) can be expressed by

$$f(t) = \frac{c_0}{2} + \sum_{n=1}^{\infty} c_n \cos(n\omega_1 t + \varphi_n), \tag{2.5}$$

where $c_0 = a_0$, $\varphi_0 = 0$ [1–3],

$$c_n = \sqrt{a_n^2 + b_n^2}, \tag{2.6}$$

$$\varphi_n = -\tan^{-1} \frac{b_n}{a_n}, \tag{2.7}$$

$$\begin{aligned} a_n &= c_n \cos \varphi_n, \\ b_n &= -c_n \sin \varphi_n. \end{aligned} \tag{2.8}$$

Alternatively [4–6], he used sinusoidal functions to express (2.1). That is [1–3],

$$f(t)=\frac{d_0}{2}+\sum_{n=1}^{\infty}d_n\sin(n\omega_1 t+\theta_n), \tag{2.9}$$

where $d_0=a_0$, $\theta_0=0$,

$$d_n=\sqrt{a_n^2+b_n^2}, \tag{2.10}$$

$$\theta_n=\tan^{-1}\frac{a_n}{b_n}, \tag{2.11}$$

$$\begin{aligned} a_n &= d_n\sin\theta_n, \\ b_n &= d_n\cos\theta_n. \end{aligned} \tag{2.12}$$

Either c_n or d_n is an amplitude of a sinusoidal wave with frequency $n\omega_1$. On the other side, φ_n or θ_n represents an initial phase of a sinusoidal function with frequency $n\omega_1$. Note that they appear in form in terms of the argument n. However, in nature, they are functions of frequency in the discrete form of $n\omega_1$. Thus, we may write

$$c_n=c(n\omega_1),\ d_n=d(n\omega_1),\ \varphi_n=\varphi(n\omega_1),\ \theta_n=\theta(n\omega_1). \tag{2.13}$$

We call $c(n\omega_1)$ or $d(n\omega_1)$ amplitude-frequency line spectrum function of $f(t)$. By line spectrum, we mean that the frequency argument is discrete. An amplitude-frequency line spectrum function is usually called amplitude-frequency spectrum function or simply spectrum function or spectrum for short. On the other hand, $\varphi(n\omega_1)$ or $\theta(n\omega_1)$ is called phase-frequency line spectrum function or phase spectrum for short. Figure 2.2 shows illustrations of $c(n\omega_1)$ and $\varphi(n\omega_1)$.

Using Euler's formula, one writes Eq. (2.5) by

$$f(t)=\frac{c_0}{2}+\sum_{n=1}^{\infty}\frac{c_n}{2}\left[e^{i(n\omega_1 t+\varphi_n)}+e^{-i(n\omega_1 t+\varphi_n)}\right]. \tag{2.14}$$

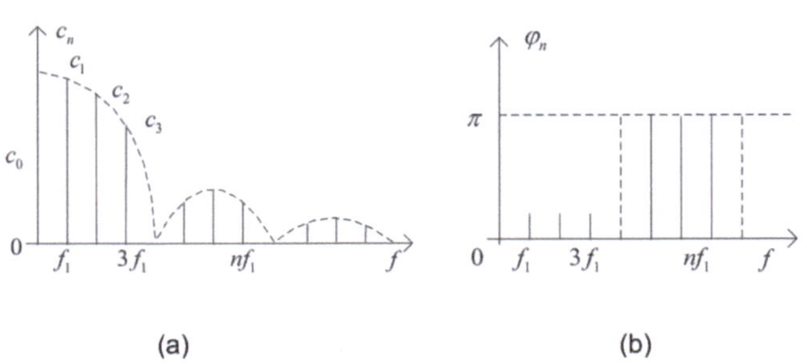

FIGURE 2.2 Spectrum illustrations. (a) Amplitude-frequency spectrum $c(n\omega_1)$. (b) Phase-frequency spectrum $\varphi(n\omega_1)$.

Because

$$\varphi_n = -\varphi_{-n}, \tag{2.15}$$

we have

$$\sum_{n=1}^{\infty} \frac{c_n}{2} e^{-i(n\omega_1 t+\varphi_n)} = \sum_{n=-\infty}^{-1} \frac{c_n}{2} e^{i(n\omega_1 t+\varphi_n)}. \tag{2.16}$$

With the above, Eq. (2.14) may be written by

$$f(t) = \sum_{n=-\infty}^{\infty} \frac{c_n}{2} e^{i(n\omega_1 t+\varphi_n)} = \sum_{n=-\infty}^{\infty} \frac{c_n e^{i\varphi_n}}{2} e^{in\omega_1 t}. \tag{2.17}$$

Denote by

$$F(n\omega_1) = \frac{c_n e^{i\varphi_n}}{2}. \tag{2.18}$$

Then, Eq. (2.17) is expressed by

$$f(t) = \sum_{n=-\infty}^{\infty} F(n\omega_1) e^{in\omega_1 t}. \tag{2.19}$$

One calls $F(n\omega_1)$ spectrum function. It is complex. Its amplitude is

$$|F(n\omega_1)| = \frac{c_n}{2}. \tag{2.20}$$

Its phase is φ_n.

Amplitude spectrum $|F(n\omega_1)|$ does not equal to the true amplitude of harmonic functions but rather one-half. Figure 2.3 shows illustrations of $|F(n\omega_1)|$ and $\varphi(n\omega_1)$.

Example 2.1: Rectangular Pulse Series

Denote by $u(t)$ the unit step function. Let $f_1(t)$ be a single pulse with the pulse width τ and pulse height E. Then,

$$f_1(t) = E\left[u\left(t+\frac{\tau}{2}\right) - u\left(t-\frac{\tau}{2}\right)\right] = \begin{cases} E, -\frac{\tau}{2} < t < \frac{\tau}{2}, \\ 0. \end{cases} \tag{2.21}$$

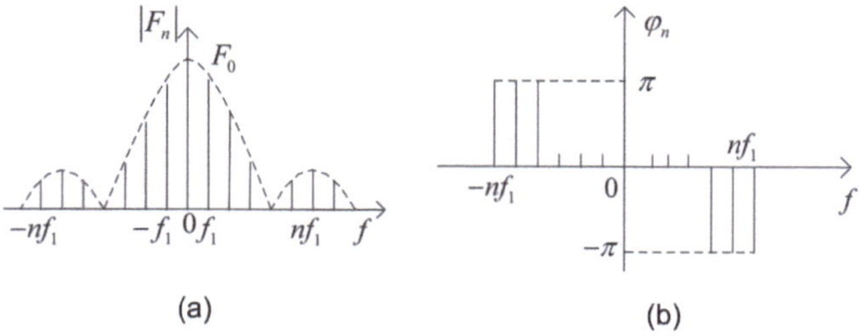

FIGURE 2.3 Spectrum illustrations. (a) Amplitude-frequency spectrum $|F(n\omega_1)|$. (b) Phase-frequency spectrum $\varphi(n\omega_1)$.

We now extend $f_1(t)$ to be a periodic function with period T_1. Then,

$$f(t) = f_1(t + nT_1), \quad n = \ldots, -2, -1, 0, 1, 2, \ldots \tag{2.22}$$

Figure 2.4 indicates its picture.

Extending it with the Fourier series yields

$$f(t) = \frac{E\tau}{T_1} + \frac{2E\tau}{T_1} \sum_{n=1}^{\infty} \mathrm{Sa}\left(\frac{n\pi\tau}{T_1}\right) \cos(n\omega_1 t), \tag{2.23}$$

where $\mathrm{Sa}(x) = \frac{\sin x}{x}$. Its spectrum is given by

$$\begin{aligned} c_0 &= \frac{E\tau}{T_1}, \\ c_n &= \frac{2E\tau}{T_1} \mathrm{Sa}\left(\frac{n\pi\tau}{T_1}\right). \end{aligned} \tag{2.24}$$

Figure 2.5 illustrates its amplitude spectrum and phase.

There are two properties of rectangular pulse series.

1. The first zero crossing of $c(n\omega_1)$ takes place at $f = \frac{1}{\tau}$ Hz. The majority of signal power is contained in $\left[0, \frac{1}{\tau}\right]$. Thus, the frequency bandwidth of $f(t)$ is taken as $f_B = \frac{1}{\tau}$ in the sense of the first zero crossing.
2. The frequency bandwidth $1/\tau$ is inversely proportional to the pulse width τ. Thus, if the pulse width τ decreases, the first zero crossing moves up in frequency and f_B increases. Contrarily, the first zero crossing moves toward the origin in frequency and f_B decreases when the pulse width τ increases.

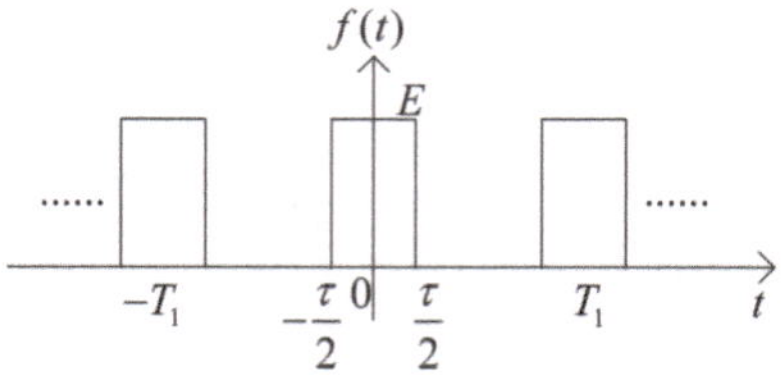

FIGURE 2.4 Rectangular pulse series.

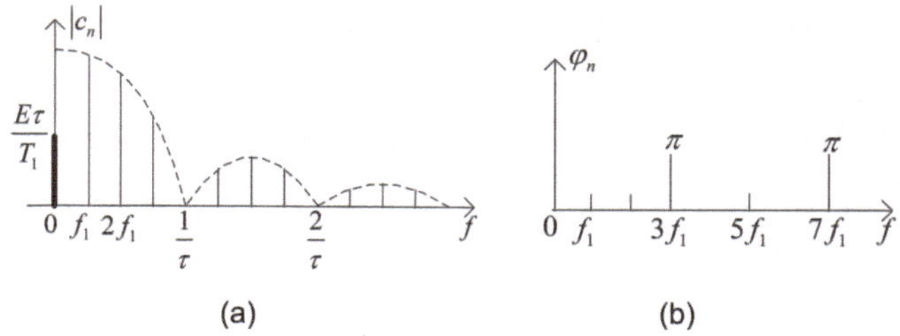

FIGURE 2.5 Spectrum illustrations of rectangular pulse series. (a) Amplitude-frequency spectrum $|c(n\omega_1)|$. (b) Phase-frequency spectrum $\varphi(n\omega_1)$.

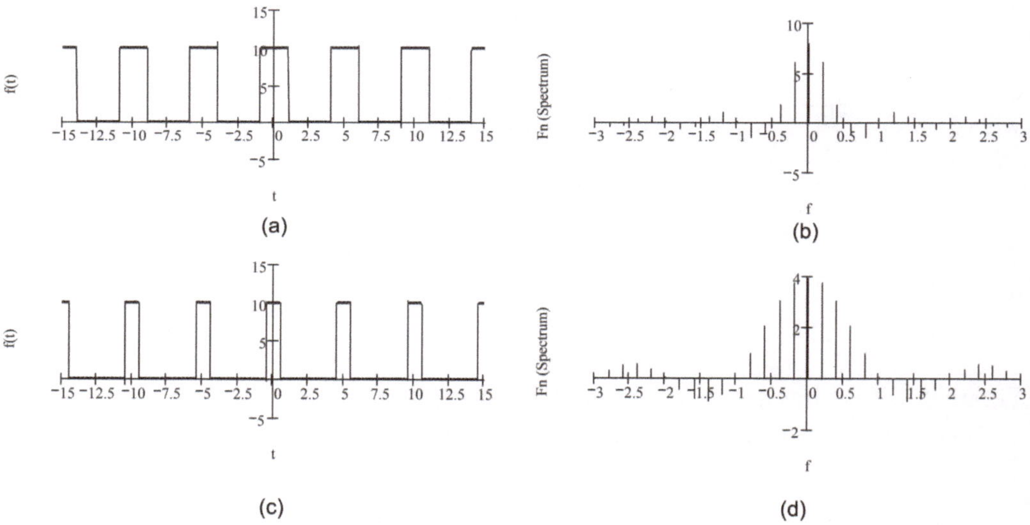

FIGURE 2.6 Observations of first zero crossings versus pulse width. (a) Rectangular pulse series with $\tau = 2$. (b) Spectrum with first zero crossing being 0.5. (c) Rectangular pulse series with $\tau = 1$. (d) Spectrum with first zero crossing $\tau = 1$.

Example 2.2

Let $E = 10$ and $T_1 = 5$. Observe the first crossings of $F(n\omega_1)$ of rectangular pulse series for the pulse width being $\tau = 2$ and 1. Figure 2.6 gives the demonstrations for the above property 2).

With the help of Example 2.2, we address two general properties of signals.

P1. The shorter the duration of a time signal, the wider the frequency bandwidth, and vice versa. Figure 2.6 is a case about P1.

P2. For a signal $f(t)=\Sigma F(n\omega_1)\exp(in\omega_1 t)$, components with low frequencies are located at the areas where $f(t)$ slowly varies in time, e.g., flat parts of $f(t)$. Conversely, components with high frequencies are concentrated in the regions where $f(t)$ varies violently in time, e.g., regions near jump edges of a rectangular pulse series.

Example 2.3

With rectangular pulse series, demonstrate P2 with the plots in Figure 2.7. Denote by $S(t, N)$ the finite summation of a rectangular pulse series. Then,

$$S(t,N)=\frac{E\tau}{T_1}+\frac{2E\tau}{T_1}\sum_{n=1}^{N}\mathrm{Sa}\left(\frac{n\pi\tau}{T_1}\right)\cos(n\omega_1 t). \tag{2.25}$$

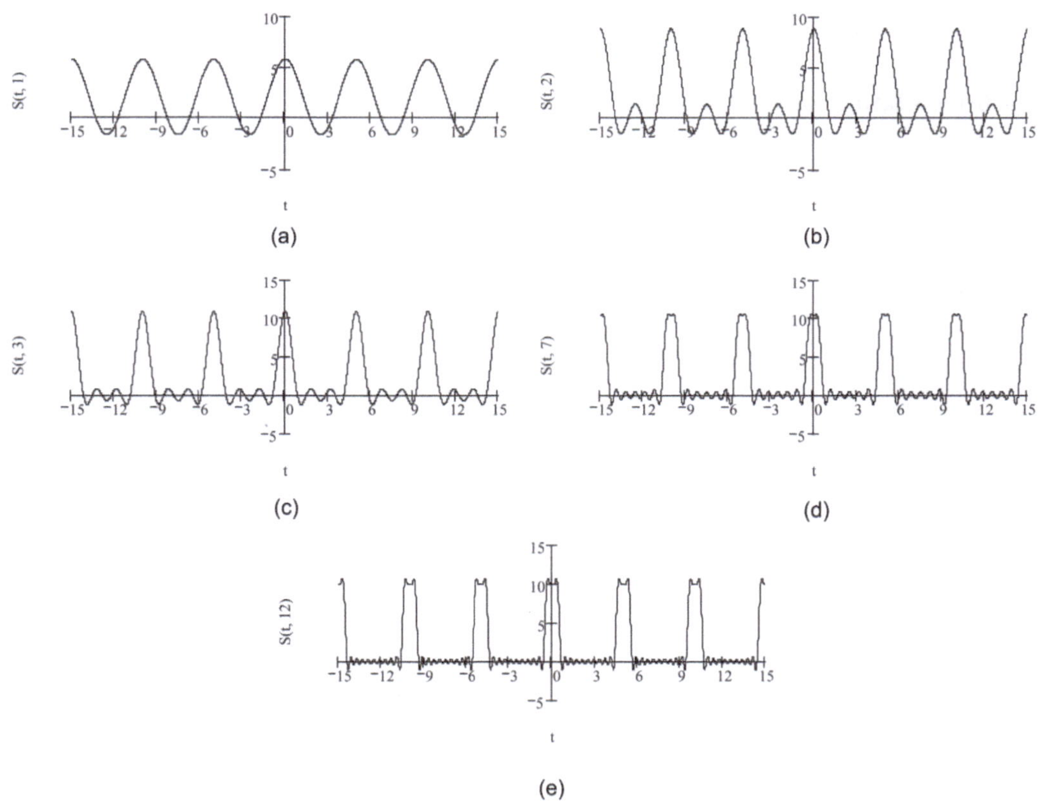

FIGURE 2.7 Demonstrations of finite summation of the Fourier series of a rectangular pulse series. (a) Pulse series with one low-frequency component $S(t, 1)$. (b) Rectangular pulse series with the summation of two low-frequency components $S(t, 2)$. (c) Rectangular pulse series with the summation of three low-frequency components $S(t, 3)$. (d) Rectangular pulse series with the summation of seven frequency components $S(t, 7)$. (e) Rectangular pulse series with the summation of 12 frequency components $S(t, 12)$.

Plots in Figure 2.7 indicate that low-frequency components of a signal contribute to the flat parts where the signal varies slowly in time while high-frequency components contribute to the sharp parts where the signal varies quickly in time, such as jump edges.

We now discuss other two periodic signals that are commonly used in vibration engineering. Periodic sawing pulse series *f(t)* is shown in Figure 2.8a, where *E* is the pulse height and T_1 the period. Its Fourier series is given by

$$f(t)=\frac{E}{\pi}\sum_{n=1}^{\infty}(-1)^{n+1}\frac{1}{n}\sin(n\omega_1 t). \tag{2.26}$$

Its spectrum $c(n\omega_1)$ is

$$\begin{aligned} &c(0)=0, \\ &c(n\omega_1)=(-1)^{n+1}\frac{E}{\pi n}, \quad (n=1,2,\ldots). \end{aligned} \tag{2.27}$$

Its finite summation is expressed by

$$S_1(t,N)=\frac{E}{\pi}\sum_{n=1}^{N}(-1)^{n+1}\frac{1}{n}\sin(n\omega_1 t). \tag{2.28}$$

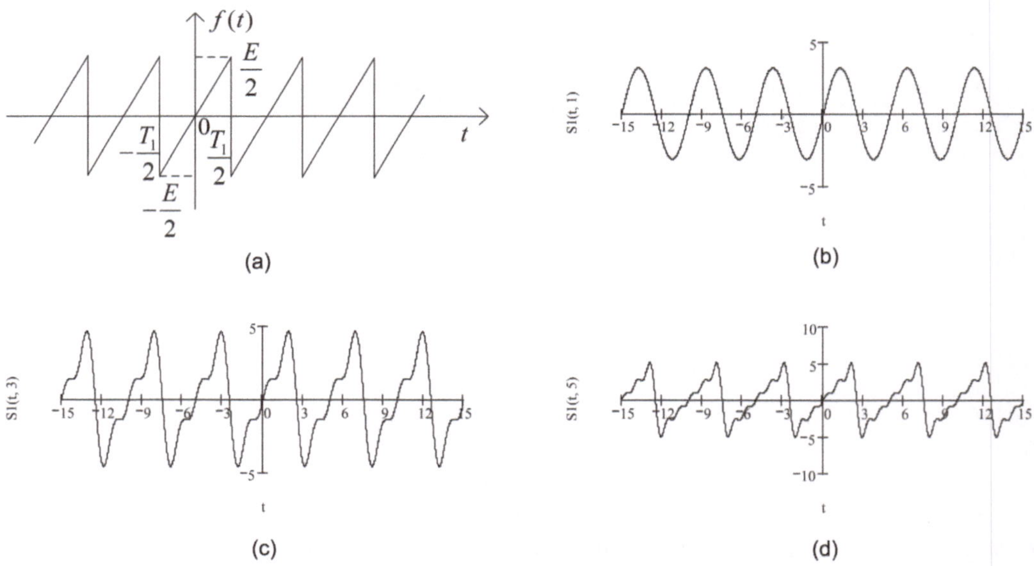

FIGURE 2.8 Periodic sawing pulse series and its finite summation. (a) Time wave. (b) Pulse series with one low frequency component *S*1(*t*, 1). (c) Pulse series with the summation of three low-frequency components *S*1(*t*, 3). (d) Pulse series with the summation of five low-frequency components *S*1(*t*, 5).

Figure 2.8 shows the illustrations of $S_1(t, N)$ for $N=1, 3, 5$.

A series in Figure 2.9a is called periodic triangular pulse series. Its Fourier series is given by

$$f(t)=\frac{E}{2}+\frac{4E}{\pi^2}\sum_{n=1}^{\infty}\frac{1}{n^2}\sin^2\left(\frac{n\pi}{2}\right)\cos(n\omega_1 t). \tag{2.29}$$

Its spectrum $c(n\omega_1)$ is

$$c(0)=\frac{E}{2},\quad c(n\omega_1)=\frac{4E}{\pi^2 n^2}\sin^2\left(\frac{n\pi}{2}\right). \tag{2.30}$$

Its finite summation is expressed by

$$S_2(t,N)=\frac{E}{2}+\frac{4E}{\pi^2}\sum_{n=1}^{N}\frac{1}{n^2}\sin^2\left(\frac{n\pi}{2}\right)\cos(n\omega_1 t). \tag{2.31}$$

Let $E=5$ and $T_1=5$. Then, its spectrum is shown in Figure 2.9b. Figures 2.9c and d show the curves of $S_2(t, 1)$ and $S_2(t, 3)$, respectively.

2.2 PARSEVAL'S RULE

Denote by $\overline{f}^2$ the average power of $f(t)$ in a period of time. It is given by

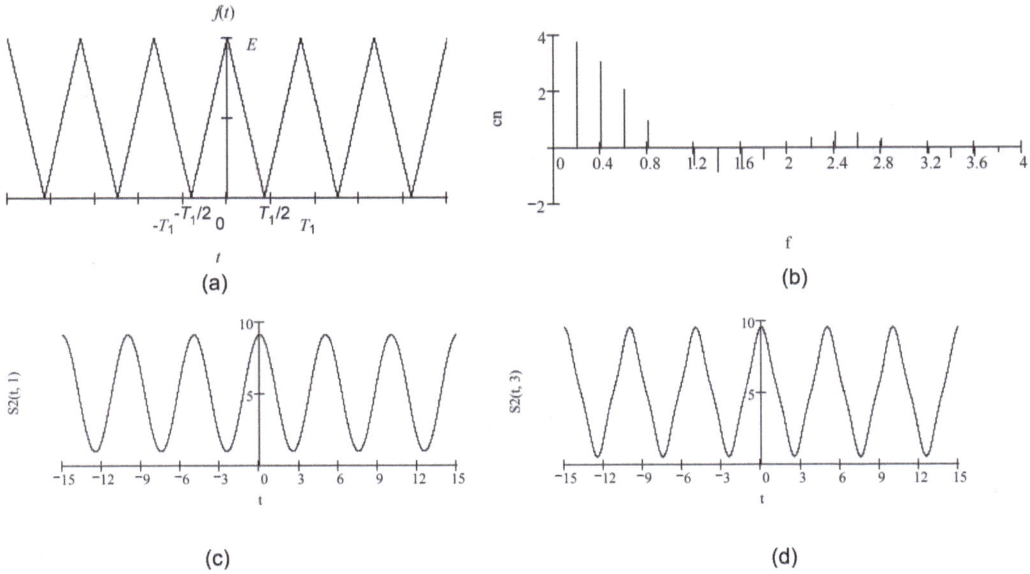

FIGURE 2.9 Periodic triangular pulse series. (a) Time wave. (b) Spectrum. (c) Pulse series with low one frequency component $S2(t, 1)$. (d) Pulse series with the summation of three low-frequency components $S2(t, 3)$.

$$\overline{f}^2 = \frac{1}{T_1}\int_{t_0}^{t_0+T_1} f^2(t)dt. \tag{2.32}$$

Then, it equals to the summation of average powers of all harmonic waves expressed by

$$\overline{f}^2 = \left(\frac{a_0}{2}\right)^2 + \frac{1}{2}\sum_{n=1}^{\infty}\left(a_n^2 + b_n^2\right) = c_0^2 + \frac{1}{2}\sum_{n=1}^{\infty} c_n^2 = \sum_{n=1}^{\infty}|F_n|^2. \tag{2.33}$$

This result is called the Parseval' rule [2,3]. In fact, it reflects the energy conservation law in mathematics. Physically, the energy represented in the time domain as that in Eq. (2.32) must equal to that in the frequency domain as described by Eq. (2.33).

2.3 RESPONSES

When a periodic force excites a vibration system, see Figure 2.10, we may ignore the static loading c_0. We now consider the nth harmonic response $x_n(t)$ that is excited by nth harmonic driven signal $c_n\cos(n\omega_1 t + \varphi_n)$, see Figure 2.11.

Let $\gamma_n = \frac{n\omega_1}{\omega_n}$. Then,

$$x_n(t) = \frac{1}{k}\frac{1}{1-\gamma_n^2 + i2\varsigma\gamma_n} c_n\cos(n\omega_1 t + \varphi_n). \tag{2.34}$$

Owing to superposition, the total response is therefore given by

$$x(t) = \sum_{n=1}^{\infty} x_n(t) = \frac{1}{k}\sum_{n=1}^{\infty}\frac{c_n\cos(n\omega_1 t + \varphi_n)}{1-\gamma_n^2 + i2\varsigma\gamma_n}. \tag{2.35}$$

Note that there is a frequency bandwidth of a vibration system. Thus, if the frequencies of a harmonic force are within the bandwidth of a vibration system, that is,

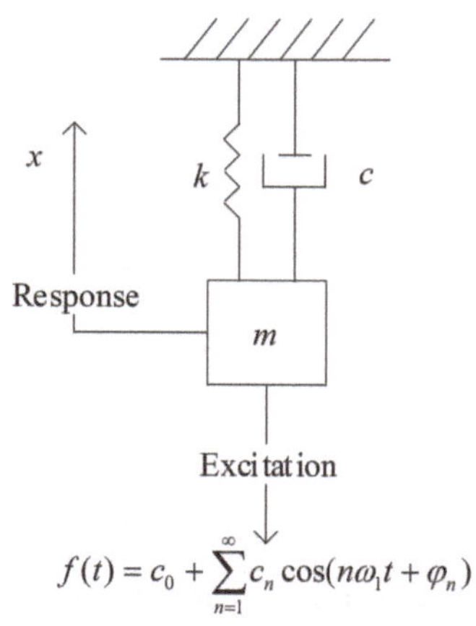

FIGURE 2.10 A periodic signal excites a vibration system.

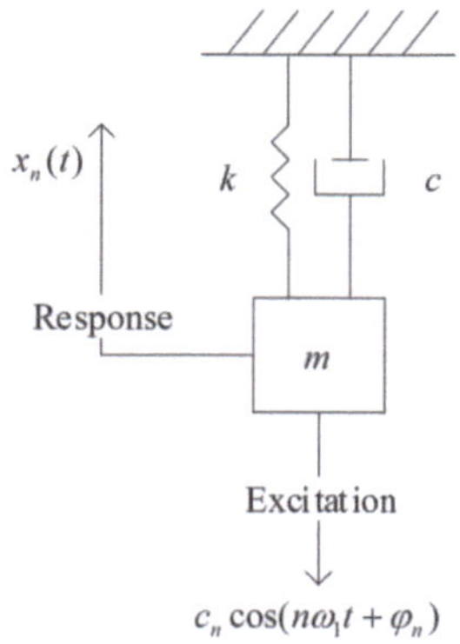

FIGURE 2.11 The *n*th harmonic signal excites a vibration system.

$$\frac{m\omega_1}{2\pi}\in\left(f_n-\mathrm{B}_{0.707},f_n+\mathrm{B}_{0.707}\right), \tag{2.36}$$

they can pass through the vibration system. All other harmonic waves in the driven signal $f(t)$ are suppressed by the vibration system.

Denote by

$$\frac{n_1\omega_1}{2\pi}=f_n-\mathrm{B}_{0.707},\quad \frac{n_2\omega_1}{2\pi}=f_n+\mathrm{B}_{0.707}. \tag{2.37}$$

Then, only those for $f\in\left(\frac{n_1\omega_1}{2\pi},\frac{n_2\omega_1}{2\pi}\right)$ are considered in a driven signal. Denote by $f_{0.707}(t)$ a signal that is made up of the components with $f\in\left(\frac{n_1\omega_1}{2\pi},\frac{n_2\omega_1}{2\pi}\right)$. Then,

$$f_{0.707}(t)=\sum_{n=n_1}^{n=n_2}c_n\cos\left(n\omega_1 t+\varphi_n\right). \tag{2.38}$$

The point I would like to stress on is that $f_{0.707}(t)$ can arrive at the output of a vibration system and contribute to vibrating a structure. Others with $f\notin\left(\frac{n_1\omega_1}{2\pi},\frac{n_2\omega_1}{2\pi}\right)$ cannot from a view of engineering.

From a view of energy, we say that only energy

$$\overline{f_{0.707}^2}=\frac{1}{T_1}\int_{t_0}^{t_0+T_1}f_{0.707}^2(t)dt \tag{2.39}$$

is active in a vibration system. Note that $\overline{f_{0.707}^2}$ is just a part of energy of driven force $f(t)$. Denote the rest energy is denoted by ΔE

$$\Delta E=\overline{f^2}-\overline{f_{0.707}^2}=\left(c_0^2+\frac{1}{2}\sum_{n=1}^{\infty}c_n^2\right)-\frac{1}{2}\sum_{n=n_1}^{n=n_2}c_n^2. \tag{2.40}$$

The point is that $\overline{f}^{2}_{0.707}$ contributes to vibrations but ΔE does not. It is consumed by a structure in other mechanical forms, such as low- and high-cycle fatigue.

2.4 DISCUSSIONS

In mathematics, a periodic function $f(t)$ has harmonic frequencies $n\omega_1$ ($n=1, 2,\ldots$) as can be seen from Eq. (2.1). In practice, one just needs to consider the finite summation $S(t, N)$. As a matter of fact, $N \to \infty$ is physically not realizable either in numeric computations or laboratory testing. Furthermore, what we should consider about harmonic frequencies of $f(t)$ is those within $B_{0.707}$ of a vibration system. Quantitatively, when simply considering the base frequency of $f(t)$, that is, ω_1, we have grasped the principal part of $f(t)$ in a laboratory vibration test. Test machine producing a harmonic excitation is obviously cheap.

2.5 EXERCISES

2.1. Consider the Fourier series of a periodic function $f(t)$ with the period T_1 in the form

$$f(t)=\frac{c_0}{2}+\sum_{n=1}^{\infty} c_n \cos(n\omega_1 t+\varphi_n),$$

where $\omega_1=\frac{2\pi}{T_1}$, $\varphi_0=0$, $c_0=\frac{2}{T_1}\int_{-T_1/2}^{T_1/2} f(t)dt$, $\varphi_n=-\tan^{-1}\frac{b_n}{a_n}$, $c_n=\sqrt{a_n^2+b_n^2}$,

Prove that the Fourier series of $f'(t)$ may be given by

$$f'(t)=-\omega_1\sum_{n=1}^{\infty} nc_n \sin(n\omega_1 t+\varphi_n).$$

2.2. Consider the Fourier series of a periodic function $f(t)$ with the period T_1 in the form

$$f(t)=\sum_{n=-\infty}^{\infty} F(n\omega_1)e^{in\omega_1 t},$$

where $F(n\omega_1)=\frac{c_n e^{i\varphi_n}}{2}$ and $\omega_1=\frac{2\pi}{T_1}$. Prove that the Fourier series of $f'(t)$ may be expressed by

$$f'(t)=in\omega_1\sum_{n=-\infty}^{\infty} F(n\omega_1)e^{in\omega_1 t}.$$

2.3. Give the proof of the Parseval' rule expressed by Eq. (2.33).

2.4. Let $f(t)$ be an even periodic function with the period T_1. Prove that

$$f(t)=\frac{a_0}{2}+\sum_{n=1}^{\infty}a_n\cos n\omega_1 t,$$

where $\omega_1=\frac{2\pi}{T_1}$ and $a_n=\frac{2}{T_1}\int_{-T_1/2}^{T_1/2}f(t)\cos n\omega_1 t dt$.

2.5. Let $f(t)$ be an odd periodic function with the period T_1. Prove that

$$f(t)=\sum_{n=1}^{\infty}b_n\sin n\omega_1 t,$$

where $\omega_1=\frac{2\pi}{T_1}$ and $b_n=\frac{2}{T_1}\int_{-T_1/2}^{T_1/2}f(t)\sin n\omega_1 t dt$.

REFERENCES

1. C. M. Harris, *Shock and Vibration Handbook*, 5th Ed., McGraw-Hill, New York, 2002.
2. R. A. Gabel and R. A. Roberts, *Signals and Linear Systems*, John Wiley & Sons, Hoboken, NJ, 1973.
3. J. L. Zheng, Q. Y. Ying, and W. L. Yang, *Signals and Systems*, 2nd Ed., vol. 1, Higher Education Press, Beijing, 2000. In Chinese.
4. J. Fourier, *The Analytic Theory of Heat*, Cambridge University Press, London, 1878. Translated by A. Freeman, M. A., p. 204.
5. J. Fourier, *The Analytic Theory of Heat*, Nover Publications, Inc., New York, 1955. Translated by A. Freeman, M. A., p. 204.
6. J. Fourier, *The Analytic Theory of Heat*, Higher Education, Beijing, 2016. Translated by A. Freeman, M. A., p. 162.

CHAPTER 3

Fourier Transform and Spectra

When a mass-spring system is driven by a deterministically non-periodic force (Figure 3.1), we need the Fourier transform and the related knowledge, such as impulse function. This chapter discusses the preliminaries for the issue of responses driven by deterministically non-periodic forces. The response issue will be addressed in Chapter 4.

3.1 BASICS OF FOURIER TRANSFORM

For a function $f(t)$ in the time domain, $t \in (-\infty, \infty)$, its Fourier transform, denoted by $F(\omega)$, is defined by

$$F(\omega)=\int_{-\infty}^{\infty} f(t)e^{-i\omega t}\,dt \tag{3.1}$$

where $\omega = 2\pi f$ is the angular frequency and f is the frequency. One usually uses F as the operator of the Fourier transform. $F(\omega)$ is a function in the frequency domain for $\omega \in (-\infty, \infty)$.

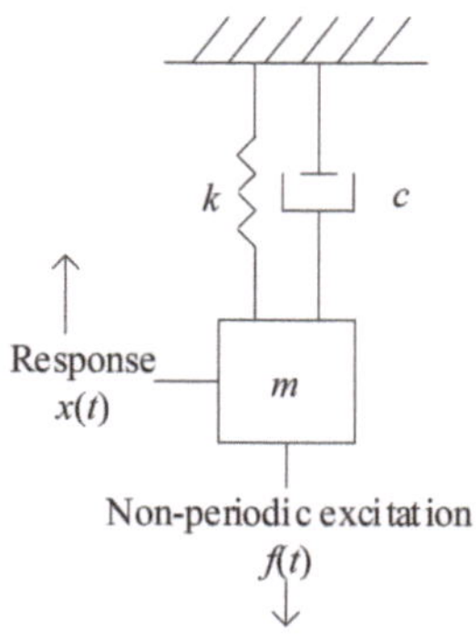

FIGURE 3.1 Mass-spring system under the excitation of non-periodic force $f(t)$.

DOI: 10.1201/9781003460947-4

Since it reveals frequency components of $f(t)$, $F(\omega)$ is called the frequency spectrum function (spectrum for short) of $f(t)$.

For a function $F(\omega)$ for $\omega \in (-\infty, \infty)$, it corresponds to a time function $f(t)$ expressed by

$$f(t)=\frac{1}{2\pi}\int_{-\infty}^{\infty} F(\omega)e^{i\omega t}\,d\omega. \tag{3.2}$$

The above is the inverse Fourier transform of $F(\omega)$. Denote by F^{-1} the operator of the inverse Fourier transform. Two functions, $f(t)$ and $F(\omega)$, constitute a pair of Fourier transform. The pair may be expressed by

$$f(t)\leftrightarrow F(\omega). \tag{3.3}$$

For two functions $f_1(t)$ and $f_2(t)$, one says that $[f_1(t)-f_2(t)]$ is a null function if

$$\int_{-\infty}^{\infty}\left[f_1(t)-f_2(t)\right]e^{-i\omega t}\,dt=0. \tag{3.4}$$

Suppose $f_1(t) \leftrightarrow F_1(\omega)$ and $f_2(t) \leftrightarrow F_2(\omega)$. If $[f_1(t)-f_2(t)]$ is a null function, we say that $F_1(\omega)=F_2(\omega)$ even if $f_1(t)\neq f_2(t)$. The equality in Fourier transform is in that sense.

Example 3.1: Null Function. Consider the unit step function in two different forms:

$$u_1(t)=\begin{cases}1, & t>0\\ 0, & t\le 0\end{cases},\quad u_2(t)=\begin{cases}1, & t\ge 0\\ 0, & t<0\end{cases}. \tag{3.5}$$

Since

$$u_2(t)-u_1(t)=\begin{cases}1, & t=0\\ 0, & \text{otherwise}\end{cases}, \tag{3.6}$$

we have

$$\int_{-\infty}^{\infty}\left[u_2(t)-u_1(t)\right]e^{-i\omega t}\,dt=0. \tag{3.7}$$

Thus, $u_2(t)-u_1(t)$ is a null function. Accordingly, $\mathrm{F}[u_1(t)]=\mathrm{F}[u_2(t)]$.

The Fourier transform is an one-to-one mapping. For $f_1(t) \leftrightarrow F_1(\omega)$ and $f_2(t) \leftrightarrow F_2(\omega)$, $F_1(\omega)=F_2(\omega)$ if $[f_1(t)-f_2(t)]$ is a null function.

Mathematically, the position of $f(t)$ is the same as that of $F(\omega)$. Both contain the same information. For instance, if $f(t)$ is white noise, $F(\omega)$ is white noise too (Robinson [1]). The difference between the two lies in the fact that $f(t)$ is in the time domain while $F(\omega)$ in the frequency domain.

Function $F(\omega)$ is complex in general. When it is expressed by

$$F(\omega)=\int_{-\infty}^{\infty} f(t)e^{-i\omega t}\,dt=\int_{-\infty}^{\infty} f(t)(\cos\omega t-i\sin\omega t)dt, \tag{3.8}$$

we can write it by

$$F(\omega)=\mathrm{Re}[F(\omega)]+i\,\mathrm{Im}[F(\omega)]=A(\omega)+iB(\omega), \tag{3.9}$$

where

$$A(\omega)=\int_{-\infty}^{\infty} f(t)\cos\omega t dt, \tag{3.10}$$

$$B(\omega)=-\int_{-\infty}^{\infty} f(t)\sin\omega t dt. \tag{3.11}$$

In polar coordinate, $F(\omega)$ is expressed by

$$F(\omega)=|F(\omega)|e^{i\varphi(\omega)}, \tag{3.12}$$

where $|F(\omega)|=\sqrt{A^2(\omega)+B^2(\omega)}$ and $\varphi(\omega)=\arctan\dfrac{B(\omega)}{A(\omega)}$.

Example 3.2. Consider the Fourier transform of a single-side exponentially decayed function in the form

$$f(t)=Ee^{-at}u(t),\quad a>0, \tag{3.13}$$

where $u(t)$ is the unit step function and E magnitude. Its Fourier transform is given by

$$F(\omega)=\int_{-\infty}^{\infty} Ee^{-at}u(t)e^{-i\omega t}\,dt=\int_{0}^{\infty} Ee^{-at}e^{-i\omega t}\,dt=\frac{E}{a+i\omega}. \tag{3.14}$$

Its amplitude-frequency spectrum is

$$|F(\omega)|=\frac{E}{\sqrt{a^2+\omega^2}}. \tag{3.15}$$

Its phase is given by

$$\varphi(\omega) = -\arctan\frac{a}{\omega}. \tag{3.16}$$

We use this example to explain the concept of principal energy and frequency band of a signal. As

$$|F(\omega)|_{\max} = |F(0)| = \frac{E}{a}, \tag{3.17}$$

and $|F(a)| = \frac{1}{\sqrt{2}}|F(0)|$. The half-power point of this signal is located at $\omega = a$. Thus, the frequency band of this signal is (0, a) in terms of ω. The principal energy of this signal contains the frequency components for $\omega \in (0, a)$. Those for $\omega \in (0, a)$ should be considered and others for $\omega > a$ may be ignored in vibration responses. Figure 3.2 indicates the plots of $f(t)$ and its Fourier transform.

Example 3.3. Consider the Fourier transform of a two-sided exponentially decayed function in the form

$$f(t) = Ee^{-|a|t}, \quad a > 0. \tag{3.18}$$

Its Fourier transform is given by

$$F(\omega) = \int_{-\infty}^{\infty} Ee^{-a|t|}e^{-i\omega t}dt = \frac{2aE}{a^2 + \omega^2}. \tag{3.19}$$

Its phase spectrum is zero. Figure 3.3 indicates the time wave and spectrum.

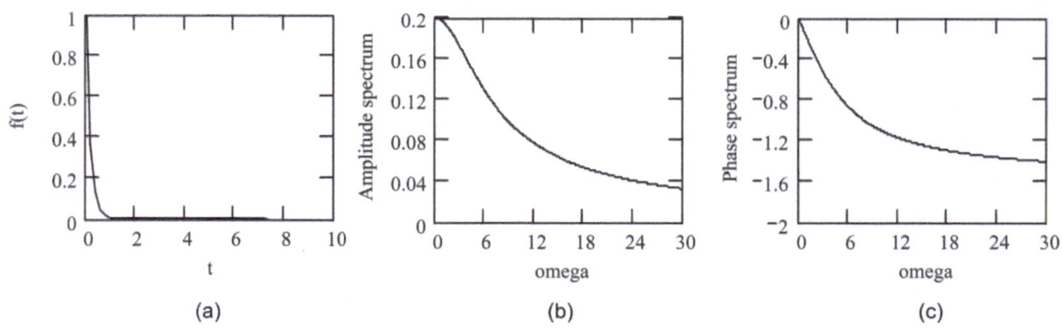

FIGURE 3.2 Single-side exponentially decayed function and its spectrum for $a = 5$, $E = 1$. (a) $f(t)$ in Eq. (3.13). (b) $|F(\omega)|$ in Eq. (3.15). (c) $\varphi(\omega)$ in Eq. (3.16).

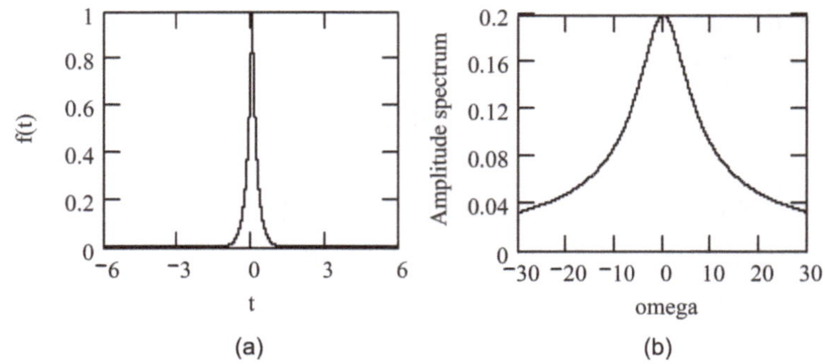

FIGURE 3.3 Two-sided exponentially decayed function and its spectrum with $a=5$, $E=1$. (a) $f(t)$ in Eq. (3.18). (b) $|F(\omega)|$ in Eq. (3.19).

For this signal, $|F(\omega)|_{\max}=|F(0)|=\frac{2E}{a}$. Denote by ω_c its half-power point. Then, we have $|F(\omega_c)|^2=\frac{1}{2}|F(0)|^2$ when $\omega_c=\sqrt{\sqrt{2}-1}a$. Thus, the frequency band that contains the principal energy of this signal is $(0, \omega_c)$.

Example 3.4. Study the Fourier transform of the single rectangular pulse function given by

$$f(t)=\begin{cases} E, & -\frac{\tau}{2}<t<\frac{\tau}{2} \\ 0, & \text{elsewhere} \end{cases}. \tag{3.20}$$

Its Fourier transform is given by

$$F(\omega)=\int_{-\infty}^{\infty} f(t)e^{-i\omega t}\,dt=2E\int_{0}^{\frac{\tau}{2}}\cos\omega t dt=E\tau \mathrm{Sa}\left(\frac{\omega\tau}{2}\right). \tag{3.21}$$

Figure 3.4 shows its plot and spectrum.

The frequency band B of this single is usually bounded by the first zero crossing $1/\tau$ in terms of f. That is, $B=(0, 1/\tau)$. It is inversely proportional to the pulse width τ. Note that one does not use half-power point to define B for this signal in engineering.

3.2 IMPULSE (DELTA) FUNCTION

Suppose that ϕ is a function that is continuous and rapidly decays on (a, b) and its derivatives of all orders exist. Then, it is called a test function on (a, b), where a may be minus infinity and b may be infinity. The set of all test functions is denoted by **D**. Now, we say that whether a function $f(t)$ can be defined or not does not rely on whether the value of $f(t)$ is finite or not. Instead, it simply depends on whether the operation $\langle f(t), \phi(t)\rangle$ is convergent

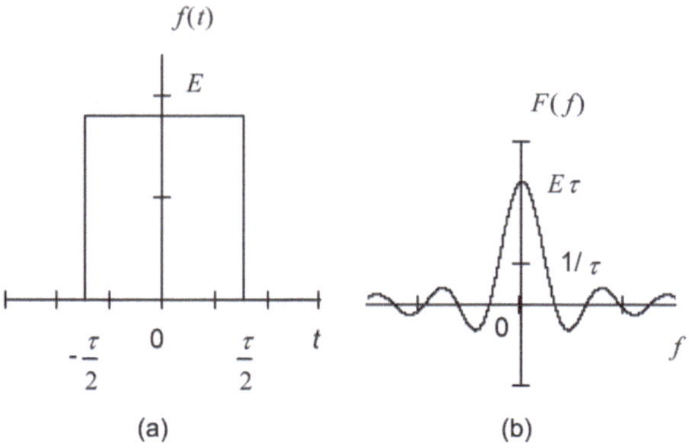

FIGURE 3.4 Single pulse and its spectrum. (a) Single pulse function. (b) Spectrum.

or not. A commonly used operation is integral. Thus, $\langle f(t),\phi(t)\rangle = \int_a^b f(t)\phi(t)dt$. Therefore, if that integral is convergent, we say that $f(t)$ is defined as a function. When a function is defined in that way, it is called a generalized function (Gelfand and Vilenkin [2]). A generalized function $f(t)$ is also called a Schwartz distribution.

Several properties of generalized functions are listed below.

Lemma 3.1. Any generalized function is infinitely differentiable.

Lemma 3.2. There exists the Fourier transform for any generalized function.

Lemma 3.3. The limit of the sequence of a generalized function is unique with possible differences in a null function.

Lemma 3.4. Product of generalized functions is undefined.

Lemma 3.5. Limit operation and integral one are interchangeable for any generalized function.

Lemma 3.6. If, for any $\phi \in \mathbf{D}$, there exists

$$\lim_{\nu\to\infty}\int_{-\infty}^{\infty} f_\nu(t)\phi(t)dt = \int_{-\infty}^{\infty} f(t)\phi(t)dt, \tag{3.22}$$

for $f_1, f_2, \ldots, f_\nu, \ldots$, which is a sequence of a generalized function, the sequence $\{f_\nu\}$ is convergent to f.

Lemma 3.7. The derivative of a generalized function f is another generalized function given by

$$\int_{-\infty}^{\infty} \frac{df(t)}{dt}\phi(t)dt = -\int_{-\infty}^{\infty} f(t)\frac{d\phi(t)}{dt}dt. \tag{3.23}$$

Because of infinitely differentiable, one has

$$\int_{-\infty}^{\infty} f^{(n)}(t)\phi(t)\,dt = (-1)^n \int_{-\infty}^{\infty} f(t)\phi^{(n)}(t)\,dt. \tag{3.24}$$

Lemma 3.8. The Fourier kernel $e^{-i\omega t}$ is a test function (Papoulis [3]).

Example 3.5. Denote by $\delta(t)$ the delta function. For all $\phi \in \mathbf{D}$, it is defined by

$$\int_{-\infty}^{\infty} \delta(t)\phi(t)\,dt = \phi(0). \tag{3.25}$$

Example 3.6. Denote by $\delta(t)$ the delta function. For all $\phi \in \mathbf{D}$, following Lemma 3.7, its derivative of order n is defined by

$$\int_{-\infty}^{\infty} \delta^{(n)}(t)\phi(t)\,dt = (-1)^{(n)} \int_{-\infty}^{\infty} \delta(t)\phi^{(n)}(t)\,dt = (-1)^{(n)}\phi^{(n)}(0). \tag{3.26}$$

There is a relationship between $\delta(t)$ and $u(t)$ in the form (Xia and Yan [4])

$$\delta(t) = \frac{du(t)}{dt}. \tag{3.27}$$

The integral relationship between $\delta(t)$ and $u(t)$ is expressed by

$$\int_{-\infty}^{t} \delta(t)\,dt = u(t). \tag{3.28}$$

From the definition of $u(t)$ and according to the above, we see that $\delta(t)$ has the property given by

$$\delta(t) = 0 \quad (t \neq 0). \tag{3.29}$$

Example 3.7. $\delta(t)$ is an even function.
Proof. According to Eq. (3.25), we have

$$\int_{-\infty}^{\infty} \delta(-t)\phi(t)\,dt \overset{t=-u}{=} -\int_{\infty}^{-\infty} \delta(u)\phi(-u)\,du = \int_{-\infty}^{\infty} \delta(u)\phi(-u)\,du = \phi(0). \tag{3.30}$$

Thus,

$$\delta(t)=\delta(-t). \tag{3.31}$$

The proof is complete.

Example 3.8. $\delta'(t)$ is an odd function.
Proof. Based on Eq. (3.26), we have

$$\int_{-\infty}^{\infty}\delta'(t)\phi(t)dt=-\int_{-\infty}^{\infty}\delta(t)\phi'(t)dt=-\phi'(0). \tag{3.32}$$

On the other hand,

$$\int_{-\infty}^{\infty}\delta'(-t)\phi(t)dt\overset{t=-u}{=}-\int_{\infty}^{-\infty}\delta(u)\phi'(-u)du=\int_{-\infty}^{\infty}\delta(u)\phi'(-u)du=\phi'(0). \tag{3.33}$$

Therefore, we have

$$\delta'(-t)=-\delta'(t). \tag{3.34}$$

This finishes the proof.

Example 3.9. $t\delta(t)=0$.
Proof. Based on Eq. (3.25), we have

$$\int_{-\infty}^{\infty}t\delta(t)\phi(t)dt=\int_{-\infty}^{\infty}\delta(t)[t\phi(t)]dt=\left[t\phi(t)\right]_{t=0}=0. \tag{3.35}$$

This completes the proof.

I emphasize that $\delta(t)$ at $t=0$ is undefined. From the point of view of generalized functions, we do not care whether $\delta(0)$ is defined or not. Due to Eq. (3.27), $\delta(0)$ is never a number. Taking into account the right hand on Eq. (3.25), we see that $\delta(0)$ just takes ∞ as limit. Accordingly, there is no picture for $\delta(0)$. One usually demonstrates it with an arrow, as shown in Figure 3.5.

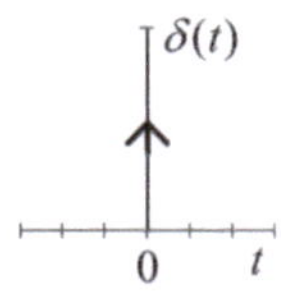

FIGURE 3.5 Illustration of $\delta(t)$.

Example 3.10. The Fourier transform of $\delta(t)$ is given by

$$\int_{-\infty}^{\infty} \delta(t) e^{-i\omega t} dt = e^{-i\omega t}\Big|_{t=0} = 1. \tag{3.36}$$

The above exhibits a particular property of $\delta(t)$. To be precise, its spectrum is constant over $\omega \in (-\infty, \infty)$. Figure 3.6 shows the plot of F[$\delta(t)$]. This property produces a useful technique for vibration test. I shall mention this late in this chapter.

The aforementioned $\delta(t)$ is often called Dirac-delta function, see e.g., [3] and Robinson [1], to memorialize the famous physicist P. A. M. Dirac [5]. However, mathematicians, such as Fourier, Cauchy, Poisson, et al., had introduced the function $\delta(t)$ to study issues of divergent series and integrals before the age of Dirac, see Laugwitz [6], Hoskins [7], Lützen [8]. Thus, this book uses the term of delta function with the prefix Dirac omitted. In mechanical engineering, $\delta(t)$ is usually termed the impulse function.

Proposition 3.1. Using $\delta(t)$, we have

$$\int_{-\infty}^{\infty} \cos \omega t \, d\omega = 2\pi\delta(t), \tag{3.37}$$

and

$$\int_{-\infty}^{\infty} \sin \omega t \, d\omega = 0. \tag{3.38}$$

Proof. Doing the inverse Fourier transform of 1 yields

$$\frac{1}{2\pi}\int_{-\infty}^{\infty} e^{-i\omega t} \, d\omega = \delta(t). \tag{3.39}$$

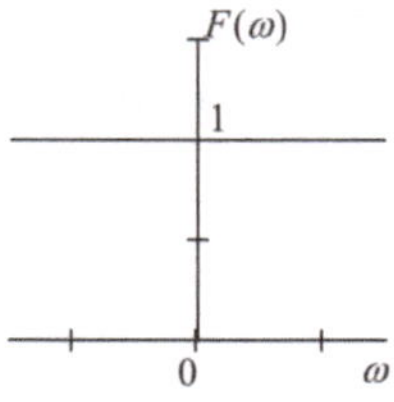

FIGURE 3.6 Fourier transform/spectrum of delta function.

From the above, we have

$$\int_{-\infty}^{\infty} \cos \omega t \, d\omega - i \int_{-\infty}^{\infty} \sin \omega t \, d\omega = 2\pi\delta(t). \tag{3.40}$$

Because $\delta(t)$ is real, we have Eqs. (3.37) and (3.38). The proof is complete.

From Proposition 3.1, we have the following corollary.

Corollary 3.1. Let $\delta(\omega)$ be the delta function in the frequency domain. Then,

$$\int_{-\infty}^{\infty} \cos \omega t \, dt = 2\pi\delta(\omega), \tag{3.41}$$

and

$$\int_{-\infty}^{\infty} \sin \omega t \, dt = 0. \tag{3.42}$$

Proof. Replacing t by ω and changing ω with t in Eqs. (3.37) and (3.38) produce Eqs. (3.41) and (3.42). This finishes the proof.

3.3 RESPONSE MEASUREMENT USING IMPULSE FUNCTION

Since $F[\delta(t)]=1$, we see that $\delta(t)$ contains all harmonic waves that are equal in amplitude for $-\infty<\omega<\infty$. Thus, if we exert a driven force in the form of $\delta(t)$ on a structure at time $t=0$, the structure actually suffers from the excitation that consists of all harmonic waves that are equal in amplitude for $-\infty<\omega<\infty$. Accordingly, we can measure the frequency response of a structure with the driven force in the form of $\delta(t)$.

If a test structure is small, using a hammer to hit it may work. If a test structure is large, such as ship structure, a specifically designed machine hammer is needed.

For a structure of very large size, such as a real ship, one way to shock a ship is to drive the ship to rush into a dock to exert an impulse on a test ship. Alternatively, one ship purposely rushes into a test ship to collide with an impulse force for measuring its impulse response. Casting anchor is also a technique to measure the impulse response of a test ship. All those are based on the theory of $F[\delta(t)]=1$.

3.4 PRACTICAL REALIZATIONS OF IMPULSE FUNCTION

Signal generation of $\delta(t)$ in a test machine producing an impulse shock needs its mathematical models. For that reason, we mention the term "nascent impulse function".

Let $f_v(t)$ be a function with a parameter v. Then, for $a\neq 0$ and $b\neq 0$, when the following limit holds

$$\lim_{v\to 0^+}\int_a^b f_v(t)dt=\begin{cases}1, & a<0<b\\ 0, & a<b<0 \text{ and } 0<a<b\end{cases}, \tag{3.43}$$

and it is a continuous function, we say that $f_v(t)$ is a nascent impulse/delta function. An alternative definition of a nascent impulse can be expressed as follows.

Let $f_v(t)$ be a function with a parameter v. It is a nascent delta function if it is a continuous function in the form

$$\lim_{v\to 0}\int_{-\infty}^{t} f_v(t)dt=u(t)=\begin{cases}1, & t\geq 0\\ 0, & t<0\end{cases}. \tag{3.44}$$

Example 3.11. Given a rectangular function in the form

$$f_v(t)=\text{rect}(t)=\begin{cases}\frac{1}{v}, & -\frac{v}{2}<t<\frac{v}{2}\\ 0, & \text{otherwise}\end{cases} \quad v>0. \tag{3.45}$$

Because

$$\lim_{v\to 0}\int_{-\infty}^{t} \text{rect}(t)dt=u(t)=\begin{cases}1, & t>0\\ 0, & t<0\end{cases},$$

rect(t) is a nascent impulse. The impulse function can be expressed by

$$\delta(t)=\lim_{v\to 0}\text{rect}(t). \tag{3.46}$$

The plots of this nascent impulse are indicated in Figure 3.7.

Since the limit $v\to 0$ cannot be realized by any practical machine, we can only set a small value for the pulse width $v=\varepsilon$. So long as v is much smaller than the transient response time

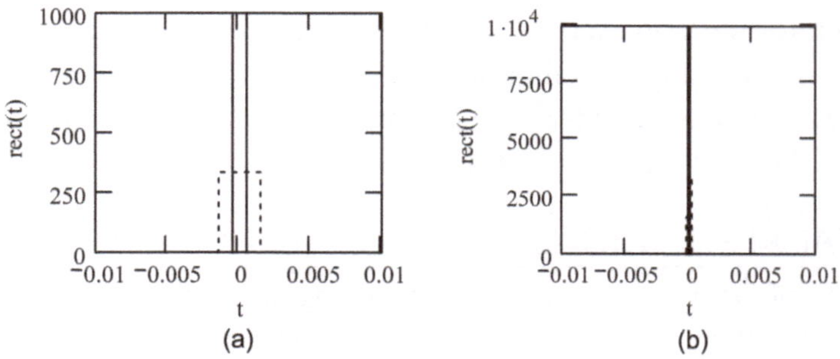

FIGURE 3.7 Plots of rect(t). (a) rect(t) when $v=0.003$ (dot), 0.001 (solid). (b) rect(t) for $v=0.0003$ (dot), 0.0001 (solid).

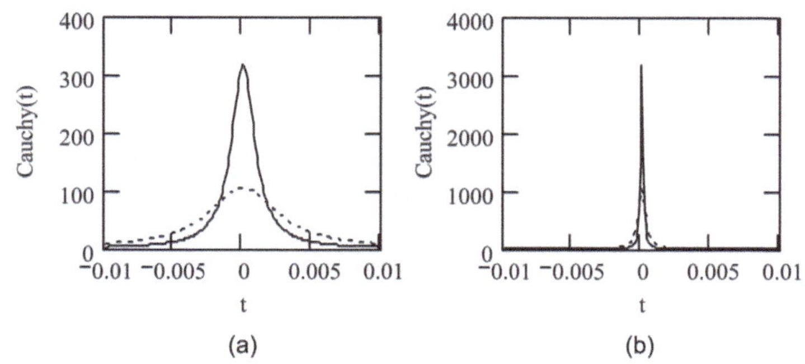

FIGURE 3.8 Plots of Cauchy(t). (a) Cauchy(t) when $v=0.003$ (dot), 0.001 (solid). (b) Cauchy(t) for $v=0.0003$ (dot), 0.0001 (solid).

of a vibration system, ε can be considered as infinitesimal in engineering. Accordingly, rect(t) may be taken as a practical impulse without the need for $v \to 0$.

Example 3.12. Given the impulse function that takes the function form of the Cauchy probability density. It is given by

$$f_v(t) = \text{Cauchy}(t) = \frac{1}{\pi}\frac{v}{t^2+v^2}. \tag{3.47}$$

Since

$$\int_a^b \text{Cauchy}(t)dt = \frac{1}{\pi}\left(\text{arctg}\frac{b}{v} - \text{arctg}\frac{a}{v}\right) = \begin{cases} 1, & a<0<b, \\ 0, & \text{elsewhere}, \end{cases} \tag{3.48}$$

we have

$$\delta(t) = \frac{1}{\pi}\lim_{v\to 0^+}\frac{v}{t^2+v^2}. \tag{3.49}$$

Thus, Cauchy(t) is a nascent impulse. Its plots are shown in Figure 3.8.

Keep in mind that $\delta(t)$ is an abstract symbol of the impulse function. Practical methods for its realizations in engineering are simply nascent impulse functions.

3.5 PROPERTIES OF FOURIER TRANSFORM

We brief some properties of $F(\omega)$. Interesting readers may refer to Papoulis [3] for details. Note that Fourier transform is a kind of linear operation. It satisfies the superposition. That is, if $f_n(t) \leftrightarrow F_n(\omega)$, $\sum_{n=0}^{\infty} a_n f_n(t) \leftrightarrow \sum_{n=0}^{\infty} a_n F_n(\omega)$.

Property 3.1. $A(\omega)$ is an even function, whereas $B(\omega)$ is odd.

Proof. According to Eq. (3.10), we have $A(\omega)=A(-\omega)$. On the other hand, In Eq. (3.11), one has $B(-\omega)=-B(\omega)$. This completes the proof.

Property 3.2. $A^2(\omega)$ and $B^2(\omega)$ are even functions.

Proof. Since $A(\omega)$ is even, $A^2(\omega)$ is even. In addition, $B^2(\omega)$ is even because $B(\omega)$ is odd.

Property 3.3. $|F(\omega)|$ is an even function while $\varphi(\omega)$ is odd.

Proof. According to Eq. (3.12), we see that $|F(\omega)|$ is even while $\varphi(\omega)$ is odd. This completes the proof.

Property 3.4. If $f(t)$ is an even function, $F(\omega)=A(\omega)$ is even. If $f(t)$ is an odd function, $F(\omega)=iB(\omega)$ is odd.

Proof is omitted.

Property 3.5: Symmetry. If $f(t) \leftrightarrow F(\omega)$, $F(t) \leftrightarrow 2\pi f(-\omega)$.

Proof. In $F(\omega)=\int_{-\infty}^{\infty} f(t)e^{-i\omega t}\,dt$, changing variable t with ω produces

$$F(t)=\int_{-\infty}^{\infty} f(\omega)e^{-i\omega t}\,d\omega=\frac{1}{2\pi}\int_{-\infty}^{\infty} 2\pi f(-\omega)e^{i(-\omega)t}\,d(-\omega).$$

Thus, $F(t) \leftrightarrow 2\pi f(-\omega)$. This completes the proof.

Property 3.5 is particularly useful for dealing with divergent integrals.

Example 3.13. Denote by rect(t) the single rectangular pulse in the form

$$\mathrm{rect}(t)=\begin{cases} E, & -\dfrac{\tau}{2}<t<\dfrac{\tau}{2} \\ 0 & \end{cases}. \tag{3.50}$$

Then,

$$\mathrm{rect}(t)\leftrightarrow\int_{-\frac{\tau}{2}}^{\frac{\tau}{2}} Ee^{-i\omega t}\,dt=E\tau\,\mathrm{Sa}\left(\frac{\omega\tau}{2}\right). \tag{3.51}$$

Using Property 3.5, one has

$$E\tau\,\mathrm{Sa}\left(\frac{t\tau}{2}\right)\leftrightarrow 2\pi\,\mathrm{rect}(\omega). \tag{3.52}$$

Example 3.14. The Fourier transform of $\cos\omega_0 t$ is given by

$$\cos\omega_0 t \leftrightarrow \pi\left[\delta(\omega+\omega_0)+\delta(\omega-\omega_0)\right]. \tag{3.53}$$

Proof. Due to $\delta(t) \leftrightarrow 1$, one has $1 \leftrightarrow 2\pi\,\delta(\omega)$ based on symmetry. Thus,

$$\cos\omega_0 t \leftrightarrow \int_{-\infty}^{\infty} \frac{e^{i\omega_0 t}+e^{-i\omega_0 t}}{2} e^{-i\omega t}\,dt = \int_{-\infty}^{\infty} \frac{e^{i(\omega_0-\omega)t}+e^{-i(\omega_0+\omega)t}}{2}\,dt$$

$$=\pi\left[\delta(\omega+\omega_0)+\delta(\omega-\omega_0)\right].$$

This finishes the proof.

Example 3.15. The Fourier transform of $\sin\omega_0 t$ is expressed by

$$\sin\omega_0 t \leftrightarrow i\pi\left[\delta(\omega+\omega_0)-\delta(\omega-\omega_0)\right]. \tag{3.54}$$

Proof. Due to

$$\sin\omega_0 t \leftrightarrow \int_{-\infty}^{\infty} \frac{e^{i\omega_0 t}-e^{-i\omega_0 t}}{2i} e^{-i\omega t}\,dt = \int_{-\infty}^{\infty} \frac{e^{i(\omega_0-\omega)t}-e^{-i(\omega_0+\omega)t}}{2i}\,dt,$$

one has Eq. (3.54). This completes the proof.

Property 3.6: Time Shifting. If $f(t) \leftrightarrow F(\omega)$, $f(t-t_0) \leftrightarrow F(\omega)e^{-i\omega t_0}$.
Proof. In fact,

$$f(t-t_0) \leftrightarrow \int_{-\infty}^{\infty} f(t-t_0)e^{-i\omega t}\,dt = \int_{-\infty}^{\infty} f(t-t_0)e^{-i\omega(t-t_0)}e^{-i\omega t_0}\,d(t-t_0)$$

$$=e^{-i\omega t_0}\int_{-\infty}^{\infty} f(u)e^{-i\omega u}\,du = F(\omega)e^{-i\omega t_0}.$$

This completes the proof.

Example 3.16. If $f(t) \leftrightarrow F(\omega)$, $f(t-t_0)+f(t+t_0) \leftrightarrow 2F(\omega)\cos\omega t_0$.
Proof. Since $f(t-t_0) \leftrightarrow F(\omega)e^{-i\omega t_0}$ and $f(t+t_0) \leftrightarrow F(\omega)e^{i\omega t_0}$, we have

$$f(t-t_0)+f(t+t_0) \leftrightarrow F(\omega)e^{-i\omega t_0} + F(\omega)e^{i\omega t_0} = 2F(\omega)\cos\omega t_0.$$

This completes the proof.

Property 3.7: Frequency Shifting. If $f(t) \leftrightarrow F(\omega)$, $f(t)e^{i\omega_0 t} \leftrightarrow F(\omega-\omega_0)$.
Proof. Due to

$$f(t)e^{i\omega_0 t} \leftrightarrow \int_{-\infty}^{\infty} f(t)e^{i\omega_0 t}e^{-i\omega t}\,dt = \int_{-\infty}^{\infty} f(t)e^{-i(\omega-\omega_0)t}\,dt = F(\omega-\omega_0),$$

the property holds. That completes the proof.

Example 3.17. If $f(t) \leftrightarrow F(\omega)$,

$$f(t)\cos\omega_0 t \leftrightarrow \frac{1}{2}\left[F(\omega-\omega_0)+F(\omega+\omega_0)\right]. \tag{3.55}$$

Proof. Since $f(t)\cos\omega_0 t=0.5\{[f(t)\exp(i\omega_0 t)]+[f(t)\exp(-i\omega_0 t)]\}$, based on Property 3.7, we have Eq. (3.55). This completes the proof.

Equation (3.55) represents the modulation theorem in electronics engineering (Zheng et al. [9]).

Example 3.18. Let $f(t)$ be a periodic signal with period T_0. Then,

$$f(t) \leftrightarrow a_0\pi\delta(\omega)+\sum_{n=1}^{\infty}\left\{\begin{matrix} a_n\pi\left[\delta(\omega+n\omega_0)+\delta(\omega-n\omega_0)\right] \\ +ib_n\pi\left[\delta(\omega+n\omega_0)-\delta(\omega-n\omega_0)\right]\end{matrix}\right\}, \tag{3.56}$$

where $\omega_0=\dfrac{2\pi}{T_0}$, a_0, a_n, and b_n are Fourier coefficients.

Proof. Extending $f(t)$ into its Fourier series yields

$$f(t)=\frac{a_0}{2}+\sum_{n=1}^{\infty}\left(a_n\cos n\omega_0 t+b_n\sin n\omega_0 t\right).$$

According to Examples 3.14 and 3.15, we have

$$\begin{aligned} &\cos n\omega_0 t \leftrightarrow \pi\left[\delta(\omega+n\omega_0)+\delta(\omega-n\omega_0)\right], \\ &\sin n\omega_0 t \leftrightarrow i\pi\left[\delta(\omega+n\omega_0)-\delta(\omega-n\omega_0)\right]. \end{aligned} \tag{3.57}$$

In addition, $\dfrac{a_0}{2} \leftrightarrow a_0\pi\delta(\omega)$. Thus, with the superposition, we have

$$f(t) \leftrightarrow \mathrm{F}\left(\frac{a_0}{2}\right) + \mathrm{F}\left(\sum_{n=1}^{\infty}(a_n \cos n\omega_0 t + b_n \sin n\omega_0 t)\right)$$

$$= a_0\pi\delta(\omega) + \sum_{n=1}^{\infty}\left[a_n\mathrm{F}(\cos n\omega_0 t) + b_n\mathrm{F}(\sin n\omega_0 t)\right]$$

$$= a_0\pi\delta(\omega) + \sum_{n=1}^{\infty}\left\{\begin{array}{l} a_n\pi\left[\delta(\omega + n\omega_0) + \delta(\omega - n\omega_0)\right] \\ +ib_n\pi\left[\delta(\omega + n\omega_0) - \delta(\omega - n\omega_0)\right] \end{array}\right\}.$$

This completes the proof.

Property 3.8: Differential. If $f(t) \leftrightarrow F(\omega)$, $f^{(n)}(t) \leftrightarrow (i\omega)^n F(\omega)$.
Proof. According to Lemma 3.7, we have

$$f^{(n)}(t) \leftrightarrow \int_{-\infty}^{\infty} f^{(n)}(t)e^{-i\omega t}\,dt = (-1)^n \int_{-\infty}^{\infty} f(t)\left(e^{-i\omega t}\right)^{(n)} dt$$

$$= (-1)^{2n}(i\omega)^n \int_{-\infty}^{\infty} f(t)e^{-i\omega t}\,dt = (i\omega)^n F(\omega).$$

This completes the proof.

Example 3.19. Let $\delta(t)$ be impulse function. Then,

$$\delta^{(n)}(t) \leftrightarrow (i\omega)^n. \tag{3.58}$$

Proof. Letting $f(t) = \delta(t)$ in Property 3.8 yields Eq. (3.58). This completes the proof.

Example 3.20. Let $\delta(\omega)$ be impulse function in the frequency domain. Then,

$$t^n \leftrightarrow 2\pi(-i)^n \delta^{(n)}(-\omega). \tag{3.59}$$

Proof. According to the property of symmetry, using Eq. (3.58), one has $(it)^n \leftrightarrow 2\pi\delta^{(n)}(-\omega)$. Moving i^n from the left to the right yields Eq. (3.59). This finishes the proof.

Example 3.21. Let $\delta(\omega)$ be impulse function in frequency domain. Then,

$$e^t \leftrightarrow \int_{-\infty}^{\infty} e^t e^{-i\omega t}\,dt = \int_{-\infty}^{\infty}\left(\sum_{n=0}^{\infty}\frac{t^n}{n!}\right)e^{-i\omega t}\,dt = 2\pi\sum_{n=0}^{\infty}\frac{(-i)^n\delta^{(n)}(-\omega)}{n!}. \tag{3.60}$$

Proof. According to Example 3.20 and the superposition property of the Fourier transform, when expanding exp(t) using its Taylor expansion produces Eq. (3.60). This completes the proof.

Example 3.22. Denote by sgn(t) the signum function in the form

$$\operatorname{sgn}(t)=\begin{cases} 1, & t\geq 0, \\ -1, & t<0. \end{cases} \tag{3.61}$$

Then,

$$\operatorname{sgn}(t)\leftrightarrow\frac{2}{i\omega}. \tag{3.62}$$

Proof. The function sgn(t) is odd. Thus, $\frac{d\operatorname{sgn}(t)}{dt}$ is even. In fact,

$$\frac{d\operatorname{sgn}(t)}{dt}=2\delta(t)\leftrightarrow 2. \tag{3.63}$$

According to the differential property, we have

$$\frac{d\operatorname{sgn}(t)}{dt}\leftrightarrow i\omega\mathrm{F}[\operatorname{sgn}(t)]=2. \tag{3.64}$$

Thus, $\mathrm{F}[\operatorname{sgn}(t)]=\frac{2}{i\omega}$. This finishes the proof.

Example 3.23. The Fourier transform of the unit step function $u(t)$ is in the form

$$u(t)\leftrightarrow\pi\delta(\omega)+\frac{1}{i\omega}. \tag{3.65}$$

Proof. Because $u(t)=\frac{1}{2}+\frac{1}{2}\operatorname{sgn}(t)$, by superposition, we have

$$u(t)\leftrightarrow\mathrm{F}\left(\frac{1}{2}\right)+\mathrm{F}\left(\frac{1}{2}\operatorname{sgn}(t)\right)=\pi\delta(\omega)+\frac{1}{i\omega}.$$

This completes the proof.

The first item $\pi\delta(\omega)$ on the right side of Eq. (3.65) stands for the average of $u(t)$ in the frequency domain. In vibrations, it implies the static loading resulting from $u(t)$. Since static loading does not interest us in vibrations, $u(t)$ and $\frac{1}{2}\operatorname{sgn}(t)$ have the same loading effects on vibrations in theory.

Mathematically, because of $u(t) \leftrightarrow \int_0^\infty u(t)e^{-i\omega t}\,dt = \int_0^\infty (\cos\omega t - i\sin\omega t)\,dt = \pi\delta(\omega) + \frac{1}{i\omega}$, we have

$$\int_0^\infty \cos\omega t\,dt = \pi\delta(\omega), \tag{3.66}$$

and

$$\int_0^\infty \sin\omega t\,dt = \frac{1}{\omega}. \tag{3.67}$$

Property 3.9: Integral. If $f(t) \leftrightarrow F(\omega)$,

$$\int_{-\infty}^{t} f(t)\,dt \leftrightarrow \pi\delta(\omega)F(0) + \frac{F(\omega)}{i\omega}. \tag{3.68}$$

Proof. Since

$$\int_{-\infty}^{t} f(t)\,dt = \left(\int_{-\infty}^{t} f(t)\,dt \right) * \delta(t),$$

with Lemma 3.10, one has

$$\int_{-\infty}^{t} f(t)\,dt = f(t) * \int_{-\infty}^{t} \delta(t)\,dt = f(t) * u(t).$$

According to Lemma 3.11, we have

$$\int_{-\infty}^{t} f(t)\,dt = f(t) * u(t) \leftrightarrow \left[\pi\delta(\omega) + \frac{1}{i\omega} \right] F(\omega) = \pi\delta(\omega)F(0) + \frac{F(\omega)}{i\omega}.$$

This completes the proof.

Property 3.10: Scaling. If $f(t) \leftrightarrow F(\omega)$,

$$f(at) \leftrightarrow \frac{1}{|a|} F\left(\frac{\omega}{a}\right). \tag{3.69}$$

Proof. As a matter of fact,

$$\int_{-\infty}^{\infty} f(at)e^{-i\omega t}\,dt = \frac{1}{a}\int_{-\infty}^{\infty} f(at)e^{-i\frac{\omega}{a}at}\,dat.$$

If $a>0, \frac{1}{a}\int_{-\infty}^{\infty} f(at)e^{-i\frac{\omega}{a}at}\,dat = \frac{1}{a}F\left(\frac{\omega}{a}\right)$. When $a<0$,

$$\frac{1}{a}\int_{-\infty}^{\infty} f(at)e^{-i\frac{\omega}{a}at}\,dat = \frac{1}{a}\int_{\infty}^{-\infty} f(u)e^{-i\frac{\omega}{a}u}\,du = -\frac{1}{a}\int_{-\infty}^{\infty} f(u)e^{-i\frac{\omega}{a}u}\,du = \frac{1}{|a|}\int_{-\infty}^{\infty} f(u)e^{-i\frac{\omega}{a}u}\,du.$$

This completes the proof.

Lemma 3.9: Parseval's Rule. If $f(t) \leftrightarrow F(\omega)$,

$$\int_{-\infty}^{\infty} |f(t)|^2\,dt = \frac{1}{2\pi}\int_{-\infty}^{\infty} |F(\omega)|^2\,d\omega. \tag{3.70}$$

Proof. Let $f(t)$ be a current that passes through a resistor with 1 Ohm. Then, $|f(t)|^2$ is the power. Accordingly, $\int_{-\infty}^{\infty} |f(t)|^2\,dt$ is the energy the resistor consumes over the time interval $(-\infty, \infty)$. Since the Fourier transform is a one-to-one mapping in mathematics, $\int_{-\infty}^{\infty} |f(t)|^2\,dt$ must equal to the energy represented in the frequency domain, that is, $\int_{-\infty}^{\infty} |F(f)|^2\,df = \frac{1}{2\pi}\int_{-\infty}^{\infty} |F(\omega)|^2\,d\omega$. According to the energy conservation law, therefore, Eq. (3.70) holds. This completes the proof.[1]

Example 3.24. Prove the following integral

$$\int_{-\infty}^{\infty} \mathrm{Sa}^2\frac{\omega}{2}\,d\omega = 2\pi. \tag{3.71}$$

Proof. Let $f(t)$ be a rectangular pulse rect(t) with the width $\tau=1$ and the magnitude $E=1$. Then,

$$\int_{-\infty}^{\infty} |f(t)|^2 dt = 1.$$

Since $f(t) \leftrightarrow \text{Sa}\frac{\omega}{2}$, according to Parseval's rule, we have Eq. (3.71). That finishes the proof.

Example 3.25. Prove the following integral.

$$\int_{-\infty}^{\infty} \left(\text{Sa}\frac{\omega\tau}{2}\right)^2 d\omega = \frac{2\pi}{\tau}. \tag{3.72}$$

Proof. Let $f(t)$ be a rectangular pulse rect(t) with the width τ and the magnitude $E = 1$. Then, $\int_{-\infty}^{\infty} |f(t)|^2 dt = \tau$. As $f(t) \leftrightarrow \tau\text{Sa}\frac{\omega\tau}{2}$, based on Parsevel's rule, we have $\frac{1}{2\pi}\int_{-\infty}^{\infty} \left(\tau\text{Sa}\frac{\omega\tau}{2}\right)^2 d\omega = \tau$.

Thus, $\int_{-\infty}^{\infty} \left(\text{Sa}\frac{\omega\tau}{2}\right)^2 d\omega = \frac{2\pi}{\tau}$. This completes the proof.

3.6 CONVOLUTION

For two functions $f_1(t)$, $f_2(t)$, $-\infty < t < \infty$, the convolution of $f_1(t)$ and $f_2(t)$ is a specific product defined by

$$f_1(t) * f_2(t) = \int_{-\infty}^{\infty} f_1(u) f_2(t-u) du. \tag{3.73}$$

In algebra, convolution follows the laws of commutative, associative, and distributive, respectively. The communitive implies that

$$f_1(t) * f_2(t) = f_2(t) * f_1(t) = \int_{-\infty}^{\infty} f_2(u) f_1(t-u) du. \tag{3.74}$$

In fact,

$$f_1(t) * f_2(t) = -\int_{-\infty}^{\infty} f_1(\tau) f_2(t-\tau) d(t-\tau) \overset{u=t-\tau}{=} -\int_{\infty}^{-\infty} f_1(u-t) f_2(u) du$$

$$= \int_{-\infty}^{\infty} f_1(t-u) f_2(u) du = f_2(t) * f_1(t).$$

The associative means

$$f_1(t)*\left[f_2(t)*f_3(t)\right]=\left[f_1(t)*f_2(t)\right]*f_3(t). \tag{3.75}$$

As a matter of fact, the right side of the above is

$$\begin{aligned}
f_1(t)*\left[f_2(t)*f_3(t)\right] &= \int_{-\infty}^{\infty}\left[\int_{-\infty}^{\infty} f_1(\lambda)f_2(\tau-\lambda)d\lambda\right]f_3(t-\tau)d\tau \\
&= \int_{-\infty}^{\infty} f_1(\lambda)\left[\int_{-\infty}^{\infty} f_2(\tau-\lambda)f_3(t-\tau)d\tau\right]d\lambda \\
&= \int_{-\infty}^{\infty} f_1(\lambda)\left[\int_{-\infty}^{\infty} f_2(\tau)f_3(t-\tau-\lambda)d\tau\right]d\lambda \\
&= f_1(t)*\left[f_2(t)*f_3(t)\right].
\end{aligned}$$

The distributive says

$$f_1(t)*\left[f_2(t)+f_3(t)\right]=\left[f_1(t)*f_2(t)\right]+\left[f_1(t)*f_3(t)\right]. \tag{3.76}$$

Note that

$$\begin{aligned}
f_1(t)*\left[f_2(t)+f_3(t)\right] &= \int_{-\infty}^{\infty} f_1(\tau)\left[f_2(t-\tau)+f_3(t-\tau)\right]d\tau \\
&= \int_{-\infty}^{\infty} f_1(\tau)f_2(t-\tau)d\tau + \int_{-\infty}^{\infty} f_1(\tau)f_3(t-\tau)d\tau \\
&= f_1(t)*f_2(t)+f_1(t)*f_3(t).
\end{aligned}$$

Lemma 3.10: Differential of Convolution. Let $g(t)=f_1(t) * f_2(t)$. If $n=l+m$ (l, $m\geq 0$), we have

$$g^{(n)}(t)=f_1^{(n)}(t)*f_2(t)=f_1(t)*f_2^{(n)}(t)=f_1^{(l)}(t)*f_2^{(m)}(t). \tag{3.77}$$

Proof. In fact,

$$g^{(n)}(t)=\frac{d^n[f_1(t)*f_2(t)]}{dt^n}=\frac{d^n}{dt^n}\int_{-\infty}^{\infty}f_1(t-\tau)f_2(\tau)d\tau$$

$$=\frac{d^l}{dt^l}\frac{d^m}{dt^m}\int_{-\infty}^{\infty}f_1(t-\tau)f_2(\tau)d\tau=\frac{d^l}{dt^l}\int_{-\infty}^{\infty}\frac{d^m}{dt^m}f_1(t-\tau)f_2(\tau)d\tau$$

$$=\frac{d^l}{dt^l}\left[\frac{d^m}{dt^m}f_1(t)*f_2(t)\right]=\frac{d^l}{dt^l}\int_{-\infty}^{\infty}\frac{d^m}{d\tau^m}f_1(\tau)f_2(t-\tau)d\tau$$

$$=\int_{-\infty}^{\infty}\frac{d^m}{d\tau^m}f_1(\tau)\frac{d^l}{dt^l}f_2(t-\tau)d\tau=\frac{d^m}{dt^m}f_1(t)*\frac{d^l}{dt^l}f_2(t).$$

This finishes the proof.

Lemma 3.11: Convolution Rule in the Time Domain. If $f_1(t)\leftrightarrow F_1(\omega)$ and $f_2(t)\leftrightarrow F_2(\omega)$, one has

$$f_1(t)*f_2(t)\leftrightarrow F_1(\omega)F_2(\omega). \tag{3.78}$$

Proof. Because

$$f_1(t)*f_2(t)\leftrightarrow\int_{-\infty}^{\infty}\left[\int_{-\infty}^{\infty}f_1(t-\tau)f_2(\tau)d\tau\right]e^{-i\omega t}dt$$

$$=\int_{-\infty}^{\infty}f_2(\tau)d\tau\int_{-\infty}^{\infty}f_1(t-\tau)e^{-i\omega t}dt=\int_{-\infty}^{\infty}f_2(\tau)e^{-i\omega\tau}d\tau F_1(\omega)$$

$$=F_1(\omega)F_2(\omega),$$

equation (3.78) holds. This completes the proof.

Example 3.26. Give the proof for the following equation.

$$\delta(t)*f(t)=f(t). \tag{3.79}$$

Proof. Let $f(t) \leftrightarrow F(\omega)$. According to Lemma 3.11 and due to $\delta(t) \leftrightarrow 1$, we have

$$\delta(t) * f(t) \leftrightarrow F(\omega). \tag{3.80}$$

Thus, Eq. (3.79) is valid. This completes the proof.

Example 3.27. Find the proof for the following equation.

$$\delta(t - t_0) * f(t) = f(t - t_0). \tag{3.81}$$

Proof. Let $f(t) \leftrightarrow F(\omega)$. According to Lemma 3.11 and due to $\delta(t - t_0) \leftrightarrow e^{-i\omega t_0}$, we obtain

$$\delta(t - t_0) * f(t) \leftrightarrow F(\omega) e^{-i\omega t_0}. \tag{3.82}$$

Note that

$$F(\omega) e^{-i\omega t_0} \leftrightarrow f(t - t_0). \tag{3.83}$$

Therefore, Eq. (3.81) holds. This completes the proof.

Lemma 3.12: Convolution Rule in the Frequency Domain. Let $f_1(t) \leftrightarrow F_1(\omega)$ and $f_2(t) \leftrightarrow F_2(\omega)$. Then,

$$f_1(t) f_2(t) \leftrightarrow \frac{1}{2\pi} F_1(\omega) * F_2(\omega). \tag{3.84}$$

Proof is omitted.

3.7 PRACTICAL CONSIDERATIONS OF SPECTRUM COMPUTATION

A practically measured signal is a finite sample of $f(t)$ with the sample size T. Without losing generality, we suppose that the finite sample is within $(-T/2, T/2)$. Let $g_T(t)$ a gate function in the form

$$g_T(t) = \begin{cases} 1, & -\frac{T}{2} < t < \frac{T}{2}, \\ 0, & \text{otherwise}, \end{cases} \quad a > 0. \tag{3.85}$$

Denote by $F(\omega)$ the true spectrum of $f(t)$. We cannot obtain $F(\omega)$ from a finite sample of $f(t)$ but its approximation in the form

$$F(\omega) \approx \int_{-\frac{T}{2}}^{\frac{T}{2}} f(t) e^{-i\omega t} dt. \tag{3.86}$$

The right side of the above equation can be expressed by

$$\int_{-\frac{T}{2}}^{\frac{T}{2}} f(t)e^{-i\omega t}dt = \int_{-\infty}^{\infty} f(t)g_T(t)e^{-i\omega t}dt. \tag{3.87}$$

Since $g_T(t) \leftrightarrow T\text{Sa}\left(\frac{\omega T}{2}\right)$, according to the convolution rule, we have

$$F(\omega) \approx \frac{1}{2\pi}F(\omega) * T\text{Sa}\left(\frac{\omega T}{2}\right) =_{\text{denoted by}} F_a(\omega). \tag{3.88}$$

The above implies that we obtain $F_a(\omega)$ instead of $F(\omega)$.

Denote the error between $F_a(\omega)$ and $F(\omega)$ by $E_a(\omega)$. It is given by

$$E_a(\omega) = F(\omega) - F_a(\omega). \tag{3.89}$$

Because

$$\lim_{T\to\infty} \frac{1}{2\pi} T\text{Sa}\left(\frac{\omega T}{2}\right) = \delta(\omega), \tag{3.90}$$

we have

$$\lim_{T\to\infty} F_a(\omega) = F(\omega). \tag{3.91}$$

In other words, for $T \to \infty$, we have $E_a(\omega) = 0$. However, $T \to \infty$ is impossible in practice. Practically, we may arrange T that is large enough such that $F_a(\omega)$ is a satisfactory approximation of $F(\omega)$, referring Li [10–12] for the descriptions of record length requirements.

3.8 TWO INTERESTING EXPRESSIONS

Equation (3.60) is an expression of the Fourier transform of e^t. Another expression of the Fourier transform of e^t is interesting. It is

$$e^t \leftrightarrow 2\pi\delta(\omega + i). \tag{3.92}$$

In fact, $1 \leftrightarrow 2\pi\,\delta(\omega)$. According to Property 3.7 (frequency shifting), we have $e^t = e^{i(-it)} \leftrightarrow 2\pi\,\delta(\omega + i)$.

From Eqs. (3.60) and (3.92), we attain another interesting expression in the form

$$\delta(\omega + i) = \sum_{n=0}^{\infty} \frac{(-i)^n \delta^{(n)}(-\omega)}{n!}. \tag{3.93}$$

3.9 A USEFUL CONVOLUTION IN VIBRATIONS

Proposition 3.2. Consider the function in the form

$$h(t)=\frac{1}{m\omega_d}e^{-\zeta\omega_n t}\sin\omega_d t u(t), \tag{3.94}$$

where $u(t)$ is the unit step function. Then,

$$h(t)*h(-t)=\frac{e^{-\zeta\omega_n t}}{(m\omega_d)^2}\left[\begin{array}{l}-\frac{\sin\omega_d t}{4}\frac{(\zeta\omega_n\sin 2\omega_d t+\omega_d\cos 2\omega_d t)}{(\zeta\omega_n)^2+(\omega_d)^2}\\ +\frac{\cos\omega_d t}{4\zeta\omega_n}-\frac{\cos\omega_d t}{4}\frac{(\zeta\omega_n\cos 2\omega_d t-\omega_d\sin 2\omega_d t)}{(\zeta\omega_n)^2+(\omega_d)^2}\end{array}\right]u(t). \tag{3.95}$$

Proof. Note that $h(-t)=-\frac{1}{m\omega_d}e^{\zeta\omega_n t}\sin\omega_d t u(-t)$. Thus,

$$\begin{aligned} h(t)*h(-t)&=-\frac{1}{(m\omega_d)^2}\int_{-\infty}^{\infty}e^{-\zeta\omega_n\tau}\sin\omega_d\tau u(\tau)e^{\zeta\omega_n(t-\tau)}\sin\omega_d(t-\tau)u(\tau-t)d\tau\\ &=-\frac{1}{(m\omega_d)^2}\int_{t}^{\infty}e^{-\zeta\omega_n\tau}\sin\omega_d\tau e^{\zeta\omega_n(t-\tau)}\sin\omega_d(t-\tau)d\tau \qquad (3.96)\\ &=-\frac{1}{(m\omega_d)^2}e^{\zeta\omega_n t}\int_{t}^{\infty}e^{-2\zeta\omega_n\tau}\sin\omega_d\tau\sin\omega_d(t-\tau)d\tau. \end{aligned}$$

Using the trigonometric functions to deal with the last item in the above equation, we have

$$h(t)*h(-t)=\frac{1}{(m\omega_d)^2}\left(\begin{array}{l}-\dfrac{e^{\zeta\omega_n t}\sin\omega_d t}{2}\displaystyle\int_t^{\infty}e^{-2\zeta\omega_n\tau}\sin 2\omega_d\tau\,d\tau\\ +\dfrac{e^{\zeta\omega_n t}\cos\omega_d t}{2}\displaystyle\int_t^{\infty}e^{-2\zeta\omega_n\tau}\,d\tau-\dfrac{e^{\zeta\omega_n t}\cos\omega_d t}{2}\displaystyle\int_t^{\infty}e^{-2\zeta\omega_n\tau}\cos 2\omega_d\tau\,d\tau\end{array}\right)$$

$$=\frac{1}{(m\omega_d)^2}\left[\begin{array}{l}-\dfrac{e^{\zeta\omega_n t}\sin\omega_d t}{2}\dfrac{2e^{-2\zeta\omega_n t}\left(\zeta\omega_n\sin 2\omega_d t+\omega_d\cos 2\omega_d t\right)}{(2\zeta\omega_n)^2+(2\omega_d)^2}\\ +\dfrac{e^{-\zeta\omega_n t}\cos\omega_d t}{4\zeta\omega_n}-\dfrac{e^{\zeta\omega_n t}\cos\omega_d t}{2}\dfrac{2e^{-2\zeta\omega_n t}\left(\zeta\omega_n\cos 2\omega_d t-\omega_d\sin 2\omega_d t\right)}{(2\zeta\omega_n)^2+(2\omega_d)^2}\end{array}\right]$$

$$=\frac{e^{-\zeta\omega_n t}}{(m\omega_d)^2}\left[\begin{array}{l}-\dfrac{\sin\omega_d t}{4}\dfrac{\left(\zeta\omega_n\sin 2\omega_d t+\omega_d\cos 2\omega_d t\right)}{(\zeta\omega_n)^2+(\omega_d)^2}\\ +\dfrac{\cos\omega_d t}{4\zeta\omega_n}-\dfrac{\cos\omega_d t}{4}\dfrac{\left(\zeta\omega_n\cos 2\omega_d t-\omega_d\sin 2\omega_d t\right)}{(\zeta\omega_n)^2+(\omega_d)^2}\end{array}\right].$$

The proof is complete.

3.10 EXERCISES

3.1. Let $f_1(t)=\begin{cases}1, & t\geq 0\\ -1, & t<0\end{cases}$ and $f_2(t)=\begin{cases}1, & t>0\\ -1, & t\leq 0\end{cases}$. Please prove that $f_1(t)-f_2(t)$ is a null function.

3.2. Prove the convolution rule in the frequency domain.

3.3. Prove that $\delta^2(t)$ is undefined.

3.4. Prove that $tu(t)\leftrightarrow i\pi\delta'(\omega)-\dfrac{1}{\omega^2}$, where $i=\sqrt{-1}$ and $u(t)$ is the unit step function.

3.5. Prove that $\cos(\omega_0 t)u(t)\leftrightarrow\dfrac{\pi}{2}\left[\delta(\omega+\omega_0)+\delta(\omega-\omega_0)\right]+\dfrac{i\omega}{\omega_0^2-\omega^2}$.

3.6. Prove that $\sin(\omega_0 t)u(t)\leftrightarrow\dfrac{i\pi}{2}\left[\delta(\omega+\omega_0)-\delta(\omega-\omega_0)\right]+\dfrac{\omega_0}{\omega_0^2-\omega^2}$.

3.7. Let $\mathrm{F}[x(t)]=X(\omega)$. Suppose that $m(\omega)$ is a function of ω. Prove that $\displaystyle\int_{-\infty}^{\infty}m(\omega)x''(t)e^{-i\omega t}d\omega=-\omega^2 m(\omega)X(\omega)$.

3.8. Let $F[x(t)] = X(\omega)$. Suppose that $c(\omega)$ is a function of ω. Prove that $F[c(\omega)x'(t)] = i\omega c(\omega)X(\omega)$.

3.9. Let $F[x(t)] = X(\omega)$. Suppose that $k(\omega)$ is a function of ω. Prove that $F[k(\omega)x(t)] = k(\omega)X(\omega)$.

3.10. Let $F[x(t)] = X(\omega)$. Let $p(t) = m(\omega)x''(t) + c(\omega)x'(t) + k(\omega)x(t)$. Prove that $F[p(t)] = [-\omega^2 m(\omega) + i\omega c(\omega) + k(\omega)]X(\omega)$.

NOTE

1 I purposely prove the well-known Parseval's rule based on the theorem of energy conservation in physics.

REFERENCES

1. E. A. Robinson, A historical perspective of spectrum estimation, *Proceedings of the IEEE*, 70(9), 1982, 885–907.
2. I. M. Gelfand and K. Vilenkin, *Generalized Functions*, vol. 1, Academic Press, New York, 1964.
3. A. Papoulis, *The Fourier Integral and Its Applications*, McGraw-Hill Inc., New York, 1962.
4. D. X. Xia and S. Z. Yan, *Real Functions and Foundation of Applied Functional Analysis*, Shanghai Science and Technology Publishing House, Beijing, China, 1982. In Chinese.
5. P. A. M. Dirac, *Quantum Mechanics*, 4th Ed., Oxford University Press, Oxford, 1958.
6. D. Laugwitz, Early delta functions and the use of infinitesimals in research, *Revue d'histoire des sciences*, 45(1) 1992, Etudes sur Cauchy (1789-1857), 115–128.
7. R. F. Hoskins, *Delta Functions: An Introduction to Generalized Functions*, Woodhead Publishing Limited, Cambridge, 2009.
8. J. Lützen, The prehistory of the theory of distributions, in: G. J. Toomer (Ed.), *Studies in the History of Mathematics and Physical Sciences*, vol. 7, Springer, New York-Berlin, 1982, pp. 112–115.
9. J. L. Zheng, Q. Y. Ying, and W. L. Yang, *Signals and Systems*, 2nd Ed., vol. 1, Higher Education Press, Beijing, 2000. In Chinese.
10. M. Li, *Multi-Fractal Traffic and Anomaly Detection in Computer Communications*, CRC Press, Boca Raton, FL, 2022.
11. M. Li, Record length requirement of long-range dependent teletraffic, *Physica A*, 472, 2017, 164–187.
12. M. Li, A method for requiring block size for spectrum measurement of ocean surface waves, *IEEE Transactions on Instrumentation and Measurement*, 55(6), 2006, 2207–2215.

CHAPTER 4

Responses Excited by Deterministically Aperiodic Forces

THIS CHAPTER CONTAINS THE author's discussions on responses to a mass-spring system under the deterministically non-periodic forces. The points addressed in this chapter are as follows:

1. Frequency transfer functions, input-output analysis method.
2. Impulse response and step one.
3. The present Theorems 4.1 and 4.2 and Corollaries 4.1 and 4.2 in this chapter suggest new ideas for dealing with vibration systems having multiple degrees of freedom in Section 4.4.
4. Problems regarding bandlimited signals are stated in Section 4.5.
5. Auto-power spectrum response is discussed in Section 4.6.
6. Autocorrelation response is explained in Section 4.7.
7. Cross-correlation response is given in Section 4.8.
8. Cross-power spectrum response is discussed in Section 4.9.
9. Mean square response is given in Section 4.10.

DOI: 10.1201/9781003460947-5

4.1 FREQUENCY TRANSFER FUNCTION

Consider a vibration equation in the time domain in the form

$$x'' + 2\omega_n \varsigma x' + \omega_n^2 x = \frac{f(t)}{m}. \tag{4.1}$$

Let $x(t) \leftrightarrow X(\omega)$ and $f(t) \leftrightarrow F(\omega)$. Doing the Fourier transform on both sides of the above yields its corresponding equation in the frequency domain in the form

$$\left[(i\omega)^2 + 2\omega_n \varsigma (i\omega) + \omega_n^2\right] X(\omega) = \frac{F(\omega)}{m}.$$

By using the Fourier transform, a differential equation in the time domain changes to be an algebraic one in the frequency domain. Usually, one writes the above by

$$\left[\left(\omega_n^2 - \omega^2\right) + i2\varsigma\omega_n\omega\right] X(\omega) = \frac{F(\omega)}{m}.$$

The ratio of $X(\omega)$ to $F(\omega)$ is defined as the frequency transfer function of a vibration system. Denote it by $H(\omega)$. Then,

$$H(\omega) = \frac{X(\omega)}{F(\omega)} = \frac{1}{m} \frac{1}{\left(\omega_n^2 - \omega^2\right) + i2\varsigma\omega_n\omega}. \tag{4.2}$$

$H(\omega)$ reflects intrinsic characteristics of a vibration system in the frequency domain. In the polar coordinate system, it is written by

$$H(\omega) = |H(\omega)| e^{i\varphi(\omega)}, \tag{4.3}$$

where $|H(\omega)|$ is called amplitude-frequency transfer function. It is given by

$$|H(\omega)| = \frac{1}{m} \frac{1}{\sqrt{\left(\omega_n^2 - \omega^2\right)^2 + (2\varsigma\omega_n\omega)^2}}. \tag{4.4}$$

Besides, $\varphi(\omega)$ is phase-frequency transfer function. It is in the form

$$\varphi(\omega) = -\arctan \frac{2\varsigma\omega_n\omega}{\omega_n^2 - \omega^2}. \tag{4.5}$$

Multiplying frequency transfer function $H(\omega)$ by excitation $F(\omega)$ produces the frequency response function in the form

$$X(\omega) = H(\omega) F(\omega). \tag{4.6}$$

A response in time is therefore obtained by

$$x(t) = \mathrm{F}^{-1} X(\omega). \tag{4.7}$$

Frequency transfer function $H(\omega)$ is also termed system transfer function that transfers input/excitation of a vibration system to output/response of the system in the frequency domain. In vibrations, such an analysis is called input-output method. For facilitating analysis, a vibration system may be characterized by $H(\omega)$, see Figure 4.1. Theoretically, the diagram in Figure 4.1 suffices to describe the relationship between the input and output of a vibration system.

Using the frequency ratio $\gamma = \dfrac{\omega}{\omega_n}$ and taking into account $\omega_n^2 = \dfrac{k}{m}$, one writes $H(\omega)$ as

$$H(\omega) = \frac{1}{k} \frac{1}{\left(1-\gamma^2\right) + i2\varsigma\gamma}. \tag{4.8}$$

In the polar coordinate system, $|H(\omega)|$ is

$$|H(\omega)| = \frac{1}{k} \frac{1}{\sqrt{\left(1-\gamma^2\right)^2 + (2\varsigma\gamma)^2}}. \tag{4.9}$$

$\varphi(\omega)$ is in the form

$$\varphi(\omega) = -\arctan \frac{2\varsigma\gamma}{1-\gamma^2}. \tag{4.10}$$

Figure 4.2 demonstrates the plots of $H(\omega)$ and $\varphi(\omega)$.

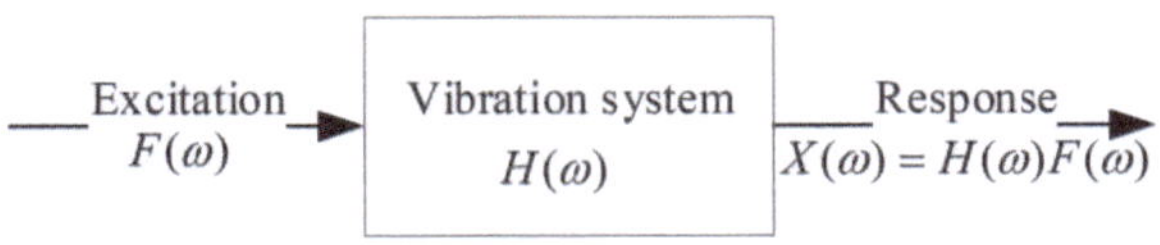

FIGURE 4.1 Input-output description of a vibration system in the frequency domain.

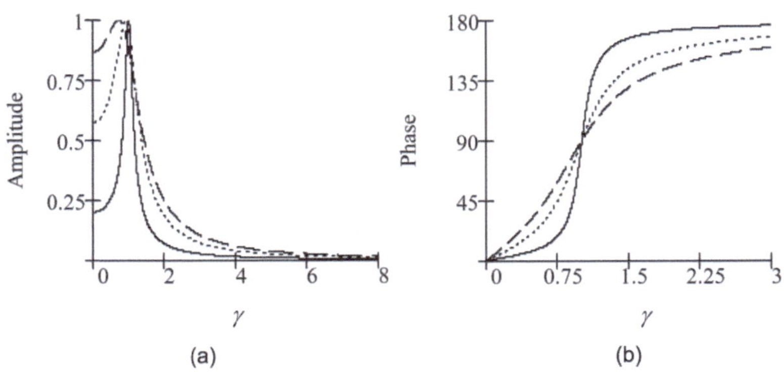

FIGURE 4.2 Demonstrations of $|H(\omega)|$ and $\varphi(\omega)$ (Deg). (a) $|H(\omega)|$ when $\zeta=0.1$ (solid), 0.3 (dot), 0.5 (dash). (b) $\varphi(\omega)$ when $\zeta=0.1$ (solid), 0.3 (dot), and 0.5 (dash).

4.2 IMPULSE RESPONSE FUNCTION

If $f(t)=\delta(t)$, $F(\omega)=1$. Accordingly, $X(\omega)=H(\omega)F(\omega)=H(\omega)$ when $f(t)=\delta(t)$. The response $\mathrm{F}^{-1}H(\omega)$ is called the impulse response function of a vibration system. Denote it by $h(t)$. Then,

$$h(t)=\mathrm{F}^{-1}H(\omega). \tag{4.11}$$

Thus, $h(t)$ and $H(\omega)$ constitute a pair of Fourier transform. An impulse response function $h(t)$ is a specific kind of zero-state response of a vibration system under the condition of $f(t)=\delta(t)$. It completely characterizes a vibration system in the time domain. We write

$$h''(t)+2\omega_n\varsigma h'(t)+\omega_n^2 h(t)=\frac{\delta(t)}{m}. \tag{4.12}$$

The expression of $h(t)$ is given by

$$h(t)=\frac{1}{m\omega_d}e^{-\zeta\omega_n t}\sin\omega_d t, \quad t\geq 0. \tag{4.13}$$

The plot of $h(t)$ is indicated in Figure 4.3.

Based on the convolution rule in the time domain, from Eq. (4.6), one has

$$x(t)=h(t)*f(t). \tag{4.14}$$

In the time domain, therefore, the input-output method can be regarded as follows. Response $x(t)$ equals to the convolution of system's impulse response $h(t)$ and excitation $f(t)$. Figure 4.4 gives the diagram.

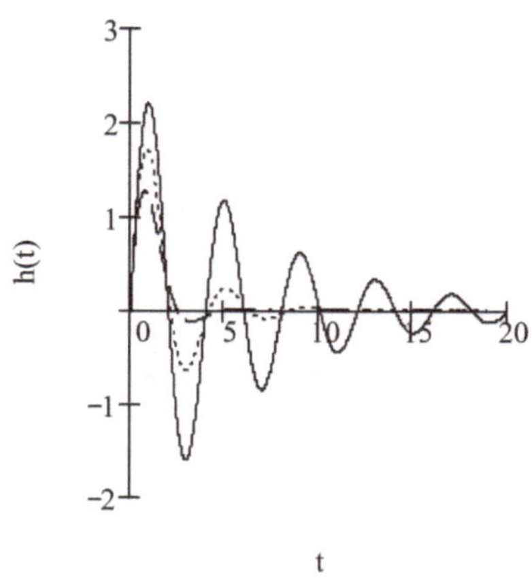

FIGURE 4.3 Impulse response with $m=0.25$ and $\omega_n=\pi/2$ when $\zeta=0.1$ (solid), 0.3 (dot), 0.6 (dash).

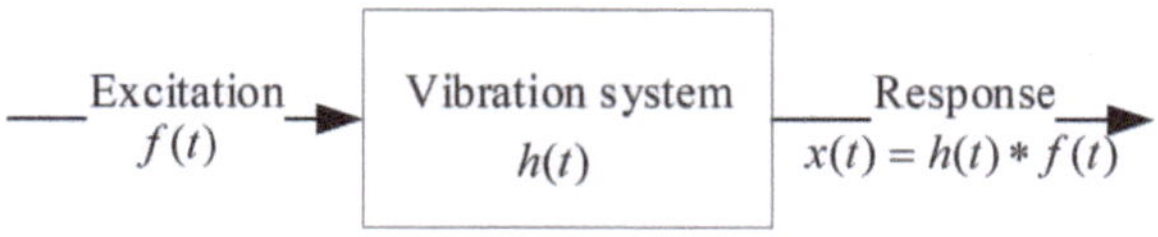

FIGURE 4.4 Input-output description of a vibration system in the time domain.

4.3 STEP RESPONSE FUNCTION

When the excitation force $f(t)=u(t)$, the response $x(t)$ is called the unit step response function. Denote it by $g(t)$. Then, one has

$$g''(t)+2\omega_n\varsigma g'(t)+\omega_n^2 g(t)=\frac{u(t)}{m}. \tag{4.15}$$

Note that $g(t)$ is a kind of zero-state response under the condition of $f(t)=u(t)$. It is related to $h(t)$ by

$$g(t)=\int_0^t h(t)dt. \tag{4.16}$$

Thus, one has

$$g(t)=\frac{1}{k}\left[1-\frac{e^{-\zeta\omega_n t}}{\sqrt{1-\zeta^2}}\cos(\omega_d t-\phi)\right],$$
$$\phi=\arctan\frac{\zeta}{\sqrt{1-\zeta^2}}. \tag{4.17}$$

Figure 4.5 gives the illustration of $g(t)$.

When using step response, we may express the relationship between input and output by

$$x(t)=h(t)*f(t)=g'(t)*f(t). \tag{4.18}$$

According to the differential property of convolution, we have

$$x(t)=g(t)*f'(t). \tag{4.19}$$

In vibrations, rise time, which is denoted by t_r, is a parameter to describe the dynamic response property of a vibration system. It may be defined by

$$t_r=t_2-t_1, \tag{4.20}$$

where $g(t_2)$ is the maximum of $g(t)$ while $g(t_1)$ is the minimum of $g(t)$, see Figure 4.6.

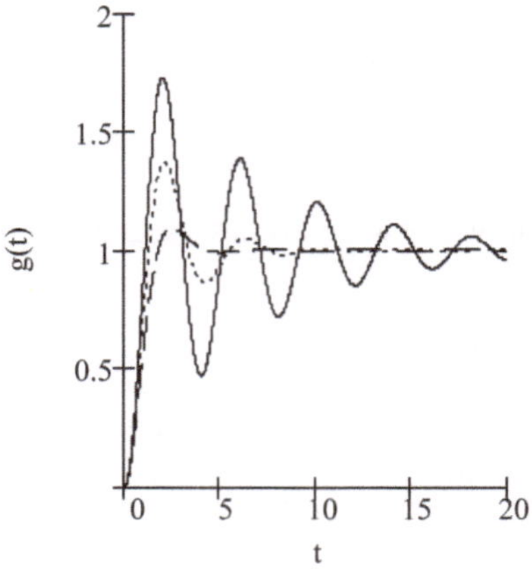

FIGURE 4.5 Unit step response with $m=0.25$ and $\omega_n=\pi/2$ when $\zeta=0.1$ (solid), 0.3 (dot), and 0.6 (dash).

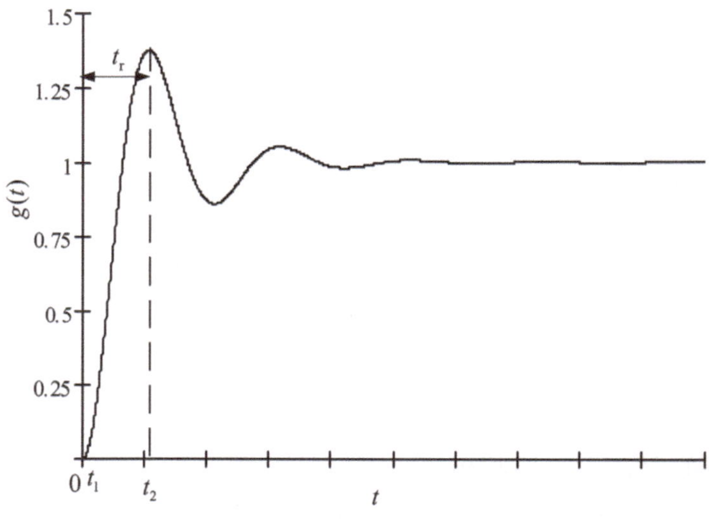

FIGURE 4.6 Illustration of rise time.

Let f_c be the cutoff frequency of a vibration system. According to the theory of linear systems (Zheng et al. [1, p. 290], Li et al. [2], Li [3]), one has

$$t_r = \frac{1}{f_c}. \tag{4.21}$$

The above means that rise time t_r is inversely proportional to the frequency band of a vibration system. In fact, it is a delay of the step response $g(t)$ in comparison with the excitation $u(t)$. Thus, the wider the bandwidth of a vibration system, the faster the output $g(t)$ responds to input $u(t)$.

4.4 APPLICATIONS OF IMPULSE RESPONSE AND FREQUENCY TRANSFER FUNCTIONS

This subsection is about my study of a vibration system that consists of several vibration subsystems in series or in parallel. The present results may yet be useful for vibration systems with multiple degrees of freedom.

4.4.1 Vibration Systems in Series

Let us consider a frequency transfer function and impulse response of a vibration system that consists of two subsystems in series as indicated in Figure 4.7.

The first subsystem, denoted by $h_1(t)$, is expressed by

$$x_1'' + 2\omega_{n1}\varsigma_1 x_1' + \omega_{n1}^2 x_1 = \frac{f_1(t)}{m_1}. \tag{4.22}$$

Since the output, $x_1(t)$, of the subsystem $h_1(t)$, is the excitation of the second subsystem $h_2(t)$, the motion equation of the second subsystem is expressed by

$$x_2'' + 2\omega_{n2}\varsigma_2 x_2' + \omega_{n2}^2 x_2 = \frac{x_1(t)}{m_2}. \tag{4.23}$$

Theorem 4.1. Let $h(t)$ be the impulse response of the total vibration system in Figure 4.7. Then,

$$h(t) = h_1(t) * h_2(t), \tag{4.24}$$

where

$$h_1(t) = \frac{1}{m_1\omega_{d1}} e^{-\varsigma_1\omega_{n1}t} \sin\omega_{d1}t, \quad t \geq 0, \tag{4.25}$$

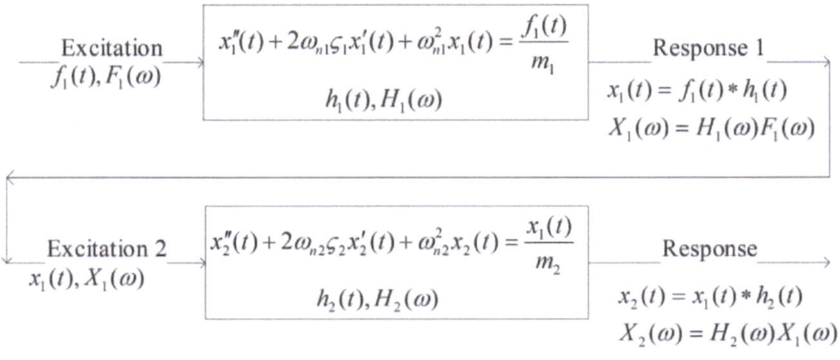

FIGURE 4.7 Two vibration subsystems in series.

where

$$\omega_{d1} = \sqrt{1-\zeta_1^2}\,\omega_{n1}, \quad |\zeta_1| < 1, \tag{4.26}$$

is the damped natural angular frequency of the first subsystem with small damping. In Eq. (4.24),

$$h_2(t) = \frac{1}{m_2\omega_{d2}} e^{-\zeta_2\omega_{n2}t} \sin\omega_{d2}t, \quad t \geq 0, \tag{4.27}$$

where

$$\omega_{d2} = \sqrt{1-\zeta_2^2}\,\omega_{n2}, \quad |\zeta_2| < 1, \tag{4.28}$$

is the damped natural angular frequency of the second subsystem with small damping.

Proof. Using convolution, we have

$$x_1(t) = f_1(t) * h_1(t). \tag{4.29}$$

Moreover, we have

$$x_2(t) = x_1(t) * h_2(t) = f_1(t) * h_1(t) * h_2(t). \tag{4.30}$$

The above means that Eq. (4.24) holds. The proof is finished.

Theorem 4.1 says that if two systems $h_1(t)$ and $h_2(t)$ are in series, the total impulse response $h(t)$ is the convolution of $h_1(t)$ and $h_2(t)$. Based on that, we have the following corollary.

Corollary 4.1. Let $H(\omega)$ be the frequency transfer function of the total vibration system in Figure 4.7. Then,

$$H(\omega) = H_1(\omega)H_2(\omega), \tag{4.31}$$

where

$$\begin{aligned} H_1(\omega) &= \frac{1}{m_1}\frac{1}{\left(\omega_{n1}^2-\omega^2\right)+i2\varsigma_1\omega_{n1}\omega}, \\ H_2(\omega) &= \frac{1}{m_2}\frac{1}{\left(\omega_{n2}^2-\omega^2\right)+i2\varsigma_2\omega_{n2}\omega}. \end{aligned} \tag{4.32}$$

Proof. Using the convolution rule in the time domain for Eq. (4.24) yields Eq. (4.31). Thus, Corollary 4.1 holds. This finishes the proof.

In the polar system,

$$H_1(\omega) = |H_1(\omega)| \arg H_1(\omega) = \frac{1}{m_1}\frac{1}{\sqrt{\left(\omega_{n1}^2-\omega^2\right)^2+\left(2\varsigma_1\omega_{n1}\omega\right)^2}} e^{i\varphi_1(\omega)}, \tag{4.33}$$

where

$$\varphi_1(\omega) = -\arctan\frac{2\varsigma_1\omega_{n1}\omega}{\omega_{n1}^2 - \omega^2}. \tag{4.34}$$

On the other hand,

$$H_2(\omega) = |H_2(\omega)| \arg H_2(\omega) = \frac{1}{m_2}\frac{1}{\sqrt{\left(\omega_{n2}^2 - \omega^2\right)^2 + \left(2\varsigma_2\omega_{n2}\omega\right)^2}} e^{i\varphi_2(\omega)}, \tag{4.35}$$

where

$$\varphi_2(\omega) = -\arctan\frac{2\varsigma_2\omega_{n2}\omega}{\omega_{n2}^2 - \omega^2}. \tag{4.36}$$

Thus, we write

$$H(\omega) = |H(\omega)| e^{i\varphi(\omega)}, \tag{4.37}$$

where

$$|H(\omega)| = |H_1(\omega)||H_2(\omega)| = \frac{1}{m_1 m_2}\frac{1}{\sqrt{\left(\omega_{n1}^2 - \omega^2\right)^2 + (2\varsigma_1\omega_{n1}\omega)^2}\sqrt{\left(\omega_{n2}^2 - \omega^2\right)^2 + (2\varsigma_2\omega_{n2}\omega)^2}}, \tag{4.38}$$

and

$$\varphi(\omega) = \varphi_1(\omega) + \varphi_2(\omega). \tag{4.39}$$

Note that there are two natural frequencies in $H(\omega)$. We write them by

$$\omega_{n1} \leq \omega_{n2}. \tag{4.40}$$

Theorem 4.1 can be extended to the general case of N subsystems in series as shown in Figure 4.8.

Theorem 4.2. Let $h(t)$ be the impulse response function of a vibration system that consists of N subsystems in series as indicated in Figure 4.8. Then,

$$h(t) = h_1(t) * \ldots h_j(t) * \ldots\ h_N(t), \tag{4.41}$$

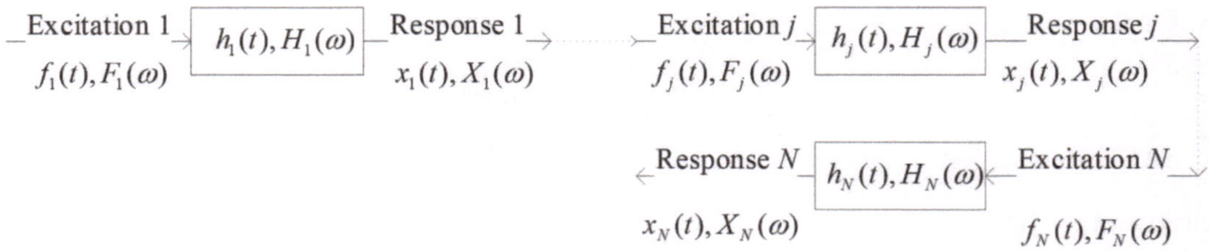

FIGURE 4.8 N vibration subsystems in series.

where

$$h_j(t)=\frac{1}{m_j\omega_{dj}}e^{-\zeta_j\omega_{nj}t}\sin\omega_{dj}t,\quad t\geq 0, \tag{4.42}$$

where, in the case of small damping,

$$\omega_{dj}=\sqrt{1-\zeta_j^2}\,\omega_{nj},\quad |\zeta_j|<1. \tag{4.43}$$

Proof. The jth subsystem follows

$$x_j''+2\omega_{nj}\varsigma_j x_j'+\omega_{nj}^2 x_j=\frac{f_j(t)}{m_j},\quad j=1,2,\ldots,N. \tag{4.44}$$

For the jth subsystem, we have

$$x_j(t)=f_j(t)*h_j(t), \tag{4.45}$$

where

$$f_j(t)=x_{j-1}(t). \tag{4.46}$$

Since

$$x_{j-1}(t)=f_{j-1}(t)*h_{j-1}(t), \tag{4.47}$$

we have

$$x_N(t)=f_N(t)*h_N(t)=f_{N-1}(t)*h_N(t)*h_{N-1}(t)*\ldots=f_1(t)*h_N(t)*h_{N-1}(t)*\ldots h_1(t). \tag{4.48}$$

This finishes the proof.

Corollary 4.2. Let $H(\omega)$ be the frequency transfer function of the total system in Figure 4.8. Then,

$$H(\omega)=H_1(\omega)H_2(\omega)\ldots H_N(\omega)=\prod_{j=1}^{N}H_j(\omega), \tag{4.49}$$

where

$$H_j(\omega)=\frac{1}{m_j}\frac{1}{\left(\omega_{nj}^2-\omega^2\right)+i2\varsigma_j\omega_{nj}\omega}. \tag{4.50}$$

Proof. Applying the convolution rule in the time domain to Eq. (4.48) results in Eq. (4.49). The proof is complete.

From Corollary 4.2, we have

$$H(\omega)=\prod_{j=1}^{N}\frac{1}{m_j}\frac{1}{\left(\omega_{nj}^2-\omega^2\right)+i2\varsigma_j\omega_{nj}\omega}. \tag{4.51}$$

The amplitude of $H(\omega)$ is given by

$$|H(\omega)|=\prod_{j=1}^{N}\frac{1}{m_j}\frac{1}{\sqrt{\left(\omega_{nj}^2-\omega^2\right)^2+\left(2\varsigma_j\omega_{nj}\omega\right)^2}}. \tag{4.52}$$

The phase of $H(\omega)$ is expressed by

$$\varphi(\omega)=\sum_{j=1}^{N}\varphi_j(\omega), \tag{4.53}$$

where

$$\varphi_j(\omega)=-\arctan\frac{2\varsigma_j\omega_{nj}\omega}{\omega_{nj}^2-\omega^2}. \tag{4.54}$$

Note that there are N natural frequencies in such a vibration system. We write them by

$$\omega_{n1}\le\omega_{n2}\le\ldots\le\omega_{nN}. \tag{4.55}$$

Example 4.1. Suppose there are three vibration subsystems in series. Their frequency transfer functions are given by

$$H_1(\omega)=\frac{1}{m_1}\frac{1}{\left(\omega_{n1}^2-\omega^2\right)+i2\varsigma_1\omega_{n1}\omega},\quad H_2(\omega)=\frac{1}{m_2}\frac{1}{\left(\omega_{n2}^2-\omega^2\right)+i2\varsigma_2\omega_{n2}\omega},$$

$$H_3(\omega)=\frac{1}{m_3}\frac{1}{\left(\omega_{n3}^2-\omega^2\right)+i2\varsigma_3\omega_{n3}\omega}.$$

Let $m_1=m_2=m_3=1$, $\omega_{n1}=1$, $\omega_{n2}=3$, $\omega_{n3}=6$, $\zeta_1=0.05$, $\zeta_2=0.015$, $\zeta_3=0.001$. Figure 4.9 shows their plots of amplitudes.

The total frequency transfer function is given by

$$H(\omega)=H_1(\omega)H_2(\omega)H_3(\omega). \tag{4.56}$$

The plot of $|H(\omega)|$ is indicated in Figure 4.10.

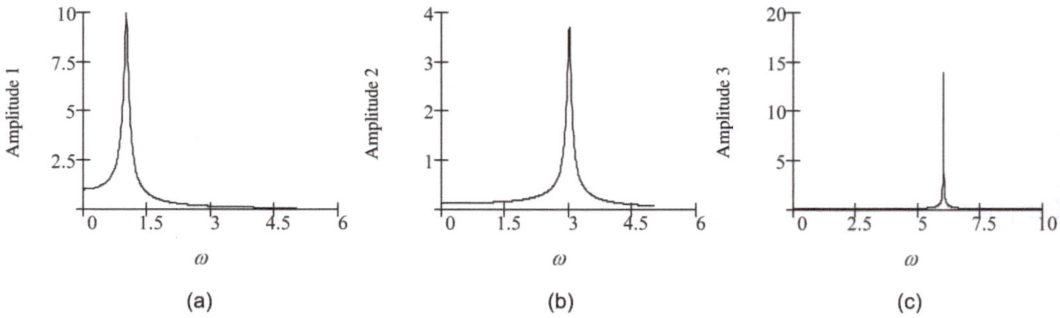

FIGURE 4.9 Plots of amplitudes of frequency transfer functions. (a) $|H_1(\omega)|$. (b) $|H_2(\omega)|$. (c) $|H_3(\omega)|$.

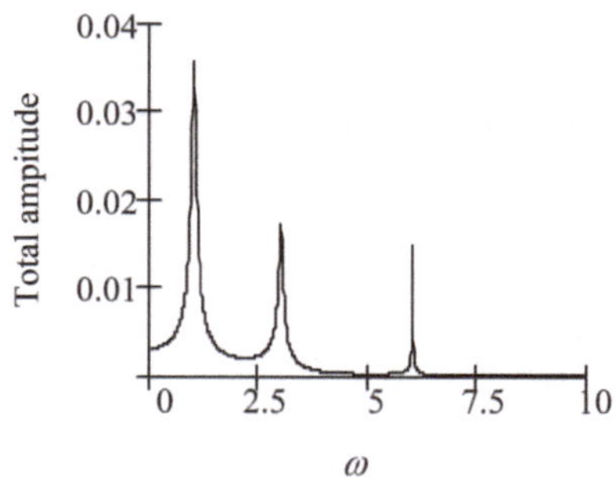

FIGURE 4.10 Plot of total amplitude of frequency transfer function $|H(\omega)|=|H_1(\omega)H_2(\omega)H_3(\omega)|$.

4.4.2 Vibration Systems in Parallel

Suppose that a vibration system consists of N subsystems in parallel. In that case, we have

$$H(\omega)=\sum_{j=1}^{N} H_j(\omega). \tag{4.57}$$

Figure 4.11 indicates a vibration system consisting of two subsystems.

Example 4.2. Suppose there are three vibration subsystems in parallel. Each subsystem is the same as that in Example 4.1. Then, $H(\omega)=H_1(\omega)+H_2(\omega)+H_3(\omega)$. Figure 4.12 gives the plot of $|H(\omega)|$.

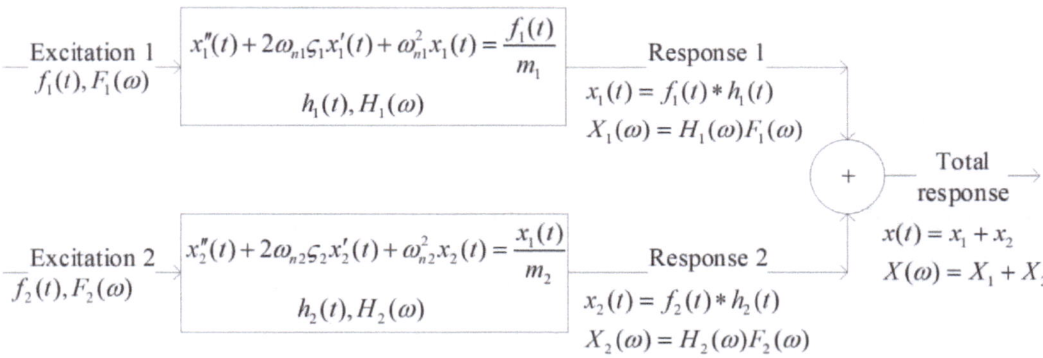

FIGURE 4.11 A vibration system with two subsystems in parallel. $H(\omega)=H_1(\omega)+H_2(\omega)$.

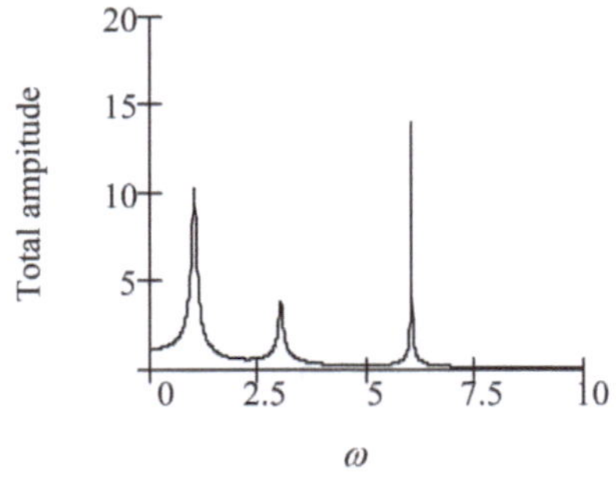

FIGURE 4.12 Plot of total amplitude of frequency transfer function $|H(\omega)|=|H_1(\omega)+H_2(\omega)+H_3(\omega)|$.

4.5 PROBLEMS ABOUT BANDLIMITED SIGNALS

Denote by $X(\omega)$ the spectrum of $x(t)$. Then, $x(t)$ is bandlimited to $(-\omega_c, \omega_c)$ if $X(\omega)=0$ for $\omega \notin (-\omega_c, \omega_c)$ (Papoulis [4,5]). In addition, $x(t)$ is said to be timelimited to $(-T/2, T/2)$ when $x(t)=0$ for $t \notin (-T/2, T/2)$ (Papoulis [4,5], Li and Zhao [6]). Papoulis addressed the fact that a signal $x(t)$ cannot be both bandlimited and timelimited. To be precise, if $x(t)$ is bandlimited, it is not timelimited. Similarly, when $x(t)$ is timelimited, it is not bandlimited. In engineering, on the one side, one can only use a finite sample of $x(t)$. Thus, a practical time signal is timelimited. On the other side, a vibration system is generally bandlimited (Papoulis [4,5]). Consequently, a response signal is both timelimited and bandlimited. That is a problem encountered in engineering. An explanation may be like this.

A vibration signal is timelimited. Theoretically, it is usually not bandlimited but it is regarded as a bandlimited one in the sense of maximum effective frequency in vibration engineering (Harris [7]). Usually, as a matter of fact, an excitation signal first passes through an anti-aliasing filter for sampling (Harris [7]). Denote by $H_a(\omega)$ the frequency transfer function of an anti-aliasing filter, see Chapter 29. It may be in the form

$$H_a(\omega)=\begin{cases} 1, & -\omega_c<\omega<\omega_c, \\ 0, & \text{otherwise.} \end{cases} \tag{4.58}$$

Denote by $x_o(t)$ the output of the anti-aliasing filter. Its Fourier transform is $F_o(\omega)$. Then,

$$F_o(\omega)=H_a(\omega)X(\omega). \tag{4.59}$$

The above implies that the output of the anti-aliasing filter is of course bandlimited.

4.6 AUTO-POWER SPECTRUM RESPONSE

Denote the complex conjugates of $X(\omega)$, $H(\omega)$, and $F(\omega)$ by $X^*(\omega)$, $H^*(\omega)$, and $F^*(\omega)$, respectively. Then, based on Eq. (4.6), we have

$$X(\omega)X^*(\omega)=H(\omega)H^*(\omega)F(\omega)F^*(\omega). \tag{4.60}$$

Owing to $X(\omega)X^*(\omega)=|X(\omega)|^2$, $H(\omega)H^*(\omega)=|H(\omega)|^2$, and $F(\omega)F^*(\omega)=|F(\omega)|^2$, we rewrite the above by

$$|X(\omega)|^2=|H(\omega)|^2|F(\omega)|^2. \tag{4.61}$$

The quantity $|X(\omega)|^2$ is the auto-power spectrum density (PSD for short) function of response $x(t)$ while $|F(\omega)|^2$ is the PSD of excitation $f(t)$. Let $S_{xx}(\omega)=|X(\omega)|^2$ and $S_{ff}(\omega)=|F(\omega)|^2$. Then, we write the auto-power spectrum response by

$$S_{xx}(\omega)=|H(\omega)|^2 S_{ff}(\omega). \tag{4.62}$$

4.7 AUTOCORRELATION RESPONSE

Using the convolution rule, we have

$$\mathrm{F}^{-1}\left[\left|H(\omega)\right|^{2}\right]=h(t)*h(-t). \tag{4.63}$$

Let $r_{xx}(\tau)$ be the autocorrelation function (ACF) of $x(t)$. Let $r_{ff}(\tau)$ be the ACF of $f(t)$. According to the Wiener-Khinchin theorem (Harris [7], Papoulis [8]), doing the inverse Fourier transform on both sides of Eq. (4.62) produces the autocorrelation response in the form

$$r_{xx}(\tau)=h(\tau)*h(-\tau)*r_{ff}(\tau). \tag{4.64}$$

The above exhibits that $h(t)*h(-t)$ is a useful convolution in vibrations as previously discussed in Section 3.9 in the last chapter.

4.8 CROSS-CORRELATION RESPONSE

From Eq. (4.18), we write the shifted response by

$$x(t+\tau)=\int_{0}^{\infty}h(u)f(t+\tau-u)du. \tag{4.65}$$

Multiplying the excitation $f(t)$ on both sides of the above yields

$$f(t)x(t+\tau)=\int_{0}^{\infty}h(u)f(t)f(t+\tau-u)du. \tag{4.66}$$

Denote by E the mean operator. Let $r_{fx}(\mathrm{t})$ be the cross-correlation between the excitation $f(t)$ and the response $x(t)$. Then, $r_{fx}(\mathrm{t})=\mathrm{E}[f(t)x(t+\mathrm{t})]$. Because

$$\mathrm{E}\left[f(t)x(t+\tau)\right]=r_{fx}(\tau)=\int_{0}^{\infty}h(u)\mathrm{E}\left[f(t)f(t+\tau-u)\right]du=\int_{0}^{\infty}h(u)r_{ff}(\tau-u)du, \tag{4.67}$$

we have the cross-correlation response expressed by

$$r_{fx}(t)=h(t)*r_{ff}(t). \tag{4.68}$$

The above equation is called the Wiener–Lee relationship.

4.9 CROSS-POWER SPECTRUM RESPONSE

Let $S_{fx}(\omega)$ be the cross-power spectrum between $f(t)$ and $x(t)$. Using the convolution on Eq. (4.68) produces the cross-power spectrum response given by

$$S_{fx}(\omega)=H(\omega)S_{ff}(\omega). \tag{4.69}$$

Note that the responses discussed in Sections 4.6–4.9 are for deterministically non-periodic excitations from the point of view of correlations and power spectra.

4.10 MEAN SQUARE RESPONSE

Since the mean square response is $E[x^2(t)]$ and $r_{xx}(\tau)=E[x(t)x(t+\tau)]$, we have the mean square response in the form

$$E\left[x^2(t)\right]=r_{xx}(0). \tag{4.70}$$

4.11 EXERCISES

4.1. Suppose that a vibration system consists of three subsystems in series. Let B_1, B_2, and B_3 be the 0.707 frequency bandwidth of each subsystem, respectively. Denote the 0.707 frequency bandwidth of the total system by B. Please interpret $B=\min(B_1, B_2, B_3)$. That is, B_1 is the minimum among B_1, B_2, and B_3.

4.2. Let $g(t)$ be a pulse function. Its pulse amplitude is 1 and pulse width is τ. Please draw the plots of the following function for $n=1, 2, 3, 4, 5, 6$.

$$\{F[g(t)]\}^n=G^n(\omega)=\left(\tau\frac{\sin\frac{\tau\omega}{2}}{\frac{\tau\omega}{2}}\right)^n.$$

Let $B(n)$ be the 0.707 bandwidth of $G^n(\omega)$ for $\tau=1$. Please observe $B(n)$ in terms of n by drawing its plot.

4.3. Denote the frequency ratio by $\gamma=\frac{\omega}{\omega_n}$, where $\omega=\sqrt{\frac{k}{m}}$. The frequency transfer function is given by

$$H(\omega)=\frac{1}{k}\frac{1}{\left(1-\gamma^2\right)+i2\varsigma\gamma},$$

where ς is damping ratio. Denote the 0.707 frequency bandwidth of the system by $B_{0.707}$ and the 0.1 frequency bandwidth of the system by $B_{0.1}$. Define the rectangular coefficient of $H(\omega)$ by $K_{0.1}=B_{0.1}/B_{0.707}$. Show that $K_{0.1}\approx 10$.

4.4. Let

$$H_1(\omega)=\frac{1}{m_1}\frac{1}{\left(\omega_{n1}^2-\omega^2\right)+i2\varsigma_1\omega_{n1}\omega},\qquad H_2(\omega)=\frac{1}{m_2}\frac{1}{\left(\omega_{n2}^2-\omega^2\right)+i2\varsigma_2\omega_{n2}\omega},$$

where $\omega_{n1}=1$, $\omega_{n2}=1.28$, $\zeta_1=0.13$, $\zeta_2=0.08$, $m_1=m_2=1$. For a system given by

$$H(\omega)=H_1(\omega)H_2(\omega)=\frac{1}{m_1m_2}\left[\frac{1}{\left(\omega_{n1}^2-\omega^2\right)+i2\varsigma_1\omega_{n1}\omega}\right]\left[\frac{1}{\left(\omega_{n2}^2-\omega^2\right)+i2\varsigma_2\omega_{n2}\omega}\right],$$

find the rectangular coefficient $K_{0.1}$ of $H(\omega)$.

REFERENCES

1. J. L. Zheng, Q. Y. Ying, and W. L. Yang, *Signals and Systems*, 2nd Ed., vol. 1, Higher Education Press, Beijing, 2000. In Chinese.
2. M. Li, B.-H. Xu, and Y.-S. Wu, An H_2-optimal control of random loading for a laboratory fatigue test, *Journal of Testing and Evaluation*, 26(6), 1998, 619–625.
3. M. Li, Experimental stability analysis of test system for doing fatigue test under random loading, *Journal of Testing and Evaluation*, 34(4), 2006, 364–367.
4. A. Papoulis, Limits on bandlimited signals, *Proceedings of the IEEE*, 55(10), 1967, 1667–1686.
5. A. Papoulis, *Signal Analysis*, McGraw-Hill, New York, 1977.
6. M. Li and W. Zhao, On bandlimitedness and lag-limitedness of fractional Gaussian noise, *Physica A*, 392(9), 2013, 1955–1961.
7. C. M. Harris, *Shock and Vibration Handbook*, 5th Ed., McGraw-Hill, New York. 2002.
8. A. Papoulis, *The Fourier Integral and Its Applications*, McGraw-Hill, New York, 1962.

CHAPTER 5

Vibrations with Multiple Degrees-of-Freedom

VIBRATIONS DISCUSSED IN PREVIOUS chapters, except Section 4.4, are in the model of mass-spring systems with single degree-of-freedom (SDOF). That model is not only crucially useful as a vibration foundation in education but also important in practice such as when designing vibration measurement instruments. However, there are many vibration phenomena that may need a vibration model with multiple degrees-of-freedom (MDOF), such as ship hull. This chapter deals with vibrations with MDOF. An interesting novelty presented in this chapter is the function form of a general solution to a vibration system with MDOF in Theorems 5.1–5.4 and Corollaries 5.1–5.4.

5.1 DEGREES-OF-FREEDOM

Suppose that there are j coordinates in a mechanical system. If there are k restriction equations about that system, the degrees-of-freedom (DOF) of that system equals to $(j-k)$ [1].

Example 5.1. Consider a double pendulum oscillation system in Figure 5.1. It consists of two masses m_1 and m_2. They are connected with two inextensible bars l_1 and l_2. Find its DOF.

Solution: There are four coordinates x_1, x_2, y_1, and y_2 in the x-y plane. Those four satisfy the restriction equations in the following form:

$$\begin{cases} x_1^2 + y_1^2 = l_1^2, \\ x_2^2 + y_2^2 = l_2^2. \end{cases} \tag{5.1}$$

According to the definition of DOF, the DOF of this system is 2.

DOI: 10.1201/9781003460947-6

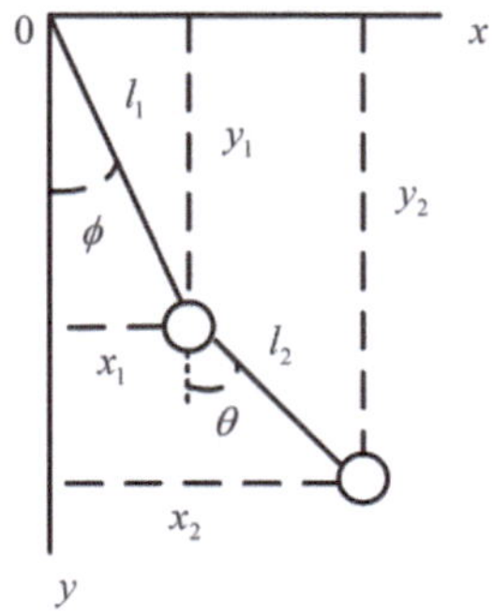

FIGURE 5.1 Double pendulum.

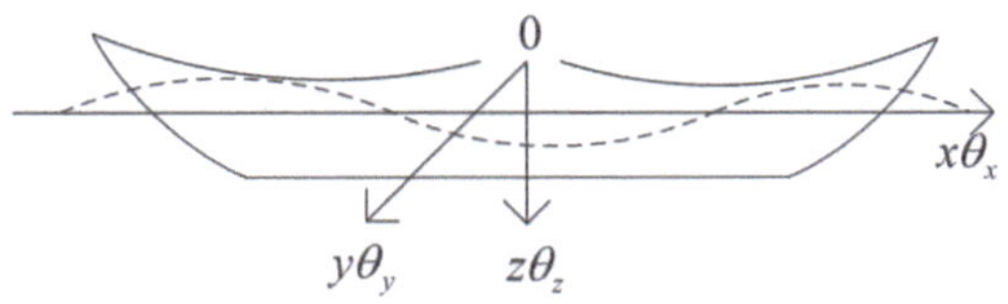

FIGURE 5.2 Illustration of a ship hull.

Since the DOF of the double pendulum system is 2, it suffices to use two coordinates to determine the positions of two pendulums. In fact, with two coordinates ϕ and θ, one has

$$\begin{cases} x_1 = l_1 \sin\phi, & y_1 = l_1 \cos\phi, \\ x_2 = l_2 \sin\theta, & y_2 = l_2 \cos\theta. \end{cases} \tag{5.2}$$

If DOF of a vibration system is necessary to be two or more than two, we say that the system is of MDOF.

There are some structures whose mass and stiffness are continuously distributed. A typically continuous vibration structure is ship hull, see Figure 5.2.

In ship engineering, a ship hull may be abstracted as a continuous beam. Its DOF is infinite. However, approximately, it may be regarded as a discrete beam. When using the theory of MDOF vibrations to deal with a discrete ship hull beam, we let it be a beam with n concentrate loadings (Figure 5.3). As long as n is large enough, an MDOF beam may be a proper approximation of a continuous one.

The concrete value of n is chosen case-by-case, relying on a ship of interest. In general, when considering the issue of ship hull beam vibration, $n=8$ may be enough.

5.2 VIBRATION EQUATIONS WITH TWO DOF

Consider a vibration system that consists of two mass-spring-damper subsystems connected in series as that in Figure 5.4. Denote by $x_1(t)$ and $x_2(t)$ the displacements of two masses in terms of the equilibrium position.

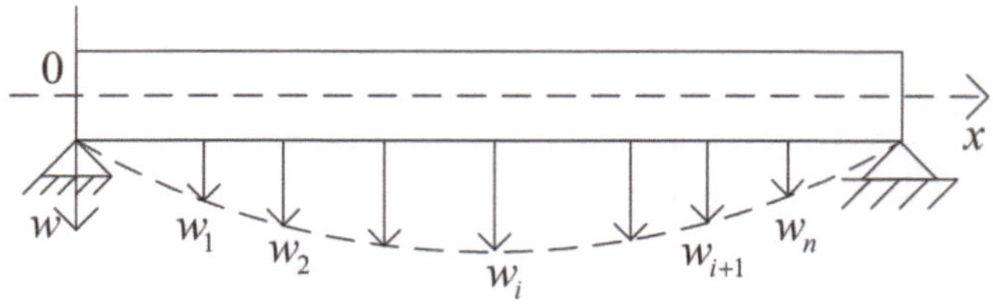

FIGURE 5.3 A discrete beam with n concentrate loadings.

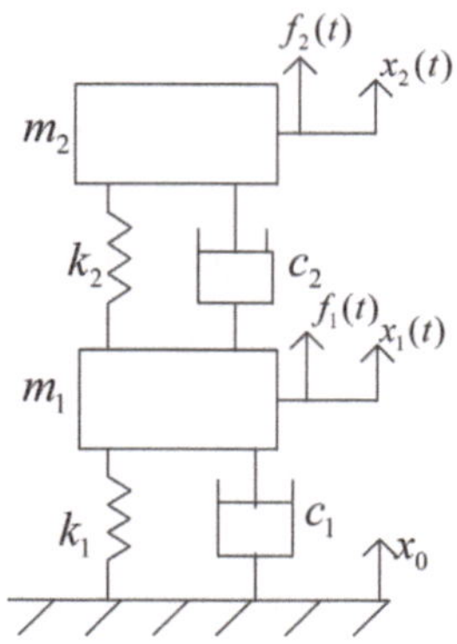

FIGURE 5.4 A system with 2 degrees-of-freedom.

Let $f_1(t)$ and $f_2(t)$ be the driven forces on two masses, respectively. When $x_0 = \dot{x}_0 = 0$, one may write the motion equations of $x_1(t)$ and $x_2(t)$ in the following form:

$$\begin{cases} m_1\ddot{x}_1 = -c_1\dot{x}_1 - k_1x_1 + c_2(\dot{x}_2 - \dot{x}_1) + k_2(x_2 - x_1) + f_1, \\ m_2\ddot{x}_2 = -c_2(\dot{x}_2 - \dot{x}_1) - k_2(x_2 - x_1) + f_2. \end{cases} \tag{5.3}$$

Rewrite the above according to the convention by

$$\begin{cases} m_1\ddot{x}_1 + (c_1 + c_2)\dot{x}_1 + (k_1 + k_2)x_1 - c_2\dot{x}_2 - k_2x_2 = f_1, \\ m_2\ddot{x}_2 + c_2\dot{x}_2 + k_2x_2 - c_2\dot{x}_1 - k_2x_1 = f_2. \end{cases} \tag{5.4}$$

On the left side of the first equation in the above for $x_1(t)$, there are two terms $-c_2(\dot{x}_2 - \dot{x}_1)$ and $-k_2(x_2 - x_1)$ that result from the second mass-spring-damper subsystem. On the other hand, on the left side of the second equation, there are two terms $-c_2\dot{x}_1$ and $-k_2x_1$ that come from the first mass-spring-damper subsystem. Therefore, it is impossible to find the solution $x_1(t)$ directly from the first motion equation in Eq. (5.4). Similarly, we cannot find the solution $x_2(t)$ simply from the second motion equation in Eq. (5.4) either. As a matter of fact, two motion equations are coupled to each other by the coupling terms.

By using the notion of matrix, one may write Eq. (5.4) by

$$\mathbf{M}\{\ddot{x}\} + \mathbf{C}\{\dot{x}\} + \mathbf{K}\{x\} = \{f\}, \tag{5.5}$$

where

$$\mathbf{M}=\begin{bmatrix} m_1 & 0 \\ 0 & m_2 \end{bmatrix},\quad \mathbf{C}=\begin{bmatrix} c_1+c_2 & -c_2 \\ -c_2 & c_2 \end{bmatrix},\quad \mathbf{K}=\begin{bmatrix} k_1+k_2 & -k_2 \\ -k_2 & k_2 \end{bmatrix},$$
$$\{x\}=\begin{Bmatrix} x_1 \\ x_2 \end{Bmatrix},\quad \{f\}=\begin{Bmatrix} f_1 \\ f_2 \end{Bmatrix}. \tag{5.6}$$

In the above, **M**, **C**, and **K** are called mass matrix, damping matrix, and stiffness matrix, respectively. Besides, $\{x\}$ and $\{f\}$ are called displacement vector and force vector, respectively.

For a structure with two DOF, one may write its motion equations in general form

$$\begin{cases} m_{11}\ddot{x}_1+m_{12}\ddot{x}_2+c_{11}\dot{x}_1+c_{12}\dot{x}_2+k_{11}x_1+k_{12}x_2=f_1, \\ m_{21}\ddot{x}_1+m_{22}\ddot{x}_2+c_{21}\dot{x}_1+c_{22}\dot{x}_2+k_{21}x_1+k_{22}x_2=f_2. \end{cases} \tag{5.7}$$

Expressing the above with Eq. (5.5) yields the general cases of **M**, **C**, and **K** in the following form:

$$\mathbf{M}=\begin{bmatrix} m_{11} & m_{12} \\ m_{21} & m_{22} \end{bmatrix},\quad \mathbf{C}=\begin{bmatrix} c_{11} & c_{12} \\ c_{21} & c_{22} \end{bmatrix},\quad \mathbf{K}=\begin{bmatrix} k_{11} & k_{12} \\ k_{21} & k_{22} \end{bmatrix}. \tag{5.8}$$

In the above, m_{11} and m_{22} are called self-mass, c_{11} and c_{22} are self-damping, k_{11} and k_{22} self-stiffness. On the other hand, m_{12} or m_{21} are called mutual mass, c_{12} or c_{21} mutual-damping, and k_{12} or k_{21} mutual stiffness. In the field, m_{12} or m_{21} is sometimes called mass influence coefficient, c_{12} or c_{21} is damping influence coefficient, k_{12} or k_{21} stiffness influence coefficient [2,3].

5.3 FREE RESPONSE TO DAMPING FREE VIBRATION SYSTEMS WITH TWO DOF

If $\mathbf{C}=[0]$ and $\{f\}=\{0\}$, we have

$$\begin{cases} m_{11}\ddot{x}_1+m_{12}\ddot{x}_2+k_{11}x_1+k_{12}x_2=0, \\ m_{21}\ddot{x}_1+m_{22}\ddot{x}_2+k_{21}x_1+k_{22}x_2=0. \end{cases} \tag{5.9}$$

We now study the general solution $\{x\}$ to the above.

Theorem 5.1. Let $X_1(\omega)$ and $X_2(\omega)$ be the Fourier transforms of $x_1(t)$ and $x_2(t)$, respectively. Then,

$$\begin{cases} \left(k_{11}x_1-\omega^2 m_{11}\right)X_1(\omega)+\left(k_{12}-\omega^2 m_{12}\right)X_2(\omega)=0, \\ \left(k_{21}x_1-\omega^2 m_{21}\right)X_1(\omega)+\left(k_{22}-\omega^2 m_{22}\right)X_2(\omega)=0. \end{cases} \tag{5.10}$$

Proof. Performing the Fourier transform on both sides of Eq. (5.9) produces

$$\begin{cases} -\omega^2 m_{11} X_1(\omega) - \omega^2 m_{12} X_2(\omega) + k_{11} X_1(\omega) + k_{12} X_2(\omega) = 0, \\ -\omega^2 m_{21} X_1(\omega) - \omega^2 m_{22} X_2(\omega) + k_{21} X_1(\omega) + k_{22} X_2(\omega) = 0. \end{cases}$$

Managing the above yields Eq. (5.10). The proof is finished. □

Equation (5.10) is homogeneous. The necessary and sufficient condition for $\begin{Bmatrix} X_1(\omega) \\ X_2(\omega) \end{Bmatrix} \neq \begin{Bmatrix} 0 \\ 0 \end{Bmatrix}$ is that Eq. (5.10) satisfies

$$D(\omega^2) = \begin{vmatrix} k_{11} - \omega^2 m_{11} & k_{12} - \omega^2 m_{12} \\ k_{21} - \omega^2 m_{21} & k_{22} - \omega^2 m_{22} \end{vmatrix} = 0. \quad (5.11)$$

The above is called *frequency equation* since it is an equation in terms of ω. However, it does not hold for all ω. It holds only specifically for $\omega^2 = \omega_n^2$. Thus, we may rewrite the above by $D(\omega_n^2)$ in the following form:

$$\begin{aligned} D(\omega_n^2) &= \begin{vmatrix} k_{11} - \omega_n^2 m_{11} & k_{12} - \omega_n^2 m_{12} \\ k_{21} - \omega_n^2 m_{21} & k_{22} - \omega_n^2 m_{22} \end{vmatrix} \\ &= m_{11} m_{22} \omega_n^4 - (k_{11} m_{22} + k_{22} m_{11} - k_{12} m_{21} - k_{21} m_{12}) \omega_n^2 + (k_{11} k_{22} - k_{12} k_{21}) = 0. \end{aligned} \quad (5.12)$$

The above is an algebraic equation in terms of ω_n^2. It is also called frequency equation.

Denote by ω_{n1}^2 and ω_{n2}^2 the roots of Eq. (5.12). Then,

$$\left.\begin{matrix} \omega_{n1}^2 \\ \omega_{n2}^2 \end{matrix}\right\} = \frac{1}{2} \frac{k_{11} m_{22} + k_{22} m_{11} - k_{12} m_{21} - k_{21} m_{12}}{m_{11} m_{22}} \mp \frac{1}{2} \sqrt{\left(\frac{k_{11} m_{22} + k_{22} m_{11} - k_{12} m_{21} - k_{21} m_{12}}{m_{11} m_{22}} \right)^2 - 4 \frac{k_{11} k_{22} - k_{12} k_{21}}{m_{11} m_{22}}}. \quad (5.13)$$

The above implies that there are four characteristic roots in Eq. (5.12). By ignoring two negative roots, we just utilize two positive ones: $\omega_{n1} > 0$ and $\omega_{n2} > 0$. Conventionally, we denote them according to the order of $\omega_{n1} < \omega_{n2}$. The quantity ω_{n1} is called the first-order natural angular frequency while ω_{n2} the second-order natural angular frequency as both are determined by the system itself irrelevant of driven forces and initial conditions. At $\omega = \omega_{n1}$ and $\omega = \omega_{n2}$, the system may vibrate with the excitation of nonzero initial conditions even if there are no external driven forces.

Theorem 5.2. Let $X_1(\omega)$ and $X_2(\omega)$ be the Fourier transforms of $x_1(t)$ and $x_2(t)$, respectively. Then,

$$\begin{cases} \left(k_{11}-\omega_{n1}^2 m_{11}\right)X_1(\omega_{n1})+\left(k_{12}-\omega_{n1}^2 m_{12}\right)X_2(\omega_{n1})=0, \\ \left(k_{21}-\omega_{n1}^2 m_{21}\right)X_1(\omega_{n1})+\left(k_{22}-\omega_{n1}^2 m_{22}\right)X_2(\omega_{n1})=0. \end{cases} \tag{5.14}$$

And

$$\begin{cases} \left(k_{11}-\omega_{n2}^2 m_{11}\right)X_1(\omega_{n2})+\left(k_{12}-\omega_{n2}^2 m_{12}\right)X_2(\omega_{n2})=0, \\ \left(k_{21}-\omega_{n2}^2 m_{21}\right)X_1(\omega_{n2})+\left(k_{22}-\omega_{n2}^2 m_{22}\right)X_2(\omega_{n2})=0. \end{cases} \tag{5.15}$$

Proof. Because nonzero solutions to Eq. (5.10) exist only at $\omega=\omega_{n1}$ and $\omega=\omega_{n2}$, we may write Eq. (5.10) by two systems of equations above. This completes the proof. □

Either Eq. (5.14) or (5.15) is homogeneous. Therefore, we cannot find $X_1(\omega)$ and $X_2(\omega)$ from either Eq. (5.14) or (5.15). Nevertheless, we are able to find the ratio X_2/X_1 from either Eq. (5.14) or (5.15).

Theorem 5.3. Let $X_1(\omega)$ and $X_2(\omega)$ be the Fourier transforms of $x_1(t)$ and $x_2(t)$, respectively. Then, at $\omega=\omega_{n1}$, we have

$$\frac{X_2(\omega_{n1})}{X_1(\omega_{n1})}=-\frac{k_{11}-\omega_{n1}^2 m_{11}}{k_{12}-\omega_{n1}^2 m_{12}}=-\frac{k_{21}-\omega_{n1}^2 m_{21}}{k_{22}-\omega_{n1}^2 m_{22}}. \tag{5.16}$$

Proof. Consider Eq. (5.14) by

$$\begin{cases} \left(k_{11}-\omega_{n1}^2 m_{11}\right)+\left(k_{12}-\omega_{n1}^2 m_{12}\right)\dfrac{X_2(\omega_{n1})}{X_1(\omega_{n1})}=0, \\ \left(k_{21}-\omega_{n1}^2 m_{21}\right)+\left(k_{22}-\omega_{n1}^2 m_{22}\right)\dfrac{X_2(\omega_{n1})}{X_1(\omega_{n1})}=0. \end{cases} \tag{5.17}$$

We obtain the ratio expressed by Eq. (5.16). The proof is complete. □

Since the ratio in Eq. (5.16) is for ω_{n1}, we use the symbol $\beta^{(1)}$ to denote it. It is in the form

$$\beta^{(1)}=\frac{X_2^{(1)}(\omega_{n1})}{X_1^{(1)}(\omega_{n1})}=-\frac{k_{11}-\omega_{n1}^2 m_{11}}{k_{12}-\omega_{n1}^2 m_{12}}=-\frac{k_{21}-\omega_{n1}^2 m_{21}}{k_{22}-\omega_{n1}^2 m_{22}}. \tag{5.18}$$

The ratio $\beta^{(1)}$ is called the vibration mode of first order. The superscript (1) means that the ratio is based on ω_{n1}.

Corollary 5.1. Let $X_1(\omega)$ and $X_2(\omega)$ be the Fourier transforms of $x_1(t)$ and $x_2(t)$, respectively. Then, at $\omega=\omega_{n1}$, we have

$$X_2^{(1)}(\omega_{n1})=\beta^{(1)}X_1^{(1)}(\omega_{n1}). \tag{5.19}$$

Proof. From Eq. (5.18), we have Eq. (5.19). This finishes the proof. □

Corollary 5.2. Let $X_1^{(1)}(\omega_{n1})$ and $X_2^{(1)}(\omega_{n1})$ be the Fourier transforms of $x_1^{(1)}(t)$ and $x_2^{(1)}(t)$, respectively. Then, we have

$$x_2^{(1)}(t)=\beta^{(1)}x_1^{(1)}(t). \tag{5.20}$$

Proof. Doing the inverse Fourier transform on both sides of Eq. (5.19) yields Eq. (5.20). The proof is finished. □

Theorem 5.4. Let $X_1(\omega)$ and $X_2(\omega)$ be the Fourier transforms of $x_1(t)$ and $x_2(t)$, respectively. At $\omega=\omega_{n2}$, we use the symbol $\beta^{(2)}$ to denote the ratio $\frac{X_2(\omega_{n2})}{X_1(\omega_{n2})}$. Then, we have

$$\beta^{(2)}=\frac{X_2^{(2)}(\omega_{n2})}{X_1^{(2)}(\omega_{n2})}=-\frac{k_{11}-\omega_{n2}^2 m_{11}}{k_{12}-\omega_{n2}^2 m_{12}}=-\frac{k_{21}-\omega_{n2}^2 m_{21}}{k_{22}-\omega_{n2}^2 m_{22}}. \tag{5.21}$$

Proof. From Eq. (5.15), we have

$$\begin{cases}\left(k_{11}-\omega_{n2}^2 m_{11}\right)+\left(k_{12}-\omega_{n2}^2 m_{12}\right)\dfrac{X_2(\omega_{n2})}{X_1(\omega_{n2})}=0,\\[2ex]\left(k_{21}-\omega_{n2}^2 m_{21}\right)+\left(k_{22}-\omega_{n2}^2 m_{22}\right)\dfrac{X_2(\omega_{n2})}{X_1(\omega_{n2})}=0.\end{cases} \tag{5.22}$$

Since Eq. (5.22) is for ω_{n2}, we use another symbol $\beta^{(2)}$ to express $\frac{X_2(\omega_{n2})}{X_1(\omega_{n2})}$. From Eq. (5.22), we have Eq. (5.21). The proof is complete. □

The ratio $\beta^{(2)}$ is termed the vibration mode of second order.

Corollary 5.3. Let $X_1(\omega)$ and $X_2(\omega)$ be the Fourier transforms of $x_1(t)$ and $x_2(t)$, respectively. At $\omega=\omega_{n2}$, we have

$$X_2^{(2)}(\omega_{n2})=\beta^{(2)}X_1^{(2)}(\omega_{n2}). \tag{5.23}$$

Proof. From Eq. (5.21), we have Eq. (5.23). The proof is complete. □

Corollary 5.4. For ω_{n2}, we denote $x_1(t)$ and $x_2(t)$ by $x_1^{(2)}(t)$ and $x_2^{(2)}(t)$, respectively. Then,

$$x_2^{(2)}(t) = \beta^{(2)} x_1^{(2)}(t). \tag{5.24}$$

Proof. Doing the inverse Fourier transform on both hands of Eq. (5.23) results in Eq. (5.24). This finishes the proof. □

A particular thing worth noting is that the present Corollaries 5.1–5.4 imply that there are infinite solutions to the system of equations with 2 DOF. In fact, there are no restrictions on $x_1(t)$ and $x_2(t)$ in Theorems 5.1–5.4 and Corollaries 5.1–5.4 in mathematics. Precisely, either $x_1(t)$ or $x_2(t)$ may be any function, including harmonic one. I use an example below to show this particular thing.

Example 5.2. Let $x_1(t) = e^{-a|t|}$ for $a > 0$. Let

$$\beta^{(1)} = -\frac{k_{11} - \omega_{n1}^2 m_{11}}{k_{12} - \omega_{n1}^2 m_{12}} = -\frac{k_{21} - \omega_{n1}^2 m_{21}}{k_{22} - \omega_{n1}^2 m_{22}}. \tag{5.25}$$

Let $x_2(t) = \beta^{(1)} e^{-a|t|}$. Then, $x_1(t)$ and $x_2(t)$ are the solutions to the motion equations (5.9).

Proof. Since

$$X_1(\omega) = \int_{-\infty}^{\infty} x_1(t) e^{-i\omega t}\, dt = \int_{-\infty}^{\infty} e^{-a|t|} e^{-i\omega t} dt = \frac{2a}{a^2 + \omega^2}, \tag{5.26}$$

we have

$$X_2(\omega) = \beta^{(1)} X_1(\omega) = -\frac{k_{11} - \omega_{n1}^2 m_{11}}{k_{12} - \omega_{n1}^2 m_{12}} \frac{2a}{a^2 + \omega^2} = -\frac{k_{21} - \omega_{n1}^2 m_{21}}{k_{22} - \omega_{n1}^2 m_{22}} \frac{2a}{a^2 + \omega^2}. \tag{5.27}$$

Replacing $X_1(\omega)$ and $X_2(\omega)$ in Eq. (5.10) by those in Eqs. (5.26) and (5.27) results in

$$\begin{cases} \left(k_{11} - \omega_{n1}^2 m_{11}\right) \dfrac{2a}{a^2 + \omega^2} + \left(k_{12} - \omega_{n1}^2 m_{12}\right)\left(-\dfrac{k_{11} - \omega_{n1}^2 m_{11}}{k_{12} - \omega_{n1}^2 m_{12}} \dfrac{2a}{a^2 + \omega^2}\right) = 0, \\ \left(k_{21} - \omega_{n1}^2 m_{21}\right) \dfrac{2a}{a^2 + \omega^2} + \left(k_{22} - \omega_{n1}^2 m_{22}\right)\left(-\dfrac{k_{21} - \omega_{n1}^2 m_{21}}{k_{22} - \omega_{n1}^2 m_{22}} \dfrac{2a}{a^2 + \omega^2}\right) = 0. \end{cases}$$

Thus, $x_1(t)$ and $x_2(t)$ are the solutions to Eq. (5.9). The proof is finished. □

The solutions to the motion equations of (5.9) are mathematically sound for $x_1(t)$ or $x_2(t)$ being any function. However, a function that is mathematically sound may be physically improper. In engineering, Eq. (5.9) stands for a vibration system with 2 DOF without damping. Thus, we have to put restrictions on solutions simply from the point of view of engineering instead of mathematics. An obvious restriction is the energy conservation rule. To be precise, when initial conditions excite the responses to Eq. (5.9), the responses must oscillate without attenuation due to being damping free. Since the solutions discussed in Example 5.2 have the properties of $\lim_{t\to\infty} x_1(t)=0$ and $\lim_{t\to\infty} x_2(t)=0$, they are correct in mathematics but inappropriate in engineering since they are against the rule of energy conservation!

In order to meet the energy conservation rule in physics, particular solutions to Eq. (5.9) should be or must be found, instead of assuming sinusoidal functions given by

$$\begin{aligned} x_1(t) &= A_1 \sin(\omega_n t+\theta), \\ x_2(t) &= A_2 \sin(\omega_n t+\theta), \end{aligned} \tag{5.28}$$

where $A_1 \neq A_2$. Substituting the above into Eq. (5.9) produces

$$\begin{cases} \left(k_{11}-\omega_n^2 m_{11}\right)A_1 \sin(\omega_n t+\theta)+\left(k_{12}-\omega_n^2 m_{12}\right)A_2 \sin(\omega_n t+\theta)=0, \\ \left(k_{21}-\omega_n^2 m_{21}\right)A_1 \sin(\omega_n t+\theta)+\left(k_{22}-\omega_n^2 m_{22}\right)A_2 \sin(\omega_n t+\theta)=0. \end{cases} \tag{5.29}$$

Since $\sin(\omega_n t+\theta)\neq 0$ for all t, the above implies

$$\begin{cases} \left(k_{11}-\omega_n^2 m_{11}\right)A_1+\left(k_{12}-\omega_n^2 m_{12}\right)A_2=0, \\ \left(k_{21}-\omega_n^2 m_{21}\right)A_1+\left(k_{22}-\omega_n^2 m_{22}\right)A_2=0. \end{cases} \tag{5.30}$$

The necessary and sufficient condition for the nonzero solutions of A_1 and A_2 is Eq. (5.12). Thus, we have

$$\begin{cases} \left(k_{11}-\omega_{n1}^2 m_{11}\right)A_1+\left(k_{12}-\omega_{n1}^2 m_{12}\right)A_2=0, \\ \left(k_{21}-\omega_{n1}^2 m_{21}\right)A_1+\left(k_{22}-\omega_{n1}^2 m_{22}\right)A_2=0, \end{cases} \tag{5.31}$$

and

$$\begin{cases} \left(k_{11}-\omega_{n2}^2 m_{11}\right)A_1+\left(k_{12}-\omega_{n2}^2 m_{12}\right)A_2=0, \\ \left(k_{21}-\omega_{n2}^2 m_{21}\right)A_1+\left(k_{22}-\omega_{n2}^2 m_{22}\right)A_2=0. \end{cases} \tag{5.32}$$

Because either Eq. (5.31) or (5.32) is homogeneous, we cannot find A_1 and A_2 from the above. However, one can find the ratio of A_2 to A_1 in terms of ω_{n1}^2 or ω_{n2}^2.

When using Eq. (5.31) to find A_2/A_1 for ω_{n1}^2, we write Eq. (5.31) by

$$\begin{cases} \left(k_{11}-\omega_{n1}^2 m_{11}\right)+\left(k_{12}-\omega_{n1}^2 m_{12}\right)\dfrac{A_2}{A_1}=0, \\ \left(k_{21}-\omega_{n1}^2 m_{21}\right)+\left(k_{22}-\omega_{n1}^2 m_{22}\right)\dfrac{A_2}{A_1}=0. \end{cases} \tag{5.33}$$

From the above, we have

$$\frac{A_2}{A_1} = -\frac{k_{11} - \omega_{n1}^2 m_{11}}{k_{12} - \omega_{n1}^2 m_{12}} = -\frac{k_{21} - \omega_{n1}^2 m_{21}}{k_{22} - \omega_{n1}^2 m_{22}}. \tag{5.34}$$

Again, we use the symbol $\beta^{(1)}$ to denote the ratio and add superscript (1) on A_2 and A_1. Thus,

$$\beta^{(1)} = \frac{A_2^{(1)}}{A_1^{(1)}} = -\frac{k_{11} - \omega_{n1}^2 m_{11}}{k_{12} - \omega_{n1}^2 m_{12}} = -\frac{k_{21} - \omega_{n1}^2 m_{21}}{k_{22} - \omega_{n1}^2 m_{22}}. \tag{5.35}$$

The relationship between A_2 and A_1 at ω_{n1} is given by

$$A_2^{(1)} = \beta^{(1)} A_1^{(1)}. \tag{5.36}$$

Therefore, we write the solutions to Eq. (5.9) at ω_{n1} by

$$\begin{aligned} x_1^{(1)}(t) &= A_1^{(1)} \sin(\omega_{n1} t + \theta_1), \\ x_2^{(1)}(t) &= A_2^{(1)} \sin(\omega_{n1} t + \theta_1) = \beta^{(1)} A_1^{(1)} \sin(\omega_{n1} t + \theta_1). \end{aligned} \tag{5.37}$$

Similarly, at ω_{n2}^2, we write Eq. (5.32) by

$$\begin{cases} \left(k_{11} - \omega_{n2}^2 m_{11}\right) + \left(k_{12} - \omega_{n2}^2 m_{12}\right) \dfrac{A_2}{A_1} = 0, \\ \left(k_{21} - \omega_{n2}^2 m_{21}\right) + \left(k_{22} - \omega_{n2}^2 m_{22}\right) \dfrac{A_2}{A_1} = 0. \end{cases} \tag{5.38}$$

Use the symbol $\beta^{(2)}$ to denote the ratio and add superscript (2) on A_2 and A_1, yielding

$$\beta^{(2)} = \frac{A_2^{(2)}}{A_1^{(2)}} = -\frac{k_{11} - \omega_{n2}^2 m_{11}}{k_{12} - \omega_{n2}^2 m_{12}} = -\frac{k_{21} - \omega_{n2}^2 m_{21}}{k_{22} - \omega_{n2}^2 m_{22}}. \tag{5.39}$$

Consequently, one has

$$A_2^{(2)} = \beta^{(2)} A_1^{(2)}. \tag{5.40}$$

The solutions to Eq. (5.9) for $\omega = \omega_{n2}^2$ are given by

$$\begin{aligned} x_1^{(2)}(t) &= A_1^{(2)} \sin(\omega_{n2} t + \theta_2), \\ x_2^{(2)}(t) &= A_2^{(2)} \sin(\omega_{n2} t + \theta_2) = \beta^{(2)} A_1^{(2)} \sin(\omega_{n2} t + \theta_2). \end{aligned} \tag{5.41}$$

The ratio $\beta^{(1)}$ is called the first-order mode of vibration while $\beta^{(2)}$ the second-order mode of vibration. Denote by $\beta^{(i)}$ for $i = 1, 2$. Then, $\beta^{(i)}$ is expressed by

$$\beta^{(i)} = \frac{A_2^{(i)}}{A_1^{(i)}} = -\frac{k_{11} - \omega_{ni}^2 m_{11}}{k_{12} - \omega_{ni}^2 m_{12}} = -\frac{k_{21} - \omega_{ni}^2 m_{21}}{k_{22} - \omega_{ni}^2 m_{22}}, \quad i = 1, 2. \tag{5.42}$$

The quantity $\beta^{(i)}$ is called the principal mode of vibration with respect to the natural frequencies $\omega_{ni}\,(i = 1, 2)$. By principal mode, we mean that a mode is not specifically for 1 or 2 but in general. From Eq. (5.42), one has

$$A_2^{(i)} = \beta^{(i)} A_1^{(i)}, \quad i = 1, 2. \tag{5.43}$$

Without confusions, the following

$$\begin{Bmatrix} A_1^{(1)} \\ A_2^{(1)} \end{Bmatrix} = \begin{Bmatrix} A_1^{(1)} \\ \beta^{(1)} A_1^{(1)} \end{Bmatrix} = \begin{Bmatrix} 1 \\ \beta^{(1)} \end{Bmatrix} A_1^{(1)}, \tag{5.44}$$

is also termed the first-order mode of vibration while

$$\begin{Bmatrix} A_1^{(2)} \\ A_2^{(2)} \end{Bmatrix} = \begin{Bmatrix} A_1^{(2)} \\ \beta^{(2)} A_1^{(2)} \end{Bmatrix} = \begin{Bmatrix} 1 \\ \beta^{(2)} \end{Bmatrix} A_1^{(2)}, \tag{5.45}$$

is the second-order mode of vibration. Besides, one calls

$$\begin{Bmatrix} A_1^{(i)} \\ A_2^{(i)} \end{Bmatrix} = \begin{Bmatrix} A_1^{(i)} \\ \beta^{(i)} A_1^{(i)} \end{Bmatrix} = \begin{Bmatrix} 1 \\ \beta^{(i)} \end{Bmatrix} A_1^{(i)}, \quad i = 1, 2 \tag{5.46}$$

the principal mode of vibration with respect to $\omega_{ni}\,(i = 1, 2)$. With the notion of principal mode of vibration, the solutions at $\omega_{ni}\,(i = 1, 2)$ are given by

$$\begin{aligned} x_1^{(i)}(t) &= A_1^{(i)} \sin(\omega_{ni} t + \theta_i), \\ x_2^{(i)}(t) &= A_2^{(i)} \sin(\omega_{ni} t + \theta_i) = \beta^{(i)} A_1^{(i)} \sin(\omega_{ni} t + \theta_i), \quad i = 1, 2. \end{aligned} \tag{5.47}$$

The above may be expressed by

$$\begin{Bmatrix} x_1^{(i)}(t) \\ x_2^{(i)}(t) \end{Bmatrix} = \begin{Bmatrix} A_1^{(i)} \\ A_2^{(i)} \end{Bmatrix} \sin(\omega_{ni} t + \theta_i) = \begin{Bmatrix} 1 \\ \beta^{(i)} \end{Bmatrix} A_1^{(i)} \sin(\omega_{ni} t + \theta_i), \quad i = 1, 2. \tag{5.48}$$

In the field, $\begin{Bmatrix} 1 \\ \beta^{(i)} \end{Bmatrix}$ is called the normal model of vibration. Accordingly, $\begin{Bmatrix} 1 \\ \beta^{(1)} \end{Bmatrix}$ is termed the first normal mode of vibration while $\begin{Bmatrix} 1 \\ \beta^{(2)} \end{Bmatrix}$ the second normal mode of vibration. The general solution is expressed by

$$\left\{\begin{array}{c} x_1(t) \\ x_2(t) \end{array}\right\} = \sum_{i=1}^{2} \left\{\begin{array}{c} x_1^{(i)}(t) \\ x_2^{(i)}(t) \end{array}\right\} = \left\{\begin{array}{c} 1 \\ \beta^{(1)} \end{array}\right\} A_1^{(1)} \sin(\omega_{n1} t + \theta_1) + \left\{\begin{array}{c} 1 \\ \beta^{(2)} \end{array}\right\} A_1^{(2)} \sin(\omega_{n2} t + \theta_2). \tag{5.49}$$

The above exhibits that the free response to a vibration system with 2 DOF without damping is non-harmonic in general.

Four parameters, $A_1^{(1)}, A_1^{(2)}, \theta_1$, and θ_2, can be determined with four initial conditions, namely, $x_1(0)=x_{10}$, $x_2(0)=x_{20}$, $\dot{x}_1(0)=v_{10}$, $\dot{x}_2(0)=v_{20}$. Since

$$\left\{\begin{array}{l} x_1(0) = x_{10} = A_1^{(1)} \sin\theta_1 + A_1^{(2)} \sin\theta_2, \\ x_2(0) = x_{20} = \beta^{(1)} A_1^{(1)} \sin\theta_1 + \beta^{(2)} A_1^{(2)} \sin\theta_2, \\ \dot{x}_1(0) = v_{10} = A_1^{(1)} \omega_{n1} \cos\theta_1 + A_1^{(2)} \omega_{n2} \cos\theta_2, \\ \dot{x}_2(0) = v_{20} = \beta^{(1)} A_1^{(1)} \omega_{n1} \cos\theta_1 + \beta^{(2)} A_1^{(2)} \omega_{n2} \cos\theta_2. \end{array}\right. \tag{5.50}$$

$A_1^{(1)}, A_1^{(2)}, \theta_1$, and θ_2 are given by

$$\left\{\begin{array}{l} A_1^{(1)} = \dfrac{1}{\beta^{(2)} - \beta^{(1)}} \sqrt{\left[\beta^{(2)} x_{10} - x_{20}\right]^2 + \left[\dfrac{\beta^{(2)} v_{10} - v_{20}}{\omega_{n1}}\right]^2}, \\ A_1^{(2)} = \dfrac{1}{\beta^{(2)} - \beta^{(1)}} \sqrt{\left[-\beta^{(1)} x_{10} + x_{20}\right]^2 + \left[\dfrac{-\beta^{(1)} v_{10} + v_{20}}{\omega_{n2}}\right]^2}, \\ \theta_1 = \arctan \dfrac{\beta^{(2)} v_{10} - v_{20}}{\omega_{n1}\left(\beta^{(2)} x_{10} - x_{20}\right)}, \\ \theta_2 = \arctan \dfrac{\beta^{(1)} v_{10} - v_{20}}{\omega_{n2}\left(\beta^{(1)} x_{10} - x_{20}\right)}. \end{array}\right. \tag{5.51}$$

Example 5.3. Consider a vibration system in Figure 5.5.

The motion equations are in the form

$$\left\{\begin{array}{l} m_1 \ddot{x}_1 + (k_1 + k_2) x_1 - k_2 x_2 = 0, \\ m_2 \ddot{x}_2 - k_2 x_1 + (k_2 + k_3) x_2 = 0. \end{array}\right. \tag{5.52}$$

Suppose that $m_1=m_2=m$, $k_1=k_2=k_3=k$. Assume that $x_1(0)=\dot{x}_1(0)=\dot{x}_2(0)=0$, $x_2(0)=x_{20}$. Find the free response to the system.

Solution: With $m_1=m_2=m$, $k_1=k_2=k_3=k$, we have $m_{11}=m_{22}=m$, $m_{12}=m_{21}=0$, $k_{11}=k_{22}=2k$, $k_{12}=k_{21}=k$. Thus, we have

$$\omega_{n1} = \sqrt{\frac{k}{m}}, \quad \omega_{n2} = \sqrt{\frac{3k}{m}}. \tag{5.53}$$

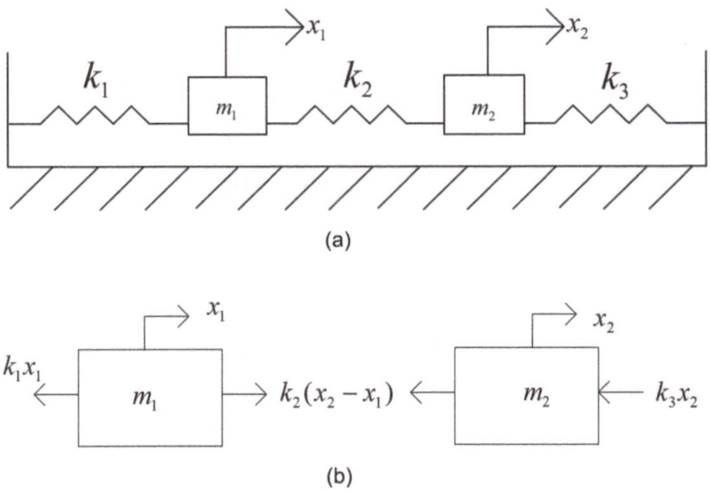

FIGURE 5.5 Damping free system with 2 DOF. (a) System. (b) Force diagram.

In addition,

$$\beta^{(1)}=1,\ \ \beta^{(2)}=-1. \tag{5.54}$$

Let

$$A_1^{(i)}=1. \tag{5.55}$$

Then,

$$A_2^{(i)}=\beta^{(i)}A_1^{(i)}=\beta^{(i)},\quad i=1,2. \tag{5.56}$$

Figure 5.6 gives the plots of the vibration modes.

Using Eq. (5.51), we have

$$\begin{Bmatrix} x_1(t) \\ x_2(t) \end{Bmatrix}=\begin{Bmatrix} \dfrac{x_{20}}{2}(\cos\omega_{n1}t+\cos\omega_{n2}t) \\ \dfrac{x_{20}}{2}(\cos\omega_{n1}t-\cos\omega_{n2}t) \end{Bmatrix}. \tag{5.57}$$

We demonstrate the plots of $x_1(t)$ and $x_2(t)$ in Figure 5.7.

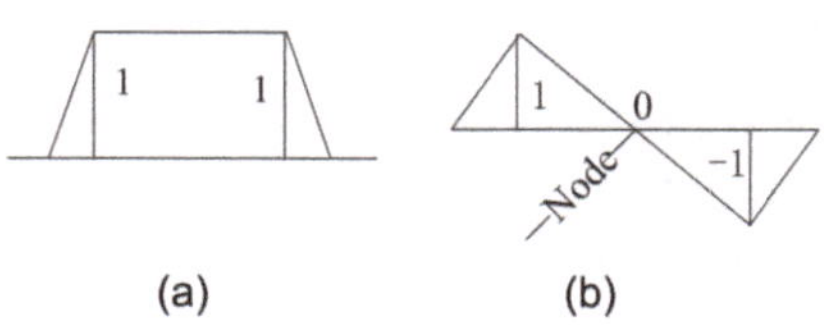

FIGURE 5.6 Illustrations of intrinsic vibrations modes. (a) First-order intrinsic vibration mode. (b) Second-order intrinsic vibration mode.

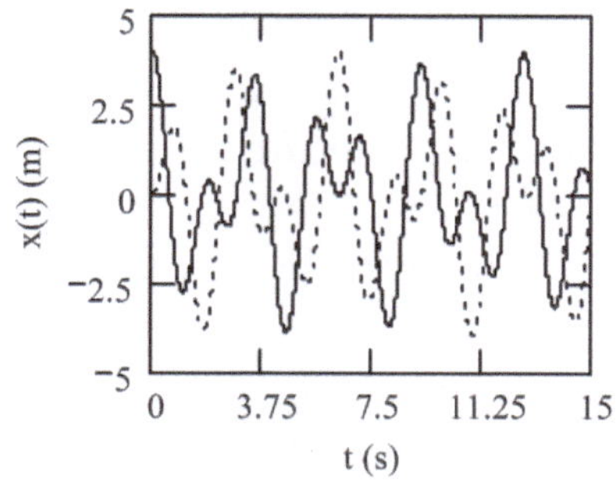

FIGURE 5.7 Plots of $x_1(t)$ (solid) and $x_2(t)$ (dot) with $x_2(0)=4\,\text{m}$, $\omega_{n1}=2\,\text{rad/s}$, $\omega_{n2}=2\sqrt{3}\,\text{rad/s}$.

5.4 FREE RESPONSE TO DAMPING FREE VIBRATION SYSTEMS WITH MDOF

Consider an MDOF vibration system with damping free. Its general motion equations are given by

$$\begin{cases} m_{11}\ddot{q}_1 + m_{12}\ddot{q}_2 + \cdots + m_{1n}\ddot{q}_n + k_{11}q_1 + k_{12}q_2 + \cdots k_{1n}q_n = 0, \\ m_{21}\ddot{q}_1 + m_{22}\ddot{q}_2 + \cdots + m_{2n}\ddot{q}_n + k_{21}q_1 + k_{22}q_2 + \cdots k_{2n}q_n = 0, \\ \vdots \\ m_{n1}\ddot{q}_1 + m_{n2}\ddot{q}_2 + \cdots + m_{nn}\ddot{q}_n + k_{n1}q_1 + k_{n2}q_2 + \cdots k_{nn}q_n = 0, \end{cases} \tag{5.58}$$

where $q_i(t)$ represents generalized displacement in the ith generalized coordinate. The above may be expressed by matrices in the form

$$\mathbf{M}_{n\times n}\ddot{\mathbf{q}}_{n\times 1} + \mathbf{K}_{n\times n}\mathbf{q}_{n\times 1} = \mathbf{0}_{n\times 1}, \tag{5.59}$$

where $\mathbf{M}_{n\times n}$ and $\mathbf{K}_{n\times n}$ are mass matrix and stiffness one. Besides, $\mathbf{q}_{n\times 1}$ and $\mathbf{0}_{n\times 1}$ are generalized displacement vector and zero driven one, respectively. They are given by

$$\mathbf{M}_{n\times n} = \begin{bmatrix} m_{11} & m_{12} & \dots & m_{1n} \\ m_{21} & m_{22} & \dots & m_{2n} \\ \vdots & \vdots & \dots & \vdots \\ m_{n1} & m_{n2} & \dots & m_{nn} \end{bmatrix}, \quad \mathbf{K}_{n\times n} = \begin{bmatrix} k_{11} & k_{12} & \dots & k_{1n} \\ k_{21} & k_{22} & \dots & k_{2n} \\ \vdots & \vdots & \dots & \vdots \\ k_{n1} & k_{n2} & \dots & k_{nn} \end{bmatrix},$$

$$\mathbf{q}_{n\times 1} = \begin{Bmatrix} q_1 \\ q_2 \\ \vdots \\ q_n \end{Bmatrix}, \quad \ddot{\mathbf{q}}_{n\times 1} = \begin{Bmatrix} \ddot{q}_1 \\ \ddot{q}_2 \\ \vdots \\ \ddot{q}_n \end{Bmatrix}, \quad \mathbf{0}_{n\times 1} = \begin{Bmatrix} 0 \\ 0 \\ \vdots \\ 0 \end{Bmatrix}. \tag{5.60}$$

Without confusions, we often omit the subscripts $n\times n$ and $n\times 1$.

The mass matrix **M** is real and symmetric. The element m_{lm} in **M** stands for the mutual mass between the lth mass and mth one. Owing to symmetry, $m_{lm}=m_{ml}$. The stiffness matrix **K** is also real and symmetric. Thus, $k_{lm}=k_{ml}$, where k_{lm} is the mutual stiffness between the lth spring and mth one.

Similar to what was previously discussed for vibrations with 2 DOF, it can be proved that there are infinite kinds of special solutions to Eq. (5.59), mathematically speaking. However, when taking into account the energy conservation rule for free vibrations with damping free in engineering, every generalized displacement $q_i(t)$ should move harmonically. Thus, $q_i(t)$ should be given by

$$q_i(t)=A_i\sin(\omega t+\theta). \tag{5.61}$$

In matrix,

$$\mathbf{q}=\begin{Bmatrix} q_1 \\ \vdots \\ q_n \end{Bmatrix}=\mathbf{A}\sin(\omega t+\theta)=\begin{Bmatrix} A_1 \\ \vdots \\ A_n \end{Bmatrix}\sin(\omega t+\theta). \tag{5.62}$$

In the above, **A**, ω, and θ are unknown.

Since

$$\begin{aligned} \dot{q}_i(t)&=\omega A_i\cos(\omega t+\theta), \\ \ddot{q}_i(t)&=-\omega^2 A_i\sin(\omega t+\theta)=-\omega^2 q_i(t), \end{aligned} \tag{5.63}$$

when substituting Eq. (5.62) into Eq. (5.59), one obtains

$$-\omega^2\left[m_{ij}\right]_{n\times n}\{q_i(t)\}_{n\times 1}+\left[k_{ij}\right]_{n\times n}\{q_i(t)\}_{n\times 1}=\{0\}_{n\times 1}. \tag{5.64}$$

The above implies

$$\begin{aligned} &\left[k_{ij}\right]_{n\times n}\left[q_i(t)\right]_{n\times 1}-\omega^2\left[m_{ij}\right]_{n\times n}\left[q_i(t)\right]_{n\times 1} \\ &=\left[k_{ij}\right]_{n\times n}\{A_i\}_{n\times 1}\sin(\omega t+\theta)-\omega^2\left[M_{ij}\right]_{n\times n}\{A_i\}_{n\times 1}\sin(\omega t+\theta) \\ &=\left\{\left[k_{ij}\right]_{n\times n}-\omega^2\left[m_{ij}\right]_{n\times n}\right\}\{A_i\}_{n\times 1}\sin(\omega t+\theta)=\{0\}_{n\times 1}. \end{aligned} \tag{5.65}$$

As $\sin(\omega t+\theta)\neq 0$ for all t, we must have

$$\left\{\left[k_{ij}\right]_{n\times n}-\omega^2\left[m_{ij}\right]_{n\times n}\right\}\{A_i\}_{n\times 1}=[0]_{n\times 1}. \tag{5.66}$$

Denote by $D\left(\omega^2\right)$ the determinant of the left side on the above. The sufficient and necessary condition that there exist nonzero solutions to the above is

$$D\left(\omega^2\right)=\left|\left[k_{ij}\right]_{n\times n}-\omega^2\left[m_{ij}\right]_{n\times n}\right|=0. \tag{5.67}$$

The above is called the frequency equation since it determines the values of ω that satisfy the above. In mathematics, it may be called characteristic equation.

According to the algebraic fundamental theorem, there are n roots to the frequency equation (5.67) in terms of ω^2. Since **M** and **K** are real and symmetry, the n roots with respect to ω^2 are positive. In terms of ω, consequently, there are $2n$ roots. Half of them are positive, and half are negative. Ignore negative roots and order positive roots by

$$\omega_1<\omega_2<\cdots<\omega_n. \tag{5.68}$$

The positive roots are natural angular frequencies of a vibration system with n DOF.

Suppose that ω_r $(r=1, 2, \ldots, n)$ satisfies $D(\omega^2)$. Then, we directly write it by

$$D\left(\omega_r^2\right)=\left|\left[k_{ij}\right]_{n\times n}-\omega_r^2\left[m_{ij}\right]_{n\times n}\right|=0. \tag{5.69}$$

The above exhibits that ω_r is related to **M** and **K** irrelevant of external driven forces and initial conditions. Expanding the above yields

$$\begin{cases}\left(k_{11}-\omega_r^2m_{11}\right)A_1^{(r)}+\left(k_{12}-\omega_r^2m_{12}\right)A_2^{(r)}+\cdots+\left(k_{1n}-\omega_r^2m_{1n}\right)A_n^{(r)}=0,\\ \left(k_{21}-\omega_r^2m_{21}\right)A_1^{(r)}+\left(k_{22}-\omega_r^2m_{22}\right)A_2^{(r)}+\cdots+\left(k_{2n}-\omega_r^2m_{2n}\right)A_n^{(r)}=0,\\ \vdots\\ \left(k_{n1}-\omega_r^2m_{n1}\right)A_1^{(r)}+\left(k_{n2}-\omega_r^2m_{n2}\right)A_2^{(r)}+\cdots+\left(k_{nn}-\omega_r^2m_{nn}\right)A_n^{(r)}=0,\end{cases} \tag{5.70}$$

where the superscript (r) means that those amplitudes are obtained at $\omega=\omega_r$ $(r=1, 2, \ldots, n)$. Because $r=1, 2,\ldots, n$, there are n systems of equations that correspond to a vibration system with n DOF. Since Eq. (5.70) is homogeneous, there are only $n-1$ independent equations. Thus, we cannot find actual A_r $(r=1, 2,\ldots, n)$ from Eq. (5.70). What we can do is to find the amplitude ratio. Select $A_n^{(r)}$ as a free variable.[1] Then, the above is expressed by

$$\begin{cases}\left(k_{11}-\omega_r^2m_{11}\right)A_1^{(r)}+\left(k_{12}-\omega_r^2m_{12}\right)A_2^{(r)}+\cdots+\left(k_{1n-1}-\omega_r^2m_{1n-1}\right)A_{n-1}^{(r)}=-\left(k_{1n}-\omega_r^2m_{1n}\right)A_n^{(r)},\\ \left(k_{21}-\omega_r^2m_{21}\right)A_1^{(r)}+\left(k_{22}-\omega_r^2m_{22}\right)A_2^{(r)}+\cdots+\left(k_{2n-1}-\omega_r^2m_{2n-1}\right)A_{n-1}^{(r)}=-\left(k_{2n}-\omega_r^2m_{2n}\right)A_n^{(r)},\\ \vdots\\ \left(k_{n1}-\omega_r^2m_{n1}\right)A_1^{(r)}+\left(k_{n2}-\omega_r^2m_{n2}\right)A_2^{(r)}+\cdots+\left(k_{nn-1}-\omega_r^2m_{nn-1}\right)A_{n-1}^{(r)}=-\left(k_{nn}-\omega_r^2m_{nn}\right)A_n^{(r)}.\end{cases} \tag{5.71}$$

Dividing both sides by $A_n^{(r)}$ on the above yields $n-1$ independent equations in the form

$$\begin{cases} \left(k_{11}-\omega_r^2 m_{11}\right)\dfrac{A_1^{(r)}}{A_n^{(r)}}+\left(k_{12}-\omega_r^2 m_{12}\right)\dfrac{A_2^{(r)}}{A_n^{(r)}}+\cdots+\left(k_{1n-1}-\omega_r^2 m_{1n-1}\right)\dfrac{A_{n-1}^{(r)}}{A_n^{(r)}}=-\left(k_{1n}-\omega_r^2 m_{1n}\right), \\ \left(k_{21}-\omega_r^2 m_{21}\right)\dfrac{A_1^{(r)}}{A_n^{(r)}}+\left(k_{22}-\omega_r^2 m_{22}\right)\dfrac{A_2^{(r)}}{A_n^{(r)}}+\cdots+\left(k_{2n-1}-\omega_r^2 m_{2n-1}\right)\dfrac{A_{n-1}^{(r)}}{A_n^{(r)}}=-\left(k_{2n}-\omega_r^2 m_{2n}\right), \\ \vdots \\ \left(k_{n1}-\omega_r^2 m_{n1}\right)\dfrac{A_1^{(r)}}{A_n^{(r)}}+\left(k_{n2}-\omega_r^2 m_{n2}\right)\dfrac{A_2^{(r)}}{A_n^{(r)}}+\cdots+\left(k_{nn-1}-\omega_r^2 m_{nn-1}\right)\dfrac{A_{n-1}^{(r)}}{A_n^{(r)}}=-\left(k_{nn}-\omega_r^2 m_{nn}\right). \end{cases} \tag{5.72}$$

Denote by $\rho_j^{(r)}$ the algebraic cofactor of the element of the first row and the jth column, i.e., $\left(k_{1j}-\omega_r^2 m_{1j}\right)$, of the determinant $D\left(\omega_r^2\right)=\left|\mathbf{K}-\omega_r^2\mathbf{M}\right|$. Then, we obtain the amplitude ratio given by

$$\frac{A_1^{(r)}}{\rho_1^{(r)}}=\frac{A_2^{(r)}}{\rho_2^{(r)}}=\cdots=\frac{A_n^{(r)}}{\rho_n^{(r)}}=p^{(r)}. \tag{5.73}$$

The ratio $p^{(r)}$ is a constant. It reflects the principal vibration of the system at $\omega=\omega_r$ $(r=1, 2,\ldots, n)$. Therefore, $A_n^{(r)}$ may be expressed by $p^{(r)}$ with the proportional coefficient $\rho_j^{(r)}$ in the form

$$A_j^{(r)}=\rho_j^{(r)}p^{(r)}. \tag{5.74}$$

The vector

$$\boldsymbol{\rho}^{(r)}=\begin{Bmatrix} \rho_1^{(r)} \\ \rho_2^{(r)} \\ \vdots \\ \rho_n^{(r)} \end{Bmatrix} \tag{5.75}$$

is called the intrinsically principal vibration mode of a vibration system. By principal vibration mode, we mean that r for $\omega=\omega_r$ is not specified but in general, $r=1, 2,\ldots, n$. The amplitude vector

$$\mathbf{A}^{(r)}=\begin{Bmatrix} A_1^{(r)} \\ A_2^{(r)} \\ \vdots \\ A_n^{(r)} \end{Bmatrix}=\boldsymbol{\rho}^{(r)}p^{(r)} \tag{5.76}$$

is also called the intrinsically principal vibration mode of a vibration system. As a matter of fact, $C\mathbf{A}^{(r)}$ is also the intrinsically principal vibration mode of a vibration system, where C is any positive constant.

Now, we summarize the properties of a vibration system with n DOF as follows.

- There are n positive natural frequencies ω_r ($r=1, 2,\ldots, n$) in a vibration system with n DOF.
- There are n intrinsic vibration modes $\boldsymbol{\rho}^{(r)} = \begin{Bmatrix} \rho_1^{(r)} \\ \rho_2^{(r)} \\ \vdots \\ \rho_n^{(r)} \end{Bmatrix}$ ($r=1, 2,\ldots, n$) for a vibration system with n DOF.
- There are $r-1$ nodes in a vibration mode of order r.

Since the motion equations of free vibrations are homogeneous, just from a view of simplifying computations, one is allowed to multiply a special solution $\mathbf{A}^{(r)}\sin(\omega_r t+\theta)$ by a constant $\dfrac{C^{(r)}}{p^{(r)}}$ so that

$$\mathbf{U}^{(r)} = \frac{C^{(r)}}{p^{(r)}}\mathbf{A}^{(r)} = C^{(r)}\boldsymbol{\rho}^{(r)}. \tag{5.77}$$

Since $\mathbf{U}^{(r)}\sin(\omega_r t+\theta) = C^{(r)}\boldsymbol{\rho}^{(r)}\sin(\omega_r t+\theta)$ is still a special solution at $\omega=\omega_r$ ($r=1, 2,\ldots, n$), we may express the generalized coordinate by

$$\mathbf{q}^{(r)} = C^{(r)}\boldsymbol{\rho}^{(r)}\sin(\omega_r t+\theta) = \mathbf{U}^{(r)}\sin(\omega_r t+\theta), \tag{5.78}$$

where

$$\mathbf{U}^{(r)} = \begin{Bmatrix} u_1^{(r)} \\ u_2^{(r)} \\ \vdots \\ u_n^{(r)} \end{Bmatrix}. \tag{5.79}$$

For facilitating computations, therefore, one may purposely select $C^{(r)}$ such that

$$\sum_{j=1}^{n}\sum_{k=1}^{n} m_{jk}u_j^{(r)}u_k^{(r)} = \mathbf{U}^{(r)T}\mathbf{M}\mathbf{U}^{(r)} = 1, \tag{5.80}$$

where $\mathbf{U}^{(r)T}$ is the transposed matrix of $\mathbf{U}^{(r)}$. In fact, when selecting

$$C^{(r)} = \frac{1}{\sqrt{\boldsymbol{\rho}^{(r)T}\mathbf{M}\boldsymbol{\rho}^{(r)}}}, \tag{5.81}$$

we have

$$\mathbf{U}^{(r)} = \frac{\boldsymbol{\rho}^{(r)}}{\sqrt{\boldsymbol{\rho}^{(r)T}\mathbf{M}\boldsymbol{\rho}^{(r)}}}. \tag{5.82}$$

If no confusions occur, we may regard $\mathbf{A}^{(r)}$ as a normalized amplitude vector for $\omega = \omega_r$. The normalized $\mathbf{A}^{(r)}$ or $\boldsymbol{\rho}^{(r)}$ is called principal vibration mode or normal vibration mode.

Example 5.4. Determine the natural frequencies, intrinsic vibration modes, normal modes of vibration of the system in Figure 5.8, where $m=1$, $k=1$.

Solution: First, we write the mass matrix and the stiffness matrix as follows.

$$\mathbf{M} = \begin{bmatrix} 1 & 0 & 0 \\ 0 & 1 & 0 \\ 0 & 0 & 1 \end{bmatrix}, \quad \mathbf{K} = \begin{bmatrix} 2 & -1 & 0 \\ -1 & 2 & -1 \\ 0 & -1 & 2 \end{bmatrix}. \tag{5.83}$$

Second, express the motion equation by

$$\mathbf{M}\ddot{x}(t) + \mathbf{K}x(t) = \mathbf{0}. \tag{5.84}$$

Third, denote the special solution by $x_i(t) = A_i \sin(\omega t + \theta)$. Write the characteristic equation by

$$-\omega_n^2 \mathbf{MA} + \mathbf{KA} = 0. \tag{5.85}$$

Expanding the above yields the frequency equation in the form

$$D(\omega_n^2) = \begin{vmatrix} 2-\omega_n^2 & -1 & 0 \\ -1 & 2-\omega_n^2 & -1 \\ 0 & -1 & 2-\omega_n^2 \end{vmatrix} = (2-\omega_n^2)(\omega_n^4 - 4\omega_n^2 + 2) = 0. \tag{5.86}$$

From the above, we obtain three natural frequencies given by

$$\omega_{n1}^2 = 2-\sqrt{2}, \quad \omega_{n2}^2 = 2, \quad \omega_{n3}^2 = 2+\sqrt{2}. \tag{5.87}$$

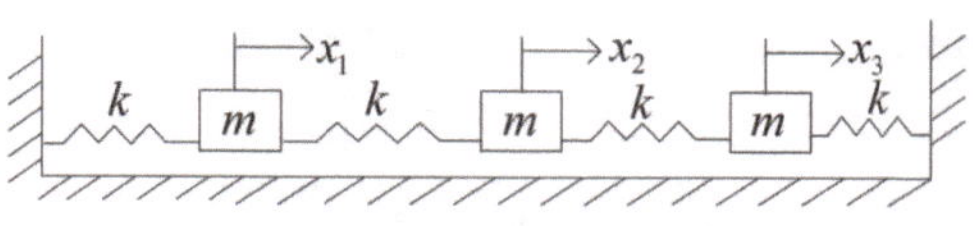

FIGURE 5.8 Damping free system with 3 DOF.

Fourth, find the vibration mode for each natural frequency one by one. Substituting ω_{n1} into Eq. (5.85) produces

$$\begin{bmatrix} \sqrt{2} & -1 & 0 \\ -1 & \sqrt{2} & -1 \\ 0 & -1 & \sqrt{2} \end{bmatrix} \begin{bmatrix} A_1^{(1)} \\ A_2^{(1)} \\ A_3^{(1)} \end{bmatrix} = \begin{bmatrix} 0 \\ 0 \\ 0 \end{bmatrix}. \tag{5.88}$$

By letting $A_1^{(1)} = 1$, the above equivalently becomes

$$\begin{bmatrix} \sqrt{2} & -1 & 0 \\ -1 & \sqrt{2} & -1 \\ 0 & -1 & \sqrt{2} \end{bmatrix} \begin{bmatrix} 1 \\ A_2^{(1)} \\ A_3^{(1)} \end{bmatrix} = \begin{bmatrix} 0 \\ 0 \\ 0 \end{bmatrix}. \tag{5.89}$$

Hence, we have $\begin{bmatrix} A_2^{(1)} \\ A_3^{(1)} \end{bmatrix} = \begin{bmatrix} \sqrt{2} \\ 1 \end{bmatrix}$. Consequently, we have the vibration mode in terms of ω_{n1} in the form

$$\mathbf{A}^{(1)} = \begin{bmatrix} 1 \\ \sqrt{2} \\ 1 \end{bmatrix}. \tag{5.90}$$

Similarly, substituting ω_{n2} into Eq. (5.85) results in

$$\begin{bmatrix} 0 & -1 & 0 \\ -1 & 0 & -1 \\ 0 & -1 & 0 \end{bmatrix} \begin{bmatrix} A_1^{(2)} \\ A_2^{(2)} \\ A_3^{(2)} \end{bmatrix} \overset{\text{Let } A_1^{(2)}=1}{=} \begin{bmatrix} 0 & -1 & 0 \\ -1 & 0 & -1 \\ 0 & -1 & 0 \end{bmatrix} \begin{bmatrix} 1 \\ A_2^{(2)} \\ A_3^{(2)} \end{bmatrix} = \begin{bmatrix} 0 \\ 0 \\ 0 \end{bmatrix}. \tag{5.91}$$

Thus, with respect to ω_{n2}, we have the vibration mode

$$\mathbf{A}^{(2)} = \begin{bmatrix} 1 \\ A_2^{(2)} \\ A_3^{(2)} \end{bmatrix} = \begin{bmatrix} 1 \\ 0 \\ -1 \end{bmatrix}. \tag{5.92}$$

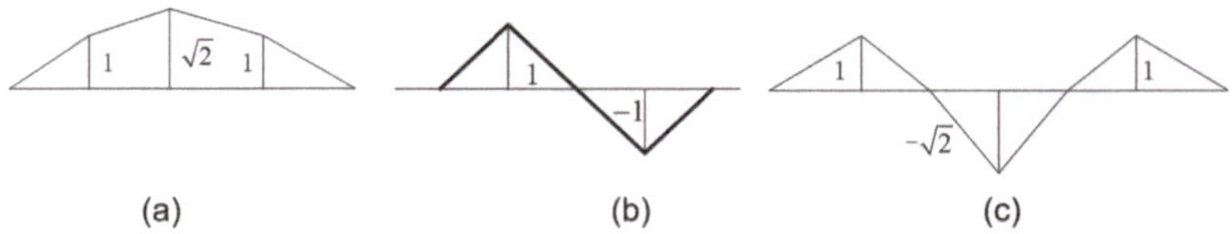

FIGURE 5.9 Vibration modes. (a) First-order intrinsic vibration mode. (b) Second-order intrinsic vibration mode. (c) Third-order intrinsic vibration mode.

In similar way, for ω_{n3}, we have

$$\begin{bmatrix} -\sqrt{2} & -1 & 0 \\ -1 & -\sqrt{2} & -1 \\ 0 & -1 & -\sqrt{2} \end{bmatrix} \begin{bmatrix} A_1^{(3)} \\ A_2^{(3)} \\ A_3^{(3)} \end{bmatrix} \overset{\text{Let } A_1^{(3)}=1}{=} \begin{bmatrix} -\sqrt{2} & -1 & 0 \\ -1 & -\sqrt{2} & -1 \\ 0 & -1 & -\sqrt{2} \end{bmatrix} \begin{bmatrix} 1 \\ A_2^{(3)} \\ A_3^{(3)} \end{bmatrix} = \begin{bmatrix} 0 \\ 0 \\ 0 \end{bmatrix}. \tag{5.93}$$

Thus, the vibration mode for ω_{n3} is expressed by

$$\mathbf{A}^{(3)} = \begin{bmatrix} 1 \\ A_2^{(3)} \\ A_3^{(3)} \end{bmatrix} = \begin{bmatrix} 1 \\ -\sqrt{2} \\ 1 \end{bmatrix}. \tag{5.94}$$

Figure 5.9 shows the plots of vibration modes.

5.5 MODAL ANALYSIS

Modal analysis is a tool for studying vibrations with MDOF. When it is used, concepts of modal mass, modal stiffness is used. In the aspect of modal analysis theory, orthogonality is essential.

5.5.1 Orthogonality

Orthogonality is an abstract concept in mathematics. Denote by $\langle f(t), g(t)\rangle$ an abstract operation between $f(t)$ and $g(t)$. If $\langle f(t), g(t)\rangle = 0$, we say that $f(t)$ is orthogonal to $g(t)$ and vice versa. When $f(t)$ is orthogonal to $g(t)$, we write $f(t) \perp g(t)$. If $f(t) \perp g(t)$, one says that the projection of $f(t)$ on $g(t)$ is zero and vice versa.

Example 5.5. Let $\vec{x}$ and $\vec{y}$ be two vectors with the same length. Denote by φ the intersection angle between $\vec{x}$ and $\vec{y}$. Then, the dot product between the two is given by

$$\langle \vec{x}, \vec{y}\rangle = \vec{x}\cdot\vec{y} = |\vec{x}||\vec{y}|\cos\varphi. \tag{5.95}$$

If $\varphi = \pi/2$, $\langle \vec{x}, \vec{y}\rangle = 0$. In this case, $\vec{x}$ is orthogonal to $\vec{y}$.

Example 5.6. Let $\lambda = \frac{2\pi}{T}$. Define the operation

$$\langle \cos n\lambda t, \sin m\lambda t \rangle = \int_{-\frac{T}{2}}^{\frac{T}{2}} \cos n\lambda t \sin m\lambda t \, dt. \tag{5.96}$$

If $m \neq n$, $\langle \cos n\lambda t, \sin m\lambda t \rangle = 0$. Thus, the condition for $\cos n\lambda t$ to be orthogonal to $\sin m\lambda t$ is $m \neq n$.

Lemma 5.1: Orthogonality of Mass Matrix. Let $\rho^{(r)}$ and $\rho^{(s)}$ be the vibration modes for the natural frequencies of order r for ω_r and order s for ω_s, respectively. Then,

$$\rho^{(r)T}\mathbf{M}\rho^{(s)} = \begin{cases} 0, & r \neq s, \\ m_r, & r = s, \end{cases} \tag{5.97}$$

where m_r is with the unit of mass. It is called modal mass.

Proof. Substituting the special solution $\mathbf{q}_{n\times 1} = \mathbf{A}_{n\times 1}\sin(\omega t + \theta)$ into the following matrix motion equation

$$\mathbf{M}_{n\times n}\ddot{\mathbf{q}}_{n\times 1} + \mathbf{K}_{n\times n}\mathbf{q}_{n\times 1} = \mathbf{0}$$

yields

$$\left(\mathbf{K}_{n\times n} - \omega^2\mathbf{M}_{n\times n}\right)\mathbf{A}_{n\times 1} = \mathbf{0}.$$

For the natural frequency ω_r, due to $\mathbf{A}^{(r)} = \rho^{(r)}p^{(r)}$, we have

$$\left(\mathbf{K} - \omega_r^2\mathbf{M}\right)\rho^{(r)} = \mathbf{0}. \tag{5.98}$$

Besides, for ω_s, we have

$$\left(\mathbf{K} - \omega_s^2\mathbf{M}\right)\rho^{(s)} = \mathbf{0}. \tag{5.99}$$

Since

$$\rho^{(s)} \times \mathbf{0} = 0, \tag{5.100}$$

multiplying the left side of Eq. (5.98) by $\rho^{(s)T}$ produces

$$\rho^{(s)T}\left(\mathbf{K} - \omega_r^2\mathbf{M}\right)\rho^{(r)} = \rho^{(s)T}\mathbf{K}\rho^{(r)} - \omega_r^2\rho^{(s)T}\mathbf{M}\rho^{(r)} = 0. \tag{5.101}$$

On the other hand, multiplying the left side of Eq. (5.99) by $\boldsymbol{\rho}^{(r)T}$ results in

$$\boldsymbol{\rho}^{(r)T}\mathbf{K}\boldsymbol{\rho}^{(s)} - \omega_s^2\boldsymbol{\rho}^{(r)T}\mathbf{M}\boldsymbol{\rho}^{(s)} = 0. \tag{5.102}$$

Subtracting Eq. (5.101) from Eq. (5.102) yields

$$\boldsymbol{\rho}^{(s)T}\mathbf{K}\boldsymbol{\rho}^{(r)} - \boldsymbol{\rho}^{(r)T}\mathbf{K}\boldsymbol{\rho}^{(s)} + \omega_s^2\boldsymbol{\rho}^{(r)T}\mathbf{M}\boldsymbol{\rho}^{(s)} - \omega_r^2\boldsymbol{\rho}^{(s)T}\mathbf{M}\boldsymbol{\rho}^{(r)} = 0. \tag{5.103}$$

Because **K** is symmetric, one has

$$\boldsymbol{\rho}^{(s)T}\mathbf{K}\boldsymbol{\rho}^{(r)} - \boldsymbol{\rho}^{(r)T}\mathbf{K}\boldsymbol{\rho}^{(s)} = 0. \tag{5.104}$$

Thus, Eq. (5.103) means

$$\omega_s^2\boldsymbol{\rho}^{(r)T}\mathbf{M}\boldsymbol{\rho}^{(s)} - \omega_r^2\boldsymbol{\rho}^{(s)T}\mathbf{M}\boldsymbol{\rho}^{(r)} = 0. \tag{5.105}$$

Since **M** is also symmetric, the following holds

$$\boldsymbol{\rho}^{(r)T}\mathbf{M}\boldsymbol{\rho}^{(s)} = \boldsymbol{\rho}^{(s)T}\mathbf{M}\boldsymbol{\rho}^{(r)}. \tag{5.106}$$

Due to Eqs. (5.106), (5.105) may be written by

$$\left(\omega_s^2 - \omega_r^2\right)\boldsymbol{\rho}^{(s)T}\mathbf{M}\boldsymbol{\rho}^{(r)} = 0. \tag{5.107}$$

Owing to

$$\omega_s^2 \neq \omega_r^2, \quad r \neq s, \tag{5.108}$$

equation (5.107) implies

$$\boldsymbol{\rho}^{(s)T}\mathbf{M}\boldsymbol{\rho}^{(r)} = 0, \quad r \neq s. \tag{5.109}$$

If $r=s$, $\boldsymbol{\rho}^{(r)T}\mathbf{M}\boldsymbol{\rho}^{(r)}$=constant, which is with the unit of mass **M**. It is denoted by m_r. Thus, Lemma 5.1 holds. This finishes the proof. □

Lemma 5.2: Orthogonality of Stiffness Matrix. Let $\boldsymbol{\rho}^{(r)}$ and $\boldsymbol{\rho}^{(s)}$ be the vibration modes for the natural frequencies of order r for ω_r and order s for ω_s, respectively. Then,

$$\boldsymbol{\rho}^{(r)T}\mathbf{K}\boldsymbol{\rho}^{(s)} = \begin{cases} 0, & r \neq s, \\ k_r, & r = s, \end{cases} \tag{5.110}$$

where k_r is with the unit of stiffness. It is called modal stiffness.

Proof. In (5.102), if $r \neq s$, $\boldsymbol{\rho}^{(r)T}\mathbf{M}\boldsymbol{\rho}^{(s)}=0$. Thus, $\boldsymbol{\rho}^{(r)T}\mathbf{K}\boldsymbol{\rho}^{(s)}=0$ for $r \neq s$. When $r=s$, $\boldsymbol{\rho}^{(r)T}\mathbf{K}\boldsymbol{\rho}^{(s)}$ equals to a constant, which is with the unit of **K**. It is denoted by k_r. The proof is complete. □

Lemma 5.1 says that vibration modes of inertia force in terms of different natural frequencies are orthogonal to each other. On the other hand, Lemma 5.2 means that vibration modes of restoration force with respect to different natural frequencies are also orthogonal to each other. Thus, n vibration modes consist of an n dimensional orthogonal set.

5.5.2 Modal Mass Matrix and Modal Stiffness Matrix

Vectors of vibration modes consist of a vibration mode matrix in the form

$$\boldsymbol{\rho}=\left[\boldsymbol{\rho}^{(1)},\boldsymbol{\rho}^{(2)},\ldots,\boldsymbol{\rho}^{(r)},\ldots,\boldsymbol{\rho}^{(n)}\right]=\begin{bmatrix} \rho_1^{(1)} & \rho_1^{(2)} & \cdots & \rho_1^{(r)} & \cdots & \rho_1^{(n)} \\ \rho_2^{(1)} & \rho_2^{(2)} & \cdots & \rho_2^{(r)} & \cdots & \rho_2^{(n)} \\ \vdots & \vdots & \ddots & \vdots & \cdots & \vdots \\ \vdots & \vdots & \cdots & \ddots & \cdots & \vdots \\ \vdots & \vdots & \cdots & \vdots & \ddots & \vdots \\ \rho_n^{(1)} & \rho_n^{(2)} & \cdots & \rho_n^{(r)} & \cdots & \rho_n^{(n)} \end{bmatrix}. \tag{5.111}$$

According to Lemma 5.1, one has

$$\boldsymbol{\rho}^T\mathbf{M}\boldsymbol{\rho}=\begin{bmatrix} m_1 & 0 & \cdots & 0 & \cdots & 0 \\ 0 & m_2 & \cdots & 0 & \cdots & 0 \\ \vdots & \vdots & \ddots & \vdots & \vdots & \vdots \\ 0 & 0 & \cdots & m_r & \vdots & \vdots \\ \vdots & \vdots & \vdots & \vdots & \ddots & \vdots \\ 0 & 0 & \cdots & 0 & \cdots & m_n \end{bmatrix}. \tag{5.112}$$

The diagonal matrix on the right side is called modal matrix, which we denote by $\mathbf{M}_m$. Thus,

$$\mathbf{M}_m=\begin{bmatrix} m_1 & 0 & \cdots & 0 & \cdots & 0 \\ 0 & m_2 & \cdots & 0 & \cdots & 0 \\ \vdots & \vdots & \ddots & \vdots & \vdots & \vdots \\ 0 & 0 & \cdots & m_r & \vdots & \vdots \\ \vdots & \vdots & \vdots & \vdots & \ddots & \vdots \\ 0 & 0 & \cdots & 0 & \cdots & m_n \end{bmatrix}. \tag{5.113}$$

Denote by $\mathbf{K}_m$ the modal stiffness matrix. Based on Lemma 5.2, we have

$$\boldsymbol{\rho}^T\mathbf{K}\boldsymbol{\rho} = \begin{bmatrix} k_1 & 0 & \cdots & 0 & \cdots & 0 \\ 0 & k_2 & \cdots & 0 & \cdots & 0 \\ \vdots & \vdots & \ddots & \vdots & \vdots & \vdots \\ 0 & 0 & \cdots & k_r & \vdots & \vdots \\ \vdots & \vdots & \vdots & \vdots & \ddots & \vdots \\ 0 & 0 & \cdots & 0 & \cdots & k_n \end{bmatrix} = \mathbf{K}_m. \tag{5.114}$$

Because $\boldsymbol{\rho}^{(s)}$ multiplied by a constant is still a vibration mode of order s, we may purposely constitute a vibration mode by $\frac{\boldsymbol{\rho}^{(r)}}{\sqrt{m_r}}$. Thus, according to Eq. (5.97), we have

$$\frac{\boldsymbol{\rho}^{(r)T}}{\sqrt{m_r}}\mathbf{M}\frac{\boldsymbol{\rho}^{(s)}}{\sqrt{m_r}} = \begin{cases} 0, & r \neq s, \\ 1, & r = s. \end{cases} \tag{5.115}$$

Let $u^{(r)} = \frac{\boldsymbol{\rho}^{(r)}}{\sqrt{m_r}}$. Denote by $\mathbf{U} = \left[u^{(1)}, u^{(2)}, \ldots, u^{(r)}, \ldots, u^{(n)}\right]$. Let $\mathbf{M}_{m1}$ be the normalized modal mass matrix. Then,

$$\mathbf{U}^T\mathbf{M}\mathbf{U} = \mathbf{M}_{m1} = \begin{bmatrix} 1 & 0 & \cdots & 0 & \cdots & 0 \\ 0 & 1 & \cdots & 0 & \cdots & 0 \\ \vdots & \vdots & \ddots & \vdots & \vdots & \vdots \\ 0 & 0 & \cdots & 1 & \vdots & \vdots \\ \vdots & \vdots & \vdots & \vdots & \ddots & \vdots \\ 0 & 0 & \cdots & 0 & \cdots & 1 \end{bmatrix}. \tag{5.116}$$

For $\mathbf{K}$, we have

$$\frac{\boldsymbol{\rho}^{(r)T}}{\sqrt{m_r}}\mathbf{K}\frac{\boldsymbol{\rho}^{(s)}}{\sqrt{m_r}} = \begin{cases} 0, & r \neq s \\ \frac{k_r}{m_r}, & r = s \end{cases} = \begin{cases} 0, & r \neq s, \\ \omega_{nr}^2, & r = s. \end{cases} \tag{5.117}$$

When using the vibration modal matrix of $\mathbf{U}$, we have the normalized modal stiffness matrix, which is denoted by $\mathbf{K}_{m1}$, in the form

$$\mathbf{U}^T\mathbf{K}\mathbf{U} = \mathbf{K}_{m1} = \begin{bmatrix} \omega_{n1}^2 & 0 & \cdots & 0 & \cdots & 0 \\ 0 & \omega_{n2}^2 & \cdots & 0 & \cdots & 0 \\ \vdots & \vdots & \ddots & \vdots & \vdots & \vdots \\ 0 & 0 & \cdots & \omega_{nr}^2 & \vdots & \vdots \\ \vdots & \vdots & \vdots & \vdots & \ddots & \vdots \\ 0 & 0 & \cdots & 0 & \cdots & \omega_{nn}^2 \end{bmatrix}. \tag{5.118}$$

5.5.3 Decoupling

Using normalized modal mass matrix and normalized modal stiffness one, the motion matrix equation can be expressed as

$$\mathbf{M}_{m1n\times n}\ddot{\mathbf{q}}_{n\times 1}+\mathbf{K}_{m1n\times n}\mathbf{q}_{n\times 1}=\begin{bmatrix}1 & & \\ & \ddots & \\ & & 1\end{bmatrix}_{n\times n}\left[\ddot{q}_r\right]_{n\times 1}+\begin{bmatrix}\omega_1^2 & & \\ & \ddots & \\ & & \omega_n^2\end{bmatrix}_{n\times n}\left[q_r\right]_{n\times 1}=\left[0\right]_{n\times 1}. \tag{5.119}$$

Thus, we have a set of n vibration systems with SDOF in the form

$$\ddot{q}_r+\omega_r^2 q_r=0,\quad r=1,2,\ldots,n. \tag{5.120}$$

For each generalized displacement, we have its solution in the form

$$q_r(t)=q_r(0)\cos\omega_{nr}t+\frac{\dot{q}_r(0)}{\omega_{nr}}\sin\omega_{nr}t. \tag{5.121}$$

The decoupled motion equations (5.119) mean that a vibration system with n DOF can be changed to a set of n independent vibration system with SDOF. Some literature, e.g., Nakagawa and Ringo [3], considers that a successful decoupling is just an ideal of modal analysis and may not be easily realized in engineering. However, in my opinion, any vibration system with n DOF in engineering can be decoupled into n independent vibration system with SDOF. That is in nature the most attractive point of modal analysis theory. As a matter of fact, once ω_r is obtained, a set of motion equations with SDOF is ready, see Eqs. (5.119) and (5.120). Thus, natural frequency ω_r plays an important role in modal analysis.

From the above, we see that the vector of generalized displacements is given by

$$\mathbf{x}(t)=\sum_{r=1}^{n}\boldsymbol{\rho}^{(r)}q_r(t). \tag{5.122}$$

Expanding the above produces

$$\begin{Bmatrix}x_1(t)\\x_2(t)\\\vdots\\x_r(t)\\\vdots\\x_n(t)\end{Bmatrix}=\begin{Bmatrix}\rho_1^{(1)}(t)\\\rho_2^{(1)}(t)\\\vdots\\\rho_r^{(1)}(t)\\\vdots\\\rho_n^{(1)}(t)\end{Bmatrix}q_1(t)+\begin{Bmatrix}\rho_1^{(2)}(t)\\\rho_2^{(2)}(t)\\\vdots\\\rho_r^{(2)}(t)\\\vdots\\\rho_n^{(2)}(t)\end{Bmatrix}q_2(t)+\cdots+\begin{Bmatrix}\rho_1^{(r)}(t)\\\rho_2^{(r)}(t)\\\vdots\\\rho_r^{(r)}(t)\\\vdots\\\rho_n^{(r)}(t)\end{Bmatrix}q_r(t)+\cdots+\begin{Bmatrix}\rho_1^{(n)}(t)\\\rho_2^{(n)}(t)\\\vdots\\\rho_r^{(n)}(t)\\\vdots\\\rho_n^{(n)}(t)\end{Bmatrix}q_n(t).$$

Example 5.7. Write the modal motion equations of the vibration system with 3 DOF as that in Example 5.4.

Solution: In Example 5.4, three natural frequencies were

$$\omega_{n1}^2 = 2-\sqrt{2},\ \omega_{n2}^2 = 2,\ \omega_{n3}^2 = 2+\sqrt{2}.$$

According to Eq. (1.6), we have

$$\begin{cases} q_1(t) = q_1(0)\cos\omega_{n1}t + \dfrac{\dot{q}_1(0)}{\omega_{n1}}\sin\omega_{n1}t = q_1(0)\cos\sqrt{2-\sqrt{2}}t + \dfrac{\dot{q}_1(0)}{\sqrt{2-\sqrt{2}}}\sin\sqrt{2-\sqrt{2}}t, \\ q_2(t) = q_2(0)\cos\sqrt{2}t + \dfrac{\dot{q}_2(0)}{\sqrt{2}}\sin\sqrt{2}t, \\ q_3(t) = q_3(0)\cos\sqrt{2+\sqrt{2}}t + \dfrac{\dot{q}_3(0)}{\sqrt{2+\sqrt{2}}}\sin\sqrt{2+\sqrt{2}}t. \end{cases} \tag{5.123}$$

Because

$$\mathbf{A} = \left[\mathbf{A}^{(1)}, \mathbf{A}^{(2)}, \mathbf{A}^{(3)}\right] = \begin{bmatrix} 1 & 1 & 1 \\ \sqrt{2} & 0 & -\sqrt{2} \\ 1 & -1 & 1 \end{bmatrix},$$

we have

$$\begin{Bmatrix} x_1(t) \\ x_2(t) \\ x_3(t) \end{Bmatrix} = \begin{Bmatrix} 1 \\ \sqrt{2} \\ 1 \end{Bmatrix} q_1(t) + \begin{Bmatrix} 1 \\ 0 \\ -1 \end{Bmatrix} q_2(t) + \begin{Bmatrix} 1 \\ -\sqrt{2} \\ 1 \end{Bmatrix} q_3(t). \tag{5.124}$$

Figure 5.10 shows the plots of responses $x_1(t)$, $x_2(t)$, and $x_3(t)$.

The previous discussions were modal analysis of free response to a vibration system with MDOF. Modal analysis for forced response to a damped vibration system with MDOF may refer to Yao [4].

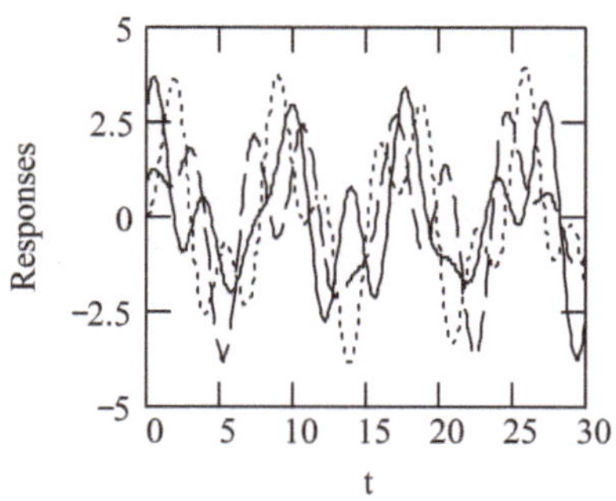

FIGURE 5.10 Responses $x_1(t)$ (solid), $x_2(t)$ (dot), $x_3(t)$ (dash) with $q_1(0) = q_2(0) = q_3(0) = 1$, $\dot{q}_1(0) = \dot{q}_2(0) = \dot{q}_3(0) = 1$.

5.6 FREQUENCY TRANSFER MATRIX OF DAMPED VIBRATION SYSTEMS WITH MDOF

Frequency response is essential to vibration analysis. Consider a vibration system with n DOF. Its general motion equations are in the form

$$\begin{cases} m_{11}\ddot{q}_1+m_{12}\ddot{q}_2+\cdots+m_{1n}\ddot{q}_n+c_{11}\dot{q}_1+c_{12}\dot{q}_2+\cdots+c_{1n}\dot{q}_n+k_{11}q_1+k_{12}q_2+\cdots k_{1n}q_n=f_1(t), \\ m_{21}\ddot{q}_1+m_{22}\ddot{q}_2+\cdots+m_{2n}\ddot{q}_n+c_{21}\dot{q}_1+c_{22}\dot{q}_2+\cdots+c_{2n}\dot{q}_n+k_{21}q_1+k_{22}q_2+\cdots k_{2n}q_n=f_2(t), \\ \vdots \\ m_{n1}\ddot{q}_1+m_{n2}\ddot{q}_2+\cdots+m_{nn}\ddot{q}_n+c_{n1}\dot{q}_1+c_{n2}\dot{q}_2+\cdots+c_{nn}\dot{q}_n+k_{n1}q_1+k_{n2}q_2+\cdots k_{nn}q_n=f_n(t), \end{cases} \tag{5.125}$$

where $q_i(t)$ represents the generalized displacement in the ith generalized coordinate and $f_i(t)$ the driven force in that coordinate. Using matrices, the above with initial conditions may be written by

$$\begin{cases} \mathbf{M}\mathbf{q}''+\mathbf{C}\mathbf{q}'+\mathbf{K}\mathbf{q}=\mathbf{f}, \\ \mathbf{q}(0)=\{q_r(0)\}_{n\times 1}=\{a_r\}_{n\times 1}, \dot{\mathbf{q}}(0)=\{\dot{q}_r(0)\}_{n\times 1}=\{b_r\}_{n\times 1}, \end{cases} \tag{5.126}$$

where **M**, **C**, and **K** are mass matrix, damping matrix, and stiffness one, respectively. **f** is the driven force vector and **q** is the generalized displacement vector.

$$\mathbf{M}=\begin{bmatrix} m_{11} & m_{12} & \dots & m_{1n} \\ m_{21} & m_{22} & \dots & m_{2n} \\ \vdots & \vdots & \dots & \vdots \\ m_{n1} & m_{n2} & \dots & m_{nn} \end{bmatrix}, \quad \mathbf{C}=\begin{bmatrix} c_{11} & c_{12} & \dots & c_{1n} \\ c_{21} & c_{22} & \dots & c_{2n} \\ \vdots & \vdots & \dots & \vdots \\ c_{n1} & c_{n2} & \dots & c_{nn} \end{bmatrix},$$

$$\mathbf{K}=\begin{bmatrix} k_{11} & k_{12} & \dots & k_{1n} \\ k_{21} & k_{22} & \dots & k_{2n} \\ \vdots & \vdots & \dots & \vdots \\ k_{n1} & k_{n2} & \dots & k_{nn} \end{bmatrix}, \tag{5.127}$$

$$\mathbf{q}=\begin{bmatrix} q_1 \\ q_2 \\ \vdots \\ q_n \end{bmatrix}, \quad \dot{\mathbf{q}}=\begin{bmatrix} \dot{q}_1 \\ \dot{q}_2 \\ \vdots \\ \dot{q}_n \end{bmatrix}, \quad \ddot{\mathbf{q}}=\begin{bmatrix} \ddot{q}_1 \\ \ddot{q}_2 \\ \vdots \\ \ddot{q}_n \end{bmatrix}, \quad \mathbf{f}=\begin{bmatrix} f_1(t) \\ f_2(t) \\ \vdots \\ f_n(t) \end{bmatrix}.$$

When ignoring initial conditions, we study stationary response. In this case, we use the Fourier transform. Denote by

$$\begin{aligned} & q_r(t)\leftrightarrow Q_r(\omega),\ \dot{q}_r(t)\leftrightarrow i\omega Q_r(\omega), \\ & \ddot{q}_r(t)\leftrightarrow (i\omega)^2 Q_r(\omega)=-\omega^2 Q_r(\omega),\ f_r(t)\leftrightarrow F_r(\omega), \\ & \mathbf{q}(t)\leftrightarrow \mathbf{Q}(\omega), \mathbf{f}(t)\leftrightarrow \mathbf{F}(\omega). \end{aligned} \tag{5.128}$$

Doing the Fourier transform on both sides of Eq. (5.126) produces

$$\left(\mathbf{K}-\omega^2\mathbf{M}+i\omega\mathbf{C}\right)\mathbf{Q}=\mathbf{F}. \tag{5.129}$$

Denote frequency transfer matrix by $\mathbf{H}(\omega)$. It is defined by

$$\mathbf{H}(\omega)=\left(\mathbf{K}-\omega^2\mathbf{M}+i\omega\mathbf{C}\right)^{-1}. \tag{5.130}$$

Then, in frequency domain, the generalized displacement matrix is given by

$$\mathbf{Q}(\omega)=\mathbf{H}(\omega)\mathbf{F}(\omega). \tag{5.131}$$

When expanding the right hand of Eq. (5.130), we have

$$\mathbf{H}(\omega)=\left(\begin{array}{l}\begin{bmatrix} k_{11} & k_{12} & \dots & k_{1n} \\ k_{21} & k_{22} & \dots & k_{2n} \\ \vdots & \vdots & \dots & \vdots \\ k_{n1} & k_{n2} & \dots & k_{nn} \end{bmatrix} - \omega^2 \begin{bmatrix} m_{11} & m_{12} & \dots & m_{1n} \\ m_{21} & m_{22} & \dots & m_{2n} \\ \vdots & \vdots & \dots & \vdots \\ m_{n1} & m_{n2} & \dots & m_{nn} \end{bmatrix} \\ +i\omega \begin{bmatrix} c_{11} & c_{12} & \dots & c_{1n} \\ c_{21} & c_{22} & \dots & c_{2n} \\ \vdots & \vdots & \dots & \vdots \\ c_{n1} & c_{n2} & \dots & c_{nn} \end{bmatrix}\end{array}\right)^{-1}. \tag{5.132}$$

That may be expressed by

$$\mathbf{H}=\left[H_{rs}(\omega)\right]=\begin{bmatrix} H_{11}(\omega) & H_{12}(\omega) & \cdots & H_{1n}(\omega) \\ H_{21}(\omega) & H_{22}(\omega) & \cdots & H_{2n}(\omega) \\ \vdots & \vdots & \ddots & \vdots \\ H_{n1}(\omega) & H_{n2}(\omega) & \cdots & H_{nn}(\omega) \end{bmatrix}. \tag{5.133}$$

In the above, $H_{rr}(\omega)$ is the self-frequency transfer function while $H_{rs}(\omega)$ for $r\neq s$ is the mutual frequency transfer function. Expanding Eq. (5.131) results in

$$\begin{bmatrix} Q_1(\omega) \\ Q_2(\omega) \\ \vdots \\ Q_n(\omega) \end{bmatrix} = \begin{bmatrix} H_{11}(\omega) & H_{12}(\omega) & \cdots & H_{1n}(\omega) \\ H_{21}(\omega) & H_{22}(\omega) & \cdots & H_{2n}(\omega) \\ \vdots & \vdots & \ddots & \vdots \\ H_{n1}(\omega) & H_{n2}(\omega) & \cdots & H_{nn}(\omega) \end{bmatrix} \begin{bmatrix} F_1(\omega) \\ F_2(\omega) \\ \vdots \\ F_n(\omega) \end{bmatrix}$$

$$= \begin{bmatrix} H_{11}(\omega)F_1(\omega) + H_{12}(\omega)F_2(\omega) + \cdots + H_{1n}(\omega)F_n(\omega) \\ H_{21}(\omega)F_1(\omega) + H_{22}(\omega)F_2(\omega) + \cdots + H_{2n}(\omega)F_n(\omega) \\ \vdots \\ H_{n1}(\omega)F_1(\omega) + H_{n2}(\omega)F_2(\omega) + \cdots + H_{nn}(\omega)F_n(\omega) \end{bmatrix} = \begin{bmatrix} \sum_{k=1}^{n} H_{1k}(\omega)F_k(\omega) \\ \sum_{k=1}^{n} H_{2k}(\omega)F_k(\omega) \\ \vdots \\ \sum_{k=1}^{n} H_{nk}(\omega)F_k(\omega) \end{bmatrix}. \tag{5.134}$$

Denote by **h** impulse response matrix. Then,

$$\mathbf{h} = [h_{rs}(t)] = F^{-1}\left[H_{rs}(\omega)\right] = \begin{bmatrix} h_{11}(t) & h_{12}(t) & \cdots & h_{1n}(t) \\ h_{21}(t) & h_{22}(t) & \cdots & h_{2n}(t) \\ \vdots & \vdots & \ddots & \vdots \\ h_{n1}(t) & h_{n2}(t) & \cdots & h_{nn}(t) \end{bmatrix}. \tag{5.135}$$

Thus, generalized displacement vector in the time domain is given by

$$\mathbf{q} = \mathbf{h} * \mathbf{f} = \begin{bmatrix} h_{11}(t) & h_{12}(t) & \cdots & h_{1n}(t) \\ h_{21}(t) & h_{22}(t) & \cdots & h_{2n}(t) \\ \vdots & \vdots & \ddots & \vdots \\ h_{n1}(t) & h_{n2}(t) & \cdots & h_{nn}(t) \end{bmatrix} * \begin{bmatrix} f_1(t) \\ f_2(t) \\ \vdots \\ f_n(t) \end{bmatrix}$$

$$= \begin{bmatrix} h_{11}(t) * f_1(t) + h_{12}(t) * f_2(t) + \cdots + h_{1n}(t) * f_n(t) \\ h_{21}(t) * f_1(t) + h_{22}(t) * f_2(t) + \cdots + h_{2n}(t) * f_n(t) \\ \vdots \\ h_{n1}(t) * f_1(t) + h_{n2}(t) * f_2(t) + \cdots + h_{nn}(t) * f_n(t) \end{bmatrix} = \begin{bmatrix} \sum_{k=1}^{n} h_{1k}(t) * f_k(t) \\ \sum_{k=1}^{n} h_{2k}(t) * f_k(t) \\ \vdots \\ \sum_{k=1}^{n} h_{nk}(t) * f_k(t) \end{bmatrix}. \tag{5.136}$$

Example 5.8. Find the frequency transfer matrix of a vibration system of 2 DOF in the form

$$\begin{cases} m_{11}\ddot{q}_1 + m_{12}\ddot{q}_2 + c_{11}\dot{q}_1 + c_{12}\dot{q}_2 + k_{11}q_1 + k_{12}q_2 = f_1(t), \\ m_{21}\ddot{q}_1 + m_{22}\ddot{q}_2 + c_{21}\dot{q}_1 + c_{22}\dot{q}_2 + k_{21}q_1 + k_{22}q_2 = f_2(t). \end{cases} \tag{5.137}$$

Solution: According to Eq. (5.130), we have

$$\begin{aligned} \begin{bmatrix} H_{11}(\omega) & H_{12}(\omega) \\ H_{21}(\omega) & H_{22}(\omega) \end{bmatrix} &= \begin{bmatrix} k_{11} - \omega^2 m_{11} + i\omega c_{11} & k_{12} - \omega^2 m_{12} + i\omega c_{12} \\ k_{21} - \omega^2 m_{21} + i\omega c_{21} & k_{22} - \omega^2 m_{22} + i\omega c_{22} \end{bmatrix}^{-1} \\ &= \frac{1}{ad - bc} \begin{bmatrix} k_{22} - \omega^2 m_{22} + i\omega c_{22} & -\left(k_{12} - \omega^2 m_{12} + i\omega c_{12}\right) \\ -\left(k_{21} - \omega^2 m_{21} + i\omega c_{21}\right) & k_{11} - \omega^2 m_{11} + i\omega c_{11} \end{bmatrix}, \end{aligned} \tag{5.138}$$

where

$$\begin{aligned} a &= k_{11} - \omega^2 m_{11} + i\omega c_{11}, \quad b = k_{12} - \omega^2 m_{12} + i\omega c_{12}, \\ c &= k_{21} - \omega^2 m_{21} + i\omega c_{21}, \quad d = k_{22} - \omega^2 m_{22} + i\omega c_{22}. \end{aligned} \tag{5.139}$$

Thus, we have

$$\begin{aligned} H_{11}(\omega) &= \frac{k_{22} - \omega^2 m_{22} + i\omega c_{22}}{ad - bc}, \quad H_{22}(\omega) = \frac{k_{11} - \omega^2 m_{11} + i\omega c_{11}}{ad - bc}, \\ H_{12}(\omega) &= -\frac{k_{12} - \omega^2 m_{12} + i\omega c_{12}}{ad - bc}, \quad H_{21}(\omega) = -\frac{k_{21} - \omega^2 m_{21} + i\omega c_{21}}{ad - bc}. \end{aligned} \tag{5.140}$$

Finally, we note that symbolic computing is useful to deal with the computation of either **H** or **h**. In the aspect to symbolic computing, the software called Maple may be encouraged (Garvan [5]). Other software, such as Mathematica (Cap [6]), is also usable.

5.7 DISCUSSIONS

For an MDOF system described by

$$\mathbf{M}_{n\times n}\mathbf{q}'' + \mathbf{K}_{n\times n}\mathbf{q} = \begin{bmatrix} m_{11} & m_{12} & \cdots & m_{1n} \\ m_{21} & m_{22} & \cdots & m_{2n} \\ \vdots & \vdots & \cdots & \vdots \\ m_{n1} & m_{n2} & \cdots & m_{nn} \end{bmatrix} \mathbf{q}'' + \begin{bmatrix} k_{11} & k_{12} & \cdots & k_{1n} \\ k_{21} & k_{22} & \cdots & k_{2n} \\ \vdots & \vdots & \cdots & \vdots \\ k_{n1} & k_{n2} & \cdots & k_{nn} \end{bmatrix} \mathbf{q} = [0]_{n\times 1}, \tag{5.141}$$

by using modal analysis, we have its equivalent equation given by

$$\mathbf{M}_{m1n\times n}\mathbf{q}'' + \mathbf{K}_{m1n\times n}\mathbf{q} = \begin{bmatrix} 1 & & \\ & \ddots & \\ & & 1 \end{bmatrix}_{n\times n} \mathbf{q}'' + \begin{bmatrix} \omega_1^2 & & \\ & \ddots & \\ & & \omega_n^2 \end{bmatrix}_{n\times n} \mathbf{q} = [0]_{n\times 1}. \quad (5.142)$$

In the above two equations, obviously, $\mathbf{M}_{m1n\times n} \neq \mathbf{M}_{n\times n}$ and $\mathbf{K}_{m1n\times n} \neq \mathbf{K}_{n\times n}$. However, natural frequencies of both equations are the same. So are the vibration modes of two equations. The key point of modal analysis is to decouple an original MDOF system into a set of SDOF systems for investigating its natural frequencies and vibration modes.

5.8 EXERCISES

5.1. Following Example 5.8, please find the inverse Fourier transform of $\begin{bmatrix} H_{11}(\omega) & H_{12}(\omega) \\ H_{21}(\omega) & H_{22}(\omega) \end{bmatrix}$. That is, $\begin{bmatrix} h_{11}(t) & h_{12}(t) \\ h_{21}(t) & h_{22}(t) \end{bmatrix}$. (Hint: Use Maple)

5.2. For a system with 3 DOF in the form

$$\begin{cases} m_{11}\ddot{q}_1 + m_{12}\ddot{q}_2 + m_{13}\ddot{q}_3 + c_{11}\dot{q}_1 + c_{12}\dot{q}_2 + c_{13}\dot{q}_3 + k_{11}q_1 + k_{12}q_2 + k_{13}q_3 = f_1(t), \\ m_{21}\ddot{q}_1 + m_{22}\ddot{q}_2 + m_{23}\ddot{q}_3 + c_{21}\dot{q}_1 + c_{22}\dot{q}_2 + c_{23}\dot{q}_3 + k_{21}q_1 + k_{22}q_2 + k_{23}q_3 = f_2(t), \\ m_{31}\ddot{q}_1 + m_{32}\ddot{q}_2 + m_{33}\ddot{q}_3 + c_{31}\dot{q}_1 + c_{32}\dot{q}_2 + c_{33}\dot{q}_3 + k_{31}q_1 + k_{32}q_2 + k_{33}q_3 = f_3(t). \end{cases}$$

Please find its frequency transfer matrix given by

$$\begin{bmatrix} H_{11}(\omega) & H_{12}(\omega) & H_{13}(\omega) \\ H_{21}(\omega) & H_{22}(\omega) & H_{23}(\omega) \\ H_{31}(\omega) & H_{32}(\omega) & H_{33}(\omega) \end{bmatrix}$$

$$= \begin{bmatrix} k_{11} - \omega^2 m_{11} + i\omega c_{11} & k_{12} - \omega^2 m_{12} + i\omega c_{12} & k_{13} - \omega^2 m_{13} + i\omega c_{13} \\ k_{21} - \omega^2 m_{21} + i\omega c_{21} & k_{22} - \omega^2 m_{22} + i\omega c_{22} & k_{23} - \omega^2 m_{23} + i\omega c_{23} \\ k_{31} - \omega^2 m_{31} + i\omega c_{31} & k_{32} - \omega^2 m_{32} + i\omega c_{32} & k_{33} - \omega^2 m_{33} + i\omega c_{33} \end{bmatrix}^{-1}.$$

(Hint: Use Maple)

5.3. Suppose that a vibration system consists of I subsystems that are in series. Each subsystem is with n DOF. Denote by $\mathbf{H}_i$ the frequency transfer matrix of the ith vibration subsystem, where $i = 1, 2, \ldots, I$ and

$$\mathbf{H}_i=\begin{bmatrix} H_{i,11}(\omega) & H_{i,12}(\omega) & \cdots & H_{i,1n}(\omega) \\ H_{i,21}(\omega) & H_{i,22}(\omega) & \cdots & H_{i,2n}(\omega) \\ \vdots & \vdots & \ddots & \vdots \\ H_{i,n1}(\omega) & H_{i,n2}(\omega) & \cdots & H_{i,nn}(\omega) \end{bmatrix}.$$

Let **H** be the frequency transfer matrix of the total vibration system. Please prove that

$$\mathbf{H}=\begin{bmatrix} H_{11}(\omega) & H_{12}(\omega) & \cdots & H_{1n}(\omega) \\ H_{21}(\omega) & H_{22}(\omega) & \cdots & H_{2n}(\omega) \\ \vdots & \vdots & \ddots & \vdots \\ H_{n1}(\omega) & H_{n2}(\omega) & \cdots & H_{nn}(\omega) \end{bmatrix}=\prod_{i=1}^{I}\begin{bmatrix} H_{i,11}(\omega) & H_{i,12}(\omega) & \cdots & H_{i,1n}(\omega) \\ H_{i,21}(\omega) & H_{i,22}(\omega) & \cdots & H_{i,2n}(\omega) \\ \vdots & \vdots & \ddots & \vdots \\ H_{i,n1}(\omega) & H_{i,n2}(\omega) & \cdots & H_{i,nn}(\omega) \end{bmatrix}.$$

(Hint: Use Theorems 4.1 and 4.2, Corollaries 4.1 and 4.2)

5.4. Denote by **H** the frequency transfer matrix of a decoupled vibration system with n DOF. Please prove that

$$\mathbf{H}=\begin{bmatrix} H_{11}(\omega) & 0 & \cdots & 0 \\ 0 & H_{22}(\omega) & \cdots & 0 \\ \vdots & \vdots & \ddots & \vdots \\ 0 & 0 & \cdots & H_{nn}(\omega) \end{bmatrix}.$$

(Hint: In Eq. (5.130), **M, C,** and **K** are diagonal).

5.5. Denote by **h** the impulse response matrix of a decoupled vibration system with n DOF. Please prove that

$$\mathbf{h}=\begin{bmatrix} h_{11}(t) & 0 & \cdots & 0 \\ 0 & h_{22}(t) & \cdots & 0 \\ \vdots & \vdots & \ddots & \vdots \\ 0 & 0 & \cdots & h_{nn}(t) \end{bmatrix}.$$

NOTE

1 A free variable is selected arbitrarily. Any A_r $(r = 1, 2, \ldots, n)$ may serve as a free variable.

REFERENCES

1. S. P. Timoshenko, *Mechanical Vibrations*, D. Van Nostrand Company, Inc., New York, 1955.
2. O. M. Palley, Г. B. Bahizov, and E. Я. Voroneysk, *Handbook of Ship Structural Mechanics*, National Defense Industry Publishing House, Beijing, China, 2002. In Chinese. Translated from Russian by B. H. Xu, X. Xu, and M. Q. Xu.
3. K. Nakagawa and M. Ringo, *Engineering Vibrations*, Shanghai Science and Technology Publishing House, Shanghai, China, 1981. In Chinese. Translated from Japanese by S. R. Xia
4. X. L. Yao, *Ship Hull Vibrations*, Harbin Engineering University Press, Harbin, China, 2004. In Chinese.
5. F. Garvan, *The Maple Book*, Chapman and Hall/CRC Press, New York/Boca Raton, FL, 2001.
6. F. F. Cap, *Mathematical Methods in Physics and Engineering with Mathematica*, CRC Press, Boca Raton, FL, 2019.

CHAPTER 6

Vibrations of Distributed Systems and Euler-Bernoulli Beam

A DISTRIBUTED VIBRATION SYSTEM IS continuous. By distributed or continuous, one means that vibrations parameters, such as mass, damping, and stiffness, are continuously distributed over the whole vibration system.

Denote by $u(x, t)$ displacement at position x and time t on a beam with l length. Due to $x \in (0, l)$, there are infinite independent displacements on the beam of l length. Thus, a distributed system has infinite degrees-of-freedom (DOF). Consequently, analysis methods for systems with multiple degrees-of-freedom (MDOF) cease to work for distributed ones. This chapter discusses longitudinal vibrations and lateral ones of beams from a view of ship hull vibrations.

6.1 LONGITUDINAL VIBRATIONS OF UNIFORM CIRCULAR BEAMS

Consider a uniform beam of l length (Figure 6.1). Its cross section is A and its material density ρ. Denote by E its elastic modulus. Suppose the beam is under the excitation of axial force $P(t)$. Then, it suffers from longitudinal vibration. Denote by $F(x, t)$ an axial force at cross section x at time t.

Using an infinitesimal segment of beam dx, we have the force diagram about dx as shown in Figure 6.1b. For dx, we have

$$\frac{\partial F}{\partial x} = \rho A \frac{\partial^2 u}{\partial t^2}. \tag{6.1}$$

DOI: 10.1201/9781003460947-7

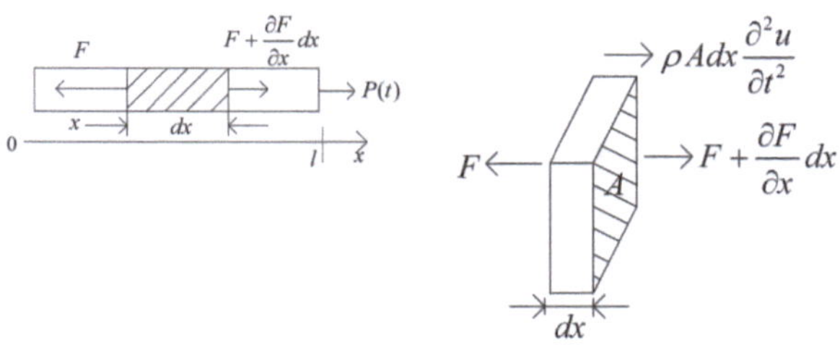

FIGURE 6.1 Illustration of longitudinal vibration of beam. (a) Beam of length l. (b) Force Diagram for an infinitesimal part of the beam.

Because of longitudinal vibration, the Hooke's law is used to express

$$\frac{F}{AE}=\frac{\partial u}{\partial x}. \tag{6.2}$$

Substituting Eq. (6.2) into (6.1) yields

$$\frac{\partial^2 u}{\partial t^2}=\frac{E}{\rho}\frac{\partial^2 u}{\partial x^2}. \tag{6.3}$$

The above is a free vibration equation. Let a be a propagation speed of stress wave in beam. It is given by

$$a=\sqrt{\frac{E}{\rho}}. \tag{6.4}$$

With Eq. (6.4), Eq. (6.3) is written by

$$\frac{\partial^2 u}{\partial t^2}=a^2\frac{\partial^2 u}{\partial x^2}. \tag{6.5}$$

When considering initial conditions and boundary ones, one has

$$\begin{cases} \dfrac{\partial^2 u}{\partial t^2}=a^2\dfrac{\partial^2 u}{\partial x^2}, \\ u(0,t)=\Psi_1(t), u(l,t)=\Psi_2(t), \\ u(x,t_0)=f(x), \left.\dfrac{\partial u}{\partial t}\right|_{t=t_0}=g(x), \end{cases} \tag{6.6}$$

where the boundary conditions $u(0,t)=\Psi_1(t)$ and $u(l,t)=\Psi_2(t)$ imply free-free beam. Usually, an initial time t_0 is set to 0 for simplicity.

6.1.1 Free Response

The above is a partial differential equation (PDF) without external excitations. Thus, $u(x, t)$ is a free response. That may generally be difficult to be obtained by methods in ordinary differential equations, such as methods of characteristic roots. However, $u(x, t)$ happens to be in the form

$$u(x,t)=\varphi(x)p(t). \tag{6.7}$$

Solving method of a PDE based on the above is called variable separation method.

Substituting the above into Eq. (6.5) yields

$$\frac{\varphi''(x)}{\varphi(x)}=\frac{p''(t)}{a^2p(t)}. \tag{6.8}$$

Since the left side on the above is a function of x while the right is a function in terms of t, the above ratio must be a constant. Denote the ratio by $-\lambda$. Then,

$$\frac{\varphi''(x)}{\varphi(x)}=\frac{p''(t)}{a^2p(t)}=-\lambda. \tag{6.9}$$

The above expression implies two ordinary differential equations given by

$$p''(t)+\lambda a^2p(t)=0, \tag{6.10}$$

$$\begin{cases} \varphi''(x)+\lambda\varphi(x)=0, \\ \varphi(0)=\varphi(l)=0, \end{cases} \tag{6.11}$$

where the boundary conditions are supposed to be $\varphi(0)=\varphi(l)=0$, implying a beam with fixed ends. That does not lose generality in our discussions.

Now we consider whether $\lambda>0$ or not. Denote by $h_p(t)$ an impulse response to Eq. (6.10). Then, we have

$$h_p''(t)+\lambda a^2h_p(t)=\delta(t). \tag{6.12}$$

Since

$$h_p(t)=\frac{1}{\sqrt{\lambda}a}\sin\left(\sqrt{\lambda}at\right), \tag{6.13}$$

$\lambda=0$ is trivial. In addition, a physically realizable vibration system never produces $h_p(t)$ for $\lambda<0$. Thus, $\lambda>0$ has to be required.

Consider the solution to Eq. (6.11), i.e., $\varphi(x)$. The characteristic equation of Eq. (6.11) is given by

$$s^2+\lambda=0. \tag{6.14}$$

Due to $\lambda>0$, one has

$$s_{1,2}=\pm\sqrt{-\lambda}=\pm i\sqrt{\lambda}. \tag{6.15}$$

A general solution to Eq. (6.11) is given by

$$\varphi(x)=C\cos\sqrt{\lambda}x+D\sin\sqrt{\lambda}x. \tag{6.16}$$

Owing to $\varphi(0)=\varphi(l)=0$, we have

$$\begin{cases} C=0, \\ D\sin\sqrt{\lambda}l=0. \end{cases} \tag{6.17}$$

Since $D=0$ is trivial for Eq. (6.17), we have

$$\sin\sqrt{\lambda}l=0. \tag{6.18}$$

Because the above holds for

$$\sqrt{\lambda}l=j\pi,\quad j=1,2,\ldots, \tag{6.19}$$

we have

$$\lambda=\frac{j^2\pi^2}{l^2},\quad j=1,2,\ldots \tag{6.20}$$

Thus, $\varphi(x)$ is given by

$$\varphi(x)=D\sin\frac{j\pi x}{l}\triangleq\varphi_j(x). \tag{6.21}$$

In mathematics, λ is called the eigenvalue and $\varphi_j(x)$ is termed eigenfunction of Eq. (6.11). In engineering vibrations, it is often called the jth order vibration mode.

Note 6.1: There are infinite vibration modes since $j=1, 2, \ldots$

Under the boundary conditions of $\varphi(0)=\varphi(l)=0$, we write $\varphi_j(x)$ by

$$\varphi_j(x)\sim\sin\frac{j\pi x}{l},\quad j=1,2,.. \tag{6.22}$$

Figure 6.2 shows the plots for $j=1, 2, 3$ with $l=10$.

With respect to $p(t)$, we have

$$p''(t)+\frac{j^2\pi^2}{l^2}a^2p(t)=0. \tag{6.23}$$

Let

$$\mu_j=\frac{j\pi}{l}. \tag{6.24}$$

Denoting

$$\omega_{nj}=a\mu_j=\frac{j\pi}{l}\sqrt{\frac{E}{\rho}} \tag{6.25}$$

yields

$$p''(t)+\omega_{nj}^2p(t)=0. \tag{6.26}$$

The quantity ω_{nj} takes the place of natural angular frequency of the jth order vibration mode. From Eq. (6.26), one has

$$p(t)=A\cos\omega_{nj}t+B\sin\omega_{nj}t, \tag{6.27}$$

where A and B are constants determined by initial conditions.

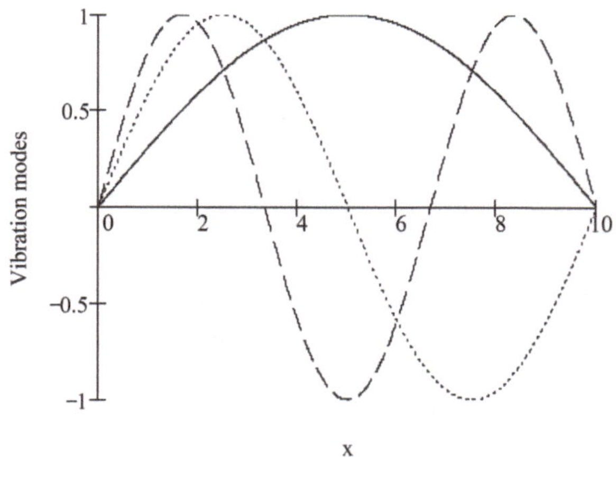

FIGURE 6.2 Illustrations of vibrations modes of fixed-fixed beam with $l=10$: $\varphi_1(x)$ (solid), $\varphi_2(x)$ (dot), $\varphi_3(x)$ (dash).

Note 6.2: There are infinite natural frequencies because of $j = 1, 2, \ldots$
For the jth order vibration mode, one has

$$p_j(t) = A_j \cos\omega_{nj}t + B_j \sin\omega_{nj}t. \tag{6.28}$$

For the jth order natural frequency, we have the jth order free response in the form

$$u_j(x,t) = \varphi_j(x)p_j(t) = \sin\frac{j\pi x}{l}\left(A_j \cos\omega_{nj}t + B_j \sin\omega_{nj}t\right). \tag{6.29}$$

Owing to the linearity of a wave equation (6.5), the general solution of total free response is given by

$$\begin{aligned} u(x,t) &= \sum_{j=1}^{\infty} u_j(x,t) = \sum_{j=1}^{\infty} \varphi_j(x)p_j(t) \\ &= \sum_{j=1}^{\infty} \sin\frac{j\pi x}{l}\left(A_j \cos\omega_{nj}t + B_j \sin\omega_{nj}t\right). \end{aligned} \tag{6.30}$$

Taking into account initial conditions $u(x, 0) = f(x)$ and $u_t(x, 0) = g(x)$, one has

$$\begin{cases} f(x) = \displaystyle\sum_{j=1}^{\infty} A_j \sin\frac{j\pi x}{l}, \\ g(x) = \displaystyle\sum_{j=1}^{\infty} B_j\omega_{nj} \sin\frac{j\pi x}{l}, \end{cases} \quad 0 < x < l. \tag{6.31}$$

Using the orthogonality of trigonometric functions, one has

$$\begin{cases} A_j = \dfrac{2}{l}\displaystyle\int_0^l f(x)\sin\frac{j\pi x}{l}\,dx, \\ B_j = \dfrac{2}{l\omega_{nj}}\displaystyle\int_0^l g(x)\sin\frac{j\pi x}{l}\,dx. \end{cases} \tag{6.32}$$

Note 6.3: Since $\lim_{j\to\infty} A_j = 0$ and $\lim_{j\to\infty} B_j = 0$ in general, eigenvibrations with high eigenvibration frequencies can be ignored in engineering.

Note 6.4: In vibrations, $\varphi(x)$ is called vibration mode while $p(t)$ is termed coordinate.

Example 6.1. Let

$$\begin{cases} \dfrac{\partial^2 u}{\partial t^2} = a^2 \dfrac{\partial^2 u}{\partial x^2}, \\ u(0,t) = 0,\ u(l,t) = 0, \\ u(x,0) = f(x) = x(l-x), \left.\dfrac{\partial u}{\partial t}\right|_{t=0} = g(x) = 0. \end{cases} \tag{6.33}$$

Find the solution $u(x, t)$.

Solution: According to Eq. (6.32), we have $B_j = 0$ and

$$A_j = \frac{2}{l}\int_0^l f(x)\sin\frac{j\pi x}{l}dx = \frac{l^3}{j\pi}(-1)^{j+1} + \left(\frac{l}{j\pi}\right)^2\left[(-1)^j - 1\right]. \tag{6.34}$$

Thus,

$$u(x,t) = \sum_{j=1}^{\infty} A_j \sin\frac{j\pi x}{l}\cos\frac{j\pi a t}{l} = \sum_{j=1}^{\infty}\left\{\frac{l^3}{j\pi}(-1)^{j+1} + \left(\frac{l}{j\pi}\right)^2\left[(-1)^j - 1\right]\right\}\sin\frac{j\pi x}{l}\cos\omega_{nj}t. \tag{6.35}$$

Example 6.2. Find the general solution $u(x, t)$ to a free-free beam.

Solution: Due to free-free beam, for longitudinal vibrations, axial forces on its both ends are zero. That is,

$$u_x(0,t) = u_x(l,t) = 0. \tag{6.36}$$

Thus, the motion equation is in the form

$$\begin{cases} \dfrac{\partial^2 u}{\partial t^2} = a^2 \dfrac{\partial^2 u}{\partial x^2}, \\ u(x,0) = f(x), u_t(x,0) = g(x), \\ \left.\dfrac{\partial u(x,t)}{\partial x}\right|_{x=0} = \left.\dfrac{\partial u(x,t)}{\partial x}\right|_{x=l} = 0. \end{cases} \tag{6.37}$$

Since $u_x(0,t) = 0$ and $u_x(l,t) = 0$ mean $\varphi'(0) = \varphi'(l) = 0$, we have

$$\begin{cases} \varphi''(x)+\lambda\varphi(x)=0, \\ \varphi'(0)=\varphi'(l)=0. \end{cases} \tag{6.38}$$

As $\varphi'(x)=-C\sqrt{\lambda}\sin\sqrt{\lambda}x+D\sqrt{\lambda}\cos\sqrt{\lambda}x$, $\varphi'(0)=0$ results in $D=0$. Besides, $\varphi'(l)=0$ yields

$$C\sin\sqrt{\lambda}l=0. \tag{6.39}$$

Because $C=0$ is trivial, the above implies

$$\lambda=\frac{j^2\pi^2}{l^2}, \quad j=1,2,\ldots$$

The natural frequencies are given by Eq. (6.25). Vibration mode function $\varphi(x)$ is in the form

$$\varphi(x)=C\cos\frac{j\pi x}{l}\triangleq\varphi_j(x). \tag{6.40}$$

Figure 6.3 gives the plots of $\varphi(x)$.

The general solution is given by

$$\begin{aligned} u(x,t)&=\sum_{j=1}^{\infty}u_j(x,t)=\sum_{j=1}^{\infty}\varphi_j(x)p_j(t) \\ &=\sum_{j=1}^{\infty}\cos\frac{j\pi x}{l}\left(A_j\cos\omega_{nj}t+B_j\sin\omega_{nj}t\right). \end{aligned} \tag{6.41}$$

Note 6.5: The set of vibration mode functions $\{\varphi_j(x)\}$ is an orthogonal space because

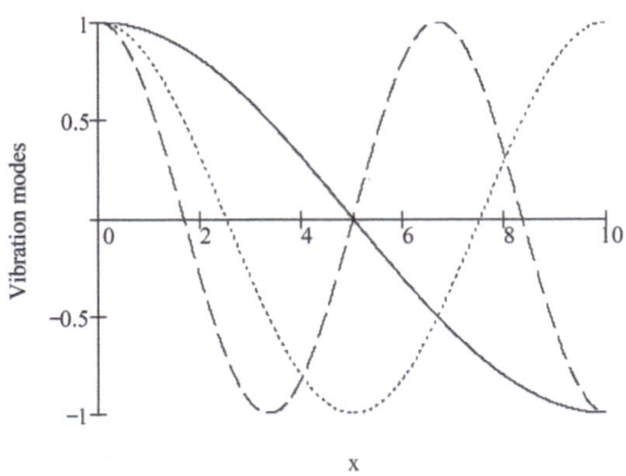

FIGURE 6.3 Illustrations of vibrations modes of free-free beam with l = 10: $\varphi_1(x)$ (solid), $\varphi_2(x)$ (dot), $\varphi_3(x)$ (dash).

$$\int_0^l \varphi_m(x)\varphi_n(x)dx = 0, \quad m \neq n. \tag{6.42}$$

Thus, intrinsic vibration modes are orthogonal. However, $\{\varphi_j(x)\}$ is not completed.

Note 6.6: The set of coordinates $\{p_j(t)\}$ is an orthogonal space because

$$\int_0^{T_n} p_m(t)p_n(t)dt = 0, \quad m \neq n, \tag{6.43}$$

where T_n is the base period of $\cos\omega_{nj}t$ and $\sin\omega_{nj}t$. Clearly, $\{p_j(t)\}$ is orthogonal and completed because $\{\cos\omega_{nj}t, \sin\omega_{nj}t\}$ is a completed orthogonal space, taking into account

$$p_j(t) = A_j\cos\omega_{nj}t + B_j\sin\omega_{nj}t. \tag{6.44}$$

Note 6.7: In engineering, we can only use the finite summation of $u_j(x,t)$. That is,

$$u(x,t) \approx \sum_{j=1}^{N} u_j(x,t) = \sum_{j=1}^{N} \varphi_j(x)p_j(t). \tag{6.45}$$

Note 6.8: $u_j(x,t) = \varphi_j(x)p_j(t)$ designates the value $\varphi_j(x)$ on the coordinate $p_j(t)$.

Note 6.9: Because $\lim_{j\to\infty} p_j(t) = 0$, $\lim_{j\to\infty} u_j(x,t) = 0$. Consequently, the higher the natural frequency ω_{nj}, the smaller the coordinate $p_j(t)$.

Two things are worth noting. The first thing is that a ship is a typical free-free beam (Palley et al. [1]). From the above, we see that any vibration mode of longitudinal vibration of a ship gets its maximum at its stem and stern, also see Figure 6.3. The second thing is any vibration mode of longitudinal vibration of free-free beam is a sinusoidal function, see Eq. (6.40).

6.1.2 Forced Response

Let $F(x, t)$ be a driven force. Denote by

$$f(x,t) = \frac{F(x,t)}{\rho}. \tag{6.46}$$

Then, damping free response to a longitudinal vibration of a beam is expressed by a wave equation with excitation $f(x, t)$ in the form

$$\begin{cases} \dfrac{\partial^2 u(x,t)}{\partial t^2} - a^2\dfrac{\partial^2 u(x,t)}{\partial x^2} = f(x,t), \\ u(0,t) = 0, \quad u(l,t) = 0, \\ u(x,0) = a(x), \quad \left.\dfrac{\partial u(x,t)}{\partial t}\right|_{t=0} = b(x). \end{cases} \tag{6.47}$$

Since the above is a linear equation, response $u(x, t)$ may be expressed by zero input response plus zero state response. Denote by $u_{zi}(x, t)$ and $u_{zs}(x, t)$ zero input response and zero state one, respectively. Then,

$$u(x,t)=u_{zi}(x,t)+u_{zs}(x,t). \tag{6.48}$$

For $u_{zi}(x, t)$, we have

$$\begin{cases} \dfrac{\partial^2 u_{zi}(x,t)}{\partial t^2}-a^2\dfrac{\partial^2 u_{zi}(x,t)}{\partial x^2}=0, \\ u_{zi}(0,t)=0, \quad u_{zi}(l,t)=0, \\ u_{zi}(x,0)=a(x), \quad \left.\dfrac{\partial u_{zi}(x,t)}{\partial t}\right|_{t=0}=b(x). \end{cases} \tag{6.49}$$

Our discussion in Section 6.1.1 is for $u_{zi}(x, t)$. Thus, we only explain $u_{zs}(x, t)$ in this subsection. To be precise, we consider a PDE in the form

$$\begin{cases} \dfrac{\partial^2 u_{zs}(x,t)}{\partial t^2}-a^2\dfrac{\partial^2 u_{zs}(x,t)}{\partial x^2}=f(x,t), \\ u_{zs}(0,t)=0, \quad u_{zs}(l,t)=0, \\ u_{zs}(x,0)=0, \quad \left.\dfrac{\partial u_{zs}(x,t)}{\partial t}\right|_{t=0}=0. \end{cases} \tag{6.50}$$

Without confusion, we omit the subscript of $u_{zs}(x, t)$ in what follows. Thus, we consider

$$\begin{cases} \dfrac{\partial^2 u(x,t)}{\partial t^2}-a^2\dfrac{\partial^2 u(x,t)}{\partial x^2}=f(x,t), \\ u(0,t)=0, \quad u(l,t)=0, \\ u(x,0)=0, \quad \left.\dfrac{\partial u(x,t)}{\partial t}\right|_{t=0}=0. \end{cases} \tag{6.51}$$

Several common methods are briefed below.

6.1.2.1 Impulse Method

External force on a beam with unit length is given by

$$F(x,t)=\rho(x,t). \tag{6.52}$$

At time $t=\tau$, an instantaneously external force is expressed by

$$\rho f(x,t)\delta(t-\tau)dx.$$

According to the sampling property of delta function, one has

$$F(x,t)=\rho f(x,t)=\int_0^t \rho f(x,\tau)\delta(t-\tau)d\tau. \tag{6.53}$$

Thus, due to superposition, we first obtain the response that is denoted by $v(x, t)$ under the excitation $\rho f(x,t)\delta(t-\tau)$. Precisely,

$$\begin{cases} \dfrac{\partial^2 v(x,t)}{\partial t^2}-a^2\dfrac{\partial^2 v(x,t)}{\partial x^2}=f(x,\tau)\delta(t-\tau), \\ v(0,t)=0, \quad v(l,t)=0, \\ v(x,0)=0, \quad \left.\dfrac{\partial v(x,t)}{\partial t}\right|_{t=0}=0. \end{cases} \tag{6.54}$$

Then, total response $u(x, t)$ is obtained by the integral

$$u(x,t)=\int_0^t v(x,t;\tau)d\tau. \tag{6.55}$$

Since

$$\begin{cases} \rho f(x,t)\delta(t-\tau)\neq 0 \quad (t=\tau-0, t=\tau+0), \\ 0, \text{ otherwise}, \end{cases} \tag{6.56}$$

one has

$$\int_{\tau-0}^{\tau+0}\frac{\partial^2 v(x,t)}{\partial t^2}dt-a^2\int_{\tau-0}^{\tau+0}\frac{\partial^2 v(x,t)}{\partial x^2}dt=\int_{\tau-0}^{\tau+0} f(x,\tau)\delta(t-\tau)dt=f(x,\tau). \tag{6.57}$$

Note that

$$\int_{\tau-0}^{\tau+0}\frac{\partial^2 v(x,t)}{\partial x^2}dt=0. \tag{6.58}$$

In fact, when letting $v(x, t)=g(x)h(t)$, one has

$$\frac{\partial^2 v(x,t)}{\partial x^2}=g''(x)h(t). \tag{6.59}$$

Substituting the above into the left side of Eq. (6.58) produces

$$\begin{aligned}\int_{\tau-0}^{\tau+0}\frac{\partial^2 v(x,t)}{\partial x^2}dt &= g''(x)\int_{\tau-0}^{\tau+0}h(t)dt = g''(x)\int_{\tau-0}^{\tau+0}d\left[\int_0^t h(t)dt\right]\\ &= g''(x)\left[\int_0^t h(u)du\bigg|_{t=\tau+0} - \int_0^t h(u)du\bigg|_{t=\tau-0}\right]=0.\end{aligned} \tag{6.60}$$

Besides, for the first item on the left side of Eq. (6.57), we have

$$\int_{\tau-0}^{\tau+0}\frac{\partial^2 v(x,t)}{\partial t^2}dt = \frac{\partial v(x,t)}{\partial t}\bigg|_{\tau-0}^{\tau+0} = \frac{\partial v(x,t)}{\partial t}\bigg|_{t=\tau+0} - \frac{\partial v(x,t)}{\partial t}\bigg|_{t=\tau-0} = \frac{\partial v(x,t)}{\partial t}\bigg|_{t=\tau+0}. \tag{6.61}$$

Thus,

$$\frac{\partial v(x,t)}{\partial t}\bigg|_{t=\tau+0} = f(x,\tau). \tag{6.62}$$

Therefore, Eq. (6.54) becomes

$$\begin{cases}\dfrac{\partial^2 v(x,t)}{\partial t^2} - a^2\dfrac{\partial^2 v(x,t)}{\partial x^2} = 0,\\ v(0,t)=0,\quad v(l,t)=0,\\ v(x,t)\big|_{t=\tau-0}=0,\quad \dfrac{\partial v(x,t)}{\partial t}\bigg|_{t=\tau+0} = f(x,\tau).\end{cases} \tag{6.63}$$

The above says that excitation $\rho f(x,t)\delta(t-\tau)$ in Eq. (6.54) changes to an initial condition in Eq. (6.62). Having solved $v(x, t)$, one can obtain the total response $u(x, t)$ by Eq. (6.55).

What discussed previously is called impulse method for solving the response to a damping free wave equation with excitation $f(x, t)$. This method consists of four steps.

1. Write $u(x, t)$ by $v(x, t)$ and let excitation item be zero.
2. Write $\dfrac{\partial u(x,t)}{\partial t}\bigg|_{t=0} = 0$ by $\dfrac{\partial v(x,t)}{\partial t}\bigg|_{t=\tau+0} = f(x,\tau)$.
3. Find free response $v(x, t)$.
4. Obtain total forced response $u(x, t)$ by Eq. (6.55).

Example 6.3. Find $u(x, t)$ to the following equation

$$\begin{cases} \dfrac{\partial^2 u(x,t)}{\partial t^2} - a^2 \dfrac{\partial^2 u(x,t)}{\partial x^2} = A\sin\dfrac{\pi x}{l}\sin\omega t, \\ u(0,t)=0, \quad u(l,t)=0, \qquad (0<x<l). \\ u(x,0)=0, \quad \left.\dfrac{\partial u(x,t)}{\partial t}\right|_{t=0}=0, \end{cases} \tag{6.64}$$

Solution: Using the impulse method, we have

$$\begin{cases} \dfrac{\partial^2 v(x,t)}{\partial t^2} - a^2 \dfrac{\partial^2 v(x,t)}{\partial x^2} = 0, \\ v(0,t)=0, \quad v(l,t)=0, \qquad (0<x<l). \\ v|_{t=\tau+0}=0, \quad v_t|_{t=\tau+0} = A\sin\dfrac{\pi x}{l}\sin\omega\tau, \end{cases} \tag{6.65}$$

The above equation is for a fixed-fixed beam. Thus, its vibration mode function is $\sin\dfrac{j\pi x}{l}$ so that

$$\left.\sin\frac{\pi x}{l}\right|_{x=0}=0, \tag{6.66}$$

and

$$v(l,t)=\left.\sin\frac{\pi x}{l}\right|_{x=l}=0. \tag{6.67}$$

Thus,

$$v(0,t)=v(l,t)=0. \tag{6.68}$$

Accordingly, the general solution is in the form

$$v(x,t;\tau)=\sum_{j=1}^{\infty} p_j(t;\tau)\sin\frac{j\pi x}{l}. \tag{6.69}$$

Thus, we have

$$\sum_{j=1}^{\infty}\left[p_j''(t;\tau)+\frac{j^2\pi^2a^2}{l^2}p_j(t;\tau)\right]\sin\frac{j\pi x}{l}=0. \tag{6.70}$$

That corresponds to an ordinary differential equation given by

$$p_j''(t;\tau)+\frac{j^2\pi^2a^2}{l^2}p_j(t;\tau)=0. \tag{6.71}$$

Because

$$p_j(t;\tau)=A_j(\tau)\cos\frac{j\pi a(t-\tau)}{l}+B_j(\tau)\sin\frac{j\pi a(t-\tau)}{l}, \tag{6.72}$$

we have

$$\begin{aligned} v(x,t;\tau)&=\sum_{j=1}^{\infty}p_j(t;\tau)\sin\frac{j\pi x}{l} \\ &=\sum_{j=1}^{\infty}\left[A_j(\tau)\cos\frac{j\pi a(t-\tau)}{l}+B_j(\tau)\sin\frac{j\pi a(t-\tau)}{l}\right]\sin\frac{j\pi x}{l}. \end{aligned} \tag{6.73}$$

Note that the initial conditions for $v(x, t; \tau)$ are at $t=\tau$ instead of $t=0$. Since

$$v(x,t;\tau)\big|_{t=\tau}=0, \tag{6.74}$$

we have

$$A_j(\tau)=0. \tag{6.75}$$

In addition, as

$$v_t(x,t;\tau)\big|_{t=\tau}=A\sin\frac{\pi x}{l}\sin\omega\tau, \tag{6.76}$$

we have

$$B_j(\tau)=0(j\neq 1),\ \ B_1(\tau)=\frac{Al}{\pi a}\sin\omega\tau. \tag{6.77}$$

Consequently, $v(x, t; \tau)$ is in the form

$$v(x,t;\tau)=\frac{Al}{\pi a}\sin\omega\tau\sin\frac{\pi a(t-\tau)}{l}\sin\frac{\pi x}{l}. \tag{6.78}$$

Finally, doing the integral of $v(x, t; \tau)$ yields the total forced response

$$\begin{aligned} u(x,t) &= \int_0^t v(x,t;\tau)d\tau = \frac{Al}{\pi a}\sin\frac{\pi x}{l}\int_0^t \sin\omega\tau \sin\frac{\pi a(t-\tau)}{l}d\tau \\ &= \frac{Al}{\pi a}\sin\frac{\pi x}{l}\frac{1}{\omega^2-\left(\frac{\pi a}{l}\right)^2}\left(\omega\sin\frac{\pi a}{l}t - \frac{\pi a}{l}\sin\omega t\right), \quad t\geq 0. \end{aligned} \tag{6.79}$$

6.1.2.2 Green Function Method

Another commonly used method in solving a wave equation is the Green function method. Its key point is to decompose excitation $f(x, t)$ by two delta functions in the form

$$f(x,t) = \int_{\tau=0}^{t}\int_{\xi=0}^{l} f(\xi,\tau)\delta(x-\xi)\delta(t-\tau)d\xi d\tau. \tag{6.80}$$

The above holds because

$$\begin{aligned} &\int_{\tau=0}^{t} f(\xi,\tau)\delta(t-\tau)d\tau = f(\xi,t), \\ &\int_{\xi=0}^{l} f(\xi,t)\delta(x-\xi)d\xi = f(x,t). \end{aligned} \tag{6.81}$$

Thus, for any excitation $f(x, t)$, one may deal with its forced response by considering the solution to the following equation with a unified excitation $\delta(x-\xi)\delta(t-\tau)$

$$\begin{cases} G_{tt}(x,t) - a^2G_{xx}(x,t) = \delta(x-\xi)\delta(t-\tau), \\ G(0,t)=0, \ \ G(l,t)=0, \\ G(x,0)=0, \ \ G_t(x,0)=0, \end{cases} \tag{6.82}$$

where $G(x, t)$ is called Green function. Once it is obtained, we have the forced response $u(x, t)$ under the excitation $f(x, t)$ in the form

$$u(x,t) = \int_{\tau=0}^{t}\int_{\xi=0}^{l} f(\xi,\tau)G(x,t;\xi,\tau)d\xi d\tau. \tag{6.83}$$

For solving $G(x, t)$ in Eq. (6.77), we may use the impulse method to write it by

$$\begin{cases} G_{tt}(x,t) - a^2G_{xx}(x,t) = 0, \\ G(0,t)=0, \ \ G(l,t)=0, \\ G(x,\tau+0)=0, \ \ G_t(x,\tau+0)=\delta(x-\xi). \end{cases} \tag{6.84}$$

In other words, the excitation $\delta(x-\xi)\delta(t-\tau)$ changes to be non-zero initial conditions at $t=\tau$, where

$$\delta(x-\xi)=\frac{1}{l}+\sum_{j=1}^{\infty}\cos\frac{j\pi\xi}{l}\sin\frac{j\pi x}{l}. \tag{6.85}$$

6.2 LATERAL VIBRATIONS OF BEAMS

6.2.1 Euler-Bernoulli Beam

If a beam suffers from pure lateral vibration or transversal vibration or bending vibration as indicated in Figure 6.4, it is called Euler-Bernoulli beam or Bernoulli beam for short.

For a micro-segmentation in Figure 6.4, at time t and position x, $\rho(x)$ is material density, $A(x)$ is cross-sectional area, E is elastic modulus, $I(x)$ is cross-sectional moment of inertia, $EI(x)$ is bending rigidity, $Q(x, t)$ is shear force, $M(x, t)$ is bending moment, $q(x, t)$ is driven force, and $w(x, t)$ is deflection.

When a force moment at $x+dx$ reaches its equilibrium status, one has

$$M(x,t)+\frac{\partial M(x,t)}{\partial x}dx-M(x,t)+Q(x,t)dx-q\frac{(dx)^2}{2}=0. \tag{6.86}$$

Ignoring high-order infinitesimal, we have

$$\frac{\partial M(x,t)}{\partial x}=-Q(x,t). \tag{6.87}$$

Based on material mechanics, bending moment $M(x, t)$ relates to deflection $w(x, t)$ by

$$EI(x)\frac{\partial^2 w(x,t)}{\partial x^2}=M(x,t).$$

When micro-segmentation dx is along the direction of w, therefore, we have

$$(\rho A dx)\frac{\partial^2 w(x,t)}{\partial t^2}=Q+\frac{\partial Q}{\partial x}dx-Q+qdx=\frac{\partial Q}{\partial x}dx+qdx, \tag{6.88}$$

where $(\rho A dx)\frac{\partial^2 w(x,t)}{\partial t^2}$ is inertial force and $(\rho A dx)$ is mass. The driven force is $Q_x(x, t)dx+qdx$. Using Eqs. (6.87) and (6.88) produces

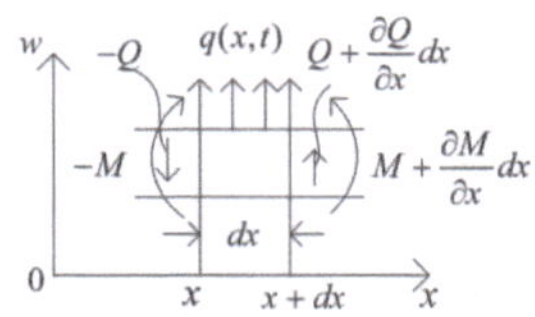

FIGURE 6.4 Diagram of force on micro-segment of a beam in lateral vibration.

$$\rho A(x)\frac{\partial^2 w(x,t)}{\partial t^2}+\frac{\partial^2}{\partial x^2}\left[EI(x)\frac{\partial^2 w(x,t)}{\partial x^2}\right]=q(x,t). \tag{6.89}$$

The above is the motion equation of the famous Euler-Bernoulli beam.

Since cross-section area $A(x)$ and cross-sectional moment of inertia $I(x)$ vary with position x, it may be difficult to solve the damping free motion equation of Euler-Bernoulli beam even if material density $\rho(x)$ is a constant. To make a thing a step further, we let $A(x)$ and $I(x)$ be constants. In that case, Eq. (6.89) is simplified by

$$EI\frac{\partial^4 w(x,t)}{\partial x^4}+\rho A\frac{\partial^2 w(x,t)}{\partial t^2}=q(x,t). \tag{6.90}$$

The above is the simplified motion equation of Euler-Bernoulli beam. Finding its analytic solution is available.

6.2.2 Free Response to Simplified Euler-Bernoulli Beam

In the following equation

$$EI\frac{\partial^4 w}{\partial x^4}+\rho A\frac{\partial^2 w}{\partial t^2}=0, \tag{6.91}$$

using $w(x, t)=\varphi(x)p(t)$ yields

$$\frac{EI}{\rho A\varphi(x)}\frac{d^4\varphi(x)}{dx^4}=-\frac{1}{p(t)}\frac{d^2 p(t)}{dt^2}. \tag{6.92}$$

Since the left side of the above is a function of x while the right is a function of t, they are equal if and only if both are equal to the same constant. Denote by ω_n^2 that constant. Then, we have two ordinary differential equations. One is for vibration mode $\varphi(x)$ in the form

$$EI\frac{d^4\varphi(x)}{dx^4}-\rho A\omega_n^2\varphi(x)=0. \tag{6.93}$$

The other is in terms of coordinate $p(t)$ in the form

$$\frac{d^2 p(t)}{dt^2}+\omega_n^2 p(t)=0. \tag{6.94}$$

Rewrite vibration mode equation by

$$\frac{d^4\varphi(x)}{dx^4}-\mu^4\varphi(x)=0, \tag{6.95}$$

where

$$\mu^4 = \frac{\rho A}{EI}\omega_n^2, \tag{6.96}$$

or

$$\omega_n = \mu^2\sqrt{\frac{EI}{\rho A}}. \tag{6.97}$$

Note 6.10: The parameter μ or ω_n is determined by system itself irrelevant of external environments.

The characteristic equation of Eq. (6.95) is given by

$$s^4 - \mu^4 = 0. \tag{6.98}$$

Four characteristic roots are expressed by

$$s_{1,2} = \pm i\mu, \quad s_{3,4} = \pm\mu. \tag{6.99}$$

Therefore, $\varphi(x)$ is given by

$$\varphi(x) = B_1 \sin\mu x + B_2 \cos\mu x + B_3 \mathrm{sh}\mu x + B_4 \mathrm{ch}\mu x, \tag{6.100}$$

where four coefficients B_1, B_2, B_3, and B_4 are determined by boundary conditions. Table 6.1 indicates three sets of boundary conditions. We study $\varphi(x)$ with a simple supported beam and free-free one below.

For a simply supported beam, we have

$$\varphi(0) = \left(B_1 \sin\mu x + B_2 \cos\mu x + B_3 \mathrm{sh}\mu x + B_4 \mathrm{ch}\mu x\right)_{x=0} = B_2 + B_4 = 0. \tag{6.101}$$

In addition,

$$\varphi''(0) = \mu^2\left(-B_2 + B_4\right) = 0. \tag{6.102}$$

TABLE 6.1 Boundary Conditions

Simply supported beam	$\varphi(0)=0$ (zero deflection); $\varphi''(0)=0$ (zero bending moment)	$\varphi(l)=0$ (zero deflection); $\varphi''(l)=0$ (zero bending moment)
Free-free beam	$\varphi''(0)=0$ (zero bending moment); $\varphi'''(0)=0$ (zero shear force)	$\varphi''(l)=0$ (zero bending moment); $\varphi'''(l)=0$ (zero shear force)
Beam with both fixed ends	$\varphi(0)=0$ (zero deflection); $\varphi'(0)=0$ (zero beam angle)	$\varphi(l)=0$ (zero deflection); $\varphi'(l)=0$ (zero beam angle)

Equations (6.101) and (6.102) imply

$$B_2 = B_4 = 0. \tag{6.103}$$

Thus, $\varphi(x)$ is expressed by

$$\varphi(x) = B_1 \sin \mu x + B_3 \mathrm{sh}\mu x. \tag{6.104}$$

Substituting $\varphi(l) = 0$ and $\varphi''(l) = 0$ into the above produces

$$\begin{aligned} B_1 \sin \mu l + B_3 \mathrm{sh}\mu l &= 0, \\ -B_1 \sin \mu l + B_3 \mathrm{sh}\mu l &= 0. \end{aligned} \tag{6.105}$$

Note that $B_1 = B_3 = 0$ is trivial. For non-zero B_1 and B_3, we must have

$$\begin{vmatrix} \sin \mu l & \mathrm{sh}\mu l \\ -\sin \mu l & \mathrm{sh}\mu l \end{vmatrix} = 0. \tag{6.106}$$

The above is called frequency equation of a simply supported Euler-Bernoulli beam. Since

$$\mathrm{sh}\mu l \neq 0, \tag{6.107}$$

we must have

$$\sin \mu l = 0. \tag{6.108}$$

Denote by

$$\mu l \triangleq \mu_j l = j\pi, \quad j = 1, 2, \ldots \tag{6.109}$$

Then,

$$\mu_j = \frac{j\pi}{l}, \quad j = 1, 2, \ldots \tag{6.110}$$

The jth order natural angular frequency is given by

$$\omega_{nj} = \mu_j^2 \sqrt{\frac{EI}{\rho A}} = \left(\frac{j\pi}{l}\right)^2 \sqrt{\frac{EI}{\rho A}}, \quad j = 1, 2, \ldots \tag{6.111}$$

Therefore, considering $B_3 = 0$, the jth order vibration mode is expressed by

$$\varphi_j(x) = B_{1j} \sin \frac{j\pi x}{l}, \quad j = 1, 2, \ldots \tag{6.112}$$

For simplicity, we use the symbol $C_j = B_{1j}$. Thus, the above is rewritten by

$$\varphi_j(x) = C_j \sin\frac{j\pi x}{l}, \quad j = 1,2,\ldots \tag{6.113}$$

Note that the general solution in terms of $p(t)$ is given by

$$p(t) = A_1 \cos\omega_n t + A_2 \sin\omega_n t = p\sin(\omega_n t + \theta). \tag{6.114}$$

The jth coordinate $p_j(t)$ is in the form

$$p_j(t) = p_j \sin(\omega_{nj} t + \theta_j). \tag{6.115}$$

Thus, there are infinite natural frequencies, vibration modes, and coordinates. Plots of vibration modes of a simple supported Euler-Bernoulli beam with $C_j = 1$ are the same as those in Figure 6.2.

The jth order normal vibration of a simply supported Euler-Bernoulli beam is expressed by

$$w_j(x,t) = \varphi_j(x)p_j(t) = \varphi_j(x)p_j \sin(\omega_{nj} t + \theta_j). \tag{6.116}$$

The total free response to a simple supported Euler-Bernoulli beam is in the form

$$w(x,t) = \sum_{j=1}^{\infty} w_j(x,t) = \sum_{j=1}^{\infty} \varphi_j(x)p_j \sin(\omega_{nj} t + \theta_j) = \sum_{j=1}^{\infty} C_j \sin\frac{j\pi x}{l} p_j \sin(\omega_{nj} t + \theta_j). \tag{6.117}$$

Note 6.11: For each natural frequency ω_{nj}, its corresponding free response $w_j(x, t)$ is harmonic (Figure 6.5). However, total free response $w(x, t)$ is non-harmonic (Figure 6.6).

To study a free-free Euler-Bernoulli beam, we first write

$$\varphi''(x) = \mu^2(-B_1 \sin\mu x - B_2 \cos\mu x + B_3 \text{sh}\mu x + B_4 \text{ch}\mu x), \tag{6.118}$$

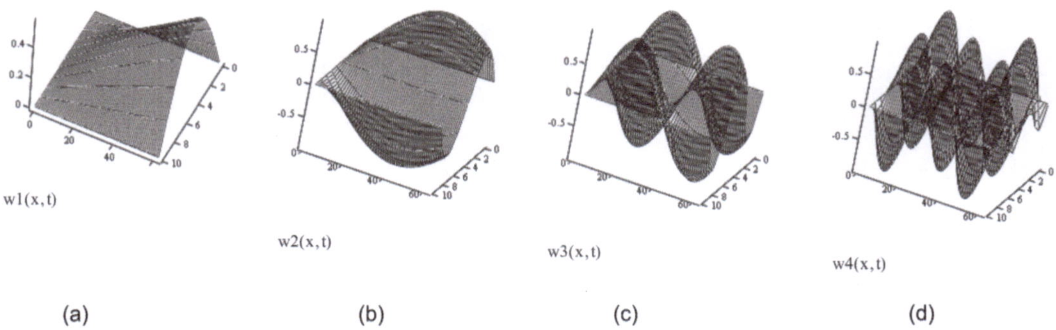

FIGURE 6.5 Free response $w_j(x, t)$ to a simply supported Euler-Bernoulli beam with $l = 10$ and $t = 1, 2, \ldots, 60$. (a) w1(x, t) = $w_1(x, t)$. (b) w2(x, t) = $w_2(x, t)$. (c) w3(x, t) = $w_3(x, t)$. (d) w4(x, t) = $w_4(x, t)$.

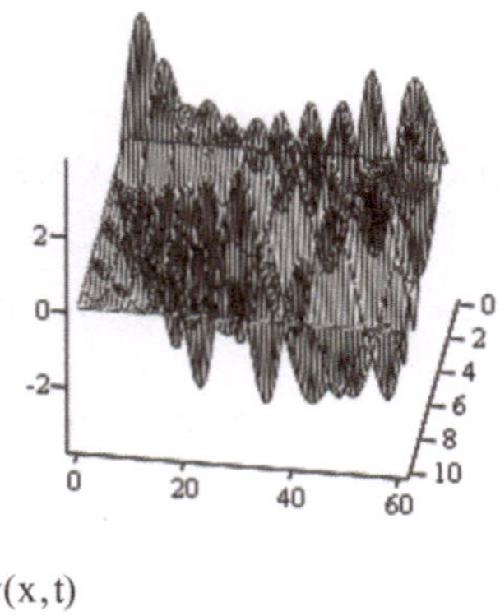

FIGURE 6.6 Free response $w(x,t) \approx \sum_{j=1}^{8} w_j(x,t)$ to a simply supported Euler-Bernoulli beam with $l=10$ and $t=1, 2, \ldots, 60$.

and

$$\varphi'''(x) = \mu^3\left(-B_1 \cos \mu x + B_2 \sin \mu x + B_3 \mathrm{ch}\mu x + B_4 \mathrm{sh}\mu x\right). \tag{6.119}$$

Substituting $\varphi''(0)=0$ and $\varphi'''(0)=0$ into the above yields

$$\varphi''(0) = -B_2 + B_4 = 0, \tag{6.120}$$

and

$$\varphi'''(0) = -B_1 + B_3 = 0. \tag{6.121}$$

Thus, we have

$$B_2 = B_4, \quad B_1 = B_3. \tag{6.122}$$

Therefore, $\varphi(x)$ can be expressed by

$$\varphi(x) = B_1\left(\sin \mu x + \mathrm{sh}\mu x\right) + B_2\left(\cos \mu x + \mathrm{ch}\mu x\right). \tag{6.123}$$

With the above, Eqs. (6.118) and (6.119) may be simplified by

$$\varphi''(x) = \mu^2\left[B_1\left(\mathrm{sh}\mu x - \sin \mu x\right) + B_2\left(\mathrm{ch}\mu x - \cos \mu x\right)\right], \tag{6.124}$$

and

$$\varphi'''(x) = \mu^3\left[B_1\left(\mathrm{ch}\mu x - \cos \mu x\right) + B_2\left(\mathrm{sh}\mu x + \sin \mu x\right)\right]. \tag{6.125}$$

Substituting $\varphi''(l)=0$ and $\varphi'''(l)=0$ into the above produces

$$B_1(\text{sh}\mu l-\sin\mu l)+B_2(\text{ch}\mu l-\cos\mu l)=0, \tag{6.126}$$

and

$$B_1(\text{ch}\mu l-\cos\mu l)+B_2(\text{sh}\mu l+\sin\mu l)=0. \tag{6.127}$$

Finding non-zero B_1 and B_2 requires

$$\begin{vmatrix} \text{sh}\mu l-\sin\mu l & \text{ch}\mu l-\cos\mu l \\ \text{ch}\mu l-\cos\mu l & \text{sh}\mu l+\sin\mu l \end{vmatrix}=0. \tag{6.128}$$

Expanding the above determinant results in the frequency equation in the form

$$\cos\mu l=\frac{1}{\text{ch}\mu l}. \tag{6.129}$$

There does not exist analytic solution to the above frequency equation. Numerically, one may attain its solution as that shown in Figure 6.7.

The numeric results for $\cos(\mu l)=1/\text{ch}(\mu l)$ are given by

$$\frac{\mu_j l}{(2j+1)\dfrac{\pi}{2}}, \quad j=1,2,\ldots \tag{6.130}$$

Thus, we have natural frequencies given by

$$\omega_{nj}=\mu_j^2\sqrt{\frac{EI}{\rho A}}=\left(\frac{(2j+1)\pi}{2l}\right)^2\sqrt{\frac{EI}{\rho A}}, \quad j=1,2,\ldots \tag{6.131}$$

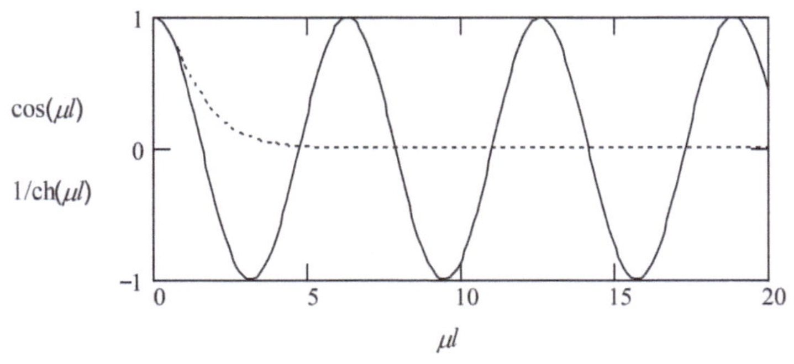

FIGURE 6.7 Illustration for numeric solution to frequency equation for μl. Solid: $\cos(\mu l)$. Dot: $1/\text{ch}(\mu l)$.

Substituting $\mu_j l$ into Eq. (6.126) yields the ratio expressed by

$$\frac{B_1}{B_2}=-\frac{\mathrm{ch}\mu_j l-\cos\mu_j l}{\mathrm{sh}\mu_j l-\sin\mu_j l}. \tag{6.132}$$

Note that a vibration mode $\varphi_j(x)$ multiplied by a constant is still a vibration mode. Thus, $\frac{\varphi(x)}{B_2}$ is a vibration mode. Taking into account $\varphi(x)$ in Eq. (6.123), we have

$$\frac{\varphi(x)}{B_2}=\frac{B_1}{B_2}(\sin\mu x+\mathrm{sh}\mu x)+(\cos\mu x+\mathrm{ch}\mu x). \tag{6.133}$$

Denoting it by $\varphi_j(x)$ yields

$$\varphi_j(x)=\frac{\varphi(x)}{B_2}=\frac{\mathrm{ch}\mu_j l-\cos\mu_j l}{\sin\mu_j l-\mathrm{sh}\mu_j l}(\sin\mu_j x+\mathrm{sh}\mu_j x)+(\cos\mu_j x+\mathrm{ch}\mu_j x). \tag{6.134}$$

Figure 6.8 indicates the plots of $\varphi_j(x)$.

Free response to a free-free Euler-Bernoulli beam is therefore given by

$$\begin{aligned} w(x,t)&=\sum_{j=1}^{\infty}w_j(x,t)=\sum_{j=1}^{\infty}\varphi_j(x)p_j\sin(\omega_{nj}t+\theta_j)\\ &=\sum_{j=1}^{\infty}\left[\frac{\mathrm{ch}\mu_j l-\cos\mu_j l}{\sin\mu_j l-\mathrm{sh}\mu_j l}(\sin\mu_j x+\mathrm{sh}\mu_j x)+(\cos\mu_j x+\mathrm{ch}\mu_j x)\right]p_j\sin(\omega_{nj}t+\theta_j). \end{aligned} \tag{6.135}$$

Note that the particularity of a free-free Euler-Bernoulli beam is that its vibration mode function $\varphi_j(x)$ is non-harmonic. It is the combination of sinusoidal functions with hyperbolic ones, see Eq. (6.134). Since a ship may be abstracted as a free-free beam (Palley et al. [1]), any vibration mode of bending vibration of a ship reaches its maximum at its stem and stern as indicated by Figure 6.8.

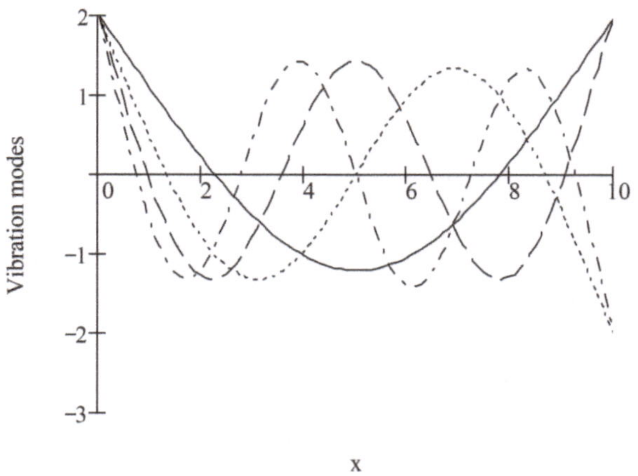

FIGURE 6.8 Plots of vibration modes for a free-free Euler-Bernoulli beam with $l=10$: $\varphi_1(x)$ (solid), $\varphi_2(x)$ (dot), $\varphi_3(x)$ (dash), $\varphi_4(x)$ (dadot).

6.2.3 Orthogonality of Normal Vibration Modes

There is orthogonality for $\varphi_j(x)$. In fact, for the rth and sth order vibration modes, one has

$$EI\frac{d^4\varphi_r(x)}{dx^4} = \rho A\omega_{nr}^2\varphi_r(x), \tag{6.136}$$

and

$$EI\frac{d^4\varphi_s(x)}{dx^4} = \rho A\omega_{ns}^2\varphi_s(x). \tag{6.137}$$

Multiplying $\varphi_s(x)$ or $\varphi_r(x)$ on both sides of Eq. (6.136) or (6.137) results in

$$EI\varphi_s(x)\frac{d^4\varphi_r(x)}{dx^4} = \rho A\omega_{nr}^2\varphi_r(x)\varphi_s(x), \tag{6.138}$$

and

$$EI\varphi_r(x)\frac{d^4\varphi_s(x)}{dx^4} = \rho A\omega_{ns}^2\varphi_s(x)\varphi_r(x). \tag{6.139}$$

Subtracting Eq. (6.138) from Eq. (6.139) produces

$$EI\varphi_r(x)\frac{d^4\varphi_s(x)}{dx^4} - EI\varphi_s(x)\frac{d^4\varphi_r(x)}{dx^4} = \rho A\omega_{ns}^2\varphi_s(x)\varphi_r(x) - \rho A\omega_{nr}^2\varphi_r(x)\varphi_s(x). \tag{6.140}$$

Doing integral from 0 to l on both sides of the above yields

$$\begin{aligned}\frac{\rho A}{EI}\left(\omega_{nr}^2 - \omega_{ns}^2\right)\int_0^l \varphi_r(x)\varphi_s(x)dx &= \int_0^l \left[\varphi_s(x)\frac{d^4\varphi_r(x)}{dx^4} - \varphi_r(x)\frac{d^4\varphi_s(x)}{dx^4}\right]dx \\ &= \varphi_s(x)\frac{d^3\varphi_r(x)}{dx^3}\bigg|_0^l - \varphi_r(x)\frac{d^3\varphi_s(x)}{dx^3}\bigg|_0^l \\ &\quad - \frac{d\varphi_s(x)}{dx}\frac{d^2\varphi_r(x)}{dx^2}\bigg|_0^l + \frac{d\varphi_r(x)}{dx}\frac{d^2\varphi_s(x)}{dx^2}\bigg|_0^l \\ &= 0\end{aligned} \tag{6.141}$$

for any supporting conditions of beam ends. Since $\omega_{nr} \neq \omega_{ns}$ for $r \neq s$, the above implies

$$\int_0^l \varphi_r(x)\varphi_s(x)dx = 0 \quad (r \neq s). \tag{6.142}$$

When $r=s$, on the other hand,

$$\int_0^l \varphi_r^2(x)dx = \text{const} \quad (r=s). \tag{6.143}$$

Equation (6.142) implies that $\varphi_s(x)$ is orthogonal to $\varphi_r(x)$ if $r \neq s$. That is called orthogonality of normal vibration modes.

One specifically calls the following the rth order generalized mass and denotes it by M_r in the form

$$\rho A \int_0^l \varphi_r^2(x)dx = M_r. \tag{6.144}$$

6.2.4 Forced Response

Suppose that a forced response corresponds to normal vibration modes in the following expression

$$w(x,t) = \sum_{j=1}^{\infty} w_j(x,t) = \sum_{j=1}^{\infty} \varphi_j(x)p_j(t), \tag{6.145}$$

where $p_j(t)$ is such that (6.90), i.e., the one below, holds

$$EI\frac{\partial^4 w(x,t)}{\partial x^4} + \rho A \frac{\partial^2 w(x,t)}{\partial t^2} = q(x,t).$$

Substituting Eq. (6.145) into the above yields

$$EI\sum_{j=1}^{\infty} \frac{d^4\varphi_j(x)}{dx^4} p_j(t) + \rho A \sum_{j=1}^{\infty} \varphi_j(x)\frac{d^2 p_j(t)}{dt^2} = q(x,t). \tag{6.146}$$

Since vibration modes are normal, we may use $EI\frac{d^4\varphi(x)}{dx^4} = \rho A\omega_n^2\varphi(x)$ in Eq. (6.93) to express the above by

$$\rho A\sum_{j=1}^{\infty} \varphi_j(x)\frac{d^2 p_j(t)}{dt^2} + \rho A\sum_{j=1}^{\infty} \omega_{nj}^2\varphi_j(x)p_j(t) = q(x,t). \tag{6.147}$$

Multiplying $\varphi_m(x)$ on both sides of the above yields

$$\rho A\sum_{j=1}^{\infty} \varphi_j(x)\varphi_m(x)\frac{d^2 p_j(t)}{dt^2} + \rho A\sum_{j=1}^{\infty} \omega_{nj}^2\varphi_j(x)\varphi_m(x)p_j(t) = q(x,t)\varphi_m(x). \tag{6.148}$$

Doing the integral from 0 to l in terms of x on both sides of the above equation produces

$$\rho A\sum_{j=1}^{\infty}\int_0^l \varphi_j(x)\varphi_m(x)dx\frac{d^2p_j(t)}{dt^2}+\rho A\sum_{j=1}^{\infty}\omega_{nj}^2\int_0^l \varphi_j(x)\varphi_m(x)dx\,p_j(t)=\int_0^l q(x,t)\varphi_m(x)dx. \quad (6.149)$$

Considering the orthogonality expressed by Eqs. (6.142) and (6.144), we have

$$M_j\frac{d^2p_j(t)}{dt^2}+M_j\omega_{nj}^2p_j(t)=\int_0^l q(x,t)\varphi_j(x)dx, \quad (6.150)$$

where $M_j\omega_{nj}^2$ is defined as the jth order generalized stiffness in the form

$$K_j=M_j\omega_{nj}^2. \quad (6.151)$$

Define the jth order generalized force $Q_j(t)$ by

$$Q_j(t)=\int_0^l q(x,t)\varphi_j(x)dx. \quad (6.152)$$

Then, Eq. (6.150) is expressed by

$$\frac{d^2p_j(t)}{dt^2}+\omega_{nj}^2p_j(t)=\frac{Q_j(t)}{M_j}. \quad (6.153)$$

The above is a vibration equation of single degree-of-freedom for the jth order vibration mode. Its general solution is in the form

$$p_j(t)=A_j\cos\omega_{nj}t+B_j\sin\omega_{nj}t+\frac{1}{M_j\omega_{nj}}\int_0^t Q_j(t)\sin\omega_{nj}(t-\tau)d\tau. \quad (6.154)$$

Thus, the forced response to an Euler-Bernoulli beam is given by

$$\begin{aligned} w(x,t)&=\sum_{j=1}^{\infty}p_j(t)\varphi_j(x)\\ &=\sum_{j=1}^{\infty}\varphi_j(x)\left[A_j\cos\omega_{nj}t+B_j\sin\omega_{nj}t+\frac{1}{M_j\omega_{nj}}\int_0^t Q_j(t)\sin\omega_{nj}(t-\tau)d\tau\right], \end{aligned} \quad (6.155)$$

where A_j and B_j are determined by initial conditions.

Example 6.4. Suppose that a force $q(x, t)$ exerts on a simple supported beam with EI=constant and ρA=constant as shown in Figure 6.9. The force is in the form

$$q(x,t)=q_0\delta\left(x-\frac{l}{2}\right)u(t). \tag{6.156}$$

When the beam is with zero initial conditions, find the forced response.

Solution: The jth order normal vibration mode is given by

$$\varphi_j(x)=\sin\left(\frac{j\pi x}{l}\right). \tag{6.157}$$

The natural frequency is

$$\omega_{nj}=\left(\frac{j\pi}{l}\right)^2\sqrt{\frac{EI}{\rho A}}. \tag{6.158}$$

The jth order generalized mass is

$$M_j=\int_0^l \rho A\varphi_j^2(x)dx=\rho A\int_0^l \sin^2(j\pi x/l)dx=\frac{\rho Al}{2}. \tag{6.159}$$

The jth order generalized force is expressed by

$$Q_j(t)=\int_0^l \varphi_j(x)q(x,t)dx=\int_0^l \sin(j\pi x/l)q_0\delta(x-l/2)u(t)dx=q_0\sin(j\pi/2)u(t). \tag{6.160}$$

The jth order coordinate is given by

$$p_j(t)=\frac{1}{M_j\omega_{nj}^2}\int_0^l Q_j(\tau)\sin\omega_{nj}(t-\tau)d\tau=\frac{2q_0\sin\frac{j\pi}{2}}{\rho Al\omega_{nj}^2}(1-\cos\omega_{nj}t),\quad t>0. \tag{6.161}$$

Thus, the response is expressed by

$$w(x,t)=\sum_{j=1}^{\infty}\varphi_j(x)p_j(t)=\sum_{j=1}^{\infty}\frac{2q_0\sin(j\pi/2)}{\rho Al\omega_{nj}^2}(1-\cos\omega_{nj}t)\sin(j\pi x/l),\quad t>0. \tag{6.162}$$

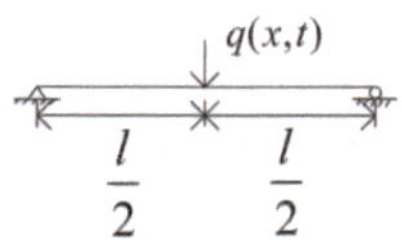

FIGURE 6.9 Simply supported beam with external lateral loading $q(x, t)$.

6.2.5 Forced Response to a Damped Euler-Bernoulli Beam

Conventionally, damping is classified into two categories. One is external damping that is conventionally viscous damping. The other is internal damping.

Denote by $c(x)$ an external damping at position x. Then, the external damping force is given by

$$-c(x)\frac{\partial w(x,t)}{\partial t}. \tag{6.163}$$

According to the Voigt assumption on materials about internal damping, the stress σ is given by

$$\sigma = E\varepsilon(x,t)+c_s\frac{\partial \varepsilon(x,t)}{\partial t}, \tag{6.164}$$

where $\varepsilon(x,t)$ is strain at time t and position x while c_s is an internal damping. In the above, one calls

$$\sigma_D = c_s\frac{\partial \varepsilon(x,t)}{\partial t}, \tag{6.165}$$

damping stress. Suppose $\varepsilon(x,t)$ is linearly distributed along cross-section. Then, strain at a cross-section at the location z away from the neutral axis is given by

$$\varepsilon = z\frac{\partial^2 w}{\partial x^2}. \tag{6.166}$$

Let $b(z)$ be the width of a beam at cross-section x and the height of the cross-section is z. Then, the bending moment is expressed by

$$\begin{aligned} M &= \int \sigma b(z)zdz = \int (E\varepsilon + c_s\dot{\varepsilon})b(z)zdz = \int\left(Ez\frac{\partial^2 w}{\partial x^2}+c_s z\frac{\partial^3 w}{\partial x^2 \partial t}\right)b(z)zdz \\ &= EI\frac{\partial^2 w}{\partial x^2}+c_s I\frac{\partial^3 w}{\partial x^2 \partial t}, \end{aligned} \tag{6.167}$$

where I is the rotary inertia given by

$$I = \int b(z)z^2\,dz. \tag{6.168}$$

Therefore, the bending moment consists of two items, namely, elastic force moment $EI\frac{\partial^2 w}{\partial x^2}$ and non-elastic damping force moment $c_s I\frac{\partial^3 w}{\partial x^2 \partial t}$. When considering both external damping and internal one, the equation becomes

$$\frac{\partial^2}{\partial x^2}\left(EI\frac{\partial^2 w}{\partial x^2}+c_s I\frac{\partial^3 w}{\partial x^2 \partial t}\right)+\rho A\frac{\partial^2 w}{\partial t^2}+c\frac{\partial w}{\partial t}=q(x,t). \tag{6.169}$$

The above is linear. Using $w(x,t)=\sum_{j=1}^{\infty}\varphi_j(x)p_j(t)$, one has

$$\sum_{j=1}^{\infty}\rho A\varphi_j(x)\frac{d^2 p_j(t)}{dt^2}+\sum_{j=1}^{\infty}c\varphi_j(x)\frac{dp_j(t)}{dt}+\sum_{j=1}^{\infty}\frac{d^2}{dx^2}\left[c_s I\frac{d^2\varphi_j(x)}{dx^2}\right]\frac{dp_j(t)}{dt}$$
$$+\sum_{j=1}^{\infty}\frac{d^2}{dx^2}\left[EI\frac{d^2\varphi_j(x)}{dx^2}\right]p_j(t)=q(x,t). \tag{6.170}$$

Multiplying $\varphi_m(x)$ on both sides of the above produces

$$\sum_{j=1}^{\infty}\rho A\varphi_j(x)\varphi_m(x)\frac{d^2 p_j(t)}{dt^2}+\sum_{j=1}^{\infty}c\varphi_j(x)\varphi_m(x)\frac{dp_j(t)}{dt}+\sum_{j=1}^{\infty}\varphi_m(x)\frac{d^2}{dx^2}\left[c_s I\frac{d^2\varphi_j(x)}{dx^2}\right]\frac{dp_j(t)}{dt}$$
$$+\sum_{j=1}^{\infty}\varphi_m(x)\frac{d^2}{dx^2}\left[EI\frac{d^2\varphi_j(x)}{dx^2}\right]p_j(t)=\varphi_m(x)q(x,t). \tag{6.171}$$

Performing the integral from 0 to l in terms of x on both sides of the above equation yields

$$\sum_{j=1}^{\infty}\rho A\int_0^l\varphi_j(x)\varphi_m(x)dx\frac{d^2 p_j(t)}{dt^2}+\sum_{j=1}^{\infty}c\int_0^l\varphi_j(x)\varphi_m(x)dx\frac{dp_j(t)}{dt}$$
$$+\sum_{j=1}^{\infty}\int_0^l\varphi_m(x)\frac{d^2}{dx^2}\left[c_s I\frac{d^2\varphi_j(x)}{dx^2}\right]dx\frac{dp_j(t)}{dt}$$
$$+\sum_{j=1}^{\infty}\int_0^l\varphi_m(x)\frac{d^2}{dx^2}\left[EI\frac{d^2\varphi_j(x)}{dx^2}\right]dxp_j(t) \tag{6.172}$$
$$=\int_0^l\varphi_m(x)q(x,t)dx.$$

Due to the orthogonality, one has

$$M_j\frac{d^2p_j(t)}{dt^2}+\frac{M_jc}{\rho A}\frac{dp_j(t)}{dt}+\sum_{j=1}^{\infty}\int_0^l\varphi_m(x)\frac{d^2}{dx^2}\left[c_sI\frac{d^2\varphi_j(x)}{dx^2}\right]dx\frac{dp_j(t)}{dt}$$
$$+\sum_{j=1}^{\infty}\int_0^l\varphi_m(x)\frac{d^2}{dx^2}\left[EI\frac{d^2\varphi_j(x)}{dx^2}\right]dxp_j(t)=Q_j(t). \tag{6.173}$$

Note that

$$\frac{d^2}{dx^2}\left[EI\frac{d^2\varphi_j(x)}{dx^2}\right]=\rho A\omega_{nj}^2\varphi_j(x). \tag{6.174}$$

Thus, Eq. (6.173) becomes

$$M_j\frac{d^2p_j(t)}{dt^2}+\frac{M_jc}{\rho A}\frac{dp_j(t)}{dt}+\sum_{j=1}^{\infty}\int_0^l\varphi_m(x)\frac{d^2}{dx^2}\left[c_sI\frac{d^2\varphi_j(x)}{dx^2}\right]dx\frac{dp_j(t)}{dt}$$
$$+M_j\omega_{nj}^2p_j(t)=Q_j(t). \tag{6.175}$$

There is a difficulty in the above expression to deal with the third item on the left side in the aspect of getting rid of Σ due to c_s. To overcome the difficulty [1], people purposely let

$$c=\rho A\alpha, \tag{6.176}$$

where α is a coefficient with the unit of time since ρA is with the unit of mass. Moreover, one assumes that

$$c_s=E\beta, \tag{6.177}$$

where β is a coefficient with the unit of frequency as E is with the unit [N/m]. The above assumptions regarding c and c_s constitute the famous Rayleigh damping assumption.

After substituting Eqs. (6.176) and (6.177) into Eq. (6.175) and taking into account the orthogonality, Σ on the third item on the left side of Eq. (6.175) is smoothed away. In doing so, one has

$$M_j\frac{d^2p_j(t)}{dt^2}+M_j\alpha\frac{dp_j(t)}{dt}+M_j\omega_{nj}^2\beta\frac{dp_j(t)}{dt}+M_j\omega_{nj}^2p_j(t)=Q_j(t). \tag{6.178}$$

Therefore, for the jth order forced vibration of Euler-Bernoulli beam, we have its corresponding motion equation in the form of single degree-of-freedom

$$\frac{d^2p_j(t)}{dt^2}+\left(\alpha+\omega_{nj}^2\beta\right)\frac{dp_j(t)}{dt}+\omega_{nj}^2p_j(t)=\frac{Q_j(t)}{M_j}. \tag{6.179}$$

Denote by

$$f_j(t)=\frac{Q_j(t)}{M_j}. \tag{6.180}$$

Write

$$\varsigma_j=\frac{\alpha}{2\omega_{nj}}+\frac{\beta\omega_{nj}}{2}. \tag{6.181}$$

Then, Eq. (6.179) becomes

$$\frac{d^2p_j(t)}{dt^2}+2\varsigma_j\omega_{nj}\frac{dp_j(t)}{dt}+\omega_{nj}^2p_j(t)=f_j(t). \tag{6.182}$$

The general solution to the above is given by

$$p_j(t)=e^{-\varsigma_j\omega_{nj}t}\left(a_j\sin\omega_{nj}t+b_j\cos\omega_{nj}t\right)+\frac{1}{\omega_{dj}}\int_0^t f_j(\tau)e^{-\varsigma_j\omega_{nj}(t-\tau)}\sin\omega_{nj}(t-\tau)d\tau, \tag{6.183}$$

where ω_{dj} is damped natural frequency for the jth vibration mode of Euler-Bernoulli beam in the form

$$\omega_{dj}=\omega_{nj}\sqrt{1-\varsigma_j^2}. \tag{6.184}$$

The coefficients a_j and b_j are determined by initial conditions.

Denote the ratio of vibration frequency to the frequency of jth order vibration mode by $\gamma_j=\frac{\omega}{\omega_{nj}}$. Then, when $q(x, t)=Q(x)\sin\omega t$, total response is given by

$$w(x,t)=\sum_{j=1}^{\infty}\varphi_j(x)p_j(t)=\sum_{j=1}^{\infty}\frac{Q_j}{M_j\omega_{nj}^2}\frac{\varphi_j(x)}{\sqrt{\left(1-\gamma_j^2\right)^2+4\zeta_j^2\gamma_j^2}}\sin\left(\omega t-\theta_j\right), \tag{6.185}$$

where $\theta_j=\frac{2\varsigma_j\gamma_j}{1-\gamma_j^2}$.

Note that the amplitude of $\sin\left(\omega t-\theta_j\right)$ in Eq. (6.185) is inversely proportional to $\omega_{nj.}^2$ Thus,

$$\lim_{\omega_{nj}\to\infty}p_j(t)=0. \tag{6.186}$$

Therefore, high vibration modes may be neglected from a view of vibration engineering. In other words, the frequency components with small j contribute to vibration while those with large j contribute other mechanical performance, such as high-frequency fatigue instead of vibration.

Denote by $h_j(t)$ the impulse response to Eq. (6.182). Then,

$$h_j(t) = \frac{1}{\omega_{dj} M_j} e^{-\varsigma_j \omega_{nj} t} \sin \omega_{nj} t. \tag{6.187}$$

Let $H_j(\omega)$ be the frequency transfer function of Eq. (6.164). Then,

$$H_j(\omega) = \frac{1}{K_j} \frac{1}{\left(1-\gamma_j^2\right)+i2\varsigma_j\gamma_j}, \tag{6.188}$$

where K_j is generalized stiffness of jth order vibration mode of Euler-Bernoulli beam. The above discussions show that a distributed system of Euler-Bernoulli beam may be identified as a system that consists of j subsystems with single degree-of-freedom in parallel as indicated in Figure 6.10.

6.2.6 Timoshenko Beam

Vibrations in an Euler-Bernoulli beam are purely transversal. In some cases, things on a structure are the combination of transversal or lateral vibration with rotary one. For instance, a ship moves obliquely across wave peaks, see Figure 6.11a, combined motion of lateral vibration with rotary one may be true (Figure 6.11b). For such a motion, the model

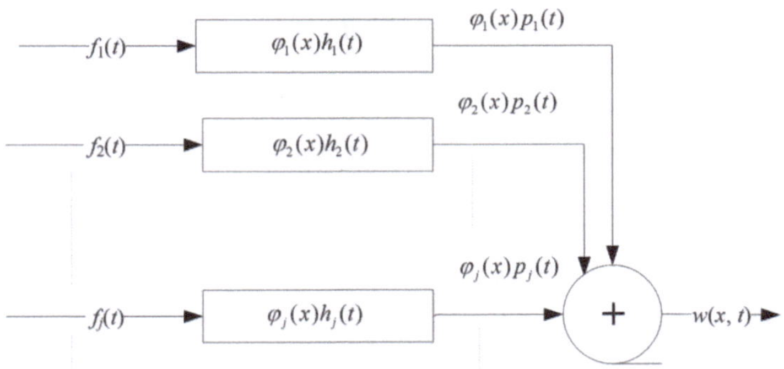

FIGURE 6.10 Diagram of forced response $w(x, t)$.

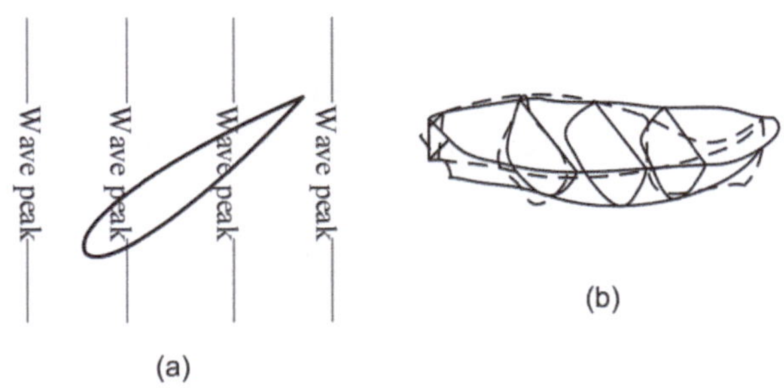

FIGURE 6.11 Ship hull beam in lateral vibration and rotary one. (a) Moving obliquely across wave peaks. (b) Combination of transversal or lateral vibration with rotary one.

of Euler-Bernoulli beam may not be enough if the length of a beam is not large enough. In that case, the model Timoshenko beam may be considered.

Symbols

E: Elastic modulus
I: Rotary inertia
ρ: Material density
$w(x, t)$: Deflection
q: Beam angle
M: Bending moment
N: Shear force
G: Shear elastic modulus
A: Section area
A_s: Section shear area, usually, $A=A_s$
k': Constant relying on section shape
r: Rotary radius
$F(x, t)$: Disturbing force.

The analysis of a Timoshenko beam refers to Timoshenko [2], Searle [3], and Palley et al. [1]. Here, we directly give the motion equation of a Timoshenko beam. It is given by

$$\underbrace{EI\frac{\partial^4 w}{\partial x^4}-\left(F-\rho A\frac{\partial^2 w}{\partial t^2}\right)}_{\text{Basic expression}}-\underbrace{\rho Ar^2\frac{\partial^4 w}{\partial x^2\partial t^2}}_{\text{Rotary inertia item}}+\underbrace{\frac{EI}{k'AG}\frac{\partial^2}{\partial x^2}\left(F-\rho A\frac{\partial^2 w}{\partial t^2}\right)}_{\text{Shear deformation item}}$$

$$-\underbrace{\frac{\rho Ar^2}{k'A_sG}\frac{\partial^2}{\partial t^2}\left(F-\rho A\frac{\partial^2 w}{\partial t^2}\right)}_{\text{Item of shear coupled with rotary inertia}}=0. \tag{6.189}$$

The first term on the left side in the above formula is basic for a Timoshenko beam. That is actually the motion equation of an Euler-Bernoulli beam. Thus, we note that the model of an Euler-Bernoulli beam is a principal part of the model of a Timoshenko beam. The second term is caused by rotary inertia, the third is produced by shear deformation, and the fourth results from the combination of shear force with rotary inertia.

If $F=0$, one has the motion equation for free response in the form

$$\underbrace{EI\frac{\partial^4 w}{\partial x^4}+\rho A\frac{\partial^2 w}{\partial t^2}}_{\text{Basic expression}}-\underbrace{\rho Ar^2\frac{\partial^4 w}{\partial x^2\partial t^2}}_{\text{Rotary inertia item}}-\underbrace{\frac{EI}{k'AG}\frac{\partial^2}{\partial x^2}\left(\rho A\frac{\partial^2 w}{\partial t^2}\right)}_{\text{Shear deformation item}}+\underbrace{\frac{\rho Ar^2}{k'A_sG}\frac{\partial^2}{\partial t^2}\left(\rho A\frac{\partial^2 w}{\partial t^2}\right)}_{\text{Item of shear coupled with rotary inertia}}=0. \tag{6.190}$$

When ρA = constant, $Ar^2 = I$, and $A = A_s$, the above becomes

$$\left(EI\frac{\partial^4 w}{\partial x^4}+\rho A\frac{\partial^2 w}{\partial t^2}\right)-\rho I\frac{\partial^4 w}{\partial x^2\partial t^2}-\frac{\rho EI}{k'G}\frac{\partial^2}{\partial x^2}\frac{\partial^2 w}{\partial t^2}+\frac{\rho^2 I}{k'G}\frac{\partial^4 w}{\partial t^4}$$
$$=\left(EI\frac{\partial^4 w}{\partial x^4}+\rho A\frac{\partial^2 w}{\partial t^2}\right)-\rho I\frac{\partial^4 w}{\partial x^2\partial t^2}\left(1+\frac{E}{k'G}\right)+\frac{\rho^2 I}{k'G}\frac{\partial^4 w}{\partial t^4}=0. \tag{6.191}$$

Suppose that a Timoshenko beam is simply supported. Then, $\varphi_j(x)\sim\sin\frac{j\pi x}{l}$. Suppose

$$w_j(x,t)=\sin\frac{j\pi x}{l}\cos(\omega_{nj}t). \tag{6.192}$$

Substituting the above into Eq. (6.191) produces

$$EI\left(\frac{j\pi}{l}\right)^4-\rho A\omega_{nj}^2-\rho I\left(1+\frac{E}{k'G}\right)\left(\frac{j\pi}{l}\right)^2\omega_{nj}^2+\frac{\rho^2 I\omega_{nj}^4}{k'G}=0. \tag{6.193}$$

Since $\frac{\rho^2 I\omega_{nj}^4}{k'G}$ is much smaller than other items, the above may approximately be written by

$$EI\left(\frac{j\pi}{l}\right)^4-\rho A\omega_{nj}^2-\rho I\left(1+\frac{E}{k'G}\right)\left(\frac{j\pi}{l}\right)^2\omega_{nj}^2\approx 0. \tag{6.194}$$

The above yields

$$\omega_{nj}^2\approx\frac{EI}{\rho A}\left(\frac{j\pi}{l}\right)^4\left[1-\frac{I}{A}\left(1+\frac{E}{k'G}\right)\left(\frac{j\pi}{l}\right)^2\right]. \tag{6.195}$$

For the convenience of comparison, we denote by $\omega_{nj,\text{Tim}}$ and $\omega_{nj,\text{Bernoulli}}$ the jth natural frequency of a Timoshenko beam and Euler-Bernoulli one, respectively. Since

$$\omega_{nj,\text{Bernoulli}}^2\approx\frac{EI}{\rho A}\left(\frac{j\pi}{l}\right)^4, \tag{6.196}$$

we have

$$\omega_{nj,\text{Tim}}^2\approx\omega_{nj,\text{Bernoulli}}^2\left[1-\frac{I}{A}\left(1+\frac{E}{k'G}\right)\left(\frac{j\pi}{l}\right)^2\right]. \tag{6.197}$$

Because

$$\left[1-\frac{I}{A}\left(1+\frac{E}{k'G}\right)\left(\frac{j\pi}{l}\right)^2\right]<1, \tag{6.198}$$

we see that

$$\omega_{nj,\text{Tim}}^2 \le \omega_{nj,\text{Bernoulli}}^2. \tag{6.199}$$

The difference between the two is given by

$$\Delta = \omega_{nj,\text{Bernoulli}}^2 - \omega_{nj,\text{Tim}}^2 \approx \frac{EI^2}{\rho A^2}\left(\frac{j\pi}{l}\right)^6\left(1+\frac{E}{k'G}\right). \tag{6.200}$$

The above exhibits that Δ is proportional to j^6 but inversely proportional to l^6. Thus, we see that shear and or rotary effects are mainly on high-order vibration modes. However, the coordinates of high-order vibration modes are small so that they may be neglected in vibration engineering. Therefore, for a structure with large l, such as conventional warship hull, we have

$$\omega_{nj,\text{Tim}}^2 \approx \omega_{nj,\text{Bernoulli}}^2. \tag{6.201}$$

In general, the theory of Euler-Bernoulli beam is suitable for long-shallow beams while that of Timoshenko beam for short-deep ones. Thus, on one hand, Euler-Bernoulli beam is a main vibration model of warships if they are in the sense of long-shallow beams because they are usually with large l and vibration modes with large j may be ignored. On the other hand, when a ship is abstracted as a short-deep beam, such as large opening merchant ships, rotary effects should be taken into account since l is not large enough.

6.3 PROBLEMS WORTH STUDYING

6.3.1 About Rayleigh Damping

Experimental vibration study of a real ship hull exhibits that mass, damping, and stiffness are varying in terms of vibration frequency (Jin and Xia [4, Chapter 4]). In previous discussions, however, only variable damping with frequency is utilized, see Eq. (6.181), based on the Rayleigh assumption of damping. How to explain the Rayleigh damping assumption in mathematics and or physics remains an open problem. In addition, what is a unified approach to give explanations of frequency varying mass and stiffness is another open problem. The author tries to give explanations to those problems from a view of fractional vibrations in Chapters 8–14, 16–27.

6.3.2 About Frequency Dependent Mass, Damping, and Stiffness

A number of structures with frequency-dependent mass or damping or stiffness are described in Harris [5]. In fact, frequency-dependent mass or damping or stiffness attracts the interest of researchers. Thus, the Euler-Bernoulli beam discussed in this chapter is not enough. From the point of view of fractional calculus, we shall present a number of results of frequency-dependent mass/damping/stiffness for Euler-Bernoulli beams in Chapters 18–27.

6.4 EXERCISES

6.1. Let C_{Ray} be the Rayleigh damping. Denote by m and k the mass and stiffness of a vibrator. Rayleigh's assumption says that C_{Ray} is in the form

$$C_{\text{Ray}} = Am + Bk,$$

where A and B are the mass-proportional and the stiffness-proportional constants, respectively. Provide its mathematical explanation. (Hint: Refer to Chapter 14)

6.2. Denote by C_{Ray1} a variation of C_{Ray}. It is given by $C_{\text{Ray1}} = Am$. Study its mathematical explanation. (Hint: Refer to Chapter 14)

6.3. Denote by C_{Ray2} another variation of C_{Ray}. It is expressed by $C_{\text{Ray2}} = Bk$. Investigate its mathematical explanation. (Hint: Refer to Chapter 14)

6.4. Consider the longitudinal vibration of a damped beam of l length. The motion equation is in the form

$$\frac{\partial^2 u}{\partial t^2} + 2\zeta \frac{\partial u}{\partial t} = a^2 \frac{\partial^2 u}{\partial x^2}.$$

Its boundary conditions are $u(0, t) = u(l, t) = 0$. Find its free response $u(x, t)$.

6.5. Consider the longitudinal vibration of a damped free-free beam of l length. The motion equation is given by

$$\frac{\partial^2 u}{\partial t^2} + 2\zeta \frac{\partial u}{\partial t} = a^2 \frac{\partial^2 u}{\partial x^2}.$$

Find its free response $u(x, t)$.

REFERENCES

1. O. M. Palley, Г. B. Bahizov, and E. Я. Voroneysk, *Handbook of Ship Structural Mechanics*, National Defense Industry Publishing House, Beijing, China, 2002. In Chinese. Translated from Russian by B. H. Xu, X. Xu, and M. Q. Xu.
2. S. P. Timoshenko, On the transverse vibrations of bars of uniform cross-section, *Philosophical Magazine (ser. 6)*, 43(253), 1922, 125–131.
3. J. H. C. Searle, The effect of rotatory inertia on the vibration of bars, *Philosophical Magazine (ser. 6)*, 14(79), 1907, 35–60.
4. X. D. Jin and L. J. Xia, *Ship Hull Vibration*, The Press of Shanghai Jiaotong University, Shanghai, China, 2011. In Chinese.
5. C. M. Harris, *Shock and Vibration Handbook*, 5th Ed., McGraw-Hill, New York, 2002.

II

Fractional Vibrations

CHAPTER 7

Six Classes of Fractional Vibrations

THE PRELIMINARIES FOR FRACTIONAL calculus are in Section 7.1. The classification of fractional vibrators is discussed in Section 7.2 for the Chapters 8–27. The present novelty is given in Section 7.3 about the representations of a fractional derivative by using derivatives of integer orders with variable coefficients from a view of convolution. In Section 7.4, we give two representations of fractional derivative of delta function from a view of integral.

7.1 FRACTIONAL DERIVATIVES AND INTEGRALS

Let $y(t)$ be a function differentiable any times for $t \in \mathbf{R}$ (the set of real numbers). Then,

$$\frac{dy(t)}{dt} = \lim_{\varepsilon \to 0} \frac{y(t) - y(t-\varepsilon)}{\varepsilon}. \tag{7.1}$$

For a positive integer n, one has

$$y^{(n)}(t) = \frac{d^n y(t)}{dt^n} = \lim_{\varepsilon \to 0} \frac{y^{(n-1)}(t) - y^{(n-1)}(t-\varepsilon)}{\varepsilon}. \tag{7.2}$$

Expanding the right side of Eq. (7.2) produces

$$\frac{d^n y(t)}{dt^n} = \lim_{\varepsilon \to 0} \frac{y(t) - \begin{pmatrix} n \\ 1 \end{pmatrix} y(t-\varepsilon) + \begin{pmatrix} n \\ 2 \end{pmatrix} y(t-2\varepsilon) - \cdots + (-1)^n \begin{pmatrix} n \\ n \end{pmatrix} y(t-n\varepsilon)}{\varepsilon^n}, \tag{7.3}$$

where

$$\begin{pmatrix} n \\ m \end{pmatrix} = \frac{n(n-1)\ldots(n-m+1)}{m!}. \tag{7.4}$$

DOI: 10.1201/9781003460947-9

When using the notion D^n for the derivative of $y(t)$ of order n, $y^{(n)}(t)$ is expressed by Mikusinski [1]

$$y^{(n)}(t) = D^n y(t). \tag{7.5}$$

Similarly, the n times integral of $y(t)$ may be expressed by

$$y^{(-n)}(t) = D^{-n} y(t). \tag{7.6}$$

When considering the fractional derivative of $y(t)$ of order $v>0$, one has (Letnikov [2, p. 109], [3])

$$y^{(\upsilon)}(t) = \frac{d^{\upsilon}y(t)}{dt^{\upsilon}} = \lim_{\varepsilon\to 0}\frac{y(t) - \begin{pmatrix} \upsilon \\ 1 \end{pmatrix} y(t-\varepsilon) + \begin{pmatrix} \upsilon \\ 2 \end{pmatrix} y(t-2\varepsilon) - \cdots + (-1)^n \begin{pmatrix} \upsilon \\ n \end{pmatrix} y(t-n\varepsilon)}{\varepsilon^{\upsilon}}. \tag{7.7}$$

A well-known example about Eq. (7.7) is $\frac{d^{\frac{1}{2}}t}{dt^{\frac{1}{2}}} = \frac{\sqrt{4t}}{\pi}$ (Letnikov [2], Leibniz [4], Cajori [5, p. 419]).

According to Letnikov [3, p. 8], one has

$$\begin{aligned} \left[\lim_{h\to 0}\frac{d^{-\upsilon}y(t)}{dt^{-\upsilon}}\right]_{t_0}^{t} &= \lim_{h\to 0}\sum_{r=0}^{n} h^{\upsilon}\begin{pmatrix} v \\ r \end{pmatrix} y(t-rh) \\ &= \frac{1}{1\cdot 2\cdots(\upsilon-1)}\int_{t_0}^{t}(t-u)^{\upsilon-1}f(u)du = \frac{1}{\Gamma(\upsilon)}\int_{t_0}^{t}(t-u)^{\upsilon-1}f(u)du. \end{aligned} \tag{7.8}$$

The above is well known today as the fractional integral of Riemann–Liouville type. Nowadays, it is rare to use the original limit like Eqs. (7.7) and (7.8) to describe a fractional derivative and integral. Instead, one simply utilizes the definitions of fractional derivatives and integrals, such as the Riemann–Liouville's (Liouville [6], Ross [7]), the Weyl's (Weyl [8], Raina and Koul [9]). There are other types of definitions of fractional derivatives and integrals, see e.g., [10–21], but we mention Riemann-Liouville's definition and use Weyl's in this book. The main contents of the theory of fractional calculus, such as the mean-value theorem, the Leibniz rule, Taylor's series, refer to, e.g., Klafter et al. [10], Fabian [22], Osler [23–26], Kleinz and Osler [27], Davis [28], Ross [29].

Denote by ${}_0D_x^{-\alpha}$ the fractional Riemann–Liouville (RL) integral operator of order $\alpha>0$. Then (Miller and Ross [30, p. 10]), for a function $x(t)$, $t \in \mathbf{R}$, one has

$${}_0D_x^{-\alpha}x(t) = \frac{1}{\Gamma(\alpha)}\int_0^t x(u)(t-u)^{\alpha-1}du. \tag{7.9}$$

Thus, the symbol ${}_0D_x^{\alpha}$ stands for the RL fractional derivative operator. Hence (Lavoie et al. [31, Eq. (5.1)]),

$$ {}_0D_x^{\alpha}x(t)=\frac{1}{\Gamma(-\alpha)}\int_0^t \frac{x(u)}{(t-u)^{\alpha+1}}du. \tag{7.10} $$

Following Weyl [8], the Weyl fractional integral operator is denoted by ${}_tW_{\infty}^{-\alpha}$ in the form (Miller and Ross [30, p. 13])

$$ {}_tW_{\infty}^{-\alpha}x(t)=\frac{1}{\Gamma(\alpha)}\int_t^{\infty} x(u)(u-t)^{\alpha-1}du. \tag{7.11} $$

When using the Weyl fractional derivative operator ${}_tW_{-\infty}^{\alpha}$, one has (Lavoie et al. [31, p. 245], Rain and Koul [9]),

$$ {}_tW_{-\infty}^{\alpha}x(t)=\frac{1}{\Gamma(-\alpha)}\int_{-\infty}^t \frac{x(u)}{(t-u)^{\alpha+1}}du. \tag{7.12} $$

In general, different derivative operators may yield different results for a specific function. An example in this regard is the fractional derivatives of the elementary function e^{at} $(a<0)$. To be precise (Lavoie et al. [31, p. 246]), on the one hand,

$$ {}_0D_x^{\alpha}e^{at}=\frac{t^{-a}}{\Gamma(1-\alpha)}\,{}_1F_1(1;1-\alpha;\alpha t), \tag{7.13} $$

where ${}_1F_1(\cdot)$ is the hypergeometric function. On the other hand,

$$ {}_tW_{-\infty}^{\alpha}e^{at}=a^{\alpha}e^{at}. \tag{7.14} $$

In this monograph, we use the fractional derivative of Weyl's unless otherwise stated. The notation $\frac{d^{\alpha}}{dt^{\alpha}}$ or $x^{(\alpha)}(t)$ is in the sense of Weyl.

Example 7.1. The fractional derivative of cos ωt is given by Miller and Ross [30, p. 248]

$$ \frac{d^{v}\cos\omega t}{dt^{v}}=\omega^{v}\cos\left(\omega t-\frac{\pi v}{2}\right)=\omega^{v}\left(\cos\omega t\cos\frac{\pi v}{2}+\sin\omega t\sin\frac{\pi v}{2}\right). \tag{7.15} $$

Example 7.2. The fractional derivative of sin ωt is in the form (Miller and Ross [30, p. 248])

$$ \frac{d^{v}\sin\omega t}{dt^{v}}=\omega^{v}\sin\left(\omega t-\frac{\pi v}{2}\right)=\omega^{v}\left(\sin\omega t\cos\frac{\pi v}{2}-\cos\omega t\sin\frac{\pi v}{2}\right). \tag{7.16} $$

Note that Eqs. (7.15) and (7.16) are in form consistent with those described by Fourier in Ref. [32, p. 437].

Example 7.3. The fractional derivative of $t^{-\lambda}$ is given by Miller and Ross [30, p. 249]

$$\frac{d^{\nu}t^{-\lambda}}{dt^{\nu}}=\frac{\Gamma(\nu+\lambda)}{\Gamma(\lambda)}t^{-\nu-\lambda}. \tag{7.17}$$

Example 7.4. The fractional derivative of t^{λ} is in the form (Miller and Ross [30, p. 249])

$$\frac{d^{\nu}t^{\lambda}}{dt^{\nu}}=\frac{\Gamma(\nu-\lambda)}{\Gamma(-\lambda)}t^{-\nu+\lambda}. \tag{7.18}$$

Example 7.5. Let $X(\omega)$ be the Fourier transform of $x(t)$. Then (Uchaikin [21, p. 230], Lavoie et al. [31, Eq. (7.1)], Uchaikin [33, 34]), the Fourier transform of $x^{(\nu)}(t)$ is given by

$$x^{(\nu)}(t)\leftrightarrow\int_{-\infty}^{\infty}x^{(\nu)}(t)e^{-i\omega t}\,dt=(i\omega)^{\nu}X(\omega), \tag{7.19}$$

where $i=\sqrt{-1}$.

7.2 FRACTIONAL VIBRATORS AND CLASSIFICATION

7.2.1 Fractional Vibrators in Class I

Following Li [35], a fractional vibrator that takes the following motion equation is termed a fractional vibrator in class I.

$$m\frac{d^{\alpha}y(t)}{dt^{\alpha}}+ky(t)=f(t). \tag{7.20}$$

This class of vibrators contains fractional inertia force $m\frac{d^{\alpha}y(t)}{dt^{\alpha}}$, see e.g., Duan [36, Eq. (3)], Zurigat [37, Eq. (16)], Blaszczyk and Ciesielski [38, Eq. (1)], Blaszczyk et al. [39, Eq. (10)], Al-rabtah et al. [40, Eq. (3.1)], Drozdov [41, Eq. (9)], Stanislavsky [42], Achar et al. ([43, Eq. (1)], [44, Eq. (9)], [45, Eq. (2)]), Tofighi [46, Eq. (2)], Ryabov and Puzenko [47, Eq. (1)], Ahmad and Elwakil [48, Eq. (1)], Uchaikin [34, Chap. 7], Duan et al. [49, Eq. (4.2)], Blaszczyk [50], Tavazoei [51], Sandev and Tomovski [52, Eq. (36)], Singh et al. [53].

Note 7.1: Conventionally, the range of α for vibrators in class I is $1<\alpha\leq 2$, see [35–53]. In this monograph, we make a development so that $1<\alpha<3$.

7.2.2 Fractional Vibrators in Class II

When the motion equation of a fractional vibrator is in the form

$$m\frac{d^{2}y(t)}{dt^{2}}+c\frac{d^{\beta}y(t)}{dt^{\beta}}+ky(t)=f(t), \tag{7.21}$$

it is called a fractional vibrator/oscillator in class II [35].

The class II consists of vibrators only with fractional friction term $c\frac{d^{\beta}y(t)}{dt^{\beta}}$, see e.g., Lin et al. [54, Eq.(2)], Duan [55, Eq. (31)], Alkhaldi et al. [56, Eq. (1a)], Dai et al. [57, Eq. (1)], Ren et al. [58, Eq. (1)], Xu et al. [59, Eq. (1)], He et al. [60, Eq. (4)], Leung et al. [61, Eq. (2)], Chen et al. [62, Eq. (1)], Deü and Matignon [63, Eq. (1)], Drăgănescu et al. [64, Eq. (4)], Rossikhin and Shitikova [65, Eq. (3)], Xie and Lin [66, Eq. (1)], Ren et al. [67, Eq. (1)], Yuan et al. [68, Eq. (8)], Dai et al. [69, Eq. (1)], Lin et al. [70, Eq. (1)], Naranjani et al. [71, Eq. (1)], He et al. [72], Matteo et al. [73], Tomovski and Sandev [74, Eq. (44)].

Note 7.2: Traditionally, the fractional order β is for $0<\beta\leq 1$, see [35] and [54–74]. However, we make a step further for $0<\beta<2$ in this monograph.

7.2.3 Fractional Vibrators in Class III

When the vibration motion equation of a fractional vibrator is expressed by

$$m\frac{d^{\alpha}y(t)}{dt^{\alpha}}+c\frac{d^{\beta}y(t)}{dt^{\beta}}+ky(t)=f(t), \tag{7.22}$$

the vibrator is called a fractional one in class III [35].

Class III is with both fractional inertia force $m\frac{d^{\alpha}y(t)}{dt^{\alpha}}$ and fractional friction $c\frac{d^{\beta}y(t)}{dt^{\beta}}$, see e.g., Gomez-Aguilar [75, Eq. (10)], Tian et al. [76], Berman and Cederbaum [77], Coronel-Escamilla et al. [78, Eq. (12)].

Note 7.3: Conventionally, in Refs. [35,54–74], fractional vibrators of Eq. (7.22) are for $1<\alpha\leq 2$ and $0<\beta\leq 1$. In this book, we show our developments for $1<\alpha<3$ and $0<\beta<2$.

7.2.4 Fractional Vibrators in Class IV

We say that a fractional vibrator is in class IV if its motion equation is in the form

$$m\frac{d^{\alpha}y(t)}{dt^{\alpha}}+k\frac{d^{\lambda}y(t)}{dt^{\lambda}}=f(t), \quad 1<\alpha<3,\ 0\leq\lambda<1. \tag{7.23}$$

A fractional vibrator in class IV includes fractional acceleration and fractional displacement. Reports about it are rarely seen. We shall put the stress on the solution to Eq. (7.23) and its analysis because it may serve as a mathematical explanation of the standard expression of the Rayleigh damping.

Equation (7.23) appears damping free in form but we shall reveal that it intrinsically contains a Rayleigh-type damping.

7.2.5 Fractional Vibrators in Class V

We call a fractional vibrator of class V if its motion equation is given by

$$m\frac{d^{2}y(t)}{dt^{2}}+k\frac{d^{\lambda}y(t)}{dt^{\lambda}}=f(t), \quad 0\leq\lambda<1. \tag{7.24}$$

Mathematically, Eq. (7.24) is a special case for $\alpha=2$ in Eq. (7.23). I take it as a class mainly from the point of view of vibration mechanics because the intrinsic damping in Eq. (7.24) is a variation of the Rayleigh damping, see Chapter 14.

7.2.6 Fractional Vibrators in Class VI

Introduce one more term for a fractional vibrator of class VI. It is defined by

$$m\frac{d^{\alpha}y(t)}{dt^{\alpha}}+c\frac{d^{\beta}y(t)}{dt^{\beta}}+k\frac{d^{\lambda}y(t)}{dt^{\lambda}}=f(t), \qquad 1<\alpha<3,\ 0<\beta<2,\ 0\leq\lambda<1. \tag{7.25}$$

Note that reports about Eqs. (7.23), (7.24), and (7.25) are rarely seen, to the best of my knowledge.

7.3 FRACTIONAL DERIVATIVES: A CONVOLUTION VIEW

I would like to propose several theorems with respect to representations of fractional derivative from the point of view of convolution.

Theorem 7.1. Let $y^{(\alpha)}(t)$ be the fractional differential of $y(t)$ of order α. Then,

$$y^{(\alpha)}(t)=y(t)*\delta^{(\alpha)}(t). \tag{7.26}$$

Proof. Let $Y(\omega)$ be the Fourier transform of $y(t)$. According to Example 7.5, the Fourier transform of $y^{(\alpha)}(t)$ is

$$y^{(\alpha)}(t)\leftrightarrow(i\omega)^{\alpha}Y(\omega). \tag{7.27}$$

Besides, the Fourier transform of the right side of Eq. (7.26) is, according to the convolution rule, in the form

$$y(t)*\delta^{(\alpha)}(t)\leftrightarrow Y(\omega)(i\omega)^{\alpha}. \tag{7.28}$$

From Eqs. (7.27) and (7.28), we see that Theorem 7.1 holds. The proof is complete. □

Note 7.4: Theorem 7.1 means that the fractional derivative of $y(t)$ of order α may be expressed by $y(t)$ convoluted with the fractional derivative of a delta function of order α.

Note 7.5: Theorem 7.1 also means that $y^{(\alpha)}(t)$ may be expressed by $y(t)$ convoluted with a variable coefficient $\delta^{(\alpha)}(t)$.

Theorem 7.2. Let $y^{(\alpha)}(t)$ be the fractional differential of $y(t)$ of order α. For a positive integer n, we have

$$y^{(\alpha)}(t)=y^{(n)}(t)*\delta^{(\alpha-n)}(t). \tag{7.29}$$

Proof. Let $Y(\omega)$ be the Fourier transform of $y(t)$. Because

$$y^{(n)}(t)\leftrightarrow(i\omega)^{n}Y(\omega), \tag{7.30}$$

and

$$\delta^{(\alpha-n)}(t) \leftrightarrow (i\omega)^{\alpha-n}. \tag{7.31}$$

Based on the convolution rule, for the right side of Eq. (7.29), we have

$$y^{(n)}(t) * \delta^{(\alpha-n)}(t) \leftrightarrow (i\omega)^n (i\omega)^{\alpha-n} Y(\omega) = (i\omega)^\alpha Y(\omega). \tag{7.32}$$

Thus, Theorem 7.2 holds. The proof is finished. □

Note 7.6: Theorem 7.2 says that the fractional derivative of $y(t)$ of order α may be expressed by an integer derivative $y^{(n)}(t)$ convoluted with the fractional derivative of a delta function of order $\alpha - n$.

Note 7.7: Theorem 7.2 also says that $y^{(\alpha)}(t)$ equals to an integer derivative $y^{(n)}(t)$ convoluted with a variable coefficient $\delta^{(\alpha-n)}(t)$.

For the complex number i^α, we just take its principal value by

$$i^\alpha = \cos\frac{\alpha\pi}{2} + i\sin\frac{\alpha\pi}{2}. \tag{7.33}$$

Theorem 7.3. Let $y^{(\alpha)}(t)$ be the fractional differential of $y(t)$ of order α. Then, we have

$$y^{(\alpha)}(t) = -\omega^{\alpha-2}\cos\frac{\alpha\pi}{2} y''(t) + \omega^{\alpha-1}\sin\frac{\alpha\pi}{2} y'(t). \tag{7.34}$$

Proof. Let $Y(\omega)$ be the Fourier transform of $y(t)$. Doing the Fourier transform of the right side on the above produces

$$\begin{aligned}
&\mathrm{F}\left[-\omega^{\alpha-2}\cos\frac{\alpha\pi}{2} y''(t) + \omega^{\alpha-1}\sin\frac{\alpha\pi}{2} y'(t)\right] \\
&= \mathrm{F}\left[-\omega^{\alpha-2}\cos\frac{\alpha\pi}{2} y''(t)\right] + \mathrm{F}\left[\omega^{\alpha-1}\sin\frac{\alpha\pi}{2} y'(t)\right] \\
&= -\omega^{\alpha-2}\cos\frac{\alpha\pi}{2}\mathrm{F}\left[y''(t)\right] + \omega^{\alpha-1}\sin\frac{\alpha\pi}{2}\mathrm{F}\left[y'(t)\right] \\
&= \left[-\omega^{\alpha-2}\cos\frac{\alpha\pi}{2}(i\omega)^2 + \omega^{\alpha-1}\sin\frac{\alpha\pi}{2} i\omega\right] Y(\omega) \\
&= \left[\omega^{\alpha}\cos\frac{\alpha\pi}{2} + i\omega^{\alpha}\sin\frac{\alpha\pi}{2}\right] Y(\omega) \\
&= (i\omega)^\alpha Y(\omega).
\end{aligned} \tag{7.35}$$

According to Eq. (7.26), we see that Theorem 7.3 is sound. This finishes the proof. □

Another way to prove Eq. (7.34) may be given as follows. In fact, due to Theorem 7.1, we have

$$(i\omega)^{\alpha} Y(\omega) \leftrightarrow \delta^{(\alpha)}(t) * y(t) = y^{(\alpha)}(t). \tag{7.36}$$

Note 7.8: Theorem 7.3 exhibits that the fractional derivative of $y(t)$ of order α may be expressed by a linear combination of integer derivatives $y''(t)$ and $y'(t)$ multiplied by variable coefficients $-\omega^{\alpha-2}\cos\frac{\alpha\pi}{2}$ and $\omega^{\alpha-1}\sin\frac{\alpha\pi}{2}$.

Note that numerical computation of fractional integrals and derivatives is an interesting field in fractional calculus and its applications, see e.g., Li and Zeng [79], Li and Cai [80], Heydari [81,82], Heydari et al. [83], just mentioning a few. The research stated in this monograph, however, is in analytical in style.

7.4 FRACTIONAL DERIVATIVE OF DELTA FUNCTION: AN INTEGRAL VIEW

Because $\delta(t) \longleftrightarrow 1$, we have

$$\frac{1}{2\pi}\int_{-\infty}^{\infty} e^{i\omega t}\, d\omega = \frac{1}{2\pi}\int_{-\infty}^{\infty} \left(\cos\omega t + i\sin\omega t\right) d\omega = \delta(t). \tag{7.37}$$

As $\delta(t)$ is real, one has

$$\int_{-\infty}^{\infty} \cos\omega t d\omega = 2\pi\delta(t). \tag{7.38}$$

Thus,

$$\delta(t) = \frac{1}{\pi}\int_{0}^{\infty} \cos\omega t d\omega. \tag{7.39}$$

Theorem 7.4. The fractional derivative of $\delta(t)$ of order $v \geq 0$ is given by

$$\delta^{(v)}(t) = \frac{1}{\pi}\int_{0}^{\infty} \frac{d^{v}\cos\omega t}{dt^{v}} d\omega. \tag{7.40}$$

Proof. According to Eq. (7.39), we have

$$\delta^{(v)}(t) = \frac{d^{v}\delta(t)}{dt^{v}} = \frac{1}{\pi}\frac{d^{v}}{dt^{v}}\int_{0}^{\infty} \cos\omega t d\omega. \tag{7.41}$$

Considering Lemma 2.3 in Li [84], we have

$$\frac{d^v}{dt^v}\int_0^\infty \cos\omega t d\omega = \int_0^\infty \frac{d^v \cos\omega t}{dt^v} d\omega. \tag{7.42}$$

Thus, the theorem holds. □

Theorem 7.5. The fractional derivative of $\delta(t)$ of order $v \geq 0$ can be written by

$$\delta^{(v)}(t) = \frac{1}{\pi}\int_0^\infty \omega^v \cos\left(\omega t - \frac{\pi v}{2}\right) d\omega. \tag{7.43}$$

Proof. According to Example 7.1, we see that the above is true. □

7.5 EXERCISES

7.1. Let $X(\omega)$ be the Fourier transform of $x(t)$. Prove that $x^{(\alpha)}(t) \leftrightarrow (i\omega)^\alpha X(\omega)$.

7.2. Let $\delta(t)$ be the delta function. Prove that $\delta^{(\alpha)}(t) \leftrightarrow (i\omega)^\alpha$.

7.3. Let $X(\omega)$ and $Y(\omega)$ be the Fourier transform of $x(t)$ and $y(t)$, respectively. Prove that $x^{(\alpha)}(t) * y^{(\beta)}(t) \leftrightarrow (i\omega)^{\alpha+\beta} X(\omega) Y(\omega)$.

7.4. Let $X(\omega)$ be the Fourier transform of $x(t)$. Let $\delta(t)$ be the delta function. Prove that $x^{(\alpha)}(t) * \delta^{(\beta)}(t) \leftrightarrow (i\omega)^{\alpha+\beta} X(\omega)$.

7.5. Let $\delta(t)$ be the delta function. Prove that $\delta^{(\alpha)}(t) * \delta^{(\beta)}(t) \leftrightarrow (i\omega)^{\alpha+\beta}$.

REFERENCES

1. J. Mikusinski, *Operational Calculus*, Pergamon Press, Oxford, 1959.
2. A. V. Letnikov, Historical development of the theory of differentiation of fractional order, *Moskow Mathematicheskii Sbornik*, 3(2), 1868, 85–119. In Russian.
3. A. V. Letnikov, Theory of differentiation of fractional order, *Moskow Mathematicheskii Sbornik*, 3(1), 1868, 1–68. In Russian.
4. G. W. Leibniz, Letter from Hanover, Germany to G. F. A. L'Hospital, September 30, 1695, in Math. Schriften, 1849 reprinted in 1962, Hildesheim, Germany (Olms Verlag), 2, pp. 301–302 (https://www.olms.com/).
5. F. Cajori, Leibniz, the master-builder of mathematical notations, *Isis*, 7(3), 1925, 412–429.
6. J. Liouville, Memoire sur le theoreme des complementaires, *Journal für die Reine und Angewandte Mathematik*, 11, 1834, 1–19.
7. B. Ross, *Fractional Calculus and Its Applications*, Lecture Notes in Mathematics, vol. 457, Springer-Verlag, New York, 1975.
8. H. Weyl, Bemerkungen zum Begriff des Differentialquotienten gebrochener Ordnung, *Vierteljahrsschrift der Naturforschenden Gesellschaft in Zürich*, 62, 1917, 296–302.

9. R. K. Raina and C. L. Koul, On Weyl Fractional Calculus, *Proceedings of the American Mathematical Society*, 73(2), 1979, 188–192.
10. J. Klafter, S. C. Lim, and R. Metzler, *Fractional Dynamics: Recent Advances*, World Scientific, Singapore, 2012.
11. J. Sabatier and C. Farge, Comments on the description and initialization of fractional partial differential equations using Riemann-Liouville's and Caputo's definitions, *Journal of Computational and Applied Mathematics*, 339, 2018, 30–39.
12. Y. A. Rossikhin and M. V. Shitikova, Application of fractional operators to the analysis of damped vibrations of viscoelastic single-mass systems, *Journal of Sound and Vibration*, 199(4), 1997, 567–586.
13. M. V. Shitikova, Fractional operator viscoelastic models in dynamic problems of mechanics of solids: A review, *Mechanics of Solids*, 57(1), 2022, 1–33.
14. Y. A. Rossikhin, Reflections on two parallel ways in progress of fractional calculus in mechanics of solids, *Applied Mechanics Reviews*, 63(1), 2010, 010701.
15. M. V. Shitikova, The fractional derivative expansion method in nonlinear dynamic analysis of structures, *Nonlinear Dynamics*, 99(1), 2020, 109–122.
16. G.-H. Gao, Z.-Z. Sun, and H. Wei Zhang, A new fractional numerical differentiation formula to approximate the Caputo fractional derivative and its applications, *Journal of Computational Physics*, 259, 2014, 33–50.
17. M. Mazandarani and M. Najariyan, Type-2 fuzzy fractional derivatives, *Communications in Nonlinear Science and Numerical Simulation*, 19(7), 2014, 2354–2372.
18. Z. Zheng, W. Zhao, and H. Dai, A new definition of fractional derivative, *International Journal of Non-Linear Mechanics*, 108, 2019, 1–6.
19. T. M. Atanackovic, S. Pilipovic, B. Stankovic, and D. Zorica, *Fractional Calculus with Applications in Mechanics*, ISTE Ltd and John Wiley & Sons, London/Hoboken, NJ, 2014.
20. M. Caputo and M. Fabrizio, The kernel of the distributed order fractional derivatives with an application to complex materials, *Fractal and Fractional*, 1(1), 2017, 11 p.
21. V. V. Uchaikin, *Fractional Derivatives for Physicists and Engineers*, vol. I, Springer, Berlin, Heidelberg, 2013.
22. W. Fabian, Fractional calculus, *The Mathematical Gazette*, 20(238), 1936, 88–92.
23. T. J. Osler, The integral analog of the Leibniz rule, *Mathematics of Computation*, 26(120), 1972, 903–915.
24. T. J. Osler, Fractional derivatives and Leibniz rule, *The American Mathematical Monthly*, 78(6), 1971, 645–649.
25. T. J. Osler, Leibniz rule for fractional derivatives generalized and an application to infinite series, *SIAM Journal of Applied Mathematics*, 18(3), 1970, 658–674.
26. T. J. Osler, An integral analogue of Taylor's series and its use in computing Fourier transforms, *Mathematics of Computation*, 26(118), 1972, 449–460.
27. M. Kleinz and T. J. Osler, A child's garden of fractional derivatives, *The College Mathematics Journal*, 31(2), 2000, 82–88.
28. H. T. Davis, The application of fractional operators to functional equations, *American Journal of Mathematics*, 49(1), 1927, 123–142.
29. B. Ross, Fractional calculus, *Mathematics Magazine*, 50(3), 1977, 115–122.
30. K. S. Miller and B. Ross, *An Introduction to the Fractional Calculus and Fractional Differential Equations*, John Wiley & Sons, Hoboken, NJ, 1993.
31. J. L. Lavoie, T. J. Osler, and R. Tremblay, Fractional derivatives and special functions, *SIAM Review*, 18(2), 1976, 240–268.
32. J. Fourier, *The Analytic Theory of Heat*, Cambridge University Press, London, 1878. Translated by A. Freeman, M. A.
33. V. V. Uchaikin, Relaxation processes and fractional differential equations, *International Journal of Theoretical Physics*, 42(1), 2003, 121–134.

34. V. V. Uchaikin, *Fractional Derivatives for Physicists and Engineers*, vol. II, Springer, Berlin, 2013.
35. M. Li, Three classes of fractional oscillators, *Symmetry-Basel*, 10(2), 2018, 91 p.
36. J.-S. Duan, The periodic solution of fractional oscillation equation with periodic input, *Advances in Mathematical Physics*, 46, 2013, 46–53.
37. M. Zurigat, Solving fractional oscillators using Laplace homotopy analysis method, *Annals of the University of Craiova, Mathematics and Computer Science Series*, 38(4), 2011, 1-11.
38. T. Blaszczyk and M. Ciesielski, Fractional oscillator equation: Transformation into integral equation and numerical solution, *Applied Mathematics and Computation*, 257, 2015, 428–435.
39. T. Blaszczyk, M. Ciesielski, M. Klimek, and J. Leszczynski, Numerical solution of fractional oscillator equation, *Applied Mathematics and Computation*, 218(6), 2011, 2480–2488.
40. A. Al-Rabtah, V. S. Ertürk, and S. Momani, Solutions of a fractional oscillator by using differential transform method, *Computers & Mathematics with Applications*, 59(3), 2010, 1356–1362.
41. A. D. Drozdov, Fractional oscillator driven by a Gaussian noise, *Physica A*, 376, 2007, 237–245.
42. A. A. Stanislavsky, Fractional oscillator, *Physical Review E*, 70(5), 2004, 051103(6 pp).
43. B. N. N. Achar, J. W. Hanneken, and T. Clarke, Damping characteristics of a fractional oscillator, *Physica A*, 339(3–4), 2004, 311–319.
44. B. N. N. Achar, J. W. Hanneken, and T. Clarke, Response characteristics of a fractional oscillator, *Physica A*, 309(3–4), 2002, 275–288.
45. B. N. N. Achar, J. W. Hanneken, T. Enck, and T. Clarke, Dynamics of the fractional oscillator, *Physica A*, 297(3–4), 2001, 361–367.
46. A. Tofighi, The intrinsic damping of the fractional oscillator, *Physica A*, 329(1–2), 2003, 29–34.
47. Y. E. Ryabov and A. Puzenko, Damped oscillations in view of the fractional oscillator equation, *Physical Review B*, 66(18), 2002, 184201.
48. W. E. Ahmad and A. S. R. Elwakil, Fractional-order Wien-bridge oscillator, *Electronics Letters*, 37(18), 2001, 1110–1112.
49. J.-S. Duan, Z. Wang, Y.-L. Liu, and X. Qiu, Eigenvalue problems for fractional ordinary differential equations, *Chaos, Solitons & Fractals*, 46, 2013, 46–53.
50. T. Blaszczyk, A numerical solution of a fractional oscillator equation in a non-resisting medium with natural boundary conditions, *Romanian Reports in Physics*, 67(2), 2015, 350–358.
51. M. S. Tavazoei, Reduction of oscillations via fractional order pre-filtering, *Signal Processing*, 107, 2015, 407–414.
52. T. Sandev and Z. Tomovski, The general time fractional wave equation for a vibrating string, *Journal of Physics A: Mathematical and Theoretical*, 43(5), 2010, 055204 (12pp).
53. H. Singh, H. M. Srivastava, and D. Kumar, A reliable numerical algorithm for the fractional vibration equation, *Chaos, Solitons & Fractals*, 103, 2017, 131–138.
54. L.-F. Lin, C. Chen, S.-C. Zhong, and H.-Q. Wang, Stochastic resonance in a fractional oscillator with random mass and random frequency, *Journal of Statistical Physics*, 160(2), 2015, 497–511.
55. J.-S. Duan, A modified fractional derivative and its application to fractional vibration equation, *Applied Mathematics & Information Sciences*, 10(5), 2016, 1863–1869.
56. H. S. Alkhaldi, I. M. Abu-Alshaikh, and A. N. Al-Rabadi, Vibration control of fractionally-damped beam subjected to a moving vehicle and attached to fractionally-damped multi-absorbers, *Advances in Mathematical Physics*, 2013, 2013, 12pp.
57. H. Dai, Z. Zheng, and W. Wang, On generalized fractional vibration equation, *Chaos, Solitons & Fractals*, 95, 2017, 48–51.
58. R. Ren, M. Luo, and K. Deng, Stochastic resonance in a fractional oscillator driven by multiplicative quadratic noise, *Journal of Statistical Mechanics: Theory and Experiment*, 2017, 2017, 023210.

59. Y. Xu, Y. Li, D. Liu, W. Jia, and H. Huang, Responses of Duffing oscillator with fractional damping and random phase, *Nonlinear Dynamics*, 74(3), 2013, 745–753.
60. G. He, Y. Tian, and Y. Wang, Stochastic resonance in a fractional oscillator with random damping strength and random spring stiffness, *Journal of Statistical Mechanics: Theory and Experiment*, 2013, 2013, P09026.
61. A. Y. T. Leung, Z. Guo, and H. X. Yang, Fractional derivative and time delay damper characteristics in Duffing-van der Pol oscillators, *Communications in Nonlinear Science and Numerical Simulation*, 18(10), 2013, 2900–2915.
62. L. C. Chen, Q. Q. Zhuang, and W. Q. Zhu, Response of SDOF nonlinear oscillators with lightly fractional derivative damping under real noise excitations, *The European Physical Journal of Special Topics*, 193(1), 2011, 81–92.
63. J.-F. Deü and D. Matignon, Simulation of fractionally damped mechanical systems by means of a Newmark-diffusive scheme, *Computers & Mathematics with Applications*, 59(5), 2010, 1745–1753.
64. G. E. Drăgănescu, L. Bereteu, A. Ercuţa, and G. Luca, Anharmonic vibrations of a nano-sized oscillator with fractional damping, *Communications in Nonlinear Science and Numerical Simulation*, 15(4), 2010, 922–926.
65. Y. A. Rossikhin and M. V. Shitikova, New approach for the analysis of damped vibrations of fractional oscillators, *Shock and Vibration*, 16(4), 2009, 365–387.
66. F. Xie and X. Lin, Asymptotic solution of the van der Pol oscillator with small fractional damping, *Physica Scripta*, T136, 2009, 014033.
67. R. Ren, M. Luo, and K. Deng, Stochastic resonance in a fractional oscillator subjected to multiplicative trichotomous noise, *Nonlinear Dynamics*, 90(1), 2017, 379–390.
68. J. Yuan, Y. Zhang, J. Liu, B. Shi, M. Gai, and S. Yang, Mechanical energy and equivalent differential equations of motion for single-degree-of-freedom fractional oscillators, *Journal of Sound and Vibration*, 397, 2017, 192–203.
69. H. Dai, Z. Zheng, and W. Wang, On generalized fractional vibration equation, *Chaos, Solitons & Fractals*, 95, 2017, 48–51.
70. L.-F. Lin, C. Chen, and H.-Q. Wang, Trichotomous noise induced stochastic resonance in a fractional oscillator with random damping and random frequency, *Journal of Statistical Mechanics: Theory and Experiment*, 2016, 2016, 023201.
71. Y. Naranjani, Y. Sardahi, Y.-Q. Chen, and J.-Q. Sun, Multi-objective optimization of distributed-order fractional damping, *Communications in Nonlinear Science and Numerical Simulation*, 24(1–3), 2015, 159–168.
72. G. He, Y. Tian, and Y. Wang, Stochastic resonance in a fractional oscillator with random damping strength and random spring stiffness, *Journal of Statistical Mechanics: Theory and Experiment*, 2013, 2013, P09026.
73. A. Di Matteo, P. D. Spanos, and A. Pirrotta, Approximate survival probability determination of hysteretic systems with fractional derivative elements, *Probabilistic Engineering Mechanics*, 54, 2018, 138–146.
74. Ž. Tomovski and T. Sandev, Effects of a fractional friction with power-law memory kernel on string vibrations, *Computers & Mathematics with Applications*, 62(3), 2011, 1554–1561.
75. J. F. Gomez-Aguilar, J. J. Rosales-Garcia, J. J. Bernal-Alvarado, T. Cordova-Fraga, and R. Guzman-Cabrera, Fractional mechanical oscillators, *Revista Mexicana de Fisica*, 58(4), 2012, 348–352.
76. Y. Tian, L.-F. Zhong, G.-T. He, T. Yu, M.-K. Luo, and H. E. Stanley, The resonant behavior in the oscillator with double fractional-order damping under the action of nonlinear multiplicative noise, *Physica A*, 490, 2018, 845–856.
77. M. Berman and L. S. Cederbaum, Fractional driven-damped oscillator and its general closed form exact solution, *Physica A*, 505, 2018, 744–762.

78. A. Coronel-Escamilla, J. F. Gómez-Aguilar, D. Baleanu, T. Córdova-Fraga, R. F. Escobar-Jiménez, V. H. Olivares-Peregrino, and M. M. Al Qurashi, Bateman-Feshbach Tikochinsky and Caldirola-Kanai oscillators with new fractional differentiation, *Entropy*, 19(2) 2017, 55 pp.
79. C. P. Li and F. H. Zeng, *Numerical Methods for Fractional Calculus*, Chapman and Hall/CRC Press, Boca Raton, FL, 2015.
80. C. P. Li and M. Cai, *Theory and Numerical Approximations of Fractional Integrals and Derivatives*, SIAM, Philadelphia, 2019.
81. M. H. Heydari, A computational method for a class of systems of nonlinear variable-order fractional quadratic integral equations, *Applied Numerical Mathematics*, 153, 2020, 164–178.
82. M. H. Heydari, Numerical solution of nonlinear 2D optimal control problems generated by Atangana-Riemann-Liouville fractal-fractional derivative, *Applied Numerical Mathematics*, 150, 2020, 507–518.
83. M. H. Heydari, M. R. Hooshmandasl, C. Cattani, F. M. Maalek Ghaini, An efficient computational method for solving nonlinear stochastic Itô integral equations: Application for stochastic problems in physics, *Journal of Computational Physics*, 283, 2015, 148–168.
84. M. Li, Integral representation of fractional derivative of delta function, *Fractal and Fractional*, 4(3), 2020, 47.

CHAPTER 8

Fractional Vibrations of Class I

THE GOAL OF THIS chapter is to discuss the following things.

1. Present an equivalent motion equation of a fractional vibrator in class I in Theorem 8.1.
2. Introduce the concepts of quasiequivalent mass, quasiequivalent damping, quasiequivalent damping ratio, quasiequivalent damping free natural frequency, quasiequivalent damped natural frequency, quasiequivalent frequency ratio, quasiequivalent Q factor, and quasiequivalent logarithmic decrement for a fractional vibrator in class I. Their expressions are proposed in Theorems 8.2, 8.3, 8.6–8.9, 8.14, and 8.15, respectively.
3. Reveal that the quasiequivalent mass of a fractional vibrator in class I may vary from 0 to ∞, relying on vibration frequency ω and fractional order α in Corollaries 8.1 and 8.2.
4. Reveal that the quasiequivalent damping of a fractional vibrator of class I type may vary from $-\infty$ to ∞, depending on ω and α (Corollaries 8.6 and 8.7).
5. Bring forward the conditions for a fractional vibrator of class I type to be self-vibrated or non-stable in Theorems 8.4, 8.5, Corollaries 8.5, and 8.9.
6. Put forward the analytic expressions of free response, impulse response, step response, and frequency transfer function of a fractional vibrator in class I in Theorems 8.10–8.13.
7. In passing, we present the representations of generalized Mittag-Leffler functions by using elementary functions in Theorems 8.16–8.20.

DOI: 10.1201/9781003460947-10

8.1 EQUIVALENT MOTION EQUATION OF CLASS I FRACTIONAL VIBRATOR

The equivalent motion equation of a fractional vibrator in class I is described in Theorem 8.1.

Theorem 8.1. For the motion equation of a fractional vibrator in class I,

$$m\frac{d^{\alpha}x(t)}{dt^{\alpha}}+kx(t)=0, \quad 1<\alpha<3,$$

its equivalent motion equation is expressed by

$$-m\omega^{\alpha-2}\cos\frac{\alpha\pi}{2}\frac{d^2x(t)}{dt^2}+m\omega^{\alpha-1}\sin\frac{\alpha\pi}{2}\frac{dx(t)}{dt}+kx(t)=0, \quad 1<\alpha<3. \tag{8.1}$$

Proof. According to Theorem 7.3, the Fourier transform of the first two items on the left side on the above is given by

$$\left[-m\omega^{\alpha-2}\cos\frac{\alpha\pi}{2}\frac{d^2x(t)}{dt^2}+m\omega^{\alpha-1}\sin\frac{\alpha\pi}{2}\frac{dx(t)}{dt}\right]\leftrightarrow m(i\omega)^{\alpha}X(\omega), \tag{8.2}$$

where $X(\omega)$ is the Fourier transform of $x(t)$. Thus,

$$\begin{aligned}&\mathrm{F}^{-1}\mathrm{F}\left[-m\omega^{\alpha-2}\cos\frac{\alpha\pi}{2}\frac{d^2x(t)}{dt^2}+m\omega^{\alpha-1}\sin\frac{\alpha\pi}{2}\frac{dx(t)}{dt}+kx(t)\right]\\&=\mathrm{F}^{-1}\left[m(i\omega)^{\alpha}X(\omega)+kX(\omega)\right]=m\frac{d^{\alpha}x(t)}{dt^{\alpha}}+kx(t).\end{aligned} \tag{8.3}$$

The proof is finished. □

8.2 QUASIEQUIVALENT VIBRATION PARAMETERS OF CLASS I FRACTIONAL VIBRATOR

8.2.1 Quasiequivalent Mass in Fractional Vibrators of Class I

Denote by m_{eq1} the quasiequivalent mass for a fractional vibrator of class I. We present its expression in Theorem 8.2.

Theorem 8.2. The quasiequivalent mass for a fractional vibrator of class I is expressed by

$$m_{\mathrm{eq1}}=-m\omega^{\alpha-2}\cos\frac{\alpha\pi}{2}, \quad 1<\alpha<3. \tag{8.4}$$

Proof. From Eq. (8.1), we see that Eq. (8.4) holds. This finishes the proof. □

Note that the unit of the primary mass m is kg. Since the unit of m_{eq1} is kg rad$^{\alpha-2}$ or kg Hz$^{\alpha-2}$ instead of kg, we introduce a term quasiequivalent mass to describe m_{eq1} for facilitating the discussions.

Since m_{eq1} is associated with ω, α, and m, it may be written by

$$m_{eq1} = m_{eq1}(\omega, \alpha, m) = -m\omega^{\alpha-2}\cos\frac{\alpha\pi}{2}, \quad 1 < \alpha < 3. \tag{8.5}$$

Quasiequivalent mass $m_{eq1}(\omega, \alpha, m) \geq 0$ for $1 < \alpha < 3$. It reduces to the primary mass m when $\alpha = 2$. It has the asymptotic properties described in Corollaries 8.1 and 8.2.

Corollary 8.1: Asymptotic Property for $\omega \to \infty$. If $1 < \alpha < 2$,

$$\lim_{\omega\to\infty} m_{eq1}(\omega, \alpha, m) = 0. \tag{8.6}$$

When $2 < \alpha < 3$,

$$\lim_{\omega\to\infty} m_{eq1}(\omega, \alpha, m) = \infty. \tag{8.7}$$

Proof. Since $\lim_{\omega\to\infty} \omega^{\alpha-2} = 0$ for $1 < \alpha < 2$, we have Eq. (8.6). On the other hand, because $\lim_{\omega\to\infty} \omega^{\alpha-2} = \infty$ if $2 < \alpha < 3$, we have Eq. (8.7). The proof is finished. □

Figure 8.1 shows the plots of $m_{eq1}(\omega, \alpha, m)$.

In vibrations, $\omega \to \infty$ implies high order vibrations modes and coordinates of high order vibrations modes approach zero in general. Thus, $\omega \to \infty$ may usually be trivial in vibrations. However, a fractional vibrator in class I has properties substantially different from an ordinary vibrator in mass. When $1 < \alpha < 2$, $m_{eq1}(\omega, \alpha, m)$ disappears for $\omega \to \infty$. On the other hand, m_{eq1} becomes ∞ for $\omega \to \infty$ if $2 < \alpha < 3$.

Corollary 8.2. When $1 < \alpha < 2$,

$$\lim_{\omega\to 0} m_{eq1}(\omega, \alpha, m) = \infty. \tag{8.8}$$

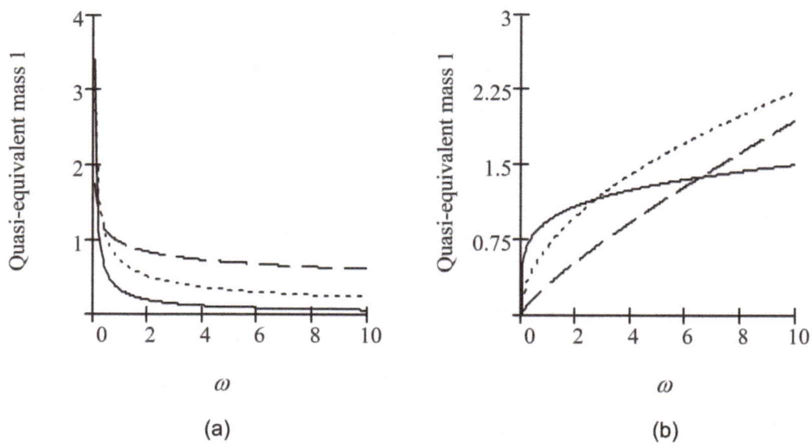

FIGURE 8.1 Plots of m_{eq1}. (a) m_{eq1} for $\alpha = 1.2$ (solid), 1.5 (dot), 1.8 (dash). (b) m_{eq1} for $\alpha = 2.2$ (solid), 2.5 (dot), 2.8 (dash). Note that "Quasi-equivalent mass 1" in the above figures means m_{eq1}.

If $2<\alpha<3$,

$$\lim_{\omega\to 0} m_{eq1}(\omega,\alpha,m)=0. \tag{8.9}$$

Proof. As $\lim_{\omega\to 0}\omega^{\alpha-2}=\infty$ for $1<\alpha<2$, we see that Eq. (8.8) holds. On the other side, since $\lim_{\omega\to 0}\omega^{\alpha-2}=0$ if $2<\alpha<3$, Eq. (8.9) holds. This completes the proof. □

It is known that $\omega=0$ is trivial in vibrations because it implies static loading. Nevertheless, the point I would like to stress on is that fractional inertia force $m\frac{d^\alpha x(t)}{dt^\alpha}$ may cause $m_{eq1}(0, \alpha, m)=0$ for $2<\alpha<3$ and $m_{eq1}(0, \alpha, m)=\infty$ for $1<\alpha<2$.

In summary, $0<m_{eq1}<\infty$. Ordinary vibrations cannot describe such phenomena of mass.

8.2.2 Quasiequivalent Damping in Fractional Vibrators of Class I

Let c_{eq1} be the quasiequivalent damping for a fractional vibrator of class I. Its expression is presented below.

Theorem 8.3. The quasiequivalent damping for a fractional vibrator of class I is given by

$$c_{eq1}=m\omega^{\alpha-1}\sin\frac{\alpha\pi}{2}, \quad 1<\alpha<3. \tag{8.10}$$

Proof. From the second item on the left side of Eq. (8.1), we see that Eq. (8.10) is valid. The proof is finished. □

As the standard unit of damping c is $N\times m^{-1}\times s$ and that of c_{eq1} is kg $(rad/s)^{\alpha-1}$ or kg $Hz^{\alpha-1}$, we use the prefix quasi to describe c_{eq1}. It is produced by fractional inertia force $m\frac{d^\alpha x(t)}{dt^\alpha}$.

Corollary 8.3. When $\alpha=2$, c_{eq1} vanishes.

Proof. Since $c_{eq1}\big|_{\alpha=2}=0$, Corollary 8.3 holds. The proof is finished. □

We now address the properties of c_{eq1}.

Corollary 8.4. When $1<\alpha<2$,

$$c_{eq1}\geq 0. \tag{8.11}$$

Proof. Since $\sin\frac{\alpha\pi}{2}>0$ for $1<\alpha<2$, the above holds. This finishes the proof. □

In passing, we say that c_{eq1} is proportional to the primary mass m. That is just a mathematical explanation of a variation of the Rayleigh damping assumption. Details about it refer to Chapter 14.

Theorem 8.4: Negative Damping. If $2<\alpha<3$ and $\omega \neq 0$,

$$c_{\text{eq1}} < 0. \tag{8.12}$$

Proof. Due to $\sin\frac{\alpha\pi}{2}<0$ if $2<\alpha<3$, we have Eq. (8.12). The proof is finished. □

Theorem 8.5. When $2<\alpha<3$ and $\omega \neq 0$, a fractional vibrator in class I is self-vibrated.

Proof. According to Theorem 8.4, $c_{\text{eq1}}<0$ when $2<\alpha<3$ and $\omega \neq 0$. Since c_{eq1} is negative, a fractional vibrator in class I is self-vibrated. The proof is complete. □

Corollary 8.5. If $2<\alpha<3$ and $\omega \neq 0$, a fractional vibrator in class I is not stable.

Proof. Since a fractional vibrator in class I is self-vibrated for $2<\alpha<3$ and $\omega \neq 0$, it is non-stable. The proof is finished. □

We now write c_{eq1} by $c_{\text{eq1}}(\omega, \alpha, m)$ for describing its asymptotic properties.

Corollary 8.6. When $1<\alpha<2$,

$$\lim_{\omega \to 0} c_{\text{eq1}}(\omega, \alpha, m) = 0. \tag{8.13}$$

In addition,

$$\lim_{\omega \to \infty} c_{\text{eq1}}(\omega, \alpha, m) = \infty. \tag{8.14}$$

Proof. Note that $\sin\frac{\alpha\pi}{2}>0$ for $1<\alpha<2$. Besides, $\lim_{\omega \to 0} \omega^{\alpha-1}=0$. Thus, Eq. (8.13) holds. Moreover, $\lim_{\omega \to \infty} \omega^{\alpha-1}=\infty$. Hence, Eq. (8.14) is valid. This finishes the proof. □

Corollary 8.7. If $2<\alpha<3$,

$$\lim_{\omega \to 0} c_{\text{eq1}}(\omega, \alpha, m) = 0. \tag{8.15}$$

In addition,

$$\lim_{\omega \to \infty} c_{\text{eq1}}(\omega, \alpha, m) = -\infty. \tag{8.16}$$

Proof. Since $\lim_{\omega \to 0} \omega^{\alpha-1}=0$ for $2<\alpha<3$, we have Eq. (8.15). Besides, $\sin\frac{\alpha\pi}{2}<0$ and $\lim_{\omega \to \infty} \omega^{\alpha-1}=\infty$ for $2<\alpha<3$, Eq. (8.16) holds. Hence, Corollary 8.7 is valid. This completes the proof. □

Properties in Eqs. (8.15) and (8.16) cannot be described by fractional vibrators in Ref. [1], letting along ordinary vibrations. As a summary, $-\infty<c_{\text{eq1}}<\infty$. Figure 8.2 indicates the illustrations of c_{eq1}.

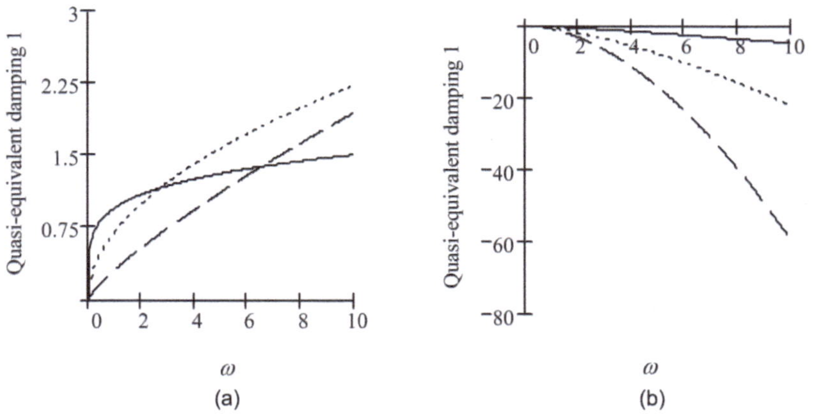

FIGURE 8.2 Illustrations of c_{eq1}. (a) c_{eq1} for $\alpha=1.2$ (solid), 1.5 (dot), 1.8 (dash). (b) c_{eq1} for $\alpha=2.2$ (solid), 2.5 (dot), 2.8 (dash). Note that "Quasi-equivalent damping 1" in the above figures stands for c_{eq1}.

8.2.3 Quasiequivalent Damping Ratio in Fractional Vibrators of Class I

Denote by ζ_{eq1} the quasiequivalent damping ratio for a fractional vibrator of class I. It is defined by

$$\varsigma_{eq1} = \frac{c_{eq1}}{2\sqrt{m_{eq1}k}}. \tag{8.17}$$

Its expression is proposed below.

Theorem 8.6. The quasiequivalent damping ratio for a fractional vibrator of class I is given by

$$\varsigma_{eq1} = \varsigma_{eq1}(\omega,\alpha) = \frac{\omega^{\frac{\alpha}{2}} \sin\frac{\alpha\pi}{2}}{2\omega_n\sqrt{-\cos\frac{\alpha\pi}{2}}}, \quad 1<\alpha<3, \tag{8.18}$$

where

$$\omega_n = \sqrt{\frac{k}{m}} \tag{8.19}$$

is the conventional natural angular frequency with damping free.

Proof. Substituting m_{eq1} and c_{eq1} into Eqs. (8.17), (8.18) results. The proof is finished. □

Note that the unit of ζ_{eq1} is $(\text{rad/s})^{0.5\alpha-1}$ instead of dimensionless. Hence, it is a quasi-equivalent damping ratio. When $\alpha=2$, it disappears.

I explain the asymptotic properties of ζ_{eq1} with the following theorems and corollaries.

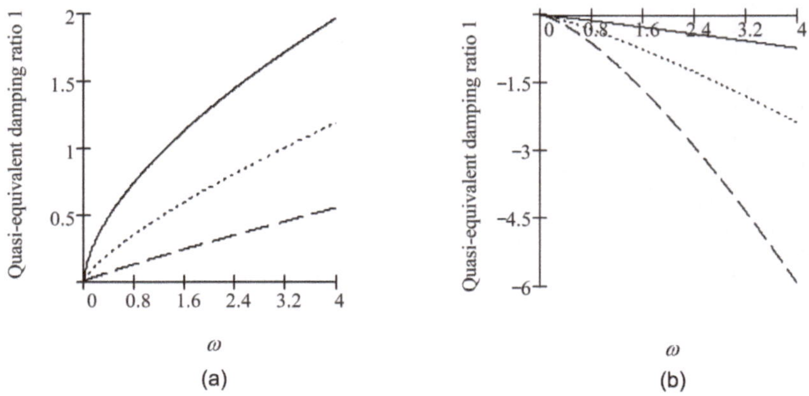

FIGURE 8.3 Plots of ζ_{eq1} with $\omega_n = 1$. (a) ζ_{eq1} for $\alpha = 1.2$ (solid), 1.5 (dot), 1.8 (dash). (b) ζ_{eq1} for $\alpha = 2.2$ (solid), 2.5 (dot), 2.8 (dash). Note that "Quasi-equivalent damping ratio 1" in the above figures stands for ζ_{eq1}.

Corollary 8.8. When $1 < \alpha < 2$, we have

$$\varsigma_{\text{eq1}} \geq 0. \tag{8.20}$$

Proof. Since $c_{\text{eq1}} > 0$ for $1 < \alpha < 2$, the above is valid. This finishes the proof. □

Corollary 8.9. If $2 < \alpha < 3$ and $\omega \neq 0$,

$$\varsigma_{\text{eq1}} < 0. \tag{8.21}$$

Proof. Owing to $c_{\text{eq1}} < 0$ if $2 < \alpha < 3$, the above holds. The proof is finished. □

Several plots of ζ_{eq1} are illustrated in Figure 8.3.

In class I, a fractional vibrator appears damping free in form. However, fractional inertia force $m\frac{d^{\alpha}x(t)}{dt^{\alpha}}$ produces c_{eq1} and accordingly ζ_{eq1}. They may be positive, negative or zero, relying on α and ω.

8.2.4 Quasiequivalent Damping Free Natural Frequency in Fractional Vibrators of Class I

The conventional natural frequency with damping free is expressed by Eq. (8.19). For a fractional vibrator in class I, we introduce the term of quasiequivalent damping-free natural frequency. Let it be ω_{eqn1}. Define it by

$$\omega_{\text{eqn1}} = \sqrt{\frac{k}{m_{\text{eq1}}}}. \tag{8.22}$$

Theorem 8.7. The quasiequivalent damping-free natural frequency for a fractional vibrator in class I is given by

$$\omega_{\text{eqn1}} = \frac{\omega_n}{\sqrt{-\omega^{\alpha-2}\cos\frac{\alpha\pi}{2}}}, \qquad 1<\alpha<3. \tag{8.23}$$

Proof. Substituting m_{eq1} into Eq. (8.22) yields

$$\omega_{\text{eqn1}} = \omega_{\text{eqn1}}(\omega,\alpha) = \sqrt{\frac{k}{-m\omega^{\alpha-2}\cos\frac{\alpha\pi}{2}}} = \sqrt{\frac{1}{-\omega^{\alpha-2}\cos\frac{\alpha\pi}{2}}}\sqrt{\frac{k}{m}} = \frac{\omega_n}{\sqrt{-\omega^{\alpha-2}\cos\frac{\alpha\pi}{2}}}. \tag{8.24}$$

The proof is finished. □

Note that we call ω_{eqn1} as "quasiequivalent damping free natural frequency" just because it takes the form of Eq. (8.22). In fact, it is not a constant but a function of ω and α. It reduces to ω_n=const if and only if $\alpha=2$. In addition, the unit of ω_{eqn1} is $(\text{rad/s})^{2-0.5\alpha}$ instead of rad/s. Its unit degenerates to rad/s when $\alpha=2$. As a matter of fact, $\omega_{\text{eqn1}}(\omega,2)=\omega_n$. Several plots of $\omega_{\text{eqn1}}(\omega, \alpha)$ are shown in Figure 8.4.

8.2.5 Quasiequivalent Damped Natural Frequency in Fractional Vibrators of Class I

Let ω_{eqd1} be the quasiequivalent damped natural frequency for a fractional vibrator of class I. Since $\zeta_{\text{eq1}}>1$ is trivial in vibrations, we restrict $|\varsigma_{\text{eq1}}|\le 1$ in what follows. Define

$$\omega_{\text{eqd1}} = \omega_{\text{eqn1}}\sqrt{1-\varsigma_{\text{eq1}}^2}, \qquad |\varsigma_{\text{eq1}}|\le 1. \tag{8.25}$$

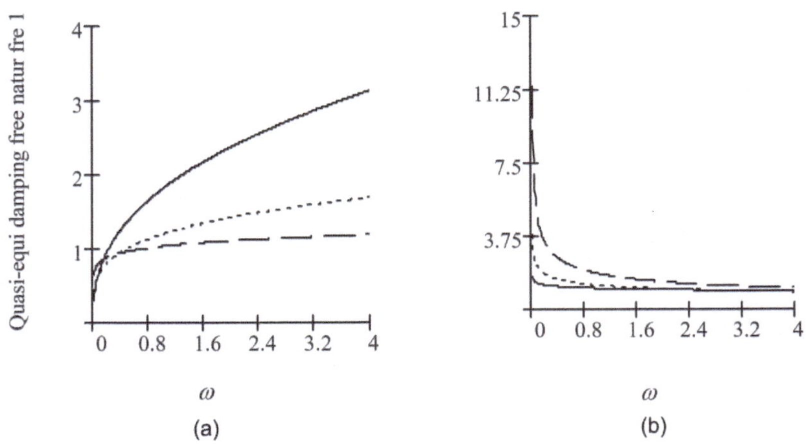

FIGURE 8.4 Plots of $\omega_{\text{eqn1}}(\omega, \alpha)$ with $\omega_n=1$. (a) ω_{eqn1} for $\alpha=1.2$ (solid), 1.5 (dot), 1.8 (dash). (b) ω_{eqn1} for $\alpha=2.2$ (solid), 2.5 (dot), 2.8 (dash). Note that "Quasi-equivalent damping free natural frequency 1" in the above figures implies ω_{eqn1}.

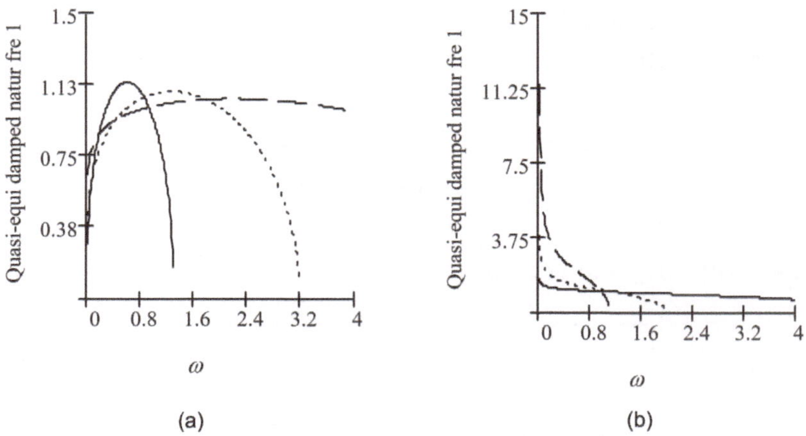

FIGURE 8.5 Illustrations of $\omega_{\mathrm{eqd1}}(\omega, \alpha)$ with $\omega_n=1$. (a) ω_{eqd1} for $\alpha=1.2$ (solid), 1.5 (dot), 1.8 (dash). (b) ω_{eqd1} for $\alpha=2.2$ (solid), 2.5 (dot), 2.8 (dash). Note that "Quasi-equivalent damped natural frequency 1" in the above figures means ω_{eqd1}.

Theorem 8.8. The quantity ω_{eqd1} is expressed by

$$\omega_{\mathrm{eqd1}}=\omega_{\mathrm{eqd1}}(\omega,\alpha)=\frac{\omega_n}{\sqrt{-\omega^{\alpha-2}\cos\frac{\alpha\pi}{2}}}\sqrt{1-\frac{\omega^{\alpha}\sin^2\frac{\alpha\pi}{2}}{4\omega_n^2\left|\cos\frac{\alpha\pi}{2}\right|}}. \tag{8.26}$$

Proof. Substituting ζ_{eq1} and ω_{eqn1} into Eq. (8.25) results in Eq. (8.26). The proof is complete. □

Several illustrations of $\omega_{\mathrm{eqd1}}(\omega, \alpha)$ are indicated in Figure 8.5.

8.2.6 Quasiequivalent Frequency Ratio in Fractional Vibrators of Class I

Let γ_{eq1} be the quasiequivalent frequency ratio of a fractional vibrator of class I. It is defined by

$$\gamma_{\mathrm{eq1}}=\frac{\omega}{\omega_{\mathrm{eqn1}}}. \tag{8.27}$$

We present its expression below.

Theorem 8.9. The quantity γ_{eq1} is in the form

$$\gamma_{\mathrm{eq1}}=\gamma_{\mathrm{eq1}}(\omega,\alpha)=\frac{\omega\sqrt{\omega^{\alpha-2}\left|\cos\frac{\alpha\pi}{2}\right|}}{\omega_n}. \tag{8.28}$$

Proof. Substituting ω_{eqn1} into Eq. (8.27) produces Eq. (8.28). This finishes the proof. □

The unit of γ_{eq1} is not dimensionless but $(\mathrm{rad/s})^{0.5\alpha-1}$. It reduces to being dimensionless if and only if $\alpha=2$. In fact, $\gamma_{\mathrm{eq1}}(\omega,2)=\gamma=\frac{\omega}{\omega_n}$. Thus, we introduce the prefix quasi to describe it. Its plots are shown in Figure 8.6.

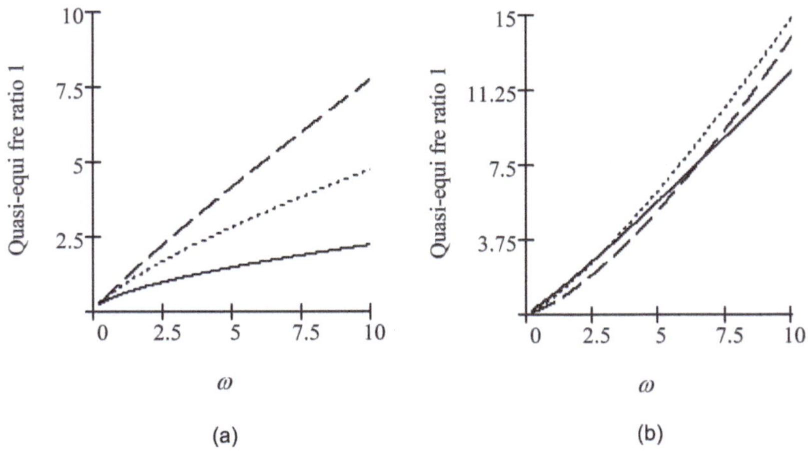

FIGURE 8.6 Plots of $\gamma_{\mathrm{eq1}}(\omega, \alpha)$ with $\omega_n = 1$. (a) γ_{eq1} for $\alpha = 1.2$ (solid), 1.5 (dot), 1.8 (dash). (b) γ_{eq1} for $\alpha = 2.2$ (solid), 2.5 (dot), 2.8 (dash). Note that "Quasi-equivalent frequency ratio 1" in the above figures implies γ_{eq1}.

8.3 FREE RESPONSE OF CLASS I FRACTIONAL VIBRATOR

Denote by $x_1(t)$ the free response of a fractional vibrator in class I. It is a solution to the fractional differential equation given by

$$\begin{cases} m\dfrac{d^\alpha x_1(t)}{dt^\alpha} + kx_1(t) = 0, \\ x_1(0) = x_{10}, x_1'(0) = v_{10}, \end{cases} \quad 1 < \alpha < 3. \tag{8.29}$$

We give its solution by Theorem 8.10.

Theorem 8.10. The solution to the above equation is in the form

$$x_1(t) = e^{-\varsigma_{\mathrm{eq1}}\omega_{\mathrm{eqn1}}t}\left(x_{10}\cos\omega_{\mathrm{eqd1}}t + \frac{v_{10} + \varsigma_{\mathrm{eq1}}\omega_{\mathrm{eqn1}}x_{10}}{\omega_{\mathrm{eqd1}}}\sin\omega_{\mathrm{eqd1}}t\right), \quad t \geq 0. \tag{8.30}$$

Proof. According to Theorem 8.1, the solution to the above equation equals to the solution to the following differential equation given by

$$\begin{cases} m_{\mathrm{eq1}}\dfrac{d^2 x_1(t)}{dt^2} + c_{\mathrm{eq1}}\dfrac{dx_1(t)}{dt} + kx_1(t) = 0, \\ x_1(0) = x_{10}, x_1'(0) = v_{10}. \end{cases} \tag{8.31}$$

Since Eq. (8.31) is a standard vibration equation, its solution is given by Eq. (8.30). The proof is finished. □

Substituting ζ_{eq1}, ω_{eqn1}, and ω_{eqd1} into Eq. (8.30) yields

$$x_1(t) = e^{-\frac{\omega^{\frac{\alpha}{2}}\sin\frac{\alpha\pi}{2}}{2\sqrt{\left|\cos\frac{\alpha\pi}{2}\right|}}\frac{t}{\sqrt{\omega^{\alpha-2}\left|\cos\frac{\alpha\pi}{2}\right|}}}\left[\begin{array}{l} x_{10}\cos\left(\frac{\omega_n}{\sqrt{\omega^{\alpha-2}\left|\cos\frac{\alpha\pi}{2}\right|}}\sqrt{1-\frac{\omega^{\alpha}\sin^2\frac{\alpha\pi}{2}}{4\omega_n^2\left|\cos\frac{\alpha\pi}{2}\right|}}\,t\right) \\ +\frac{v_{10}+\frac{\omega^{\frac{\alpha}{2}}\sin\frac{\alpha\pi}{2}}{2\sqrt{\left|\cos\frac{\alpha\pi}{2}\right|}}\frac{x_{10}}{\sqrt{\omega^{\alpha-2}\left|\cos\frac{\alpha\pi}{2}\right|}}}{\frac{\omega_n}{\sqrt{\omega^{\alpha-2}\left|\cos\frac{\alpha\pi}{2}\right|}}\sqrt{1-\frac{\omega^{\alpha}\sin^2\frac{\alpha\pi}{2}}{4\omega_n^2\left|\cos\frac{\alpha\pi}{2}\right|}}} \\ \sin\left(\frac{\omega_n}{\sqrt{\omega^{\alpha-2}\left|\cos\frac{\alpha\pi}{2}\right|}}\sqrt{1-\frac{\omega^{\alpha}\sin^2\frac{\alpha\pi}{2}}{4\omega_n^2\left|\cos\frac{\alpha\pi}{2}\right|}}\,t\right) \end{array}\right]. \quad (8.32)$$

Figure 8.7 shows several plots of $x_1(t)$.

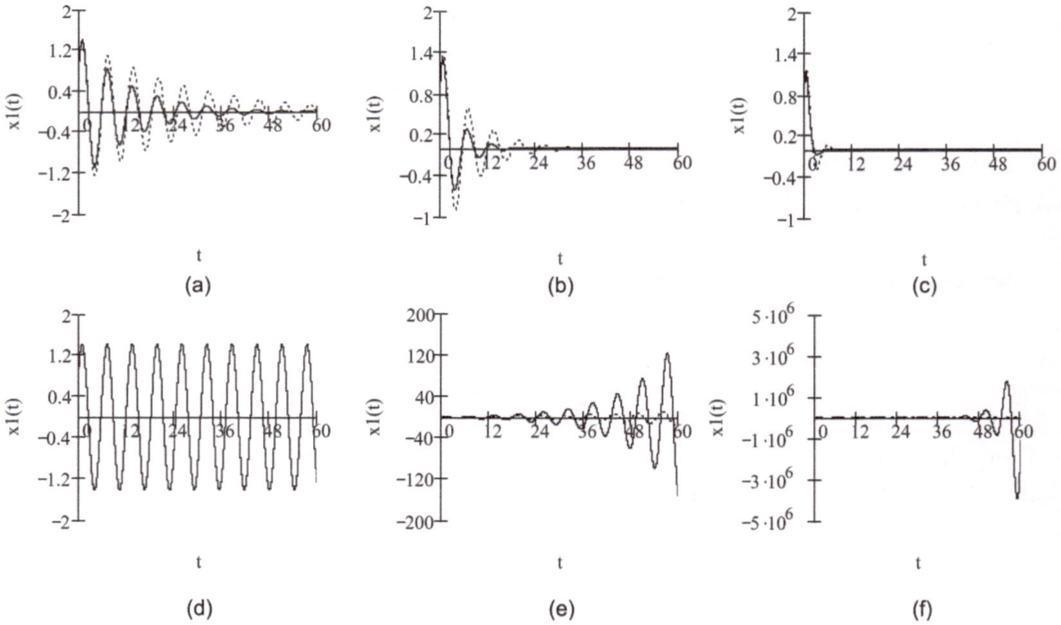

FIGURE 8.7 Illustrations of free response $x_1(t)$ for $x_{10}=1$, $v_{10}=1$, and $\omega_n=1$. (a) $x_1(t)$ when $\alpha=1.9$. Solid line: $\omega=1$ ($\zeta_{\text{eq1}}=0.08$). Dot line: $\omega=0.7$ ($\zeta_{\text{eq1}}=0.04$). (b) $x_1(t)$ when $\alpha=1.6$. Solid line: $\omega=1$ ($\zeta_{\text{eq1}}=0.33$). Dot line: $\omega=0.7$ ($\zeta_{\text{eq1}}=0.16$). (c) $x_1(t)$ when $\alpha=1.3$. Solid line: $\omega=1$ ($\zeta_{\text{eq1}}=0.66$). Dot line: $\omega=0.7$ ($\zeta_{\text{eq1}}=0.42$). (d) $x_1(t)$ when $\alpha=2$. Solid line: $\omega=1$ ($\zeta_{\text{eq1}}=0$). Dot line: $\omega=0.7$ ($\zeta_{\text{eq1}}=0$). (e) $x_1(t)$ when $\alpha=2.1$. Solid line: $\omega=1$ ($\zeta_{\text{eq1}}=-0.08$). Dot line: $\omega=0.7$ ($\zeta_{\text{eq1}}=-0.04$) (self-vibration). (f) $x_1(t)$ when$\alpha=2.3$. Solid line: $\omega=1$ ($\zeta_{\text{eq1}}=-0.24$). Dot line: $\omega=0.7$ ($\zeta_{\text{eq1}}=-0.11$) (self-vibration).

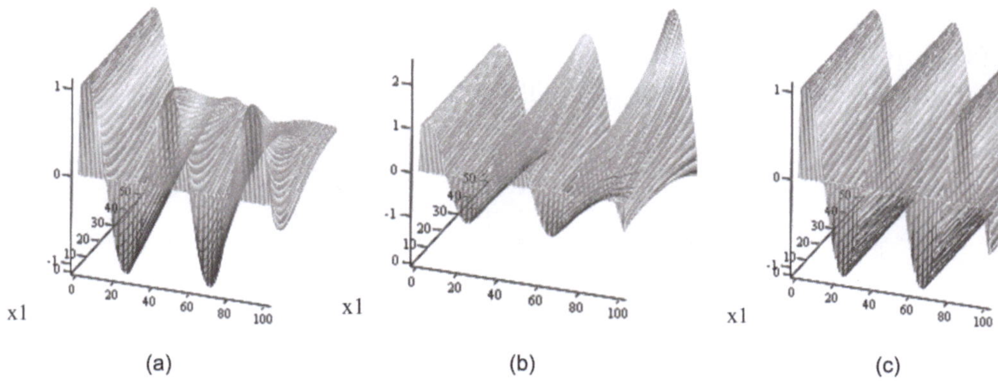

FIGURE 8.8 Illustrations of free response $x_1(t, \omega)$ in t-ω plane for $t=0, 1, \ldots, 60$ and $\omega=1, 2, \ldots, 50$ with $x_{10}=1$, $v_{10}=1$, and $\omega_n=3$. (a) $x_1(t, \omega)$ when $\alpha=1.95$ ($0.008 \le \zeta_{eq1} \le 17.464$). (b) $x_1(t, \omega)$ when $\alpha=2.01$ ($-0.003 \le \zeta_{eq1} \le -6.806$) (self-vibration). (c) $x_1(t, \omega)$ when $\alpha=2$ ($\zeta_{eq1}=0$).

Note that free response $x_1(t)$ is actually a time-frequency distribution on a time-frequency plane since $m_{eq1}(\omega, \alpha, m)$ and $c_{eq1}(\omega, \alpha, m)$ are all functions of ω. Thus, $x_1(t)$ may be expressed by

$$x_1(t)=x_1(t,\omega). \tag{8.33}$$

Figure 8.8 gives several illustrations of $x_1(t, \omega)$.

8.4 IMPULSE RESPONSE OF CLASS I FRACTIONAL VIBRATOR

Let $h_1(t)$ be the impulse response of a class I fractional vibrator. It is a solution to the fractional differential equation given by

$$\begin{cases} m\dfrac{d^\alpha h_1(t)}{dt^\alpha}+kh_1(t)=\delta(t), & 1<\alpha<3. \\ h_1(0)=0, h_1'(0)=0, & \end{cases} \tag{8.34}$$

We propose Theorem 8.11 to express $h_1(t)$.

Theorem 8.11. The solution to the above equation is given by

$$h_1(t)=\frac{e^{-\varsigma_{eq1}\omega_{eqn1}t}}{m_{eq1}\omega_{eqd1}}\sin\omega_{eqd1}t, \quad t\ge 0. \tag{8.35}$$

Proof. According to Theorem 8.1, the solution to the above equation is equivalent to the solution to the differential equation given by

$$\begin{cases} m_{eq1}\dfrac{d^2 h_1(t)}{dt^2}+c_{eq1}\dfrac{dh_1(t)}{dt}+kh_1(t)=\delta(t), \\ h_1(0)=0, h_1'(0)=0. \end{cases} \tag{8.36}$$

Note that Eq. (8.36) is a standard vibration equation. Its solution is accordingly expressed by Eq. (8.35). This finishes the proof. □

Substituting m_{eq1}, ζ_{eq1}, ω_{eqn1}, and ω_{eqd1} into Eq. (8.35) results in

$$h_1(t)=\frac{e^{-\frac{\omega\sin\frac{\alpha\pi}{2}}{2\left|\cos\frac{\alpha\pi}{2}\right|}t}\sin\left(\frac{\omega_n}{\sqrt{\omega^{\alpha-2}\left|\cos\frac{\alpha\pi}{2}\right|}}\sqrt{1-\frac{\omega^{2\alpha}\sin^2\frac{\alpha\pi}{2}}{4\omega_n^2\left|\cos\frac{\alpha\pi}{2}\right|}}\,t\right)}{m\omega_n\sqrt{\omega^{\alpha-2}\left|\cos\frac{\alpha\pi}{2}\right|}\sqrt{1-\frac{\omega^{2\alpha}\sin^2\frac{\alpha\pi}{2}}{4\omega_n^2\left|\cos\frac{\alpha\pi}{2}\right|}}},\quad t\geq 0. \tag{8.37}$$

The plots of $h_1(t)$ are shown in Figure 8.9.

Again, $h_1(t)$ is a time-frequency distribution, and it may be in the form

$$h_1(t)=h_1(t,\omega). \tag{8.38}$$

Several plots of $h_1(t,\omega)$ are indicated in Figure 8.10.

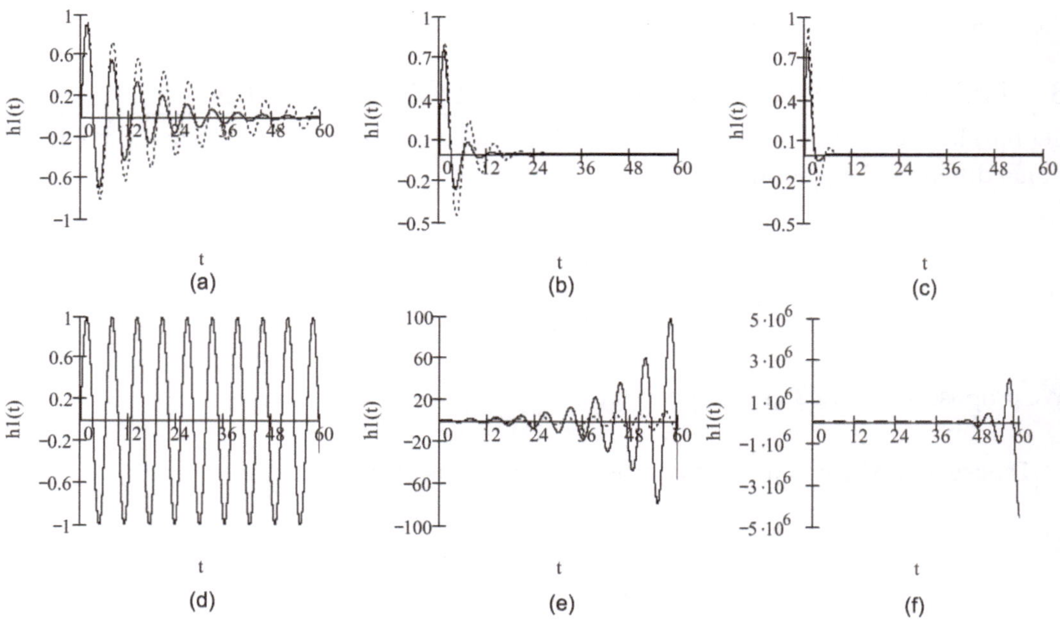

FIGURE 8.9 Plots of impulse response $h_1(t)$ for $m=1$ and $\omega_n=1$. (a) $h_1(t)$ when $\alpha=1.9$. Solid line: $\omega=1$ ($\zeta_{eq1}=0.08$). Dot line: $\omega=0.7$ ($\zeta_{eq1}=0.04$). (b) $h_1(t)$ when $\alpha=1.6$. Solid line: $\omega=1$ ($\zeta_{eq1}=0.33$). Dot line: $\omega=0.7$ ($\zeta_{eq1}=0.16$). (c) $h_1(t)$ when $\alpha=1.3$. Solid line: $\omega=1$ ($\zeta_{eq1}=0.66$). Dot line: $\omega=0.7$ ($\zeta_{eq1}=0.42$). (d) $h_1(t)$ when $\alpha=2$. Solid line: $\omega=1$ ($\zeta_{eq1}=0$). Dot line: $\omega=0.7$ ($\zeta_{eq1}=0$). (e) $h_1(t)$ when $\alpha=2.1$. Solid line: $\omega=1$ ($\zeta_{eq1}=-0.08$). Dot line: $\omega=0.7$ ($\zeta_{eq1}=-0.04$) (self-vibration). (f) $h_1(t)$ when $\alpha=2.3$. Solid line: $\omega=1$ ($\zeta_{eq1}=-0.24$). Dot line: $\omega=0.7$ ($\zeta_{eq1}=-0.11$) (self-vibration).

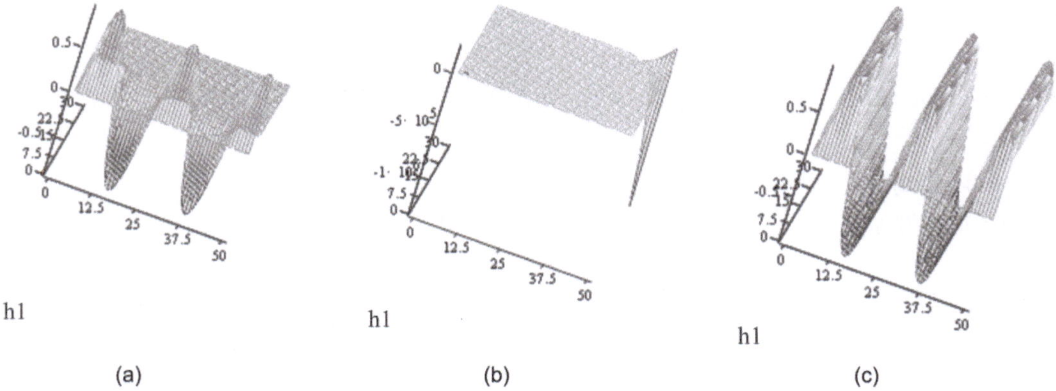

FIGURE 8.10 Plots of impulse response $h_1(t, \omega)$ in t-ω plane for $t=0, 1, \ldots, 60$ and $\omega=1, 2, \ldots, 30$, with $m=1$ and $\omega_n=0.3$. (a) $h_1(t, \omega)$ when $\alpha=1.95$ ($0.131 \le \zeta_{\mathrm{eq1}} \le 99.437$). (b) $h_1(t, \omega)$ when $\alpha=2.01$ ($-24.378 \le \zeta_{\mathrm{eq1}} \le -0.026$) (self-vibration). (c) $h_1(t, \omega)$ when $\alpha=2$ ($\zeta_{\mathrm{eq1}}=0$).

8.5 STEP RESPONSE OF CLASS I FRACTIONAL VIBRATOR

Denote by $g_1(t)$ the unit step response of a class I fractional vibrator. It is a solution to

$$\begin{cases} m\dfrac{d^{\alpha} g_1(t)}{dt^{\alpha}} + kg_1(t) = u(t), \\ g_1(0) = 0, g_1'(0) = 0, \end{cases} \qquad 1 < \alpha < 3. \tag{8.39}$$

The present Theorem 8.12 describes it.

Theorem 8.12. The solution to the above equation is expressed by

$$g_1(t) = \int_0^t h_1(\tau)\,d\tau = \frac{1}{k}\left[1 - \frac{e^{-\varsigma_{\mathrm{eq1}}\omega_{\mathrm{eqn1}}t}\,\omega_{\mathrm{eqn1}}}{\omega_{\mathrm{eqd1}}}\cos\left(\omega_{\mathrm{eqd1}}t - \phi_1\right)\right], \qquad t \ge 0, \tag{8.40}$$

where

$$\phi_1 = \tan^{-1}\frac{\varsigma_{\mathrm{eq1}}}{\sqrt{1-\zeta_{\mathrm{eq1}}^2}}. \tag{8.41}$$

Proof. Following Theorem 8.1, the solution to Eq. (8.39) equals to the solution to the following differential equation given by

$$\begin{cases} m_{\mathrm{eq1}}\dfrac{d^2 g_1(t)}{dt^2} + c_{\mathrm{eq}}\dfrac{dg_1(t)}{dt} + kg_1(t) = u(t), \\ g_1(0) = 0, \quad g_1'(0) = 0. \end{cases} \tag{8.42}$$

The above is a standard vibration equation of order 2. Since

$$u(t) = \int_{-\infty}^{t} \delta(t)\,dt, \tag{8.43}$$

equation (8.40) is valid. The proof is finished. □

Substituting ζ_{eq1}, ω_{eqn1}, and ω_{eqd1} into Eq. (8.40) produces

$$g_1(t) = \frac{1}{k}\left[1 - \frac{1}{\sqrt{1-\zeta_{eq1}^2}} e^{-\frac{\omega \sin\frac{\alpha\pi}{2}}{2\left|\cos\frac{\alpha\pi}{2}\right|}t} \cos\left(\frac{\omega_n}{\sqrt{-\omega^{\alpha-2}\cos\frac{\alpha\pi}{2}}}\sqrt{1-\frac{\omega^{\alpha}\sin^2\frac{\alpha\pi}{2}}{4\omega_n^2\left|\cos\frac{\alpha\pi}{2}\right|}}t - \phi_1\right)\right]. \tag{8.44}$$

Figure 8.11 shows several illustrations of $g_1(t)$.

Step response $g_1(t)$ is also a time-frequency distribution. Thus,

$$g_1(t) = g_1(t,\omega). \tag{8.45}$$

Several illustrations of $g_1(t, \omega)$ are shown in Figure 8.12.

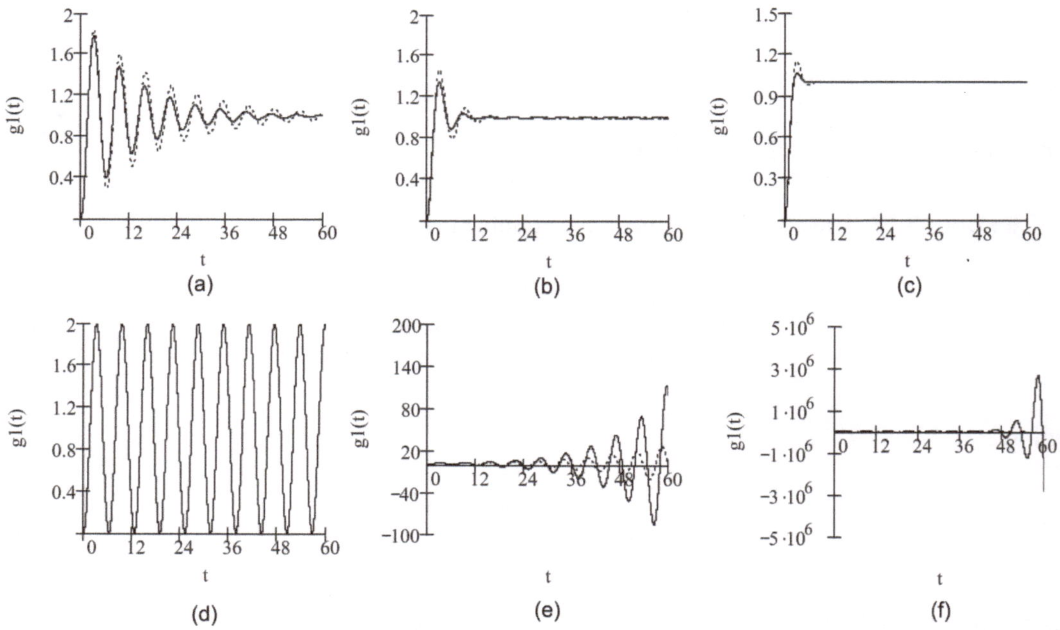

FIGURE 8.11 Illustrations of step response $g_1(t)$ for $k=1$ and $\omega_n=1$. (a) $g_1(t)$ when $\alpha=1.9$. Solid line: $\omega=1$ ($\zeta_{eq1}=0.08$). Dot line: $\omega=0.7$ ($\zeta_{eq1}=0.04$). (b) $g_1(t)$ when $\alpha=1.6$. Solid line: $\omega=1$ ($\zeta_{eq1}=0.33$). Dot line: $\omega=0.7$ ($\zeta_{eq1}=0.16$). (c) $g_1(t)$ when $\alpha=1.3$. Solid line: $\omega=1$ ($\zeta_{eq1}=0.66$). Dot line: $\omega=0.7$ ($\zeta_{eq1}=0.42$). (d) $g_1(t)$ when $\alpha=2$. Solid line: $\omega=1$ ($\zeta_{eq1}=0$). Dot line: $\omega=0.7$ ($\zeta_{eq1}=0$). (e) $g_1(t)$ when $\alpha=2.1$. Solid line: $\omega=1$ ($\zeta_{eq1}=-0.08$). Dot line: $\omega=0.7$ ($\zeta_{eq1}=-0.04$) (self-vibration). (f) $g_1(t)$ when $\alpha=2.3$. Solid line: $\omega=1$ ($\zeta_{eq1}=-0.24$). Dot line: $\omega=0.7$ ($\zeta_{eq1}=-0.11$) (self-vibration).

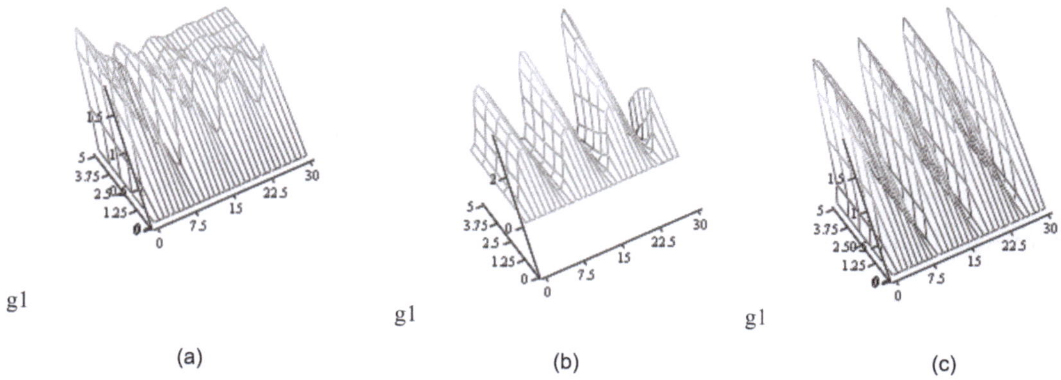

FIGURE 8.12 Illustrations of step response $g_1(t, \omega)$ in t-ω plane for $t=0, 1, \ldots, 30$ and $\omega=1, 2, \ldots, 5$, with $k=1$ and $\omega_n=7$. (a) $g_1(t, \omega)$ when $\alpha=1.95$ ($0.006 \le \zeta_{\text{eq1}} \le 0.027$). (b) $g_1(t, \omega)$ when $\alpha=2.01$ ($-0.006 \le \zeta_{\text{eq1}} \le -0.001$) (self-vibration). (c) $g_1(t, \omega)$ when $\alpha=2$ ($\zeta_{\text{eq1}}=0$).

8.6 FREQUENCY TRANSFER FUNCTION OF CLASS I FRACTIONAL VIBRATOR

Theorem 8.13 expresses the frequency transfer function of a fractional vibrator of class I.

Theorem 8.13. Let $H_1(\omega)$ be the frequency transfer function of a fractional vibrator of class I. Then,

$$H_1(\omega) = \frac{1}{k\left(1 - \frac{\omega^\alpha}{\omega_n^2}\left|\cos\frac{\alpha\pi}{2}\right| + i\frac{\omega^\alpha}{\omega_n^2}\sin\frac{\alpha\pi}{2}\right)}. \tag{8.46}$$

Proof. $H_1(\omega)$ is the Fourier transform of $h_1(t)$. Doing the Fourier transform on both sides of Eq. (8.34) produces

$$\left[m(i\omega)^\alpha + k\right]H_1(\omega) = 1. \tag{8.47}$$

The above implies

$$H_1(\omega) = \frac{1}{m\left[(i\omega)^\alpha + \omega_n^2\right]} = \frac{1}{k\left(1 - \frac{\omega^\alpha}{\omega_n^2}\left|\cos\frac{\alpha\pi}{2}\right| + i\frac{\omega^\alpha}{\omega_n^2}\sin\frac{\alpha\pi}{2}\right)}.$$

The proof is complete. □

Another expression of $H_1(\omega)$ is given by

$$H_1(\omega) = \frac{1}{k\left(1 - \gamma_{\text{eq1}}^2 + i2\varsigma_{\text{eq1}}\gamma_{\text{eq1}}\right)}. \tag{8.48}$$

The amplitude-frequency response is given by

$$\left|H_1(\omega)\right| = \frac{1/k}{\sqrt{\left(1-\frac{\omega^\alpha}{\omega_n^2}\left|\cos\frac{\alpha\pi}{2}\right|\right)^2 + \left(\frac{\omega^\alpha}{\omega_n^2}\sin\frac{\alpha\pi}{2}\right)^2}}. \tag{8.49}$$

Equivalently, that is expressed by

$$\left|H_1(\omega)\right| = \frac{1/k}{\sqrt{\left(1-\gamma_{\text{eq1}}^2\right)^2 + \left(2\varsigma_{\text{eq1}}\gamma_{\text{eq1}}\right)^2}}. \tag{8.50}$$

The phase is given by

$$\varphi_1(\omega) = \tan^{-1}\frac{\omega^\alpha \sin\frac{\alpha\pi}{2}}{\omega_n^2 - \omega^\alpha\left|\cos\frac{\alpha\pi}{2}\right|}. \tag{8.51}$$

For the purpose of computation of $\varphi_1(\omega)$ with digital computers, we use the expression in the form

$$\varphi_1(\omega) = \cos^{-1}\frac{1-\gamma_{\text{eq1}}^2}{\sqrt{\left(1-\gamma_{\text{eq1}}^2\right)^2 + \left(2\varsigma_{\text{eq1}}\gamma_{\text{eq1}}\right)^2}}. \tag{8.52}$$

Figures 8.13 and 8.14 show illustrations of $\left|H_1(\omega)\right|$. Plots of phase-frequency response function are in Figures 8.15–8.17.

When $\alpha=2$, a fractional vibrator reduces to a conventional one. Figure 8.17 indicates its plots about $\left|H_1(\omega)\right|$ and $\varphi_1(\omega)$ for $\alpha=2$.

8.7 QUASIEQUIVALENT LOGARITHMIC DECREMENT AND Q FACTOR OF CLASS I FRACTIONAL VIBRATOR

8.7.1 Quasiequivalent Logarithmic Decrement in Fractional Vibrators of Class I

We present a theorem below to express the quasiequivalent logarithmic decrement of $x_1(t)$.

Theorem 8.14. Let t_i and t_{i+1} be two successive time points of free response $x_1(t)$, where $x_1(t)$ reaches its successive peak values of $x_1(t_i)$ and $x_1(t_{i+1})$, respectively. Let Δ_{eq1} be the quasiequivalent logarithmic decrement of $x_1(t)$. Then,

$$\Delta_{\text{eq1}} = \Delta_{\text{eq1}}(\omega,\alpha) = \frac{\pi}{\sqrt{1-\left(\frac{\omega^{\frac{\alpha}{2}}\sin\frac{\alpha\pi}{2}}{2\omega_n\sqrt{-\cos\frac{\alpha\pi}{2}}}\right)^2}}\frac{\omega^{\frac{\alpha}{2}}\sin\frac{\alpha\pi}{2}}{\omega_n\sqrt{-\cos\frac{\alpha\pi}{2}}}, \quad 1<\alpha<3. \tag{8.53}$$

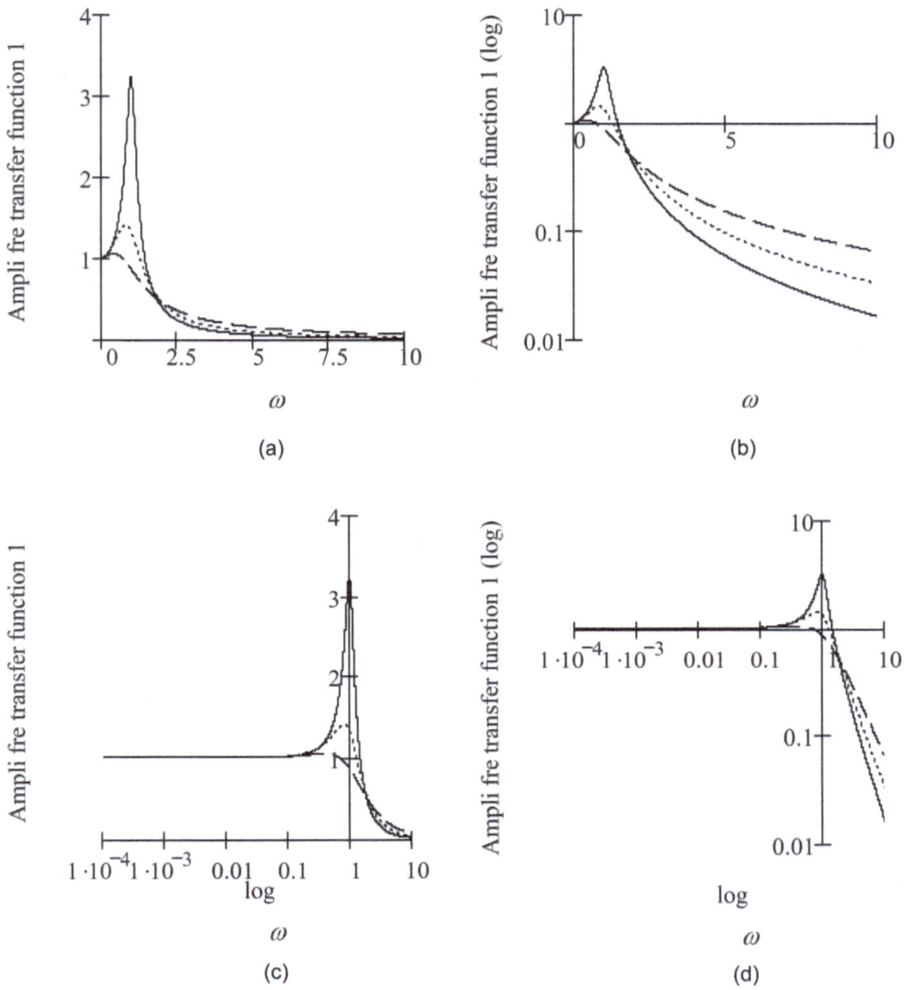

FIGURE 8.13 Illustrations of amplitude-frequency response $|H_1(\omega)|$ for $\omega_n = 1$, $m=1$, and $k=1$ with $\alpha=1.8$ (solid), 1.5 (dot), 1.2 (dash). (a) $|H_1(\omega)|$. (b) $|H_1(\omega)|$ in log-scale. (c) $|H_1(\omega)|$ with ω in log-scale. (d) $|H_1(\omega)|$ in log-log scale.

Proof. According to the definition of logarithmic decrement, we have

$$\Delta_{\text{eq1}} = \ln\frac{x_1(t_i)}{x_1(t_{i+1})} = \frac{2\pi\varsigma_{\text{eq1}}}{\sqrt{1-\varsigma_{\text{eq1}}^2}}. \tag{8.54}$$

Substituting ζ_{eq1} into the above yields Eq. (8.53). The proof is complete. □

Note that conventional logarithmic decrement is dimensionless. However, Δ_{eq1} is not. Hence, the prefix quasi is utilized. In addition, conventionally, logarithmic decrement is a constant. Nonetheless, $\Delta_{\text{eq1}}(\omega,\alpha)$ is a function of ω and α. When $\alpha=2$, $\Delta_{\text{eq1}}(\omega,2)=0$. Its plots are shown in Figure 8.18. Note that negative Δ_{eq1} in Figure 8.18b means self-vibration.

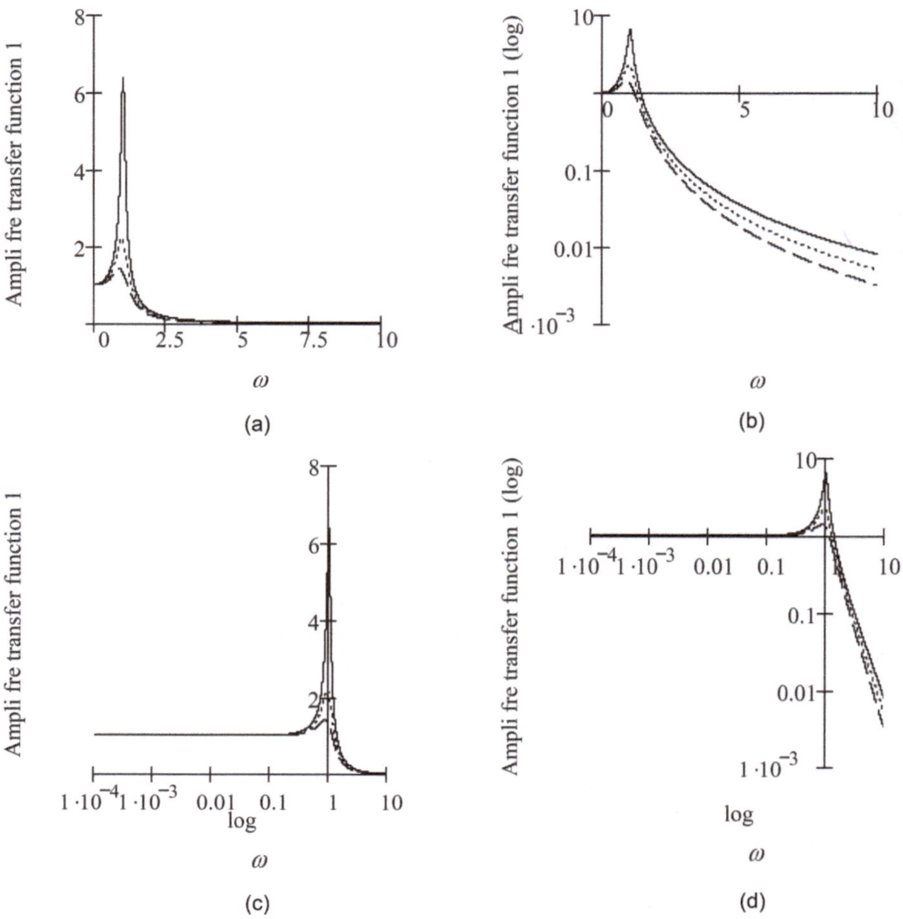

FIGURE 8.14 Illustrations of frequency transfer function $|H_1(\omega)|$ for $\omega_n=1$, $m=1$, and $k=1$ with $\alpha=2.1$ (solid), 2.3 (dot), 2.5 (dash). (a) $|H_1(\omega)|$. (b) $|H_1(\omega)|$ in log-scale. (c) $|H_1(\omega)|$ with ω in log-scale. (d) $|H_1(\omega)|$ in log-log scale.

8.7.2 Quasiequivalent Q Factor in Fractional Vibrators of Class I

The expression of Q factor of a fractional vibrator of class I is proposed below.

Theorem 8.15. Let Q_{eq1} be the quasiequivalent Q factor of a fractional vibrator of class I. Then,

$$Q_{\text{eq1}}=Q_{\text{eq1}}(\omega,\alpha)=\frac{\omega_n\sqrt{-\cos\frac{\alpha\pi}{2}}}{\omega^{\frac{\alpha}{2}}\sin\frac{\alpha\pi}{2}},\quad 1<\alpha<3. \tag{8.55}$$

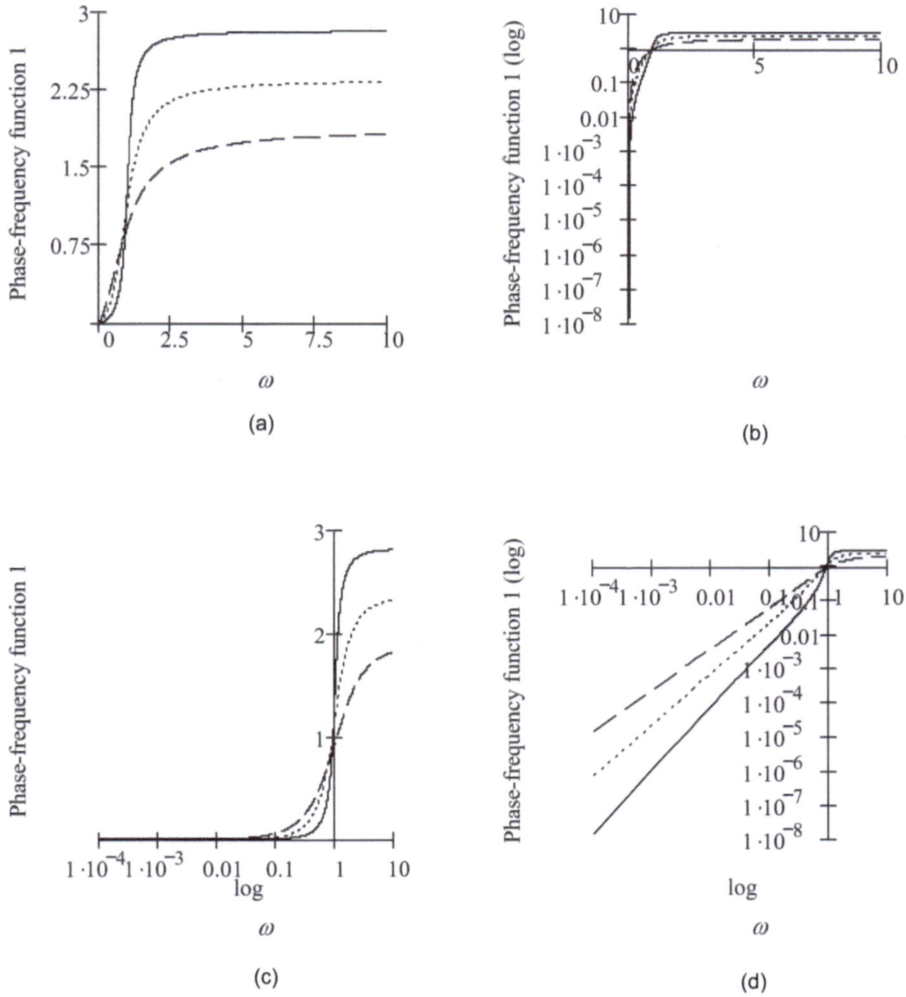

FIGURE 8.15 Plots of phase-frequency response function $\varphi_1(\omega)$ for $\omega_n=1$, $k=1$, and $m=1$ with $\alpha=1.8$ (solid), 1.5 (dot), 1.2 (dash). (a) $\varphi_1(\omega)$. (b) $\varphi_1(\omega)$ in log-scale. (c) $\varphi_1(\omega)$ with ω in log-scale. (d) $\varphi_1(\omega)$ in log-log scale.

Proof. According to the definition of Q factor, we have

$$Q_{\mathrm{eq1}} = \frac{1}{2\varsigma_{\mathrm{eq1}}}. \tag{8.56}$$

Substituting ζ_{eq1} into the above results in Eq. (8.55). This completes the proof. □

Note that Q_{eq1} is usually not a constant as that of a conventional one. Instead, it is a function with respect to ω and α. Besides, conventional Q is dimensionless but Q_{eq1} is not. If $\alpha=2$, $Q_{\mathrm{eq1}}(\omega,2)=\infty$. Figure 8.19 gives its plots, where negative Q_{eq1} implies self-vibration.

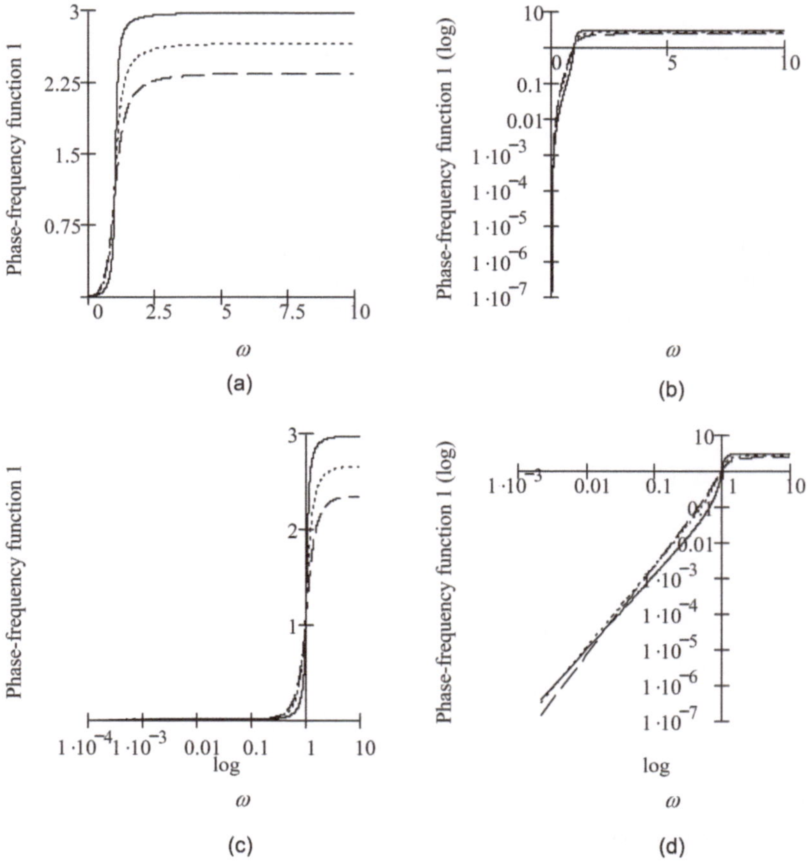

FIGURE 8.16 Plots of phase-frequency response function $\varphi_1(\omega)$ for $\omega_n=1$, $k=1$, and $m=1$ with $\alpha=2.1$ (solid), 2.3 (dot), 2.5 (dash). (a) $\varphi_1(\omega)$. (b) $\varphi_1(\omega)$ in log-scale. (c) $\varphi_1(\omega)$ with ω in log-scale. (d) $\varphi_1(\omega)$ in log-log scale.

8.8 GENERALIZED MITTAG-LEFFLER FUNCTIONS: A VIEW OF ELEMENTARY FUNCTIONS

8.8.1 Preliminaries

Following Achar et al. [2–4], Uchaikin [5, Chapter 7]), the free response, using the Mittage-Leffler functions, is expressed by

$$x_1(t)=x_{10}E_{\alpha,1}\left[-(\omega_n t)^\alpha\right]+v_{10}tE_{\alpha,2}\left[-(\omega_n t)^\alpha\right], \tag{8.57}$$

where $E_{a,\,b}(z)$ is the generalized Mittag-Leffler function given by

$$E_{a,b}(z)=\sum_{k=0}^{\infty}\frac{z^k}{\Gamma(ak+b)}, \quad a,b\in C,\ \mathrm{Re}(a)>0,\ \mathrm{Re}(b)>0. \tag{8.58}$$

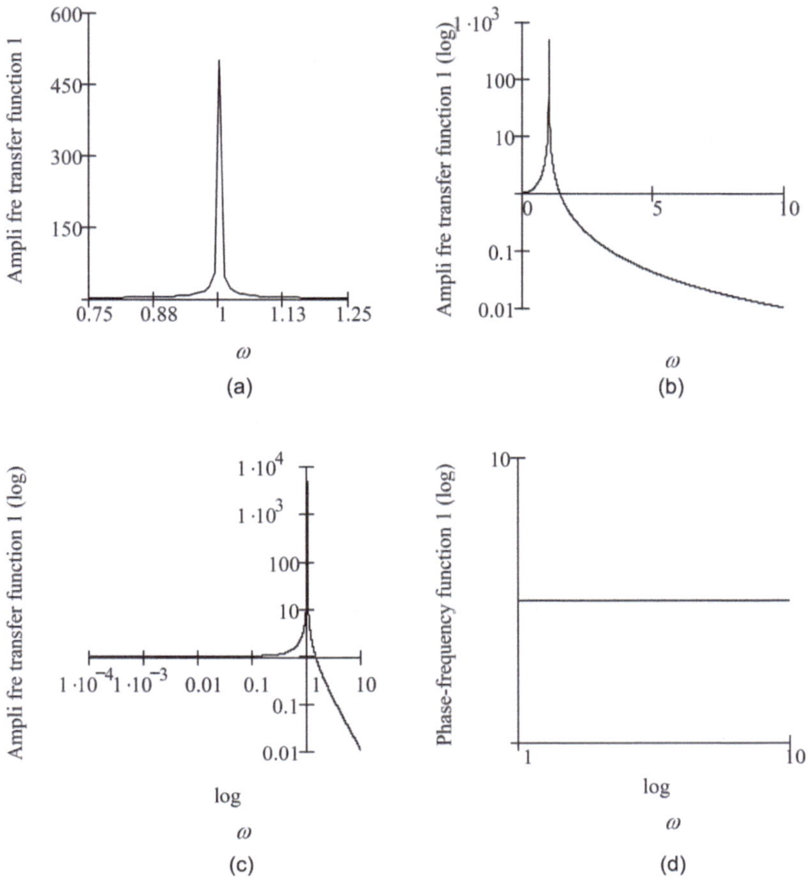

FIGURE 8.17 Plots of $|H_1(\omega)|$ and $\varphi_1(\omega)$ for $\omega_n = 1$, $k=1$, and $m=1$, and $\alpha=2$. (a) $|H_1(\omega)|$. (b) $|H_1(\omega)|$ in log-scale. (c) $|H_1(\omega)|$ in log-log scale. (d) $\varphi_1(\omega)$ in log-log scale.

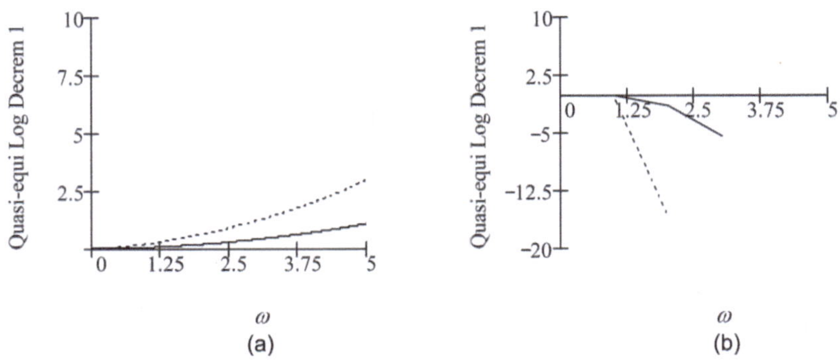

FIGURE 8.18 Plots of equivalent logarithmic decrement Δ_{eq1} of free response of a fractional vibrator of class I for $\omega_n = 10$. (a) Δ_{eq1} when $\alpha=1.9$ (solid), 1.6 (dot). (b) Δ_{eq1} when $\alpha=2.5$ (solid), 2.9 (dot). Note that "Quasi-equivalent logarithmic decrement 1" in the above figures stands for Δ_{eq1}.

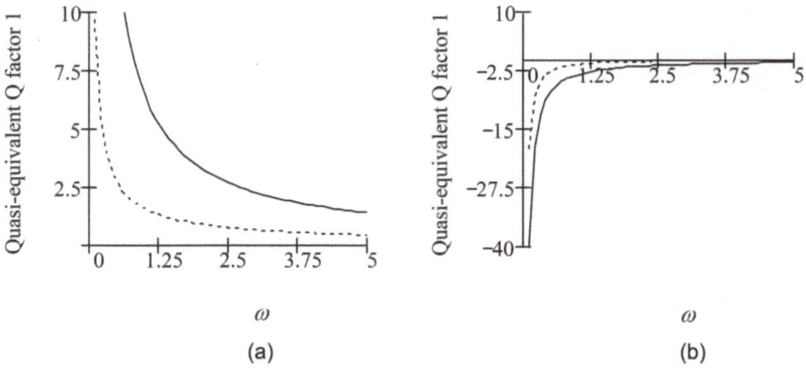

FIGURE 8.19 Plots of quasiequivalent Q factor Q_{eq1} of a fractional vibrator of class I for $\omega_n = 1$. (a) Q_{eq1} when $\alpha = 1.9$ (solid), 1.6 (dot). (b) Q_{eq1} when $\alpha = 2.2$ (solid), 2.6 (dot) (self-vibration). Note that "Quasi-equivalent Q factor 1" in the above figures stands for Q_{eq1}.

The Mittag-Leffler function denoted by $E_a(t)$ is in the form

$$E_a(z) = \sum_{k=0}^{\infty} \frac{z^k}{\Gamma(ak+1)}, \quad a \in C, \text{Re}(a) > 0, \tag{8.59}$$

referring Mathai and Haubold [6], or Gorenflo et al. [7], or Erdelyi et al. [8] for the Mittag-Leffler functions.

The impulse response to a fractional oscillator in Class I. Then (Uchaikin [5, Chapter 7]),

$$h_1(t) = t^{\alpha-1} E_{\alpha,\alpha}\left[-(\omega_n t)^{\alpha}\right]. \tag{8.60}$$

The step response to a fractional oscillator of Class I type is expressed by

$$g_1(t) = t^{\alpha} E_{\alpha,\alpha+1}\left[-(\omega_n t)^{\alpha}\right]. \tag{8.61}$$

8.8.2 Results

Theorem 8.16. The generalized Mittag-Leffler function in the form Eq. (8.57) can be expressed by the one in Eq. (8.32) for $1 < \alpha \leq 2$, $t \geq 0$. That is,

$$x_1(t)=x_{10}E_{\alpha,1}\left[-(\omega_n t)^\alpha\right]+v_{10}tE_{\alpha,2}\left[-(\omega_n t)^\alpha\right]$$

$$=e^{-\frac{\omega\sin\frac{\alpha\pi}{2}}{2\left|\cos\frac{\alpha\pi}{2}\right|}t}\left[\begin{array}{l} x_{10}\cos\left(\dfrac{\omega_n}{\sqrt{\omega^{\alpha-2}\left|\cos\frac{\alpha\pi}{2}\right|}}\sqrt{1-\dfrac{\omega^\alpha\sin^2\frac{\alpha\pi}{2}}{4\omega_n^2\left|\cos\frac{\alpha\pi}{2}\right|}}t\right) \\ +\left[\dfrac{v_{10}+\dfrac{\omega^{\frac{\alpha}{2}}\sin\frac{\alpha\pi}{2}}{2\omega_n\sqrt{\left|\cos\frac{\alpha\pi}{2}\right|}}x_{10}}{\sqrt{1-\dfrac{\omega^\alpha\sin^2\frac{\alpha\pi}{2}}{4\omega_n^2\left|\cos\frac{\alpha\pi}{2}\right|}}}\right]\sin\left(\dfrac{\omega_n}{\sqrt{\omega^{\alpha-2}\left|\cos\frac{\alpha\pi}{2}\right|}}\sqrt{1-\dfrac{\omega^\alpha\sin^2\frac{\alpha\pi}{2}}{4\omega_n^2\left|\cos\frac{\alpha\pi}{2}\right|}}t\right)\end{array}\right]. \tag{8.62}$$

Proof. Since Eqs. (8.32) and (8.57) are the solutions to the same equation for $1<\alpha\leq 2$, $t\geq 0$, they are equal. □

Theorem 8.17. The generalized Mittag-Leffler function given by

$$x_1(t)=x_{10}E_{\alpha,1}\left[-(\omega_n t)^\alpha\right], \tag{8.63}$$

can be expressed by the elementary functions, for $1<\alpha\leq 2$, $t\geq 0$, in the form

$$x_1(t)=x_{10}E_{\alpha,1}\left[-(\omega_n t)^\alpha\right]$$

$$=e^{-\frac{\omega\sin\frac{\alpha\pi}{2}}{2\left|\cos\frac{\alpha\pi}{2}\right|}t}\left[\begin{array}{l} x_{10}\cos\left(\dfrac{\omega_n}{\sqrt{\omega^{\alpha-2}\left|\cos\frac{\alpha\pi}{2}\right|}}\sqrt{1-\dfrac{\omega^\alpha\sin^2\frac{\alpha\pi}{2}}{4\omega_n^2\left|\cos\frac{\alpha\pi}{2}\right|}}t\right) \\ +\left(\dfrac{\dfrac{\omega^{\frac{\alpha}{2}}\sin\frac{\alpha\pi}{2}}{2\omega_n\sqrt{\left|\cos\frac{\alpha\pi}{2}\right|}}x_{10}}{\sqrt{1-\dfrac{\omega^\alpha\sin^2\frac{\alpha\pi}{2}}{4\omega_n^2\left|\cos\frac{\alpha\pi}{2}\right|}}}\right)\sin\left(\dfrac{\omega_n}{\sqrt{\omega^{\alpha-2}\left|\cos\frac{\alpha\pi}{2}\right|}}\sqrt{1-\dfrac{\omega^\alpha\sin^2\frac{\alpha\pi}{2}}{4\omega_n^2\left|\cos\frac{\alpha\pi}{2}\right|}}t\right)\end{array}\right]. \tag{8.64}$$

Proof. If $v_{10}=0$ in Eqs. (8.32) and (8.62), (8.64) results. The proof completes. □

If $x_{10}=0$ in Eqs. (8.57) and (8.62), we obtain another result as follows.

Theorem 8.18. The generalized Mittag-Leffler function expressed by

$$x_1(t)=v_{10}tE_{\alpha,2}\left[-(\omega_n t)^\alpha\right], \tag{8.65}$$

can be represented, for $1<\alpha\le 2$, $t\ge 0$, by the elementary functions in the form

$$x_1(t)=v_{10}tE_{\alpha,2}\left[-(\omega_n t)^\alpha\right]$$
$$=e^{-\frac{\omega\sin\frac{\alpha\pi}{2}}{2\left|\cos\frac{\alpha\pi}{2}\right|}t}\left[\frac{v_{10}\sin\left(\frac{\omega_n}{\sqrt{\omega^{\alpha-2}\left|\cos\frac{\alpha\pi}{2}\right|}}\sqrt{1-\frac{\omega^\alpha\sin^2\frac{\alpha\pi}{2}}{4\omega_n^2\left|\cos\frac{\alpha\pi}{2}\right|}}\,t\right)}{\sqrt{1-\frac{\omega^\alpha\sin^2\frac{\alpha\pi}{2}}{4\omega_n^2\left|\cos\frac{\alpha\pi}{2}\right|}}}\right]. \tag{8.66}$$

Proof. When $x_{10}=0$ in Eqs. (8.62), (8.66) result. The proof finishes. □

Theorem 8.19. For $1<\alpha\le 2$, $t\ge 0$, the generalized Mittag-Leffler function in the form Eq. (8.60) can be expressed by the elementary functions in Theorem 8.11 in the form

$$t^{\alpha-1}E_{\alpha,\alpha}\left[-(\omega_n t)^\alpha\right]=\frac{e^{-\frac{\omega\sin\frac{\alpha\pi}{2}}{2\left|\cos\frac{\alpha\pi}{2}\right|}t}\sin\frac{\omega_n\sqrt{1-\frac{\omega^{2\alpha}\sin^2\frac{\alpha\pi}{2}}{4\omega_n^2\left|\cos\frac{\alpha\pi}{2}\right|}}}{\sqrt{\omega^{\alpha-2}\left|\cos\frac{\alpha\pi}{2}\right|}}t}{m\omega_n\sqrt{\omega^{\alpha-2}\left|\cos\frac{\alpha\pi}{2}\right|}\sqrt{1-\frac{\omega^{2\alpha}\sin^2\frac{\alpha\pi}{2}}{4\omega_n^2\left|\cos\frac{\alpha\pi}{2}\right|}}}. \tag{8.67}$$

Proof. As Eqs. (8.60) and (8.37) are the solutions to the same equation for $1<\alpha\le 2$, $t\ge 0$, they are equal. This finishes the proof. □

Theorem 8.20. For $1<\alpha\le 2$, $t\ge 0$, the generalized Mittag-Leffler function expressed by Eq. (8.61) can be represented by using the elementary functions described in Theorem 8.12. To be precise, we have

$$t^{\alpha}E_{\alpha,\alpha+1}\left[-(\omega_n t)^{\alpha}\right]$$
$$=\frac{1}{k}\left[1-\frac{e^{-\frac{\omega\sin\frac{\alpha\pi}{2}}{2\left|\cos\frac{\alpha\pi}{2}\right|}t}\cos\left(\frac{\omega_n}{\sqrt{-\omega^{\alpha-2}\cos\frac{\alpha\pi}{2}}}\sqrt{1-\frac{\omega^{\alpha}\sin^2\frac{\alpha\pi}{2}}{4\omega_n^2\left|\cos\frac{\alpha\pi}{2}\right|}}t-\phi_1\right)}{\sqrt{1-\left(\frac{\omega^{\frac{\alpha}{2}}\sin\frac{\alpha\pi}{2}}{2\omega_n\sqrt{-\cos\frac{\alpha\pi}{2}}}\right)^2}}\right], \tag{8.68}$$

where ϕ_1 is given by Eq. (8.41).

Proof. When $1<\alpha\leq 2$, $t\geq 0$, from Eqs. (8.40) and (8.61), we see that both sides of the above equation are the solutions to the same equation. Thus, the above holds. This completes the proof. □

8.9 VIEWING C_{EQ1} WITH RAYLEIGH DAMPING

Denote by c_r assumed by Rayleigh. It is in the form $c_r=am+bk$, where m and k are primary mass and stiffness, respectively, a is proportional to vibration frequency ω while b is inversely proportional to ω (Jin and Xia [9], Harris [10], Palley et al. [11]). From Eq. (8.10), we see that c_{eq1} stands for a variation of c_r. It is proportional to ω with $a=\omega^{\alpha-1}\sin\frac{\alpha\pi}{2}$ and $b=0$.

8.10 CONCLUDING REMARK

The present results in this chapter cannot be observed in ordinary vibrators. They are novel in the field.

8.11 EXERCISES

8.1. Prove that $\frac{d^{\alpha}x(t)}{dt^{\alpha}}$ for $1<\alpha<2$ or $2<\alpha<3$ does not stand for a standard acceleration, where $x(t)$ is a displacement.

8.2. If the unit of m is kg, prove that $m\frac{d^{\alpha}x(t)}{dt^{\alpha}}$ does not designate a standard inertia force for $1<\alpha<2$ or $2<\alpha<3$, where $x(t)$ is a displacement.

8.3. Find the ranges of ω and α for $0<\zeta_{eq1}<1$.

8.4. Prove that ζ_{eq1} is not dimensionless when $1<\alpha<2$ or $2<\alpha<3$.

8.5. Prove that ω_{eqn1} is not a standard angular frequency unless $\alpha=2$.

8.6. Write the expression of sinusoidal response to a class I fractional vibrator. (Hint: Refer to [1]).

REFERENCES

1. M. Li, Three classes of fractional oscillators, *Symmetry-Basel*, 10(2), 2018, 91 p.
2. B. N. N. Achar, J. W. Hanneken, and T. Clarke, Damping characteristics of a fractional oscillator, *Physica A*, 339(3–4), 2004, 311–319.
3. B. N. N. Achar, J. W. Hanneken, and T. Clarke, Response characteristics of a fractional oscillator, *Physica A*, 309(3–4), 2002, 275–288.
4. B. N. N. Achar, J. W. Hanneken, T. Enck, and T. Clarke, Dynamics of the fractional oscillator, *Physica A*, 297(3–4), 2001, 361–367.
5. V. V. Uchaikin, *Fractional Derivatives for Physicists and Engineers*, vol. II, Springer, Berlin, 2013.
6. A. M. Mathai and H. J. Haubold, *Special Functions for Applied Scientists*, Springer, Berlin, 2008.
7. R. Gorenflo, A. A. Kilbas, F. Mainardi, and S. V. Rogosin, *Mittag-Leffler Functions: Related Topics and Applications, Theory and Applications*, Springer, Berlin, 2014.
8. A. Erdelyi, W. Magnus, F. Oberhettinger, and F. G. Tricomi, *Higher Transcendental Functions*, vol. I, McGraw Hill, New York, 1955.
9. X. D. Jin and L. J. Xia, *Ship Hull Vibration*, The Press of Shanghai Jiaotong University, Shanghai, China, 2011. In Chinese.
10. C. M. Harris, *Shock and Vibration Handbook*, 5th Ed., McGraw-Hill, New York, 2002.
11. O. M. Palley, Г. B. Bahizov, and E. Я. Voroneysk, *Handbook of Ship Structural Mechanics*, National Defense Industry Publishing House, Beijing, China, 2002. In Chinese. Translated from Russian by B. H. Xu, X. Xu, and M. Q. Xu.

CHAPTER 9

Fractional Vibrations of Class II

THIS CHAPTER AIMS TO address the research regarding class II fractional vibrators in seven aspects. (1) Put forward the expression of an equivalent motion equation of a fractional vibrator of class II type in Theorem 9.1. (2) Bring on the concepts about the terms of quasiequivalent mass, quasiequivalent damping, quasiequivalent damping ratio, quasiequivalent damping free natural frequency, quasiequivalent damped natural frequency, quasiequivalent frequency ratio, quasiequivalent logarithmic decrement, and quasiequivalent Q factor of a fractional vibrator in class II. Their analytic expressions are presented in Theorems 9.2, 9.3, 9.6–9.9, 9.14, and 9.15, respectively. (3) Reveal that the quasiequivalent mass of a fractional vibrator in class II is in the range of $(-\infty, m)$, relying on vibration frequency ω and the fractional order β in Corollaries 9.1 and 9.2. (4) The condition for the equivalent mass being negative regarding a fractional vibrator in class II is proposed in Corollary 9.3. (5) We present Corollaries 9.4–9.7 to show that the quasiequivalent damping of a fractional vibrator in class II is positive definite. (6) Put forward a result that a fractional vibrator in class II is stable (Theorem 9.5). (7) Present a set of analytic expressions to represent free response, impulse response, step response, and frequency transfer function of a fractional vibrator of class II type in Theorems 9.10–9.13, respectively.

9.1 EQUIVALENT MOTION EQUATION OF CLASS II FRACTIONAL VIBRATOR

We address a fractional vibrator of class II starting with its equivalent motion equation in Theorem 9.1.

Theorem 9.1. For a fractional vibrator of class II

$$m\frac{d^2y(t)}{dt^2}+c\frac{d^\beta y(t)}{dt^\beta}+ky(t)=0, \quad 0<\beta<2, \tag{9.1}$$

DOI: 10.1201/9781003460947-11

its equivalent motion equation is in the form

$$\left(m-c\omega^{\beta-2}\cos\frac{\beta\pi}{2}\right)\frac{d^2x_2(t)}{dt^2}+\left(c\omega^{\beta-1}\sin\frac{\beta\pi}{2}\right)\frac{dx_2(t)}{dt}+kx_2(t)=0,\quad 0<\beta<2. \tag{9.2}$$

Proof. Doing the Fourier transform of the left side of Eq. (9.1) produces

$$\begin{aligned}\left[m\frac{d^2y(t)}{dt^2}+c\frac{d^\beta y(t)}{dt^\beta}+ky(t)\right]&\leftrightarrow\left[-m\omega^2+c(i\omega)^\beta+k\right]Y(\omega)\\&=\left[-m\omega^2+c\left(\cos\frac{\beta\pi}{2}+i\sin\frac{\beta\pi}{2}\right)\omega^\beta+k\right]Y(\omega)\\&=\left[-m\omega^2+c\omega^\beta\cos\frac{\beta\pi}{2}+ic\omega^\beta\sin\frac{\beta\pi}{2}+k\right]Y(\omega),\end{aligned} \tag{9.3}$$

where $Y(\omega)$ the Fourier transform of $y(t)$. On the other hand, if doing the Fourier transform of the left side of Eq. (9.2), we have

$$\begin{aligned}&\left[\left(m-c\omega^{\beta-2}\cos\frac{\beta\pi}{2}\right)\frac{d^2x_2(t)}{dt^2}+\left(c\omega^{\beta-1}\sin\frac{\beta\pi}{2}\right)\frac{dx_2(t)}{dt}+kx_2(t)\right]\\&\leftrightarrow\left[-m\omega^2+c\omega^\beta\cos\frac{\beta\pi}{2}+ic\omega^\beta\sin\frac{\beta\pi}{2}+k\right]X_2(\omega),\end{aligned} \tag{9.4}$$

where $X_2(\omega)$ is the Fourier transform of $x_2(t)$. Comparing the right side of Eq. (9.3) to that of Eq. (9.4) results in

$$Y(\omega)=X_2(\omega). \tag{9.5}$$

Thus,

$$y(t)=x_2(t). \tag{9.6}$$

Therefore, Theorem 9.1 holds. The proof is finished. □

9.2 QUASIEQUIVALENT VIBRATION PARAMETERS OF CLASS II FRACTIONAL VIBRATOR

9.2.1 Quasiequivalent Mass of Fractional Vibrators of Class II

Let m_{eq2} be the quasiequivalent mass of a class II fractional vibrator. We present the following theorem to describe its expression.

Theorem 9.2. The expression of m_{eq2} is in the form

$$m_{eq2} = m_{eq2}(\omega, \beta) = m - c\omega^{\beta-2}\cos\frac{\beta\pi}{2}, \quad 0 < \beta < 2. \tag{9.7}$$

Proof. From Eq. (9.2), we see that $\left(m - c\omega^{\beta-2}\cos\frac{\beta\pi}{2}\right)\frac{d^2x_2(t)}{dt^2}$ is inertia force according to Newton's second law. Hence, Theorem 9.2 is valid. The proof is finished. □

Note that the unit of m_{eq2} is not kg unless $\beta = 1$ or $\omega \to \infty$. When $\beta = 1$ or $\omega \to \infty$, $m_{eq2} = m$. Thus, we use the prefix quasi for m_{eq2}.

The quantity m_{eq2} relates to ω and β. It consists of two parts. One is the primary mass m and the other $-c\omega^{\beta-2}\cos\frac{\beta\pi}{2}$. The latter is contributed by fractional friction force $c\frac{d^\beta y(t)}{dt^\beta}$. In other words, fractional friction force may produce a certain added mass. Taking into account $\frac{c}{m} = 2\varsigma\omega_n$, we rewrite Eq. (9.7) by

$$m_{eq2} = m\left(1 - 2\varsigma\omega_n\omega^{\beta-2}\cos\frac{\beta\pi}{2}\right). \tag{9.8}$$

Now, we discuss interesting asymptotic properties regarding m_{eq2}.

Corollary 9.1: Asymptotic Property for $\omega \to \infty$. For m_{eq2}, we have

$$\lim_{\omega\to\infty} m_{eq2}(\omega, \beta, m) = m. \tag{9.9}$$

Proof. In Eq. (9.8), we have $\lim_{\omega\to\infty}\omega^{\beta-2} = 0$, which yields the above. The proof is finished. □

Corollary 9.2: Asymptotic Property for $\omega \to 0$. For m_{eq2}, we have

$$\lim_{\omega\to 0} m_{eq2}(\omega, \beta, m) = -\infty. \tag{9.10}$$

Proof. As $0 < \beta < 2$, we have $\lim_{\omega\to 0}\omega^{\beta-2} = \infty$ in Eq. (9.9), which results in the above. The proof ends. □

Corollary 9.3: Negative Mass. The quantity m_{eq2} is negative if

$$2\varsigma\omega_n\omega^{\beta-2}\cos\frac{\beta\pi}{2} > 1. \tag{9.11}$$

Proof. When Eq. (9.11) holds, we have $\left(1 - 2\varsigma\omega_n\omega^{\beta-2}\cos\frac{\beta\pi}{2}\right) < 0$. Thus, $m_{eq2} < 0$ if Eq. (9.11) holds. The proof is finished. □

Note that fractional vibrations in class II provide a model of negative mass. The range of m_{eq2} is $(-\infty, m)$. That is a particularly interesting phenomenon.

Figure 9.1 gives several plots of m_{eq2}.

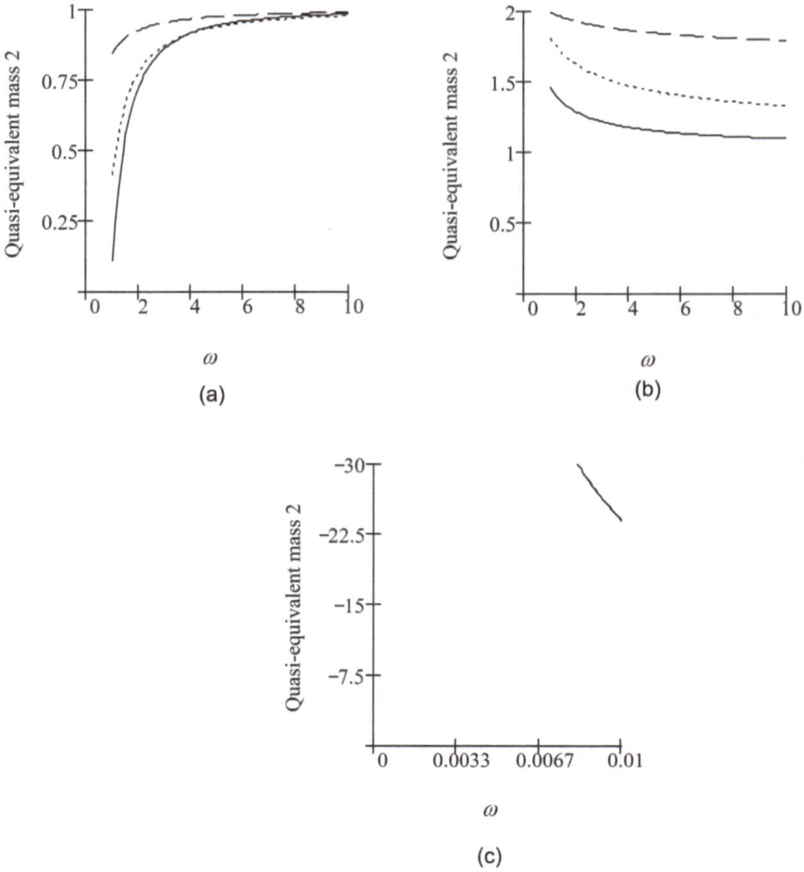

FIGURE 9.1 Plots of m_{eq2} with $m=1$, $c=1$, and $\omega_n=1$. (a) m_{eq2} for $\omega \geq 1$ and $\beta=0.3$ (solid), $\beta=0.6$ (dot), 0.9 (dash). (b) m_{eq2} for $\omega \geq 1$ and $\beta=1.3$ (solid), 1.6 (dot), 1.9 (dash). (c) m_{eq2} for $0.0001 \leq \omega \leq 0.01$ (negative mass). Note that "Quasi-equivalent mass 2" in the above figures stands for m_{eq2}.

9.2.2 Quasiequivalent Damping in Fractional Vibrators of Class II

Denote by c_{eq2} the quasiequivalent damping for a fractional vibrator of class II type. The proposed theorem below describes its expression.

Theorem 9.3. c_{eq2} is expressed by

$$c_{eq2} = c_{eq2}(\omega, \beta) = c\omega^{\beta-1} \sin\frac{\beta\pi}{2}, \quad 0 < \beta < 2. \tag{9.12}$$

Proof. The friction force in Eq. (9.2) is $\left(c\omega^{\beta-1} \sin\frac{\beta\pi}{2}\right)\frac{dx_2(t)}{dt}$. Thus, the above holds. The proof is finished. □

We call c_{eq2} quasiequivalent damping because its unit is $\mathrm{N} \times \mathrm{m}^{-1} \times \mathrm{s} \times (\mathrm{rad/s})^{\beta-1}$ instead of standard unit of damping. It reduces to the primary damping c when $\beta=1$.

Corollary 9.4. c_{eq2} is positive definite for $0<\beta<2$, i.e.,

$$c_{eq2} \geq 0. \tag{9.13}$$

Proof. Due to $\sin\frac{\beta\pi}{2}>0$ if $0<\beta<2$, the above holds. The proof is finished. □

Theorem 9.4. $\beta=2$ is not allowed for $c_{eq2}\neq 0$.

Proof. If $\beta=2$, $c_{eq2}=0$. Thus, Theorem 9.4 results. The proof is finished. □

As a matter of fact, if $\beta=2$, $\frac{d^{\beta}y(t)}{dt^{\beta}}$ becomes acceleration instead of fractional velocity. Note that fractional friction force vanishes if $c_{eq2}=0$. In that case, a class II fractional vibrator reduces to a standard vibrator of order 2 without damping.

Theorem 9.5. A fractional vibrator of class II type is stable if $\omega\neq 0$.

Proof. Since $c_{eq2}\geq 0$, a fractional vibrator of class II type is never self-vibrated. Accordingly, it is stable. The proof is finished. □

Corollary 9.5. Asymptotically, if $0<\beta<1$,

$$\lim_{\omega\to 0} c_{eq2}(\omega,\beta)=\infty. \tag{9.14}$$

In addition,

$$\lim_{\omega\to \infty} c_{eq2}(\omega,\beta)=0. \tag{9.15}$$

Proof. As $\sin\frac{\beta\pi}{2}>0$ for $0<\beta<1$ and $\lim_{\omega\to 0}\omega^{\beta-1}=\infty$, Eq. (9.14) is sound. Since $\lim_{\omega\to\infty}\omega^{\beta-1}=0$, Eq. (9.15) is valid. This completes the proof. □

Corollary 9.6. When $1<\beta<2$,

$$\lim_{\omega\to 0} c_{eq2}(\omega,\beta)=0. \tag{9.16}$$

In addition,

$$\lim_{\omega\to \infty} c_{eq2}(\omega,\beta)=\infty. \tag{9.17}$$

Proof. Since $\lim_{\omega\to 0}\omega^{\beta-1}=0$ for $1<\beta<2$, Eq. (9.16) is valid. Additionally, $\lim_{\omega\to\infty}\omega^{\beta-1}=\infty$ for $1<\beta<2$. Thus, Eq. (9.17) holds. The proof is complete. □

As a whole, the range of c_{eq2} is $(0, \infty)$. Its illustrations are shown in Figure 9.2.

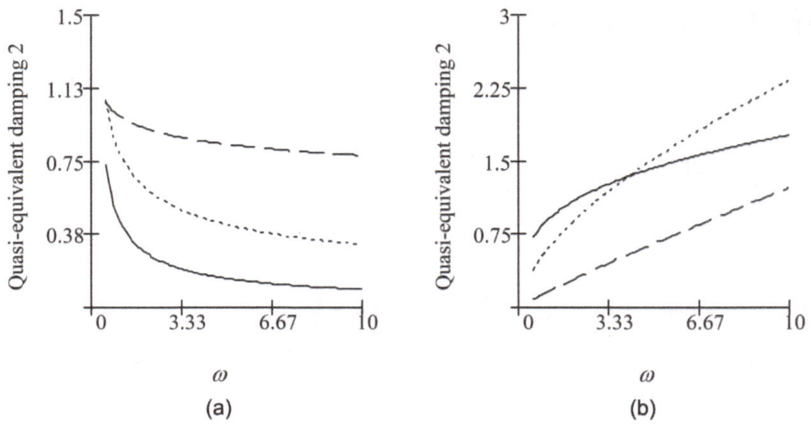

FIGURE 9.2 Plots of c_{eq2}. (a) c_{eq2} when $\beta=0.3$ (solid), 0.6 (dot), 0.9 (dash). (b) c_{eq2} when $\beta=1.3$ (solid), 1.6 (dot), 1.9 (dash). Note that "Quasi-equivalent damping 2" in the above figures stands for c_{eq2}.

9.2.3 Quasiequivalent Damping Ratio in Fractional Vibrators of Class II

Let ζ_{eq2} be the quasiequivalent damping ratio for a fractional vibrator of class II. Define it by

$$\varsigma_{eq2} = \frac{c_{eq2}}{2\sqrt{m_{eq2}k}}. \tag{9.18}$$

The presented theorem below is for its expression.

Theorem 9.6. ζ_{eq2} is given by

$$\varsigma_{eq2} = \varsigma_{eq2}(\omega, \beta) = \frac{\varsigma \omega^{\beta-1} \sin\frac{\beta\pi}{2}}{\sqrt{1 - \frac{c}{m}\omega^{\beta-2}\cos\frac{\beta\pi}{2}}}, \tag{9.19}$$

where $\varsigma = \frac{c}{2\sqrt{mk}}$.

Proof. Substituting m_{eq2} and c_{eq2} into Eq. (9.18) yields Eq. (9.19). The proof completes. □

Since ζ_{eq2} is not dimensionless, the prefix quasi is used for it. It is dimensionless if and only if $\beta=1$.

The asymptotic properties of ζ_{eq2} are presented below.

Corollary 9.7. When $0<\beta<2$, we have

$$\varsigma_{eq2} \geq 0. \tag{9.20}$$

Proof. Since $c_{eq2}>0$ for $0<\beta<2$, the above holds. The proof is finished. □

Figure 9.3 shows several plots of ζ_{eq2}.

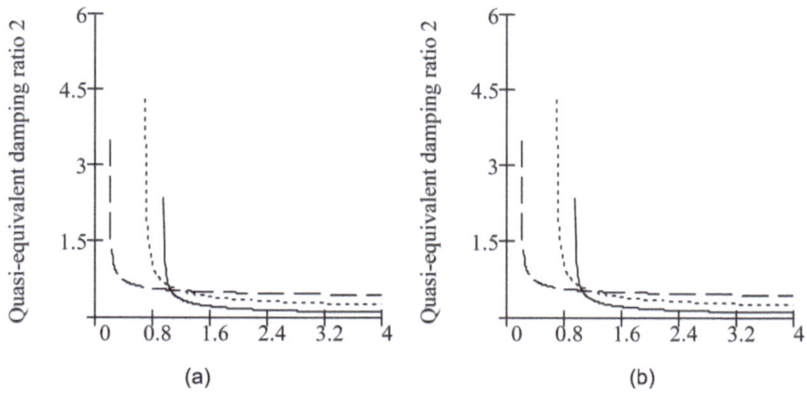

FIGURE 9.3 Plots of ζ_{eq2} with $m=1$, $c=1$, and $k=1$. (a) ζ_{eq2} for $\beta=0.3$ (solid), 0.6 (dot), 0.9 (dash). (b) ζ_{eq2} for $\beta=1.3$ (solid), 1.6 (dot), 1.9 (dash). Note that "Quasi-equivalent damping ratio 2" in the above figures stands for ζ_{eq2}.

9.2.4 Quasiequivalent Damping Free Natural Frequency of Fractional Vibrators of Class II

Let ω_{eqn2} be the quasiequivalent damping-free natural frequency for a fractional vibrator of class II. It is defined by

$$\omega_{eqn2} = \sqrt{\frac{k}{m_{eq2}}}. \tag{9.21}$$

Theorem 9.7. ω_{eqn2} is expressed by

$$\omega_{eqn2} = \omega_{eqn2}(\omega, \beta) = \frac{\omega_n}{\sqrt{1 - \frac{c}{m}\omega^{\beta-2}\cos\frac{\beta\pi}{2}}}. \tag{9.22}$$

Proof. Substituting m_{eq2} into Eq. (9.21) produces

$$\omega_{eqn2} = \sqrt{\frac{k}{m - c\omega^{\beta-2}\cos\frac{\beta\pi}{2}}} = \sqrt{\frac{k}{m\left(1 - \frac{c}{m}\omega^{\beta-2}\cos\frac{\beta\pi}{2}\right)}} = \frac{\omega_n}{\sqrt{1 - \frac{c}{m}\omega^{\beta-2}\cos\frac{\beta\pi}{2}}}.$$

This finishes the proof. □

Note that we describe ω_{eqn2} with the prefix quasi because its unit is not rad/s. Its unit is rad/s if and only $\beta=1$ or $\omega \to \infty$. As a matter of fact,

$$\omega_{eqn2}\big|_{\beta=1} = \omega_n. \tag{9.23}$$

Besides,

$$\lim_{\omega\to\infty} \omega_{eqn2} = \omega_n. \tag{9.24}$$

Several plots of $\omega_{eqn2}(\omega, \beta)$ are indicated in Figure 9.4.

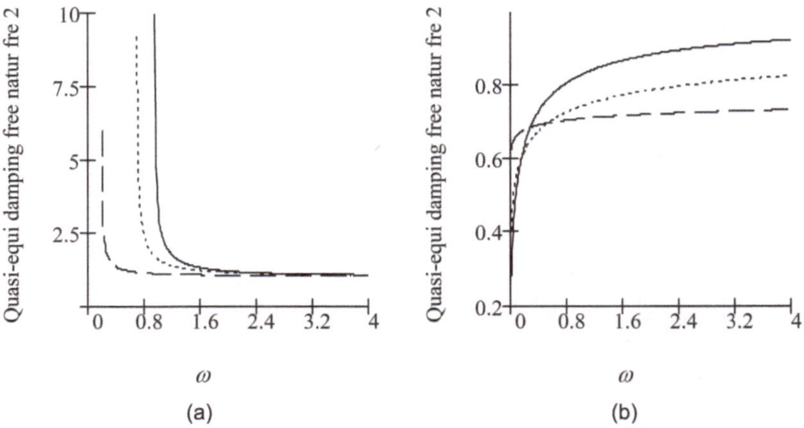

FIGURE 9.4 Plots of $\omega_{eqn2}(\omega, \beta)$ with $\omega_n=1$, $m=1$, and $c=1$. (a) ω_{eqn2} for $\beta=0.3$ (solid), 0.6 (dot), 0.9 (dash). (b) ω_{eqn2} for $\beta=1.3$ (solid), 1.6 (dot), 1.9 (dash). Note that "Quasi-equivalent damping free natural frequency 2" in the above figures stands for ω_{eqn2}.

9.2.5 Quasiequivalent Damped Natural Frequency of Fractional Vibrators of Class II

Without losing generality, small damping $|\zeta_{eq2}| \le 1$ is assumed in what follows. We introduce the term "quasiequivalent damped natural frequency" for a fractional vibrator of class II. It is denoted by ω_{eqd2} and defined by

$$\omega_{eqd2} = \omega_{eqn2}\sqrt{1-\zeta_{eq2}^2}, \quad |\zeta_{eq2}| \le 1. \tag{9.25}$$

Theorem 9.8. ω_{eqd2} is expressed by

$$\omega_{eqd2} = \omega_{eqd2}(\omega, \beta) = \frac{\omega_n}{\sqrt{1-\frac{c}{m}\omega^{\beta-2}\cos\frac{\beta\pi}{2}}}\sqrt{1-\frac{\varsigma^2\omega^{2(\beta-1)}\sin^2\frac{\beta\pi}{2}}{1-\frac{c}{m}\omega^{\beta-2}\cos\frac{\beta\pi}{2}}}. \tag{9.26}$$

Proof. Substituting ζ_{eq2} and ω_{eqn2} into Eq. (9.25) yields Eq. (9.26). The proof is complete. □

The unit of ω_{eqd2} is not rad/s unless $\beta=1$ or $\omega \to \infty$. Illustrations of ω_{eqd2} are shown in Figure 9.5.

9.2.6 Quasiequivalent Frequency Ratio of Fractional Vibrators of Class II

We introduce a quantity called "quasiequivalent frequency ratio" for a fractional vibrator of class II. It is denoted by γ_{eq2} and defined by

$$\gamma_{eq2} = \frac{\omega}{\omega_{eqn2}}. \tag{9.27}$$

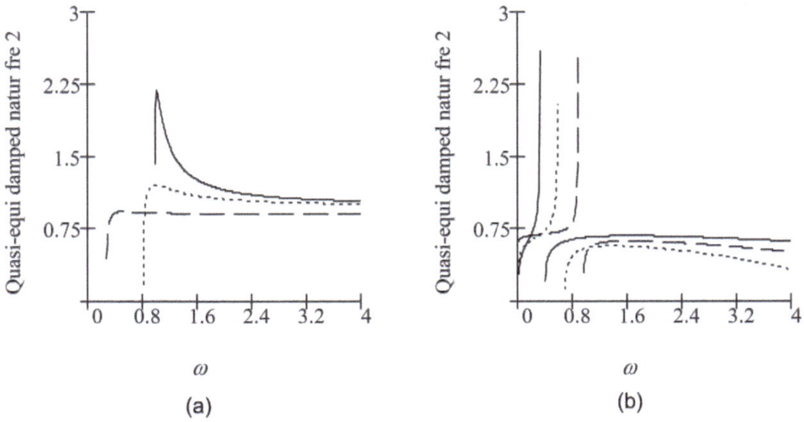

FIGURE 9.5 Illustrations of ω_{eqd2} with $m=1$, $c=1$, and $k=1$. (a) ω_{eqd2} for $\beta=0.3$ (solid), 0.6 (dot), 0.9 (dash). (b) ω_{eqd2} for $\beta=1.3$ (solid), 1.6 (dot), 1.9 (dash). Note that "Quasi-equivalent damped natural frequency 2" in the above figures stands for ω_{eqd2}.

Theorem 9.9. The parameter γ_{eq2} is given by

$$\gamma_{\mathrm{eq2}}=\gamma_{\mathrm{eq2}}(\omega,\beta)=\gamma\sqrt{1-\frac{c}{m}\omega^{\beta-2}\cos\frac{\beta\pi}{2}}, \tag{9.28}$$

where $\gamma=\dfrac{\omega}{\omega_n}$.

Proof. Substituting ω_{eqn2} into Eq. (9.27) results in Eq. (9.28). This completes the proof. □

Note that the unit of γ_{eq2} is not dimensionless. It reduces to being dimensionless when $\beta=1$. In fact, $\gamma_{\mathrm{eq2}}(\omega,1)=\gamma$. Hence, comes the meaning of "quasi". Figure 9.6 shows several plots of γ_{eq2}.

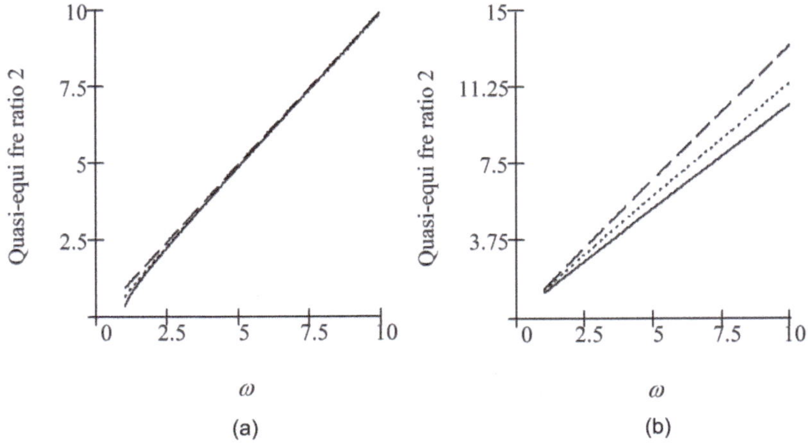

FIGURE 9.6 Plots of $\gamma_{\mathrm{eq2}}(\omega,\beta)$ with $\omega_n=1$, $m=1$, and $c=1$. (a) γ_{eq2} for $\beta=0.3$ (solid), 0.6 (dot), 0.9 (dash). (b) γ_{eq2} for $\beta=1.3$ (solid), 1.6 (dot), 1.9 (dash). Note that "Quasi-equivalent frequency ratio 2" in the above figures stands for γ_{eq2}.

9.3 FREE RESPONSE OF CLASS II FRACTIONAL VIBRATOR

Let $x_2(t)$ be the free response of a fractional vibrator of class II. It is a solution to the fractional differential equation given by

$$\begin{cases} m\dfrac{d^2x_2(t)}{dt^2}+c\dfrac{d^{\beta}x_2(t)}{dt^{\beta}}+kx_2(t)=0, \\ x_2(0)=x_{20}, x_2'(0)=v_{20}, \end{cases} \quad 0<\beta<2. \tag{9.29}$$

Theorem 9.10 gives its solution.

Theorem 9.10. The solution to the aforementioned equation is in the form

$$x_2(t)=e^{-\varsigma_{\mathrm{eq2}}\omega_{\mathrm{eqn2}}t}\left(x_{20}\cos\omega_{\mathrm{eqd2}}t+\frac{v_{20}+\varsigma_{\mathrm{eq2}}\omega_{\mathrm{eqn2}}x_{20}}{\omega_{\mathrm{eqd2}}}\sin\omega_{\mathrm{eqd2}}t\right), \quad t\geq 0. \tag{9.30}$$

Proof. According to Theorem 9.1, the solution to Eq. (9.29) is equivalent to the solution to the differential equation given by

$$\begin{cases} m_{\mathrm{eq2}}\dfrac{d^2x_2(t)}{dt^2}+c_{\mathrm{eq2}}\dfrac{dx_2(t)}{dt}+kx_2(t)=0, \\ x_2(0)=x_{20}, x_2'(0)=v_{20}, \end{cases} \quad 0<\beta<2. \tag{9.31}$$

As the above is a standard vibration equation, its solution is given by Eq. (9.30). The proof is complete. □

After substituting $\zeta_{\mathrm{eq2}}(\omega,\beta)$ and $\omega_{\mathrm{eqd2}}(\omega,\beta)$ into Eq. (9.30), we have

$$x_2(t)=e^{-\frac{\varsigma\omega_n\omega^{\beta-1}\sin\frac{\beta\pi}{2}}{1-\frac{c}{m}\omega^{\beta-2}\cos\frac{\beta\pi}{2}}t}\left[\begin{array}{l} x_{20}\cos\left(\dfrac{\omega_n}{\sqrt{\left(1-\frac{c}{m}\omega^{\beta-2}\cos\frac{\beta\pi}{2}\right)}}\sqrt{1-\dfrac{c^2\omega^{2(\beta-1)}\sin^2\frac{\beta\pi}{2}}{4\left(m-c\omega^{\beta-2}\cos\frac{\beta\pi}{2}\right)k}}\,t\right) \\ +\dfrac{v_{20}+\dfrac{c\omega^{\beta-1}\sin\frac{\beta\pi}{2}}{2\left(m-c\omega^{\beta-2}\cos\frac{\beta\pi}{2}\right)}x_{20}}{\omega_n\sqrt{1-\dfrac{c^2\omega^{2(\beta-1)}\sin^2\frac{\beta\pi}{2}}{4\left(m-c\omega^{\beta-2}\cos\frac{\beta\pi}{2}\right)k}}} \\ \sin\left(\dfrac{\omega_n}{\sqrt{\left(1-\frac{c}{m}\omega^{\beta-2}\cos\frac{\beta\pi}{2}\right)}}\sqrt{1-\dfrac{c^2\omega^{2(\beta-1)}\sin^2\frac{\beta\pi}{2}}{4\left(m-c\omega^{\beta-2}\cos\frac{\beta\pi}{2}\right)k}}\,t\right) \end{array}\right]. \tag{9.32}$$

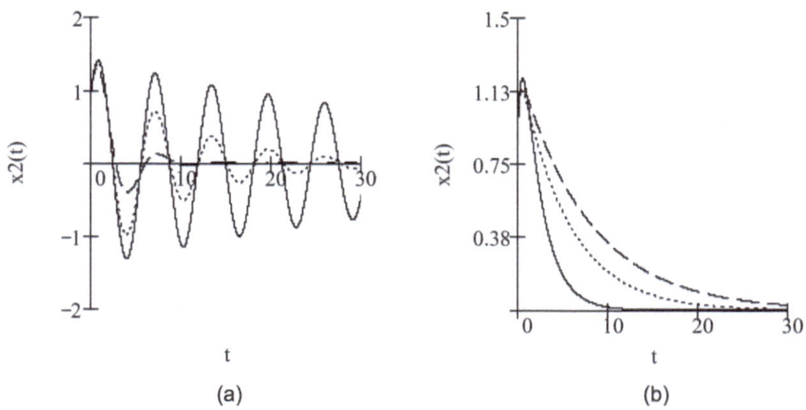

FIGURE 9.7 Illustrations of free response $x_2(t)$ for $x_{20}=1$, $v_{20}=1$, $m=1$, $c=1$, $k=1$, and $\omega=30$. (a) $x_2(t)$ when $\beta=0.3$ (solid), 0.6 (dot), 0.9 (dash). (b) $x_2(t)$ when $\beta=1.3$ (solid), 1.6 (dot), 1.9 (dash).

Several plots of $x_2(t)$ are shown in Figure 9.7.

In general, $x_2(t)$ is a time-frequency distribution on a time-frequency plane since m_{eq2} and c_{eq2} are functions of ω. Thus, it may be expressed by

$$x_2(t)=x_2(t,\omega). \tag{9.33}$$

9.4 IMPULSE RESPONSE OF CLASS II FRACTIONAL VIBRATOR

Denote by $h_2(t)$ the impulse response of a fractional vibrator of class II. It is a solution to the following fractional differential equation

$$m\frac{d^2h_2(t)}{dt^2}+c\frac{d^\beta h_2(t)}{dt^\beta}+kh_2(t)=\delta(t), \quad 0<\beta<2 \tag{9.34}$$

with zero initial conditions. The proposed theorem below expresses $h_2(t)$.

Theorem 9.11. The solution to the aforementioned equation is given by

$$h_2(t)=e^{-\varsigma_{\mathrm{eq2}}\omega_{\mathrm{eqn2}}t}\frac{1}{m_{\mathrm{eq2}}\omega_{\mathrm{eqd2}}}\sin\omega_{\mathrm{eqd2}}t, \quad t\geq 0. \tag{9.35}$$

Proof. The solution to Eq. (9.34) is equivalent to the solution to the following equation

$$m_{\mathrm{eq2}}\frac{d^2h_2(t)}{dt^2}+c_{\mathrm{eq2}}\frac{dh_2(t)}{dt}+kh_2(t)=\delta(t), \quad 0<\beta<2. \tag{9.36}$$

Since Eq. (9.36) is a standard vibration equation, its solution is given by Eq. (9.35). The proof is finished. □

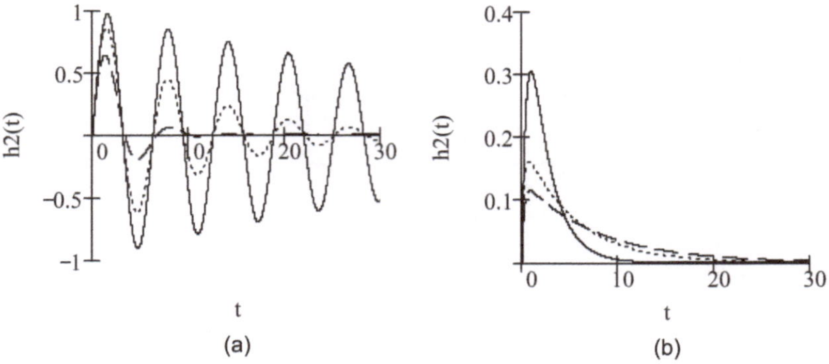

FIGURE 9.8 Illustrations of $h_2(t)$ for $m=1$, $c=1$, $k=1$, and $\omega=30$. (a) $h_2(t)$ for $\beta=0.3$ (solid), 0.6 (dot), 0.9 (dash). (b) $h_2(t)$ for $\beta=1.3$ (solid), 1.6 (dot), $\beta=1.9$ (dash).

When substituting m_{eq2}, ζ_{eq2}, ω_{eqn2}, and ω_{eqd2} into Eq. (9.35), we have

$$h_2(t)=\frac{e^{-\frac{\varsigma\omega_n\omega^{\beta-1}\sin\frac{\beta\pi}{2}}{1-\frac{c}{m}\omega^{\beta-2}\cos\frac{\beta\pi}{2}}t}\sin\dfrac{\omega_n\sqrt{1-\dfrac{\varsigma^2\omega^{2(\beta-1)}\sin^2\frac{\beta\pi}{2}}{1-\frac{c}{m}\omega^{\beta-2}\cos\frac{\beta\pi}{2}}}}{\sqrt{1-\frac{c}{m}\omega^{\beta-2}\cos\frac{\beta\pi}{2}}}t}{\omega_n m\sqrt{1-\frac{c}{m}\omega^{\beta-2}\cos\frac{\beta\pi}{2}}\sqrt{1-\dfrac{\varsigma^2\omega^{2(\beta-1)}\sin^2\frac{\beta\pi}{2}}{1-\frac{c}{m}\omega^{\beta-2}\cos\frac{\beta\pi}{2}}}},\quad t\geq 0. \tag{9.37}$$

Several illustrations of $h_2(t)$ are indicated in Figure 9.8.

$h_2(t)$ is a time-frequency distribution. It may be written by

$$h_2(t)=h_2(t,\omega). \tag{9.38}$$

9.5 UNIT STEP RESPONSE OF CLASS II FRACTIONAL VIBRATOR

Let $g_2(t)$ be the unit step response of a fractional vibrator of class II. It is the solution to

$$m\frac{d^2g_2(t)}{dt^2}+c\frac{d^\beta g_2(t)}{dt^\beta}+kg_2(t)=u(t),\quad 0<\beta<2 \tag{9.39}$$

with zero initial conditions. The following present theorem gives the solution.

Theorem 9.12. The solution to the above equation is expressed by

$$g_2(t)=\frac{1}{k}\left[1-\frac{e^{-\varsigma_{\text{eq2}}\omega_{\text{eqn2}}t}}{\sqrt{1-\varsigma_{\text{eq2}}^2}}\cos\left(\omega_{\text{eqd2}}t-\phi_2\right)\right],\quad t\geq 0, \tag{9.40}$$

where

$$\phi_2 = \tan^{-1}\frac{\varsigma_{\mathrm{eq2}}}{\sqrt{1-\varsigma_{\mathrm{eq2}}^2}}. \tag{9.41}$$

Proof. The solution to Eq. (9.39) equals to the solution to the following equation

$$m_{\mathrm{eq2}}\frac{d^2 g_2(t)}{dt^2}+c_{\mathrm{eq2}}\frac{dg_2(t)}{dt}+kg_2(t)=u(t), \quad 0<\beta<2. \tag{9.42}$$

The above is a standard vibration equation. Its unit step response is given by Eq. (9.40). This finishes the proof. □

If substituting $\zeta_{\mathrm{eq2}}(\omega,\beta)$, $\omega_{\mathrm{eqn2}}(\omega,\beta)$, and $\omega_{\mathrm{eqd2}}(\omega,\beta)$ into Eq. (9.40), we have

$$g_2(t)=\frac{1}{k}\left[1-\frac{e^{-\frac{\varsigma\omega_n\omega^{\beta-1}\sin\frac{\beta\pi}{2}}{1-\frac{c}{m}\omega^{\beta-2}\cos\frac{\beta\pi}{2}}t}}{\sqrt{1-\frac{\varsigma^2\omega^{2(\beta-1)}\sin^2\frac{\beta\pi}{2}}{1-\frac{c}{m}\omega^{\beta-2}\cos\frac{\beta\pi}{2}}}}\cos\left(\frac{\omega_n}{\sqrt{\left(1-\frac{c}{m}\omega^{\beta-2}\cos\frac{\beta\pi}{2}\right)}}\sqrt{1-\frac{\varsigma^2\omega^{2(\beta-1)}\sin^2\frac{\beta\pi}{2}}{1-\frac{c}{m}\omega^{\beta-2}\cos\frac{\beta\pi}{2}}}\,t-\phi_2\right)\right], \tag{9.43}$$

where

$$\phi_2=\tan^{-1}\frac{\dfrac{\varsigma\omega^{\beta-1}\sin\frac{\beta\pi}{2}}{\sqrt{1-\frac{c}{m}\omega^{\beta-2}\cos\frac{\beta\pi}{2}}}}{\sqrt{1-\dfrac{\varsigma^2\omega^{2(\beta-1)}\sin^2\frac{\beta\pi}{2}}{1-\frac{c}{m}\omega^{\beta-2}\cos\frac{\beta\pi}{2}}}}. \tag{9.44}$$

Illustrations of $g_2(t)$ are presented in Figure 9.9.

Step response $g_2(t)$ may be expressed by

$$g_2(t)=g_2(t,\omega). \tag{9.45}$$

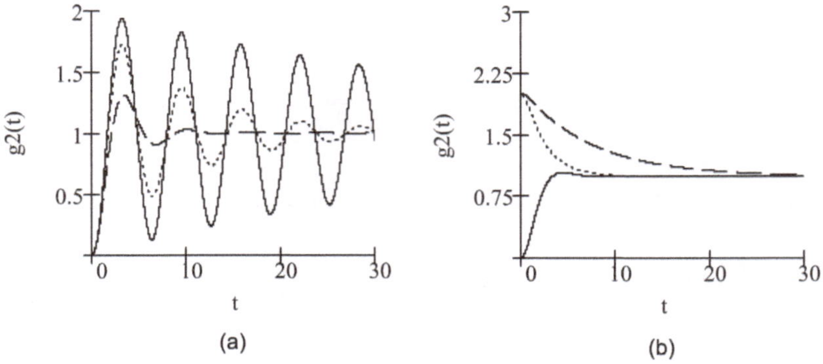

FIGURE 9.9 Illustrations of step response $g_2(t)$ for $m=1$, $c=1$, $k=1$, and $\omega=30$. (a) $g_2(t)$ for $\beta=0.3$ (solid), 0.6 (dot), 0.9 (dash). (b) $g_2(t)$ for $\beta=1.1$ (solid), 1.3 (dot), 1.5 (dash).\

9.6 FREQUENCY TRANSFER FUNCTION OF CLASS II FRACTIONAL VIBRATOR

We propose a theorem below to express the frequency transfer function of a fractional vibrator of class II.

Theorem 9.13. Denote by $H_2(\omega)$ the frequency transfer function of a fractional vibrator of class II. It is expressed by

$$H_2(\omega)=\frac{1/k}{1-\gamma^2\left(1-\frac{c}{m}\omega^{\beta-2}\cos\frac{\beta\pi}{2}\right)+i\frac{2\varsigma\omega^{\beta}\sin\frac{\beta\pi}{2}}{\omega_n}}, \tag{9.46}$$

where $\gamma=\frac{\omega}{\omega_n}$.

Proof. Denote by $H_2(\omega)$ the Fourier transform of $h_2(t)$, where $h_2(t)$ is the solution of the following equation

$$\left(m-c\omega^{\beta-2}\cos\frac{\beta\pi}{2}\right)\frac{d^2h_2(t)}{dt^2}+\left(c\omega^{\beta-1}\sin\frac{\beta\pi}{2}\right)\frac{dh_2(t)}{dt}+kh_2(t)=\delta(t).$$

Doing the Fourier transform on both sides of the above yields

$$\left[\left(m-c\omega^{\beta-2}\cos\frac{\beta\pi}{2}\right)(i\omega)^2+\left(c\omega^{\beta-1}\sin\frac{\beta\pi}{2}\right)(i\omega)+k\right]H_2(\omega)=1.$$

Dividing both sides of the above equation by m produces

$$\left[\left(1-\frac{c}{m}\omega^{\beta-2}\cos\frac{\beta\pi}{2}\right)(i\omega)^2+\left(\frac{c}{m}\omega^{\beta-1}\sin\frac{\beta\pi}{2}\right)(i\omega)+\frac{k}{m}\right]H_2(\omega)=\frac{1}{m}.$$

Thus, we have

$$\left[\left(\frac{c}{m}\omega^{\beta-2}\cos\frac{\beta\pi}{2}-1\right)\omega^2+i\left(\frac{c}{m}\omega^{\beta-1}\sin\frac{\beta\pi}{2}\right)\omega+\omega_n^2\right]H_2(\omega)=\frac{1}{m}.$$

Dividing both sides of the above by ω_n yields

$$\left[\left(\frac{c}{m}\omega^{\beta-2}\cos\frac{\beta\pi}{2}-1\right)\frac{\omega^2}{\omega_n^2}+i\left(2\varsigma\omega_n\omega^{\beta-1}\sin\frac{\beta\pi}{2}\right)\frac{\omega}{\omega_n^2}+1\right]H_2(\omega)=\frac{1}{m\omega_n^2}=\frac{1}{k}.$$

Simplifying the aforementioned results in

$$\left[\left(\frac{c}{m}\omega^{\beta-2}\cos\frac{\beta\pi}{2}-1\right)\gamma^2+i\frac{2\varsigma\omega^{\beta}\sin\frac{\beta\pi}{2}}{\omega_n}+1\right]H_2(\omega)=\frac{1}{k}.$$

The above is consistent with Eq. (9.46). The proof is complete. □

Since $\frac{c}{m}=2\varsigma\omega_n$, Eq. (9.46) may be rewritten by

$$H_2(\omega)=\frac{1/k}{1-\gamma^2\left(1-2\varsigma\omega_n\omega^{\beta-2}\cos\frac{\beta\pi}{2}\right)+i\frac{2\varsigma\omega^{\beta}\sin\frac{\beta\pi}{2}}{\omega_n}}. \tag{9.47}$$

Another expression of $H_2(\omega)$ is given by

$$H_2(\omega)=\frac{1}{k\left(1-\gamma_{\mathrm{eq2}}^2+i2\varsigma_{\mathrm{eq2}}\gamma_{\mathrm{eq2}}\right)}. \tag{9.48}$$

The amplitude-frequency response function is in the form

$$|H_2(\omega)|=\frac{1/k}{\sqrt{\left[1-\gamma^2\left(1-\frac{c}{m}\omega^{\beta-2}\cos\frac{\beta\pi}{2}\right)\right]^2+\left(\frac{2\varsigma\omega^{\beta}\sin\frac{\beta\pi}{2}}{\omega_n}\right)^2}}, \tag{9.49}$$

which equals to

$$|H_2(\omega)|=\frac{1/k}{\sqrt{\left(1-\gamma_{\mathrm{eq2}}^2\right)^2+\left(2\varsigma_{\mathrm{eq2}}\gamma_{\mathrm{eq2}}\right)^2}}. \tag{9.50}$$

The phase-frequency response function is expressed by

$$\varphi_2(\omega)=\tan^{-1}\frac{\dfrac{2\varsigma\omega^{\beta}\sin\dfrac{\beta\pi}{2}}{\omega_n}}{1-\gamma^2\left(1-\dfrac{c}{m}\omega^{\beta-2}\cos\dfrac{\beta\pi}{2}\right)}. \tag{9.51}$$

For facilitating the computation of $\varphi_2(\omega)$, we use

$$\varphi_2(\omega)=\cos^{-1}\frac{1-\gamma_{\text{eq2}}^2}{\sqrt{\left(1-\gamma_{\text{eq2}}^2\right)^2+\left(2\varsigma_{\text{eq2}}\gamma_{\text{eq2}}\right)^2}}. \tag{9.52}$$

Figures 9.10 and 9.11 show plots of $|H_2(\omega)|$. Plots of $\varphi_2(\omega)$ are in Figures 9.12 and 9.13.

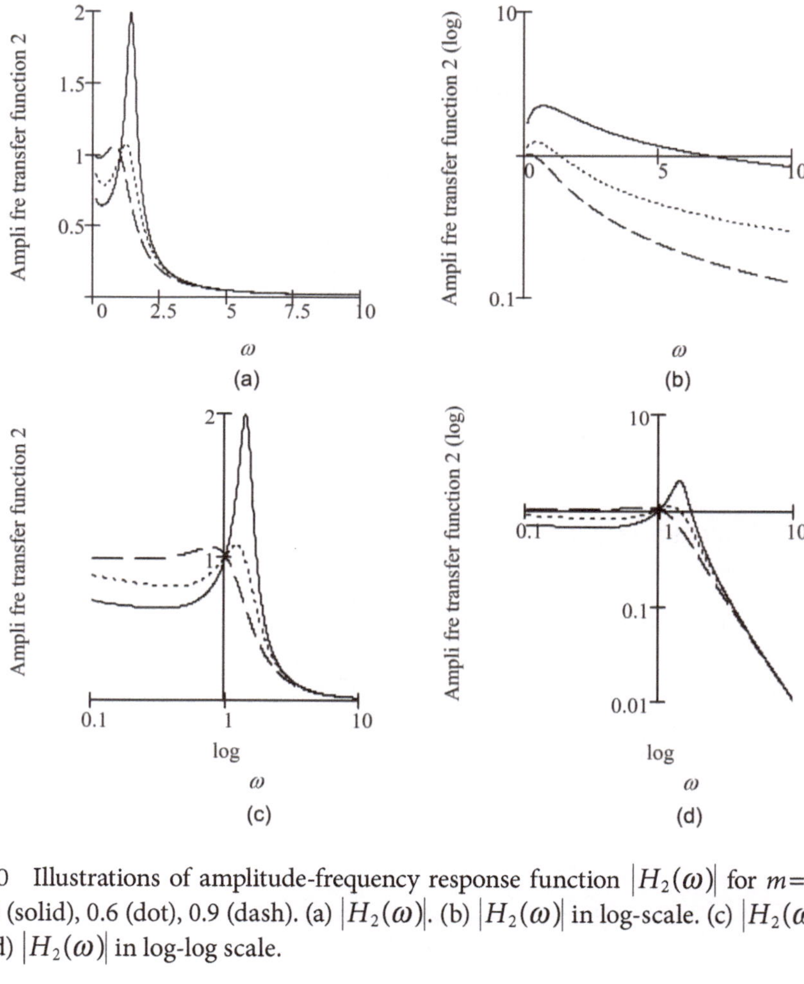

FIGURE 9.10 Illustrations of amplitude-frequency response function $|H_2(\omega)|$ for $m=1$, $c=1$, $k=1$ with $\beta=0.3$ (solid), 0.6 (dot), 0.9 (dash). (a) $|H_2(\omega)|$. (b) $|H_2(\omega)|$ in log-scale. (c) $|H_2(\omega)|$ with ω in log-scale. (d) $|H_2(\omega)|$ in log-log scale.

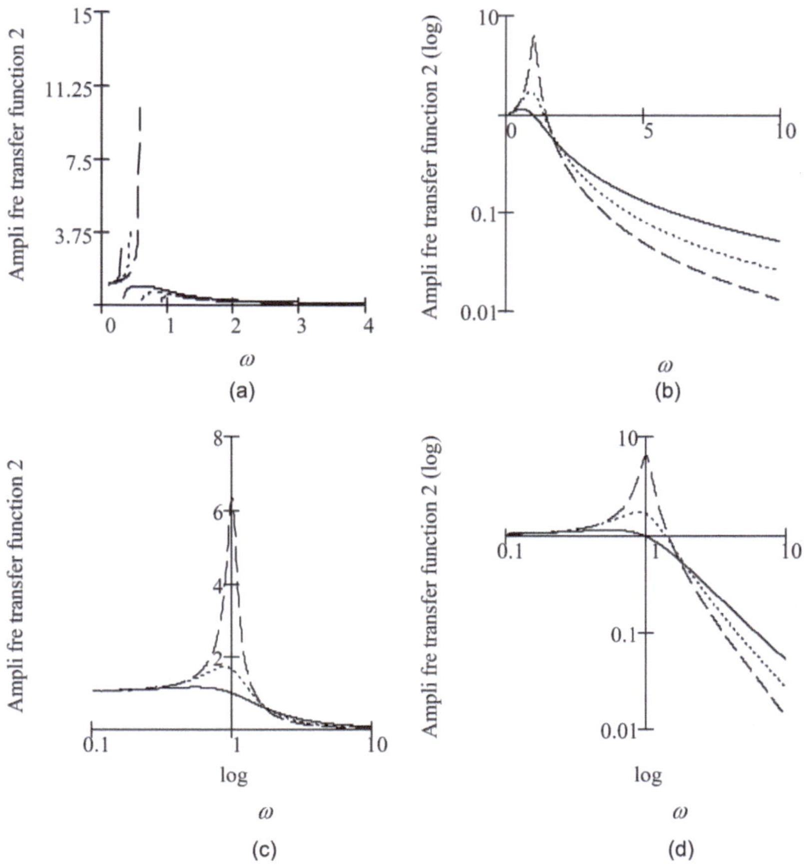

FIGURE 9.11 Illustrations of $|H_2(\omega)|$ for $m=1$, $c=1$, $k=1$ with $\beta=1.3$ (solid), 1.6 (dot), 1.9 (dash). (a) $|H_2(\omega)|$. (b) $|H_2(\omega)|$ in log-scale. (c) $|H_2(\omega)|$ with ω in log-scale. (d) $|H_2(\omega)|$ in log-log scale.

9.7 QUASIEQUIVALENT LOGARITHMIC DECREMENT AND Q FACTOR OF CLASS II FRACTIONAL VIBRATOR

9.7.1 Quasiequivalent Logarithmic Decrement of Fractional Vibrators of Class II

The following theorem expresses the quasiequivalent logarithmic decrement of $x_2(t)$.

Theorem 9.14. Suppose that t_i and t_{i+1} are two successive time points of fractional free response $x_2(t)$, where $x_2(t)$ reaches its successive peak values of $x_2(t_i)$ and $x_2(t_{i+1})$, respectively. Denote by Δ_{eq2} the quasiequivalent logarithmic decrement of $x_2(t)$. Then,

$$\Delta_{\text{eq2}} = \Delta_{\text{eq2}}(\omega,\beta) = \frac{2\pi}{\sqrt{1-\left(\dfrac{\varsigma\omega^{\beta-1}\sin\dfrac{\beta\pi}{2}}{\sqrt{1-\dfrac{c}{m}\omega^{\beta-2}\cos\dfrac{\beta\pi}{2}}}\right)^2}}\frac{\varsigma\omega^{\beta-1}\sin\dfrac{\beta\pi}{2}}{\sqrt{1-\dfrac{c}{m}\omega^{\beta-2}\cos\dfrac{\beta\pi}{2}}}, \quad 0<\beta<2. \tag{9.53}$$

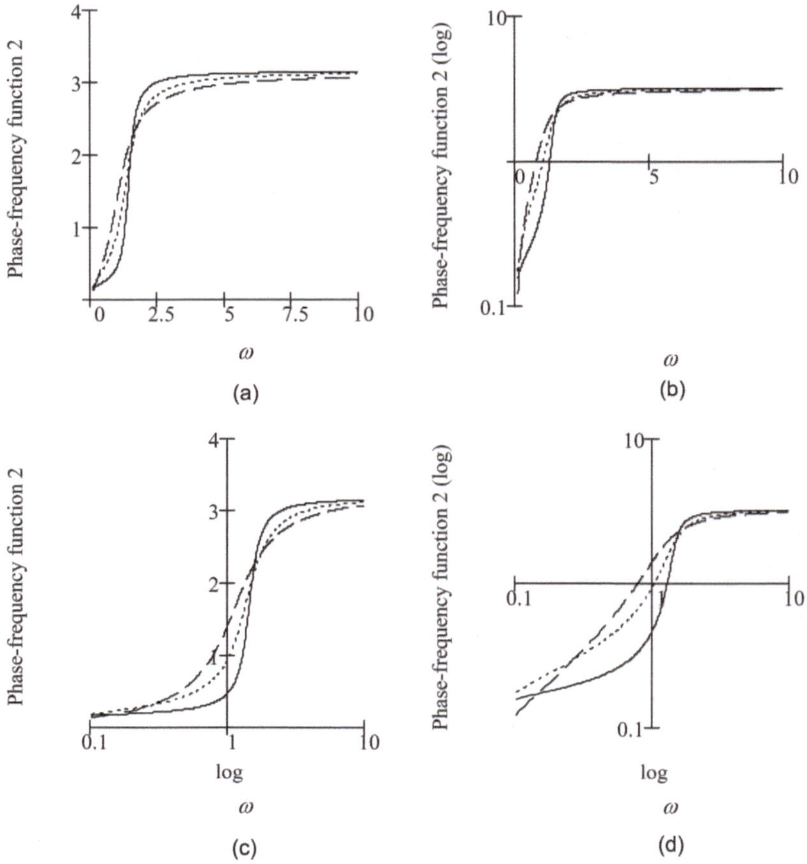

FIGURE 9.12 Plots of $\varphi_2(\omega)$ for $m=1$, $c=1$, $k=1$ with $\beta=0.3$ (solid), 0.6 (dot), 0.9 (dash). (a) $\varphi_2(\omega)$. (b) $\varphi_2(\omega)$ in log-scale. (c) $\varphi_2(\omega)$ with ω in log-scale. (d) $\varphi_2(\omega)$ in log-log scale.

Proof. Following the definition of logarithmic decrement, we have

$$\Delta_{\text{eq}2} = \ln \frac{x_2(t_i)}{x_2(t_{i+1})} = \frac{2\pi\varsigma_{\text{eq}2}}{\sqrt{1-\varsigma_{\text{eq}2}^2}}. \tag{9.54}$$

When substituting $\zeta_{\text{eq}2}$ into the above, we have Eq. (9.53). The proof is finished. □

Note that $\Delta_{\text{eq}2}$ is not dimensionless as that of a conventional one. Hence, we call it quasi-equivalent logarithmic decrement. Besides, it is in general not a constant but a function of ω and β. If $\beta=1$, it reduces to conventional logarithmic decrement. That is, $\Delta_{\text{eq}2}\big|_{\beta=1} = \frac{2\pi\varsigma}{\sqrt{1-\varsigma^2}}$. Several plots of $\Delta_{\text{eq}2}$ are shown in Figure 9.14.

9.7.2 Quasiequivalent *Q* Factor of Fractional Vibrators of Class II

Theorem 9.15 is about the representation of the quasiequivalent *Q* factor of a fractional vibrator of class II.

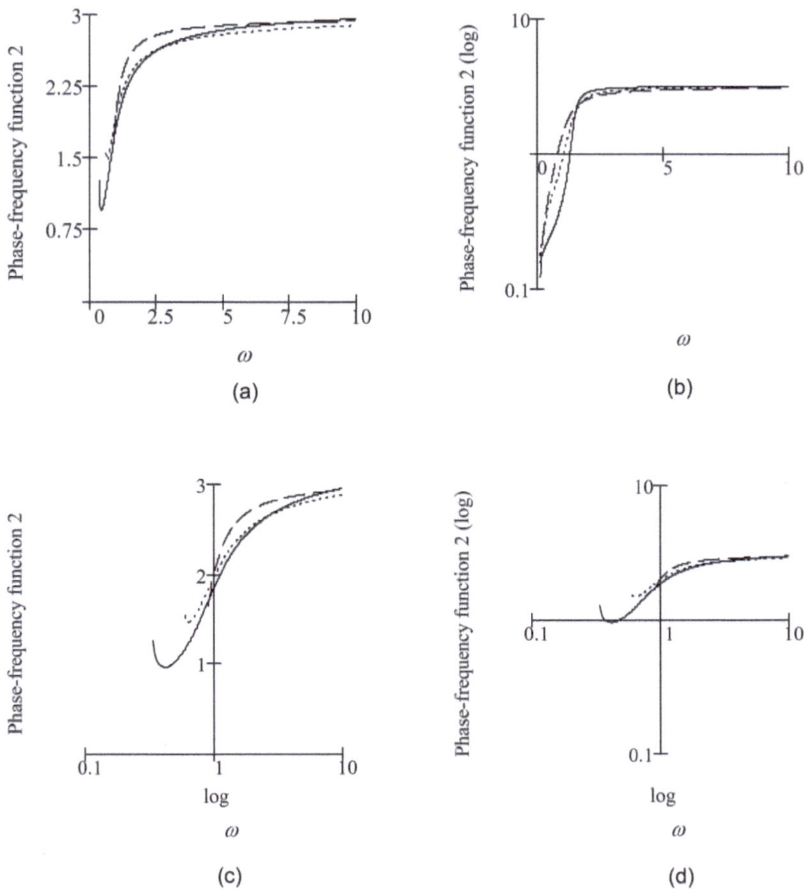

FIGURE 9.13 Plots of $\varphi_2(\omega)$ for $m=1$, $c=1$, $k=1$ with $\beta=1.3$ (solid), 1.6 (dot), 1.9 (dash). (a) $\varphi_2(\omega)$. (b) $\varphi_2(\omega)$ in log-scale. (c) $\varphi_2(\omega)$ with ω in log-scale. (d) $\varphi_2(\omega)$ in log-log scale.

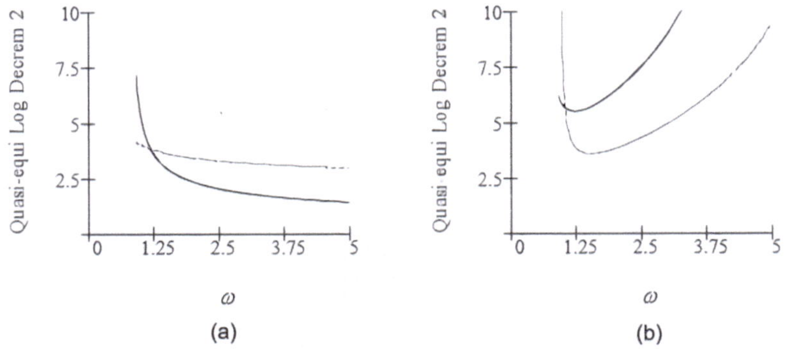

FIGURE 9.14 Plots of quasiequivalent logarithmic decrement Δ_{eq2} of $x_2(t)$ for $m=1$, $c=1$, and $k=1$. (a) Δ_{eq2} for $\beta=0.6$ (solid), 0.9 (dot). (b) Δ_{eq2} for $\beta=1.6$ (solid), 1.9 (dot). Note that "Quasi-equivalent logarithmic decrement 2" in the above figures means Δ_{eq2}.

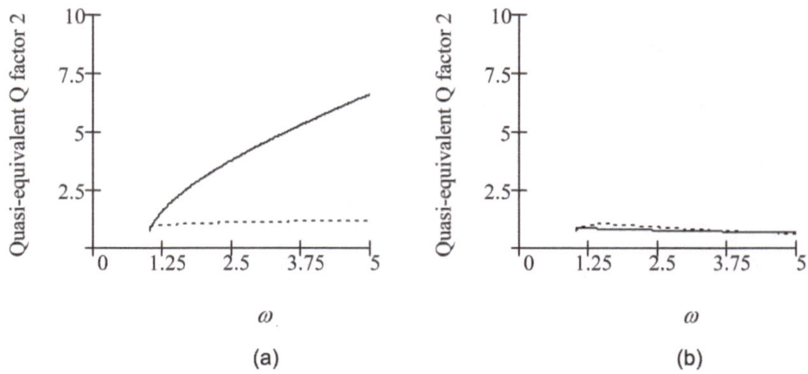

FIGURE 9.15 Plots of quasiequivalent Q factor Q_{eq2} for $m=1$, $c=1$, and $k=1$. (a) Q_{eq2} when $\beta=0.3$ (solid), 0.9 (dot). (b) Q_{eq2} when $\beta=1.3$ (solid), 1.9 (dot). Note that "Quasi-equivalent Q factor 2" in the above figures implies Q_{eq2}.

Theorem 9.15. Let Q_{eq2} be the quasiequivalent Q factor of class II fractional vibrator. Then,

$$Q_{eq2} = Q_{eq2}(\omega, \beta) = \frac{\sqrt{1 - \frac{c}{m}\omega^{\beta-2}\cos\frac{\beta\pi}{2}}}{2\varsigma\omega^{\beta-1}\sin\frac{\beta\pi}{2}}, \quad 0 < \beta < 2. \tag{9.55}$$

Proof. Following the definition of Q factor, Q_{eq2} is defined by

$$Q_{eq2} = \frac{1}{2\varsigma_{eq2}}. \tag{9.56}$$

Substituting ζ_{eq2} into the above produces Eq. (9.55). This completes the proof. □

Q_{eq2} is generally not a constant but a function of ω and β. If $\beta=1$, it reduces to conventional Q factor $\frac{1}{2\varsigma}$. In addition, it is not dimensionless unless $\beta=1$. Hence, the prefix "quasi" is used. Its plots are shown in Figure 9.15.

9.8 NOTE ON C_{EQ2}

Let c_r be the Rayleigh damping. It is given by $c_r = am + bk$, where m and k are primary mass and stiffness, respectively, a is proportional to vibration frequency ω while b is inversely proportional to ω (Rayleigh [1], Jin and Xia [2], Harris [3], Palley et al. [4]). For this class of fractional vibrators, $c_{eq2} = c\omega^{\beta-1}\sin\frac{\beta\pi}{2}$, where c is the primary damping, see Eq. (9.12). Two things are worth noting. On the one hand, we see that c_{eq2} is proportional to ω when $1<\beta<2$ while it is inversely proportional to ω if $0<\beta<1$. On the other hand, c_{eq2} does not relate to m and k.

9.9 CONCLUDING REMARK

The present results in this chapter are new in the field. They may not be investigated in ordinary vibrators.

9.10 EXERCISES

9.1. Prove that $\frac{d^{\beta}x(t)}{dt^{\beta}}$ for $0<\beta<1$ or $1<\beta<2$ is not a standard velocity, where $x(t)$ is a displacement.

9.2. If the unit of c is N m^{-1}s, prove that $c\frac{d^{\beta}x(t)}{dt^{\beta}}$ does not designate a standard friction force for $0<\beta<1$ or $1<\beta<2$, where $x(t)$ is a displacement.

9.3. Find the ranges of ω and β for $0<\zeta_{\text{eq2}}<1$.

9.4. Find the ranges of ω and β with specific m and c such that $m_{\text{eq2}}<0$.

9.5. Prove that ω_{eqn2} is not a standard angular frequency if $\beta\neq 1$.

9.6. Write the expression of sinusoidal response to a class II fractional vibrator. (Hint: Refer to [5]).

REFERENCES

1. J. W. Strutt and 3rd B. Rayleigh, *The Theory of Sound*, vol. 1, Macmillan & Co., Ltd., London, 1877.
2. X. D. Jin and L. J. Xia, *Ship Hull Vibration*, The Press of Shanghai Jiaotong University, Shanghai, China, 2011. In Chinese.
3. C. M. Harris, *Shock and Vibration Handbook*, 5th Ed., McGraw-Hill, New York, 2002.
4. O. M. Palley, Г. B. Bahizov, and E. Я. Voroneysk, *Handbook of Ship Structural Mechanics*, National Defense Industry Publishing House, Beijing, China, 2002. In Chinese. Translated from Russian by B. H. Xu, X. Xu, and M. Q. Xu.
5. M. Li, Three classes of fractional oscillators, *Symmetry-Basel*, 10(2), 2018, 91 p.

CHAPTER 10

Class III Fractional Vibrations

THE MAIN HIGHLIGHTS PRESENTED in this chapter are listed below.

1. An equivalent motion equation of a fractional vibrator in class III is presented in Theorem 10.1.
2. The terminologies in terms of quasiequivalent mass, quasiequivalent damping, quasiequivalent damping ratio, quasiequivalent damping free natural frequency, quasiequivalent damped natural frequency, quasiequivalent frequency ratio, quasi-equivalent logarithmic decrement, and quasiequivalent Q factor for a fractional vibrator in class III are introduced. Their analytical representations are proposed in Theorems 10.2, 10.3, 10.6–10.9, 10.14, and 10.15, respectively.
3. The phenomenon that the quasiequivalent mass of a fractional vibrator in class III may vary from $-\infty$ to ∞ is revealed in Corollaries 10.1–10.3.
4. The quasiequivalent damping of a fractional vibrator in class III may vary from $-\infty$ to ∞; this is brought to light in Corollary 10.4.
5. The conditions for a fractional vibrator in class III to be self-vibrated are put forward in Theorem 10.5.
6. The analytical representations of free response, impulse response, step response, and frequency transfer function of a fractional vibrator in class III are presented in Theorems 10.10–10.13.

10.1 CLASS III EQUIVALENT MOTION EQUATION

We present Theorem 10.1 to address an equivalent motion equation of a class III fractional vibrator.

DOI: 10.1201/9781003460947-12

Theorem 10.1. A fractional vibrator in class III expressed by

$$m\frac{d^{\alpha}x_3(t)}{dt^{\alpha}}+c\frac{d^{\beta}x_3(t)}{dt^{\beta}}+kx_3(t)=0, \quad 1<\alpha<3,\ 0<\beta<2 \tag{10.1}$$

can be equivalently expressed by

$$\begin{aligned}&-\left(m\omega^{\alpha-2}\cos\frac{\alpha\pi}{2}+c\omega^{\beta-2}\cos\frac{\beta\pi}{2}\right)\frac{d^2x_3(t)}{dt^2}\\&+\left(m\omega^{\alpha-1}\sin\frac{\alpha\pi}{2}+c\omega^{\beta-1}\sin\frac{\beta\pi}{2}\right)\frac{dx_3(t)}{dt}+kx_3(t)=0, \quad 1<\alpha<3,\ 0<\beta<2.\end{aligned} \tag{10.2}$$

Proof. Denote by $X_3(\omega)$ the Fourier transform of $x_3(t)$. Consider the principal values of i^{α} and i^{β}. Doing the Fourier transform of the left side of Eq. (10.1) yields

$$\begin{aligned}&\left[m\frac{d^{\alpha}x_3(t)}{dt^{\alpha}}+c\frac{d^{\beta}x_3(t)}{dt^{\beta}}+kx_3(t)\right]\leftrightarrow\left[m(i\omega)^{\alpha}+c(i\omega)^{\beta}+k\right]X_3(\omega)\\&=\left[m\left(\cos\frac{\alpha\pi}{2}+i\sin\frac{\alpha\pi}{2}\right)\omega^{\alpha}+c\left(\cos\frac{\beta\pi}{2}+i\sin\frac{\beta\pi}{2}\right)\omega^{\beta}+k\right]X_3(\omega)\\&=\left[m\omega^{\alpha}\cos\frac{\alpha\pi}{2}+c\omega^{\beta}\cos\frac{\beta\pi}{2}+k+i\left(m\omega^{\alpha}\sin\frac{\alpha\pi}{2}+c\omega^{\beta}\sin\frac{\beta\pi}{2}\right)\right]X_3(\omega).\end{aligned} \tag{10.3}$$

On the other side, doing the Fourier transform of the left side of Eq. (10.2) results in

$$\begin{aligned}&\left[\begin{array}{l}-\left(m\omega^{\alpha-2}\cos\frac{\alpha\pi}{2}+c\omega^{\beta-2}\cos\frac{\beta\pi}{2}\right)\frac{d^2x_3(t)}{dt^2}\\+\left(m\omega^{\alpha-1}\sin\frac{\alpha\pi}{2}+c\omega^{\beta-1}\sin\frac{\beta\pi}{2}\right)\frac{dx_3(t)}{dt}+kx_3(t)\end{array}\right]\leftrightarrow\\&=-\left(m\omega^{\alpha-2}\cos\frac{\alpha\pi}{2}+c\omega^{\beta-2}\cos\frac{\beta\pi}{2}\right)(i\omega)^2X_3(\omega)\\&+\left(m\omega^{\alpha-1}\sin\frac{\alpha\pi}{2}+c\omega^{\beta-1}\sin\frac{\beta\pi}{2}\right)(i\omega)X_3(\omega)+kX_3(\omega)\\&=\left[m\omega^{\alpha}\cos\frac{\alpha\pi}{2}+c\omega^{\beta}\cos\frac{\beta\pi}{2}+k+i\left(m\omega^{\alpha}\sin\frac{\alpha\pi}{2}+c\omega^{\beta}\sin\frac{\beta\pi}{2}\right)\right]X_3(\omega).\end{aligned} \tag{10.4}$$

By comparing the right side of Eq. (10.3) to that of Eq. (10.4), we see that Theorem 10.1 holds. This finishes the proof. □

10.2 CLASS III QUASIEQUIVALENT VIBRATION PARAMETERS

10.2.1 Quasiequivalent Mass of Class III Fractional Vibrators

Denote by m_{eq3} the quasiequivalent mass for a fractional vibrator in class III. We propose its expression with the theorem below.

Theorem 10.2. The quasiequivalent mass m_{eq3} is in the form

$$m_{eq3} = -\left(m\omega^{\alpha-2}\cos\frac{\alpha\pi}{2} + c\omega^{\beta-2}\cos\frac{\beta\pi}{2} \right). \tag{10.5}$$

Proof. Since the above is the coefficient of the inertia force in Eq. (10.2), the above holds according to Newton's second law. The proof is finished. □

The unit of m_{eq3} is not kg unless $\alpha=2$ and $\beta=1$. If $\alpha=2$ and $\beta=1$, m_{eq3} reduces to the primary mass m. Thus, the prefix "quasi" is adopted.

Note that m_{eq3} is associated with ω, α, and β. It consists of two parts. One is contributed by fractional acceleration, which is denoted by m_{eq31}

$$m_{eq31} = -m\omega^{\alpha-2}\cos\frac{\alpha\pi}{2}. \tag{10.6}$$

The other results from fractional velocity and it is denoted by m_{eq32} where

$$m_{eq32} = -c\omega^{\beta-2}\cos\frac{\beta\pi}{2}. \tag{10.7}$$

When considering $\frac{c}{m} = 2\varsigma\omega_n$, rewrite Eq. (10.5) as

$$m_{eq3} = m_{eq3}(\omega,\alpha,\beta) = -\left(\omega^{\alpha-2}\cos\frac{\alpha\pi}{2} + 2\varsigma\omega_n\omega^{\beta-2}\cos\frac{\beta\pi}{2} \right). \tag{10.8}$$

Now, we show that the asymptotic properties of m_{eq3} are interesting.

Corollary 10.1: Asymptotic Property for $\omega \to \infty$. For m_{eq3}, we have

$$\lim_{\omega\to\infty} m_{eq3}(\omega,\alpha,\beta) = \begin{cases} \infty, & 2<\alpha<3, \\ 0, & 1<\alpha<2. \end{cases} \tag{10.9}$$

Proof. On the one hand, $\lim_{\omega\to\infty}\omega^{\alpha-2} = \infty$ and $\cos\frac{\alpha\pi}{2} < 0$ for $2<\alpha<3$. On the other hand, $\lim_{\omega\to\infty}\omega^{\alpha-2} = 0$ for $1<\alpha<2$. Besides, $\lim_{\omega\to\infty}\omega^{\beta-2} = 0$ for $1<\beta<2$. Thus, the above is valid. This finishes the proof. □

Corollary 10.2: Asymptotic Property for $\omega \to 0$. For m_{eq3} with $1<\beta<2$, we have

$$\lim_{\omega \to 0} m_{eq3}(\omega,\alpha,\beta)=\infty. \tag{10.10}$$

Proof. As $\lim_{\omega \to 0} \omega^{\beta-2}=\infty$ and $\cos\frac{\beta\pi}{2}<0$ for $1<\beta<2$, the above holds. The proof ends. □

Corollary 10.3: Negative Mass. If $0<\beta<1$, m_{eq3} is negative when $\omega \to 0$.

Proof. Since $\cos\frac{\beta\pi}{2}>0$ for $0<\beta<1$, $m_{eq3}<0$ if

$$m\omega^{\alpha-2}\cos\frac{\alpha\pi}{2}<c\omega^{\beta-2}\cos\frac{\beta\pi}{2}. \tag{10.11}$$

The proof is finished. □

Note that m_{eq3} is relatively complicated due to two parameters α and β.

Demonstrations of m_{eq3} are given in Figure 10.1.

10.2.2 Quasiequivalent Damping of Fractional Vibrators of Class III

Let c_{eq3} be the quasiequivalent damping for a class III fractional vibrator. Theorem 10.3 presents its expression.

Theorem 10.3. The quantity c_{eq3} is given by

$$c_{eq3}=c_{eq3}(\omega,\alpha,\beta)=m\omega^{\alpha-1}\sin\frac{\alpha\pi}{2}+c\omega^{\beta-1}\sin\frac{\beta\pi}{2}. \tag{10.12}$$

Proof. As the above is the coefficient of $\frac{dx_3(t)}{dt}$ in Eq. (10.2), $c_{eq3}\frac{dx_3(t)}{dt}$ takes the position of friction force. Thus, the above is valid. This finishes the proof. □

Note that c_{eq3} is termed "quasi" because its unit in general is not $\mathrm{N}\times\mathrm{m}^{-1}\times\mathrm{s}$ unless $\alpha=2$ and $\beta=1$. We propose the following corollaries and theorems for its interesting asymptotic properties.

Corollary 10.4. When $\omega \to \infty$,

$$c_{eq3}=\begin{cases} \infty, & 1<\alpha<2, 0<\beta<2, \\ -\infty, & 2<\alpha<3, 0<\beta<1. \end{cases} \tag{10.13}$$

Proof. Since $\sin\frac{\beta\pi}{2}>0$ if $0<\beta<2$ and $\sin\frac{\alpha\pi}{2}>0$ when $1<\alpha<2$, $\lim_{\omega\to\infty} c_{eq3}(\omega,\alpha,\beta)=\infty$ for $1<\alpha<2$ and $0<\beta<2$. On the other side, because $\lim_{\omega\to\infty}\omega^{\beta-1}=0$ if $0<\beta<1$ and $\sin\frac{\alpha\pi}{2}<0$ when $2<\alpha<3$, $\lim_{\omega\to\infty} c_{eq3}(\omega,\alpha,\beta)=-\infty$ and negative damping when $2<\alpha<3$ and $0<\beta<1$. Thus, (10.13) holds. The proof is complete. □

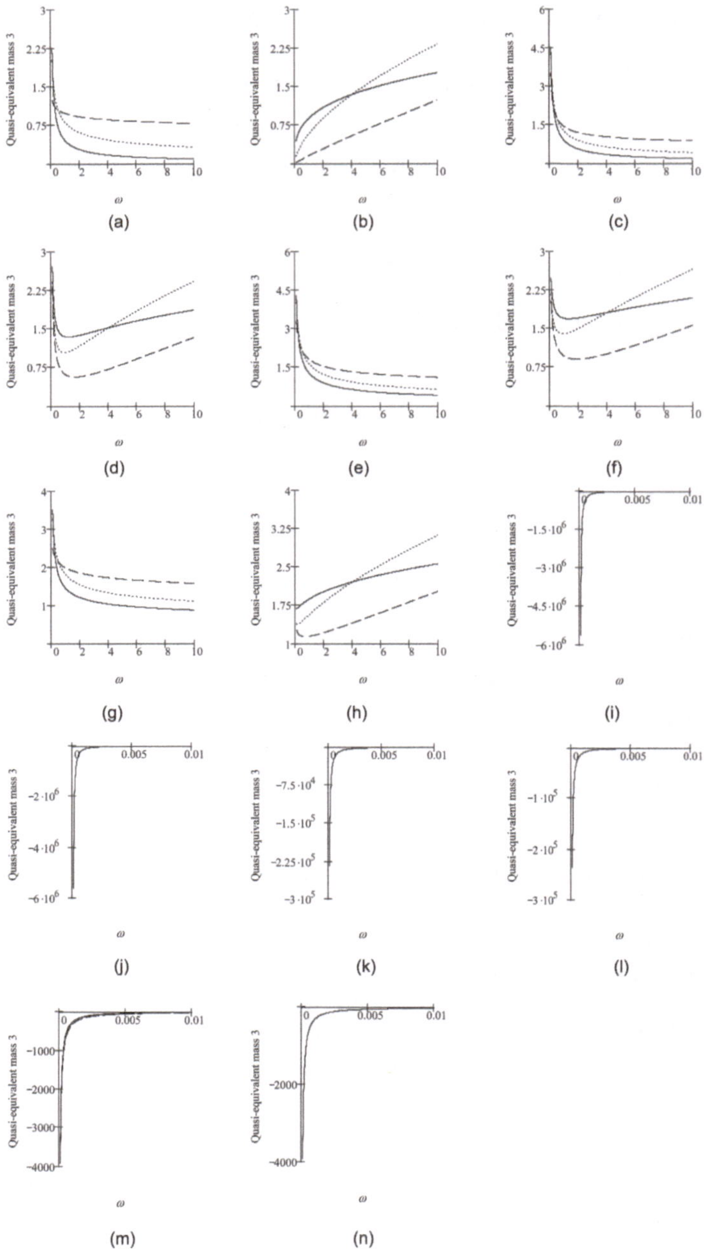

FIGURE 10.1 Plots of m_{eq3} with $m=1$, $c=1$, and $k=1$. (a) m_{eq3} for $\beta=1$ and $\alpha=1.3$ (solid), 1.6 (dot), 1.9 (dash). (b) m_{eq3} for $\beta=1$ and $\alpha=2.3$ (solid), 2.6 (dot), 2.9 (dash). (c) m_{eq3} for $\beta=1.3$ and $\alpha=1.3$ (solid), 1.6 (dot), 1.9 (dash). (d) m_{eq3} for $\beta=1.3$ and $\alpha=2.3$ (solid), 2.6 (dot), 2.9 (dash). (e) m_{eq3} for $\beta=1.6$ and $\alpha=1.3$ (solid), 1.6 (dot), 1.9 (dash). (f) m_{eq3} for $\beta=1.6$ and $\alpha=2.3$ (solid), 2.6 (dot), 2.9 (dash). (g) m_{eq3} for $\beta=1.9$ and $\alpha=1.3$ (solid), 1.6 (dot), 1.9 (dash). (h) m_{eq3} for $\beta=1.9$ and $\alpha=2.3$ (solid), 2.6 (dot), 2.9 (dash). (i) Observing negative m_{eq3} when $\beta=0.3$ and $\alpha=1.3$ (solid), 1.6 (dot), 1.9 (dash). (j) Observing negative m_{eq3} when $\beta=0.6$ and $\alpha=2.3$ (solid), 2.6 (dot), 2.9 (dash). (k) Observing negative m_{eq3} when $\beta=0.6$ and $\alpha=1.3$ (solid), 1.6 (dot), 1.9 (dash). (l) Observing negative m_{eq3} when $\beta=0.6$ and $\alpha=2.3$ (solid), 2.6 (dot), 2.9 (dash). (m) Observing negative m_{eq3} when $\beta=0.9$ and $\alpha=1.3$ (solid), 1.6 (dot), 1.9 (dash). (n) Observing negative m_{eq3} when $\beta=0.9$ and $\alpha=2.3$ (solid), 2.6 (dot), 2.9 (dash). Note that "Quasi-equivalent mass 3" in the above figures means m_{eq3}.

Theorem 10.4. c_{eq3} is proportional to m and c.

Proof. From Eq. (10.12), we see that Theorem 10.4 holds. The proof is finished. □

Theorem 10.5. A fractional vibrator of class III may be self-vibrated and accordingly non-stable if $2<\alpha<3$.

Proof. $\sin\frac{\alpha\pi}{2}<0$ when $2<\alpha<3$. Thus, c_{eq3} may be negative if $2<\alpha<3$. Thus, a vibrator may be self-vibrated and non-stable in that case. This finishes the proof. □

Corollary 10.5. When $1<\alpha<3$ and $1<\beta<2$,

$$\lim_{\omega\to 0} c_{eq3}(\omega,\alpha,\beta)=0. \tag{10.14}$$

In addition, if $1<\alpha<3$ and $0<\beta<1$,

$$\lim_{\omega\to 0} c_{eq3}(\omega,\alpha,\beta)=\infty. \tag{10.15}$$

Proof. Since $\lim_{\omega\to 0}\omega^{\alpha-1}=0$ for $1<\alpha<3$ and $\lim_{\omega\to 0}\omega^{\beta-1}=0$ for $1<\beta<2$, Eq. (10.14) holds. In addition, $\lim_{\omega\to 0}\omega^{\beta-1}=\infty$ if $0<\beta<1$ and $\lim_{\omega\to 0}\omega^{\alpha-1}=0$ for $1<\alpha<3$. Thus, Eq. (10.15) is valid. This finishes the proof. □

In summary, $c_{eq3}\in(-\infty,\infty)$.

Corollary 10.6. If $1<\alpha<3$ and $0<\beta<1$, in the case of large ω,

$$c_{eq3}\propto m\omega^{\alpha-1}\sin\frac{\alpha\pi}{2}. \tag{10.16}$$

On the contrary, for $1<\alpha<3$ and $0<\beta<1$, if ω is small enough,

$$c_{eq3}\propto c\omega^{\beta-1}\sin\frac{\beta\pi}{2}. \tag{10.17}$$

Proof. If $1<\alpha<3$, $0<\beta<1$ and ω is sufficiently large, $\omega^{\beta-1}\approx 0$. Hence, Eq. (10.16) holds. On the other hand, if ω is small enough, $\omega^{\alpha-1}\approx 0$. Thus, Eq. (10.17) is valid. This finishes the proof. □

The plots of c_{eq3} are indicated in Figure 10.2.

10.2.3 Quasiequivalent Damping Ratio of Fractional Vibrators of Class III

Denote by ζ_{eq3} the quasiequivalent damping ratio for a fractional vibrator of class III. It is defined by

$$\varsigma_{eq3}=\frac{c_{eq3}}{2\sqrt{m_{eq3}k}}. \tag{10.18}$$

Theorem 10.6 proposes its expression.

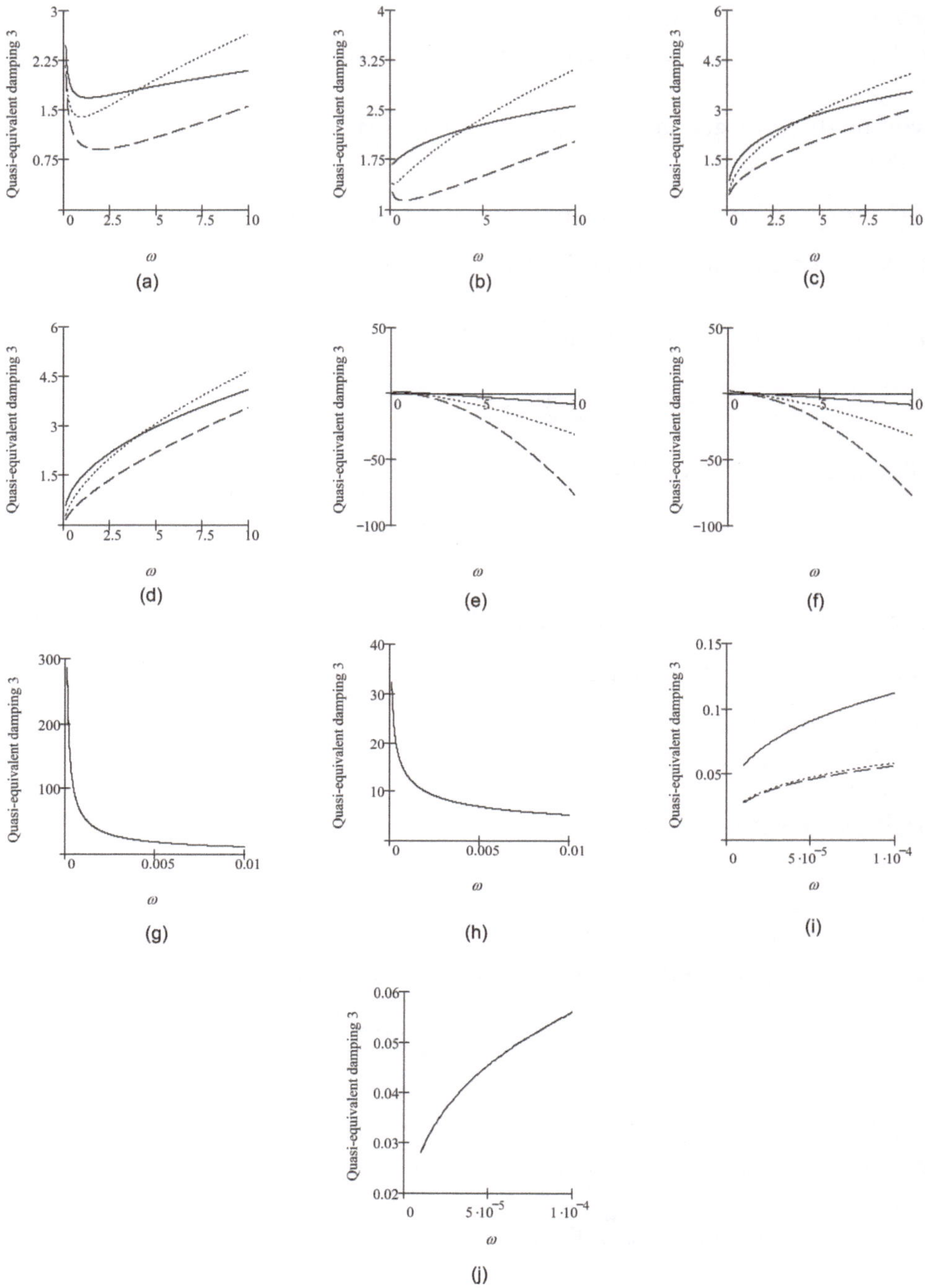

FIGURE 10.2 Plots of c_{eq3}. (a) c_{eq3} for $\beta=0.6$ and $\alpha=1.3$ (solid), 1.6 (dot), 1.9 (dash). (b) c_{eq3} for $\beta=0.9$ and $\alpha=1.3$ (solid), 1.6 (dot), 1.9 (dash). (c) c_{eq3} for $\beta=1.3$ and $\alpha=1.3$ (solid), 1.6 (dot), 1.9 (dash). (d) c_{eq3} for $\beta=1.6$ and $\alpha=1.3$ (solid), 1.6 (dot), 1.9 (dash). (e) c_{eq3} for $\beta=0.9$ and $\alpha=2.3$ (solid), 2.6 (dot), 2.9 (dash) (negative damping, see Corollary 10.4). (f) c_{eq3} for $\beta=0.6$ and $\alpha=2.3$ (solid), 2.6 (dot), 2.9 (dash) (negative damping). (g) c_{eq3} for $\beta=0.3$ and $\alpha=2.3$ (solid), 2.6 (dot), 2.9 (dash) (infinite damping at $\omega\approx0$, see Corollary 10.5). (h) c_{eq3} for $\beta=0.6$ and $\alpha=2.3$ (solid), 2.6 (dot), 2.9 (dash) (infinite damping at $\omega\approx0$). (i) c_{eq3} for $\beta=1.3$ and $\alpha=1.3$ (solid), 1.6 (dot), 1.9 (dash) (zero damping at $\omega\approx0$ for Corollary 10.5). (j) c_{eq3} for $\beta=1.3$ and $\alpha=2.3$ (solid), 2.6 (dot), 2.9 (dash) (zero damping at $\omega\approx0$, see Corollary 10.5). Note that "Quasi-equivalent damping 3" in the above figures implies c_{eq3}.

Theorem 10.6. ζ_{eq3} is expressed by

$$\varsigma_{eq3}=\varsigma_{eq3}(\omega,\alpha,\beta)=\frac{\omega^{\alpha-1}\sin\frac{\alpha\pi}{2}+2\varsigma\omega_n\omega^{\beta-1}\sin\frac{\beta\pi}{2}}{2\omega_n\sqrt{-\left(\omega^{\alpha-2}\cos\frac{\alpha\pi}{2}+2\varsigma\omega_n\omega^{\beta-2}\cos\frac{\beta\pi}{2}\right)}},\tag{10.19}$$

where $\varsigma=\frac{c}{2\sqrt{mk}}$.

Proof. Substituting m_{eq3} and c_{eq3} into Eq. (10.18) results in Eq. (10.19). This completes the proof. □

Since ζ_{eq3} is not dimensionless, we use the prefix "quasi" to term it. It is dimensionless if and only if $\alpha=2$ and $\beta=1$. As a matter of fact, $\zeta_{eq3}(\omega, 2, 1)=\zeta$.

We propose the asymptotic properties of ζ_{eq3} below.

Corollary 10.7. If $1<\alpha<2$ and $0<\beta<2$, we have

$$\zeta_{eq3}\geq 0.\tag{10.20}$$

However, when $2<\alpha<3$, it may be negative.

Proof. If $1<\alpha<2$ and $0<\beta<2$, ζ_{eq3} is positive definite. Nonetheless, $\sin\frac{\alpha\pi}{2}<0$ when $2<\alpha<3$. Thus, it may be negative if $2<\alpha<3$. This completes the proof. □

The demonstrations of ζ_{eq3} are indicated in Figure 10.3.

10.2.4 Quasiequivalent Damping Free Natural Frequency of Class III Fractional Vibrators

For a fractional vibrator of class III, we introduce the term quasiequivalent damping-free natural frequency to describe the quantity defined by

$$\omega_{eqn3}=\sqrt{\frac{k}{m_{eq3}}}.\tag{10.21}$$

Theorem 10.7. The quasiequivalent damping-free natural frequency ω_{eqn3} is expressed by

$$\omega_{eqn3}=\frac{\omega_n}{\sqrt{-\left(\omega^{\alpha-2}\cos\frac{\alpha\pi}{2}+\frac{c}{m}\omega^{\beta-2}\cos\frac{\beta\pi}{2}\right)}}.\tag{10.22}$$

Proof. If substituting m_{eq3} into Eq. (10.21), we have

$$\omega_{eqn3}=\sqrt{\frac{k}{m_{eq3}}}=\sqrt{\frac{k}{-\left(m\omega^{\alpha-2}\cos\frac{\alpha\pi}{2}+c\omega^{\beta-2}\cos\frac{\beta\pi}{2}\right)}}=\frac{\omega_n}{\sqrt{-\left(\omega^{\alpha-2}\cos\frac{\alpha\pi}{2}+\frac{c}{m}\omega^{\beta-2}\cos\frac{\beta\pi}{2}\right)}}.$$

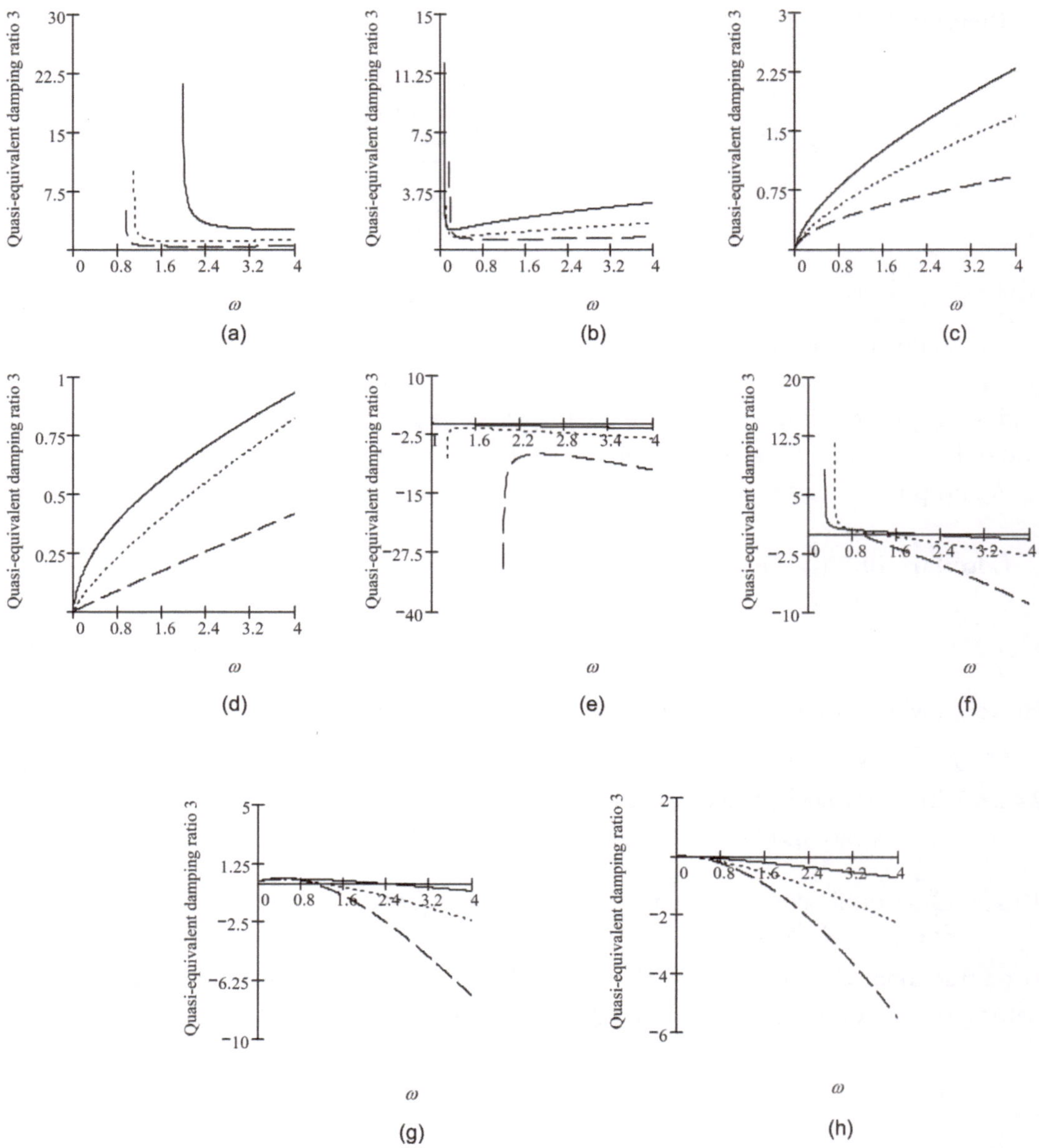

FIGURE 10.3 Plots of ζ_{eq3} with $m=1$, $c=1$, and $k=1$. (a) ζ_{eq3} for $\beta=0.3$ and $\alpha=1.3$ (solid), 1.6 (dot), 1.9 (dash). (b) ζ_{eq3} for $\beta=0.9$ and $\alpha=1.3$ (solid), 1.6 (dot), 1.9 (dash). (c) ζ_{eq3} for $\beta=1.3$ and $\alpha=1.3$ (solid), 1.6 (dot), 1.9 (dash). (d) ζ_{eq3} for $\beta=1.9$ and $\alpha=1.3$ (solid), 1.6 (dot), 1.9 (dash). (e) ζ_{eq3} for $\beta=0.3$ and $\alpha=2.3$ (solid), 2.6 (dot), 2.9 (dash) (negative damping ratio). (f) ζ_{eq3} for $\beta=0.9$ and $\alpha=2.3$ (solid), 2.6 (dot), 2.9 (dash) (negative damping ratio). (g) ζ_{eq3} for $\beta=1.3$ and $\alpha=2.3$ (solid), 2.6 (dot), 2.9 (dash) (negative damping ratio). (h) ζ_{eq3} for $\beta=1.9$ and $\alpha=2.3$ (solid), 2.6 (dot), 2.9 (dash) (negative damping ratio). Note that "Quasi-equivalent damping ratio 3" in the above figures stands for ζ_{eq3}.

The proof is complete. □

We use the prefix "quasi" to describe ω_{eqn3} because its unit is generally not rad/s. It reduces to rad/s if $\alpha=2$ and $\beta=1$. In fact,

$$\omega_{\text{eqn3}}\big|_{\alpha=2,\beta=1}=\omega_n. \tag{10.23}$$

Since ω_{eqn3} relates to ω, α, and β, it may be written by

$$\omega_{\text{eqn3}}=\omega_{\text{eqn3}}(\omega,\alpha,\beta). \tag{10.24}$$

Figure 10.4 demonstrates several plots of ω_{eqn3}.

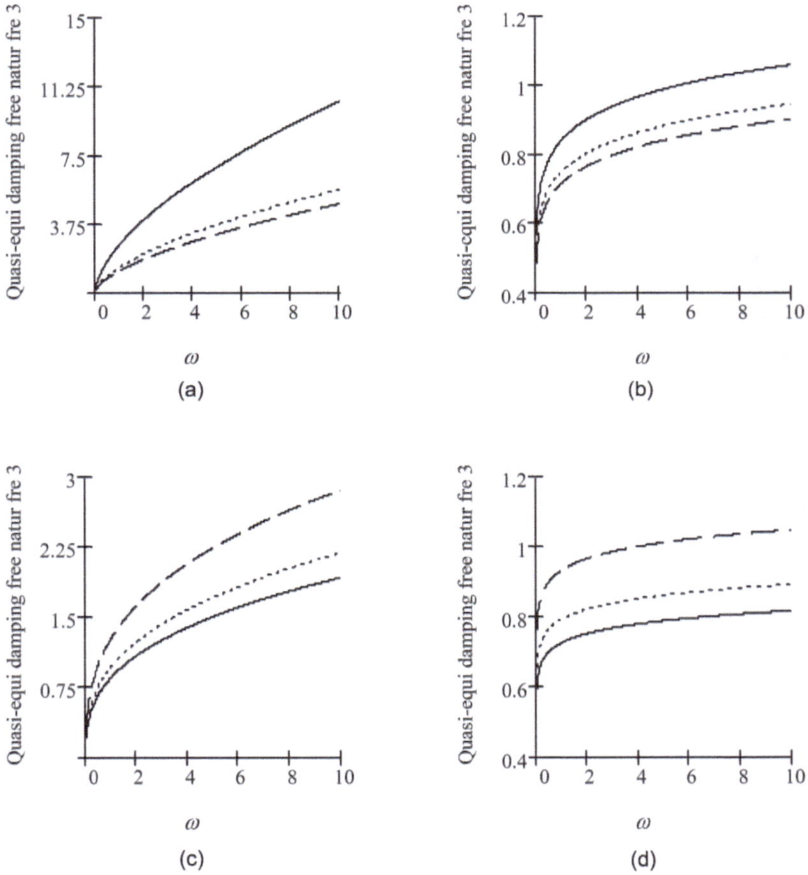

FIGURE 10.4 Illustrations of ω_{eqn3} with $\omega_n=1$, $m=1$, and $c=1$. (a) ω_{eqn3} for $\beta=0.8$ and $\alpha=1.3$ (solid), 1.6 (dot), 1.9 (dash). (b) ω_{eqn3} for $\beta=1.8$ and $\alpha=1.3$ (solid), 1.6 (dot), 1.9 (dash). (c) ω_{eqn3} for $\beta=1.3$ and $\alpha=2.3$ (solid), 2.6 (dot), 2.9 (dash). (d) ω_{eqn3} for $\beta=1.9$ and $\alpha=2.3$ (solid), 2.6 (dot), 2.9 (dash). Note that "Quasi-equivalent damping free natural frequency 3" in the above figures stands for ω_{eqn3}.

10.2.5 Quasiequivalent Damped Natural Frequency of Class III Fractional Vibrators

The term "quasiequivalent damped natural frequency" is introduced for a fractional vibrator of class III type. Denote it by ω_{eqd3}. Small damping, i.e., $|\zeta_{\text{eq3}}| \le 1$, is assumed in what follows. Define ω_{eqd3} by

$$\omega_{\text{eqd3}} = \omega_{\text{eqn3}}\sqrt{1-\varsigma_{\text{eq3}}^2}, \quad |\varsigma_{\text{eq3}}| \le 1. \tag{10.25}$$

Theorem 10.8. The expression of ω_{eqd3} is given by

$$\omega_{\text{eqd3}} = \frac{\omega_n\sqrt{1-\left[\dfrac{\left(\omega^{\alpha-1}\sin\dfrac{\alpha\pi}{2}+2\varsigma\omega_n\omega^{\beta-1}\sin\dfrac{\beta\pi}{2}\right)^2}{4\omega_n^2\left[-\left(\omega^{\alpha-2}\cos\dfrac{\alpha\pi}{2}+2\varsigma\omega_n\omega^{\beta-2}\cos\dfrac{\beta\pi}{2}\right)\right]}\right]^2}}{\sqrt{-\left(\omega^{\alpha-2}\cos\dfrac{\alpha\pi}{2}+\dfrac{c}{m}\omega^{\beta-2}\cos\dfrac{\beta\pi}{2}\right)}}. \tag{10.26}$$

Proof. Substituting ζ_{eq3} and ω_{eqn3} into Eq. (10.25) produces (10.26). This finishes the proof. □

When $\alpha=2$ and $\beta=1$, ω_{eqd3} degenerates the conventional ω_d with the unit of rad/s. In general, its unit is not rad/s. Hence, the prefix "quasi" is used. Figure 10.5 shows its illustrations.

10.2.6 Quasiequivalent Frequency Ratio of Class III Fractional Vibrators

We now introduce the term quasiequivalent frequency ratio for a fractional vibrator of class III type. It is denoted by γ_{eq3} and defined by

$$\gamma_{\text{eq3}} = \frac{\omega}{\omega_{\text{eqn3}}}. \tag{10.27}$$

Theorem 10.9. The quasiequivalent frequency ratio γ_{eq3} is expressed by

$$\gamma_{\text{eq3}} = \gamma_{\text{eq3}}(\omega,\alpha,\beta) = \gamma\sqrt{-\left(\omega^{\alpha-2}\cos\frac{\alpha\pi}{2}+\frac{c}{m}\omega^{\beta-2}\cos\frac{\beta\pi}{2}\right)}, \tag{10.28}$$

where $\gamma = \dfrac{\omega}{\omega_n}$.

Proof. When substituting ω_{eqn3} into Eq. (10.27), we have Eq. (10.28). The proof is complete. □

The prefix "quasi" is adopted to describe γ_{eq3} since it is not dimensionless. It reduces to being dimensionless if and only if $\alpha=2$ and $\beta=1$. That is, $\gamma_{\text{eq3}}(\omega, 2, 1)=\gamma$. Its plots are shown in Figure 10.6.

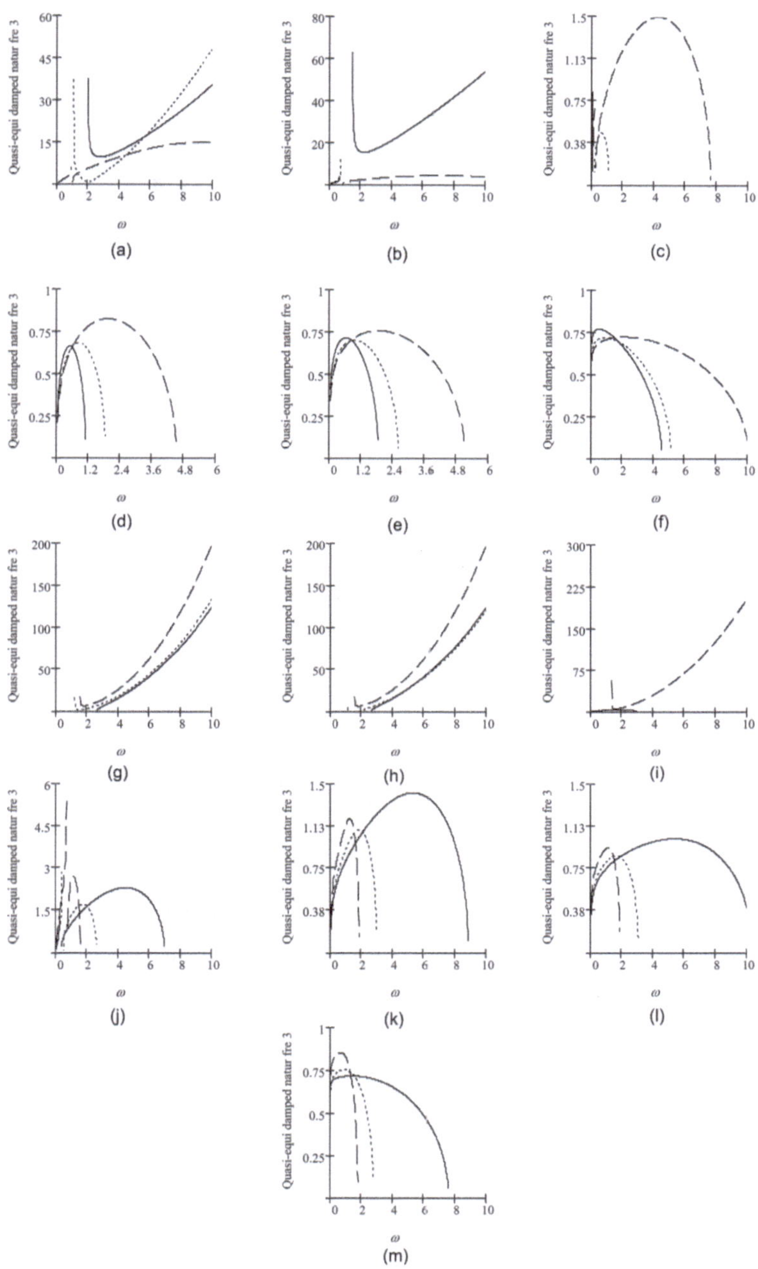

FIGURE 10.5 Illustrations of ω_{eqd3} with $m=1$, $c=1$, $k=1$, and $\omega_n=1$. (a) ω_{eqd3} for $\beta=0.3$ and $\alpha=1.3$ (solid), 1.6 (dot), 1.9 (dash). (b) ω_{eqd3} for $\beta=0.6$ and $\alpha=1.3$ (solid), 1.6 (dot), 1.9 (dash). (c) ω_{eqd3} for $\beta=0.9$ and $\alpha=1.3$ (solid), 1.6 (dot), 1.9 (dash). (d) ω_{eqd3} for $\beta=1.3$ and $\alpha=1.3$ (solid), 1.6 (dot), 1.9 (dash). (e) ω_{eqd3} for $\beta=1.6$ and $\alpha=1.3$ (solid), 1.6 (dot), 1.9 (dash). (f) ω_{eqd3} for $\beta=1.9$ and $\alpha=1.3$ (solid), 1.6 (dot), 1.9 (dash). (g) ω_{eqd3} for $\beta=0.3$ and $\alpha=2.2$ (solid), 2.5 (dot), 2.8 (dash). (h) ω_{eqd3} for $\beta=0.3$ and $\alpha=2.2$ (solid), 2.5 (dot), 2.8 (dash). (i) ω_{ecd3} for $\beta=0.6$ and $\alpha=2.4$ (solid), 2.5 (dot), 2.8 (dash). (j) ω_{eqd3} for $\beta=0.9$ and $\alpha=2.2$ (solid), 2.5 (dot), 2.8 (dash). (k) ω_{eqd3} for $\beta=1.3$ and $\alpha=2.2$ (solid), 2.5 (dot), 2.8 (dash). (l) ω_{eqd3} for $\beta=1.6$ and $\alpha=2.2$ (solid), 2.5 (dot), 2.8 (dash). (m) ω_{eqd3} for $\beta=1.9$ and $\alpha=2.2$ (solid), 2.5 (dot), 2.8 (dash). Note that "Quasi-equivalent damped natural frequency 3" in the above figures stands for ω_{eqd3}.

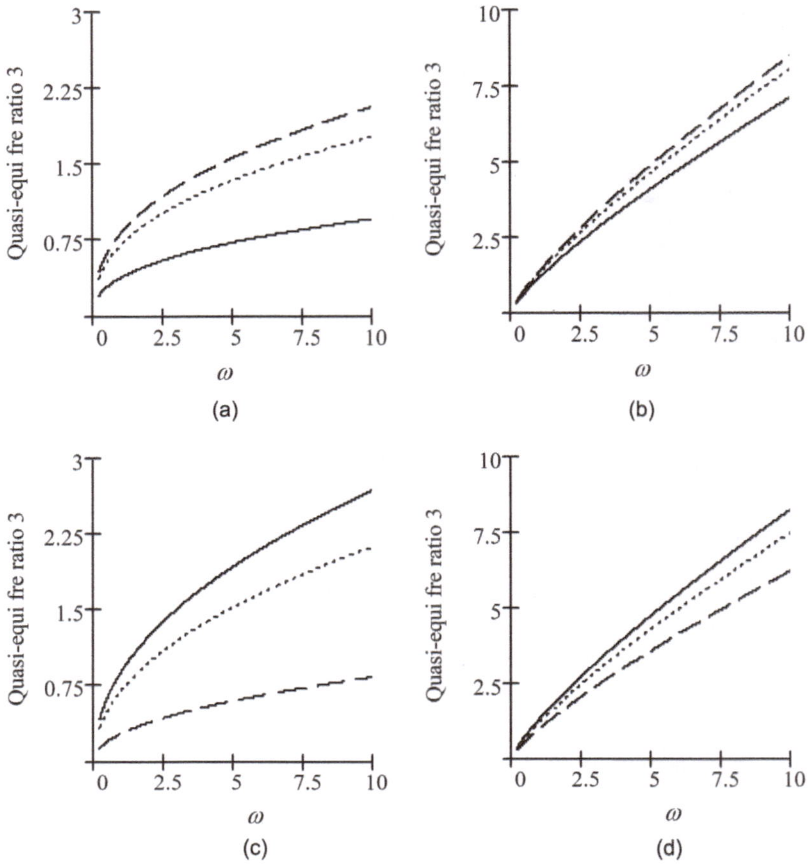

FIGURE 10.6 Plots of γ_{eq3} with $\omega_n=1$, $m=1$, and $c=1$. (a) γ_{eq3} for $\beta=0.8$ and $\alpha=1.3$ (solid), 1.6 (dot), 1.9 (dash). (b) γ_{eq3} for $\beta=1.6$ and $\alpha=1.3$ (solid), 1.6 (dot), 1.9 (dash). (c) γ_{eq3} for $\beta=0.95$ and $\alpha=2.3$ (solid), 2.6 (dot), 2.9 (dash). (d) γ_{eq3} for $\beta=1.6$ and $\alpha=2.3$ (solid), 2.6 (dot), 2.9 (dash). Note that "Quasi-equivalent frequency ratio 3" in the above figures stands for γ_{eq3}.

10.3 CLASS III FREE RESPONSE

Denote by $x_3(t)$ the free response of a fractional vibrator of class III. Then, it is a solution to the fractional differential equation in the form

$$\begin{cases} m\dfrac{d^{\alpha}x_3(t)}{dt^{\alpha}}+c\dfrac{d^{\beta}x_3(t)}{dt^{\beta}}+kx_3(t)=0, \\ x_3(0)=x_{30}, x_3'(0)=v_{30}, \end{cases} \quad 1<\alpha<3, 0<\beta<2. \tag{10.29}$$

The proposed theorem below provides its solution.

Theorem 10.10. $x_3(t)$ is given by

$$x_3(t)=e^{-\varsigma_{eq3}\omega_{eqn3}t}\left(x_{30}\cos\omega_{eqd3}t+\frac{v_{30}+\varsigma_{eq3}\omega_{eqn3}x_{30}}{\omega_{eqd3}}\sin\omega_{eqd3}t\right),\quad t\geq 0. \tag{10.30}$$

Proof. Following Theorem 10.1, the solution to Eq. (10.29) equivalently equals to the one to the equation below

$$\begin{cases} m_{eq3}\dfrac{d^2x_3(t)}{dt^2}+c_{eq3}\dfrac{dx_3(t)}{dt}+kx_3(t)=0, \\ x_3(0)=x_{30},x_3'(0)=v_{30}, \end{cases} \quad 1<\alpha<3, 0<\beta<2. \tag{10.31}$$

Because the above is a standard vibration equation of order 2, its solution is given by Eq. (10.30). This finishes the proof. □

Substituting ζ_{eq3} and ω_{eqn3} into Eq. (10.30) yields

$$x_3(t)=e^{-\frac{m\omega^{\alpha-1}\sin\frac{\alpha\pi}{2}+c\omega^{\beta-1}\sin\frac{\beta\pi}{2}}{2\left(m\omega^{\alpha-2}\left|\cos\frac{\alpha\pi}{2}\right|-c\omega^{\beta-2}\cos\frac{\beta\pi}{2}\right)}t}\left(x_{30}\cos\omega_{eqd3}t+A\sin\omega_{eqd3}t\right),\quad t\geq 0, \tag{10.32}$$

where

$$A=\frac{v_{30}+\dfrac{m\omega^{\alpha-1}\sin\frac{\alpha\pi}{2}+c\omega^{\beta-1}\sin\frac{\beta\pi}{2}}{2\left(m\omega^{\alpha-2}\left|\cos\frac{\alpha\pi}{2}\right|-c\omega^{\beta-2}\cos\frac{\beta\pi}{2}\right)}x_{30}}{\dfrac{\omega_n}{\sqrt{\left(\omega^{\alpha-2}\left|\cos\frac{\alpha\pi}{2}\right|-\frac{c}{m}\omega^{\beta-2}\cos\frac{\beta\pi}{2}\right)}}\sqrt{1-\dfrac{\left(m\omega^{\alpha-1}\sin\frac{\alpha\pi}{2}+c\omega^{\beta-1}\sin\frac{\beta\pi}{2}\right)^2}{4\left[-\left(m\omega^{\alpha-2}\cos\frac{\alpha\pi}{2}+c\omega^{\beta-2}\cos\frac{\beta\pi}{2}\right)k\right]}}}.$$

Figure 10.7 illustrates several plots of $x_3(t)$.

As can be seen from Theorem 10.10, $x_3(t)$ is a time-frequency distribution on a time-frequency plane. We may write it as

$$x_3(t)=x_3(t,\omega). \tag{10.33}$$

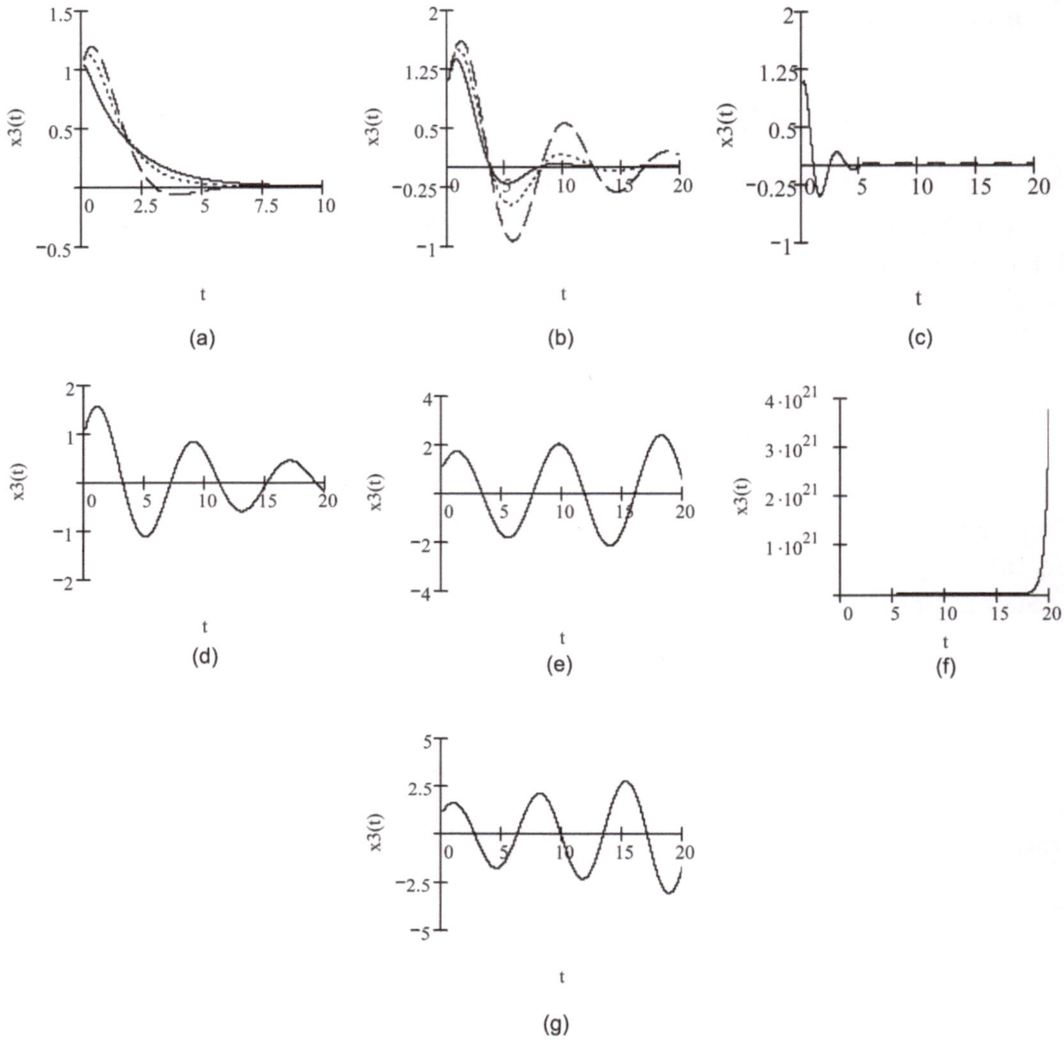

FIGURE 10.7 Illustrations of free response $x_3(t)$ for $x_{30}=1$, $v_{30}=1$, $m=1$, $c=1$, $k=1$, and $\omega=1$. (a) $x_3(t)$ when $\beta=0.8$ and $\alpha=1.3$ (solid), 1.6 (dot), 1.9 (dash). (b) $x_3(t)$ when $\beta=1.8$ and $\alpha=1.3$ (solid), 1.6 (dot), 1.9 (dash). (c) $x_3(t)$ when $\alpha=2.3$, $\beta=0.5$. (d) $x_3(t)$ when $\alpha=2.3$, $\beta=1.5$. (e) $x_3(t)$ when $\alpha=2.3$, $\beta=1.75$ (self-vibration). (f) $x_3(t)$ when $\alpha=2.6$, $\beta=0.5$ (self-vibration). (g) $\alpha=2.6$, $\beta=1.5$ (self-vibration).

10.4 CLASS III IMPULSE RESPONSE

Designate $h_3(t)$ as the impulse response of a fractional vibrator of class III. It results from the fractional differential equation in the form

$$m\frac{d^{\alpha}h_3(t)}{dt^{\alpha}}+c\frac{d^{\beta}h_3(t)}{dt^{\beta}}+kh_3(t)=\delta(t),\quad 1<\alpha<3,\ 0<\beta<2 \tag{10.34}$$

with zero initial conditions. We present a theorem below to give its analytical expression.

Theorem 10.11. The analytical expression of $h_3(t)$ is in the form

$$h_3(t) = e^{-\varsigma_{eq3}\omega_{eqn3}t}\frac{1}{m_{eq3}\omega_{eqd3}}\sin\omega_{eqd3}t, \quad t \geq 0. \tag{10.35}$$

Proof. The impulse response $h_3(t)$ in Eq. (10.34) is equivalent to the one in the following differential equation according to Theorem 10.1:

$$m_{eq3}\frac{d^2h_3(t)}{dt^2} + c_{eq3}\frac{dh_3(t)}{dt} + kh_3(t) = \delta(t), \quad 1 < \alpha < 3,\ 0 < \beta < 2. \tag{10.36}$$

As the above is a standard vibration equation, its solution is expressed by Eq. (10.35). This finishes the proof. □

Substituting m_{eq3}, ζ_{eq3}, and ω_{eqn3} into Eq. (10.35) results in

$$h_3(t) = \frac{e^{-\frac{m\omega^{\alpha-1}\sin\frac{\alpha\pi}{2} + c\omega^{\beta-1}\sin\frac{\beta\pi}{2}}{2\sqrt{-\left(m\omega^{\alpha-2}\cos\frac{\alpha\pi}{2} + c\omega^{\beta-2}\cos\frac{\beta\pi}{2}\right)k}}\omega_{eqn3}t}\sin\omega_{eqd3}t}{-\left(m\omega^{\alpha-2}\cos\frac{\alpha\pi}{2} + c\omega^{\beta-2}\cos\frac{\beta\pi}{2}\right)\omega_{eqd3}}, \quad t \geq 0. \tag{10.37}$$

Several plots of $h_3(t)$ are demonstrated in Figure 10.8.

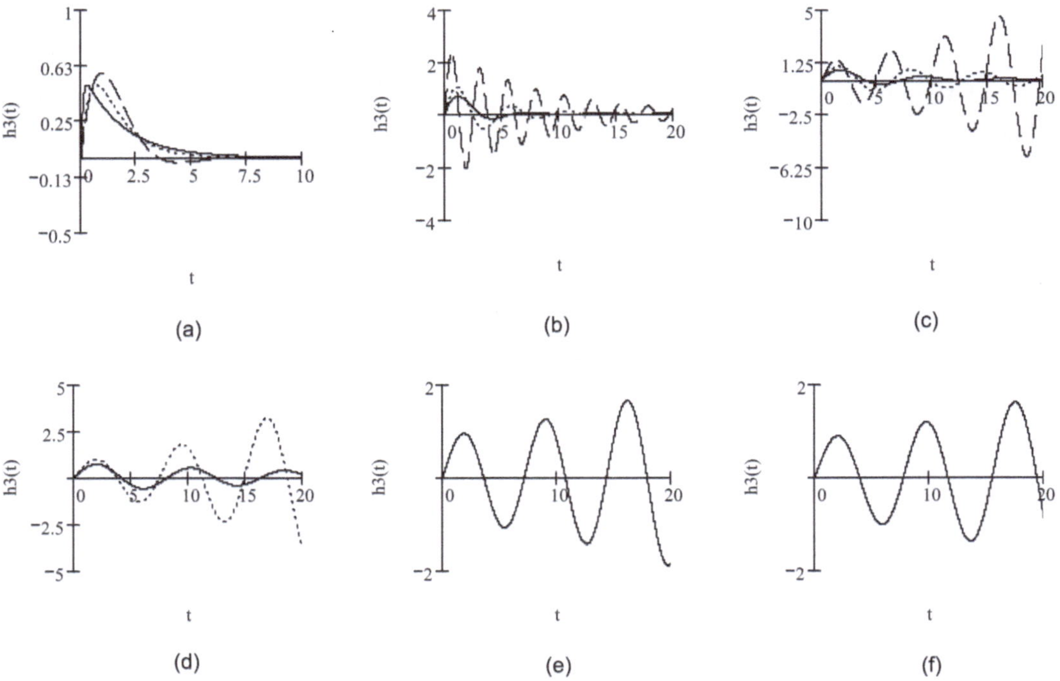

FIGURE 10.8 Plots of $h_3(t)$ for $m=1$, $c=1$, $k=1$, and $\omega=1$. (a) $h_3(t)$ when $\beta=0.8$ and $\alpha=1.3$ (solid), 1.6 (dot), 1.9 (dash). (b) $h_3(t)$ when $\beta=0.9$ and $\alpha=2.2$ (solid), 2.5 (dot), 2.8 (dash). (c) $h_3(t)$ when $\beta=1.3$ and $\alpha=2.3$ (solid), 2.6 (dot), 2.9 (dash) (self-vibration). (d) $h_3(t)$ when $\beta=1.6$ and $\alpha=2.3$ (solid), 2.6 (dot) (Self-vibration). (e) $h_3(t)$ when $\alpha=2.6$, $\beta=1.5$ (self-vibration). (f) $h_3(t)$ when $\alpha=2.5$, $\beta=1.6$ (self-vibration).

The impulse response function $h_3(t)$ may be expressed as a time-frequency plane by

$$h_3(t)=h_3(t,\omega). \tag{10.38}$$

10.5 CLASS III STEP RESPONSE

Designate $g_3(t)$ as the unit step response of a fractional vibrator of class III. It is a solution to

$$m\frac{d^\alpha g_3(t)}{dt^\alpha}+c\frac{d^\beta g_3(t)}{dt^\beta}+kg_3(t)=u(t),\quad 1<\alpha<3, 0<\beta<2 \tag{10.39}$$

with zero initial conditions. The present theorem below depicts the solution.

Theorem 10.12. The solution to the aforementioned equation is expressed by

$$g_3(t)=\frac{1}{k}\left[1-\frac{e^{-\varsigma_{eq3}\omega_{eqn3}t}}{\sqrt{1-\varsigma_{eq3}^2}}\cos\left(\omega_{eqd3}t-\phi_3\right)\right],\quad t\geq 0, \tag{10.40}$$

where

$$\phi_3=\tan^{-1}\frac{\varsigma_{eq3}}{\sqrt{1-\varsigma_{eq3}^2}}. \tag{10.41}$$

Proof. Based on Theorem 10.1, $g_3(t)$ in Eq. (10.39) equals to the solution to the equation below

$$m_{eq3}\frac{d^2 g_3(t)}{dt^2}+c_{eq3}\frac{dg_3(t)}{dt}+kg_3(t)=u(t),\quad 1<\alpha<3,\ 0<\beta<2. \tag{10.42}$$

Since the above is a standard vibration equation, $g_3(t)$ is expressed by Eq. (10.40). The proof is finished. □

When substituting ζ_{eq3} and ω_{eqn3} into Eq. (10.40), we have

$$g_3(t)=\frac{1}{k}\left[1-\frac{e^{-\frac{m\omega^{\alpha-1}\sin\frac{\alpha\pi}{2}+c\omega^{\beta-1}\sin\frac{\beta\pi}{2}}{2\sqrt{-\left(m\omega^{\alpha-2}\cos\frac{\alpha\pi}{2}+c\omega^{\beta-2}\cos\frac{\beta\pi}{2}\right)k}}\omega_{eqn3}t}}{\sqrt{1-\left[\frac{m\omega^{\alpha-1}\sin\frac{\alpha\pi}{2}+c\omega^{\beta-1}\sin\frac{\beta\pi}{2}}{2\sqrt{-\left(m\omega^{\alpha-2}\cos\frac{\alpha\pi}{2}+c\omega^{\beta-2}\cos\frac{\beta\pi}{2}\right)k}}\right]^2}}\cos\left(\omega_{eqd3}t-\phi_3\right)\right], \tag{10.43}$$

where

$$\phi_3 = \tan^{-1}\frac{\omega^{\alpha-1}\sin\frac{\alpha\pi}{2} + 2\varsigma\omega_n\omega^{\beta-1}\sin\frac{\beta\pi}{2}}{\sqrt{\left[2\omega_n\sqrt{-\left(\omega^{\alpha-2}\cos\frac{\alpha\pi}{2} + 2\varsigma\omega_n\omega^{\beta-2}\cos\frac{\beta\pi}{2}\right)}\right]^2 - \left(\omega^{\alpha-1}\sin\frac{\alpha\pi}{2} + 2\varsigma\omega_n\omega^{\beta-1}\sin\frac{\beta\pi}{2}\right)^2}}. \tag{10.44}$$

Several plots of $g_3(t)$ are illustrated in Figure 10.9.

The function $g_3(t)$ may be expressed by

$$g_3(t) = g_3(t,\omega). \tag{10.45}$$

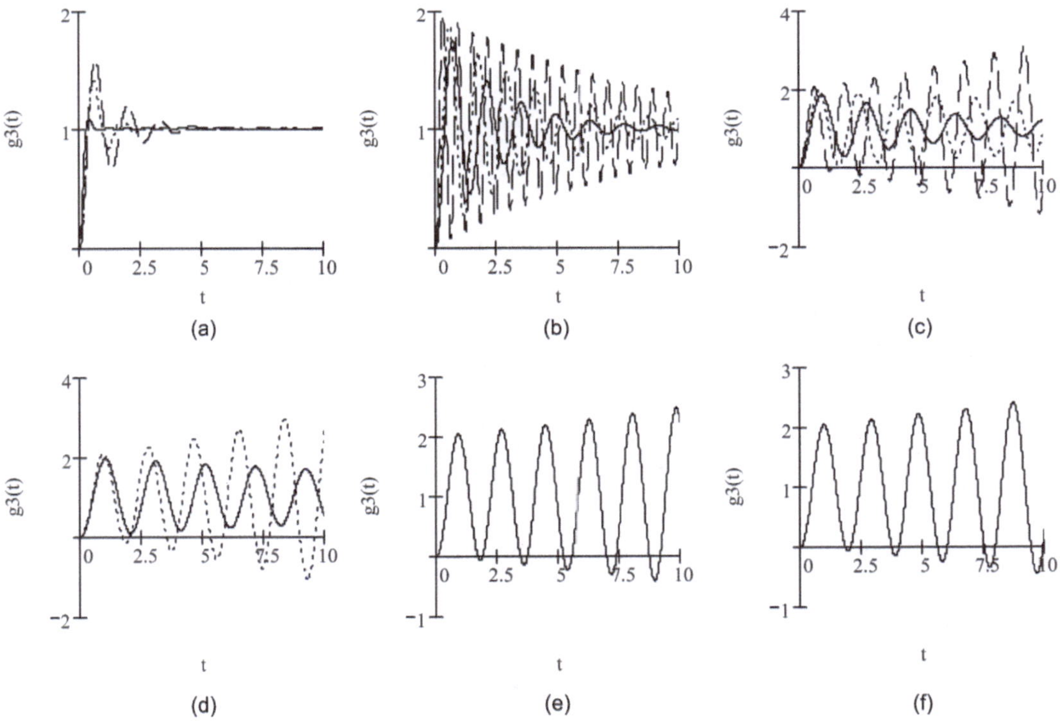

FIGURE 10.9 Plots of $g_3(t)$ for $m=1$, $c=1$, $k=16$, and $\omega=1$. (a) $g_3(t)$ when $\beta=0.8$ and $\alpha=1.3$ (solid), 1.6 (dot), 1.9 (dash). (b) $g_3(t)$ when $\beta=0.9$ and $\alpha=2.2$ (solid), 2.5 (dot), 2.8 (dash). (c) $g_3(t)$ when $\beta=1.3$ and $\alpha=2.3$ (solid), 2.6 (dot), 2.9 (dash) (self-vibration). (d) $g_3(t)$ when $\beta=1.6$ and $\alpha=2.3$ (solid), 2.6 (dot) (self-vibration). (e) $g_3(t)$ when $\alpha=2.6$, $\beta=1.5$ (self-vibration). (f) $\alpha=2.5$, $\beta=1.6$ (self-vibration).

10.6 CLASS III FREQUENCY TRANSFER FUNCTION

We use a theorem below to present the frequency transfer function of a fractional vibrator of class III.

Theorem 10.13. Let $H_3(\omega)$ be the frequency transfer function of a fractional vibrator of class III. Its expression is given by

$$H_3(\omega)=\frac{1/k}{1-\gamma^2\left(\omega^{\alpha-2}\left|\cos\frac{\alpha\pi}{2}\right|-2\varsigma\omega_n\omega^{\beta-2}\cos\frac{\beta\pi}{2}\right)+i\frac{\gamma\left(\omega^{\alpha-1}\sin\frac{\alpha\pi}{2}+2\varsigma\omega_n\omega^{\beta-1}\sin\frac{\beta\pi}{2}\right)}{\omega_n\left(\omega^{\alpha-2}\left|\cos\frac{\alpha\pi}{2}\right|-2\varsigma\omega_n\omega^{\beta-2}\cos\frac{\beta\pi}{2}\right)}},\tag{10.46}$$

where $\gamma=\frac{\omega}{\omega_n}$.

Proof. Consider the following equation

$$-\left(m\omega^{\alpha-2}\cos\frac{\alpha\pi}{2}+c\omega^{\beta-2}\cos\frac{\beta\pi}{2}\right)\frac{d^2h_3(t)}{dt^2}$$
$$+\left(m\omega^{\alpha-1}\sin\frac{\alpha\pi}{2}+c\omega^{\beta-1}\sin\frac{\beta\pi}{2}\right)\frac{dh_3(t)}{dt}+kh_3(t)=\delta(t).$$

Let $H_3(\omega)$ be the Fourier transform of $h_3(t)$. Then, we have

$$\begin{bmatrix}-\left(m\omega^{\alpha-2}\cos\frac{\alpha\pi}{2}+c\omega^{\beta-2}\cos\frac{\beta\pi}{2}\right)(i\omega)^2\\ +\left(m\omega^{\alpha-1}\sin\frac{\alpha\pi}{2}+c\omega^{\beta-1}\sin\frac{\beta\pi}{2}\right)(i\omega)+k\end{bmatrix}H_3(\omega)=1.$$

From the above, we have Eq. (10.46). This finishes the proof. □

Due to $\frac{c}{m}=2\varsigma\omega_n$, the above may be given by

$$H_3(\omega)=\frac{1/k}{1-\gamma^2\left(\omega^{\alpha-2}\left|\cos\frac{\alpha\pi}{2}\right|-\frac{c\omega^{\beta-2}\cos\frac{\beta\pi}{2}}{m}\right)+i\frac{\gamma\left(\omega^{\alpha-1}\sin\frac{\alpha\pi}{2}+\frac{m}{c}\omega^{\beta-1}\sin\frac{\beta\pi}{2}\right)}{\omega_n\left(\omega^{\alpha-2}\left|\cos\frac{\alpha\pi}{2}\right|-\frac{m}{c}\omega^{\beta-2}\cos\frac{\beta\pi}{2}\right)}}.\tag{10.47}$$

Moreover, $H_3(\omega)$ may be expressed by

$$H_3(\omega)=\frac{1}{k\left(1-\gamma_{\mathrm{eq}3}^2+i2\varsigma_{\mathrm{eq}3}\gamma_{\mathrm{eq}3}\right)}. \tag{10.48}$$

The amplitude $|H_3(\omega)|$ is given by

$$|H_3(\omega)|=\frac{1/k}{\sqrt{\left(1-\gamma_{\mathrm{eq}3}^2\right)^2+\left(2\varsigma_{\mathrm{eq}3}\gamma_{\mathrm{eq}3}\right)^2}}. \tag{10.49}$$

Expanding the above yields

$$|H_3(\omega)|=\frac{1/k}{\sqrt{\left\{\begin{array}{l}\left[1-\gamma^2\left(\omega^{\alpha-2}\left|\cos\frac{\alpha\pi}{2}\right|-2\varsigma\omega_n\omega^{\beta-2}\cos\frac{\beta\pi}{2}\right)\right]^2\\+\left[\dfrac{\gamma\left(\omega^{\alpha-1}\sin\frac{\alpha\pi}{2}+2\varsigma\omega_n\omega^{\beta-1}\sin\frac{\beta\pi}{2}\right)}{\omega_n\left(\omega^{\alpha-2}\left|\cos\frac{\alpha\pi}{2}\right|-2\varsigma\omega_n\omega^{\beta-2}\cos\frac{\beta\pi}{2}\right)}\right]^2\end{array}\right\}}}. \tag{10.50}$$

The phase-frequency response function is expressed by

$$\varphi_3(\omega)=\tan^{-1}\frac{\dfrac{\gamma\left(\omega^{\alpha-1}\sin\frac{\alpha\pi}{2}+2\varsigma\omega_n\omega^{\beta-1}\sin\frac{\beta\pi}{2}\right)}{\omega_n\left(\omega^{\alpha-2}\left|\cos\frac{\alpha\pi}{2}\right|-2\varsigma\omega_n\omega^{\beta-2}\cos\frac{\beta\pi}{2}\right)}}{1-\gamma^2\left(\omega^{\alpha-2}\left|\cos\frac{\alpha\pi}{2}\right|-2\varsigma\omega_n\omega^{\beta-2}\cos\frac{\beta\pi}{2}\right)}. \tag{10.51}$$

When computing $\varphi_3(\omega)$ using digital computers, we use the expression below

$$\varphi_3(\omega)=\cos^{-1}\frac{1-\gamma_{\mathrm{eq}3}^2}{\sqrt{\left(1-\gamma_{\mathrm{eq}3}^2\right)^2+\left(2\varsigma_{\mathrm{eq}3}\gamma_{\mathrm{eq}3}\right)^2}}. \tag{10.52}$$

Figures 10.10 and 10.11 show the plots of $|H_3(\omega)|$. Plots of $\varphi_3(\omega)$ are in Figures 10.12 and 10.13.

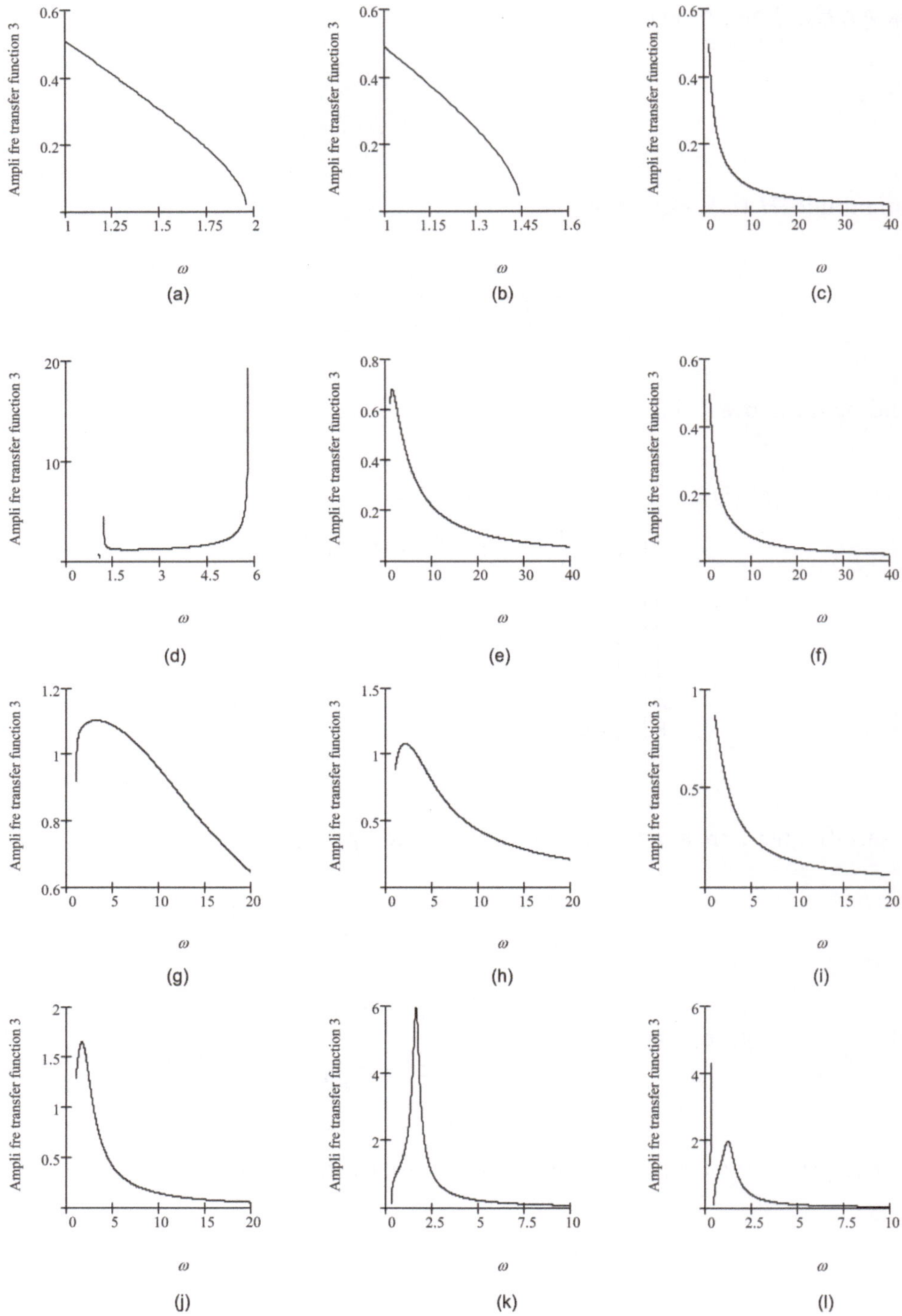

FIGURE 10.10 Plots of amplitude-frequency response $|H_3(\omega)|$ for $m=1$, $c=1$, and $k=1$ with $\omega \geq 1$. (a) $|H_3(\omega)|$ when $(\alpha, \beta)=(1.3, 0.3)$. (b) $|H_3(\omega)|$ when $(\alpha, \beta)=(1.3, 0.6)$. (c) $|H_3(\omega)|$ when $(\alpha, \beta)=(1.3, 0.9)$. (d) $|H_3(\omega)|$ when $(\alpha, \beta)=(1.6, 0.3)$. (e) $|H_3(\omega)|$ when $(\alpha, \beta)=(1.6, 0.6)$. (f) $|H_3(\omega)|$ when $(\alpha, \beta)=(1.6, 0.9)$. (g) $|H_3(\omega)|$ when $(\alpha, \beta)=(1.9, 0.3)$. (h) $|H_3(\omega)|$ when $(\alpha, \beta)=(1.9, 0.6)$. (i) $|H_3(\omega)|$ when $(\alpha, \beta)=(1.9, 0.9)$. (j) $|H_3(\omega)|$ when $(\alpha, \beta)=(2.3, 0.6)$. (k) $|H_3(\omega)|$ when $(\alpha, \beta)=(2.3, 0.9)$. (l) $|H_3(\omega)|$ when $(\alpha, \beta)=(2.6, 0.9)$.

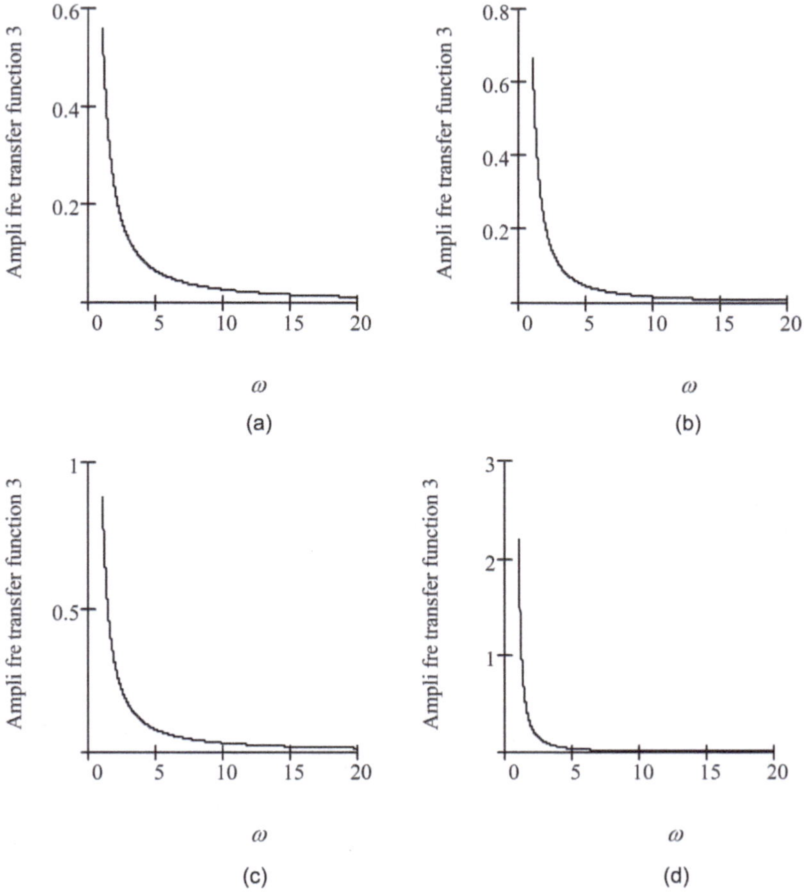

FIGURE 10.11 Plots of $|H_3(\omega)|$ for $m=1$, $c=1$, and $k=1$ with $\omega \geq 1$. (a) $|H_3(\omega)|$ when $(\alpha, \beta)=(1.3, 1.3)$. (b) $|H_3(\omega)|$ when $(\alpha, \beta)=(1.3, 1.6)$. (c) $|H_3(\omega)|$ when $(\alpha, \beta)=(1.9, 1.3)$. (d) $|H_3(\omega)|$ when $(\alpha, \beta)=(2.6, 1.6)$.

10.7 CLASS III QUASIEQUIVALENT LOGARITHMIC DECREMENT AND *Q* FACTOR

10.7.1 Quasiequivalent Logarithmic Decrement of Class III Fractional Vibrators

We now address the quasiequivalent logarithmic decrement for a class III fractional vibrator.

Theorem 10.14. Let t_i and t_{i+1} be two time points of fractional free response $x_3(t)$, where $x_3(t_i)$ and $x_3(t_{i+1})$ are successive peak values at t_i and t_{i+1}. Let Δ_{eq3} be the quasiequivalent logarithmic decrement of $x_3(t)$. Then,

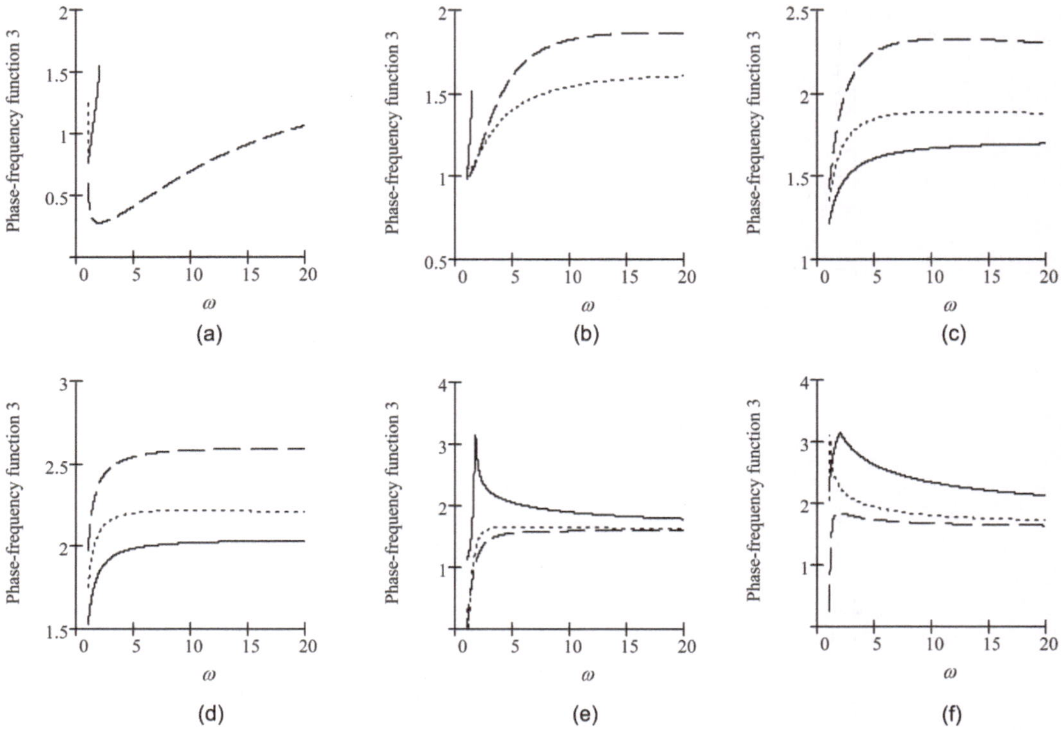

FIGURE 10.12 Illustrations of $\varphi_3(\omega)$ for $m=1$, $c=1$, and $k=1$ with $\omega \geq 1$. (a) $\varphi_3(\omega)$ when $\beta=0.3$ and $\alpha=1.3$ (solid), 1.6 (dot), 1.9 (dash). (b) $\varphi_3(\omega)$ when $\beta=0.6$ and $\alpha=1.3$ (solid), 1.6 (dot), 1.9 (dash). (c) $\varphi_3(\omega)$ when $\beta=0.9$ and $\alpha=1.3$ (solid), 1.6 (dot), 1.9 (dash). (d) $\varphi_3(\omega)$ when $\beta=1.3$ and $\alpha=1.3$ (solid), 1.6 (dot), 1.9 (dash). (e) $\varphi_3(\omega)$ when $\beta=0.9$ and $\alpha=2.3$ (solid), 2.6 (dot), 2.8 (dash). (f) $\varphi_3(\omega)$ when $\beta=1.3$ and $\alpha=2.3$ (solid), 2.6 (dot), 2.8 (dash).

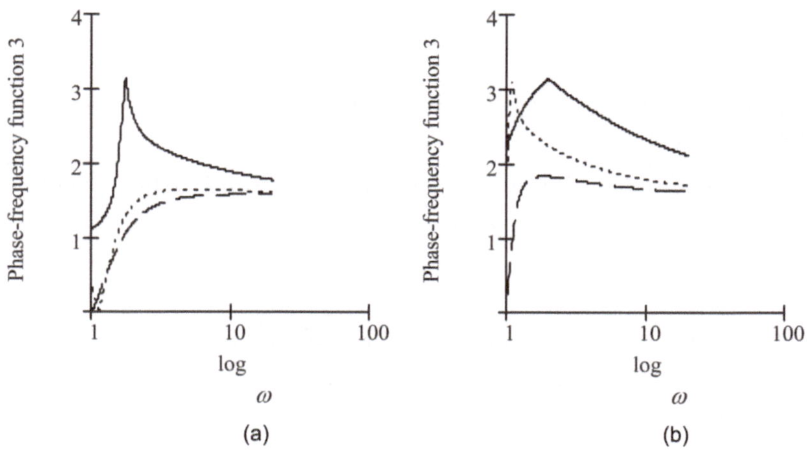

FIGURE 10.13 Plots of $\varphi_3(\omega)$ with log-scale for $m=1$, $c=1$, and $k=1$ with $\omega \geq 1$ with $\alpha=2.3$ (solid), 2.6 (dot), 2.8 (dash). (a) $\varphi_3(\omega)$ for $\beta=0.9$. (b) $\varphi_3(\omega)$ for $\beta=1.3$.

$$\Delta_{eq3}=\frac{\dfrac{\pi\left(\omega^{\alpha-1}\sin\dfrac{\alpha\pi}{2}+2\varsigma\omega_n\omega^{\beta-1}\sin\dfrac{\beta\pi}{2}\right)}{\omega_n\sqrt{-\left(\omega^{\alpha-2}\cos\dfrac{\alpha\pi}{2}+2\varsigma\omega_n\omega^{\beta-2}\cos\dfrac{\beta\pi}{2}\right)}}}{\sqrt{1-\left(\dfrac{\omega^{\alpha-1}\sin\dfrac{\alpha\pi}{2}+2\varsigma\omega_n\omega^{\beta-1}\sin\dfrac{\beta\pi}{2}}{2\omega_n\sqrt{-\left(\omega^{\alpha-2}\cos\dfrac{\alpha\pi}{2}+2\varsigma\omega_n\omega^{\beta-2}\cos\dfrac{\beta\pi}{2}\right)}}\right)^2}}. \tag{10.53}$$

Proof. According to the definition of logarithmic decrement, we have

$$\Delta_{eq3}=\Delta_{eq3}(\omega,\alpha,\beta)=\ln\frac{x_3(t_i)}{x_3(t_{i+1})}=\frac{2\pi\varsigma_{eq3}}{\sqrt{1-\varsigma_{eq3}^2}},\quad 1<\alpha<3, 0<\beta<2. \tag{10.54}$$

Substituting ς_{eq3} into the above produces Eq. (10.53). The proof is complete. □

Since Δ_{eq3} is not dimensionless, we use the prefix "quasi" to describe such an equivalent logarithmic decrement. When $\alpha=2$ and $\beta=1$, it reduces to conventional logarithmic decrement $\frac{2\pi\varsigma}{\sqrt{1-\varsigma^2}}$. The plots of Δ_{eq3} are demonstrated in Figure 10.14.

10.7.2 Quasiequivalent Q Factor of Class III Fractional Vibrators

Theorem 10.15. Let Q_{eq3} be the quasiequivalent Q factor of a fractional vibrator of class III. Then, it is expressed by

$$Q_{eq3}=Q_{eq3}(\omega,\alpha,\beta)=\frac{\omega_n\sqrt{-\left(\omega^{\alpha-2}\cos\dfrac{\alpha\pi}{2}+2\varsigma\omega_n\omega^{\beta-2}\cos\dfrac{\beta\pi}{2}\right)}}{\omega^{\alpha-1}\sin\dfrac{\alpha\pi}{2}+2\varsigma\omega_n\omega^{\beta-1}\sin\dfrac{\beta\pi}{2}},\quad 1<\alpha<3, 0<\beta<2. \tag{10.55}$$

Proof. Based on the definition of Q factor, we have

$$Q_{eq3}=\frac{1}{2\varsigma_{eq3}}. \tag{10.56}$$

Substituting ζ_{eq3} into the above results in Eq. (10.55). The proof is complete. □

Note that Q_{eq3} is not a constant but a function in terms of ω, α, and β. When $\alpha=2$ and $\beta=1$, it reduces to the conventional Q factor $\frac{1}{2\varsigma}$. In general, it is not dimensionless unless $\alpha=2$ and $\beta=1$. That is the reason we put a prefix "quasi" on it. Its plots are illustrated in Figure 10.15.

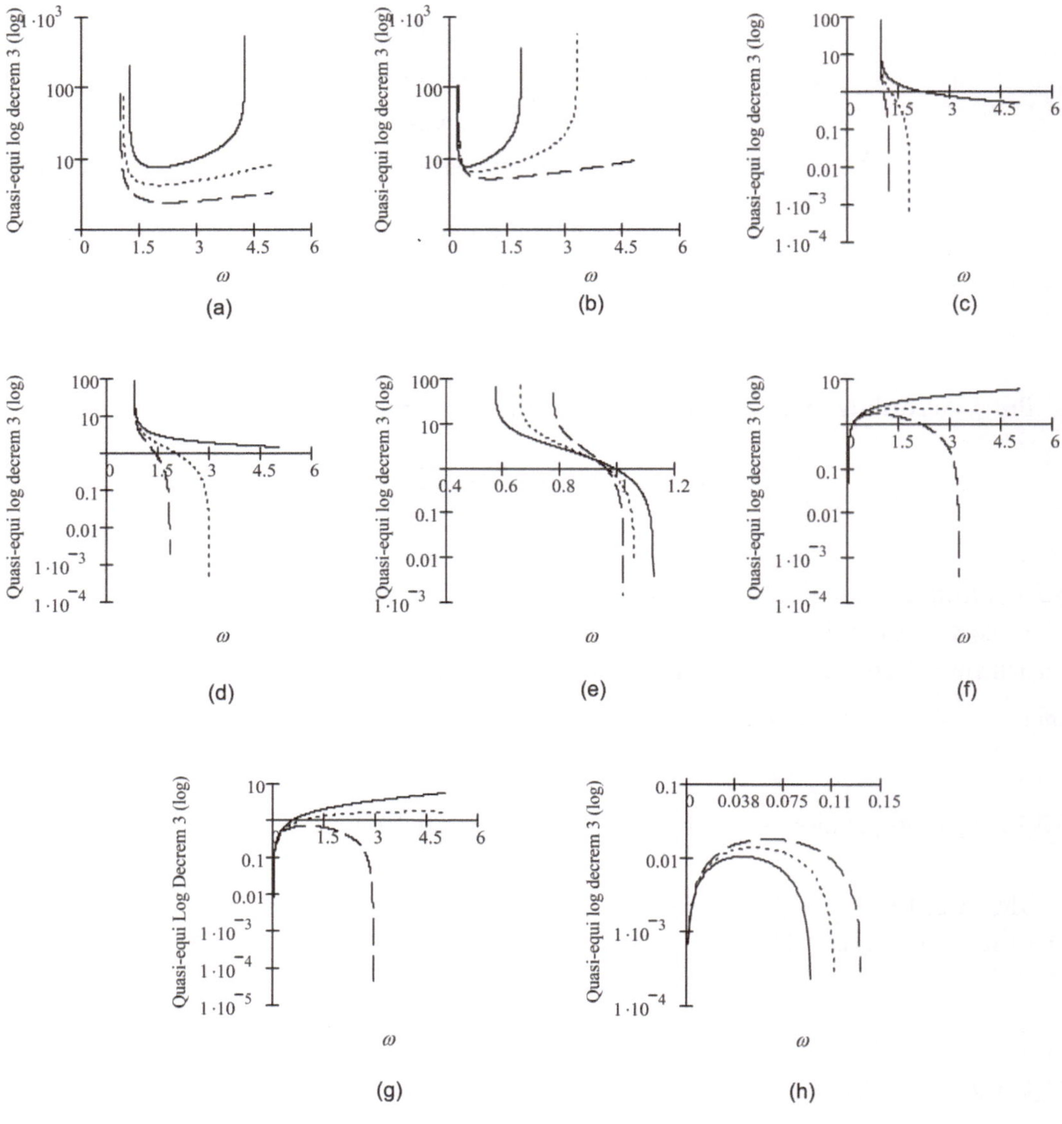

FIGURE 10.14 Plots of quasiequivalent logarithmic decrement Δ_{eq3} (log) for $m=1$, $c=1$, and $k=1$. (a) Δ_{eq3} when $\beta=0.3$ and $\alpha=1.7$ (solid), 1.8 (dot), 1.9 (dash). (b) Δ_{eq3} when $\beta=0.9$ and $\alpha=1.7$ (solid), 1.8 (dot), 1.9 (dash). (c) Δ_{eq3} when $\beta=0.3$ and $\alpha=2.0$ (solid), 2.1 (dot), 2.2 (dash). (d) Δ_{eq3} when $\beta=0.6$ and $\alpha=2.0$ (solid), 2.1 (dot), 2.2 (dash). (e) Δ_{eq3} when $\beta=0.9$ and $\alpha=2.6$ (solid), 2.7 (dot), 2.9 (dash). (f) Δ_{eq3} when $\beta=1.3$ and $\alpha=2.0$ (solid), 2.1 (dot), 2.2 (dash). (g) Δ_{eq3} when $\beta=1.6$ and $\alpha=2.0$ (solid), 2.1 (dot), 2.2 (dash). (h) Δ_{eq3} when $\beta=1.9$ and $\alpha=2.6$ (solid), 2.7 (dot), 2.9 (dash). Note that "Quasi-equivalent logarithmic decrement 3" in the above figures stands for Δ_{eq3}.

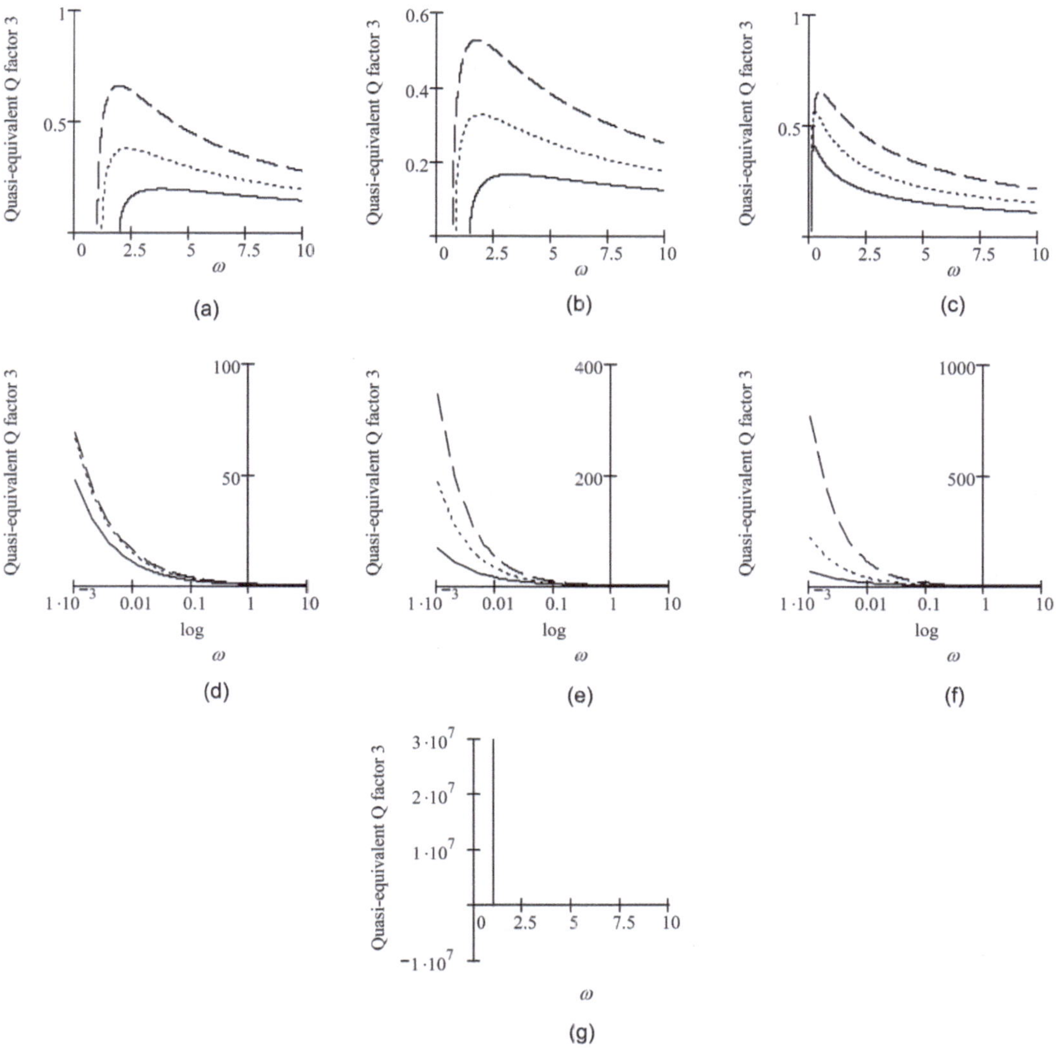

FIGURE 10.15 Plots of Q_{eq3} for $m=1$, $c=1$, and $k=1$. (a) Q_{eq3} when $\beta=0.3$ and $\alpha=1.3$ (solid), 1.5 (dot), 1.7 (dash). (b) Q_{eq3} when $\beta=0.6$ and $\alpha=1.3$ (solid), 1.5 (dot), 1.7 (dash). (c) Q_{eq3} when $\beta=0.9$ and $\alpha=1.3$ (solid), 1.5 (dot), 1.7 (dash). (d) Q_{eq3} with ω in log-scale and when $\beta=1.3$ and $\alpha=1.3$ (solid), 1.5 (dot), 1.7 (dash). (e) Q_{eq3} with ω in log-scale and for $\beta=1.6$ and $\alpha=1.3$ (solid), 1.5 (dot). (f) With ω in log-scale, Q_{eq3} when $\beta=1.9$ and $\alpha=1.3$ (solid), 1.5 (dot). (g) $\beta=0.3$ and $\alpha=2.3$ (solid), 2.5 (dot), 2.7 (dash). Note that "Quasi-equivalent Q factor 3" in the above figures stands for Q_{eq3}.

10.8 REMARK C_{EQ3} FROM A VIEW OF RAYLEIGH DAMPING

Denote by c_r the Rayleigh damping. It is given by $c_r = am + bk$, where m and k are primary mass and stiffness, respectively, a is proportional to vibration frequency ω while b is inversely proportional to ω (Jin and Xia [1], Harris [2], Palley et al. [3]). When writing c_{eq3} by

$$c_{eq3} = c_{eq31} + c_{eq32}, \tag{10.57}$$

where

$$c_{eq31} = m\omega^{\alpha-1} \sin\frac{\alpha\pi}{2}, \tag{10.58}$$

and

$$c_{eq32} = c\omega^{\beta-1} \sin\frac{\beta\pi}{2}. \tag{10.59}$$

The part c_{eq31} follows the standard form of the Rayleigh damping with respect to am but c_{eq32} appears new regarding the Rayleigh damping. As a matter of fact, $c_{eq32} = c_{eq2}$, see subsection 9.8.

10.9 CONCLUDING REMARK

The proposed results in this chapter are novel in the field. They are interesting and may not be observed in the field of ordinary vibrators.

10.10 EXERCISES

10.1. Write the expression of sinusoidal response to a fractional vibrator in class III. (Hint: Refer to [4]).

10.2. Find the condition of $m_{eq3} < 0$.

10.3. Write the condition of $c_{eq3} < 0$.

10.4. Find the condition for $0 < \zeta_{eq3} < 1$.

10.5. Prove that ζ_{eq3} is not dimensionless if $\alpha \neq 2$ or $\beta \neq 1$.

10.6. Prove that ω_{eqn3} is not a standard angular frequency when $\alpha \neq 2$ or $\beta \neq 1$.

REFERENCES

1. X. D. Jin and L. J. Xia, *Ship Hull Vibration*, The Press of Shanghai Jiaotong University, Shanghai, China, 2011. In Chinese.
2. C. M. Harris, *Shock and Vibration Handbook*, 5th Ed., McGraw-Hill, New York, 2002.
3. O. M. Palley, Г. B. Bahizov, and E. Я. Voroneysk, *Handbook of Ship Structural Mechanics*, National Defense Industry Publishing House, Beijing, China, 2002. In Chinese. Translated from Russian by B. H. Xu, X. Xu, and M. Q. Xu.
4. M. Li, Three classes of fractional oscillators, *Symmetry-Basel*, 10(2), 2018, 91 p.

CHAPTER 11

Fractional Vibrations of Class IV

THIS CHAPTER AIMS TO discuss the contributions in eight folds as follows.

1. Propose an equivalent motion equation for fractional vibrators in class IV in Theorem 11.1.
2. Introduce the quasiequivalent mass, quasiequivalent damping, quasiequivalent stiffness, quasiequivalent damping ratio, quasiequivalent damping free natural frequency, quasiequivalent damped natural frequency, quasiequivalent frequency ratio, quasiequivalent logarithmic decrement, and quasiequivalent Q factor for fractional vibrators in class IV. Their expressions are given in Theorems 11.2, 11.3, 11.6–11.10, 11.15, and 11.16, respectively.
3. Show that the quasiequivalent mass of a fractional vibrator in class IV equals to that of a fractional vibrator in class I in Theorem 11.2.
4. Reveal the phenomenon that the quasiequivalent damping of a fractional vibrator in class IV may vary from $-\infty$ to ∞ in Corollary 11.2.
5. Bring forward the condition for a fractional vibrator in class IV to be self-vibrated or non-stable in Theorem 11.5.
6. Propose the asymptotic properties of the quasiequivalent stiffness in Corollaries 11.4–11.6.
7. Present a mathematical explanation of the Rayleigh damping assumption in Corollary≈11.1.
8. Bring forward the analytic expressions of free response, impulse response, step response, frequency transfer function of a fractional vibrator of class IV in Theorems 11.11–11.14.

DOI: 10.1201/9781003460947-13

11.1 EQUIVALENT MOTION EQUATION OF CLASS IV

We propose Theorem 11.1 to explain an equivalent motion equation of a fractional vibrator of class IV.

Theorem 11.1. The motion equation of a fractional vibrator in class IV is

$$m\frac{d^{\alpha}x_4(t)}{dt^{\alpha}}+k\frac{d^{\lambda}x_4(t)}{dt^{\lambda}}=0,\quad 1<\alpha<3, 0\leq\lambda<1. \tag{11.1}$$

Its equivalent equation is given by

$$\begin{aligned}&-m\omega^{\alpha-2}\cos\frac{\alpha\pi}{2}\frac{d^2x_4(t)}{dt^2}+\left(m\omega^{\alpha-1}\sin\frac{\alpha\pi}{2}+k\omega^{\lambda-1}\sin\frac{\lambda\pi}{2}\right)\frac{dx_4(t)}{dt}\\&+k\omega^{\lambda}\cos\frac{\lambda\pi}{2}x_4(t)=0,\quad 1<\alpha<3,\ 0\leq\lambda<1.\end{aligned} \tag{11.2}$$

Proof. Let $X_4(\omega)$ be the Fourier transform of $x_4(t)$. Doing the Fourier transform of the left side of Eq. (11.1) yields

$$\begin{aligned}&\left[m\frac{d^{\alpha}x_4(t)}{dt^{\alpha}}+k\frac{d^{\lambda}x_4(t)}{dt^{\lambda}}\right]\leftrightarrow\left[m(i\omega)^{\alpha}+k(i\omega)^{\lambda}\right]X_4(\omega)\\&=\left[m\left(\cos\frac{\alpha\pi}{2}+i\sin\frac{\alpha\pi}{2}\right)\omega^{\alpha}+k\left(\cos\frac{\lambda\pi}{2}+i\sin\frac{\lambda\pi}{2}\right)\omega^{\lambda}\right]X_4(\omega)\\&=\left[m\omega^{\alpha}\cos\frac{\alpha\pi}{2}+i\left(m\omega^{\alpha}\sin\frac{\alpha\pi}{2}+k\omega^{\lambda}\sin\frac{\lambda\pi}{2}\right)+k\omega^{\lambda}\cos\frac{\lambda\pi}{2}\right]X_4(\omega).\end{aligned} \tag{11.3}$$

On the other hand, doing the Fourier transform of the left side of Eq. (11.2) produces

$$\begin{aligned}&\left[-m\omega^{\alpha-2}\cos\frac{\alpha\pi}{2}\frac{d^2x_4(t)}{dt^2}+\left(m\omega^{\alpha-1}\sin\frac{\alpha\pi}{2}+k\omega^{\lambda-1}\sin\frac{\lambda\pi}{2}\right)\frac{dx_4(t)}{dt}+k\omega^{\lambda}\cos\frac{\lambda\pi}{2}x_4(t)\right]\\&\leftrightarrow-\left(m\omega^{\alpha-2}\cos\frac{\alpha\pi}{2}\right)(i\omega)^2X_4(\omega)\\&+\left(m\omega^{\alpha-1}\sin\frac{\alpha\pi}{2}+k\omega^{\lambda-1}\sin\frac{\lambda\pi}{2}\right)(i\omega)X_4(\omega)+k\omega^{\lambda}\cos\frac{\lambda\pi}{2}X_4(\omega)\\&=\left[m\omega^{\alpha}\cos\frac{\alpha\pi}{2}+i\left(m\omega^{\alpha}\sin\frac{\alpha\pi}{2}+k\omega^{\lambda}\sin\frac{\lambda\pi}{2}\right)+k\omega^{\lambda}\cos\frac{\lambda\pi}{2}\right]X_4(\omega).\end{aligned} \tag{11.4}$$

From right sides of Eqs. (11.3) and (11.4), we see that Theorem 11.1 holds. This finishes the proof. □

11.2 QUASIEQUIVALENT VIBRATION PARAMETERS OF CLASS IV

11.2.1 Quasiequivalent Mass of Fractional Vibrators of Class IV

Let m_{eq4} be the quasiequivalent mass of a fractional vibrator of class IV. Theorem 11.2 depicts it.

Theorem 11.2. m_{eq4} equals to m_{eq1}. That is,

$$m_{eq4} = m_{eq1} = -m\omega^{\alpha-2}\cos\frac{\alpha\pi}{2}, \quad 1<\alpha<3. \tag{11.5}$$

Proof. From the item for inertia force in Eq. (11.2) and Theorem 8.2, we see that the above is valid. This finishes the proof. □

Due to Theorem 11.2, we do not discuss m_{eq4} anymore. Refer Chapter 8 for m_{eq1}.

11.2.2 Quasiequivalent Damping of Class IV Fractional Vibrators

The most interesting and attractive thing of a fractional vibrator of class IV is its damping.

Theorem 11.3. Let c_{eq4} be the quasiequivalent damping of a fractional vibrator of class IV. Then,

$$c_{eq4} = c_{eq4}(\omega,\alpha,\lambda) = m\omega^{\alpha-1}\sin\frac{\alpha\pi}{2} + k\omega^{\lambda-1}\sin\frac{\lambda\pi}{2}, \quad 1<\alpha<3,\ 0\le\lambda<1. \tag{11.6}$$

Proof: The second term on the left side of Eq. (11.2) is the friction force with the damping coefficient denoted by Eq. (11.6). The proof is completed. □

Theorem 11.4. c_{eq4} is proportional to the primary mass m and the primary stiffness k.

Proof. From Eq. (11.6), we see that c_{eq4} is proportional to m with the coefficient $\omega^{\alpha-1}\sin\frac{\alpha\pi}{2}$ and to k with the coefficient $\omega^{\lambda-1}\sin\frac{\lambda\pi}{2}$. The proof is completed. □

Corollary 11.1. The proposed c_{eq4} is consistent with the function form of the Rayleigh damping.

Proof. Note that c_{eq4} is proportional to m and k. Besides, $m\omega^{\alpha-1}\sin\frac{\alpha\pi}{2}$ is proportional to ω due to $1<\alpha<3$ while $k\omega^{\lambda-1}\sin\frac{\lambda\pi}{2}$ is inversely proportional to ω owing to $0<\lambda<1$. The damping c_{eq4} is consistent with the function form of the Rayleigh damping, referring [1–3] for the function form of the Rayleigh damping assumption. The proof is finished. □

It is worth noting that Theorem 11.4 may be taken as a mathematical explanation of the Rayleigh damping assumption.

As a matter of fact, c_{eq4} is made up of two parts, namely, $c_{eq4}=c_{eq41}+c_{eq42}$, where c_{eq41} results from fractional inertia. It is given by

$$c_{\text{eq41}} = m\omega^{\alpha-1}\sin\frac{\alpha\pi}{2}, \quad 1<\alpha<3. \tag{11.7}$$

On the other hand, c_{eq42} is contributed by fractional restoration. It is expressed by

$$c_{\text{eq42}} = k\omega^{\lambda-1}\sin\frac{\lambda\pi}{2}, \quad 0\le\lambda<1. \tag{11.8}$$

Now, we address the asymptotic properties of c_{eq4}.

Corollary 11.2: Asymptotic Properties for $\omega \to 0$ and $\omega \to \infty$. For c_{eq4}, we have

$$\lim_{\omega\to 0} c_{\text{eq4}}(\omega,\alpha,\lambda)=\infty, \quad \lambda\neq 0. \tag{11.9}$$

Besides,

$$\lim_{\omega\to\infty} c_{\text{eq4}}(\omega,\alpha,\lambda)=\begin{cases} \infty, & 1<\alpha<2, \\ -\infty, & 2<\alpha<3. \end{cases} \tag{11.10}$$

Proof. When $0<\lambda<1$, we have

$$\lim_{\omega\to 0}\omega^{\lambda-1}\sin\frac{\lambda\pi}{2}=\infty. \tag{11.11}$$

Thus, Eq. (11.9) holds. On the other hand,

$$\lim_{\omega\to\infty}\omega^{\alpha-1}\sin\frac{\alpha\pi}{2}=\begin{cases} \infty, & 1<\alpha<2, \\ -\infty, & 2<\alpha<3. \end{cases} \tag{11.12}$$

Thus, Eq. (11.10) is valid. This finishes the proof. □

Note that $c_{\text{eq4}}=0$ when $\alpha=2$ and $\lambda=0$. Figure 11.1 shows its plots.

Theorem 11.5. A fractional vibrator of class IV is self-vibrated if

$$\omega > \sqrt[\frac{1}{\alpha-1}]{\omega_n\omega^{\lambda-1}\frac{\sin\frac{\lambda\pi}{2}}{\left|\sin\frac{\alpha\pi}{2}\right|}}, \quad 2<\alpha<3,\ 0\le\lambda<1, \tag{11.13}$$

where $\omega_n=\sqrt{\frac{k}{m}}$.

Proof. If $2<\alpha<3$,

$$\sin\frac{\alpha\pi}{2}=-\left|\sin\frac{\alpha\pi}{2}\right|. \tag{11.14}$$

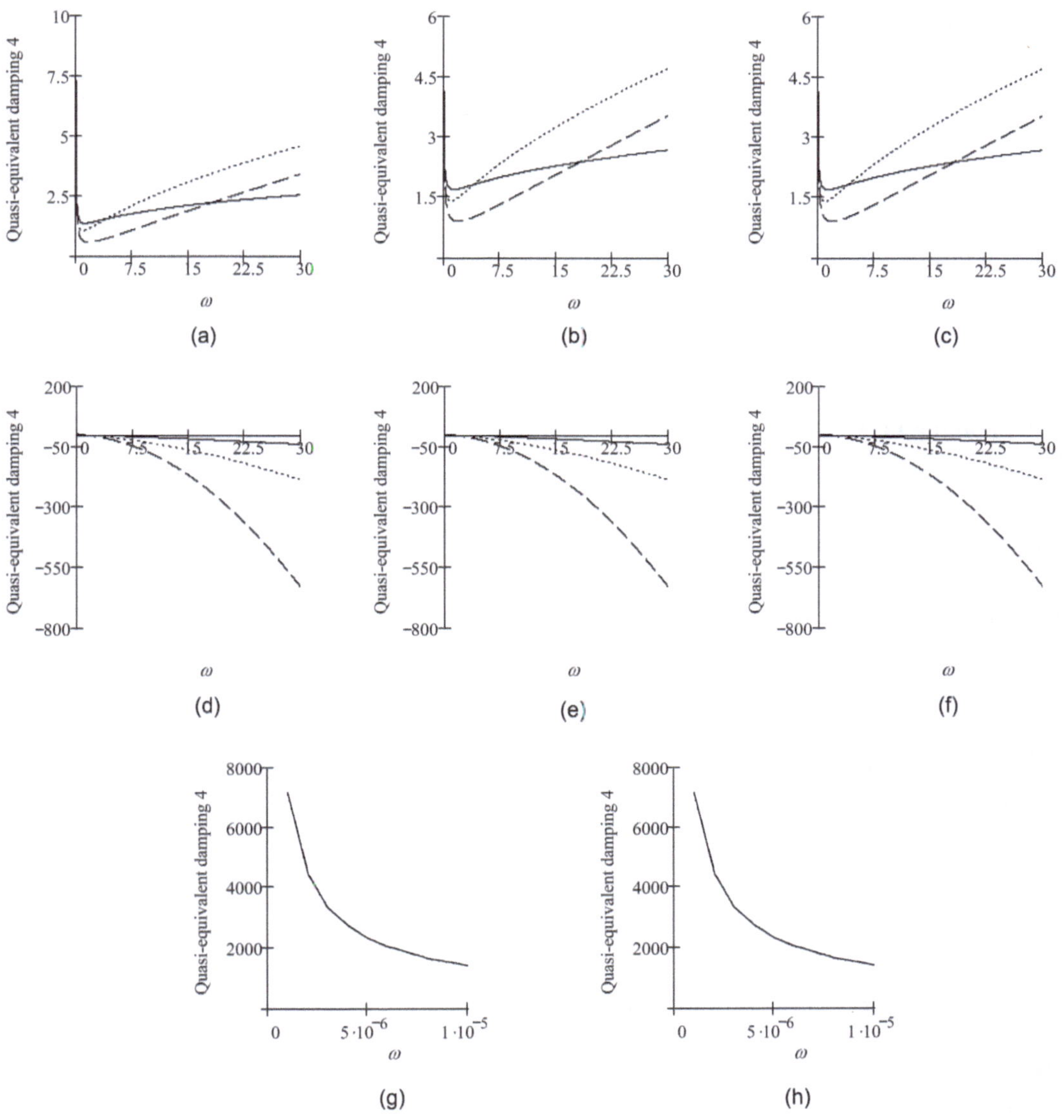

FIGURE 11.1 Plots of c_{eq4} with $m=1$ and $k=1$. (a) c_{eq4} when $\lambda=0.3$ and $\alpha=1.3$ (solid), 1.6 (dot), 1.9 (dash). (b) c_{eq4} when $\lambda=0.6$ and $\alpha=2.3$ (solid), 2.6 (dot), 2.9 (dash). (c) c_{eq4} when $\lambda=0.9$ and $\alpha=1.3$ (solid), 1.6 (dot), 1.9 (dash). (d) Negative c_{eq4} when $\lambda=0.3$ and $\alpha=2.3$ (solid), 2.6 (dot), 2.9 (dash). (e) Negative c_{eq4} when $\lambda=0.6$ and $\alpha=2.3$ (solid), 2.6 (dot), 2.9 (dash). (f) Negative damping. $\lambda=0.9$ and $\alpha=2.3$ (solid), 2.6 (dot), 2.9 (dash). (g) c_{eq4} when $\lambda=0.3$. $\alpha=1.3$ (Observing c_{eq4} at $\omega\approx 0$). (h) c_{eq4} when $\lambda=0.3$. $\alpha=2.3$ (Observing c_{eq4} at $\omega\approx 0$). Note that "Quasi-equivalent damping 4" in the above figures stands for c_{eq4}.

Let $c_{eq4}<0$. Then, when $2<\alpha<3$, we have

$$k\omega^{\lambda-1}\sin\frac{\lambda\pi}{2}<m\omega^{\alpha-1}\left|\sin\frac{\alpha\pi}{2}\right|. \tag{11.15}$$

Thus, if $\omega>\sqrt[\frac{1}{\alpha-1}]{\omega_n\omega^{\lambda-1}\frac{\sin\frac{\lambda\pi}{2}}{\left|\sin\frac{\alpha\pi}{2}\right|}}$, we have $c_{eq4}<0$ and a fractional vibrator in class IV is self-vibrated. The proof is complete. □

Corollary 11.3. A fractional vibrator in class IV is non-stable if $2<\alpha<3$ and Eq. (11.13) is satisfied.

Proof. If $2<\alpha<3$ and Eq. (11.13) is satisfied, $c_{eq4}<0$. Accordingly, the vibrator is self-vibrated and consequently non-stable. This finishes the proof. □

11.2.3 Quasiequivalent Stiffness of Fractional Vibrators of Class IV

We now present Theorem 11.6 to explain an interesting phenomenon of quasiequivalent stiffness.

Theorem 11.6. Let k_{eq4} be the quasiequivalent stiffness of a fractional vibrator of class IV. Then,

$$k_{eq4}=k_{eq4}(\omega,\lambda)=k\omega^{\lambda}\cos\frac{\lambda\pi}{2}. \tag{11.16}$$

Proof. In Eq. (11.2), restoration force is given by $k\omega^{\lambda}\cos\frac{\lambda\pi}{2}x_4(t)$. Hence, $k\omega^{\lambda}\cos\frac{\lambda\pi}{2}$ takes the position of stiffness. The proof is finished. □

Since the unit of k_{eq4} is (N/s) × (rad/s)$^{\lambda}$ instead of N/s. We use a prefix "quasi" to describe it. Its unit reduces to N/s if and only if $\lambda=0$. In that case, k_{eq4} degenerates to the primary stiffness k.

Note that $0\le k_{eq4}<\infty$.

Corollary 11.4. When $\omega\to\infty$, $k_{eq4}\to\infty$.

Proof is straightforward and omitted. □

Corollary 11.5. $k_{eq4}\to 0$ if $\omega\to 0$.

Proof is straightforward and accordingly omitted. □

Corollary 11.6. k_{eq4} is positive definite for $0<\lambda<1$.

Proof is straightforward and omitted. □

Figure 11.2 gives the plot of k_{eq4}.

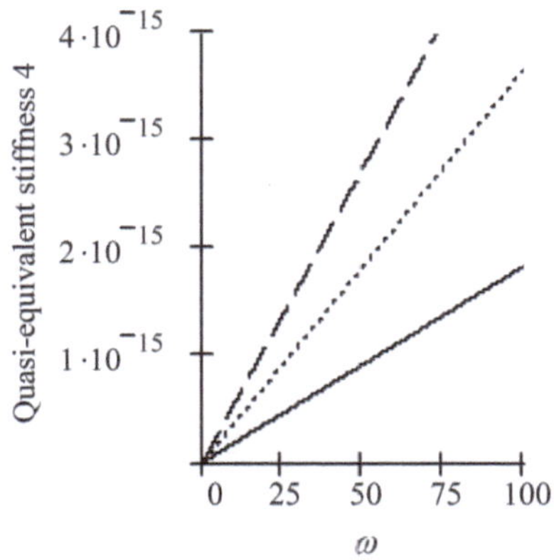

FIGURE 11.2 Plot of k_{eq4} with $k=1$ for $\lambda=0.3$ (solid), 0.6 (dot), 0.9 (dash). Note that "Quasi-equivalent stiffness 4" in the above figure means k_{eq4}.

11.2.4 Quasiequivalent Damping Ratio of Fractional Vibrators of Class IV

Designate ζ_{eq4} as the quasiequivalent damping ratio for a fractional vibrator of class IV. It is defined by

$$\varsigma_{eq4} = \frac{c_{eq4}}{2\sqrt{m_{eq4}k_{eq4}}}. \tag{11.17}$$

Theorem 11.7 depicts its representation.

Theorem 11.7. The representation of ζ_{eq4} is given by

$$\varsigma_{eq4} = \varsigma_{eq4}(\omega,\alpha,\lambda) = \frac{m\omega^{\alpha-1}\sin\frac{\alpha\pi}{2} + k\omega^{\lambda-1}\sin\frac{\lambda\pi}{2}}{2\sqrt{mk\omega^{\alpha+\lambda-2}\left|\cos\frac{\alpha\pi}{2}\right|\cos\frac{\lambda\pi}{2}}}. \tag{11.18}$$

Proof. Substituting m_{eq4}, c_{eq4}, and k_{eq4} into Eq. (11.17) results in Eq. (11.18). The proof is complete. □

As ζ_{eq4} is not dimensionless, the prefix "quasi" is used to term it. If $\alpha=2$ and $\lambda=0$, $\zeta_{eq4}=0$.

Corollary 11.7. If $c_{eq4}<0$,

$$\zeta_{eq4} < 0. \tag{11.19}$$

Proof. From Eq. (11.17), we see that Eq. (11.19) is valid. The proof ends. □

The plots of ζ_{eq4} are demonstrated in Figure 11.3.

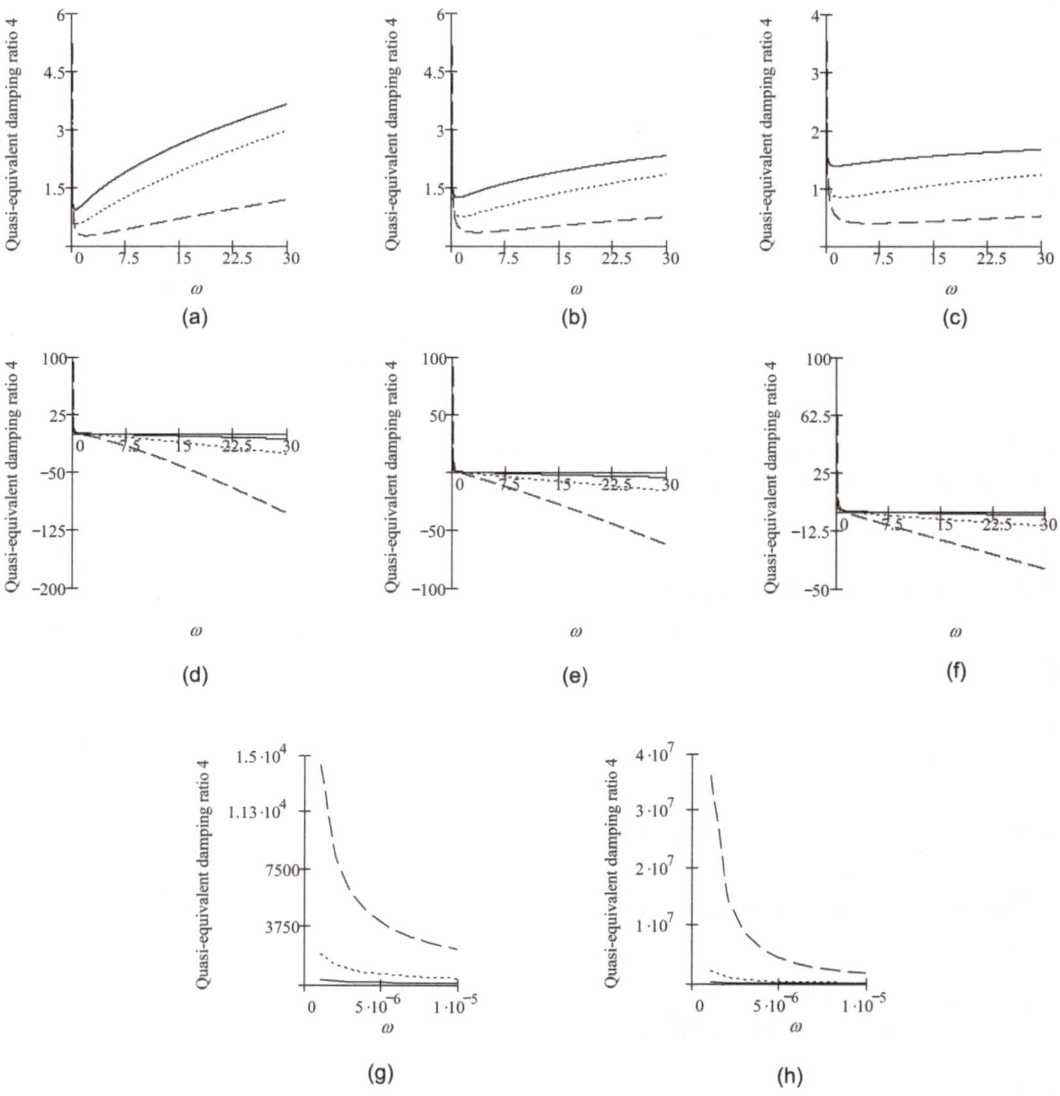

FIGURE 11.3 Plots of ζ_{eq4} with $m=1$ and $k=1$. (a) ζ_{eq4} when $\lambda=0.3$ and $\alpha=1.3$ (solid), 1.6 (dot), 1.9 (dash). (b) ζ_{eq4} when $\lambda=0.6$ and $\alpha=2.3$ (solid), 2.6 (dot), 2.9 (dash). (c) ζ_{eq4} when $\lambda=0.9$ and $\alpha=1.3$ (solid), 1.6 (dot), 1.9 (dash). (d) Negative damping ratio ζ_{eq4} when $\lambda=0.3$ and $\alpha=2.3$ (solid), 2.6 (dot), 2.9 (dash). (e) Negative damping ratio ζ_{eq4} when $\lambda=0.6$ and $\alpha=1.3$ (solid), 1.6 (dot), 1.9 (dash). (f) Negative damping ratio ζ_{eq4} when $\lambda=0.9$ and $\alpha=2.3$ (solid), 2.6 (dot), 2.9 (dash). (g) $\lambda=0.3$ and $\alpha=1.3$ (solid), 1.6 (dot), 1.9 (dash) (Observing ζ_{eq4} at $\omega\approx 0$). (h) $\lambda=0.3$ and $\alpha=2.3$ (solid), 2.6 (dot), 2.9 (dash) (Observing ζ_{eq4} at $\omega\approx 0$). Note that "Quasi-equivalent damping ratio 4" in the above figures stands for ζ_{eq4}.

11.2.5 Quasiequivalent Damping Free Natural Frequency of Fractional Vibrators of Class IV

For a fractional vibrator of class IV, we introduce a term quasiequivalent damping free natural frequency and denote it by ω_{eqn4}. Define it by

$$\omega_{\mathrm{eqn4}} = \sqrt{\frac{k_{\mathrm{eq4}}}{m_{\mathrm{eq4}}}}. \tag{11.20}$$

Theorem 11.8. The expression of ω_{eqn4} is given by

$$\omega_{\mathrm{eqn4}} = \omega_n \sqrt{\frac{\omega^{\lambda} \cos\frac{\lambda\pi}{2}}{-\omega^{\alpha-2} \cos\frac{\alpha\pi}{2}}}, \tag{11.21}$$

where $\omega_n = \sqrt{\frac{k}{m}}$.

Proof. Substituting m_{eq4} and k_{eq4} into Eq. (11.20) produces

$$\omega_{\mathrm{eqn4}} = \sqrt{\frac{k\omega^{\lambda} \cos\frac{\lambda\pi}{2}}{-m\omega^{\alpha-2} \cos\frac{\alpha\pi}{2}}} = \omega_n \sqrt{\frac{\omega^{\lambda} \cos\frac{\lambda\pi}{2}}{-\omega^{\alpha-2} \cos\frac{\alpha\pi}{2}}}.$$

This completes the proof. □

The quantity ω_{eqn4} is with the prefix "quasi" since its unit is not rad/s. Its unit deteriorates to rad/s when $\alpha=2$ and $\lambda=0$. As a matter of fact,

$$\omega_{\mathrm{eqn4}}\big|_{\alpha=2,\lambda=0} = \omega_n. \tag{11.22}$$

We write ω_{eqn4} by

$$\omega_{\mathrm{eqn4}} = \omega_{\mathrm{eqn4}}(\omega, \alpha, \lambda). \tag{11.23}$$

The plots of ω_{eqn4} are shown in Figure 11.4.

11.2.6 Quasiequivalent Damped Natural Frequency of Fractional Vibrators in Class IV

We introduce a term quasiequivalent damped natural frequency for a fractional vibrator of class IV. Let it be ω_{eqd4}. In what follows, small damping is supposed. Define it by

$$\omega_{\mathrm{eqd4}} = \omega_{\mathrm{eqn4}} \sqrt{1-\varsigma_{\mathrm{eq4}}^2}, \quad |\varsigma_{\mathrm{eq4}}| \le 1. \tag{11.24}$$

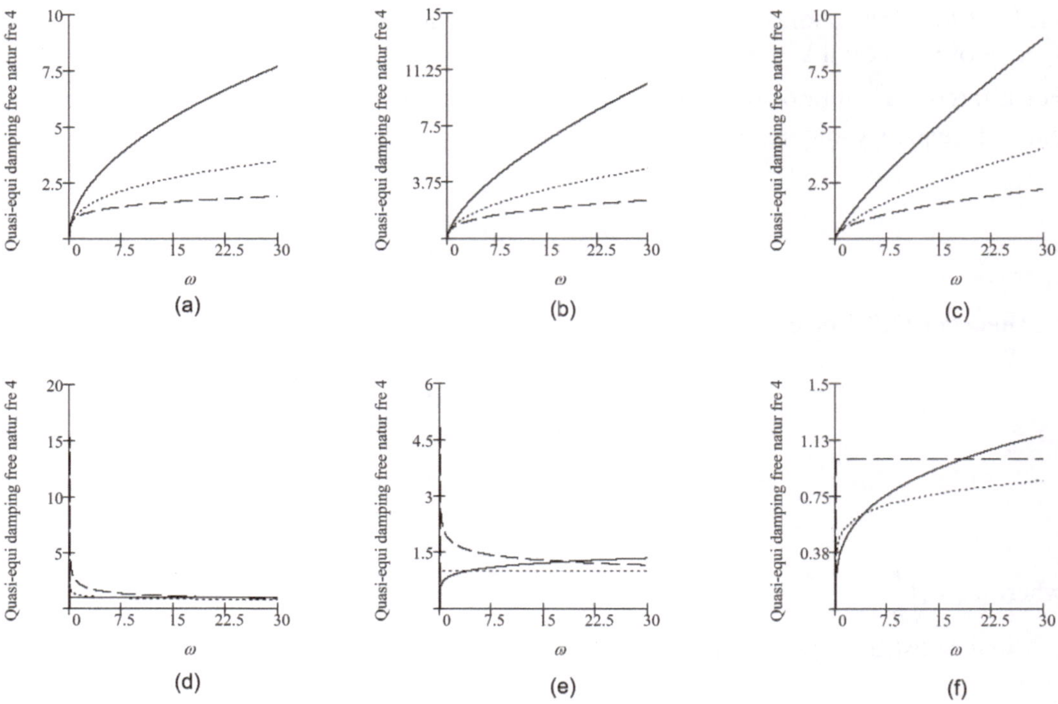

FIGURE 11.4 Plots of ω_{eqn4} with $m=1$ and $k=1$. (a) ω_{eqn4} when $\lambda=0.3$ and $\alpha=1.3$ (solid), 1.6 (dot), 1.9 (dash). (b) ω_{eqn4} when $\lambda=0.6$ and $\alpha=1.3$ (solid), 1.6 (dot), 1.9 (dash). (c) ω_{eqn4} when $\lambda=0.9$ and $\alpha=2.3$ (solid), 2.6 (dot), 2.9 (dash). (d) ω_{eqn4} when $\lambda=0.3$ and $\alpha=2.3$ (solid), 2.6 (dot), 2.9 (dash). (e) ω_{eqn4} when $\lambda=0.6$ and $\alpha=2.3$ (solid), 2.6 (dot), 2.9 (dash). (f) ω_{eqn4} when $\lambda=0.9$ and $\alpha=2.3$ (solid), 2.6 (dot), 2.9 (dash). Note that "Quasi-equivalent damping free natural frequency 4" in the above figures means ω_{eqn4}.

Theorem 11.9. The expression of ω_{eqd4} is given by

$$\omega_{\text{eqd4}} = \omega_n \sqrt{\frac{\omega^{\lambda}\cos\frac{\lambda\pi}{2}}{-\omega^{\alpha-2}\cos\frac{\alpha\pi}{2}}} \sqrt{1-\left(\frac{m\omega^{\alpha-1}\sin\frac{\alpha\pi}{2}+k\omega^{\lambda-1}\sin\frac{\lambda\pi}{2}}{2\sqrt{mk\omega^{\alpha+\lambda-2}\left|\cos\frac{\alpha\pi}{2}\right|\cos\frac{\lambda\pi}{2}}}\right)^2}. \tag{11.25}$$

Proof. Substituting ζ_{eq4} and ω_{eqn4} into Eq. (11.24) yields Eq. (11.25). The proof is finished. □

Note that the unit of ω_{eqd4} is not dimensionless. Hence, the prefix "quasi". If $\alpha=2$ and $\lambda=0$, we have

$$\omega_{\text{eqd4}}\Big|_{\alpha=2,\lambda=0} = \omega_n. \tag{11.26}$$

Figure 11.5 indicates several of its plots.

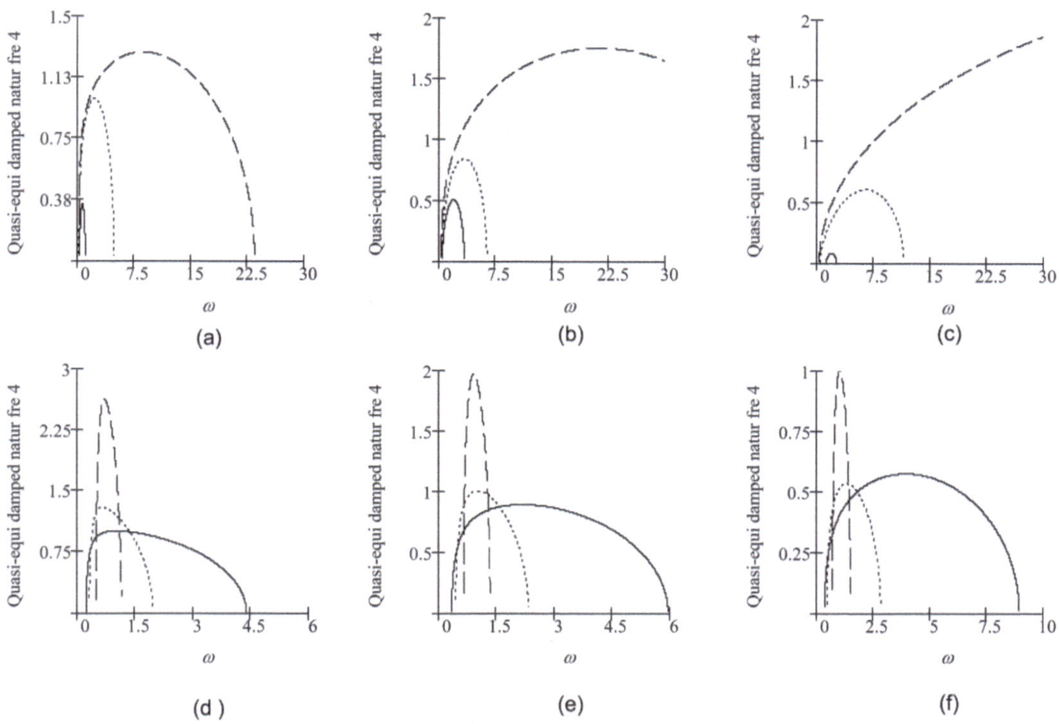

FIGURE 11.5 Plots of ω_{eqd4} with $m=1$ and $k=1$. (a) ω_{eqd4} when $\lambda=0.3$ and $\alpha=1.3$ (solid), 1.6 (dot), 1.9 (dash). (b) ω_{eqd4} when $\lambda=0.6$ and $\alpha=1.5$ (solid), 1.6 (dot), 1.9 (dash). (c) ω_{eqd4} when $\lambda=0.9$ and $\alpha=1.5$ (solid), 1.6 (dot), 1.9 (dash). (d) ω_{eqd4} when $\lambda=0.3$ and $\alpha=2.3$ (solid), 2.6 (dot), 2.9 (dash). (e) ω_{eqd4} when $\lambda=0.6$ and $\alpha=2.3$ (solid), 2.6 (dot), 2.9 (dash). (f) ω_{eqd4} when $\lambda=0.9$ and $\alpha=2.3$ (solid), 2.6 (dot), 2.9 (dash). Note that "Quasi-equivalent damped natural frequency 4" in the above figures stands for ω_{eqd4}.

11.2.7 Quasiequivalent Frequency Ratio of Fractional Vibrators of Class IV

Introduce a term quasiequivalent frequency ratio for a fractional vibrator of class IV. Denote it by γ_{eq4}. Define it by

$$\gamma_{eq4} = \frac{\omega}{\omega_{eqn4}}. \tag{11.27}$$

Theorem 11.10 is for its representation.

Theorem 11.10. The representation of γ_{eq4} is given by

$$\gamma_{eq4} = \gamma \sqrt{\frac{-\omega^{\alpha-2} \cos\frac{\alpha\pi}{2}}{\omega^{\lambda} \cos\frac{\lambda\pi}{2}}}, \tag{11.28}$$

where $\gamma = \dfrac{\omega}{\omega_n}$.

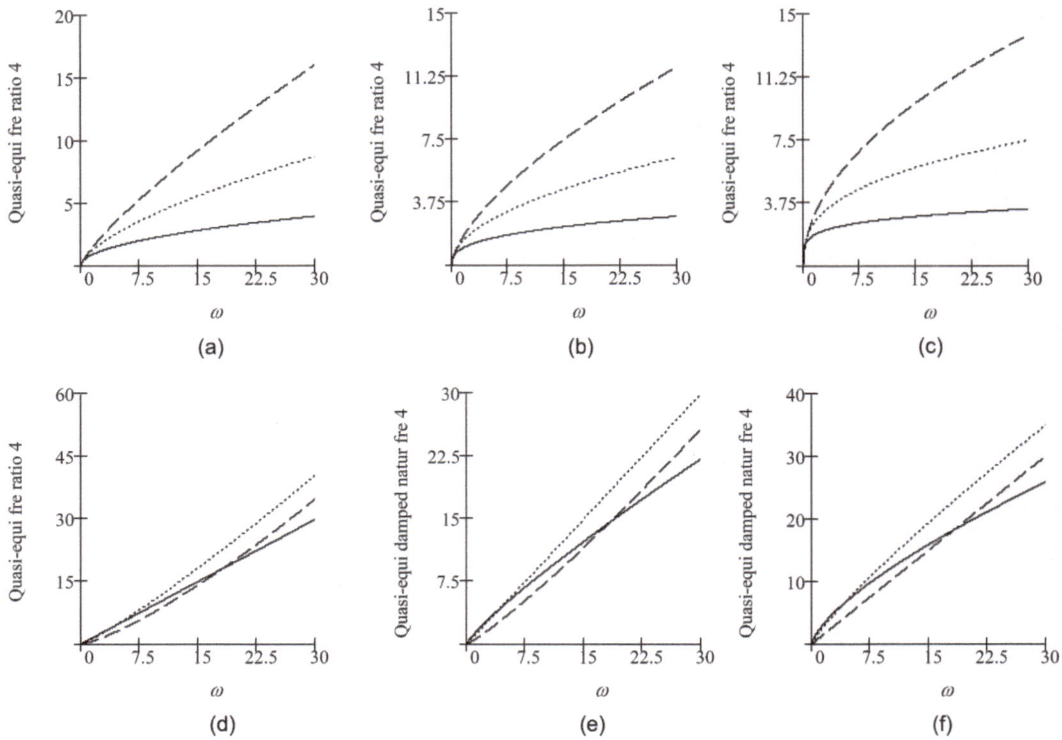

FIGURE 11.6 Plots of γ_{eq4} with $m=1$ and $k=1$. (a) γ_{eq4} when $\lambda=0.3$ and $\alpha=1.3$ (solid), 1.6 (dot), 1.9 (dash). (b) γ_{eq4} when $\lambda=0.6$ and $\alpha=1.3$ (solid), 1.6 (dot), 1.9 (dash). (c) γ_{eq4} when $\lambda=0.9$ and $\alpha=1.3$ (solid), 1.6 (dot), 1.9 (dash). (d) γ_{eq4} when $\lambda=0.3$ and $\alpha=2.3$ (solid), 2.6 (dot), 2.9 (dash). (e) γ_{eq4} when $\lambda=0.6$ and $\alpha=2.3$ (solid), 2.6 (dot), 2.9 (dash). (f) γ_{eq4} when $\lambda=0.9$ and $\alpha=2.3$ (solid), 2.6 (dot), 2.9 (dash). Note that "Quasi-equivalent frequency ratio 4" in the above figures stands for γ_{eq4}.

Proof. If substituting ω_{eqn4} into Eq. (11.27), we have Eq. (11.28). The proof is complete.□

Note that the reason we introduce the prefix quasi to describe γ_{eq4} is that its unit is not dimensionless. Its unit reduces to dimensionless if $\alpha=2$ and $\lambda=0$. Figure 11.6 shows several of its plots.

11.3 FREE RESPONSE OF CLASS IV

Designate $x_4(t)$ as the free response of a fractional vibrator of class IV. It results from the following fractional differential equation

$$\begin{cases} m\dfrac{d^{\alpha}x_4(t)}{dt^{\alpha}}+k\dfrac{d^{\lambda}x_4(t)}{dt^{\lambda}}=0, \\ x_4(0)=x_{40}, x_4'(0)=v_{40}, \end{cases} \quad 1<\alpha<3,\ 0\le\lambda<1. \tag{11.29}$$

Theorem 11.11 gives the analytical representation of $x_4(t)$.

Theorem 11.11. The representation of $x_4(t)$ is expressed by

$$x_4(t)=e^{-\varsigma_{eq4}\omega_{eqn4}t}\left(x_{40}\cos\omega_{eqd4}t+\frac{v_{40}+\varsigma_{eq4}\omega_{eqn4}x_{40}}{\omega_{eqd4}}\sin\omega_{eqd4}t\right),\quad t\geq 0. \tag{11.30}$$

Proof. Theorem 11.1 says that Eq. (11.29) equals to the one below

$$\begin{cases} m_{eq4}\dfrac{d^2x_4(t)}{dt^2}+c_{eq4}\dfrac{dx_4(t)}{dt}+k_{eq4}x_4(t)=0, \\ x_4(0)=x_{40},x_4'(0)=v_{40}, \end{cases} \tag{11.31}$$

which is a standard vibration equation. Thus, its solution is represented by Eq. (11.30). The proof is finished. □

Using ζ_{eq4} and ω_{eqn4}, we have

$$x_4(t)=e^{-\frac{m\omega^{\alpha-1}\sin\frac{\alpha\pi}{2}+k\omega^{\lambda-1}\sin\frac{\lambda\pi}{2}}{2\sqrt{mk\omega^{\alpha+\lambda-2}\left|\cos\frac{\alpha\pi}{2}\right|\cos\frac{\lambda\pi}{2}}}\sqrt{\frac{\omega^{\lambda}\cos\frac{\lambda\pi}{2}}{-\omega^{\alpha-2}\cos\frac{\alpha\pi}{2}}}\omega_n t} \tag{11.32}$$
$$\left(x_{40}\cos\omega_{eqd4}t+\frac{v_{40}+\varsigma_{eq4}\omega_{eqn4}x_{40}}{\omega_{eqd4}}\sin\omega_{eqd4}t\right),\quad t\geq 0.$$

Several illustrations of $x_4(t)$ are given in Figure 11.7

$x_4(t)$ may be expressed by

$$x_4(t)=x_4(t,\omega). \tag{11.33}$$

11.4 IMPULSE RESPONSE OF CLASS IV

Let $h_4(t)$ be the impulse response of a fractional vibrator of class IV. It is a solution to the fractional differential equation in the form

$$m\frac{d^{\alpha}h_4(t)}{dt^{\alpha}}+k\frac{d^{\lambda}h_4(t)}{dt^{\lambda}}=\delta(t),\quad 1<\alpha<3,\ 0\leq\lambda<1 \tag{11.34}$$

with zero initial conditions. Theorem 11.12 gives its analytical expression.

Theorem 11.12. $h_4(t)$ is expressed by

$$h_4(t)=e^{-\varsigma_{eq4}\omega_{eqn4}t}\frac{1}{m_{eq4}\omega_{eqd4}}\sin\omega_{eqd4}t,\quad t\geq 0. \tag{11.35}$$

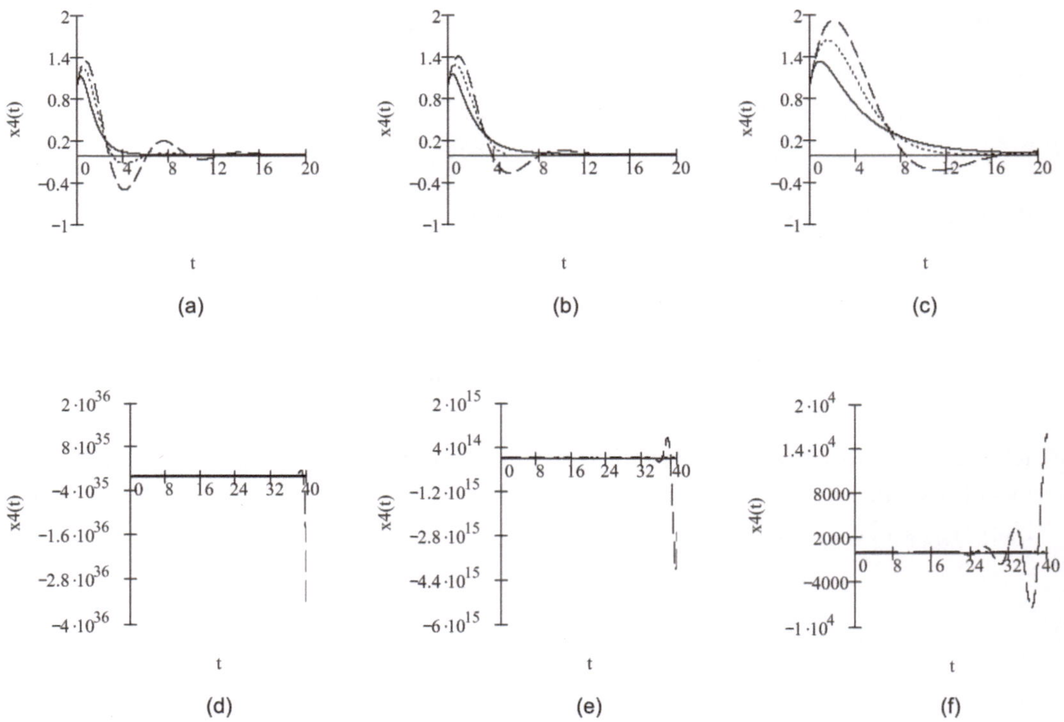

FIGURE 11.7 Illustrations of $x_4(t)$ for $x_{40}=1$, $v_{40}=1$, $m=1$, $k=1$, and $\omega=1.1$. (a) $x_4(t)$ when $\lambda=0.3$ and $\alpha=1.3$ (solid), 1.6 (dot), 1.9 (dash). (b) $x_4(t)$ when $\lambda=0.6$ and $\alpha=1.3$ (solid), 1.6 (dot), 1.9 (dash). (c) $x_4(t)$ when $\lambda=0.9$ and $\alpha=1.3$ (solid), 1.6 (dot), 1.9 (dash). (d) $x_4(t)$ when $\lambda=0.3$ and $\alpha=2.3$ (solid), 2.6 (dot), 2.9 (dash) (Self-vibration). (e) $x_4(t)$ when $\lambda=0.6$ and $\alpha=2.3$ (solid), 2.6 (dot), 2.9 (dash) (Self-vibration). (f) $x_4(t)$ when $\lambda=0.9$ and $\alpha=2.3$ (solid), 2.6 (dot), 2.9 (dash) (Self-vibration).

Proof. The impulse response $h_4(t)$ in Eq. (11.34) is equivalent to the one in the following differential equation according to Theorem 11.1

$$m_{eq4}\frac{d^2h_4(t)}{dt^2}+c_{eq4}\frac{dh_4(t)}{dt}+k_{eq4}h_4(t)=\delta(t). \tag{11.36}$$

The above is a standard vibration equation. Its solution is, therefore, given by Eq. (11.35). The proof is finished. □

If substituting ζ_{eq4} and ω_{eqn4} into Eq. (11.35), we have

$$h_4(t)=e^{-\frac{m\omega^{\alpha-1}\sin\frac{\alpha\pi}{2}+k\omega^{\lambda-1}\sin\frac{\lambda\pi}{2}}{2\sqrt{mk\omega^{\alpha+\lambda-2}\left|\cos\frac{\alpha\pi}{2}\right|\cos\frac{\lambda\pi}{2}}}\sqrt{\frac{\omega^{\lambda}\cos\frac{\lambda\pi}{2}}{-\omega^{\alpha-2}\cos\frac{\alpha\pi}{2}}}\omega_n t}\frac{1}{m_{eq4}\omega_{eqd4}}\sin\omega_{eqd4}t, \quad t\geq 0. \tag{11.37}$$

We demonstrate $h_4(t)$ with Figure 11.8.

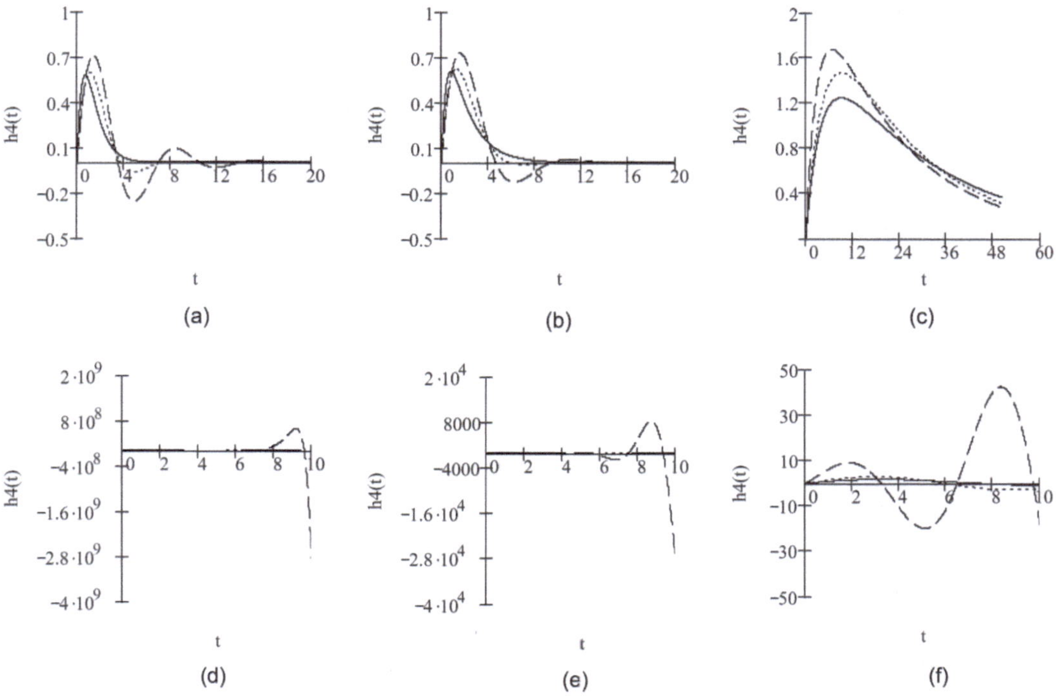

FIGURE 11.8 Plots of $h_4(t)$ for $m=1$, $k=1$, and $\omega=1.1$. (a) $h_4(t)$ when $\lambda=0.3$ and $\alpha=1.3$ (solid), 1.6 (dot), 1.9 (dash). (b) $h_4(t)$ when $\lambda=0.6$ and $\alpha=1.3$ (solid), 1.6 (dot), 1.9 (dash). (c) $h_4(t)$ when $\lambda=0.9$ and $\alpha=1.3$ (solid), 1.6 (dot), 1.9 (dash). (d) $h_4(t)$ when $\lambda=0.3$ and $\alpha=2.3$ (solid), 2.6 (dot), 2.9 (dash) (Self-vibration). (e) $h_4(t)$ when $\lambda=0.6$ and $\alpha=2.3$ (solid), 2.6 (dot), 2.9 (dash) (self-vibration). (f) $h_4(t)$ when $\lambda=0.9$ and $\alpha=2.3$ (solid), 2.6 (dot), 2.9 (dash) (self-vibration).

The function $h_4(t)$ may be taken as a time-frequency distribution. Thus, it may be written by

$$h_4(t)=h_4(t,\omega). \tag{11.38}$$

11.5 STEP RESPONSE OF CLASS IV

Denote by $g_4(t)$ the unit step response of a fractional vibrator of class IV. It is in fact a solution to

$$m_{\text{eq4}}\frac{d^{\alpha}g_4(t)}{dt^{\alpha}}+k_{\text{eq4}}\frac{d^{\lambda}g_4(t)}{dt^{\lambda}}=u(t),\quad 1<\alpha<3, 0\le\lambda<1 \tag{11.39}$$

with zero initial conditions. We present Theorem 11.13 to show the solution.

Theorem 11.13. The expression of $g_4(t)$ is given by

$$g_4(t)=\frac{1}{k_{\text{eq4}}}\left[1-\frac{e^{-\varsigma_{\text{eq4}}\omega_{\text{eqn4}}t}}{\sqrt{1-\varsigma_{\text{eq4}}^2}}\cos\left(\omega_{\text{eqd4}}t-\phi_4\right)\right],\quad t\geq 0, \tag{11.40}$$

where

$$\phi_4=\tan^{-1}\frac{\varsigma_{\text{eq4}}}{\sqrt{1-\zeta_{\text{eq4}}^2}}. \tag{11.41}$$

Proof. Note that Eq. (11.39) is equivalent to the equation given by

$$m_{\text{eq4}}\frac{d^2g_4(t)}{dt^2}+c_{\text{eq4}}\frac{dg_4(t)}{dt}+kg_4(t)=u(t). \tag{11.42}$$

As the above is a standard vibration equation of order 2, Eq. (11.40) stands for the solution to Eq. (11.42) and equivalently to Eq. (11.39). This finishes the proof. □

Eq. (11.40) can be written by

$$g_4(t)=\frac{1}{k\omega^\lambda\cos\frac{\lambda\pi}{2}}\left[1-\frac{e^{-\frac{m\omega^{\alpha-1}\sin\frac{\alpha\pi}{2}+k\omega^{\lambda-1}\sin\frac{\lambda\pi}{2}}{2\sqrt{mk\omega^{\alpha+\lambda-2}\left|\cos\frac{\alpha\pi}{2}\right|\cos\frac{\lambda\pi}{2}}}\sqrt{\frac{\omega^\lambda\cos\frac{\lambda\pi}{2}}{-\omega^{\alpha-2}\cos\frac{\alpha\pi}{2}}}\omega_n t}}{\sqrt{1-\varsigma_{\text{eq4}}^2}}\cos\left(\omega_{\text{eqd4}}t-\phi_4\right)\right],\quad t\geq 0, \tag{11.43}$$

where

$$\phi_4(\omega)=\tan^{-1}\frac{\dfrac{m\omega^{\alpha-1}\sin\frac{\alpha\pi}{2}+k\omega^{\lambda-1}\sin\frac{\lambda\pi}{2}}{2\sqrt{mk\omega^{\alpha+\lambda-2}\left|\cos\frac{\alpha\pi}{2}\right|\cos\frac{\lambda\pi}{2}}}}{\sqrt{1-\left(\dfrac{m\omega^{\alpha-1}\sin\frac{\alpha\pi}{2}+k\omega^{\lambda-1}\sin\frac{\lambda\pi}{2}}{2\sqrt{mk\omega^{\alpha+\lambda-2}\left|\cos\frac{\alpha\pi}{2}\right|\cos\frac{\lambda\pi}{2}}}\right)^2}}. \tag{11.44}$$

We show several plots of $g_4(t)$ in Figure 11.9.

The function $g_4(t)$ may be written by

$$g_4(t)=g_4(t,\omega). \tag{11.45}$$

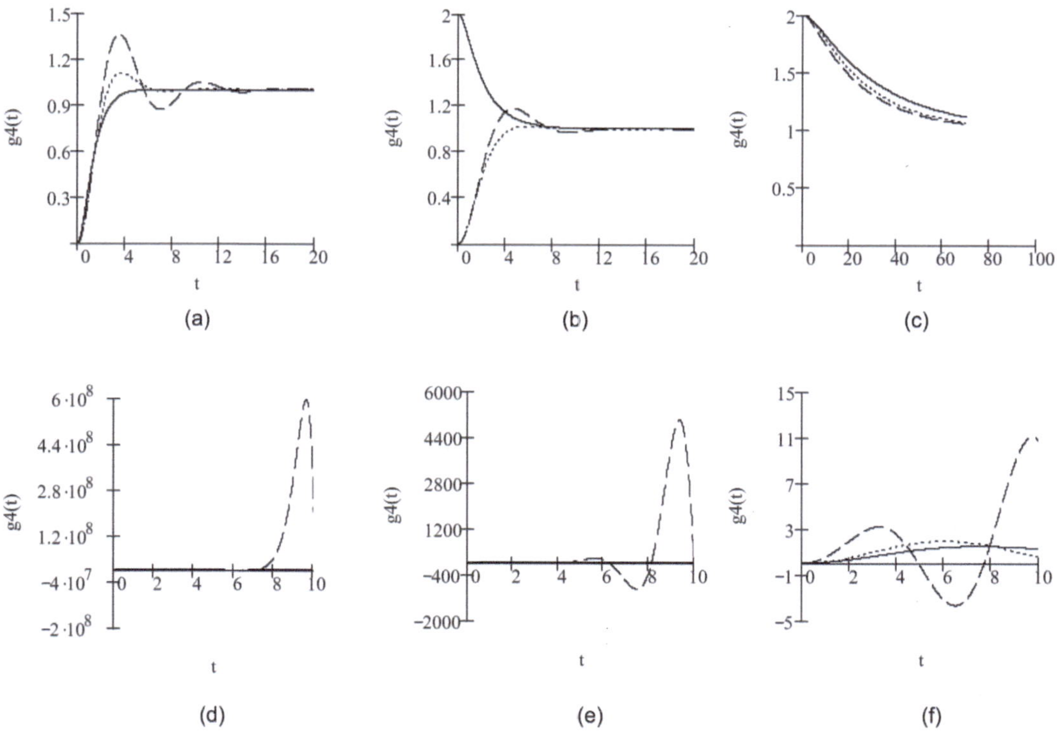

FIGURE 11.9 Plots of $g_4(t)$ for $m=1$, $k=1$, and $\omega=1$. (a) $g_4(t)$ when $\lambda=0.3$ and $\alpha=1.3$ (solid), 1.6 (dot), 1.9 (dash). (b) $g_4(t)$ when $\lambda=0.6$ and $\alpha=1.3$ (solid), 1.6 (dot), 1.9 (dash). (c) $g_4(t)$ when $\lambda=0.9$ and $\alpha=1.3$ (solid), 1.6 (dot), 1.9 (dash). (d) $g_4(t)$ when $\lambda=0.3$ and $\alpha=2.3$ (solid), 2.6 (dot), 2.9 (dash) (self-vibration). (e) $g_4(t)$ when $\lambda=0.6$ and $\alpha=2.3$ (solid), 2.6 (dot), 2.9 (dash) (Self-vibration). (f) $g_4(t)$ when $\lambda=0.9$ and $\alpha=2.3$ (solid), 2.6 (dot), 2.9 (dash) (self-vibration).

11.6 FREQUENCY TRANSFER FUNCTION OF CLASS IV

Theorem 11.14 gives the expression of frequency transfer function of a fractional vibrator of class IV.

Theorem 11.14. Denote by $H_4(\omega)$ the frequency transfer function of a fractional vibrator in class IV. It is expressed by

$$H_4(\omega)=\frac{1}{k_{eq4}\left(1-\gamma_{eq4}^2+i2\varsigma_{eq4}\gamma_{eq4}\right)}. \tag{11.46}$$

Proof. Doing the Fourier transforms on both sides of Eq. (11.36) yields

$$\left(-m_{eq4}\omega^2+ic_{eq4}\omega+k_{eq4}\right)H_4(\omega)=1.$$

Thus, we have

$$\left(-\omega^2+i\frac{c_{\mathrm{eq4}}}{m_{\mathrm{eq4}}}\omega+\frac{k_{\mathrm{eq4}}}{m_{\mathrm{eq4}}}\right)H_4(\omega)=\frac{1}{m_{\mathrm{eq4}}}.$$

Therefore,

$$\left(1-\gamma_{\mathrm{eq4}}^2+i2\varsigma_{\mathrm{eq4}}\gamma_{\mathrm{eq4}}\right)H_4(\omega)=\frac{1}{k_{\mathrm{eq4}}}.$$

From the above, we Eq. (11.46). The proof is complete. □

Substituting ζ_{eq4}, k_{eq4}, and γ_{eq4} into Eq. (11.46) produces

$$H_4(\omega)=\frac{1}{k\omega^{\lambda}\cos\frac{\lambda\pi}{2}\left(1-\gamma^2\frac{-\omega^{\alpha-2}\cos\frac{\alpha\pi}{2}}{\omega^{\lambda}\cos\frac{\lambda\pi}{2}}+i2\gamma\frac{m\omega^{\alpha-1}\sin\frac{\alpha\pi}{2}+k\omega^{\lambda-1}\sin\frac{\lambda\pi}{2}}{2\sqrt{mk\omega^{\alpha+\lambda-2}\left|\cos\frac{\alpha\pi}{2}\right|\left|\cos\frac{\lambda\pi}{2}\right|}}\sqrt{\frac{-\omega^{\alpha-2}\cos\frac{\alpha\pi}{2}}{\omega^{\lambda}\cos\frac{\lambda\pi}{2}}}\right)},\tag{11.47}$$

where $\gamma=\dfrac{\omega}{\omega_n}$.

The amplitude-frequency response function $|H_4(\omega)|$ is given by

$$|H_4(\omega)|=\frac{1}{k_{\mathrm{eq4}}}\frac{1}{\sqrt{\left(1-\gamma_{\mathrm{eq4}}^2\right)^2+\left(2\varsigma_{\mathrm{eq4}}\gamma_{\mathrm{eq4}}\right)^2}}.\tag{11.48}$$

The phase-frequency response function is expressed by

$$\varphi_4(\omega)=\cos^{-1}\frac{1-\gamma_{\mathrm{eq4}}^2}{\sqrt{\left(1-\gamma_{\mathrm{eq4}}^2\right)^2+\left(2\varsigma_{\mathrm{eq4}}\gamma_{\mathrm{eq4}}\right)^2}}.\tag{11.49}$$

We show several plots of $|H_4(\omega)|$ and $\varphi_4(\omega)$ in Figures 11.10 and 11.11.

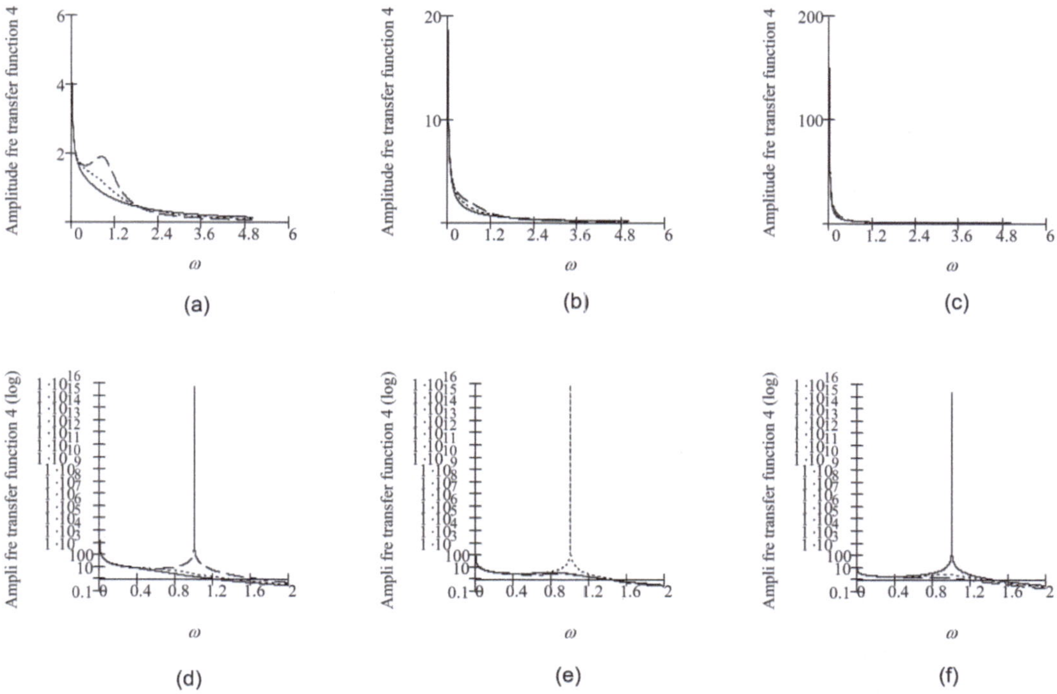

FIGURE 11.10 Plots of amplitude-frequency response $|H_4(\omega)|$ for $m=1$ and $k=1$. (a) $|H_4(\omega)|$ when $\lambda=0.3$ and $\alpha=1.3$ (solid), 1.6 (dot), 1.9 (dash). (b) $|H_4(\omega)|$ when $\lambda=0.6$ and $\alpha=1.3$ (solid), 1.6 (dot), 1.9 (dash). (c) $|H_4(\omega)|$ when $\lambda=0.9$ and $\alpha=1.3$ (solid), 1.6 (dot), 1.9 (dash). (d) $|H_4(\omega)|$ when $\lambda=0.3$ and $\alpha=2.3$ (solid), 2.6 (dot), 2.9 (dash). (e) $|H_4(\omega)|$ when $\lambda=0.6$ and $\alpha=2.3$ (solid), 2.6 (dot), 2.9 (dash). (f) $|H_4(\omega)|$ when $\lambda=0.9$ and $\alpha=2.3$ (solid), 2.6 (dot), 2.9 (dash).

11.7 QUASIEQUIVALENT LOGARITHMIC DECREMENT AND *Q* FACTOR OF CLASS IV

11.7.1 Quasiequivalent Logarithmic Decrement of Fractional Vibrators of Class IV

We present the expression of the quasiequivalent logarithmic decrement of $x_4(t)$ in this subsection.

Theorem 11.15. Denote by t_i and t_{i+1} two time points of fractional free response $x_4(t)$, where $x_4(t_i)$ and $x_4(t_{i+1})$ are successive peak values at t_i and t_{i+1}. Designate $\Delta_{\mathrm{eq}4}$ as the quasi-equivalent logarithmic decrement of $x_4(t)$. Then,

$$\Delta_{\mathrm{eq}4}=\ln\frac{x_4(t_i)}{x_4(t_{i+1})}=\frac{2\pi\dfrac{m\omega^{\alpha-1}\sin\dfrac{\alpha\pi}{2}+k\omega^{\lambda-1}\sin\dfrac{\lambda\pi}{2}}{2\sqrt{mk\omega^{\alpha+\lambda-2}\left|\cos\dfrac{\alpha\pi}{2}\right|\cos\dfrac{\lambda\pi}{2}}}}{\sqrt{1-\left(\dfrac{m\omega^{\alpha-1}\sin\dfrac{\alpha\pi}{2}+k\omega^{\lambda-1}\sin\dfrac{\lambda\pi}{2}}{2\sqrt{mk\omega^{\alpha+\lambda-2}\left|\cos\dfrac{\alpha\pi}{2}\right|\cos\dfrac{\lambda\pi}{2}}}\right)^2}},\quad 1<\alpha<3,\ 0\le\lambda<1. \quad (11.50)$$

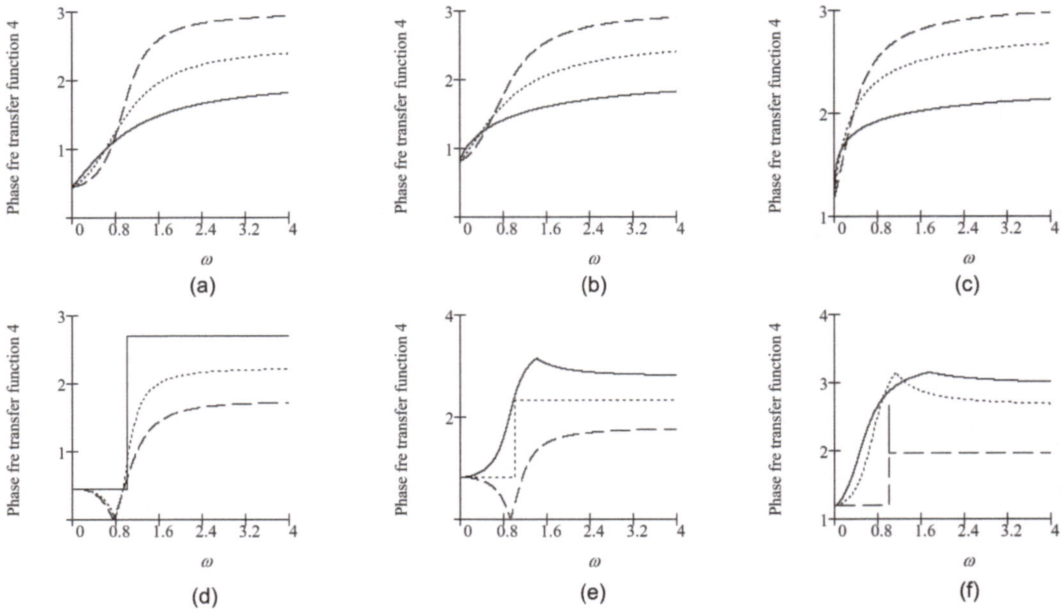

FIGURE 11.11 Plots of $\varphi_4(\omega)$ for $m=1$ and $k=1$. (a) $\varphi_4(\omega)$ when $\lambda=0.3$ and $\alpha=1.3$ (solid), 1.6 (dot), 1.9 (dash). (b) $\varphi_4(\omega)$ when $\lambda=0.6$ and $\alpha=1.3$ (solid), 1.6 (dot), 1.9 (dash). (c) $\varphi_4(\omega)$ when $\lambda=0.9$ and $\alpha=1.3$ (solid), 1.6 (dot), 1.9 (dash). (d) $\varphi_4(\omega)$ when $\lambda=0.3$ and $\alpha=2.3$ (solid), 2.6 (dot), 2.9 (dash). (e) $\varphi_4(\omega)$ when $\lambda=0.6$ and $\alpha=2.3$ (solid), 2.6 (dot), 2.9 (dash). (f) $\varphi_4(\omega)$ when $\lambda=0.9$ and $\alpha=2.3$ (solid), 2.6 (dot), 2.9 (dash).

Proof. Following to the definition of logarithmic decrement, we have

$$\Delta_{\text{eq4}} = \ln \frac{x_4(t_i)}{x_4(t_{i+1})} = \frac{2\pi\varsigma_{\text{eq4}}}{\sqrt{1-\varsigma_{\text{eq4}}^2}}. \tag{11.51}$$

Substituting ζ_{eq4} into the above yields Eq. (11.50). This completes the proof. □

Note that Δ_{eq4} is a quasiequivalent logarithmic decrement since it is not dimensionless. If $\alpha=2$ and $\lambda=0$, it reduces to zero because the vibration system reduces to a conventional damping free one. Several plots of Δ_{eq4} are shown in Figure 11.12, where negative Δ_{eq4} implies self-vibration.

11.7.2 Quasiequivalent *Q* Factor of Fractional Vibrators of Class IV

Theorem 11.16. Denote by Q_{eq4} the quasiequivalent Q factor of a fractional vibrator of class IV. It is represented by

$$Q_{\text{eq4}} = \frac{\sqrt{mk\omega^{\alpha+\lambda-2}\left|\cos\frac{\alpha\pi}{2}\right|\cos\frac{\lambda\pi}{2}}}{m\omega^{\alpha-1}\sin\frac{\alpha\pi}{2} + k\omega^{\lambda-1}\sin\frac{\lambda\pi}{2}}. \tag{11.52}$$

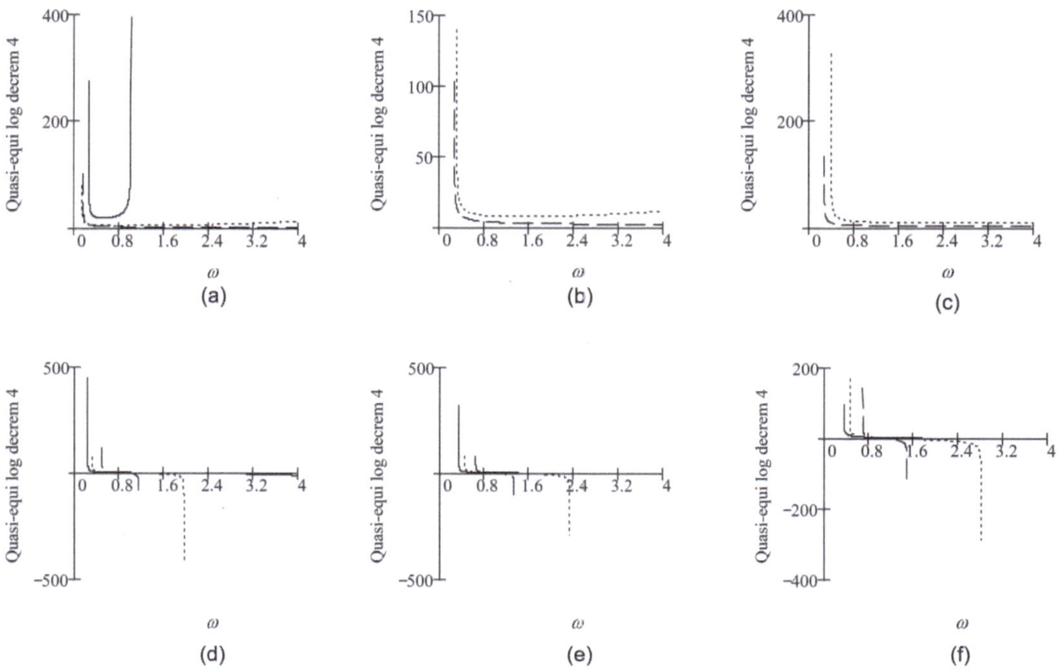

FIGURE 11.12 Plots of quasiequivalent logarithmic decrement Δ_{eq4} for $m=1$ and $k=1$. (a) Δ_{eq4} when $\lambda=0.3$ and $\alpha=1.3$ (solid), 1.6 (dot), 1.9 (dash). (b) Δ_{eq4} when $\lambda=0.6$ and $\alpha=1.3$ (solid), 1.6 (dot), 1.9 (dash). (c) Δ_{eq4} when $\lambda=0.9$ and $\alpha=1.3$ (solid), 1.6 (dot), 1.9 (dash). (d) Δ_{eq4} when $\lambda=0.3$ and $\alpha=2.3$ (solid), 2.6 (dot), 2.9 (dash). (e) Δ_{eq4} when $\lambda=0.6$ and $\alpha=2.3$ (solid), 2.6 (dot), 2.9 (dash). (f) Δ_{eq4} when $\lambda=0.9$ and $\alpha=2.3$ (solid), 2.6 (dot), 2.9 (dash). Note that "Quasi-equivalent logarithmic decrement 4" in the above figures means Δ_{eq4}.

Proof. Following the definition of Q factor, we have

$$Q_{eq4} = \frac{1}{2\varsigma_{eq4}}. \tag{11.53}$$

When substituting ζ_{eq4} into the above, we have Eq. (11.52). The proof is complete. □

Since Q_{eq4} is not a constant but a function in terms of ω, α, and λ. In addition, it is not dimensionless. $Q_{eq4}=\infty$ if $\alpha=2$ and $\lambda=0$. That is the reason we put a prefix "quasi" on it. Several of its plots are shown in Figure 11.13.

11.8 PARTICULARITY OF C_{EQ4} IN COMPARISON WITH RAYLEIGH DAMPING

Let c_r be the Rayleigh damping. It is in the form $c_r=am+bk$, where m and k are primary mass and stiffness, respectively, a is proportional to vibration frequency ω while b is inversely proportional to ω (Cruz and Miranda [2], Harris [3]). Since $c_{eq4}=c_{eq41}+c_{eq42}$, where c_{eq41} is proportional to ω with the coefficient m while c_{eq42} is inversely proportional to ω with the coefficient k, see Eqs. (11.7) and (11.8), we show that c_{eq4} is consistent with the

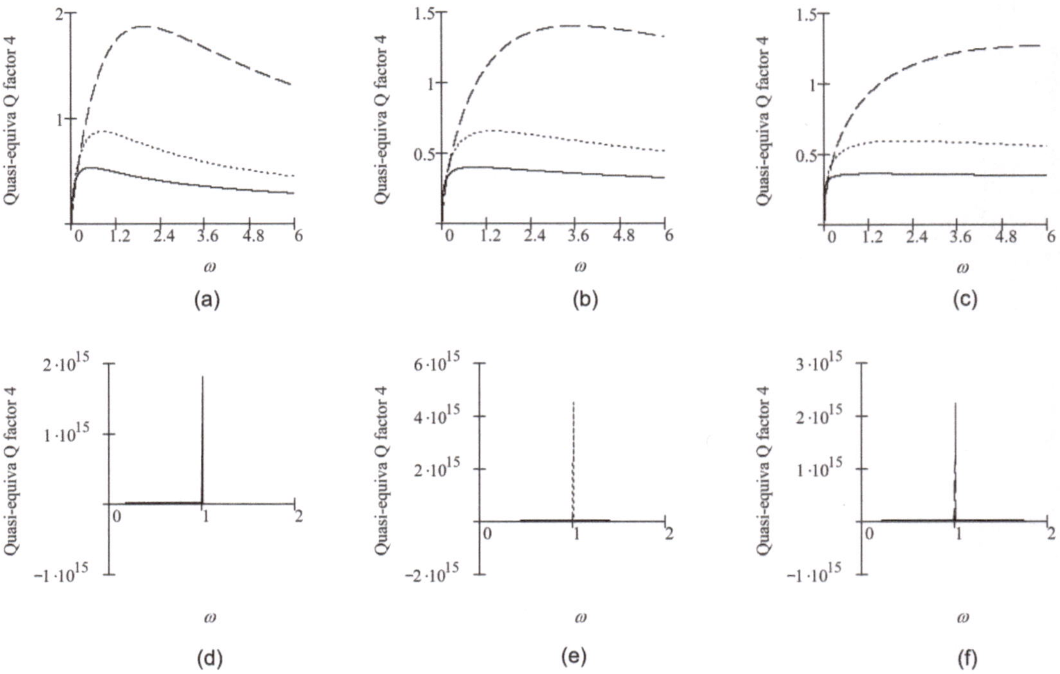

FIGURE 11.13 Plots of Q_{eq4} for $m=1$ and $k=1$. (a) Q_{eq4} when $\lambda=0.3$ and $\alpha=1.3$ (solid), 1.6 (dot), 1.9 (dash). (b) Q_{eq4} when $\lambda=0.6$ and $\alpha=1.3$ (solid), 1.6 (dot), 1.9 (dash). (c) Q_{eq4} when $\lambda=0.9$ and $\alpha=1.3$ (solid), 1.6 (dot), 1.9 (dash). (d) Q_{eq4} when $\lambda=0.3$ and $\alpha=2.3$ (solid), 2.6 (dot), 2.9 (dash). (e) Q_{eq4} when $\lambda=0.6$ and $\alpha=2.3$ (solid), 2.6 (dot), 2.9 (dash). (f) Q_{eq4} when $\lambda=0.9$ and $\alpha=2.3$ (solid), 2.6 (dot), 2.9 (dash). Note that "Quasi-equivalent Q factor 4" in the above figures implies Q_{eq4}.

Rayleigh damping in form. However, the conventional Rayleigh damping is positive while c_{eq4} revealed in this chapter may be negative, see Eq. (11.12). That is the particularity of c_{eq4} in comparison with the Rayleigh damping.

11.9 CONCLUDING REMARK

The proposed results in this chapter are novel in the field of vibrations. In addition, it is worth noting that a fractional vibrator in class IV provides us with a model to explain the Rayleigh damping assumption in mathematics. Since the Rayleigh damping assumption is a big deal in vibrations as well as material mechanics, we would like to detail its mathematical explanation in Chapter 14 further.

11.10 EXERCISES

11.1. Find the condition of $c_{eq4}<0$.

11.2. Find the condition for $0<\zeta_{eq4}<1$.

11.3. Prove that ζ_{eq4} is not dimensionless if $\alpha\neq 2$ or $\lambda\neq 0$.

11.4. Prove that k_{eq4} is positive definite.

11.5. Prove that ω_{eqn4} is not a standard angular frequency when $\alpha \neq 2$ or $\lambda \neq 0$.

11.6. Prove that γ_{eq4} is dimensionless if $\alpha = 2$ and $\lambda = 0$.

REFERENCES

1. J. W. Strutt and 3rd B. Rayleigh, *The Theory of Sound*, vol. 1, Macmillan & Co., Ltd., London, 1877.
2. C. Cruz and E. Miranda, Evaluation of the Rayleigh damping model for buildings, *Engineering Structures*, 138, 2017, 324–336.
3. C. M. Harris, *Shock and Vibration Handbook*, 5th Ed., McGraw-Hill, New York, 2002.

CHAPTER 12

Class V Fractional Vibrations

THIS CHAPTER AIMS TO explain the contributions listed below. (1) Give an equivalent motion equation for fractional vibrators in class V in Theorem 12.1. (2) Introduce the quasiequivalent damping, quasiequivalent damping ratio, quasiequivalent damping free natural frequency, quasiequivalent damped natural frequency, quasiequivalent frequency ratio, quasiequivalent logarithmic decrement, and quasiequivalent Q factor for fractional vibrators in class V. Their expressions are presented in Theorems 12.2, 12.5–12.8, 12.13, and 12.14, respectively. (3) Reveal the phenomenon that the quasiequivalent damping of a fractional vibrator in class V may vary from 0 to ∞ in Corollary 12.1. (4) Conclude that fractional vibrators in class V are stable in Theorem 12.4. (5) Put forward a mathematical explanation of a variation of the Rayleigh damping that is proportional to stiffness only in Theorem 12.3. (6) Present the analytic expressions of free response, impulse response, step response, and frequency transfer function of a fractional vibrator in class V in Theorems 12.9, 12.10–12.12, respectively.

12.1 CLASS V EQUIVALENT MOTION EQUATION

Theorem 12.1 is for an equivalent motion equation of a fractional vibrator of class V.

Theorem 12.1. The motion equation of a fractional vibrator of class V below

$$m\frac{d^2x_5(t)}{dt^2}+k\frac{d^\lambda x_5(t)}{dt^\lambda}=0,\quad 0\le\lambda<1 \tag{12.1}$$

is equivalently expressed by

$$m\frac{d^2x_5(t)}{dt^2}+k\omega^{\lambda-1}\sin\frac{\lambda\pi}{2}\frac{dx_5(t)}{dt}+k\omega^\lambda\cos\frac{\lambda\pi}{2}x_5(t)=0,\quad 0\le\lambda<1. \tag{12.2}$$

Proof. Substituting $\alpha=2$ into Eq. (11.2) yields Eq. (12.2). This completes the proof. □

Mathematically, a fractional vibrator of class V is a special case of one of class IV for $\alpha=2$. However, in vibration engineering, Eq. (12.1) stands for a class of vibrators from a

DOI: 10.1201/9781003460947-14

view of a variation of the Rayleigh damping, which is proportional to stiffness only. Thus, considering Eq. (12.1) as a class is preferred in this book.

12.2 QUASIEQUIVALENT PARAMETERS OF CLASS V FRACTIONAL VIBRATORS

12.2.1 Quasiequivalent Damping in Fractional Vibrators in Class V

Theorem 12.2 presents the expression of equivalent damping of Eq. (12.1).

Theorem 12.2. Denote by c_{eq5} the quasiequivalent damping of a fractional vibrator of class V. Then, it is given by

$$c_{eq5} = c_{eq5}(\omega, \lambda) = k\omega^{\lambda-1} \sin\frac{\lambda\pi}{2}, \quad 0 \le \lambda < 1. \tag{12.3}$$

Proof: If substituting $\alpha = 2$ into Eq. (11.8), we have Eq. (12.3). This completes the proof. □

Now, we reveal an important phenomenon of c_{eq5} with the following theorem.

Theorem 12.3. c_{eq5} is only proportional to the stiffness k.

Proof. (12.3) implies that c_{eq5} is proportional to k with the coefficient $\omega^{\lambda-1}\sin\frac{\lambda\pi}{2}$. This finishes the proof. □

Note that Theorem 12.3 may serve as a mathematical explanation of a kind of variation of the Rayleigh damping assumption, which is only proportional to stiffness; we refer to Chapter 14 for details.

Now, we present the asymptotic properties of c_{eq5} below.

Corollary 12.1: Asymptotic Properties for $\omega \to 0$ and $\omega \to \infty$. For c_{eq5}, we have

$$\lim_{\omega \to 0} c_{eq5}(\omega, \lambda) = \infty. \tag{12.4}$$

In addition,

$$\lim_{\omega \to \infty} c_{eq5}(\omega, \lambda) = 0. \tag{12.5}$$

Proof. When $0 \le \lambda < 1$, we have

$$\lim_{\omega \to 0} \omega^{\lambda-1} \sin\frac{\lambda\pi}{2} = \infty. \tag{12.6}$$

Therefore, Eq. (12.4) is valid. On the other side,

$$\lim_{\omega \to \infty} \omega^{\lambda-1} \sin\frac{\lambda\pi}{2} = 0. \tag{12.7}$$

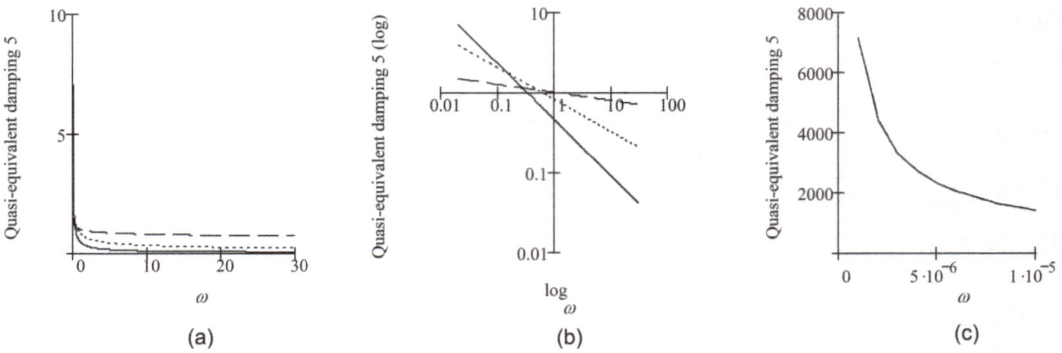

FIGURE 12.1 Illustrations of c_{eq5} with $m=1$ and $k=1$. (a) c_{eq5} when $\lambda=0.3$ (solid), 0.6 (dot), 0.9 (dash). (b) c_{eq5} in log-long scale when $\lambda=0.3$ (solid), 0.6 (dot), 0.9 (dash). (c) c_{eq5} when $\lambda=0.3$ (observing c_{eq5} at $\omega\approx 0$). Note that "Quasi-equivalent damping 5" in the above figures means c_{eq5}.

Hence, Eq. (12.5) holds. The proof is finished. □

Note that $c_{eq5}=0$ if $\lambda=0$. In fact, c_{eq5} is produced by fractional displacement. Figure 12.1 indicates several of its illustrations.

The stable property of a fractional vibrator in class V is proposed in Theorem 12.4.

Theorem 12.4. A fractional vibrator of class V is stable.

Proof. Since $c_{eq5}\geq 0$, a fractional vibrator of class V is not self-vibrated. Accordingly, it is stable. The proof is finished. □

12.2.2 Quasiequivalent Stiffness of Fractional Vibrators of Class V

Let k_{eq5} be the quasiequivalent stiffness of a fractional vibrator of class V. Then,

$$k_{eq5}=k_{eq4}. \tag{12.8}$$

Thus, we do not discuss with it anymore, referring Chapter 11 for k_{eq4}.

12.2.3 Quasiequivalent Damping Ratio of Fractional Vibrators of Class V

Denote by ζ_{eq5} the quasiequivalent damping ratio for a fractional vibrator of class V. Define it by

$$\varsigma_{eq5}=\frac{c_{eq5}}{2\sqrt{mk_{eq5}}}. \tag{12.9}$$

Theorem 12.5 depicts its representation.

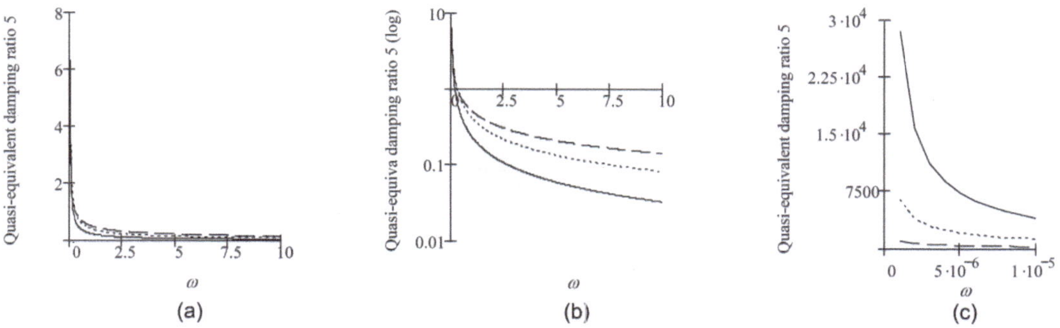

FIGURE 12.2 Plots of ζ_{eq5} with $m=1$ and $k=1$. (a) ζ_{eq5} when $\lambda=0.3$ (solid), 0.6 (dot), 0.9 (dash). (b) ζ_{eq5} in log-scale when $\lambda=0.3$ (solid), 0.6 (dot), 0.9 (dash). (c) ζ_{eq5} with ω in log-scale when $\lambda=0.3$ (solid), 0.6 (dot), 0.9 (dash) (observations at $\omega\approx0$). Note that "Quasi-equivalent damping ratio 5" in the above figures stands for ζ_{eq5}.

Theorem 12.5. The expression of ζ_{eq5} is given by

$$\varsigma_{eq5}=\varsigma_{eq5}(\omega,\lambda)=\frac{k\omega^{\lambda-1}\sin\frac{\lambda\pi}{2}}{2\sqrt{mk\omega^{\lambda}\cos\frac{\lambda\pi}{2}}}=\frac{\omega_n\omega^{\frac{\lambda}{2}-1}\sin\frac{\lambda\pi}{2}}{2\sqrt{\cos\frac{\lambda\pi}{2}}}. \tag{12.10}$$

Proof. Substituting c_{eq5} and k_{eq5} into Eq. (12.9) produces Eq. (12.10). The proof is finished. □

Note that ζ_{eq5} is not dimensionless. Hence, the prefix "quasi" is used. It is a positive quantity. If $\lambda=0$, $\zeta_{eq5}=0$. Figure 12.2 shows several of its plots.

12.2.4 Quasiequivalent Damping Free Natural Frequency of Fractional Vibrators of Class V

For a fractional vibrator of class V, its quasiequivalent damping free natural frequency is defined by

$$\omega_{eqn5}=\sqrt{\frac{k_{eq5}}{m}}. \tag{12.11}$$

I present its expression with Theorem 12.6.

Theorem 12.6. The representation of ω_{eqn5} is in the form

$$\omega_{eqn5}=\omega_n\sqrt{\omega^{\lambda}\cos\frac{\lambda\pi}{2}}, \tag{12.12}$$

where $\omega_n=\sqrt{\frac{k}{m}}$.

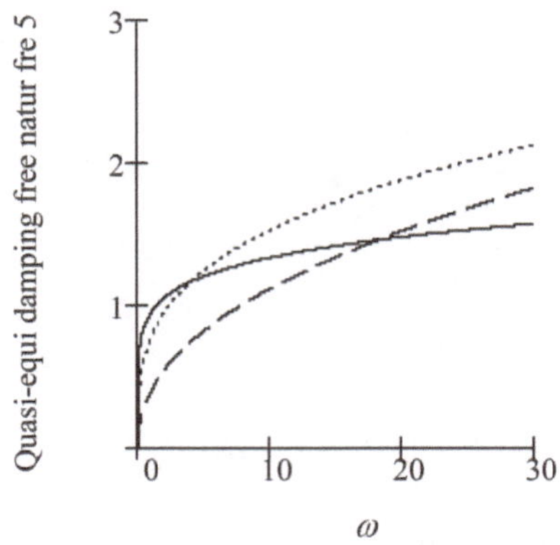

FIGURE 12.3 Plot of ω_{eqn5} with $m=1$ and $k=1$ when $\lambda=0.3$ (solid), 0.6 (dot), 0.9 (dash). Note that "Quasi-equivalent damping free natural frequency 5" in the above figure stands for ω_{eqn5}.

Proof. If substituting k_{eq5} into Eq. (12.11), we have Eq. (12.12). The proof is finished. □

The unit of ω_{eqn5} is not rad/s. Its unit deteriorates to rad/s if $\lambda=0$. Hence, the prefix "quasi". In fact,

$$\omega_{eqn5}\Big|_{\lambda=0}=\omega_n. \tag{12.13}$$

We may write ω_{eqn5} by

$$\omega_{eqn5}=\omega_{eqn5}(\omega,\lambda). \tag{12.14}$$

Figure 12.3 indicates several plots of ω_{eqn5}.

12.2.5 Quasiequivalent Damped Natural Frequency of Fractional Vibrators of Class V

Now, we introduce a term quasiequivalent damped natural frequency for a fractional vibrator of class V. Denote it by ω_{eqd5}. Unless otherwise stated, we use small damping in what follows. Define it by

$$\omega_{eqd5}=\omega_{eqn5}\sqrt{1-\varsigma_{eq5}^2},\quad |\varsigma_{eq5}|\le 1. \tag{12.15}$$

Theorem 12.7. The representation of ω_{eqd5} is expressed by

$$\omega_{eqd5}=\omega_n\sqrt{\omega^{\lambda}\cos\frac{\lambda\pi}{2}}\sqrt{1-\left(\frac{k\omega^{\lambda-1}\sin\frac{\lambda\pi}{2}}{2\sqrt{mk\omega^{\lambda}\cos\frac{\lambda\pi}{2}}}\right)^2}. \tag{12.16}$$

Proof. Letting $\alpha=2$ in Eq. (11.25) yields Eq. (12.16). The proof is finished. □

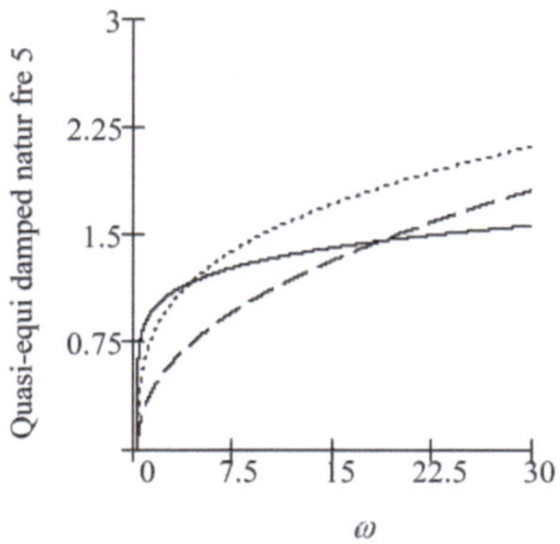

FIGURE 12.4 Plot of ω_{eqd5} with $m=1$ and $k=1$ when $\lambda=0.3$ (solid), 0.6 (dot), 0.9 (dash). Note that "Quasi-equivalent damped natural frequency 5" in the above figure stands for ω_{eqd5}.

Since the unit of ω_{eqd5} is not rad/s, we call it quasiequivalent damped natural frequency. When $\lambda=0$, we have

$$\omega_{eqd5}\big|_{\lambda=0}=\omega_n. \tag{12.17}$$

Figure 12.4 shows its several plots.

12.2.6 Quasiequivalent Frequency Ratio of Fractional Vibrators of Class V

Introduce a term quasiequivalent frequency ratio for a fractional vibrator of class V. Denote it by γ_{eq5}. Define it by

$$\gamma_{eq5}=\frac{\omega}{\omega_{eqn5}}. \tag{12.18}$$

I propose its expression with Theorem 12.8.

Theorem 12.8. γ_{eq5} is given by

$$\gamma_{eq5}=\gamma\sqrt{\frac{1}{\omega^{\lambda}\cos\frac{\lambda\pi}{2}}}, \tag{12.19}$$

where $\gamma=\frac{\omega}{\omega_n}$.

Proof. Letting $\alpha=2$ in Eq. (11.28) results in Eq. (12.19). The proof is complete. □

The unit of γ_{eq5} is not dimensionless. Its unit reduces to being dimensionless only when $\lambda=0$. We use Figure 12.5 to illustrate its plots.

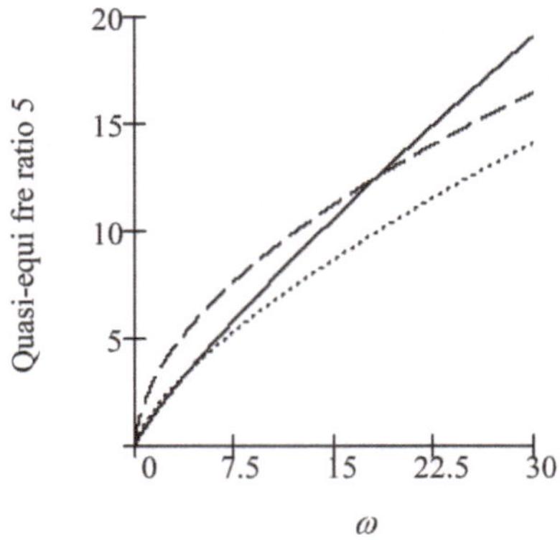

FIGURE 12.5 Plots of γ_{eq5} with $m=1$ and $k=1$ when $\lambda=0.3$ (solid), 0.6 (dot), 0.9 (dash). Note that "Quasi-equivalent frequency ratio 5" in the above figures stands for γ_{eq5}.

12.3 CLASS V FREE RESPONSE

Let $x_5(t)$ be the free response of a fractional vibrator of class V. It is the solution to the fractional differential equation below

$$\begin{cases} m\dfrac{d^2x_5(t)}{dt^2}+k\dfrac{d^\lambda x_5(t)}{dt^\lambda}=0, \\ x_5(0)=x_{50}, x_5'(0)=v_{50}, \end{cases} \quad 0\leq\lambda<1. \tag{12.20}$$

Theorem 12.9 describes its representation.

Theorem 12.9. The function $x_5(t)$ is expressed by

$$x_5(t)=e^{-\varsigma_{eq5}\omega_{eqn5}t}\left(x_{50}\cos\omega_{eqd5}t+\frac{v_{50}+\varsigma_{eq5}\omega_{eqn5}x_{50}}{\omega_{eqd5}}\sin\omega_{eqd5}t\right), \quad t\geq 0. \tag{12.21}$$

Proof. Based on Theorem 12.1, we see that Eq. (12.20) is equal to the one

$$\begin{cases} m\dfrac{d^2x_5(t)}{dt^2}+c_{eq5}\dfrac{dx_5(t)}{dt}+k_{eq5}x_5(t)=0, \\ x_5(0)=x_{50}, x_5'(0)=v_{50}, \end{cases} \quad 0\leq\lambda<1, \tag{12.22}$$

where c_{eq5} and k_{eq5} are expressed by Eqs. (12.3) and (12.8), respectively. As Eq. (12.22) is a standard vibration equation, we can immediately attain its solution as Eq. (12.21). This finishes the proof. □

Expanding Eq. (12.21) produces

$$x_5(t)=e^{-\frac{k\omega^{\lambda-1}\sin\frac{\lambda\pi}{2}}{2\sqrt{mk\omega^\lambda\cos\frac{\lambda\pi}{2}}}\sqrt{\omega^\lambda\cos\frac{\lambda\pi}{2}}\,\omega_{n}t}\left(x_{50}\cos\omega_{eqd5}t+\frac{v_{50}+\varsigma_{eq5}\omega_{eqn5}x_{50}}{\omega_{eqd5}}\sin\omega_{eqd5}t\right), \quad t\geq 0. \tag{12.23}$$

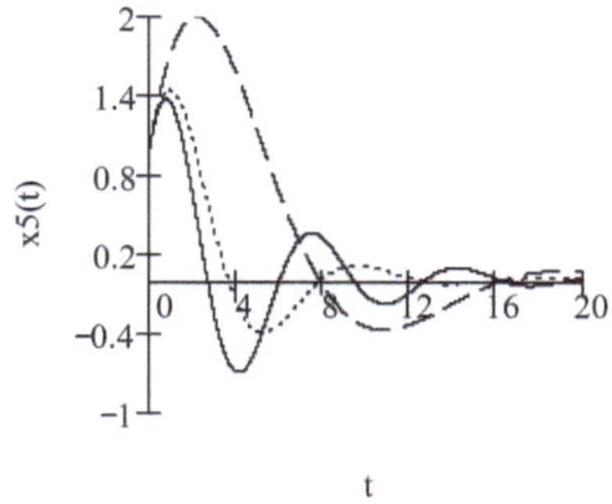

FIGURE 12.6 Plots of $x_5(t)$ for $x_{50}=1$, $v_{50}=1$, $m=1$, $k=1$, and $\omega=1.1$ when $\lambda=0.3$ (solid), 0.6 (dot), 0.9 (dash).

Several plots of $x_5(t)$ are shown in Figure 12.6

We may write $x_5(t)$ by

$$x_5(t)=x_5(t,\omega). \tag{12.24}$$

12.4 CLASS V IMPULSE RESPONSE

Denote by $h_5(t)$ the impulse response of a fractional vibrator of class V. It is a solution to the fractional differential equation given by

$$m\frac{d^2h_5(t)}{dt^2}+k\frac{d^\lambda h_5(t)}{dt^\lambda}=\delta(t), \quad 0\le\lambda<1 \tag{12.25}$$

with zero initial conditions. Theorem 12.10 gives its analytical expression.

Theorem 12.10. The expression of $h_5(t)$ is given by

$$h_5(t)=e^{-\varsigma_{\text{eq}5}\omega_{\text{eqn}5}t}\frac{1}{m\omega_{\text{eqd}5}}\sin\omega_{\text{eqd}5}t, \quad t\ge 0. \tag{12.26}$$

Proof. Since $h_5(t)$ in Eq. (12.25) is equivalently to the one in the following differential equation based on to Theorem 12.1

$$m\frac{d^2h_5(t)}{dt^2}+c_{\text{eq}5}\frac{dh_5(t)}{dt}+k_{\text{eq}5}h_5(t)=\delta(t), \quad 0\le\lambda<1, \tag{12.27}$$

$h_5(t)$ is given by Eq. (12.26) because Eq. (12.27) is a standard vibration equation. This finishes the proof. □

If we substitute $\zeta_{\text{eq}5}$ and$\omega_{\text{eqn}5}$ into Eq. (12.26), the following results

$$h_5(t)=e^{-\frac{k\omega^{\lambda-1}\sin\frac{\lambda\pi}{2}}{2\sqrt{mk\omega^\lambda\cos\frac{\lambda\pi}{2}}}\sqrt{\omega^\lambda\cos\frac{\lambda\pi}{2}}\,\omega_n t}\frac{1}{m\omega_{\text{eqd}5}}\sin\omega_{\text{eqd}5}t, \quad t\ge 0. \tag{12.28}$$

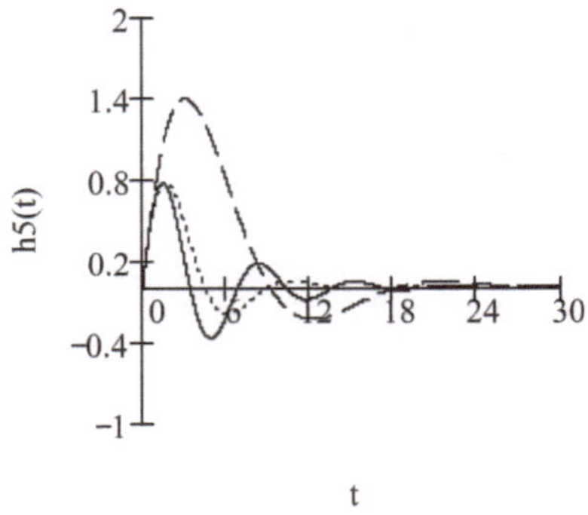

FIGURE 12.7 Illustrations of $h_5(t)$ for $m=1$, $k=1$, and $\omega=1$ when $\lambda=0.3$ (solid), 0.6 (dot), 0.9 (dash).

Several illustrations of $h_5(t)$ are given in Figure 12.7.

When taking $h_5(t)$ as a time-frequency distribution, we write

$$h_5(t)=h_5(t,\omega). \tag{12.29}$$

12.5 CLASS V STEP RESPONSE

Let $g_5(t)$ be the unit step response of a fractional vibrator of class V. Then, it is a solution to

$$m\frac{d^2 g_5(t)}{dt^2}+k_{eq5}\frac{d^\lambda g_5(t)}{dt^\lambda}=u(t), \quad 0\le\lambda<1 \tag{12.30}$$

with zero initial conditions. Theorem 12.11 gives its expression.

Theorem 12.11. $g_5(t)$ is given by

$$g_5(t)=\frac{1}{k_{eq5}}\left[1-\frac{e^{-\varsigma_{eq5}\omega_{eqn5}t}}{\sqrt{1-\varsigma_{eq5}^2}}\cos\left(\omega_{eqd5}t-\phi_5\right)\right], \quad t\ge 0, \tag{12.31}$$

where

$$\phi_5=\tan^{-1}\frac{\varsigma_{eq5}}{\sqrt{1-\zeta_{eq5}^2}}. \tag{12.32}$$

Proof. Because Eq. (12.30) equals to the one below

$$m\frac{d^2 g_5(t)}{dt^2}+c_{eq5}\frac{dg_5(t)}{dt}+kg_5(t)=u(t), \quad 0\le\lambda<1, \tag{12.33}$$

$g_5(t)$ is expressed by Eq. (12.31) as (12.33) is a standard vibration equation. This completes the proof. □

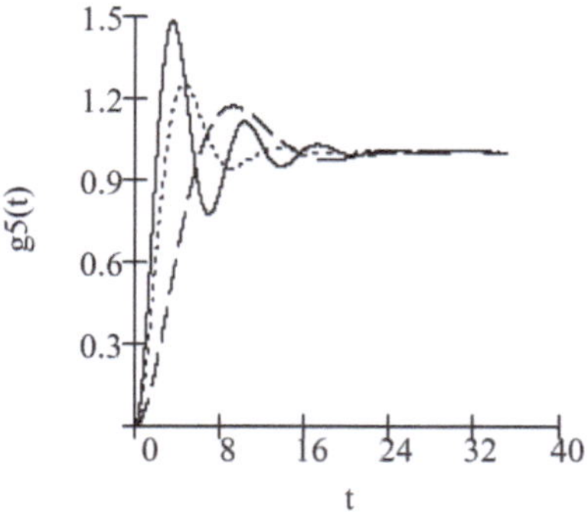

FIGURE 12.8 Plots of $g_5(t)$ for $m=1$, $k=1$, and $\omega=1$ when $\lambda=0.3$ (solid), 0.6 (dot), 0.9 (dash).

Expanding Eq. (12.31) produces

$$g_5(t)=\frac{1}{k\omega^\lambda\cos\frac{\lambda\pi}{2}}\left[1-\frac{e^{-\frac{k\omega^{\lambda-1}\sin\frac{\lambda\pi}{2}}{2\sqrt{mk\omega^\lambda\cos\frac{\lambda\pi}{2}}}\sqrt{\omega^\lambda\cos\frac{\lambda\pi}{2}}\omega_n t}}{\sqrt{1-\varsigma_{eq5}^2}}\cos\left(\omega_{eqd5}t-\phi_5\right)\right],\quad t\ge 0, \tag{12.34}$$

where

$$\phi_5(\omega)=\tan^{-1}\frac{\frac{k\omega^{\lambda-1}\sin\frac{\lambda\pi}{2}}{2\sqrt{mk\omega^\lambda\cos\frac{\lambda\pi}{2}}}}{\sqrt{1-\left(\frac{k\omega^{\lambda-1}\sin\frac{\lambda\pi}{2}}{2\sqrt{mk\omega^\lambda\cos\frac{\lambda\pi}{2}}}\right)^2}}. \tag{12.35}$$

Figure 12.8 shows several plots of $g_5(t)$.

Also $g_5(t)$may be written by

$$g_5(t)=g_5(t,\omega). \tag{12.36}$$

12.6 CLASS V FREQUENCY TRANSFER FUNCTION

Theorem 12.12 is for frequency transfer function of a fractional vibrator of class V.

Theorem 12.12. Let $H_5(\omega)$ be the frequency transfer function of a fractional vibrator of class V. Its analytical expression is given by

$$H_5(\omega)=\frac{1}{k_{eq5}\left(1-\gamma_{eq5}^2+i2\varsigma_{eq5}\gamma_{eq5}\right)}. \tag{12.37}$$

Proof. Denote by $H_5(\omega)$ the Fourier transform of $h_5(t)$, which is the solution to Eq. (12.27). Doing the Fourier transform on both sides of Eq. (12.27) produces

$$\left[m(i\omega)^2+c_{eq5}(i\omega)+k_{eq5}\right]H_5(\omega)=1.$$

The above means

$$\left[(i\omega)^2+2\varsigma_{eq5}\omega_{eqn5}(i\omega)+\omega_{eqn5}^2\right]H_5(\omega)=\frac{1}{m}.$$

Dividing both sides of the above by ω_{eqn5}^2 results in

$$\left[1-(\gamma_{eq5})^2+i2\varsigma_{eq5}\gamma_{eq5}\right]H_5(\omega)=\frac{1}{m\omega_{eqn5}^2}.$$

Since $\frac{1}{m\omega_{eqn5}^2}=\frac{k_{eq5}}{mk_{eq5}\omega_{eqn5}^2}=\frac{1}{k_{eq5}}$, the above implies

$$H_5(\omega)=\frac{1}{k_{eq5}\left(1-\gamma_{eq5}^2+i2\varsigma_{eq5}\gamma_{eq5}\right)}.$$

The above is Eq. (12.37). The proof is finished. □

If substituting ζ_{eq5}, k_{eq5}, and γ_{eq5} into Eq. (12.37), we have

$$H_5(\omega)=\frac{1}{k\omega^{\lambda}\cos\frac{\lambda\pi}{2}\left(1-\frac{\gamma^2}{\omega^{\lambda}\cos\frac{\lambda\pi}{2}}+i2\gamma\frac{k\omega^{\lambda-1}\sin\frac{\lambda\pi}{2}}{2\sqrt{mk\omega^{\lambda}\cos\frac{\lambda\pi}{2}}}\sqrt{\frac{1}{\omega^{\lambda}\cos\frac{\lambda\pi}{2}}}\right)}, \tag{12.38}$$

where $\gamma=\frac{\omega}{\omega_n}$.

The amplitude-frequency response of $H_5(\omega)$ is expressed by

$$|H_5(\omega)|=\frac{1}{k_{eq5}}\frac{1}{\sqrt{\left(1-\gamma_{eq5}^2\right)^2+\left(2\varsigma_{eq5}\gamma_{eq5}\right)^2}}. \tag{12.39}$$

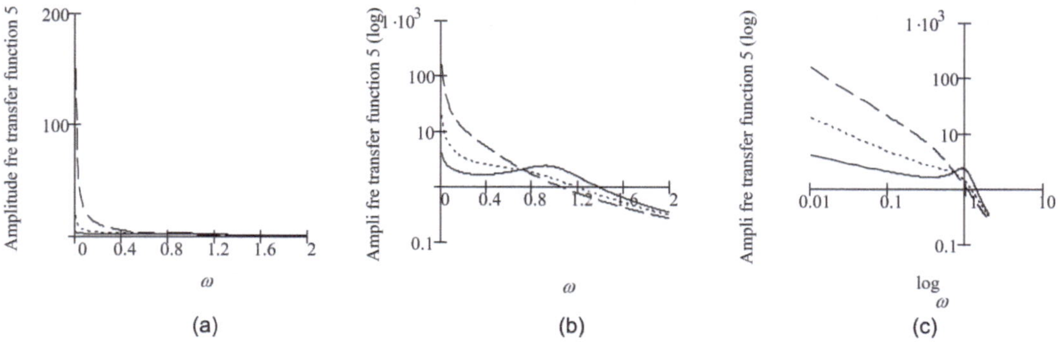

FIGURE 12.9 Plots of $|H_5(\omega)|$ for $m=1$ and $k=1$ when $\lambda=0.3$ (solid), 0.6 (dot), 0.9 (dash). (a) $|H_5(\omega)|$ in conventional coordinate. (b) $|H_5(\omega)|$ in log-scale. (c) $|H_5(\omega)|$ in log-log scale.

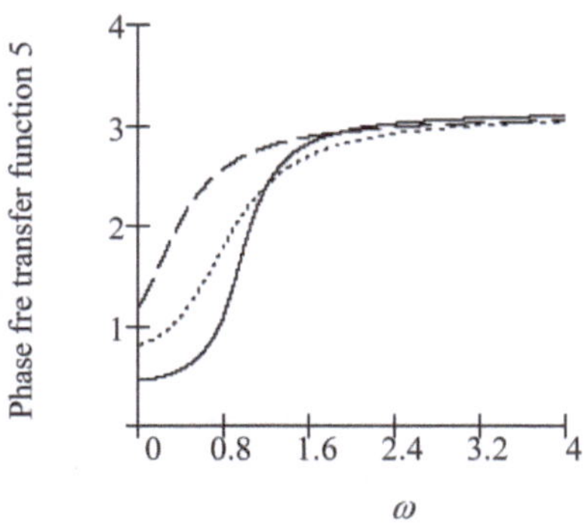

FIGURE 12.10 Plots of $\varphi_5(\omega)$ for $m = 1$ and $k = 1$ when $\lambda = 0.3$ (solid), 0.6 (dot), 0.9 (dash).

The phase-frequency response function is given by

$$\varphi_5(\omega) = \cos^{-1}\frac{1-\gamma_{eq5}^2}{\sqrt{\left(1-\gamma_{eq5}^2\right)^2+\left(2\varsigma_{eq5}\gamma_{eq5}\right)^2}}. \tag{12.40}$$

A few plots of $|H_5(\omega)|$ and $\varphi_5(\omega)$ are given in Figures 12.9 and 12.10.

12.7 CLASS V QUASIEQUIVALENT LOGARITHMIC DECREMENT AND Q FACTOR

12.7.1 Quasiequivalent Logarithmic Decrement of Fractional Vibrators of Class V

The expression of the quasiequivalent logarithmic decrement of $x_5(t)$ of a fractional vibrator of class V is proposed below.

Theorem 12.13. Let t_i and t_{i+1} be two time points of fractional free response $x_5(t)$ so that $x_5(t_i)$ and $x_5(t_{i+1})$ are successive peak values at t_i and t_{i+1}. Denote by Δ_{eq5} the quasiequivalent logarithmic decrement of $x_5(t)$. Then,

$$\Delta_{eq5} = \ln\frac{x_5(t_i)}{x_5(t_{i+1})} = \frac{2\pi\dfrac{k\omega^{\lambda-1}\sin\dfrac{\lambda\pi}{2}}{2\sqrt{mk\omega^{\lambda}\cos\dfrac{\lambda\pi}{2}}}}{\sqrt{1-\left(\dfrac{k\omega^{\lambda-1}\sin\dfrac{\lambda\pi}{2}}{2\sqrt{mk\omega^{\lambda}\cos\dfrac{\lambda\pi}{2}}}\right)^2}}, \quad 0 \le \lambda < 1. \tag{12.41}$$

Proof. Based on the definition of logarithmic decrement, we have

$$\Delta_{eq5} = \ln\frac{x_5(t_i)}{x_5(t_{i+1})} = \frac{2\pi\varsigma_{eq5}}{\sqrt{1-\varsigma_{eq5}^2}}, \quad 0 \le \lambda < 1. \tag{12.42}$$

Substituting ζ_{eq5} into the above results in Eq. (12.41). This finishes the proof. □

Δ_{eq5} is a quasiequivalent logarithmic decrement since it is not dimensionless in general. If $\lambda = 0$, it reduces to zero because the vibration system reduces to a conventional one without damping. A few illustrations of Δ_{eq5} are indicated in Figure 12.11.

12.7.2 Quasiequivalent Q Factor of Fractional Vibrators of Class V

The expression of the Q factor of a fractional vibrator of class V is presented below.

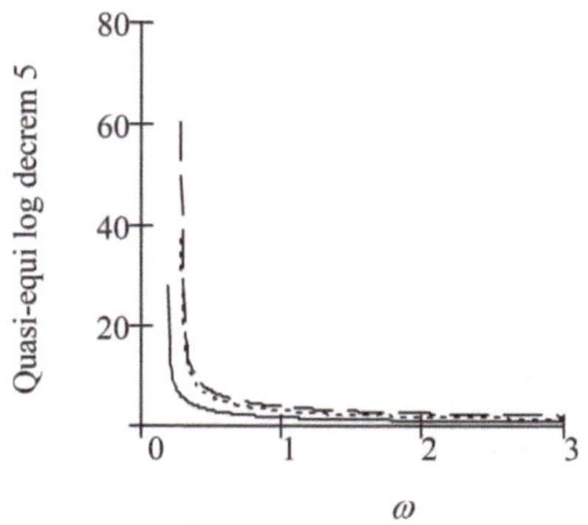

FIGURE 12.11 Illustrations of Δ_{eq5} for $m = 1$ and $k = 1$ when $\lambda = 0.3$ (solid), 0.6 (dot), 0.9 (dash). Note that "Quasi-equivalent logarithmic decrement 5" in the above figure stands for Δ_{eq5}.

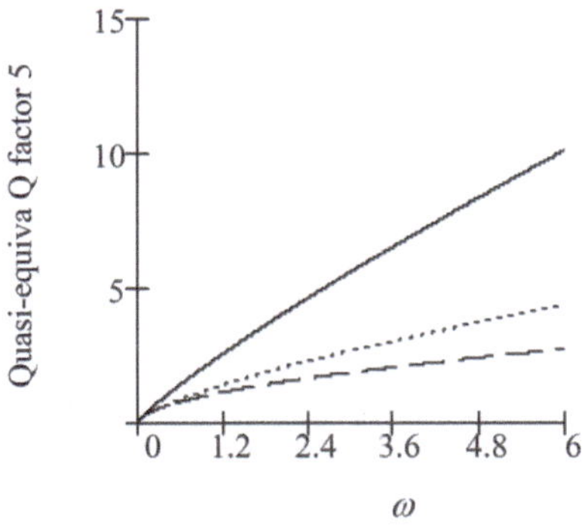

FIGURE 12.12 Illustrations of Q_{eq5} for $m=1$ and $k=1$ when $\lambda=0.3$ (solid), 0.6 (dot), 0.9 (dash). Note that "Quasi-equivalent Q factor 5" in the above figure stands for Q_{eq5}.

Theorem 12.14. Let Q_{eq5} be the quasiequivalent Q factor of a fractional vibrator of class V. It is given by

$$Q_{eq5} = \frac{\sqrt{mk\omega^{\lambda}\cos\frac{\lambda\pi}{2}}}{k\omega^{\lambda-1}\sin\frac{\lambda\pi}{2}}. \tag{12.43}$$

Proof. Based on the definition of Q factor, we have

$$Q_{eq5} = \frac{1}{2\varsigma_{eq5}}. \tag{12.44}$$

Substituting ζ_{eq5} into the above yields Eq. (12.43). The proof is finished. □

The quantity Q_{eq5} is not a constant but a function in terms of ω and λ. Besides, it is not dimensionless unless $\lambda=0$. That is the reason we use a prefix "quasi" for it. In the case of $\lambda=0$, $Q_{eq5}=\infty$. Several of its illustrations are indicated in Figure 12.12.

12.8 NOTING C_{EQ5} WITH RAYLEIGH DAMPING

When denoting the Rayleigh damping by c_r, we write $c_r=am+bk$, where m and k are primary mass and stiffness, respectively, a is proportional to ω while b is inversely proportional to ω (Jin and Xia [1], Harris [2], Palley et al. [3]). According to Eq. (12.3), we see that c_{eq5} is inversely proportional to ω with the coefficient $k\sin\frac{\lambda\pi}{2}$. Thus, it may yet be a variation of the Rayleigh damping. More about it will be addressed in Chapter 14.

12.9 CONCLUDING REMARK

Reports regarding fractional vibrators in class V are rarely seen. Thus, the present results in this chapter are new in the field. Importantly, we note that this model can be used to explain a king of variation of the Rayleigh damping.

12.10 EXERCISES

12.1. Prove that c_{eq5} is positive definite.

12.2. Prove that a fractional vibrator of class V is stable.

12.3. Find the condition for $0<\zeta_{eq5}<1$.

12.4. Prove that ζ_{eq5} is not dimensionless for $0<\lambda<1$.

12.5. Prove that ω_{eqn5} is not a standard angular frequency unless $\lambda=0$.

12.6. Prove that γ_{eq5} is not dimensionless if $0<\lambda<1$.

REFERENCES

1. X. D. Jin and L. J. Xia, *Ship Hull Vibration*, The Press of Shanghai Jiaotong University, Shanghai, China, 2011. In Chinese.
2. C. M. Harris, *Shock and Vibration Handbook*, 5th Ed., McGraw-Hill, New York, 2002.
3. O. M. Palley, Г. B. Bahizov, and E. Я. Voroneysk, *Handbook of Ship Structural Mechanics*, National Defense Industry Publishing House, Beijing, 2002. In Chinese. Translated from Russian by B. H. Xu, X. Xu, and M. Q. Xu.

CHAPTER 13

Fractional Vibrations of Class VI

THIS CHAPTER BRINGS FORWARD the results listed below:

1. An equivalent motion equation of a fractional vibrator of class VI is given in Theorem 13.1.
2. The quasiequivalent mass, quasiequivalent damping, quasiequivalent stiffness, quasiequivalent damping ratio, quasiequivalent damping free natural frequency, quasiequivalent damped natural frequency, quasiequivalent frequency ratio, quasi-equivalent logarithmic decrement, and quasiequivalent Q factor for a fractional vibrator in class VI are put forward. Their representations are presented in Theorems 13.2, 13.3, 13.6–13.10, 13.15, and 13.16, respectively.
3. The phenomenon that quasiequivalent damping of a fractional vibrator in class VI may vary from $-\infty$ to ∞ is exhibited in Corollary 13.1.
4. Conclude that a fractional vibrator in class VI may be self-vibrated in Theorem 13.5 and Corollary 13.4.
5. Present the analytical expressions of free response, impulse response, step response, and frequency transfer function of a fractional vibrator in class VI in Theorems 13.11–13.14, respectively.

13.1 EQUIVALENT MOTION EQUATION OF CLASS VI

Theorem 13.1 proposes to give an equivalent motion equation of a fractional vibrator of class VI.

DOI: 10.1201/9781003460947-15

Theorem 13.1. The motion equation of a fractional vibrator of class VI below

$$m\frac{d^{\alpha}x_6(t)}{dt^{\alpha}}+c\frac{d^{\beta}x_6(t)}{dt^{\beta}}+k\frac{d^{\lambda}x_6(t)}{dt^{\lambda}}=0,\quad 1<\alpha<3,\ 0<\beta<2,\ 0\le\lambda<1 \tag{13.1}$$

is equivalently given by

$$\begin{aligned}&-\left(m\omega^{\alpha-2}\cos\frac{\alpha\pi}{2}+c\omega^{\beta-2}\cos\frac{\beta\pi}{2}\right)\frac{d^2x_6(t)}{dt^2}\\&+\left(m\omega^{\alpha-1}\sin\frac{\alpha\pi}{2}+c\omega^{\beta-1}\sin\frac{\beta\pi}{2}+k\omega^{\lambda-1}\sin\frac{\lambda\pi}{2}\right)\frac{dx_6(t)}{dt}\\&+k\omega^{\lambda}\cos\frac{\lambda\pi}{2}x_6(t)=0,\quad 1<\alpha<3,\ 0<\beta<2,\ 0\le\lambda<1.\end{aligned} \tag{13.2}$$

Proof. Let $X_6(\omega)$ be the Fourier transform of $x_6(t)$. Doing the Fourier transform on Eq. (13.2) produces

$$\begin{aligned}&\left(m\omega^{\alpha}\cos\frac{\alpha\pi}{2}+c\omega^{\beta}\cos\frac{\beta\pi}{2}\right)X_6(\omega)+i\left(m\omega^{\alpha}\sin\frac{\alpha\pi}{2}+c\omega^{\beta}\sin\frac{\beta\pi}{2}+k\omega^{\lambda}\sin\frac{\lambda\pi}{2}\right)X_6(\omega)\\&+k\omega^{\lambda}\cos\frac{\lambda\pi}{2}X_6(\omega)=0.\end{aligned} \tag{13.3}$$

Rewrite the above by

$$\begin{aligned}&\left(m\omega^{\alpha}\cos\frac{\alpha\pi}{2}+im\omega^{\alpha}\sin\frac{\alpha\pi}{2}\right)X_6(\omega)+\left(c\omega^{\beta}\cos\frac{\beta\pi}{2}+ic\omega^{\beta}\sin\frac{\beta\pi}{2}\right)X_6(\omega)\\&+\left(k\omega^{\lambda}\cos\frac{\lambda\pi}{2}+ik\omega^{\lambda}\sin\frac{\lambda\pi}{2}\right)X_6(\omega)=0.\end{aligned} \tag{13.4}$$

Since the principal values of i^{α}, i^{β}, and i^{λ} are expressed by

$$\begin{aligned}i^{\alpha}&=\cos\frac{\alpha\pi}{2}+i\sin\frac{\alpha\pi}{2},\\i^{\beta}&=\cos\frac{\beta\pi}{2}+i\sin\frac{\beta\pi}{2},\\i^{\lambda}&=\cos\frac{\lambda\pi}{2}+i\sin\frac{\lambda\pi}{2},\end{aligned} \tag{13.5}$$

we have

$$m(i\omega)^{\alpha}X_6(\omega)+c(i\omega)^{\beta}X_6(\omega)+k(i\omega)^{\lambda}X_6(\omega)=0. \tag{13.6}$$

Doing the inverse Fourier transform of the above yields $m\frac{d^{\alpha}x_6(t)}{dt^{\alpha}}+c\frac{d^{\beta}x_6(t)}{dt^{\beta}}+k\frac{d^{\lambda}x_6(t)}{dt^{\lambda}}=0$. That is the original equation of a fractional vibrator of class VI, see Eq. (13.1). Thus, Theorem 13.1 holds. The proof is completed. □

13.2 QUASIEQUIVALENT VIBRATION PARAMETERS OF CLASS VI

13.2.1 Quasiequivalent Mass of Fractional Vibrator of Class VI

Let m_{eq6} be the quasiequivalent mass for a fractional vibrator of class VI. Theorem 13.2 depicts it.

Theorem 13.2. The expression of m_{eq6} is given by

$$m_{eq6}=m_{eq3}=-\left(m\omega^{\alpha-2}\cos\frac{\alpha\pi}{2}+c\omega^{\beta-2}\cos\frac{\beta\pi}{2}\right). \tag{13.7}$$

Proof. Taking into account Theorem 10.2, we see that the above holds. The proof is finished. □

Note that $m_{eq6}=m_{eq3}\in(-\infty,\infty)$. Refer to Chapter 10 for m_{eq3}.

13.2.2 Quasiequivalent Damping of Fractional Vibrators of Class VI

Denote by c_{eq6} the quasiequivalent damping for a fractional vibrator of class VI. Theorem 13.3 is for its analytical expression.

Theorem 13.3. c_{eq6} is expressed by

$$c_{eq6}=c_{eq6}(\omega,\alpha,\beta,\lambda)=m\omega^{\alpha-1}\sin\frac{\alpha\pi}{2}+c\omega^{\beta-1}\sin\frac{\beta\pi}{2}+k\omega^{\lambda-1}\sin\frac{\lambda\pi}{2}. \tag{13.8}$$

Proof. The above is the coefficient of $\frac{dx_6(t)}{dt}$ in Eq. (13.2). The proof is finished. □

The quantity c_{eq6} is termed with the prefix "quasi" as its unit is not $N\times m^{-1}\times s$ unless $\alpha=2$, $\beta=1$, and $\lambda=0$. If $\alpha=2$, $\beta=1$, and $\lambda=0$, c_{eq6} reduces to the primary damping c with the unit $N\times m^{-1}\times s$.

We propose the following corollaries and theorems for the asymptotic properties of c_{eq6}.

Corollary 13.1. If $\omega\to\infty$, for $0\le\lambda<1$,

$$c_{eq6}=\begin{cases}\infty, & 1<\alpha<2, 0<\beta<2,\\ -\infty, & 2<\alpha<3, 0<\beta<1.\end{cases} \tag{13.9}$$

Proof. Note that $\lim_{\omega\to\infty}\omega^{\lambda-1}=0$ if $0\le\lambda<1$. Besides, $\sin\frac{\beta\pi}{2}>0$ when $0<\beta<2$ and $\sin\frac{\alpha\pi}{2}>0$ if $1<\alpha<2$. Thus, $\lim_{\omega\to\infty}c_{eq6}=\infty$ if $1<\alpha<2$ and $0<\beta<2$. Further, since $\lim_{\omega\to\infty}\omega^{\beta-1}=0$ if $0<\beta<1$ but $\sin\frac{\alpha\pi}{2}<0$ if $2<\alpha<3$, $\lim_{\omega\to\infty}c_{eq6}=-\infty$ for $2<\alpha<3$ and $0<\beta<1$. Thus, Eq. (13.9) holds. This finishes the proof. □

The above corollary says that $-\infty<c_{eq6}<\infty$.

Theorem 13.4. c_{eq6} is proportional to m, c, and k.

The proof is straightforward from Eq. (13.8) and omitted. □

One interesting thing about c_{eq6} is that is consistent with the Rayleigh damping form if $c=0$, also see Chapter 11. When $c\neq0$, it is a variation of the Rayleigh damping.

Theorem 13.5. A fractional vibrator of class VI may be self-vibrated and accordingly non-stable if $2<\alpha<3$ and $0<\beta<1$.

Proof. Because c_{eq6} may be negative if $2<\alpha<3$ and $0<\beta<1$, a vibrator may be self-vibrated and non-stable in that case. The proof is finished. □

Corollary 13.2. Asymptotically,

$$\lim_{\omega\to0}c_{eq6}=\infty. \tag{13.10}$$

Proof. Note that $\lim_{\omega\to0}\omega^{\lambda-1}=\infty$ for $0\le\lambda<1$. The proof is finished. □

Corollary 13.3. For $1<\alpha<3$ and $0<\beta<1$, when ω is large, we have

$$c_{eq6}\propto m\omega^{\alpha-1}\sin\frac{\alpha\pi}{2}. \tag{13.11}$$

Besides, for $1<\alpha<3$ and $0<\beta<1$, if ω is small enough,

$$c_{eq6}\propto c\omega^{\beta-1}\sin\frac{\beta\pi}{2}+k\omega^{\lambda-1}\sin\frac{\lambda\pi}{2}. \tag{13.12}$$

Proof. If $1<\alpha<3$, $0<\beta<1$ and ω is sufficiently large, $\omega^{\beta-1}\approx0$ and $\omega^{\lambda-1}\approx0$. Therefore, Eq. (13.11) results. On the other side, when ω is small enough, $\omega^{\alpha-1}\approx0$. Thus, Eq. (13.12) holds. This completes the proof. □

Figure 13.1 shows a few plots of c_{eq6}.

13.2.3 Quasiequivalent Stiffness of Fractional Vibrators of Class VI

Theorem 13.6 is for the quasiequivalent stiffness of a fractional vibrator of class VI.

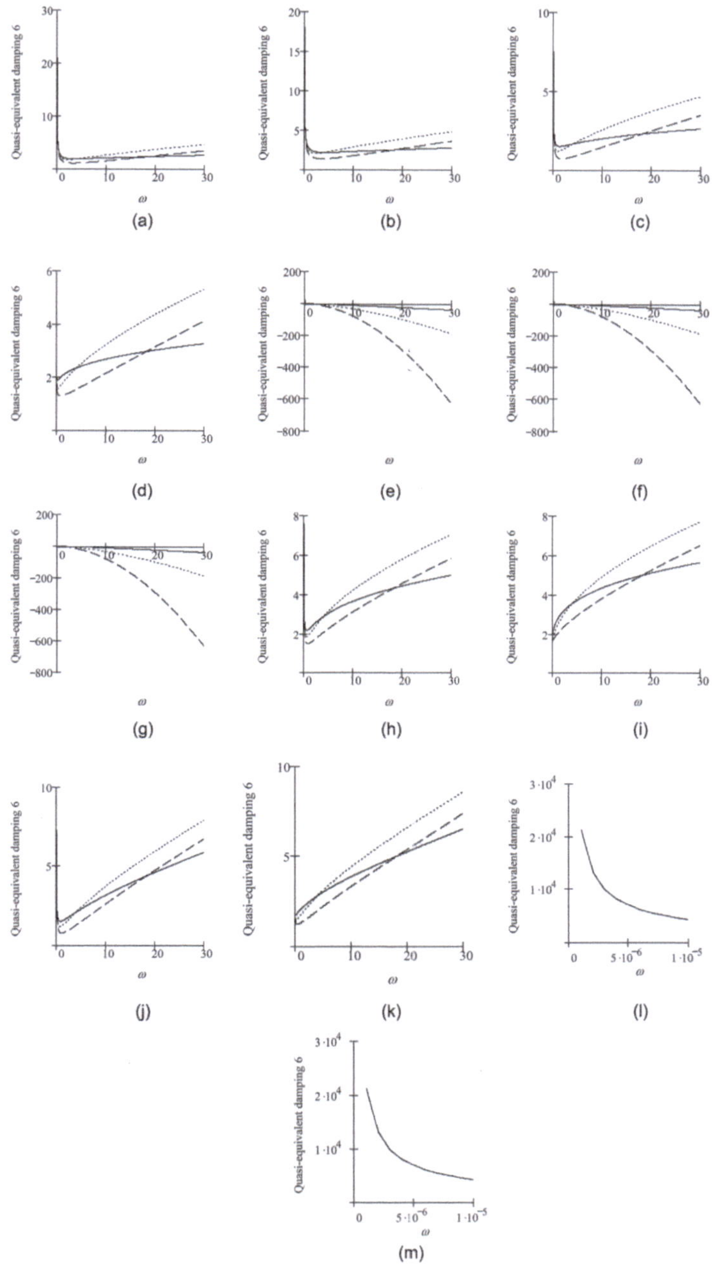

FIGURE 13.1 Plots of c_{eq6}. (a) c_{eq6} when $\beta=0.3$, $\lambda=0.3$, and $\alpha=1.3$ (solid), 1.6 (dot), 1.9 (dash). (b) c_{eq6} when $\beta=0.9$, $\lambda=0.3$, and $\alpha=1.3$ (solid), 1.6 (dot), 1.9 (dash). (c) c_{eq6} when $\beta=0.3$, $\lambda=0.9$, and $\alpha=1.3$ (solid), 1.6 (dot), 1.9 (dash). (d) c_{eq6} when $\beta=0.9$, $\lambda=0.9$, and $\alpha=1.3$ (solid), 1.6 (dot), 1.9 (dash). (e) c_{eq6} when $\beta=0.3$, $\lambda=0.3$, and $\alpha=2.3$ (solid), 2.6 (dot), 2.9 (dash) (negative damping). (f) c_{eq6} when $\beta=0.9$, $\lambda=0.3$, and $\alpha=2.3$ (solid), 2.6 (dot), 2.9 (dash) (negative damping). (g) c_{eq6} when $\beta=0.9$, $\lambda=0.9$, and $\alpha=2.3$ (solid), 2.6 (dot), 2.9 (dash) (negative damping). (h) c_{eq6} when $\beta=1.3$, $\lambda=0.3$, and $\alpha=1.3$ (solid), 1.6 (dot), 1.9 (dash). (i) c_{eq6} when $\beta=1.3$, $\lambda=0.9$, and $\alpha=1.3$ (solid), 1.6 (dot), 1.9 (dash). (j) c_{eq6} when $\beta=1.9$, $\lambda=0.3$, and $\alpha=1.3$ (solid), 1.6 (dot), 1.9 (dash). (k) c_{eq6} when $\beta=1.9$, $\lambda=0.9$, and $\alpha=1.3$ (solid), 1.6 (dot), 1.9 (dash). (l) c_{eq6} when $\beta=0.3$, $\lambda=0.3$, $\alpha=1.5$ (infinite damping at $\omega\approx0$). (m) c_{eq6} when $\beta=0.3$, $\lambda=0.3$, $\alpha=2.5$ (infinite damping at $\omega\approx0$). Note that "Quasi-equivalent damping 6" in the above figures stands for c_{eq6}.

Theorem 13.6. Let k_{eq6} be the quasiequivalent stiffness of a fractional vibrator of class VI. Then,

$$k_{eq6} = k_{eq4} = k\omega^{\lambda} \cos\frac{\lambda\pi}{2}. \tag{13.13}$$

Proof. In Eq. (13.2), restoration force is $k\omega^{\lambda} \cos\frac{\lambda\pi}{2} x_6(t)$ only. This finishes the proof, referring Chapter 11 for k_{eq4}. □

13.2.4 Quasiequivalent Damping Ratio of Fractional Vibrators in Class VI

Let ζ_{eq6} be the quasiequivalent damping ratio for a fractional vibrator of class VI. Define it by

$$\varsigma_{eq6} = \frac{c_{eq6}}{2\sqrt{m_{eq6}k_{eq6}}}. \tag{13.14}$$

Theorem 13.7 gives its expression.

Theorem 13.7. ζ_{eq6} is expressed by

$$\varsigma_{eq6} = \varsigma_{eq6}(\omega,\alpha,\beta,\lambda) = \frac{m\omega^{\alpha-1}\sin\frac{\alpha\pi}{2} + c\omega^{\beta-1}\sin\frac{\beta\pi}{2} + k\omega^{\lambda-1}\sin\frac{\lambda\pi}{2}}{2\sqrt{-\left(m\omega^{\alpha-2}\cos\frac{\alpha\pi}{2} + c\omega^{\beta-2}\cos\frac{\beta\pi}{2}\right)k\omega^{\lambda}\cos\frac{\lambda\pi}{2}}}. \tag{13.15}$$

Proof. Substituting m_{eq6}, c_{eq6}, and k_{eq6} into Eq. (13.14) yields Eq. (13.15). The proof is complete. □

Note that ζ_{eq6} is not dimensionless in general. It is dimensionless if $\alpha=2$, $\beta=1$, and $\lambda=0$. In fact, $\zeta_{eq6}(\omega, 2, 1, 0)=\zeta$.

Corollary 13.4. The quasiequivalent damping ratio ζ_{eq6} may be negative relying on the ranges of α, β, and λ.

The proof is omitted. □

Several illustrations of ζ_{eq6} are shown in Figure 13.2.

13.2.5 Quasiequivalent Damping Free Natural Frequency of Fractional Vibrators of Class VI

For a fractional vibrator of class VI, we coin a term quasiequivalent damping free natural frequency to the quantity given by

$$\omega_{eqn6} = \sqrt{\frac{k_{eq6}}{m_{eq6}}}. \tag{13.16}$$

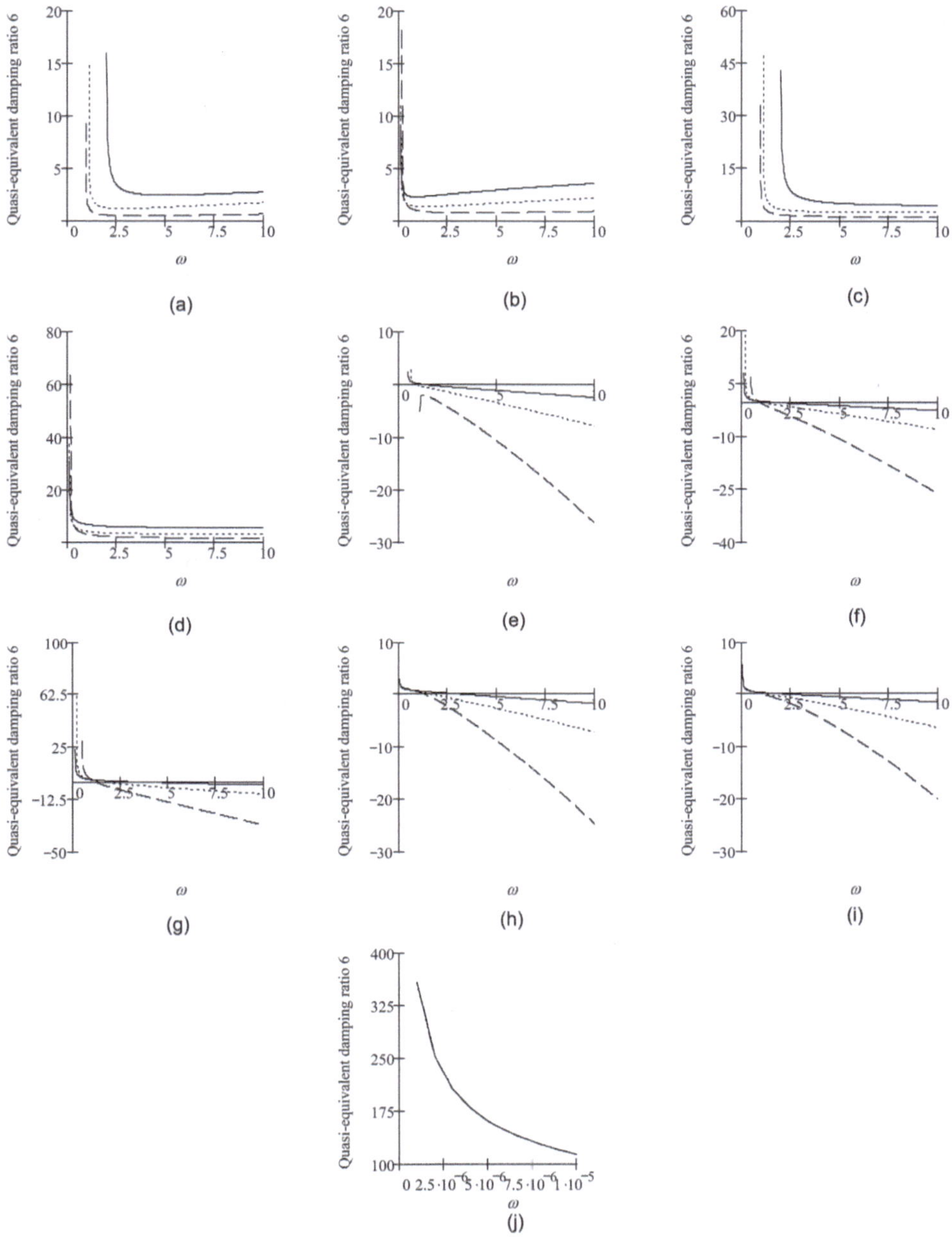

FIGURE 13.2 Illustration of ζ_{eq6} with $m=1$ and $k=1$. (a) ζ_{eq6} when $c=1$. $\beta=0.3$, $\lambda=0.3$, and $\alpha=1.3$ (solid), 1.6 (dot), 1.9 (dash). (b) ζ_{eq6} when $c=1$. $\beta=0.9$, $\lambda=0.3$, and $\alpha=1.3$ (solid), 1.6 (dot), 1.9 (dash). (c) ζ_{eq6} when $c=1$. $\beta=0.3$, $\lambda=0.9$, and $\alpha=1.3$ (solid), 1.6 (dot), 1.9 (dash). (d) ζ_{eq6} when $c=1$. $\beta=0.9$, $\lambda=0.9$, and $\alpha=1.3$ (solid), 1.6 (dot), 1.9 (dash). (e) ζ_{eq6} when $c=0.2$. $\beta=0.3$, $\lambda=0.3$, and $\alpha=2.3$ (solid), 2.6 (dot), 2.9 (dash) (negative damping ratio). (f) ζ_{eq6} when $c=0.2$. $\beta=0.9$, $\lambda=0.3$, and $\alpha=2.3$ (solid), 2.6 (dot), 2.9 (dash) (negative damping ratio). (g) ζ_{eq6} when $c=0.2$. $\beta=0.9$, $\lambda=0.9$, and $\alpha=2.3$ (solid), 2.6 (dot), 2.9 (dash) (negative damping ratio). (h) ζ_{eq6} when $c=1$. $\beta=1.3$, $\lambda=0.3$, and $\alpha=2.3$ (solid), 2.6 (dot), 2.9 (dash) (negative damping ratio). (i) ζ_{eq6} when $c=1$. $\beta=1.9$, $\lambda=0.9$, and $\alpha=2.3$ (solid), 2.6 (dot), 2.9 (dash) (negative damping ratio). (j) ζ_{eq6} when $c=1$. $\beta=1.3$, $\lambda=0.3$, and $\alpha=2.3$ (solid), 2.6 (dot), 2.9 (dash) (infinite damping ratio at $\omega\approx0$). Note that "Quasi-equivalent damping ratio 6" in the above figures implies ζ_{eq6}.

Theorem 13.8. The quasiequivalent damping free natural frequency ω_{eqn6} is represented by

$$\omega_{eqn6}=\sqrt{\frac{k\omega^{\lambda}\cos\frac{\lambda\pi}{2}}{-\left(m\omega^{\alpha-2}\cos\frac{\alpha\pi}{2}+c\omega^{\beta-2}\cos\frac{\beta\pi}{2}\right)}}. \tag{13.17}$$

Proof. Substituting m_{eq6} and k_{eq6} into Eq. (13.16) produces Eq. (13.17). The proof is finished. □

The prefix "quasi" is used to describe ω_{eqn6} as its unit is not rad/s. Its unit reduces to rad/s if $\alpha=2$, $\beta=1$, and $\lambda=0$. In that case,

$$\omega_{eqn6}\Big|_{\alpha=2,\beta=1,\lambda=0}=\omega_n. \tag{13.18}$$

As ω_{eqn6} relates to four parameters ω, α, β, and λ, it may be expressed by

$$\omega_{eqn6}=\omega_{eqn6}(\omega,\alpha,\beta,\lambda). \tag{13.19}$$

Figure 13.3 exhibits the plots of ω_{eqn6}.

13.2.6 Quasiequivalent Damped Natural Frequency of Fractional Vibrators of Class VI

By introducing a term quasiequivalent damped natural frequency for a fractional vibrator of class VI, we denote it by ω_{eqd6}. We suppose small damping here. That is, $|\zeta_{eq6}|\le 1$. It is defined by

$$\omega_{eqd6}=\omega_{eqn6}\sqrt{1-\zeta_{eq6}^{2}},\quad |\zeta_{eq6}|\le 1. \tag{13.20}$$

Theorem 13.9 is used to describe its representation.

Theorem 13.9. The representation of ω_{eqd6} is given by

$$\omega_{eqd6}=\sqrt{\frac{k\omega^{\lambda}\cos\frac{\lambda\pi}{2}}{-\left(m\omega^{\alpha-2}\cos\frac{\alpha\pi}{2}+c\omega^{\beta-2}\cos\frac{\beta\pi}{2}\right)}}\sqrt{1-\left(\frac{m\omega^{\alpha-1}\sin\frac{\alpha\pi}{2}+c\omega^{\beta-1}\sin\frac{\beta\pi}{2}+k\omega^{\lambda-1}\sin\frac{\lambda\pi}{2}}{2\sqrt{-\left(m\omega^{\alpha-2}\cos\frac{\alpha\pi}{2}+c\omega^{\beta-2}\cos\frac{\beta\pi}{2}\right)k\omega^{\lambda}\cos\frac{\lambda\pi}{2}}}\right)^{2}}. \tag{13.21}$$

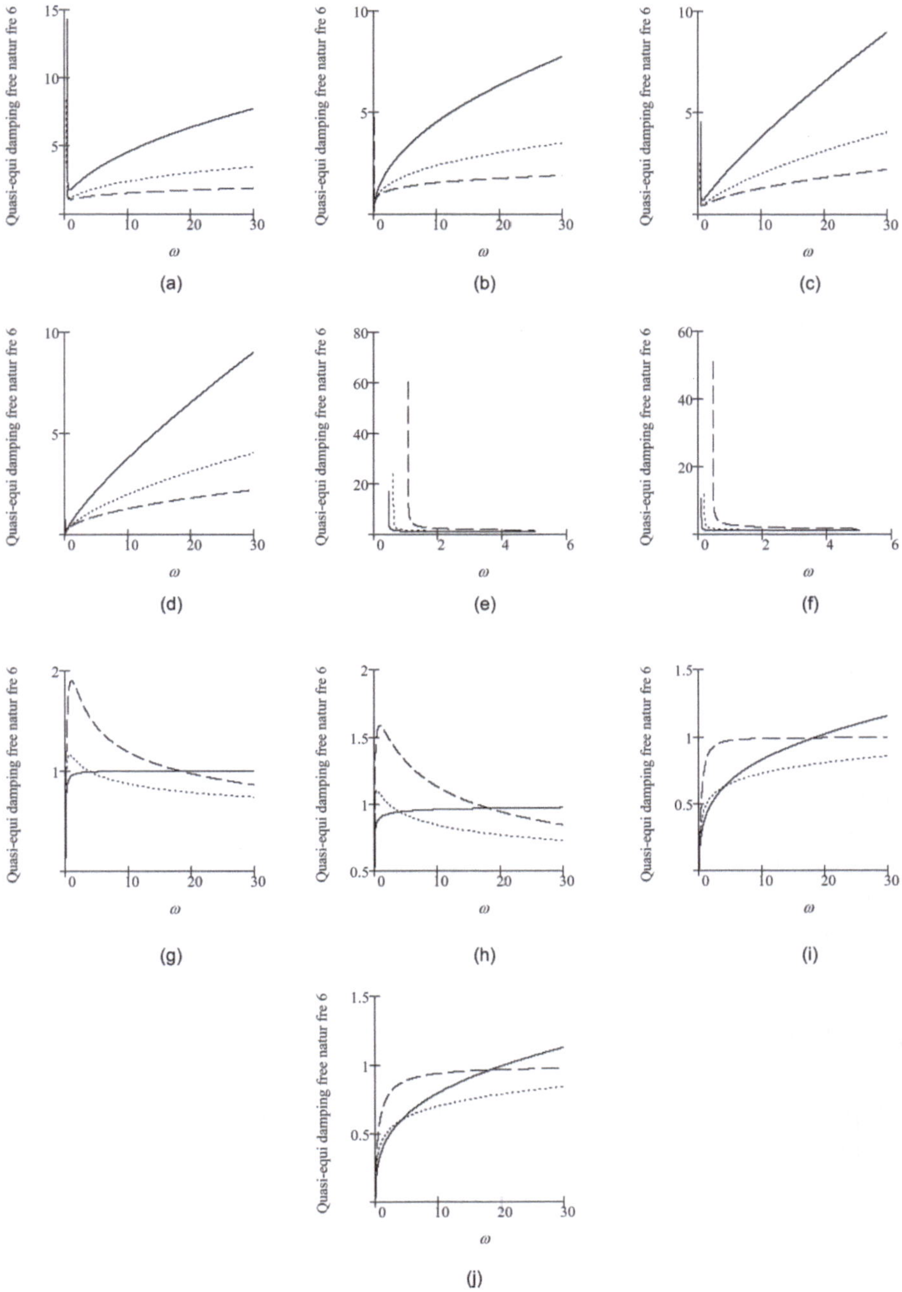

FIGURE 13.3 Plots of ω_{eqn6} with $m=1$ and $k=1$, $c=0.2$. (a) ω_{eqn6} when $\beta=0.3$, $\lambda=0.3$, and $\alpha=1.3$ (solid), 1.6 (dot), 1.9 (dash). (b) ω_{eqn6} when $\beta=0.9$, $\lambda=0.3$, and $\alpha=1.3$ (solid), 1.6 (dot), 1.9 (dash). (c) ω_{eqn6} when $\beta=0.3$, $\lambda=0.9$, and $\alpha=1.3$ (solid), 1.6 (dot), 1.9 (dash). (d) ω_{eqn6} when $\beta=0.9$, $\lambda=0.9$, and $\alpha=1.3$ (solid), 1.6 (dot), 1.9 (dash). (e) ω_{eqn6} when $\beta=0.3$, $\lambda=0.3$, and $\alpha=2.3$ (solid), 2.6 (dot), 2.9 (dash). (f) ω_{eqn6} when $\beta=0.9$, $\lambda=0.3$, and $\alpha=2.3$ (solid), 2.6 (dot), 2.9 (dash). (g) ω_{eqn6} when $\beta=1.3$, $\lambda=0.3$, and $\alpha=2.3$ (solid), 2.6 (dot), 2.9 (dash). (h) ω_{eqn6} when $\beta=1.9$, $\lambda=0.3$, and $\alpha=2.3$ (solid), 2.6 (dot), 2.9 (dash). (i) ω_{eqn6} when $\beta=1.3$, $\lambda=0.9$, and $\alpha=2.3$ (solid), 2.6 (dot), 2.9 (dash). (j) ω_{eqn6} when $\beta=1.9$, $\lambda=0.9$, and $\alpha=2.3$ (solid), 2.6 (dot), 2.9 (dash). Note that "Quasi-equivalent damping free natural frequency 6" in the above figures stands for ω_{eqn6}.

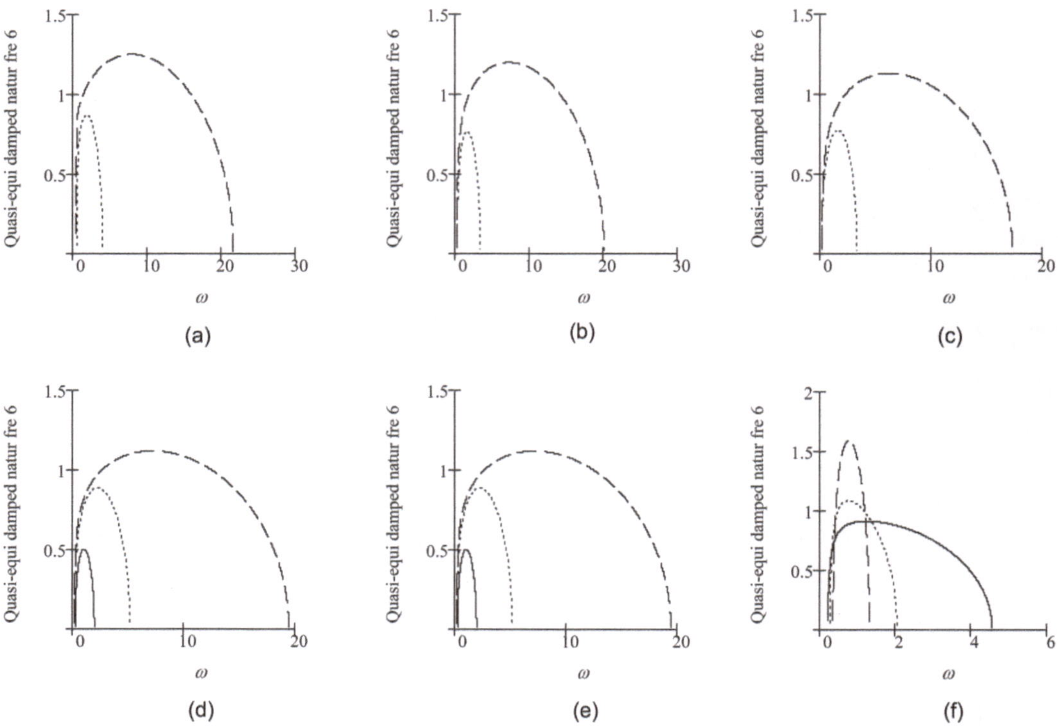

FIGURE 13.4 Plots of ω_{eqd6} with $m=1$ and $k=1$, $c=0.2$. (a) ω_{eqd6} when $\beta=0.3$, $\lambda=0.3$, and $\alpha=1.3$ (solid), 1.6 (dot), 1.9 (dash). (b) ω_{eqd6} when $\beta=0.9$, $\lambda=0.3$, and $\alpha=1.3$ (solid), 1.6 (dot), 1.9 (dash). (c) ω_{eqd6} when $\beta=1.3$, $\lambda=0.3$, and $\alpha=1.3$ (solid), 1.6 (dot), 1.9 (dash). (d) ω_{eqd6} when $\beta=1.9$, $\lambda=0.3$, and $\alpha=1.3$ (solid), 1.6 (dot), 1.9 (dash). (e) ω_{eqd6} when $\beta=0.9$, $\lambda=0.3$, and $\alpha=2.3$ (solid), 2.6 (dot), 2.9 (dash). (f) ω_{eqd6} when $\beta=1.3$, $\lambda=0.3$, and $\alpha=2.3$ (solid), 2.6 (dot), 2.9 (dash). Note that "Quasi-equivalent damped natural frequency 6" in the above figures stands for ω_{eqd6}.

Proof. If substituting ζ_{eq6} and ω_{eqn6} into Eq. (13.20), we have Eq. (13.21). The proof is finished. □

If $\alpha=2$, $\beta=1$, and $\lambda=0$, ω_{eqd6} degenerates the conventional ω_d with the unit of rad/s. Generally, its unit is not rad/s. Thus, the prefix "quasi" is adopted. Its plots are shown in Figure 13.4.

13.2.7 Quasiequivalent Frequency Ratio of Fractional Vibrators of Class VI

Introduce a term quasiequivalent frequency ratio for a fractional vibrator of class VI. Denote it by γ_{eq6} and define it by

$$\gamma_{eq6} = \frac{\omega}{\omega_{eqn6}}. \tag{13.22}$$

We present its representation with Theorem 13.10.

Theorem 13.10. The representation of γ_{eq6} is given by

$$\gamma_{eq6} = \gamma \sqrt{\frac{-\left(\omega^{\alpha-2}\cos\frac{\alpha\pi}{2} + 2\varsigma\omega_n\omega^{\beta-2}\cos\frac{\beta\pi}{2}\right)}{\omega^{\lambda}\cos\frac{\lambda\pi}{2}}}, \tag{13.23}$$

where $\gamma = \frac{\omega}{\omega_n}$.

Proof. Substituting ω_{eqn6} into Eq. (13.22) produces Eq. (13.23). The proof is finished. □

The quantity γ_{eq6} reduces to γ when $\alpha=2$, $\beta=1$, $\lambda=0$. Its plots are indicated in Figure 13.5.

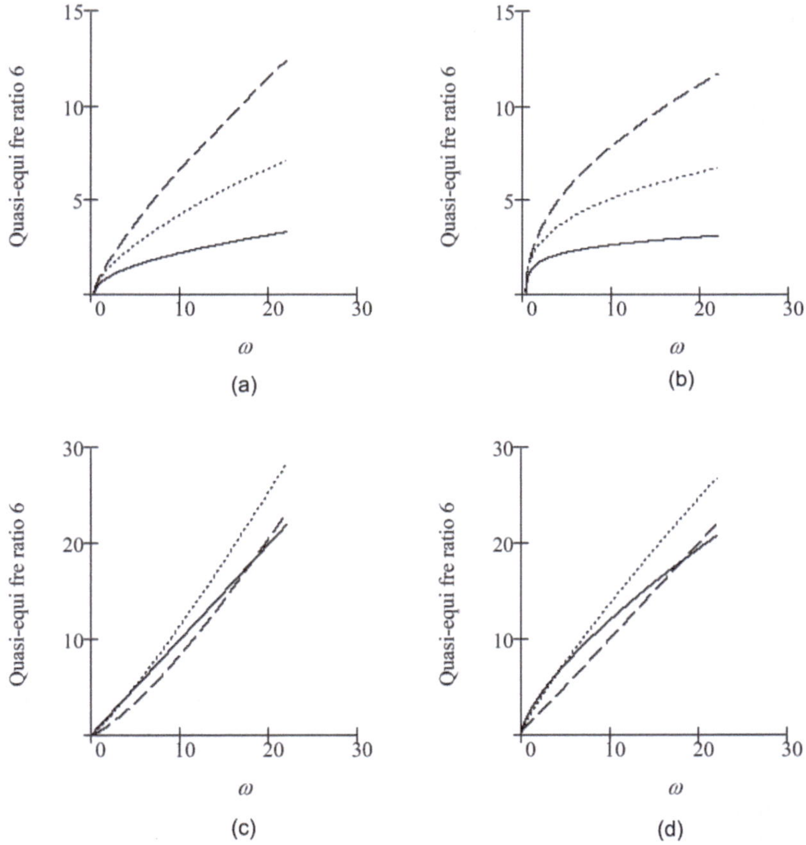

FIGURE 13.5 Plots of γ_{eq6} with $m=1$ and $k=1$, $c=0.2$. (a) γ_{eq6} when $\beta=0.3$, $\lambda=0.3$, and $\alpha=1.3$ (solid), 1.6 (dot), 1.9 (dash). (b) γ_{eq6} when $\beta=0.3$, $\lambda=0.9$, and $\alpha=1.3$ (solid), 1.6 (dot), 1.9 (dash). (c) γ_{eq6} when $\beta=0.9$, $\lambda=0.3$, and $\alpha=2.3$ (solid), 2.6 (dot), 2.9 (dash). (d) γ_{eq6} when $\beta=1.3$, $\lambda=0.9$, and $\alpha=2.3$ (solid), 2.6 (dot), 2.9 (dash). Note that "Quasi-equivalent frequency ratio 6" in the above figures stands for γ_{eq6}.

13.3 FREE RESPONSE OF CLASS VI

Let $x_6(t)$ be the free response of a fractional vibrator of class VI. It is a solution to the fractional differential equation given by

$$\begin{cases} m\dfrac{d^\alpha x_6(t)}{dt^\alpha}+c\dfrac{d^\beta x_6(t)}{dt^\beta}+k\dfrac{d^\lambda x_6(t)}{dt^\lambda}=0, \\ x_6(0)=x_{60}, x_6'(0)=v_{60}, \end{cases} \quad 1<\alpha<3,\ 0<\beta<2,\ 0\le\lambda<1. \tag{13.24}$$

Theorem 13.11 provides its solution.

Theorem 13.11. $x_6(t)$ is in the form

$$x_6(t)=e^{-\varsigma_{eq6}\omega_{eqn6}t}\left(x_{60}\cos\omega_{eqd6}t+\frac{v_{60}+\varsigma_{eq6}\omega_{eqn6}x_{60}}{\omega_{eqd6}}\sin\omega_{eqd6}t\right),\quad t\ge 0. \tag{13.25}$$

Proof. According to Theorem 13.1, Eq. (13.24) can be equivalently expressed by

$$\begin{cases} m_{eq6}\dfrac{d^2x_6(t)}{dt^2}+c_{eq6}\dfrac{dx_6(t)}{dt}+k_{eq6}x_6(t)=0, \\ x_6(0)=x_{60}, x_6'(0)=v_{60}. \end{cases}$$

Since the above is a standard vibration equation, Eq. (13.25) is its representation. The proof is finished. □

Figure 13.6 shows several plots of $x_6(t)$.

From a view of time-frequency plane, we have

$$x_6(t)=x_6(t,\omega). \tag{13.26}$$

13.4 IMPULSE RESPONSE OF CLASS VI

Theorem 13.12 is used to give the expression of impulse response of a fractional vibrator of class VI.

Theorem 13.12. Let $h_6(t)$ be the impulse response of a fractional vibrator of class VI. It is a solution to the following fractional differential equation:

$$m\frac{d^\alpha h_6(t)}{dt^\alpha}+c\frac{d^\beta h_6(t)}{dt^\beta}+k\frac{d^\lambda h_6(t)}{dt^\lambda}=\delta(t),\quad 1<\alpha<3,\ 0<\beta<2,\ 0\le\lambda<1 \tag{13.27}$$

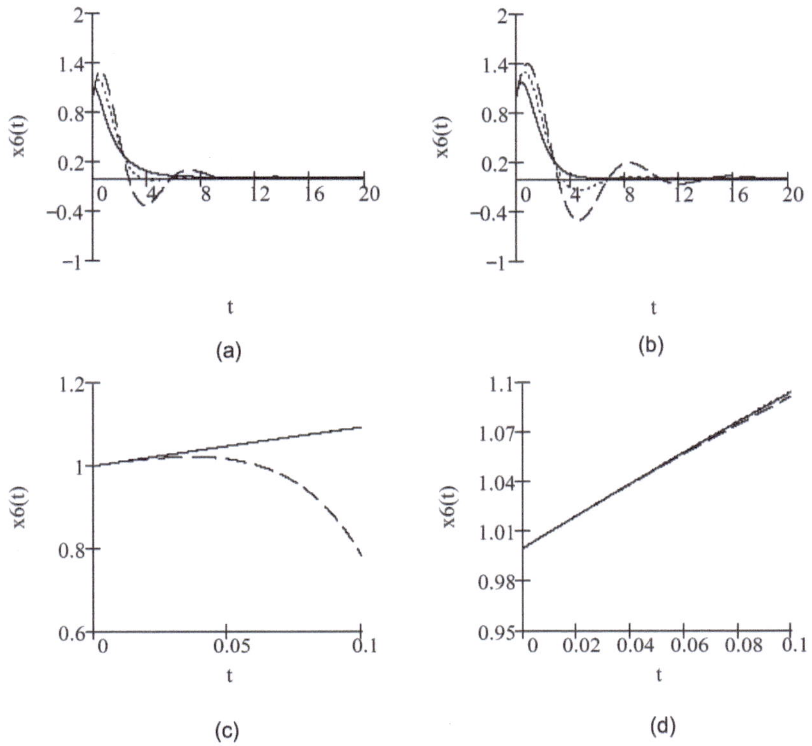

FIGURE 13.6 Plots of $x_6(t)$ for $x_{60}=1$, $v_{60}=1$, $m=1$, $k=1$, $\omega=1.1$, and $c=0.2$. (a) $x_6(t)$ when $\beta=0.3$, $\lambda=0.3$, and $\alpha=1.3$ (solid), 1.6 (dot), 1.9 (dash). (b) $x_6(t)$ when $\beta=1.9$, $\lambda=0.3$, and $\alpha=1.3$ (solid), 1.6 (dot), 1.9 (dash). (c) $x_6(t)$ when $\beta=0.3$, $\lambda=0.3$, and $\alpha=2.3$ (solid), 2.6 (dot), 2.9 (dash) (self-vibration). (d) $x_6(t)$ when $\beta=1.3$, $\lambda=0.3$, and $\alpha=2.3$ (solid), 2.6 (dot), 2.9 (dash) (self-vibration).

with zero initial conditions. The analytical expression of $h_6(t)$ is given by

$$h_6(t)=e^{-\varsigma_{eq6}\omega_{eqn6}t}\frac{1}{m_{eq6}\omega_{eqd6}}\sin\omega_{eqd6}t,\quad t\geq 0. \tag{13.28}$$

Proof. According to Theorem 13.1, Eq. (13.27) can be equivalently expressed by

$$m_{eq6}\frac{d^2h_6(t)}{dt^2}+c_{eq6}\frac{dh_6(t)}{dt}+k_{eq6}h_6(t)=\delta(t).$$

The above is a standard vibration equation. Its solution is Eq. (13.28). The proof is complete. □

Figure 13.7 shows a few plots of $h_6(t)$.

In time-frequency plan, $h_6(t)$ may be written as

$$h_6(t)=h_6(t,\omega). \tag{13.29}$$

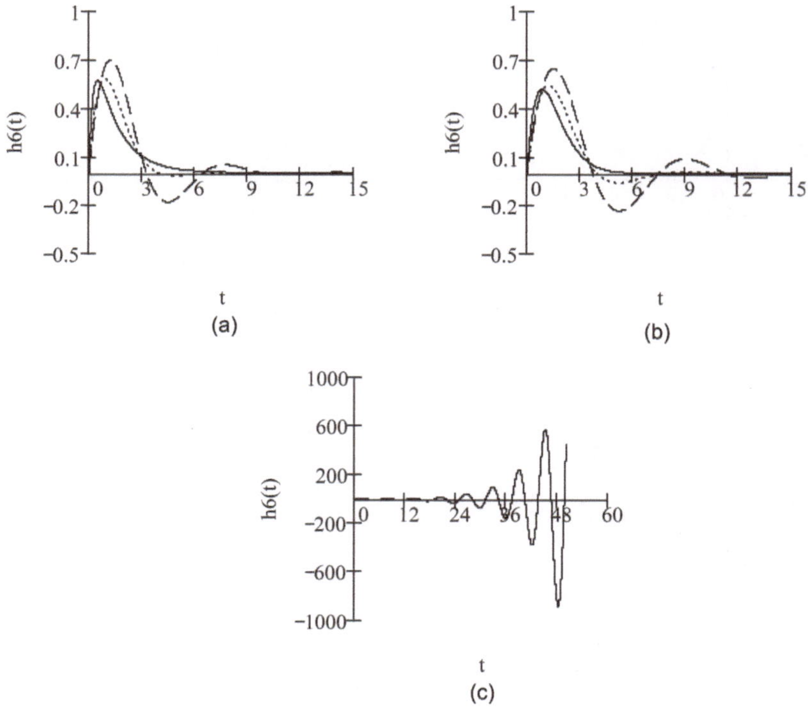

FIGURE 13.7 Plots of $h_6(t)$ for $m=1$, $k=1$, $\omega=1.1$, and $c=0.2$. (a) $h_6(t)$ when $\beta=0.3$, $\lambda=0.3$, and $\alpha=1.3$ (solid), 1.6 (dot), 1.9 (dash). (b) $h_6(t)$ when $\beta=1.9$, $\lambda=0.3$, and $\alpha=1.3$ (solid), 1.6 (dot), 1.9 (dash). (c) $h_6(t)$ when $\alpha=2.5$, $\beta=1.5$, $\lambda=0.3$ (self-vibration).

13.5 STEP RESPONSE OF CLASS VI

Theorem 13.13 gives the expression of unit step response to a fractional vibrator of class VI.

Theorem 13.13. Denote by $g_6(t)$ the unit step response of a fractional vibrator of class VI. It is a solution to

$$m\frac{d^{\alpha}g_6(t)}{dt^{\alpha}}+c\frac{d^{\beta}g_6(t)}{dt^{\beta}}+k\frac{d^{\lambda}g_6(t)}{dt^{\lambda}}=u(t),\quad 1<\alpha<3,\ 0<\beta<2,\ 0\le\lambda<1 \tag{13.30}$$

with zero initial conditions. Then

$$g_6(t)=\frac{1}{k_{eq6}}\left[1-\frac{e^{-\varsigma_{eq6}\omega_{eqn6}t}}{\sqrt{1-\varsigma_{eq6}^2}}\cos\left(\omega_{eqd6}t-\phi_6\right)\right],\quad t\ge 0, \tag{13.31}$$

where

$$\phi_6=\tan^{-1}\frac{\varsigma_{eq6}}{\sqrt{1-\zeta_{eq6}^2}}. \tag{13.32}$$

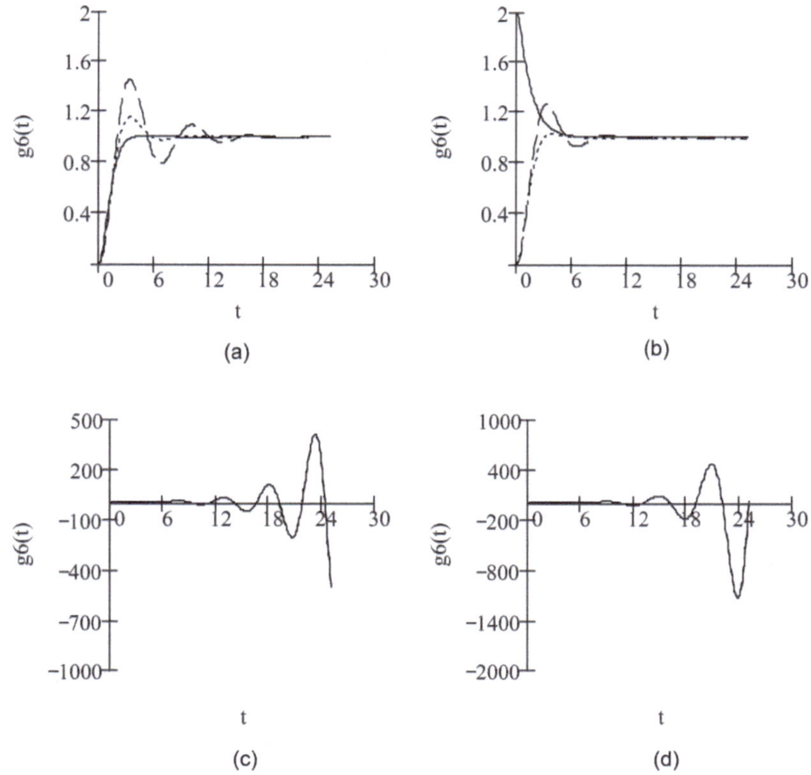

FIGURE 13.8 Plots of $g_6(t)$ for $m=1$, $k=1$, $c=0.2$, and $\omega=1.1$. (a) $g_6(t)$ when $\beta=0.3$, $\lambda=0.1$, and $\alpha=1.3$ (solid), 1.6 (dot), 1.9 (dash). (b) $g_6(t)$ when $\beta=0.3$, $\lambda=0.3$, and $\alpha=1.3$ (solid), 1.6 (dot), 1.9 (dash). (c) $g_6(t)$ when $\alpha=2.5$, $\beta=0.3$, $\lambda=0.3$ (self-vibration). (d) $g_6(t)$ when $\alpha=2.5$, $\beta=1.3$, $\lambda=0.3$ (self-vibration).

Proof. Taking into account Theorem 13.1, Eq. (13.30) is equivalently expressed by

$$m_{\text{eq6}}\frac{d^2 g_6(t)}{dt^2}+c_{\text{eq6}}\frac{dg_6(t)}{dt}+k_{\text{eq6}}g_6(t)=u(t).$$

Because the above is a standard vibration equation of order 2, its solution is given by Eq. (13.31). The proof is finished. □

Figure 13.8 show several plots of $g_6(t)$.

Also $g_6(t)$ may be written by

$$g_6(t)=g_6(t,\omega). \tag{13.33}$$

13.6 FREQUENCY TRANSFER FUNCTION OF CLASS VI

Theorem 13.14 presents the expression of frequency transfer function of a fractional vibrator of class VI.

Theorem 13.14. Let $H_6(\omega)$ be the frequency transfer function of a fractional vibrator of class VI. It is expressed by

$$H_6(\omega)=\frac{1}{k_{eq6}\left(1-\gamma_{eq6}^2+i2\varsigma_{eq6}\gamma_{eq6}\right)}. \tag{13.34}$$

Proof. Doing the Fourier transform on both sides of the following equation

$$m_{eq6}\frac{d^2h_6(t)}{dt^2}+c_{eq6}\frac{dh_6(t)}{dt}+k_{eq6}h_6(t)=\delta(t)$$

yields

$$\left[m_{eq6}(i\omega)^2+c_{eq6}(i\omega)+k_{eq6}\right]H_6(\omega)=1.$$

From the above,

$$\left[(i\omega)^2+2\varsigma_{eq6}\omega_{eqn6}(i\omega)+\omega_{eqn6}^2\right]H_6(\omega)=\frac{1}{m_{eq6}}.$$

The above implies

$$\left[1-\left(\frac{\omega}{\omega_{eqn6}}\right)^2+i2\varsigma_{eq6}\frac{\omega}{\omega_{eqn6}}\right]H_6(\omega)=\frac{1}{m_{eq6}\omega_{eqn6}^2}=\frac{k_{eq6}}{m_{eq6}k_{eq6}\omega_{eqn6}^2}=\frac{1}{k_{eq6}}.$$

Thus,

$$H_6(\omega)=\frac{1}{k_{eq6}\left(1-\gamma_{eq6}^2+i2\varsigma_{eq6}\gamma_{eq6}\right)}.$$

That is Eq. (13.34). This finishes the proof. □

The amplitude-frequency response $|H_6(\omega)|$ is expressed by

$$|H_6(\omega)|=\frac{1}{k_{eq6}}\frac{1}{\sqrt{\left(1-\gamma_{eq6}^2\right)^2+\left(2\varsigma_{eq6}\gamma_{eq6}\right)^2}}. \tag{13.35}$$

For the phase-frequency response function $\varphi_6(\omega)$, we adopt

$$\varphi_6(\omega)=\cos^{-1}\frac{1-\gamma_{eq6}^2}{\sqrt{\left(1-\gamma_{eq6}^2\right)^2+\left(2\varsigma_{eq6}\gamma_{eq6}\right)^2}}. \tag{13.36}$$

Figure 13.9 illustrates the plots of $|H_6(\omega)|$, while several plots of $\varphi_6(\omega)$ are in Figure 13.10.

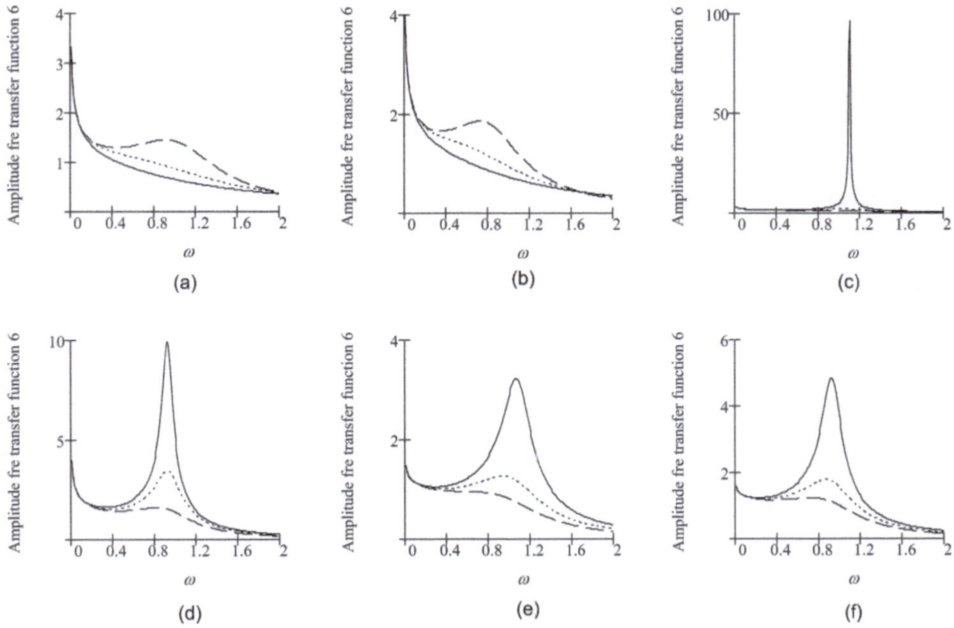

FIGURE 13.9 Plots of $|H_6(\omega)|$ for $m=1$, $k=1$, and $c=0.2$. (a) $|H_6(\omega)|$ when $\beta=0.3$, $\lambda=0.3$, and $\alpha=1.3$ (solid), 1.6 (dot), 1.9 (dash). (b) $|H_6(\omega)|$ when $\beta=1.9$, $\lambda=0.3$, and $\alpha=1.3$ (solid), 1.6 (dot), 1.9 (dash). (c) $|H_6(\omega)|$ when $\beta=0.3$, $\lambda=0.3$, and $\alpha=2.3$ (solid), 2.6 (dot), 2.9 (dash). (d) $|H_6(\omega)|$ when $\beta=1.9$, $\lambda=0.3$, and $\alpha=2.3$ (solid), 2.6 (dot), 2.9 (dash). (e) $|H_6(\omega)|$ when $\beta=0.3$, $\lambda=0.1$, and $\alpha=2.3$ (solid), 2.6 (dot), 2.9 (dash). (f) $|H_6(\omega)|$ when $\beta=1.9$, $\lambda=0.1$, and $\alpha=2.3$ (solid), 2.6 (dot), 2.9 (dash).

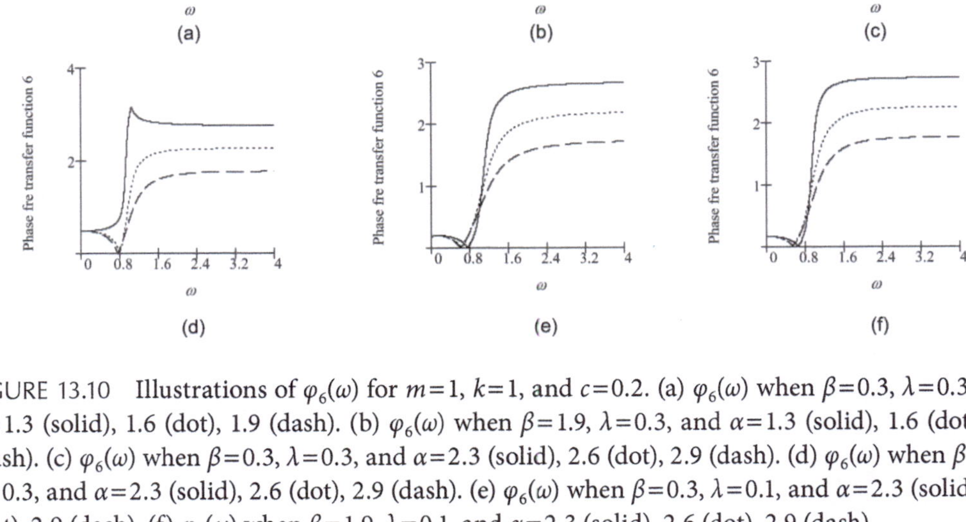

FIGURE 13.10 Illustrations of $\varphi_6(\omega)$ for $m=1$, $k=1$, and $c=0.2$. (a) $\varphi_6(\omega)$ when $\beta=0.3$, $\lambda=0.3$, and $\alpha=1.3$ (solid), 1.6 (dot), 1.9 (dash). (b) $\varphi_6(\omega)$ when $\beta=1.9$, $\lambda=0.3$, and $\alpha=1.3$ (solid), 1.6 (dot), 1.9 (dash). (c) $\varphi_6(\omega)$ when $\beta=0.3$, $\lambda=0.3$, and $\alpha=2.3$ (solid), 2.6 (dot), 2.9 (dash). (d) $\varphi_6(\omega)$ when $\beta=1.9$, $\lambda=0.3$, and $\alpha=2.3$ (solid), 2.6 (dot), 2.9 (dash). (e) $\varphi_6(\omega)$ when $\beta=0.3$, $\lambda=0.1$, and $\alpha=2.3$ (solid), 2.6 (dot), 2.9 (dash). (f) $\varphi_6(\omega)$ when $\beta=1.9$, $\lambda=0.1$, and $\alpha=2.3$ (solid), 2.6 (dot), 2.9 (dash).

13.7 QUASIEQUIVALENT LOGARITHMIC DECREMENT AND *Q* FACTOR OF CLASS VI

In this subsection, we propose the expressions of the quasiequivalent logarithmic decrement of $x_6(t)$ and Q factor of a fractional vibrator of class VI.

13.7.1 Quasiequivalent Logarithmic Decrement of Fractional Vibrator of Class VI

Theorem 13.15. Let t_i and t_{i+1} be two time points of fractional free response $x_6(t)$, where $x_6(t_i)$ and $x_6(t_{i+1})$ are successive peak values at t_i and t_{i+1}. Let Δ_{eq6} be the quasiequivalent logarithmic decrement of $x_6(t)$. Then,

$$\Delta_{eq6} = \text{In}\frac{x_6(t_i)}{x_6(t_{i+1})} = \frac{2\pi\varsigma_{eq6}}{\sqrt{1-\zeta_{eq6}^2}}. \tag{13.37}$$

Proof. According to the definition of logarithmic decrement, the above holds. The proof is finished. □

Several plots of Δ_{eq6} are given in Figure 13.11, where negative values of Δ_{eq6} imply self-vibrations.

13.7.2 Quasiequivalent *Q* Factor of Fractional Vibrator of Class VI

We present Theorem 13.16 to give the expression of the quasiequivalent Q factor of a fractional vibrator of class VI.

Theorem 13.16. Denote by Q_{eq6} the quasiequivalent Q factor of a fractional vibrator of class VI. Then,

$$Q_{eq6} = \frac{\sqrt{-\left(m\omega^{\alpha-2}\cos\frac{\alpha\pi}{2} + c\omega^{\beta-2}\cos\frac{\beta\pi}{2}\right)k\omega^{\lambda}\cos\frac{\lambda\pi}{2}}}{m\omega^{\alpha-1}\sin\frac{\alpha\pi}{2} + c\omega^{\beta-1}\sin\frac{\beta\pi}{2} + k\omega^{\lambda-1}\sin\frac{\lambda\pi}{2}}. \tag{13.38}$$

Proof. Since

$$Q_{eq6} = \frac{1}{2\varsigma_{eq6}},$$

substituting ζ_{eq6} into the above results in Eq. (13.38). The proof ends. □

A few plots of Q_{eq6} are shown in Figure 13.12.

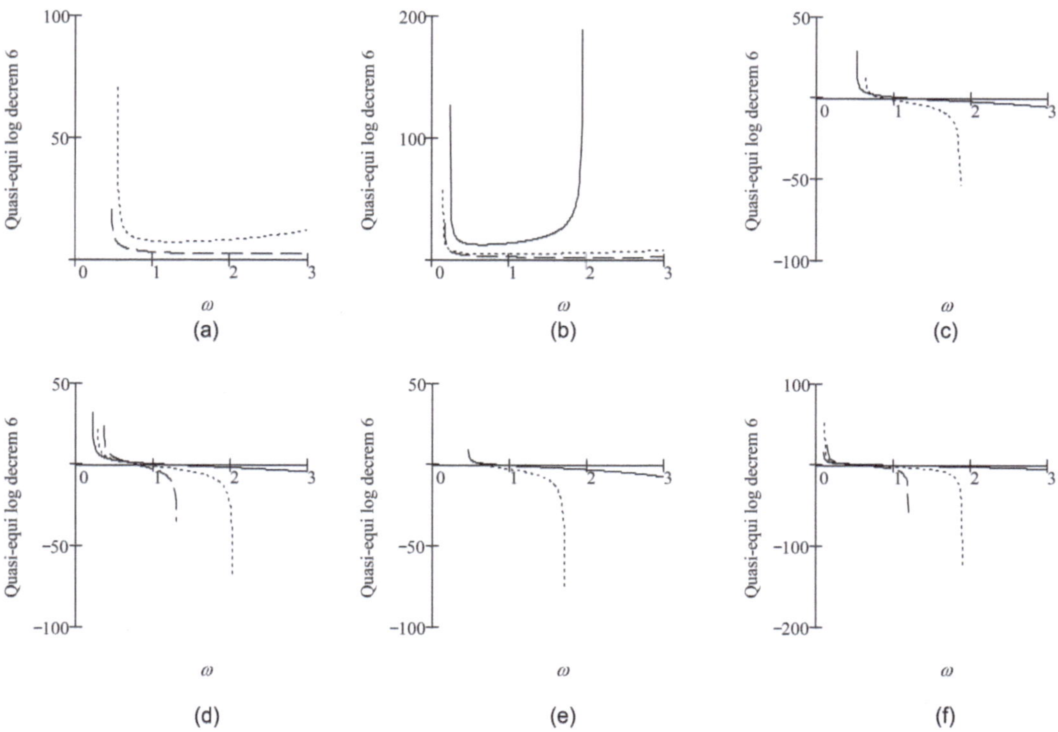

FIGURE 13.11 Plots of Δ_{eq6} for $m=1$, $k=1$, and $c=0.2$. (a) Δ_{eq6} when $\beta=0.3$, $\lambda=0.3$, and $\alpha=1.3$ (solid), 1.6 (dot), 1.9 (dash). (b) Δ_{eq6} when $\beta=1.9$, $\lambda=0.3$, and $\alpha=1.3$ (solid), 1.6 (dot), 1.9 (dash). (c) Δ_{eq6} when $\beta=0.3$, $\lambda=0.3$, and $\alpha=2.3$ (solid), 2.6 (dot), 2.9 (dash). (d) Δ_{eq6} when $\beta=1.9$, $\lambda=0.3$, and $\alpha=2.3$ (solid), 2.6 (dot), 2.9 (dash). (e) Δ_{eq6} when $\beta=0.3$, $\lambda=0.1$, and $\alpha=2.3$ (solid), 2.6 (dot), 2.9 (dash). (f) Δ_{eq6} when $\beta=1.9$, $\lambda=0.1$, and $\alpha=2.3$ (solid), 2.6 (dot), 2.9 (dash). Note that "Quasi-equivalent logarithmic decrement 6" in the above figures stands for Δ_{eq6}.

13.8 TAKING C_{EQ6} AS A GENERALIZATION OF RAYLEIGH DAMPING

The Rayleigh damping is denoted by c_r. It is in the form $c_r=am+bk$, where m and k are primary mass and stiffness, respectively, a is proportional to ω while b is inversely proportional to ω (Jin and Xia [1], Harris [2], Palley et al. [3]). We now rewrite c_{eq6} by $c_{eq6}=c_{eq61}+c_{eq62}+c_{eq63}$, where

$$c_{eq61}=m\omega^{\alpha-1}\sin\frac{\alpha\pi}{2},$$

$$c_{eq62}=c\omega^{\beta-1}\sin\frac{\beta\pi}{2},$$

$$c_{eq63}=k\omega^{\lambda-1}\sin\frac{\lambda\pi}{2}.$$

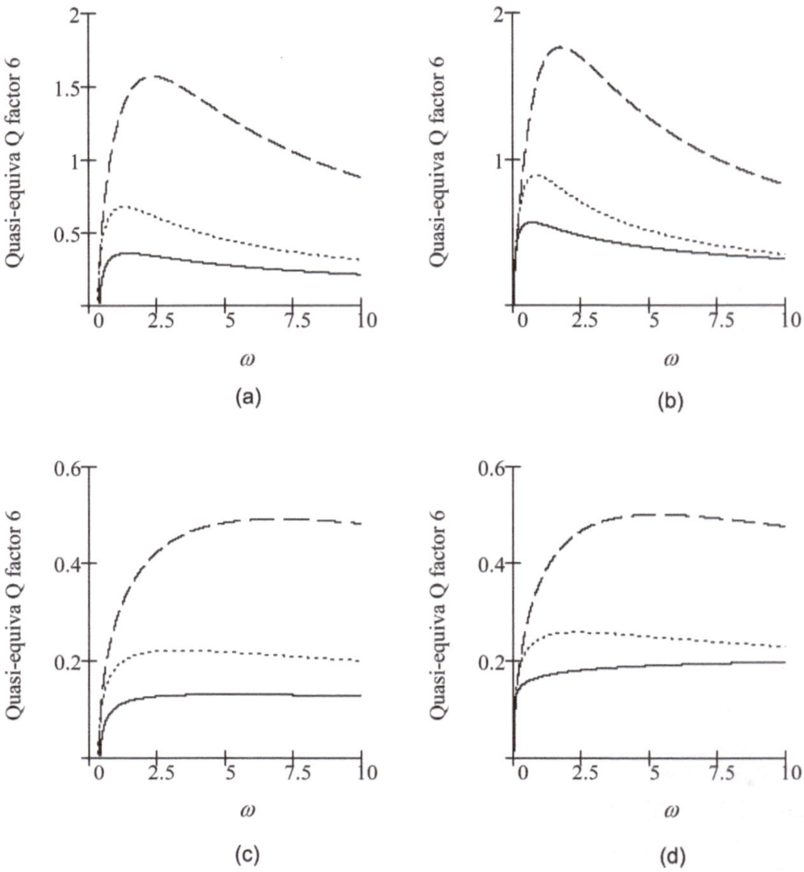

FIGURE 13.12 Plots of Q_{eq6} for $m=1$, $k=1$, and $c=0.2$ with $\alpha=1.3$ (solid), 1.6 (dot), 1.9 (dash). (a) Q_{eq6} when $\beta=0.3$, $\lambda=0.3$. (b) Q_{eq6} when $\beta=1.9$, $\lambda=0.3$. (c) Q_{eq6} when $\beta=0.3$, $\lambda=0.9$. (d) Q_{eq6} when $\beta=1.9$, $\lambda=0.1$. Note that "Quasi-equivalent Q factor 6" in the above figures stands for Q_{eq6}.

Note that $c_{eq61}+c_{eq63}=c_{eq4}$, which is in form consistent with the Rayleigh damping, see Chapter 11. However, the item c_{eq62} is new in comparison with the Rayleigh damping. Therefore, we reveal a generalization of the Rayleigh damping with c_{eq6}.

13.9 CONCLUDING REMARK

Note that references with respect to fractional vibrators of class VI are seldom reported. The results in this chapter are novel in the field. Mathematically speaking, this class covers five classes discussed in Chapters 8–12. Therefore, class VI has its importance and generality in mathematics. From a view of engineering, however, it is preferred to classify them as six. For instance, when explaining the Rayleigh damping assumption, using Class V is convenient.

13.10 EXERCISES

13.1. Find the condition of $c_{eq6}<0$.

13.2. Show that $c_{eq6}=c_{eq4}$ for $c=0$.

13.3. Show that c_{eq6} stands for three variations of the Rayleigh damping assumption for $\beta=1$, $0<\beta<1$, and $1<\beta<2$, respectively.

13.4. Find the condition for $0<\zeta_{eq6}<1$.

13.5. Prove that ω_{eqn6} is not a standard angular frequency unless $\alpha=2$, $\beta=1$, and $\lambda=0$.

13.6. Find the condition for ω_{eqd6} being real.

13.7. Prove that γ_{eq6} is dimensionless if $\alpha=2$, $\beta=1$, and $\lambda=0$.

REFERENCES

1. X. D. Jin and L. J. Xia, *Ship Hull Vibration*, The Press of Shanghai Jiaotong University, Shanghai, China, 2011. In Chinese.
2. C. M. Harris, *Shock and Vibration Handbook*, 5th Ed., McGraw-Hill, New York, 2002.
3. O. M. Palley, Г. B. Bahizov, and E. Я. Voroneysk, *Handbook of Ship Structural Mechanics*, National Defense Industry Publishing House, Beijing, China, 2002. In Chinese. Translated from Russian by B. H. Xu, X. Xu, and M. Q. Xu.

CHAPTER 14

Explanation of Rayleigh Damping Assumption Based on Fractional Vibrations

In previous Chapters 8–13, I addressed the general theory of six classes of fractional vibrators and mentioned the mathematical explanation of the Rayleigh damping assumption in passing. It is known that how to explain the Rayleigh damping assumption in either mathematics or physics remains a problem open. Therefore, I would like to specifically discuss my understanding of the mathematical explanation of the Rayleigh type damping from the point of view of fractional vibrations in this chapter. The contributions in this chapter are as follows.

1. Mathematical explanation of the standard Rayleigh damping form is addressed in Theorem 14.1.
2. In Theorem 14.2, we present a mathematical explanation of a variation of the Rayleigh damping, which is proportional to mass only.
3. In Theorem 14.3, we present a mathematical explanation of another variation of the Rayleigh damping, which is just proportional to stiffness.
4. Mathematical explanation of one more variation form of the Rayleigh damping is put forward in Theorem 14.4.

14.1 RAYLEIGH DAMPING ASSUMPTION

The famous Rayleigh damping assumption was introduced by Rayleigh [1]. Denote by c_r the damping assumed by Rayleigh. Its standard form is given by

$$c_r = am + bk, \tag{14.1}$$

DOI: 10.1201/9781003460947-16

where a and b are frequency-dependent parameters. To be precise, a is proportional to ω while b is inversely proportional to ω. In short, Rayleigh assumed that damping is proportional to m and k with the coefficients a and b. The literature regarding the study of Rayleigh damping and its applications is rich, see e.g., [2–20], simply citing a few.

In some applications, researchers found its variation or modification form given by

$$c_{r1} = am. \tag{14.2}$$

The above means that a variation of the Rayleigh damping in Eq. (14.2), which is denoted by c_{r1}, is only proportional to mass m with the coefficient a that is proportional to ω, irrelevant of stiffness k, see e.g., Filiatrault [21, Eqs. (26), (27)], Marur and Kant [22, Eq. (12)]. In other applications, scholars observed another variation form of the Rayleigh damping. Denote that by c_{r2}. It is in the form

$$c_{r2} = bk. \tag{14.3}$$

The above is just proportional to k with the coefficient b that is inversely proportional to ω, see e.g., El-Aidi and Hall [23, Eq. (5)], Takayanagi and Schnobrich [24, Eq. (18), (19)]. Moreover, Cruz and Miranda [25] investigated one more variation of Rayleigh damping in the form

$$c_{r3} = c + bk, \tag{14.4}$$

where c is a linear viscous damping.

14.2 EXPLANATION 1 WITH FRACTIONAL VIBRATORS OF CLASS IV

We first present a mathematical explanation of the standard Rayleigh damping form with the following theorem.

Theorem 14.1. The damping c_{eq4} given in Chapter 11 is consistent with the standard form of Rayleigh damping.

Proof. Taking into account Theorem 11.3, we have

$$c_{eq4} = m\omega^{\alpha-1}\sin\frac{\alpha\pi}{2} + k\omega^{\lambda-1}\sin\frac{\lambda\pi}{2}, \quad 1<\alpha<3, 0\le\lambda<1.$$

Let

$$a = \omega^{\alpha-1}\sin\frac{\alpha\pi}{2}, \quad b = \omega^{\lambda-1}\sin\frac{\lambda\pi}{2}. \tag{14.5}$$

Then, we have

$$c_{eq4} = am + bk. \tag{14.6}$$

Since a is proportional to ω while b is inversely proportional to ω, c_{eq4} is consistent with the form of the Rayleigh damping c_r in Eq. (14.1). The proof completes. □

Note 14.1: From Theorem 14.1, we see that damping in fractional vibrators of class IV type takes the standard form of Rayleigh damping. Mathematically, the standard form of Rayleigh damping may result from

$$m\frac{d^{\alpha}x(t)}{dt^{\alpha}}+k\frac{d^{\lambda}x(t)}{dt^{\lambda}}=0, \quad 1<\alpha<3, \;\; 0\leq\lambda<1. \tag{14.7}$$

14.3 EXPLANATION 2 WITH FRACTIONAL VIBRATORS OF CLASS I

Second, we propose a mathematical explanation of a variation in Eq. (14.2) of the Rayleigh damping with Theorem 14.2.

Theorem 14.2. The damping c_{eq1} described in Chapter 8 is consistent with the variation form of the Rayleigh damping in Eq. (14.2).

Proof. Consider Theorem 8.3, we have

$$c_{eq1}=m\omega^{\alpha-1}\sin\frac{\alpha\pi}{2}=am. \tag{14.8}$$

Because $a=\omega^{\alpha-1}\sin\frac{\alpha\pi}{2}$ in the above is proportional to ω, we see that Eq. (14.8) is consistent with c_{r1} in Eq. (14.2). This finishes the proof. □

Note 14.2: According to Theorem 14.2, we conclude that damping in fractional vibrators in class I has the form of c_{r1}. In mathematics, we say that the variation form of the Rayleigh damping, expressed by Eq. (14.2), may be obtained from

$$m\frac{d^{\alpha}x(t)}{dt^{\alpha}}+kx(t)=0, \quad 1<\alpha<3. \tag{14.9}$$

14.4 EXPLANATION 3 WITH FRACTIONAL VIBRATORS OF CLASS V

Third, we bring forward a mathematical explanation of a variation in Eq. (14.3) of the Rayleigh damping with Theorem 14.3.

Theorem 14.3. The damping c_{eq5} stated in Chapter 12 is in agreement with the variation form of the Rayleigh damping, which is given by Eq. (14.3).

Proof. According to Theorem 12.2, we have

$$c_{eq5}=k\omega^{\lambda-1}\sin\frac{\lambda\pi}{2}\triangleq bk. \tag{14.10}$$

As $b=\omega^{\lambda-1}\sin\frac{\lambda\pi}{2}$ is inversely proportional to ω, Eq. (14.10) is in agreement with Eq. (14.3). This completes the proof. □

Note 14.3: Theorem 14.3 says that damping in fractional vibrators of class V has the same form of the variation of the Rayleigh damping c_{r2}. Mathematically speaking, we may consider that the following fractional vibrator produces c_{r2}

$$m\frac{d^2x(t)}{dt^2}+k\frac{d^\lambda x(t)}{dt^\lambda}=0, \quad 0\le\lambda<1. \tag{14.11}$$

14.5 EXPLANATION 4 WITH FRACTIONAL VIBRATORS OF CLASS VI

Finally, we put forward a mathematical explanation of a variation of the Rayleigh damping in Eq. (14.4).

Theorem 14.4. The damping c_{eq6} explained in Chapter 13 agrees with the variation form of the Rayleigh damping in the form of Eq. (14.4).

Proof. According to Theorem 13.3, for $\alpha=2$ and $\beta=1$, we have

$$c_{\text{eq6}}\big|_{\alpha=2,\beta=1}=c+bk. \tag{14.12}$$

The proof is finished. □

Note 14.4: Following Theorem 14.4, we conclude that the damping form in fractional vibrators of class VI is consistent with that of the variation form of the Rayleigh damping c_{r3} when $\alpha=2$ and $\beta=1$. From the point of view of mathematics, we may say that the fractional vibration system below yields c_{r3} if $\alpha=2$ and $\beta=1$

$$m\frac{d^2x(t)}{dt^2}+c\frac{dx(t)}{dt}+k\frac{d^\lambda x(t)}{dt^\lambda}=0, \quad 0\le\lambda<1. \tag{14.13}$$

14.6 MORE WORK ON EXPLANATION OF RAYLEIGH DAMPING

So far, we only consider that fractional orders α, β, and λ are constants. However, when they are functions in terms of ω, we can attain a new explanation of Rayleigh damping, see Chapter 16 for details.

14.7 EXERCISES

14.1. Design a structure with the damping in the form am.

14.2. Design a structure with the damping in the form bk.

14.3. State engineering meaning of the damping $m\omega^{\alpha-1}\sin\frac{\alpha\pi}{2}$ that is produced by the fractional inertia force $m\frac{d^\alpha x(t)}{dt^\alpha}$.

14.4. State engineering meaning of the damping $k\omega^{\lambda-1}\sin\frac{\lambda\pi}{2}$ that is produced by the fractional restoration force $k\frac{d^{\lambda}x(t)}{dt^{\lambda}}$.

REFERENCES

1. J. W. Strutt and 3rd B. Rayleigh, *The Theory of Sound*, vol. 1, Macmillan & Co., Ltd., London, 1877.
2. T. Trombetti and S. Silvestri, On the modal damping ratios of shear-type structures equipped with Rayleigh damping systems, *Journal of Sound and Vibration*, 292(1–2), 2006, 21–58.
3. M. K. Poul and A. Zerva, Efficient time-domain deconvolution of seismic ground motions using the equivalent-linear method for soil-structure interaction analyses, *Soil Dynamics and Earthquake Engineering*, 112, 2018, 138–151.
4. P. Nåvik, A. Rønnquist, and S. Stichel, Identification of system damping in railway catenary wire systems from full-scale measurements, *Engineering Structures*, 113, 2016, 71–78.
5. S.-K. Park, H. W. Park, S. Shin, and H. S. Lee, Detection of abrupt structural damage induced by an earthquake using a moving time window technique, *Computers & Structures*, 86(11–12), 2008, 1253–1265.
6. M. I. Hussein, S. Biringen, O. R. Bilal, and A. Kucala, Flow stabilization by subsurface phonons, *Proceedings: Mathematical, Physical and Engineering Sciences*, 471(2177), 2015, 1–19.
7. O. Sigmund and J. S. Jensen, Systematic design of phononic band-gap materials and structures by topology optimization, *Philosophical Transactions: Mathematical, Physical and Engineering Sciences*, 361(1806), 2003, 1001–1019.
8. F. Chen, H. Tan, J. Chen, and Z. Jiao, Influences of earthquake characteristics on seismic performance of anchored sheet pile quay with barrette piles, *Journal of Coastal Research*, Special Issue No. 85: *Proceedings of 15th International Coastal Symposium*, Haeundae, Busan, 13–18 May, 2018, 701–705.
9. D. S. Battisti, A. C. Hirst, and E. S. Sarachik, Instability and predictability in coupled atmosphere-ocean models, *Philosophical Transactions of the Royal Society of London. Series A, Mathematical and Physical Sciences*, 329(1604), 1989, 237–247.
10. D. Cox, J. Kämpf, and M. Fernandes, Dispersion and connectivity of land-based discharges near the mouth of a coastal inlet, *Journal of Coastal Research*, 29(6A), 2013, 100–109.
11. F. Tisseur and K. Meerbergen, The quadratic eigenvalue problem, *SIAM Review*, 43(2), 2001, 235–286.
12. E. K.-W. Chu, T.-M. Huang, W.-W. Lin, and C.-T. Wu, Palindromic eigenvalue problems: A brief survey, *Taiwanese Journal of Mathematics*, 14(3A), 2010, 743–779.
13. J. P. Fay, S. Puria, and C. R. Steele, The discordant eardrum, *Proceedings of the National Academy of Sciences of the United States of America*, 103(52), 2006, 19743–19748.
14. H. Naderian, M. M. S. Cheung, M. Mohammadian, and E. Dragomirescu, Integrated finite strip flutter analysis of bridges, *Computers & Structures*, 212, 2019, 145–161.
15. Z. Tian, L. Huo, W. Gao, H. Li, and G. Song, Modeling of the attenuation of stress waves in concrete based on the Rayleigh damping model using time-reversal and PZT transducers, *Smart Materials and Structures*, 26(10), 2017, 105030.
16. G. Iovane and A. V. Nasedkin, Finite element dynamic analysis of anisotropic elastic solids with voids, *Computers & Structures*, 87(15–16), 2009, 981–989.
17. L. A. S. Kouris, A. Penna, and G. Magenes, Seismic damage diagnosis of a masonry building using short-term damping measurements, *Journal of Sound and Vibration*, 394, 2017, 366–391.
18. X. D. Jin and L. J. Xia, *Ship Hull Vibration*, The Press of Shanghai Jiaotong University, Shanghai, China, 2011. In Chinese.

19. C. M. Harris, *Shock and Vibration Handbook*, 5th Ed., McGraw-Hill, New York, 2002.
20. E. DeWeaver and S. Nigam, Influence of mountain ranges on the mid-latitude atmospheric response to El Niño events, *Nature*, 378(6554), 1995, 706–708.
21. A. Filiatrault, Static and dynamic analysis of timber shear walls, *Canadian Journal of Civil Engineering*, 17(4), 1990, 643–651.
22. S. R. Marur and T. Kant, A modified form of the central difference predictor scheme for damped nonlinear systems, *Computers & Structures*, 50(5), 1994, 615–618.
23. B. El-Aidi and J. F. Hall, Nonlinear earthquake response of concrete gravity dams. Part 1: Modeling, *Earthquake Engineering and Structural Dynamics*, 18(6), 1989, 837–851.
24. T. Takayanagi and W. C. Schnobrich, Non-linear analysis of coupled wall systems, *Earthquake Engineering & Structural Dynamics*, 7(1), 1979, 1–22.
25. C. Cruz and E. Miranda, Evaluation of the Rayleigh damping model for buildings, *Engineering Structures*, 138, 2017, 324–336.

CHAPTER 15

Mass

WE BRIEFLY DISCUSS MASS encountered in fractional vibrations previously addressed. The particularity about mass in fractional vibrations may be in three aspects: (1) Negative mass, which attracts the interests of scientists and engineers; (2) infinite mass, which is conventionally described from the special theory of relativity; and (3) zero mass. Writing this short chapter aims at providing readers with a theoretic issue toward engineering applications of fractional vibrations.

15.1 MASS IN THE RESTRICTED THEORY OF RELATIVITY

Let m be the primary static mass of a body. Denote by v the velocity of the body with the mass m. Following the special relativity theory of Einstein's [1], one has the equivalent mass, which is denoted by $m_{\text{Einstein eq}}$, expressed by

$$m_{\text{Einstein eq}} = \frac{m}{\sqrt{1-\frac{v^2}{C^2}}}, \tag{15.1}$$

where C is the speed of light. An obvious property of $m_{\text{Einstein eq}}$ is $m_{\text{Einstein eq}} \geq 0$ since $v \leq C$. Note that an important property is $m_{\text{Einstein eq}} \to \infty$ if $v \to C$. However, $m_{\text{Einstein eq}}$ is never zero. As a matter of fact,

$$\min\left(m_{\text{Einstein eq}}\right) = m. \tag{15.2}$$

15.2 MASS IN THE VIBRATION MOTION OF CLASS I

From Corollaries 8.1 and 8.2, we obtain

$$\begin{aligned} &\lim_{\omega\to\infty} m_{\text{eq1}}(\omega,\alpha) = 0, \quad 1<\alpha<2, \\ &\lim_{\omega\to 0} m_{\text{eq1}}(\omega,\alpha) = \infty, \quad 1<\alpha<2. \end{aligned} \tag{15.3}$$

DOI: 10.1201/9781003460947-17

On the other hand, we have

$$\begin{aligned} &\lim_{\omega\to\infty} m_{\mathrm{eq1}}(\omega,\alpha)=\infty, \quad 2<\alpha<3, \\ &\lim_{\omega\to 0} m_{\mathrm{eq1}}(\omega,\alpha)=0, \quad 2<\alpha<3. \end{aligned} \tag{15.4}$$

Thus, we have

$$\min\left(m_{\mathrm{eq1}}\right)=0, \quad \max\left(m_{\mathrm{eq1}}\right)\to\infty. \tag{15.5}$$

The range of m_{eq1} is $(0, \infty)$.

15.3 MASS IN THE VIBRATION MOTION OF CLASS II

Corollary 9.2 exhibits that $\lim_{\omega\to 0} m_{\mathrm{eq2}}(\omega,\beta,m)=-\infty$. Thus,

$$\min\left(m_{\mathrm{eq2}}\right)\to-\infty. \tag{15.6}$$

Therefore, fractional vibration motion may produce negative mass.

Note negative mass attracts the interests of researchers in either physics or engineering, see e.g., Farnes [2], Bondi [3], Mbarek and Paranjape [4], Khamehchi et al. [5], Li et al. [6], Saha and Mishra [7], Talib et al. [8], Wang [9], simply citing a few.

From the above and those discussed previously, we see that the equivalent mass of a fractional vibrator may be in the range of $(-\infty, \infty)$, in general.

15.4 EXERCISES

15.1. Show a structure with frequency-dependent mass $m(\omega)$.

15.2. State engineering meaning of the mass $\left(-c\omega^{\beta-2}\cos\frac{\beta\pi}{2}\right)$ that is produced by the fractional friction force $c\frac{d^{\beta}x(t)}{dt^{\beta}}$.

15.3. State engineering meaning of the mass $\left(-m\omega^{\alpha-2}\cos\frac{\alpha\pi}{2}\right)$ that is produced by the fractional inertia force $m\frac{d^{\alpha}x(t)}{dt^{\alpha}}$.

REFERENCES

1. A. Einstein, *Relativity, Special and General Theory*, Shanghai Science and Technology Publishing House, 1964. In Chinese. Translated from Japanese by R. Y. Yang.
2. J. S. Farnes, A unifying theory of dark energy and dark matter: Negative masses and matter creation within a modified ΛCDM framework, *Astronomy & Astrophysics*, 620, 2018, A92.
3. H. Bondi, Negative mass in general relativity, *Reviews of Modern Physics*, 9(3), 1957, 423–428.

4. S. Mbarek and M. Paranjape, Negative mass bubbles in de sitter spacetime, *Physical Review D*, 90(10), 2014, 101502.
5. M. A. Khamehchi, K. Hossain, M. E. Mossman, Y. P. Zhang, Th. Busch, M. M. Forbes, and P. Engels, Negative-mass hydrodynamics in a spin-orbit-coupled Bose-Einstein condensate, *Physical Review Letters*, 118(15), 2017, 155301.
6. Z. Li, C. Wang, and X. Wang, Modelling of elastic metamaterials with negative mass and modulus based on translational resonance, *International Journal of Solids and Structures*, 162, 2019, 271–284.
7. A. Saha and S. K. Mishra, Adaptive negative stiffness device based nonconventional tuned mass damper for seismic vibration control of tall buildings, *Soil Dynamics and Earthquake Engineering*, 126, 2019, 105767.
8. E. Talib, J.-H. Shin, and M. K. Kwak, Designing multi-input multi-output modal-space negative acceleration feedback control for vibration suppression of structures using active mass dampers, *Journal of Sound and Vibration*, 439, 2019, 77–98.
9. X. Wang, Dynamic behaviour of a metamaterial system with negative mass and modulus, *International Journal of Solids and Structures*, 51(7–8), 2014, 1534–1541.

CHAPTER 16

Vibrators with Variable-Order Fractional Forces

IN THIS CHAPTER, WE present a theory of six classes of vibrators with variable-order fractional forces of inertia, damping, and restoration. The novelty and contributions are reflected in six aspects as follows. (1) Equivalent motion equations of those variable-order fractional vibrators. (2) The analytical expressions of the effective mass, damping, and stiffness of those variable-order fractional vibrators. (3) The asymptotic properties of the effective mass, damping, and stiffness of a class of variable-order fractional vibrators. (4) The restricted effective parameters (damping ratio, damping free natural frequency, damped natural frequency, and frequency ratio) of the variable-order fractional vibrators. (5) The analytical representations of the free responses, the impulse responses, and the frequency transfer functions of those variable-order fractional vibrators. (6) Mathematical explanation of Rayleigh damping assumption based on the present theory of variable-order fractional vibrations.

16.1 BACKGROUND

Fractional vibrations attract the interest of researchers in various fields, ranging from physics to mathematics, see e.g., [1–20]. The literature for a class of fractional vibrators below is rich,

$$m\frac{d^{\alpha}x(t)}{dt^{\alpha}}+kx(t)=f(t), \tag{16.1}$$

see e.g., Uchaikin [1, Chapter 7], Duan [2, Eq. (3)], Duan et al. [3, Eq. (4.2)], Al-rabtah et al. [4, Eq. (3.1)], Zurigat [5, Eq. (16)], Blaszczyk and Ciesielski [6, Eq. (1)], Blaszczyk et al. [7, Eq. (10)], Drozdov [8, Eq. (9)], Stanislavsky [9], Achar et al. ([10, Eq. (1)], [11, Eq. (9)], [12, Eq. (2)]), Tofighi [13, Eq. (2)], Ryabov and Puzenko [14, Eq. (1)], Ahmad and Elwakil [15, Eq. (1)], Blaszczyk [16], Tavazoei [17], Sandev and Tomovski [18, Eq. (36)], Singh et al. [19], Eab and Lim [20], Li et al. [21,22], Li [23,24].

DOI: 10.1201/9781003460947-18

Following the terms used in Chapter 7, a vibrator in Eq. (16.1) is called a class I fractional one. A class I fractional vibrator is only with the fractional inertia force $mx^{(\alpha)}(t)$. In general, $1<\alpha\leq 2$ [1–22], but we extend it to $1<\alpha<3$ in our research.

If a vibrator's motion equation takes the following form, it is called a class II fractional vibrator

$$m\frac{d^2x(t)}{dt^2}+c\frac{d^{\beta}x(t)}{dt^{\beta}}+kx(t)=f(t). \tag{16.2}$$

Usually, $1<\beta\leq 1$ but we enlarge it to $1<\beta<2$ in our research.

A class II fractional vibrator is simply with the fractional damping force $cx^{(\beta)}(t)$. The reports regarding the research of Eq. (16.2) are affluent, see e.g., Li [23,24], Lin et al. [25, Eq. (2)], Duan [26, Eq. (31)], Alkhaldi et al. [27, Eq. (1a)], Dai et al. [28, Eq. (1)], Ren et al. [29, Eq. (1)], Xu et al. [30, Eq. (1)], He et al. [31, Eq. (4)], Leung et al. [32, Eq. (2)], Chen et al. [33, Eq. (1)], Deü and Matignon [34, Eq. (1)], Drăgănescu et al. [35, Eq. (4)], Rossikhin and Shitikova [36, Eq. (3)], Xie and Lin [37, Eq. (1)], Ren et al. [38, Eq. (1)], Yuan et al. [39, Eq. (8)], Dai et al. [40, Eq. (1)], Lin et al. [41, Eq. (1)], Naranjani et al. [42, Eq. (1)], Lim et al. [43], Matteo et al. [44], Tomovski and Sandev [45, Eq. (44)], Kumar et al. [46], Tian et al. [47, 48], He et al. [49, Eq. (1)], Varanis et al. [50], Pang et al. [51, Eq. (1.1)], Yu et al. [52, Eq. (2.1)], He et al. [53, Eq. (7)], Spanos and Malara [54, 55], Golmankhaneh [56], Duan et al [57, Eq. (8)].

When the motion equation of a fractional vibrator is expressed by

$$m\frac{d^{\alpha}x_3(t)}{dt^{\alpha}}+c\frac{d^{\beta}x_3(t)}{dt^{\beta}}+kx_3(t)=f(t), \tag{16.3}$$

it is called a class III fractional vibrator. A class III fractional vibrator is with both the fractional inertia force $mx^{(\alpha)}(t)$ and the fractional damping one $cx^{(\beta)}(t)$. For the reports about (16.3), we refer to Gomez-Aguilar et al. [58, Eq. (10)], Tian et al. [59], Berman and Cederbaum [60], Coronel-Escamilla et al. [61, Eq. (12)], Sene and Aguilar [62, Eq. (8)], Vishwamittar et al. [63, Eq. (1)], Ismail et al. [64, Eq. (3)], Lim et al. [65].

There are other three classes of fractional vibrators. They are expressed by Eqs. (16.4)–(16.6), respectively.

$$m\frac{d^{\alpha}x(t)}{dt^{\alpha}}+k\frac{d^{\lambda}x(t)}{dt^{\lambda}}=f(t),\quad 1<\alpha<3,\ 0\leq\lambda<1, \tag{16.4}$$

$$m\frac{d^2x(t)}{dt^2}+k\frac{d^{\lambda}x(t)}{dt^{\lambda}}=f(t),\quad 0\leq\lambda<1, \tag{16.5}$$

and

$$m\frac{d^{\alpha}x(t)}{dt^{\alpha}}+c\frac{d^{\beta}x(t)}{dt^{\beta}}+k\frac{d^{\lambda}x(t)}{dt^{\lambda}}=f(t),\quad 1<\alpha<3,\ 0<\beta<2,\ 0\leq\lambda<1. \tag{16.6}$$

The literature regarding (16.4)–(16.6) is rare.

Following Chapters 11–13, we call (16.4) the motion equation of a class IV fractional vibrator, (16.5) for a class V fractional vibrator, and (16.6) a class VI fractional one. A class IV fractional vibrator is with both the fractional inertia force $mx^{(\alpha)}(t)$ and the fractional restoration force $kx^{(\lambda)}(t)$. A class V fractional vibrator is only with the fractional restoration force $kx^{(\lambda)}(t)$. A class VI fractional vibrator contains the fractional inertia force $mx^{(\alpha)}(t)$, the fractional damping force $cx^{(\beta)}(t)$, and the fractional restoration one $kx^{(\lambda)}(t)$.

In structural vibrations, effective mass or damping or stiffness may be frequency-dependent. Considering variable fractional orders $\alpha(\omega)$: $[0,\infty) \to (1, 3)$, $\beta(\omega)$: $[0,\infty) \to (1, 2)$, and $\lambda(\omega)$: $[0,\infty) \to [0,1)$ for Eqs. (16.1)–(16.6), we have six classes of variable-order fractional vibrators respectively expressed by Eqs. (16.7)–(16.12) below.

$$m\frac{d^{\alpha(\omega)}x(t)}{dt^{\alpha(\omega)}}+kx(t)=f(t), \tag{16.7}$$

$$m\frac{d^2x(t)}{dt^2}+c\frac{d^{\beta(\omega)}x(t)}{dt^{\beta(\omega)}}+kx(t)=f(t), \tag{16.8}$$

$$m\frac{d^{\alpha(\omega)}x(t)}{dt^{\alpha(\omega)}}+c\frac{d^{\beta(\omega)}x(t)}{dt^{\beta(\omega)}}+kx(t)=f(t), \tag{16.9}$$

$$m\frac{d^{\alpha(\omega)}x(t)}{dt^{\alpha(\omega)}}+k\frac{d^{\lambda(\omega)}x(t)}{dt^{\lambda(\omega)}}=f(t), \tag{16.10}$$

$$m\frac{d^2x(t)}{dt^2}+k\frac{d^{\lambda(\omega)}x(t)}{dt^{\lambda(\omega)}}=f(t), \tag{16.11}$$

and

$$m\frac{d^{\alpha(\omega)}x(t)}{dt^{\alpha(\omega)}}+c\frac{d^{\beta(\omega)}x(t)}{dt^{\beta(\omega)}}+k\frac{d^{\lambda(\omega)}x(t)}{dt^{\lambda(\omega)}}=f(t). \tag{16.12}$$

In Eqs. (16.7)–(16.12), $m\frac{d^{\alpha(\omega)}x(t)}{dt^{\alpha(\omega)}}$, $c\frac{d^{\beta(\omega)}x(t)}{dt^{\beta(\omega)}}$, and $k\frac{d^{\lambda(\omega)}x(t)}{dt^{\lambda(\omega)}}$ designate variable-order fractional forces of inertia, damping, and restoration, respectively. The literature regarding Eqs. (16.7)–(16.12) is rarely seen.

Different classes of vibrators have their specific application areas. For instance, Eq. (16.7) implies a vibrator that is damping free in form but m moves at $\frac{d^{\alpha(\omega)}x(t)}{dt^{\alpha(\omega)}}$ instead of $x''(t)$ and its displacement is $x(t)$.

In this chapter, we aim at establishing a theory with respect to the variable-order fractional vibrators Eqs. (16.7)–(16.12). Note that Eqs. (16.7)–(16.11) are the special cases of Eq. (16.12). For example, when $c=0$ and $\lambda(\omega)=0$, Eq. (16.12) reduces to Eq. (16.7). Therefore, we detail the analysis of Eq. (16.12) in Section 16.2. The results regarding Eqs. (16.7)–(16.11) are given in Section 16.7 as the consequences of the results in Section 16.12.

The present theory consists of a set of results in six aspects as follows. (1) The equivalent equations of (16.7)–(16.12) are proposed in Sections 16.2 and 16.7. (2) The analytical expressions of the effective mass, damping, and stiffness of Eqs. (16.7)–(16.12) are presented in Sections 16.2 and 16.7. (3) The asymptotic properties of its effective mass, damping, and stiffness of Eq. (16.12) are put forward in Theorems 16.5–16.7 and Corollaries 16.2 and 16.3. (4) The analytic expressions of the restricted effective damping ratio, natural frequencies, and frequency ratio regarding Eqs. (16.7)–(16.12) are brought forward in Sections 16.4 and 16.7. (5) The closed-form solutions of free responses, impulse ones, and frequency transfer functions of Eqs. (16.7)–(16.12) are given in Sections 16.5 and 16.7. (6) A solution to the open problem of how to mathematically explain the Rayleigh damping assumption is given in Section 16.6.

The rest of this chapter is organized as follows. In Section 16.2, we propose an equivalent equation of (16.12) and the analytic expressions of the effective mass, damping, and stiffness of Eq. (16.12). In Section 16.3, we present the asymptotic properties of the effective mass, damping, and stiffness of Eq. (16.12). In Section 16.4, we bring forward the analytic expressions of the restricted effective damping ratio, restricted effective natural frequencies, and restricted effective frequency ratio of Eq. (16.12). In Section 16.5, the expressions of the free response, impulse one, and frequency transfer function regarding Eq. (16.12) are presented. In Section 16.6, we propose a mathematical explanation of the Rayleigh damping assumption. The corresponding results of Eqs. (16.7)–(16.11) are given in Section 16.7, which is followed by a summary.

16.2 EQUIVALENT MOTION EQUATION AND EFFECTIVE MASS, DAMPING, AND STIFFNESS OF FRACTIONAL VIBRATOR (16.12)

In this section, we first brief the preliminaries. Then, we present the equivalent motion equation of (16.12). Finally, we propose the analytical expressions of the effective mass, damping, and stiffness of Eq. (16.12).

16.2.1 Preliminaries

In this research, we use the Weyl fractional derivative [66–83]. Let $X(\omega)$ be the Fourier transform of $x(t)$. Using the Weyl fractional derivative, the Fourier transform of $x^{(v)}(t)$ for $v \geq 0$ is given by

$$\mathrm{F}\left[x^{(v)}(t)\right] = (i\omega)^{v} X(\omega), \tag{16.13}$$

where F the Fourier transform operator (Uchaikin [1], Miller and Ross [82], Lavoie et al. [83]).

Lemma 16.1. Let $F_1(\omega) = \mathrm{F}[f_1(t)]$ and $F_2(\omega) = \mathrm{F}[f_2(t)]$. If $\mathrm{F}[f_1(t) - f_2(t)] = 0$, $f_1(t) - f_2(t)$ is a null function. If $f_1(t) - f_2(t)$ is a null function, $f_1(t) = f_2(t)$ in the sense of $F_1(\omega) = F_2(\omega)$ (Papoulis [84], Gelfand and Vilenkin [85], Bracewell [86]). □

16.2.2 Equivalent Motion Equation of (16.12)

Theorem 16.1 gives an equivalent motion equation of (16.12).

Theorem 16.1. The motion equation of the fractional vibrator (16.12) is equivalently given by

$$-\left[m\omega^{\alpha(\omega)-2}\cos\frac{\alpha(\omega)\pi}{2}+c\omega^{\beta(\omega)-2}\cos\frac{\beta(\omega)\pi}{2}\right]\frac{d^2x(t)}{dt^2}$$
$$+\left[m\omega^{\alpha(\omega)-1}\sin\frac{\alpha(\omega)\pi}{2}+c\omega^{\beta(\omega)-1}\sin\frac{\beta(\omega)\pi}{2}+k\omega^{\lambda(\omega)-1}\sin\frac{\lambda(\omega)\pi}{2}\right]\frac{dx(t)}{dt} \quad (16.14)$$
$$+k\omega^{\lambda(\omega)}\cos\frac{\lambda(\omega)\pi}{2}x(t)=f(t), \quad 1<\alpha(\omega)<3,\ 0<\beta(\omega)<2,\ 0\le\lambda(\omega)<1.$$

Proof. Let $X(\omega)$ be the Fourier transform of $x(t)$. Rewrite Eq. (16.12) by

$$f_1(t)=m\frac{d^{\alpha(\omega)}x(t)}{dt^{\alpha(\omega)}}+c\frac{d^{\beta(\omega)}x(t)}{dt^{\beta(\omega)}}+k\frac{d^{\lambda(\omega)}x(t)}{dt^{\lambda(\omega)}}. \quad (16.15)$$

On the one hand, Eq. (16.14) is rewritten by

$$f_2(t)=-\left[m\omega^{\alpha(\omega)-2}\cos\frac{\alpha(\omega)\pi}{2}+c\omega^{\beta(\omega)-2}\cos\frac{\beta(\omega)\pi}{2}\right]\frac{d^2x(t)}{dt^2}$$
$$+\left[m\omega^{\alpha(\omega)-1}\sin\frac{\alpha(\omega)\pi}{2}+c\omega^{\beta(\omega)-1}\sin\frac{\beta(\omega)\pi}{2}+k\omega^{\lambda(\omega)-1}\sin\frac{\lambda(\omega)\pi}{2}\right]\frac{dx(t)}{dt} \quad (16.16)$$
$$+k\omega^{\lambda(\omega)}\cos\frac{\lambda(\omega)\pi}{2}x(t).$$

Doing $\mathrm{F}[f_1(t)]$ yields

$$\mathrm{F}\left[f_1(t)\right]=\left[m(i\omega)^{\alpha(\omega)}+c(i\omega)^{\beta(\omega)}+k(i\omega)^{\lambda(\omega)}\right]X(\omega). \quad (16.17)$$

On the other hand, doing $\mathrm{F}[f_2(t)]$ produces

$$\mathrm{F}\left[f_2(t)\right]=\left[m\omega^{\alpha(\omega)}\cos\frac{\alpha(\omega)\pi}{2}+c\omega^{\beta(\omega)}\cos\frac{\beta(\omega)\pi}{2}\right]X(\omega)$$
$$+i\left[m\omega^{\alpha(\omega)}\sin\frac{\alpha(\omega)\pi}{2}+c\omega^{\beta(\omega)}\sin\frac{\beta(\omega)\pi}{2}+k\omega^{\lambda(\omega)}\sin\frac{\lambda(\omega)\pi}{2}\right]X(\omega) \quad (16.18)$$
$$+k\omega^{\lambda(\omega)}\cos\frac{\lambda(\omega)\pi}{2}X(\omega),$$

where $i=\sqrt{-1}$. Rewriting the above yields

$$\begin{aligned} \mathrm{F}\left[f_2(t)\right] = & \left[m\omega^{\alpha(\omega)}\cos\frac{\alpha(\omega)\pi}{2}+im\omega^{\alpha(\omega)}\sin\frac{\alpha(\omega)\pi}{2}\right]X(\omega) \\ & +\left[c\omega^{\beta(\omega)}\cos\frac{\beta(\omega)\pi}{2}+ic\omega^{\beta(\omega)}\sin\frac{\beta(\omega)\pi}{2}\right]X(\omega) \\ & +\left[k\omega^{\lambda(\omega)}\cos\frac{\lambda(\omega)\pi}{2}+ik\omega^{\lambda(\omega)}\sin\frac{\lambda(\omega)\pi}{2}\right]X(\omega). \end{aligned} \tag{16.19}$$

Since the principal values of $i^{\alpha(\omega)}$, $i^{\beta(\omega)}$, and $i^{\lambda(\omega)}$ are given by

$$\begin{aligned} i^{\alpha(\omega)} &= \cos\frac{\alpha(\omega)\pi}{2}+i\sin\frac{\alpha(\omega)\pi}{2}, \\ i^{\beta(\omega)} &= \cos\frac{\beta(\omega)\pi}{2}+i\sin\frac{\beta(\omega)\pi}{2}, \\ i^{\lambda(\omega)} &= \cos\frac{\lambda(\omega)\pi}{2}+i\sin\frac{\lambda(\omega)\pi}{2}, \end{aligned} \tag{16.20}$$

we can write Eq. (16.19) by

$$\mathrm{F}\left[f_2(t)\right]=\left[m(i\omega)^{\alpha(\omega)}+c(i\omega)^{\beta(\omega)}+k(i\omega)^{\lambda(\omega)}\right]X(\omega). \tag{16.21}$$

From Eqs. (16.17) and (16.21), we see that $\mathrm{F}[f_1(t)]=\mathrm{F}[f_2(t)]$. Thus, $\mathrm{F}[f_1(t)-f_2(t)]=0$. Hence, $f_1(t)=f_2(t)$ as $f_1(t)-f_2(t)$ is a null function. This finishes the proof. □

16.2.3 Effective Mass, Damping, and Stiffness of Equation (16.12)

Note that, in Eq. (16.12), the unit of $mx^{(\alpha)}(t)$ is not Newtonian. Thus, the primary mass m is not an effective quantity to be an inertia measure unless $\alpha=2$. Also, the primary damping coefficient c does not effectively quantify a damping measure in Eq. (16.12) since the unit of $cx^{(\beta)}(t)$ is not Newtonian if $\beta\neq 1$. Similarly, the primary stiffness k in Eq. (16.12) is not an effective quantity to be a restoration measure as the unit of $kx^{(\lambda)}(t)$ is non-Newtonian when $\lambda\neq 0$. In vibrations, the effective mass of a vibrator is the coefficient of $x''(t)$, the effective damping is the coefficient of $x'(t)$, and the effective stiffness is that of $x(t)$ (Harris [87], Den Hartog [88], Timoshenko [89], Palley et al. [90], Nakagawa and Ringo [91]). Now, we present the effective mass, damping, and stiffness of Eq. (16.12) by the theorems below.

Theorem 16.2. Let m_{eff} be the effective mass of the fractional vibrator (16.12). Then,

$$m_{\text{eff}} = -\left[m\omega^{\alpha(\omega)-2}\cos\frac{\alpha(\omega)\pi}{2} + c\omega^{\beta(\omega)-2}\cos\frac{\beta(\omega)\pi}{2}\right]. \tag{16.22}$$

Proof. The above is the coefficient of $x''(t)$ in Eq. (16.14). Thus, m_{eff} is an inertia measure of the vibrator (16.14) and equivalently (16.12). The proof is finished. □

Though the unit of m_{eff} is not kg unless $\alpha(\omega)=2$ and $\beta(\omega)=1$, it measures the inertia of a vibrator (16.14) and equivalently (16.12). The quantity m_{eff} reduces to the primary m when $\alpha(\omega)=2$ and $\beta(\omega)=1$.

Theorem 16.3. Denote by c_{eff} the effective damping coefficient for the fractional vibrator (16.12). Then,

$$c_{\text{eff}} = m\omega^{\alpha(\omega)-1}\sin\frac{\alpha(\omega)\pi}{2} + c\omega^{\beta(\omega)-1}\sin\frac{\beta(\omega)\pi}{2} + k\omega^{\lambda(\omega)-1}\sin\frac{\lambda(\omega)\pi}{2}. \tag{16.23}$$

Proof. The above is the coefficient of $x'(t)$ in Eq. (16.14) that is equivalent to Eq. (16.12). This completes the proof. □

Although the unit of c_{eff} is not $\text{N}\times\text{m}^{-1}\times\text{s}$, it measures the damping coefficient of a vibrator (16.14) and equivalently (16.12). The quantity c_{eff} degenerates to c if $\alpha(\omega)=2$, $\beta(\omega)=1$, and $\lambda(\omega)=0$.

Theorem 16.4. Let k_{eff} be the effective stiffness of the fractional vibrator (16.12). Then,

$$k_{\text{eff}} = k\omega^{\lambda(\omega)}\cos\frac{\lambda(\omega)\pi}{2}. \tag{16.24}$$

Proof. The above is the coefficient of $x(t)$ in Eq. (16.14) which is equivalent to Eq. (16.12). The proof is finished. □

The unit of k_{eff} is not $\text{N}\times\text{m}^{-1}$, but it measures the restoration of the vibrator (16.14) and equivalently (16.12). The quantity k_{eff} reduces to the primary k if $\lambda(\omega)=0$.

16.3 ASYMPTOTIC PROPERTIES OF EFFECTIVE MASS, DAMPING, AND STIFFNESS OF FRACTIONAL VIBRATOR (16.12)

As m_{eff}, c_{eff}, and k_{eff} are the functions of vibration frequency ω, we may write them by $m_{\text{eff}}(\omega)$, $c_{\text{eff}}(\omega)$, and $k_{\text{eff}}(\omega)$.

Theorem 16.5: Asymptotic Property for $\omega \to \infty$. If $2<\alpha(\omega)<3$, $m_{\text{eff}}(\omega) \to \infty$ when $\omega \to \infty$. On the other side, $m_{\text{eff}}(\omega) \to 0$ when $1<\alpha(\omega)<2$ if $\omega \to \infty$. That is,

$$\lim_{\omega\to\infty} m_{\text{eff}}(\omega) = \begin{cases} \infty, & 2<\alpha(\omega)<3, \\ 0, & 1<\alpha(\omega)<2. \end{cases} \tag{16.25}$$

Proof. Because $\lim_{\omega\to\infty}\omega^{\alpha(\omega)-2}=\infty$ and $\cos\frac{\alpha(\omega)\pi}{2}<0$ if $2<\alpha(\omega)<3$. On the other hand, $\lim_{\omega\to\infty}\omega^{\alpha(\omega)-2}=0$ for $1<\alpha(\omega)<2$. Besides, $\lim_{\omega\to\infty}\omega^{\beta(\omega)-2}=0$ for $0<\beta(\omega)<2$. Thus, the above is valid. The proof is finished. □

Theorem 16.6: Negative Mass. If $0<\beta(\omega)<1$ and $2<\alpha(\omega)<3$, $m_{\text{eff}}\to-\infty$ if $\omega\to 0$.

Proof. Note that $\cos\frac{\beta(\omega)\pi}{2}>0$ and $\lim_{\omega\to 0}\omega^{\beta-2}=\infty$ for $0<\beta(\omega)<1$. Besides, $\lim_{\omega\to 0}\omega^{\alpha(\omega)-2}=0$ for $2<\alpha(\omega)<3$. Thus, for $0<\beta(\omega)<1$ and $2<\alpha(\omega)<3$, we have

$$\lim_{\omega\to 0} m_{\text{eff}}(\omega)=-\infty. \tag{16.26}$$

The proof is finished. □

The above implies that the range of $m_{\text{eff}}(\omega)$ is $(-\infty,\infty)$ in general. A particular thing about $m_{\text{eff}}(\omega)$ is that it may stand for a king of a new form of negative mass. Other forms of negative mass may refer to Chapter 15 and references therein. In Figure 16.1, we indicate a few plots of $m_{\text{eff}}(\omega)$.

We now consider the asymptotic property of c_{eff} for $\omega\to\infty$.

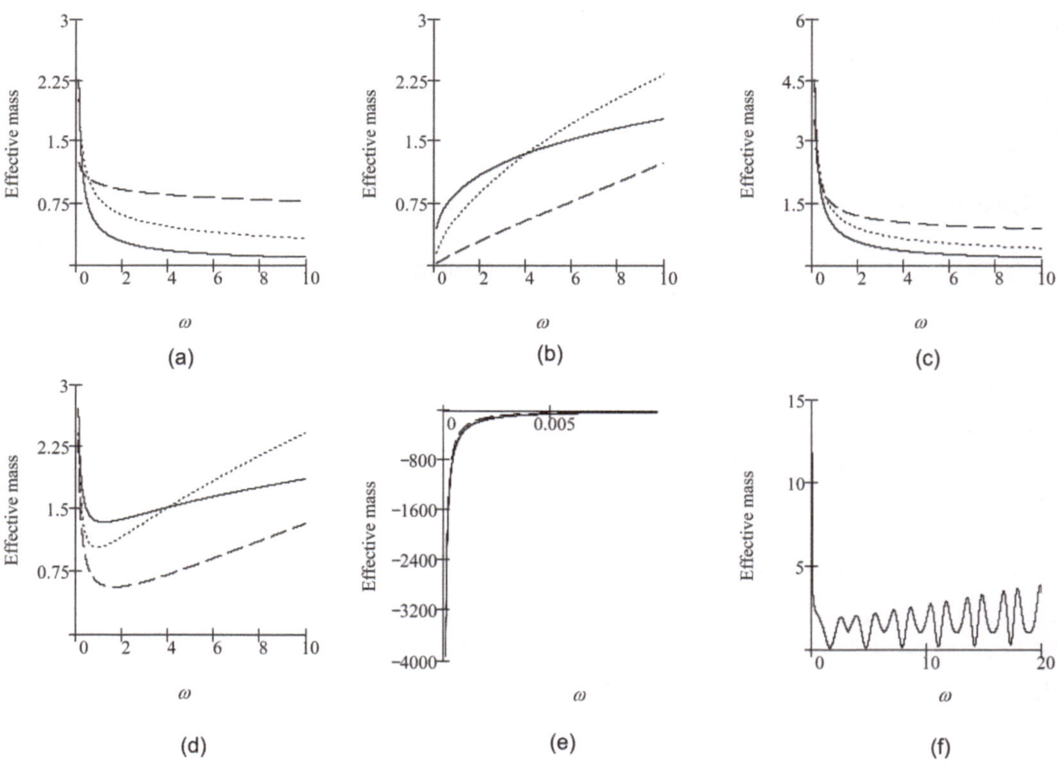

FIGURE 16.1 Plots of $m_{\text{eff}}(\omega)$ with $m=1$, $c=1$, and $k=1$. (a) $m_{\text{eff}}(\omega)$ when $\beta=1$ and $\alpha=1.3$ (solid), 1.6 (dot), 1.9 (dash). (b) $m_{\text{eff}}(\omega)$ when $\beta=1$ and $\alpha=2.3$ (solid), 2.6 (dot), 2.9 (dash). (c) $m_{\text{eff}}(\omega)$ when $\beta=1.3$ and $\alpha=1.3$ (solid), 1.6 (dot), 1.9 (dash). (d) $m_{\text{eff}}(\omega)$ when $\beta=1.3$ and $\alpha=2.3$ (solid), 2.6 (dot), 2.9 (dash). (e) Observing negative m_{eff} at small ω when $\beta=0.9$ and $\alpha=1.3$ (solid), 1.6 (dot), 1.9 (dash). (f) $m_{\text{eff}}(\omega)$ when $\alpha(\omega)=1.10+1.89|\sin(\omega)|$ and $\beta(\omega)=1+0.99|\cos(\omega)|$.

Theorem 16.7. If $\omega \to \infty$,

$$c_{\text{eff}} = \begin{cases} \infty, & 1<\alpha(\omega)<2, 0<\beta(\omega)<2, \\ -\infty, & 2<\alpha(\omega)<3, 0<\beta(\omega)<1. \end{cases} \tag{16.27}$$

Proof. Since $0 \le \lambda(\omega) < 1$, we have $\lim_{\omega\to\infty} \omega^{\lambda(\omega)-1} = 0$. In addition, because $\sin\frac{\beta(\omega)\pi}{2} > 0$ when $0<\beta(\omega)<2$ and $\sin\frac{\alpha(\omega)\pi}{2} > 0$ if $1<\alpha(\omega)<2$, $c_{\text{eff}}(\omega) \to \infty$ if $\omega \to \infty$ for $1<\alpha(\omega)<2$ and $0<\beta(\omega)<2$. Additionally, because $\lim_{\omega\to\infty} \omega^{\beta(\omega)-1} = 0$ when $0<\beta(\omega)<1$ but $\sin\frac{\alpha(\omega)\pi}{2} < 0$ if $2<\alpha(\omega)<3$, we have $c_{\text{eff}}(\omega) \to -\infty$ for $2<\alpha(\omega)<3$ and $0<\beta(\omega)<1$ when $\omega \to \infty$. The proof completes. □

The above exhibits that $-\infty < c_{\text{eff}} < \infty$.

Corollary 16.1. The fractional vibrator (16.12) may be self-vibrated and accordingly non-stable if $2<\alpha(\omega)<3$ and $0<\beta(\omega)<1$.

Proof. As c_{eff} may be negative if $2<\alpha(\omega)<3$ and $0<\beta(\omega)<1$, a vibrator (16.12) may be self-vibrated and non-stable accordingly. □

Corollary 16.2. If $1<\alpha(\omega)<3$, $0<\beta(\omega)<1$ and ω is large enough, we have

$$c_{\text{eff}} \cong m\omega^{\alpha(\omega)-1} \sin\frac{\alpha(\omega)\pi}{2}. \tag{16.28}$$

In addition, for $1<\alpha(\omega)<3$ and $0<\beta(\omega)<1$, if ω is small enough,

$$c_{\text{eff}} \cong c\omega^{\beta(\omega)-1} \sin\frac{\beta(\omega)\pi}{2} + k\omega^{\lambda(\omega)-1} \sin\frac{\lambda(\omega)\pi}{2}. \tag{16.29}$$

Proof. If $1<\alpha(\omega)<3$, $0<\beta(\omega)<1$ and ω is sufficiently large, $\omega^{\beta(\omega)-1} \approx 0$ and $\omega^{\lambda(\omega)-1} \approx 0$. Thus, Eq. (16.28) holds. On the other hand, if ω is small enough, $\omega^{\alpha(\omega)-1} \approx 0$. Thus, Eq. (16.29) results. This completes the proof. □

Figure 16.2 gives a few illustrations of c_{eff}.

Corollary 16.3. For k_{eff}, we have $k_{\text{eff}}(\omega) \ge 0$. Besides, $\lim_{\omega\to\infty} k_{\text{eff}}(\omega) = \infty$ and $\lim_{\omega\to 0} k_{\text{eff}}(\omega) = 0$.

Proof. Since $\lim_{\omega\to\infty} \omega^{\lambda(\omega)} = \infty$ and $\lim_{\omega\to 0} \omega^{\lambda(\omega)} = 0$, from Eq. (16.24), we see that the above is true. The proof is finished. □

We show its plots in Figure 16.3.

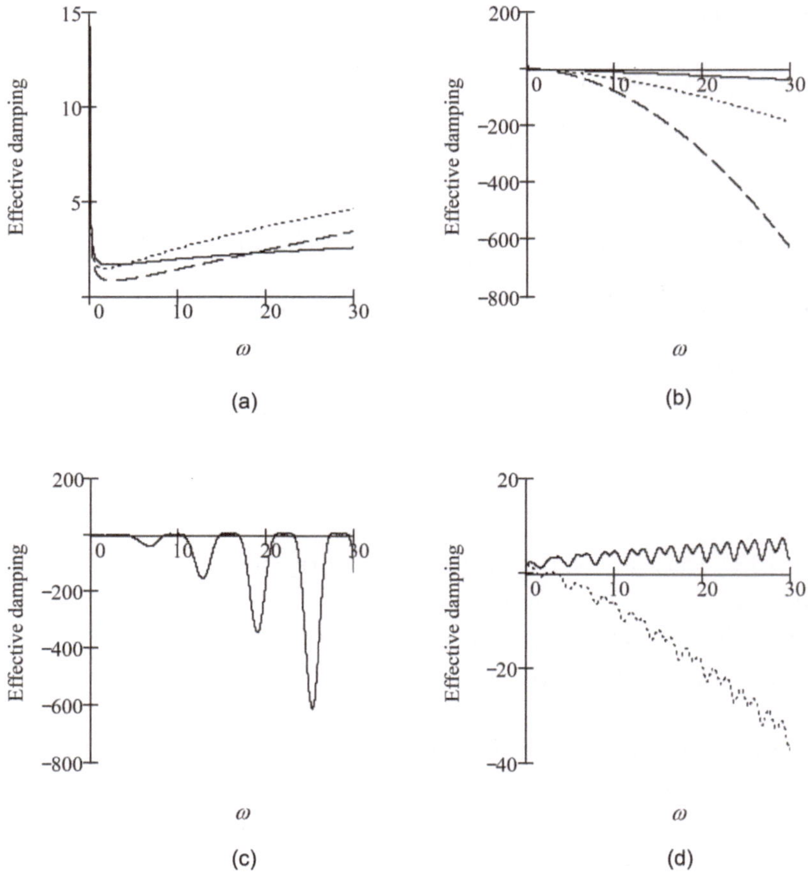

FIGURE 16.2 Plots of $c_{\text{eff}}(\omega)$ with $m=1$, $c=1$, and $k=1$. (a) $c_{\text{eff}}(\omega)$ for $\beta=0.3$, $\lambda=0.3$ for $\alpha=1.3$ (solid), 1.6 (dot), 1.9 (dash). (b) Observing negative $c_{\text{eff}}(\omega)$ for $\beta=0.3$, $\lambda=0.3$ when $\alpha=2.3$ (solid), 2.6 (dot), 2.9 (dash). (c) Observing negative $c_{\text{eff}}(\omega)$ for $\alpha(\omega)=1.10+1.89|\cos(0.5\omega)|$, $\beta(\omega)=1+0.99|\sin(\omega)|$, and $\lambda(\omega)=0.99|\sin(\omega)|$. (d) $c_{\text{eff}}(\omega)$ for $\beta(\omega)=1+0.99|\sin(\omega)|$, $\lambda(\omega)=0.99|\cos(\omega)|$ when $\alpha=1.3$ (solid), 2.3 (dot).

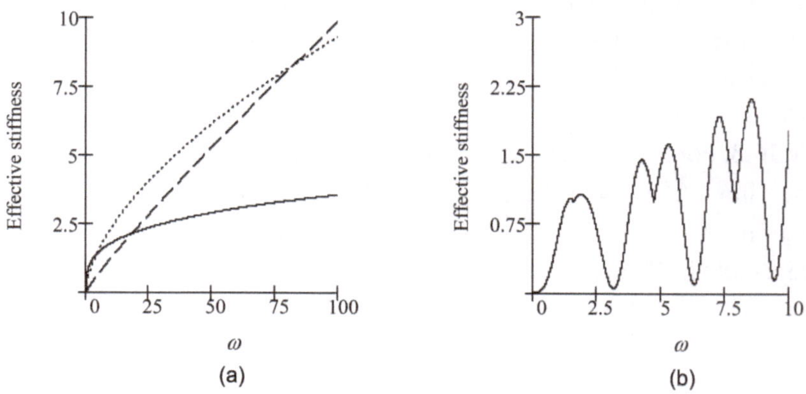

FIGURE 16.3 Plots of $k_{\text{eff}}(\omega)$ with $k=1$. (a) $k_{\text{eff}}(\omega)$ for $\lambda=0.3$ (solid), 0.6 (dot), 0.9 (dash). (b) $k_{\text{eff}}(\omega)$ when $\lambda(\omega)=0.99|\cos(\omega)|$.

16.4 RESTRICTED EFFECTIVE DAMPING RATIO, NATURAL FREQUENCIES, AND FREQUENCY RATIO OF FRACTIONAL VIBRATOR (16.12)

The effective parameters m_{eff}, c_{eff}, and k_{eff} are directly obtained from Eq. (16.14). In this section, we address the restricted effective damping ratio, natural frequencies, and frequency ratio. By restricted, we mean that additional conditions are considered for expressing those parameters.

16.4.1 Restricted Effective Damping Ratio of Equation (16.12)

As can be seen from Section 16.3, $m_{\text{eff}} \in (-\infty, \infty)$, $c_{\text{eff}} \in (-\infty, \infty)$, and $k_{\text{eff}} \in (0, \infty)$. Let ζ_{eff} be the restricted effective damping ratio (effective damping ratio for short) of the fractional vibrator (16.12). Here, we only consider ζ_{eff} in the case of $m_{\text{eff}}>0$ from a view of engineering. With that restriction, we define ζ_{eff} by

$$\varsigma_{\text{eff}} = \frac{c_{\text{eff}}}{2\sqrt{m_{\text{eff}} k_{\text{eff}}}}. \tag{16.30}$$

Theorem 16.8. The quantity ζ_{eff} is in the form

$$\varsigma_{\text{eff}} = \frac{m\omega^{\alpha(\omega)-1}\sin\frac{\alpha(\omega)\pi}{2} + c\omega^{\beta(\omega)-1}\sin\frac{\beta(\omega)\pi}{2} + k\omega^{\lambda(\omega)-1}\sin\frac{\lambda(\omega)\pi}{2}}{2\sqrt{-\left(m\omega^{\alpha(\omega)-2}\cos\frac{\alpha(\omega)\pi}{2} + c\omega^{\beta(\omega)-2}\cos\frac{\beta(\omega)\pi}{2}\right)k\omega^{\lambda(\omega)}\cos\frac{\lambda(\omega)\pi}{2}}}. \tag{16.31}$$

Proof. Substituting m_{eff}, c_{eff}, and k_{eff} into Eq. (16.30) results in Eq. (16.31). This finishes the proof. □

Although ζ_{eff} is not dimensionless in general, it effectively reflects the damping ratio of Eq. (16.14) and equivalently (16.12) in the case of $m_{\text{eff}}>0$. It is dimensionless if $\alpha(\omega)=2$, $\beta(\omega)=1$, and $\lambda(\omega)=0$. In that case, it equals to the standard damping ratio since $\varsigma_{\text{eff}}\big|_{\alpha(\omega)=2,\beta(\omega)=1,\lambda(\omega)=0} = \frac{c}{2\sqrt{mk}} = \varsigma$.

A few plots of ζ_{eff} are indicated in Figure 16.4.

16.4.2 Restricted Effective Damping Free Natural Frequency of Equation (16.12)

For the fractional vibrator (16.12), with the restriction $m_{\text{eff}}>0$ from the point of view of engineering, we coin a term restricted effective damping free natural frequency (effective damping free natural frequency in short) to the quantity defined by

$$\omega_{\text{effn}} = \sqrt{\frac{k_{\text{eff}}}{m_{\text{eff}}}}. \tag{16.32}$$

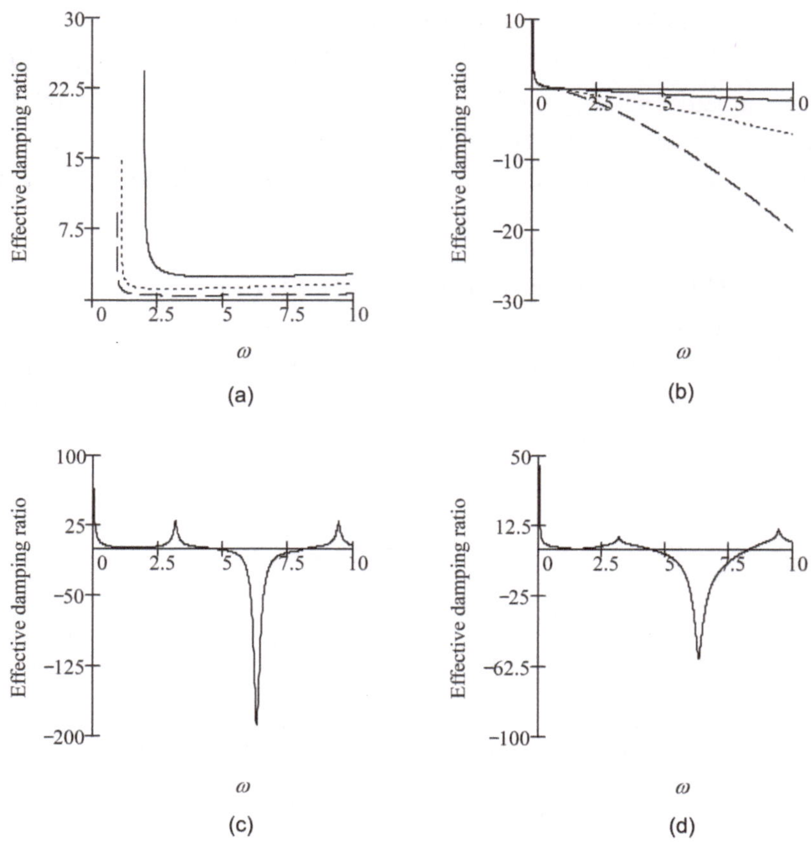

FIGURE 16.4 Plots of $\zeta_{eff}(\omega)$ for $m=1=k=c=1$. (a) $\zeta_{eff}(\omega)$ for $\beta=0.3$ and $\lambda=0.3$ when $\alpha=1.3$ (solid), 1.6 (dot), 1.9 (dash). (b) Observing negative $\zeta_{eff}(\omega)$ when $\beta=1.9$ and $\lambda=0.3$ when $\alpha=2.3$ (solid), 2.6 (dot), $\alpha=2.9$ (dash). (c) Observing negative $\zeta_{eff}(\omega)$ when $\alpha(\omega)=1.10+1.89\ |\cos(0.1\omega)|$, $\beta(\omega)=1+0.99\ |\sin(\omega)|$, and $\lambda(\omega)=0.99\ |\cos(\omega)|$. (d) Observing negative $\zeta_{eff}(\omega)$ when $\alpha(\omega)=1.10+1.89\ |\cos(0.1\omega)|$, $\beta(\omega)=1+0.99\ |\sin(\omega)|$, and $\lambda(\omega)=0.99\ |\exp(-\omega)|$.

Theorem 16.9. The quantity ω_{effn} is in the form

$$\omega_{\text{effn}} = \sqrt{\frac{k\omega^{\lambda(\omega)}\cos\frac{\lambda(\omega)\pi}{2}}{-\left(m\omega^{\alpha(\omega)-2}\cos\frac{\alpha(\omega)\pi}{2}+c\omega^{\beta(\omega)-2}\cos\frac{\beta(\omega)\pi}{2}\right)}}. \tag{16.33}$$

Proof. Substituting m_{eff} and k_{eff} into Eq. (16.32) yields Eq. (16.33). The proof is finished. □

Though the unit of ω_{effn} is not rad/s, it functionally takes the form of the conventional damping free natural frequency. Its unit reduces to rad/s if $\alpha(\omega)=2$, $\beta(\omega)=1$, and $\lambda(\omega)=0$. In fact,

$$\omega_{\text{effn}}\big|_{\alpha(\omega)=2,\beta(\omega)=1,\lambda(\omega)=0} = \omega_n = \sqrt{\frac{k}{m}}. \tag{16.34}$$

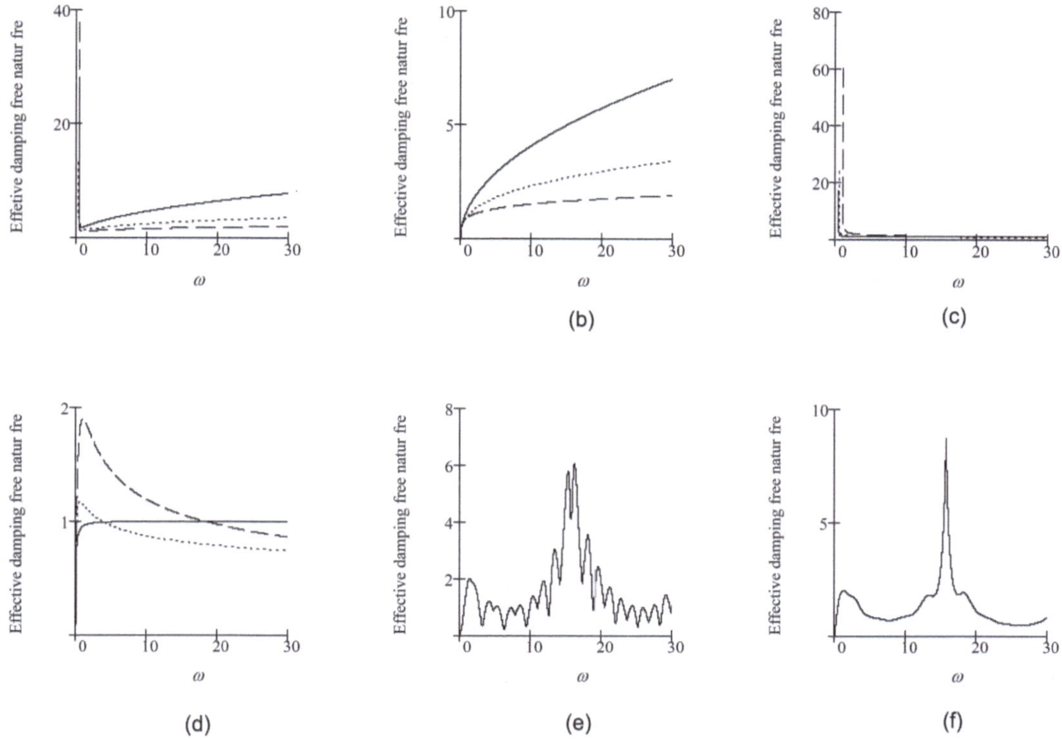

FIGURE 16.5 Plots of ω_{effn} with $m=1$ and $k=1$, $c=0.2$. (a) ω_{effn} for $\beta=0.3$ and $\lambda=0.3$ when $\alpha=1.3$ (solid), 1.6 (dot), 1.9 (dash). (b) ω_{effn} for $\beta=1.3$ and $\lambda=0.3$ when $\alpha=1.3$ (solid), 1.6 (dot), 1.9 (dash). (c) ω_{effn} for $\beta=0.3$ and $\lambda=0.3$ for $\alpha=2.3$ (solid), 2.6 (dot), 2.9 (dash). (d) ω_{effn} for $\beta=1.3$ and $\lambda=0.3$ for $\alpha=2.3$ (solid), 2.6 (dot), 2.9 (dash). (e) ω_{effn} for $\alpha(\omega)=1.10+1.89|\cos(0.1\omega)|$, $\beta(\omega)=1+0.99|\sin(\omega)|$, and $\lambda(\omega)=0.99|\cos(\omega)|$. (f) ω_{effn} for $\alpha(\omega)=1.10+1.89|\cos(0.1\omega)|$, $\beta(\omega)=1+0.99|\sin(\omega)|$, and $\lambda(\omega)=0.99|\exp(-\omega)|$.

In general, ω_{effn} is a function of ω. Thus, $\omega_{\text{effn}}=\omega_{\text{effn}}(\omega)$. Figure 16.5 illustrates a few plots of ω_{effn}. For variable fractional orders, $\omega_{\text{effn}}(\omega)$ is complicated, see Figure 16.5e and f.

16.4.3 Restricted Effective Damped Natural Frequency of Equation (16.12)

Now we introduce the term restricted effective damped natural frequency (effective damped natural frequency in short) for Eq. (16.12). Denote it by ω_{effd}. Small damping is commonly assumed in vibration engineering [87–92], [109]. Thus, ω_{effd} is practically restricted in the sense of small damping. Therefore, with the restriction $|\zeta_{\text{eff}}|\le 1$, we define it by

$$\omega_{\text{effd}}=\omega_{\text{effn}}\sqrt{1-\varsigma_{\text{eff}}^{2}}\,. \tag{16.35}$$

Theorem 16.10. The representation of ω_{effd} is in the form

$$\omega_{\text{effd}}=\sqrt{\frac{k\omega^{\lambda(\omega)}\cos\frac{\lambda(\omega)\pi}{2}}{-\left(m\omega^{\alpha(\omega)-2}\cos\frac{\alpha(\omega)\pi}{2}+c\omega^{\beta(\omega)-2}\cos\frac{\beta(\omega)\pi}{2}\right)}}\sqrt{1-\left(\frac{m\omega^{\alpha(\omega)-1}\sin\frac{\alpha(\omega)\pi}{2}+c\omega^{\beta(\omega)-1}\sin\frac{\beta(\omega)\pi}{2}+k\omega^{\lambda(\omega)-1}\sin\frac{\lambda(\omega)\pi}{2}}{2\sqrt{-\left(m\omega^{\alpha(\omega)-2}\cos\frac{\alpha(\omega)\pi}{2}+c\omega^{\beta(\omega)-2}\cos\frac{\beta(\omega)\pi}{2}\right)k\omega^{\lambda(\omega)}\cos\frac{\lambda(\omega)\pi}{2}}}\right)^2}. \tag{16.36}$$

Proof. Substituting ζ_{eff} and ω_{effn} into Eq. (16.35) yields Eq. (16.36). This finishes the proof. □

Note that if $\alpha(\omega)=2$, $\beta(\omega)=1$, and $\lambda(\omega)=0$, ω_{effd} degenerates the conventional ω_{d} with the unit of rad/s. It is generally a function of ω. Hence, $\omega_{\text{effd}}=\omega_{\text{effd}}(\omega)$.

16.4.4 Restricted Effective Frequency Ratio of Equation (16.12)

Introduce a term restricted effective frequency ratio (effective frequency ratio in short) for the fractional vibrator (16.12). Denote it by γ_{eff}. Define it by

$$\gamma_{\text{eff}}=\frac{\omega}{\omega_{\text{effn}}}. \tag{16.37}$$

Both ω_{effn} and γ_{eff} are restricted by $m_{\text{eff}}>0$. That restriction is taken from the point of view of vibration engineering.

Theorem 16.11. The representation of γ_{eff} is given by

$$\gamma_{\text{eff}}=\gamma\sqrt{\frac{-\left(\omega^{\alpha(\omega)-2}\cos\frac{\alpha(\omega)\pi}{2}+2\varsigma\omega_n\omega^{\beta(\omega)-2}\cos\frac{\beta(\omega)\pi}{2}\right)}{\omega^{\lambda(\omega)}\cos\frac{\lambda(\omega)\pi}{2}}}, \tag{16.38}$$

where $\gamma=\frac{\omega}{\omega_n}$.

Proof. Substituting ω_{effn} into Eq. (16.37) results in Eq. (16.38). The proof is finished. □

The quantity γ_{eff} reduces to γ when $\alpha(\omega)=2$, $\beta(\omega)=1$, and $\lambda(\omega)=0$. Its plots are indicated in Figure 16.6.

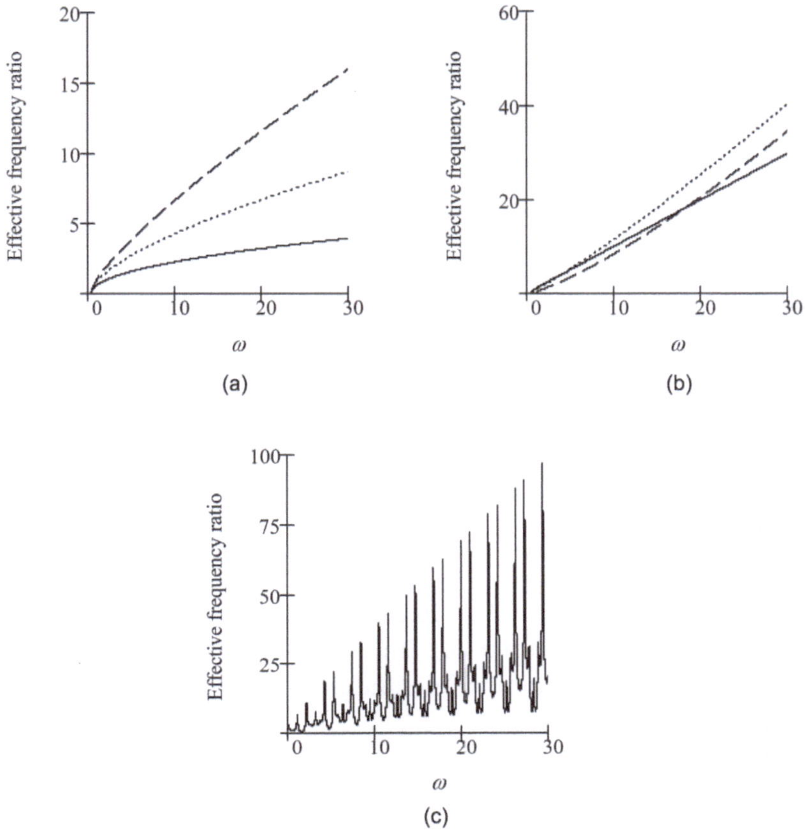

FIGURE 16.6 Plots of γ_{eff} with $m=1$, $k=1$, $c=0.2$. (a) γ_{eff} when $\beta=0.3$ and $\lambda=0.3$ for $\alpha=1.3$ (solid), 1.6 (dot), 1.9 (dash). (b) γ_{eff} when $\beta=0.3$ and $\lambda=0.3$ when $\alpha=2.3$ (solid), 2.6 (dot), 2.9 (dash). (c) γ_{eff} when $\alpha(\omega)=1.10+1.89|\cos(0.1\omega)|$, $\beta(\omega)=1+0.99|\sin(\omega)|$, and $\lambda(\omega)=0.99|\cos(\omega)|$.

16.5 RESTRICTED RESPONSES (FREE, IMPULSE) OF FRACTIONAL VIBRATOR (16.12)

Here and below, we use two restrictions, namely, $m_{\text{eff}}>0$ and $|\zeta_{\text{eff}}|\leq 1$, unless otherwise stated.

16.5.1 Equivalent Representation of Equation (16.14)

Based on $m_{\text{eff}}(\omega)$, $c_{\text{eff}}(\omega)$, and $k_{\text{eff}}(\omega)$, we rewrite Eq. (16.14) by

$$m_{\text{eff}}\frac{d^2x(t)}{dt^2}+c_{\text{eff}}\frac{dx(t)}{dt}+k_{\text{eff}}x(t)=f(t). \tag{16.39}$$

The above can be further written by

$$\frac{d^2x(t)}{dt^2}+2\varsigma_{\text{eff}}\omega_{\text{effn}}\frac{dx(t)}{dt}+\omega_{\text{effn}}^2x(t)=\frac{f(t)}{m_{\text{eff}}}. \tag{16.40}$$

16.5.2 Restricted Free Response

Let $x(t)$ be the restricted free response (free response for short) of the fractional vibrator (16.12). It is a solution to the differential equation given by

$$\begin{cases} \dfrac{d^2x(t)}{dt^2}+2\varsigma_{\text{eff}}\omega_{\text{effn}}\dfrac{dx(t)}{dt}+\omega_{\text{effn}}^2x(t)=0, & 1<\alpha(\omega)<3,\ 0<\beta(\omega)<2,\ \ 0\le\lambda(\omega)<1. \\ x(0)=x_0, x'(0)=v_0, & \end{cases} \tag{16.41}$$

The theorem proposed below provides its solution.

Theorem 16.12. The free response $x(t)$ to Eq. (16.12) is given by

$$x(t)=e^{-\varsigma_{\text{eff}}\omega_{\text{effn}}t}\left(x_0\cos\omega_{\text{effd}}t+\frac{v_0+\varsigma_{\text{eff}}\omega_{\text{effn}}x_0}{\omega_{\text{effd}}}\sin\omega_{\text{effd}}t\right),\quad t\ge 0. \tag{16.42}$$

Proof. According to Theorem 16.1, Eq. (16.14) is equivalent to Eq. (16.12). On the other hand, Eq. (16.40) is equivalent to Eq. (16.14). Since Eq. (16.41) is a standard vibration equation in form, Eq. (16.42) holds. This completes the proof. □

Figure 16.7 shows a few plots of $x(t)$.

In the time-frequency plane, we have

$$x(t)=x(t,\omega). \tag{16.43}$$

16.5.3 Restricted Impulse Response

Theorem 16.13 gives the expression of the restricted impulse response (impulse response for short) to the fractional vibrator (16.12).

Theorem 16.13. Let $h(t)$ be the impulse response to the fractional vibrator (16.12). It is a solution to the following fractional differential equation:

$$m\frac{d^{\alpha(\omega)}h(t)}{dt^{\alpha(\omega)}}+c\frac{d^{\beta(\omega)}h(t)}{dt^{\beta(\omega)}}+k\frac{d^{\lambda(\omega)}h(t)}{dt^{\lambda(\omega)}}=\delta(t),\ 1<\alpha(\omega)<3,\ 0<\beta(\omega)<2,\ 0\le\lambda(\omega)<1 \tag{16.44}$$

with zero initial conditions. The expression of $h(t)$ is given by

$$h(t)=e^{-\varsigma_{\text{eff}}\omega_{\text{effn}}t}\frac{1}{m_{\text{eff}}\omega_{\text{effd}}}\sin\omega_{\text{effd}}t,\quad t\ge 0. \tag{16.45}$$

Proof. According to Theorem 16.1, Eq. (16.44) is equivalently expressed by

$$m_{\text{eff}}\frac{d^2h(t)}{dt^2}+c_{\text{eff}}\frac{dh(t)}{dt}+k_{\text{eff}}h(t)=\delta(t).$$

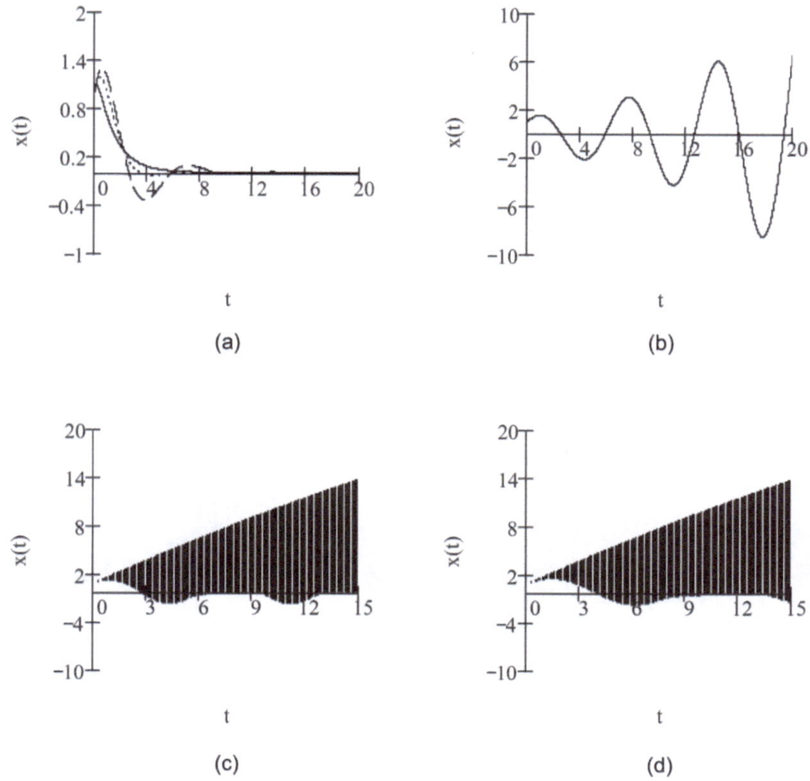

FIGURE 16.7 Plots of $x(t)$ for $x_0=1$, $v_0=1$, $m=1$ and $k=1$. (a) $x(t)$ for $\omega=1.1$, $c=0.2$, $\beta=0.3$, $\lambda=0.3$ when $\alpha=1.3$ (solid), 1.6 (dot), 1.9 (dash). (b) Self-vibration. $x(t)$ for $\omega=1.1$ and $c=0.2$ when $\beta=1.9$, $\lambda=0.3$ and $\alpha=2.4$. (c) $x(t)$ for $\omega \in (0, 1)$ and $c=0.2$ when $\alpha(\omega)=1.10+1.89|\cos(0.1\omega)|$, $\beta(\omega)=1+0.99|\sin(\omega)|$, and $\lambda(\omega)=0.99|\cos(\omega)|$. (d) $x(t)$ for $\omega \in (0, 1)$ and $c=1.2$ when $\alpha(\omega)=1.10+1.89|\cos(0.1\omega)|$, $\beta(\omega)=1+0.99|\sin(\omega)|$, and $\lambda(\omega)=0.99|\cos(\omega)|$.

The above is a standard vibration equation in form. Its solution is Eq. (16.45). The proof is finished. □

Figure 16.8 shows a few plots of $h(t)$.

In time-frequency plan, $h(t)$ may be written as

$$h(t)=h(t,\omega). \tag{16.46}$$

16.5.4 Restricted Frequency Transfer Function

Theorem 16.14. Let $H(\omega)$ be the restricted frequency transfer function (frequency transfer function in short) of the fractional vibrator (16.12). It is given by

$$H(\omega)=\frac{1}{k_{\text{eff}}\left(1-\gamma_{\text{eff}}^2+i2\varsigma_{\text{eff}}\gamma_{\text{eff}}\right)}. \tag{16.47}$$

Proof. Doing the Fourier transform of Eq. (16.45) produces the above. This finishes the proof. □

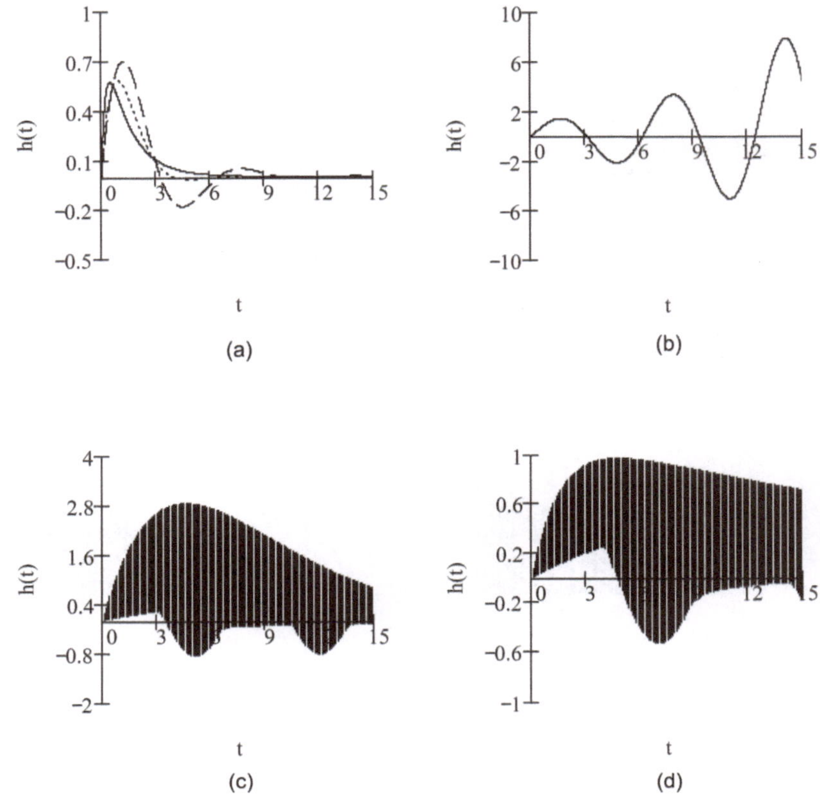

FIGURE 16.8 Plots of $h(t)$ for $m=1$ and $k=1$. (a) $h(t)$ for $\omega=1.1, c=0.2, \beta=0.3, \lambda=0.3$ when $\alpha=1.3$ (solid), 1.6 (dot), 1.9 (dash). (b) Self-vibration. $h(t)$ for $\omega=1.1, c=0.2, \alpha=2.5, \beta=1.5, \lambda=0.3$. (c) $h(t)$ for $\omega \in (0, 1)$ and $c=0.2$ when $\alpha(\omega)=1.10+1.89|\cos(0.1\omega)|$, $\beta(\omega)=1+0.99|\sin(\omega)|$, and $\lambda(\omega)=0.99|\cos(\omega)|$. (d) $h(t)$ for $\omega \in (0, 1)$ and $c=1.2$ when $\alpha(\omega)=1.10+1.89|\cos(0.1\omega)|$, $\beta(\omega)=1+0.99|\sin(\omega)|$, and $\lambda(\omega)=0.99|\cos(\omega)|$.

Using polar coordinates, we have

$$H(\omega)=|H(\omega)|e^{i\varphi(\omega)}, \tag{16.48}$$

where the amplitude-frequency response $|H(\omega)|$ is in the form

$$|H(\omega)|=\frac{1}{k_{\text{eff}}}\frac{1}{\sqrt{\left(1-\gamma_{\text{eff}}^2\right)^2+\left(2\varsigma_{\text{eff}}\gamma_{\text{eff}}\right)^2}}, \tag{16.49}$$

and the phase-frequency response function $\varphi(\omega)$ is expressed by

$$\varphi(\omega)=\cos^{-1}\frac{1-\gamma_{\text{eff}}^2}{\sqrt{\left(1-\gamma_{\text{eff}}^2\right)^2+\left(2\varsigma_{\text{eff}}\gamma_{\text{eff}}\right)^2}}. \tag{16.50}$$

Figures 16.9 and 16.10 illustrate several plots of $|H(\omega)|$ and $\varphi(\omega)$, respectively.

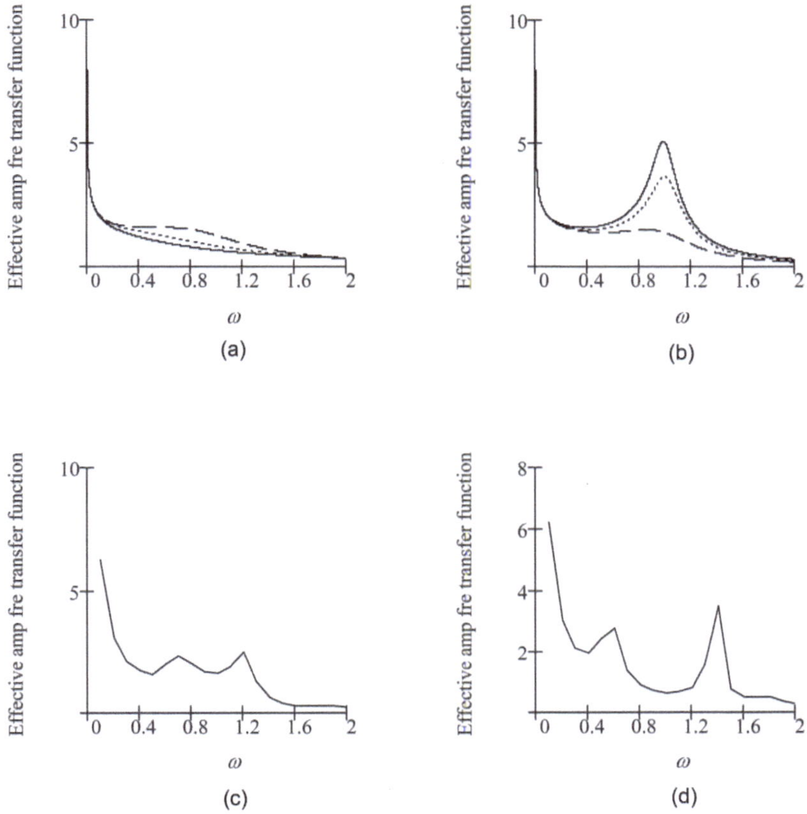

FIGURE 16.9 Plots of $|H(\omega)|$ when $m=1$ and $k=1$. (a) $|H(\omega)|$ for $\omega=1.1$, $c=0.2$, $\beta=1.3$, $\lambda=0.3$ when $\alpha=1.3$ (solid), 1.6 (dot), 1.9 (dash). (b) $|H(\omega)|$ for $\omega=1.1$, $c=0.2$, $\beta=1.3$, $\lambda=0.3$ when $\alpha=2.3$ (solid), 2.6 (dot), 2.9 (dash). (c) $|H(\omega)|$ for $\omega \in (0,2)$, $c=0.2$ when $\alpha(\omega)=1.10+1.89|\cos(\omega)|$, $\beta(\omega)=1+0.99|\sin(\omega)|$, and $\lambda(\omega)=0.99|\cos(\omega)|$. (d) $|H(\omega)|$ for $\omega \in (0, 2)$, $c=1.2$ when $\alpha(\omega)=1.10+1.89|\cos(\omega)|$, $\beta(\omega)=1+0.99|\sin(\omega)|$, and $\lambda(\omega)=0.99|\cos(\omega)|$.

16.6 MATHEMATICAL EXPLANATION OF RAYLEIGH DAMPING ASSUMPTION

Rayleigh introduced a damping assumption [92]. It has been widely used in structural mechanics, see e.g., Palley et al. [90], Trombetti and Silvestri [93], Poul and Zerva [94], Nåvik et al. [95], Park et al. [96], Hussein et al. [97], Sigmund and Jensen [98], Chen et al. [99], Battisti et al. [100], Cox et al. [101], Tisseur and Meerbergen [102], Chu et al. [103], Fay et al. [104], Naderian et al. [105], Tian et al. [106], Iovane and Nasedkin [107], Kouris et al. [108], Jin and Xia [109], just mentioning a few. In the field, it is well known that how to give a mathematical explanation of the Rayleigh damping assumption is an open problem.

In Chapter 14, the author presented a mathematical explanation of the Rayleigh damping assumption when α, β, and λ are constants. In this chapter, we further address an explanation by taking into account the functions $\alpha(\omega)$, $\beta(\omega)$, and $\lambda(\omega)$ instead of α, β, and λ being constants.

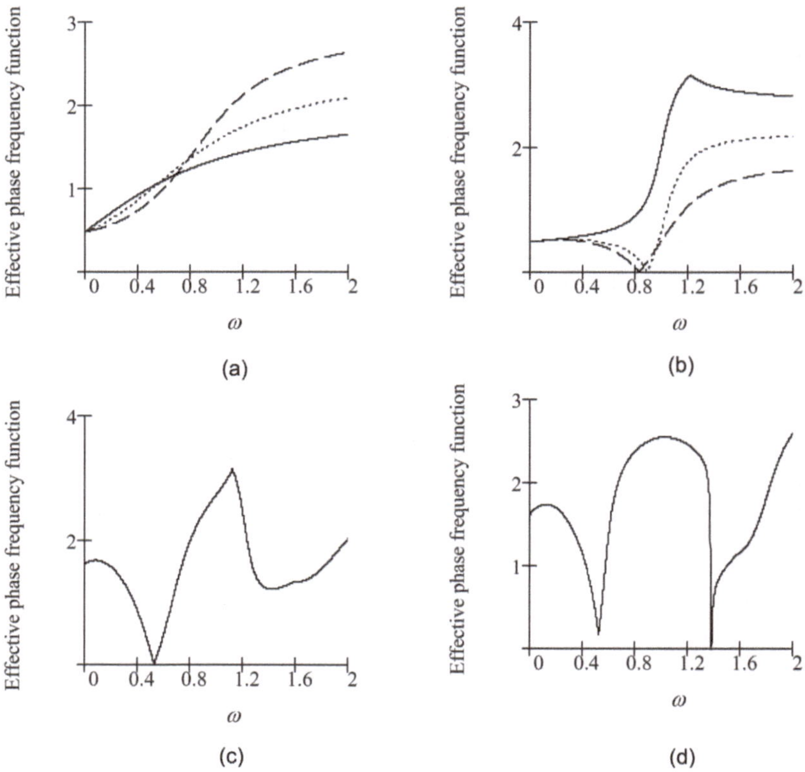

FIGURE 16.10 Plots of $\varphi(\omega)$ when $m=1$, $k=1$. (a) $\varphi(\omega)$ for $\omega=1.1$, $c=0.2$, $\beta=1.3$, $\lambda=0.3$ when $\alpha=1.3$ (solid), 1.6 (dot), 1.9 (dash). (b) $\varphi(\omega)$ for $\omega=1.1$, $c=0.2$, $\beta=1.3$, $\lambda=0.3$ when $\alpha=2.3$ (solid), 2.6 (dot), 2.9 (dash). (c) $\varphi(\omega)$ for $\omega \in (0, 2)$, $c=0.2$ when $\alpha(\omega)=1.10+1.89|\cos(\omega)|$, $\beta(\omega)=1+0.99|\sin(\omega)|$, and $\lambda(\omega)=0.99|\cos(\omega)|$. (d) $\varphi(\omega)$ for $\omega \in (0, 2)$, $c=1.2$ when $\alpha(\omega)=1.10+1.89|\cos(\omega)|$, $\beta(\omega)=1+0.99|\sin(\omega)|$, and $\lambda(\omega)=0.99|\cos(\omega)|$.

Let c_r be the Rayleigh damping. Its standard form is expressed by

$$c_r = am + bk, \tag{16.51}$$

where a and b are frequency-dependent parameters. Precisely, a is proportional to ω while b is inversely proportional to ω. With the variable-order fractional vibrator (16.12), we use its effective damping to give a more general explanation of the Rayleigh damping assumption.

Theorem 16.15. Let $c=0$ in Eq. (16.23). Then, c_{eff} in Eq. (16.23) reduces to the generalized Rayleigh damping, denoted by c_{gray}, in the form

$$c_{gray} = m\omega^{\alpha(\omega)-1} \sin\frac{\alpha(\omega)\pi}{2} + k\omega^{\lambda(\omega)-1} \sin\frac{\lambda(\omega)\pi}{2}. \tag{16.52}$$

Proof. Let $c=0$ in Eq. (16.23). Then, c_{eff} becomes the above. The above exhibits that c_{gray} is proportional to m with the coefficient a in the form

$$a = a(\omega) = \omega^{\alpha(\omega)-1} \sin\frac{\alpha(\omega)\pi}{2}. \tag{16.53}$$

Besides, c_{gray} is proportional to k with the coefficient b in the form

$$b = b(\omega) = \omega^{\lambda(\omega)-1} \sin\frac{\lambda(\omega)\pi}{2}. \tag{16.54}$$

Thus, we have

$$c_{\text{gray}} = a(\omega)m + b(\omega)k. \tag{16.55}$$

Because $a(\omega)$ is proportional to ω while $b(\omega)$ is inversely proportional to ω. Equation (16.52) is a mathematical expression that is consistent with the Rayleigh damping assumption. The proof is finished. □

Note that $a(\omega) \geq 0$ and $b(\omega) \geq 0$ for $1 < \alpha(\omega) < 2$ and $0 \leq \lambda(\omega) < 1$. Accordingly, $c_{\text{gray}} \geq 0$ in that cases of $1 < \alpha(\omega) < 2$ and $0 \leq \lambda(\omega) < 1$. On the other hand, if $2 < \alpha(\omega) < 3$ and $0 \leq \lambda(\omega) < 1$, $a(\omega)$ may be negative and $b(\omega) \geq 0$. In that case, c_{gray} may be negative. Because $\alpha(\omega)$ and $\lambda(\omega)$ are varying in terms of ω, Eq. (16.52) is a general form of the Rayleigh damping. In fact, when both $\alpha(\omega)$ and $\lambda(\omega)$ are constants, c_{gray} reduces to a specific form of the Rayleigh damping addressed in Chapter 14 in the form

$$c_r = m\omega^{\alpha-1} \sin\frac{\alpha\pi}{2} + k\omega^{\lambda-1} \sin\frac{\lambda\pi}{2}. \tag{16.56}$$

Figure 16.11 shows the plots of $a(\omega)$ and $b(\omega)$ while Figure 16.12 demonstrates the plots of c_{gray}.

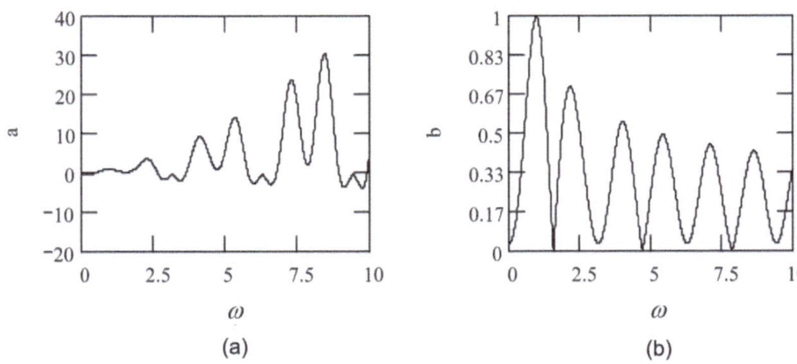

FIGURE 16.11 Plots of $a(\omega)$ and $b(\omega)$ with $m = k = 1$. (a) $a(\omega)$ with $\alpha(\omega) = 1.80 + 1.19|\sin(\omega)|$. (b) $b(\omega)$ with $\lambda(\omega) = 0.99|\cos(\omega)|$.

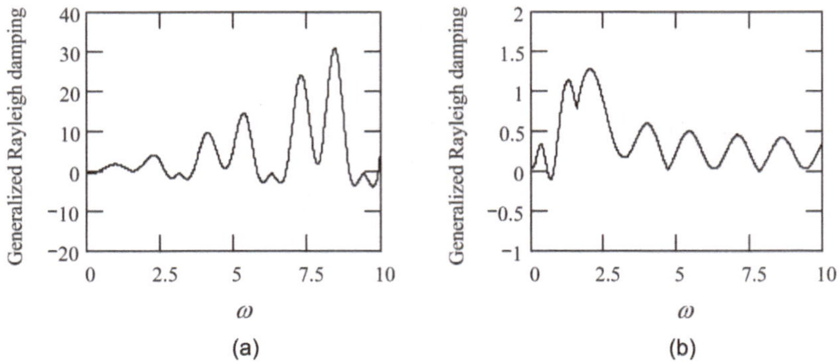

FIGURE 16.12 Plots of c_{gray} with $m=k=1$. (a) c_{gray} with $\alpha(\omega)=1.80+1.19|\sin(\omega)|$ and $\lambda(\omega)=0.99|\cos(\omega)|$. (b) c_{gray} with $\alpha(\omega)=2.99\exp(-\omega)$ and $\lambda(\omega)=0.99|\cos(\omega)|$.

16.7 RESULTS FOR THE VARIABLE-ORDER FRACTIONAL VIBRATORS (16.7)–(16.11)

We now propose the results regarding the variable-order fractional vibrators expressed by Eqs. (16.7)–(16.11). Mathematically, Eqs. (16.7)–(16.11) are the special cases of Eq. (16.12). However, from a view of engineering, studying each class from Eq. (16.7) to (16.11) is meaningful since each has its specific application area. For instance, Eq. (16.7) is for the case of damping free structures with variable-order fractional inertia force and conventional restoration one. The class (16.10) stands for the case of damping free structures with variable-order fractional inertia force and variable-order fractional restoration one. The importance of the results here lies in representing the results for each class from Eq. (16.7) to (16.11) based on the results with respect to Eq. (16.12) addressed in Sections 16.2–16.5. In what follows, x_0 and v_0 are initial conditions.

16.7.1 Results Regarding Variable-Order Fractional Vibrator (16.7)

Consider Eq. (16.7). Letting $c=0$ and $\lambda(\omega)=0$ in Eq. (16.14) yields the equivalent equation of (16.7) in the form

$$-m\omega^{\alpha(\omega)-2}\cos\frac{\alpha(\omega)\pi}{2}\frac{d^2x(t)}{dt^2}+m\omega^{\alpha(\omega)-1}\sin\frac{\alpha(\omega)\pi}{2}\frac{dx(t)}{dt}+kx(t)=f(t),\quad 1<\alpha(\omega)<3. \tag{16.57}$$

Let m_{eff1} be the effective mass of the fractional vibrator (16.7). Then, letting $c=0$ in Eq. (16.22) produces

$$m_{\text{eff1}}=-m\omega^{\alpha(\omega)-2}\cos\frac{\alpha(\omega)\pi}{2}. \tag{16.58}$$

Denote by c_{eff1} the effective damping of the fractional vibrator (16.7). Then, letting $c=0$ and $\lambda(\omega)=0$ in Eq. (16.23) results in

$$c_{\text{eff1}}=m\omega^{\alpha(\omega)-1}\sin\frac{\alpha(\omega)\pi}{2}. \tag{16.59}$$

Let ω_{effn1} be the restricted effective damping free natural frequency of Eq. (16.7) for $m_{\text{eff1}} > 0$. It is given by

$$\omega_{\text{effn1}} = \sqrt{\frac{k}{m_{\text{eff1}}}} = \sqrt{\frac{k}{-m\omega^{\alpha(\omega)-2}\cos\frac{\alpha(\omega)\pi}{2}}}. \tag{16.60}$$

Let ζ_{eff1} be the restricted effective damping ratio of Eq. (16.7) under the condition of $m_{\text{eff1}} > 0$. Then, from Eq. (16.30), we have

$$\varsigma_{\text{eff1}} = \frac{c_{\text{eff1}}}{2\sqrt{m_{\text{eff1}}k}} = \frac{m\omega^{\alpha(\omega)-1}\sin\frac{\alpha(\omega)\pi}{2}}{2\sqrt{-mk\omega^{\alpha(\omega)-2}\cos\frac{\alpha(\omega)\pi}{2}}}. \tag{16.61}$$

Consider the restricted effective damped natural frequency for Eq. (16.7) when $m_{\text{eff1}} > 0$. Denote it by ω_{effd1}. With the restriction $|\zeta_{\text{eff1}}| \le 1$, we have

$$\omega_{\text{effd1}} = \omega_{\text{effn1}}\sqrt{1-\varsigma_{\text{eff1}}^2}. \tag{16.62}$$

Let γ_{eff1} be the restricted effective frequency ratio of Eq. (16.7). It is, according to Eq. (16.37), given by

$$\gamma_{\text{eff1}} = \frac{\omega}{\omega_{\text{effn1}}}. \tag{16.63}$$

Both ω_{effn} and γ_{eff} are restricted by $m_{\text{eff}} > 0$.

Denote by $x_1(t)$ the free response to Eq. (16.7). According to Theorem 16.12, we have

$$x_1(t) = e^{-\varsigma_{\text{eff1}}\omega_{\text{effn1}}t}\left(x_0\cos\omega_{\text{effd1}}t + \frac{v_0 + \varsigma_{\text{eff1}}\omega_{\text{effn1}}x_0}{\omega_{\text{effd1}}}\sin\omega_{\text{effd1}}t\right), \quad t \ge 0. \tag{16.64}$$

Let $h_1(t)$ be the impulse response to Eq. (16.7). Based on Theorem 16.13, we have

$$h_1(t) = e^{-\varsigma_{\text{eff1}}\omega_{\text{effn1}}t}\frac{1}{m_{\text{eff1}}\omega_{\text{effd1}}}\sin\omega_{\text{effd1}}t, \quad t \ge 0. \tag{16.65}$$

Let $H_1(\omega)$ be the restricted frequency transfer function of the fractional vibrator (16.7). Following Theorem 16.14, we have

$$H_1(\omega) = \frac{1}{k\left(1-\gamma_{\text{eff1}}^2 + i2\varsigma_{\text{eff1}}\gamma_{\text{eff1}}\right)}. \tag{16.66}$$

16.7.2 Results Regarding Variable-Order Fractional Vibrator (16.8)

Taking into account (16.8), we substitute $\alpha(\omega)$ by 2 and $\lambda(\omega)$ with 0 in Eq. (16.14). Then, we have the equivalent equation of (16.8) given by

$$\left[m - c\omega^{\beta(\omega)-2}\cos\frac{\beta(\omega)\pi}{2}\right]\frac{d^2x(t)}{dt^2} + c\omega^{\beta(\omega)-1}\sin\frac{\beta(\omega)\pi}{2}\frac{dx(t)}{dt} + kx(t) = f(t), \quad 0 < \beta(\omega) < 2. \tag{16.67}$$

Write the above by

$$m_{\text{eff2}}\frac{d^2x(t)}{dt^2} + c_{\text{eff2}}\frac{dx(t)}{dt} + kx(t) = f(t), \tag{16.68}$$

where m_{eff2} and c_{eff2} are the effective mass and damping of the fractional vibrator (16.8), respectively. From Eqs. (16.67) and (16.68), we see that

$$m_{\text{eff2}} = m - c\omega^{\beta(\omega)-2}\cos\frac{\beta(\omega)\pi}{2}, \tag{16.69}$$

$$c_{\text{eff2}} = c\omega^{\beta(\omega)-1}\sin\frac{\beta(\omega)\pi}{2}. \tag{16.70}$$

Denote by ω_{effn2} the restricted effective damping free natural frequency of Eq. (16.8) when $m_{\text{eff2}} > 0$. It is

$$\omega_{\text{effn2}} = \sqrt{\frac{k}{m_{\text{eff2}}}}. \tag{16.71}$$

Let ζ_{eff2} be the restricted effective damping ratio of Eq. (16.8) in the case of $m_{\text{eff2}} > 0$. Then,

$$\varsigma_{\text{eff2}} = \frac{c_{\text{eff2}}}{2\sqrt{m_{\text{eff2}}k}}. \tag{16.72}$$

Denote by ω_{effd2} the restricted effective damped natural frequency of Eq. (16.8) for $m_{\text{eff2}} > 0$. Using the restriction $|\zeta_{\text{eff2}}| \le 1$, we have

$$\omega_{\text{effd2}} = \omega_{\text{effn2}}\sqrt{1 - \varsigma_{\text{eff2}}^2}. \tag{16.73}$$

Let γ_{eff2} be the restricted effective frequency ratio of Eq. (16.8). It is given by

$$\gamma_{\text{eff2}} = \frac{\omega}{\omega_{\text{effn2}}}. \tag{16.74}$$

Let $x_2(t)$ be the free response to Eq. (16.8). With Theorem 16.12 or from Eq. (16.68), we have

$$x_2(t)=e^{-\varsigma_{\text{eff2}}\omega_{\text{effn2}}t}\left(x_0\cos\omega_{\text{effd2}}t+\frac{v_0+\varsigma_{\text{eff2}}\omega_{\text{effn2}}x_0}{\omega_{\text{effd2}}}\sin\omega_{\text{effd2}}t\right),\quad t\geq 0. \tag{16.75}$$

Denote by $h_2(t)$ the impulse response to Eq. (16.8). Following Theorem 16.13 or from Eq. (16.68), we have

$$h_2(t)=e^{-\varsigma_{\text{eff2}}\omega_{\text{effn2}}t}\frac{1}{m_{\text{eff2}}\omega_{\text{effd2}}}\sin\omega_{\text{effd2}}t,\quad t\geq 0. \tag{16.76}$$

Let $H_2(\omega)$ be the restricted frequency transfer function of the fractional vibrator (16.8). With Theorem 16.14 or by doing the Fourier transform of $h_2(t)$, we have

$$H_2(\omega)=\frac{1}{k\left(1-\gamma_{\text{eff2}}^2+i2\varsigma_{\text{eff2}}\gamma_{\text{eff2}}\right)}. \tag{16.77}$$

16.7.3 Results Regarding Variable-Order Fractional Vibrator (16.9)

Consider Eq. (16.9). Replacing $\lambda(\omega)$ with 0 in Eq. (16.14) results in

$$\begin{aligned}&-\left[m\omega^{\alpha(\omega)-2}\cos\frac{\alpha(\omega)\pi}{2}+c\omega^{\beta(\omega)-2}\cos\frac{\beta(\omega)\pi}{2}\right]\frac{d^2x(t)}{dt^2}\\&+\left[m\omega^{\alpha(\omega)-1}\sin\frac{\alpha(\omega)\pi}{2}+c\omega^{\beta(\omega)-1}\sin\frac{\beta(\omega)\pi}{2}\right]\frac{dx(t)}{dt}\\&+kx(t)=f(t),\quad 1<\alpha(\omega)<3,\ 0<\beta(\omega)<2.\end{aligned} \tag{16.78}$$

The above is the equivalent equation of Eq. (16.9). Let m_{eff3} and c_{eff3} be the effective mass and damping of the fractional vibrator (16.9), respectively. Write the above by

$$m_{\text{eff3}}\frac{d^2x(t)}{dt^2}+c_{\text{eff3}}\frac{dx(t)}{dt}+kx(t)=f(t). \tag{16.79}$$

By comparing Eqs. (16.78) and (16.79), we have

$$m_{\text{eff3}}=-\left[m\omega^{\alpha(\omega)-2}\cos\frac{\alpha(\omega)\pi}{2}+c\omega^{\beta(\omega)-2}\cos\frac{\beta(\omega)\pi}{2}\right], \tag{16.80}$$

$$c_{\text{eff3}}=m\omega^{\alpha(\omega)-1}\sin\frac{\alpha(\omega)\pi}{2}+c\omega^{\beta(\omega)-1}\sin\frac{\beta(\omega)\pi}{2}. \tag{16.81}$$

Let ω_{effn3} be the restricted effective damping free natural frequency of Eq. (16.9) for $m_{\text{eff3}}>0$. It is given by

$$\omega_{\text{effn3}} = \sqrt{\frac{k}{m_{\text{eff3}}}}. \tag{16.82}$$

Let ζ_{eff3} be the restricted effective damping ratio of Eq. (16.9) when $m_{\text{eff3}}>0$. Then,

$$\varsigma_{\text{eff3}} = \frac{c_{\text{eff3}}}{2\sqrt{m_{\text{eff3}}k}}. \tag{16.83}$$

Denote by ω_{effd3} the restricted effective damped natural frequency of Eq. (16.9) when $m_{\text{eff3}}>0$. With $|\zeta_{\text{eff3}}|\le 1$, we have

$$\omega_{\text{effd3}} = \omega_{\text{effn3}}\sqrt{1-\varsigma_{\text{eff3}}^2}. \tag{16.84}$$

Denote by γ_{eff3} the restricted effective frequency ratio of Eq. (16.9). It is in the form

$$\gamma_{\text{eff3}} = \frac{\omega}{\omega_{\text{effn3}}}. \tag{16.85}$$

Denote $x_3(t)$ the free response to Eq. (16.9). Since Eq. (16.79) is a standard vibration equation, $x_3(t)$ is given by

$$x_3(t) = e^{-\varsigma_{\text{eff3}}\omega_{\text{effn3}}t}\left(x_0\cos\omega_{\text{effd3}}t + \frac{v_0+\varsigma_{\text{eff3}}\omega_{\text{effn3}}x_0}{\omega_{\text{effd3}}}\sin\omega_{\text{effd3}}t\right),\quad t\ge 0. \tag{16.86}$$

Denote by $h_3(t)$ the impulse response to Eq. (16.9). Following Theorem 16.13 or from Eq. (16.79), we have

$$h_3(t) = e^{-\varsigma_{\text{eff3}}\omega_{\text{effn3}}t}\frac{1}{m_{\text{eff3}}\omega_{\text{effd3}}}\sin\omega_{\text{effd3}}t,\quad t\ge 0. \tag{16.87}$$

Let $H_3(\omega)$ be the restricted frequency transfer function of the fractional vibrator (16.9). Doing the Fourier transform of $h_3(t)$ yields

$$H_3(\omega) = \frac{1}{k\left(1-\gamma_{\text{eff3}}^2+i2\varsigma_{\text{eff3}}\gamma_{\text{eff3}}\right)}. \tag{16.88}$$

16.7.4 Results Regarding Variable-Order Fractional Vibrator (16.10)

Let the equivalent equation of (16.10) be

$$m_{\text{eff4}}\frac{d^2x(t)}{dt^2} + c_{\text{eff4}}\frac{dx(t)}{dt} + k_{\text{eff4}}x(t) = f(t), \tag{16.89}$$

where m_{eff4}, c_{eff4}, and k_{eff4} are the effective mass, damping, and stiffness of the fractional vibrator (16.10), respectively. Taking into account (16.10), we replace c by 0 in Eq. (16.14) and have the equivalent equation of (16.10) expressed by

$$-m\omega^{\alpha(\omega)-2}\cos\frac{\alpha(\omega)\pi}{2}\frac{d^2x(t)}{dt^2}+\left[m\omega^{\alpha(\omega)-1}\sin\frac{\alpha(\omega)\pi}{2}+k\omega^{\lambda(\omega)-1}\sin\frac{\lambda(\omega)\pi}{2}\right]\frac{dx(t)}{dt} \tag{16.90}$$

$$+k\omega^{\lambda(\omega)}\cos\frac{\lambda(\omega)\pi}{2}x(t)=f(t),\quad 1<\alpha(\omega)<3,\ 0\le\lambda(\omega)<1.$$

By comparing Eqs. (16.89)–(16.90), we see that

$$m_{\text{eff4}}=-m\omega^{\alpha(\omega)-2}\cos\frac{\alpha(\omega)\pi}{2}, \tag{16.91}$$

$$c_{\text{eff4}}=m\omega^{\alpha(\omega)-1}\sin\frac{\alpha(\omega)\pi}{2}+k\omega^{\lambda(\omega)-1}\sin\frac{\lambda(\omega)\pi}{2}, \tag{16.92}$$

$$k_{\text{eff4}}=k\omega^{\lambda(\omega)}\cos\frac{\lambda(\omega)\pi}{2}. \tag{16.93}$$

Because Eq. (16.89) is a standard vibration equation in form, we write its free response, denoted by $x_4(t)$, by

$$x_4(t)=e^{-\varsigma_{\text{eff4}}\omega_{\text{effn4}}t}\left(x_0\cos\omega_{\text{effd4}}t+\frac{v_0+\varsigma_{\text{eff4}}\omega_{\text{effn4}}x_0}{\omega_{\text{effd4}}}\sin\omega_{\text{effd4}}t\right),\quad t\ge 0, \tag{16.94}$$

where $\varsigma_{\text{eff4}}=\dfrac{c_{\text{eff4}}}{2\sqrt{m_{\text{eff4}}k_{\text{eff4}}}}$ is the restricted effective damping ratio of Eq. (16.10), $\omega_{\text{effn4}}=\sqrt{\dfrac{k_{\text{eff4}}}{m_{\text{eff4}}}}$ is the restricted effective damping free natural frequency of Eq. (16.10), $\omega_{\text{effd4}}=\omega_{\text{effn4}}\sqrt{1-\varsigma_{\text{eff4}}^2}$ for $|\zeta_{\text{eff4}}|\le 1$ is the restricted effective damped natural frequency of Eq. (16.10). By restricted, we mean that they are in the sense of $m_{\text{eff4}}>0$.

Denote by $h_4(t)$ the impulse response to Eq. (16.10). From Theorem 16.13, we at once write it by

$$h_4(t)=e^{-\varsigma_{\text{eff4}}\omega_{\text{effn4}}t}\frac{1}{m_{\text{eff4}}\omega_{\text{effd4}}}\sin\omega_{\text{effd4}}t,\quad t\ge 0. \tag{16.95}$$

According to Theorem 16.14, the restricted frequency transfer function of the fractional vibrator (16.10), which is denoted by $H_4(\omega)$, is given by

$$H_4(\omega)=\frac{1}{k_{\text{eff4}}\left(1-\gamma_{\text{eff4}}^2+i2\varsigma_{\text{eff4}}\gamma_{\text{eff4}}\right)}, \tag{16.96}$$

where $\gamma_{\text{eff4}}=\dfrac{\omega}{\omega_{\text{effn4}}}$.

16.7.5 Results Regarding Variable-Order Fractional Vibrator (16.11)

The equivalent equation of (16.11) is expressed by

$$m\frac{d^2x(t)}{dt^2}+k\omega^{\lambda(\omega)-1}\sin\frac{\lambda(\omega)\pi}{2}\frac{dx(t)}{dt}+k\omega^{\lambda(\omega)}\cos\frac{\lambda(\omega)\pi}{2}x(t)=f(t),\ \ 0\le\lambda(\omega)<1. \tag{16.97}$$

As a matter of fact, in Eq. (16.14), letting $\alpha(\omega)=2$ and $c=0$ produces the above.

Let c_{eff5} and k_{eff5} be the effective damping and stiffness of the fractional vibrator (16.11), respectively. From the above, we have

$$c_{\mathrm{eff5}}=k\omega^{\lambda(\omega)-1}\sin\frac{\lambda(\omega)\pi}{2}, \tag{16.98}$$

$$k_{\mathrm{eff5}}=k\omega^{\lambda(\omega)}\cos\frac{\lambda(\omega)\pi}{2}. \tag{16.99}$$

Thus, (16.97) can be rewritten by

$$m\frac{d^2x(t)}{dt^2}+c_{\mathrm{eff5}}\frac{dx(t)}{dt}+k_{\mathrm{eff5}}x(t)=f(t). \tag{16.100}$$

The above in form is a standard vibration equation. Let $x_5(t)$ be the free response of Eq. (16.100). Then,

$$x_5(t)=e^{-\varsigma_{\mathrm{eff5}}\omega_{\mathrm{effn5}}t}\left(x_0\cos\omega_{\mathrm{effd5}}t+\frac{v_0+\varsigma_{\mathrm{eff5}}\omega_{\mathrm{effn5}}x_0}{\omega_{\mathrm{effd5}}}\sin\omega_{\mathrm{effd5}}t\right),\quad t\ge0, \tag{16.101}$$

where $\varsigma_{\mathrm{eff5}}=\frac{c_{\mathrm{eff5}}}{2\sqrt{mk_{\mathrm{eff5}}}}$, $\omega_{\mathrm{effn5}}=\sqrt{\frac{k_{\mathrm{eff5}}}{m}}$, and $\omega_{\mathrm{effd5}}=\omega_{\mathrm{effn5}}\sqrt{1-\varsigma_{\mathrm{eff5}}^2}$ for $|\zeta_{\mathrm{eff5}}|\le1$.

Denote by $h_5(t)$ the impulse response to Eq. (16.11). Then,

$$h_5(t)=e^{-\varsigma_{\mathrm{eff5}}\omega_{\mathrm{effn5}}t}\frac{1}{m\omega_{\mathrm{effd5}}}\sin\omega_{\mathrm{effd5}}t,\quad t\ge0. \tag{16.102}$$

Let $H_5(\omega)$ be the frequency transfer function of the fractional vibrator (16.11). Then,

$$H_5(\omega)=\frac{1}{k_{\mathrm{eff5}}\left(1-\gamma_{\mathrm{eff5}}^2+i2\varsigma_{\mathrm{eff5}}\gamma_{\mathrm{eff5}}\right)}, \tag{16.103}$$

where $\gamma_{\mathrm{eff5}}=\frac{\omega}{\omega_{\mathrm{effn5}}}$.

16.8 SUMMARY

We have derived an equivalent vibration equation of a class of variable-order fractional vibrators (16.12), which is expressed by Eq. (16.14). Based on it, we have presented the analytical expressions of its effective mass, damping, and stiffness in Section 16.2. We have brought forward the analytical expressions of its damping ratio, damping free natural frequency, damped natural frequency, and frequency ratio in Section 16.4. In Section 16.5, we have put forward the closed-form expressions of free response, impulse response, and frequency transfer function. The asymptotic properties of its effective mass, damping, and stiffness have been given in Section 16.3. In addition, we have given a mathematical explanation of the Rayleigh damping assumption when $\alpha(\omega)$ and $\lambda(\omega)$ are varying in terms of ω in Section 16.6. Besides, based on Eq. (16.14), we have proposed equivalent equations, effective vibration parameters (mass, damping, stiffness, damping ratio, damping free natural frequency, damped natural frequency, frequency ratio), as well as free responses, impulse responses, and frequency transfer functions of five classes of variable-order fractional vibrators (16.7)–(16.11) in Section 16.7. When $\alpha(\omega)$, $\beta(\omega)$, and $\lambda(\omega)$ are constants, the present theory is applicable to the fractional vibrators expressed by Eqs. (16.1)–(16.6), referring Chapters 8–13 for the details about (16.1)–(16.6).

16.9 EXERCISES

16.1. Show a vibration case with stochastically frequency varying mass.

16.2. Show a vibration case with stochastically frequency-dependent stiffness.

16.3. Show a vibration case with stochastically frequency varying damping.

16.4. Show a vibration case with stochastically frequency varying mass, damping, and stiffness.

16.5. Draw the plots of added mass $-c\omega^{\beta(\omega)-2}\cos\frac{\beta(\omega)\pi}{2}$ that is produced by the multi-fractional friction force $c\frac{d^{\beta(\omega)}y(t)}{dt^{\beta(\omega)}}$.

REFERENCES

1. V. V. Uchaikin, *Fractional Derivatives for Physicists and Engineers*, vol. II, Springer, Berlin, 2013.
2. J.-S. Duan, The periodic solution of fractional oscillation equation with periodic input, *Advances in Mathematical Physics*, 2013, 2013, 6 pp.
3. J.-S. Duan, Z. Wang, Y.-L. Liu, and X. Qiu, Eigenvalue problems for fractional ordinary differential equations, *Chaos, Solitons & Fractals*, 46, 2013, 46–53.
4. A. Al-Rabtah, V. S. Ertürk, and S. Momani, Solutions of a fractional oscillator by using differential transform method, *Computers & Mathematics with Applications*, 59(3), 2010, 1356–1362.
5. M. Zurigat, Solving fractional oscillators using Laplace homotopy analysis method, *Annals of the University of Craiova, Mathematics and Computer Science Series*, 38(4), 2011, 1–11.
6. T. Blaszczyk and M. Ciesielski, Fractional oscillator equation: Transformation into integral equation and numerical solution, *Applied Mathematics and Computation*, 257, 2015, 428–435.
7. T. Blaszczyk, M. Ciesielski, M. Klimek, and J. Leszczynski, Numerical solution of fractional oscillator equation, *Applied Mathematics and Computation*, 218(6), 2011, 2480–2488.
8. A. D. Drozdov, Fractional oscillator driven by a Gaussian noise, *Physica A*, 376, 2007, 237–245.
9. A. A. Stanislavsky, Fractional oscillator, *Physical Review E*, 70(5), 2004, 051103 (6 pp).

10. B. N. N. Achar, J. W. Hanneken, and T. Clarke, Damping characteristics of a fractional oscillator, *Physica A*, 339(3–4), 2004, 311–319.
11. B. N. N. Achar, J. W. Hanneken, and T. Clarke, Response characteristics of a fractional oscillator, *Physica A*, 309(3–4), 2002, 275–288.
12. B. N. N. Achar, J. W. Hanneken, T. Enck, and T. Clarke, Dynamics of the fractional oscillator, *Physica A*, 297(3–4), 2001, 361–367.
13. A. Tofighi, The intrinsic damping of the fractional oscillator, *Physica A*, 329(1–2), 2003, 29–34.
14. Y. E. Ryabov and A. Puzenko, Damped oscillations in view of the fractional oscillator equation, *Physical Review B*, 66(18), 2002, 184201.
15. W. E. Ahmad and A. S. R. Elwakil, Fractional-order Wien-bridge oscillator, *Electronics Letters*, 37(18), 2001, 1110–1112.
16. T. Blaszczyk, A numerical solution of a fractional oscillator equation in a non-resisting medium with natural boundary conditions, *Romanian Reports in Physics*, 67(2), 2015, 350–358.
17. M. S. Tavazoei, Reduction of oscillations via fractional order pre-filtering, *Signal Processing*, 107, 2015, 407–414.
18. T. Sandev and Z. Tomovski, The general time fractional wave equation for a vibrating string, *Journal of Physics A: Mathematical and Theoretical*, 43(5), 2010, 055204 (12 pp).
19. H. Singh, H. M. Srivastava, and D. Kumar, A reliable numerical algorithm for the fractional vibration equation, *Chaos, Solitons & Fractals*, 103, 2017, 131–138.
20. C. H. Eab and S. C. Lim, Path integral representation of fractional harmonic oscillator, *Physica A*, 371(2), 2006, 303–316.
21. M. Li, S. C. Lim, and S. Y. Chen, Exact solution of impulse response to a class of fractional oscillators and its stability, *Mathematical Problems in Engineering*, 2011, 2011, 9 pp.
22. M. Li, S. C. Lim, C. Cattani, and M. Scalia, Characteristic roots of a class of fractional oscillators, *Advances in High Energy Physics*, 2013, 2013, 7 pp.
23. M. Li, Three classes of fractional oscillators, *Symmetry-Basel*, 10(2), 2018, 91 pp.
24. M. Li, *Theory of Fractional Engineering Vibrations*, De Gruyter, Berlin/Boston, MA, 2021.
25. L.-F. Lin, C. Chen, S.-C. Zhong, and H.-Q. Wang, Stochastic resonance in a fractional oscillator with random mass and random frequency, *Journal of Statistical Physics*, 160(2), 2015, 497–511.
26. J.-S. Duan, A modified fractional derivative and its application to fractional vibration equation, *Applied Mathematics & Information Sciences*, 10(5), 2016, 1863–1869.
27. H. S. Alkhaldi, I. M. Abu-Alshaikh, and A. N. Al-Rabadi, Vibration control of fractionally-damped beam subjected to a moving vehicle and attached to fractionally-damped multi-absorbers, *Advances in Mathematical Physics*, 2013, 2013, 12 pp.
28. H. Dai, Z. Zheng, and W. Wang, On generalized fractional vibration equation, *Chaos, Solitons & Fractals*, 95, 2017, 48–51.
29. R. Ren, M. Luo, and K. Deng, Stochastic resonance in a fractional oscillator driven by multiplicative quadratic noise, *Journal of Statistical Mechanics: Theory and Experiment*, 2017, 2017, 023210.
30. Y. Xu, Y. Li, D. Liu, W. Jia, and H. Huang, Responses of Duffing oscillator with fractional damping and random phase, *Nonlinear Dynamics*, 74(3), 2013, 745–753.
31. G. He, Y. Tian, and Y. Wang, Stochastic resonance in a fractional oscillator with random damping strength and random spring stiffness, *Journal of Statistical Mechanics: Theory and Experiment*, 2013, 2013, P09026.
32. A. Y. T. Leung, Z. Guo, and H. X. Yang, Fractional derivative and time delay damper characteristics in Duffing-van der Pol oscillators, *Communications in Nonlinear Science and Numerical Simulation*, 18(10), 2013, 2900–2915.
33. L. C. Chen, Q. Q. Zhuang, and W. Q. Zhu, Response of SDOF nonlinear oscillators with lightly fractional derivative damping under real noise excitations, *The European Physical Journal of Special Topics*, 193(1), 2011, 81–92.

34. J.-F. Deü and D. Matignon, Simulation of fractionally damped mechanical systems by means of a Newmark-diffusive scheme, *Computers & Mathematics with Applications*, 59(5), 2010, 1745–1753.
35. G. E. Drăgănescu, L. Bereteu, A. Ercuța, and G. Luca, Anharmonic vibrations of a nano-sized oscillator with fractional damping, *Communications in Nonlinear Science and Numerical Simulation*, 15(4), 2010, 922–926.
36. Y. A. Rossikhin and M. V. Shitikova, New approach for the analysis of damped vibrations of fractional oscillators, *Shock and Vibration*, 16(4), 2009, 365–387.
37. F. Xie and X. Lin, Asymptotic solution of the van der Pol oscillator with small fractional damping, *Physica Scripta*, T136, 2009, 014033.
38. R. Ren, M. Luo, and K. Deng, Stochastic resonance in a fractional oscillator subjected to multiplicative trichotomous noise, *Nonlinear Dynamics*, 90(1), 2017, 379–390.
39. J. Yuan, Y. Zhang, J. Liu, B. Shi, M. Gai, and S. Yang, Mechanical energy and equivalent differential equations of motion for single-degree-of-freedom fractional oscillators, *Journal of Sound and Vibration*, 397, 2017, 192–203.
40. H. Dai, Z. Zheng, and W. Wang, On generalized fractional vibration equation, *Chaos, Solitons & Fractals*, 95, 2017, 48–51.
41. L.-F. Lin, C. Chen, and H.-Q. Wang, Trichotomous noise induced stochastic resonance in a fractional oscillator with random damping and random frequency, *Journal of Statistical Mechanics: Theory and Experiment*, 2016, 2016, 023201.
42. Y. Naranjani, Y. Sardahi, Y.-Q. Chen, and J.-Q. Sun, Multi-objective optimization of distributed-order fractional damping, *Communications in Nonlinear Science and Numerical Simulation*, 24(1–3), 2015, 159–168.
43. S. C. Lim, M. Li, and L. P. Teo, Langevin equation with two fractional orders, *Physics Letters A*, 372(42), 2008, 6309–6320.
44. A. Di Matteo, P. D. Spanos, and A. Pirrotta, Approximate survival probability determination of hysteretic systems with fractional derivative elements, *Probabilistic Engineering Mechanics*, 54, 2018, 138–146.
45. Ž. Tomovski and T. Sandev, Effects of a fractional friction with power-law memory kernel on string vibrations, *Computers & Mathematics with Applications*, 62(3), 2011, 1554–1561.
46. V. Kumar, M. Malik, and A. Debbouche, Stability and controllability analysis of fractional damped differential system with non-instantaneous impulses, *Applied Mathematics and Computation*, 391, 2021, 125633.
47. Y. Tian, G. He, Z. Liu, L. Zhong, X. Yang, H. E. Stanley, and Z. Tu, The impact of memory effect on resonance behavior in a fractional oscillator with small time delay, *Physica A*, 563, 2021, 125383.
48. Y. Tian, T. Yu, G. He, L. Zhong, and H. E. Stanley, The resonance behavior in the fractional harmonic oscillator with time delay and fluctuating mass, *Physica A*, 545, 2020, 123731.
49. L. He, X. Wu, and G. Zhang, Stochastic resonance in coupled fractional-order linear harmonic oscillators with damping fluctuation, *Physica A*, 545, 2020, 123345.
50. M. V. Varanis, A. M. Tusset, J. M. Balthazar, G. Litak, C. Oliveira, R. T. Rocha, A. Nabarrete, and V. Piccirillo, Dynamics and control of periodic and non-periodic behavior of Duffing vibrating system with fractional damping and excited by a non-ideal motor, *Journal of the Franklin Institute*, 357(4), 2020, 2067–2082.
51. D. Pang, W. Jiang, S. Liu, and D. Jun, Stability analysis for a single degree of freedom fractional oscillator, *Physica A*, 523, 2019, 498–506.
52. T. Yu, L. Zhang, Y. Ji, and L. Lai, Stochastic resonance of two coupled fractional harmonic oscillators Matteo with fluctuating mass, *Communications in Nonlinear Science and Numerical Simulation*, 72, 2019, 26–38.
53. Y. He, Y. Fu, Z. Qiao, and Y. Kang, Chaotic resonance in a fractional-order oscillator system with application to mechanical fault diagnosis, *Chaos, Solitons & Fractals*, 142, 2021, 110536.

54. P. D. Spanos and G. Malara, Nonlinear vibrations of beams and plates with fractional derivative elements subject to combined harmonic and random excitations, *Probabilistic Engineering Mechanics*, 59, 2020, 103043.
55. P. D. Spanos and G. Malara, Nonlinear random vibrations of beams with fractional derivative elements, *Journal of Engineering Mechanics*, 140(9), 2014, 04014069.
56. A. K. Golmankhaneh, On the fractal Langevin equation, *Fractal and Fractional*, 3(1), 2019, 11.
57. J.-S. Duan, C.-P. Cheng, and L. Chen, A comparison study of steady-state vibrations with single fractional-order and distributed-order derivatives, *Open Physics*, 15(1), 2017, 809–818.
58. J. F. Gomez-Aguilar, J. J. Rosales-Garcia, J. J. Bernal-Alvarado, T. Cordova-Fraga, and R. Guzman-Cabrera, Fractional mechanical oscillators, *Revista Mexicana de Fisica*, 58(4), 2012, 348–352.
59. Y. Tian, L.-F. Zhong, G.-T. He, T. Yu, M.-K. Luo, and H. E. Stanley, The resonant behavior in the oscillator with double fractional-order damping under the action of nonlinear multiplicative noise, *Physica A*, 490, 2018, 845–856.
60. M. Berman and L. S. Cederbaum, Fractional driven-damped oscillator and its general closed form exact solution, *Physica A*, 505, 2018, 744–762.
61. A. Coronel-Escamilla, J. F. Gómez-Aguilar, D. Baleanu, T. Córdova-Fraga, R. F. Escobar-Jiménez, V. H. Olivares-Peregrino, and M. M. Al Qurashi, Bateman-Feshbach Tikochinsky and Caldirola-Kanai oscillators with new fractional differentiation, *Entropy*, 19(2), 2017, 55 pp.
62. N. Sene and J. F. G. Aguilar, Fractional mass-spring-damper system described by generalized fractional order derivatives, *Fractal and Fractional*, 3(3), 2019, 39.
63. P. B. Vishwamittar, and R. Chopra, Stochastic resonance in two coupled fractional oscillators with potential and coupling parameters subjected to quadratic asymmetric dichotomous noise, *Physica A*, 561, 2021, 125148.
64. G. M. Ismail, H. R. Abdl-Rahim, A. Abdel-Aty, R. Kharabsheh, W. Alharbi, and M. Abdel-Aty, An analytical solution for fractional oscillator in a resisting medium, *Chaos, Solitons & Fractals*, 130, 2020, 109412.
65. S. C. Lim, C. H. Eab, K.-H. Mak, M. Li, and S. Y. Chen, Solving linear coupled fractional differential equations and their applications, *Mathematical Problems in Engineering*, 2012, Article ID 653939, 2012.
66. C. Cattani, H. M. Srivastava, and X.-J. Yang, *Fractional Dynamics*, De Gruyter, Berlin/Boston, MA, 2015.
67. M. D. Ortigueira and D. Valério, *Fractional Signals and Systems*, De Gruyter, Berlin/Boston, MA, 2020.
68. J. Klafter, S. C. Lim, and R. Metzler, *Fractional Dynamics: Recent Advances*, World Scientific, Singapore, 2012.
69. V. V. Uchaikin, *Fractional Derivatives for Physicists and Engineers*, vol. I, Springer, Berlin, 2013.
70. T. M. Atanackovic, S. Pilipovic, B. Stankovic, and D. Zorica, *Fractional Calculus with Applications in Mechanics*, John Wiley & Sons, Croydon, 2014.
71. C. P. Li and M. Cai, *Theory and Numerical Approximations of Fractional Integrals and Derivatives*, SIAM, Philadelphia, 2019.
72. M. Li, Generalized fractional Gaussian noise and its application to traffic modeling, *Physica A*, 579, 2021, 1236137.
73. M. Li, Multi-fractional generalized Cauchy process and its application to teletraffic, *Physica A*, 550, 2020, 123982.
74. C. Cattani and R. Spigler, Fractional dynamics, *Fractal and Fractional*, 2(2), 2018, 19.
75. M. Li, Integral representation of fractional derivative of delta function, *Fractal and Fractional*, 4(3), 2020, 47.
76. M. H. Heydari, A computational method for a class of systems of nonlinear variable-order fractional quadratic integral equations, *Applied Numerical Mathematics*, 153, 2020, 164–178.

77. M. H. Heydari, Numerical solution of nonlinear 2D optimal control problems generated by Atangana-Riemann-Liouville fractal-fractional derivative, *Applied Numerical Mathematics*, 150, 2020, 507–518.
78. M. H. Heydari, M. R. Hooshmandasl, C. Cattani, F. M. Maalek Ghaini, An efficient computational method for solving nonlinear stochastic Itô integral equations: Application for stochastic problems in physics, *Journal of Computational Physics*, 283, 2015, 148–168.
79. M. H. Heydari, M. R. Hooshmandasl, A. Shakiba, and C. Cattani, Legendre wavelets Galerkin method for solving nonlinear stochastic integral equations, *Nonlinear Dynamics*, 86(2), 2016, 1185–1202.
80. H. Weyl, Bemerkungen zum Begriff des Differentialquotienten gebrochener Ordnung, *Vierteljahrsschrift der Naturforschenden Gesellschaft in Zürich*, 62, 1917, 296–302.
81. R. K. Raina and C. L. Koul, On Weyl fractional calculus, *Proceedings of the American Mathematical Society*, 73(2), 1979, 188–192.
82. K. S. Miller and B. Ross, *An Introduction to the Fractional Calculus and Fractional Differential Equations*, John Wiley & Sons, New York, 1993.
83. J. L. Lavoie, T. J. Osler, and R. Tremblay, Fractional derivatives and special functions, *SIAM Review*, 18(2), 1976, 240–268.
84. A. Papoulis, *The Fourier Integral and Its Applications*, McGraw-Hill Inc., New York, 1962.
85. I. M. Gelfand and K. Vilenkin, *Generalized Functions*, vol. 1, Academic Press, New York, 1964.
86. R. N. Bracewell, *The Fourier Transform and Its Applications*, 2nd Ed., McGraw-Hill, New York, 1978.
87. C. M. Harris, *Shock and Vibration Handbook*, 5th Ed., McGraw-Hill, New York, 2002.
88. J. P. Den Hartog, *Mechanical Vibrations*, McGraw-Hill, New York, 1956.
89. S. P. Timoshenko, *Vibration Problems in Engineering*, 2nd Ed., Fifth Printing, D. Van Nostrand Company, Inc., New York, 1955.
90. O. M. Palley, G. B. Bahizov, and E. Я. Voroneysk, *Handbook of Ship Structural Mechanics*, National Defense Industry Publishing House, Beijing, China, 2002. In Chinese. Translated from Russian by B. H. Xu, X. Xu, and M. Q. Xu.
91. K. Nakagawa and M. Ringo, *Engineering Vibrations*, Shanghai Science and Technology Publishing House, Shanghai, China, 1981. In Chinese. Translated by S. R. Xia from Japanese.
92. J. W. Strutt and 3rd B. Rayleigh, *The Theory of Sound*, vol. 1, Macmillan & Co., Ltd., London, 1877.
93. T. Trombetti and S. Silvestri, On the modal damping ratios of shear-type structures equipped with Rayleigh damping systems, *Journal of Sound and Vibration*, 292(1–2), 2006, 21–58.
94. M. K. Poul and A. Zerva, Efficient time-domain deconvolution of seismic ground motions using the equivalent-linear method for soil-structure interaction analyses, *Soil Dynamics and Earthquake Engineering*, 112, 2018, 138–151.
95. P. Nåvik, A. Rønnquist, and S. Stichel, Identification of system damping in railway catenary wire systems from full-scale measurements, *Engineering Structures*, 113, 2016, 71–78.
96. S.-K. Park, H. W. Park, S. Shin, and H. S. Lee, Detection of abrupt structural damage induced by an earthquake using a moving time window technique, *Computers & Structures*, 86(11–12), 2008, 1253–1265.
97. M. I. Hussein, S. Biringen, O. R. Bilal, and A. Kucala, Flow stabilization by subsurface phonons, *Proceedings of the Royal Society A: Mathematical, Physical and Engineering Sciences*, 471(2177), 2015, 1–19.
98. O. Sigmund and J. S. Jensen, Systematic design of phononic band-gap materials and structures by topology optimization, *Philosophical Transactions of the Royal Society A: Mathematical, Physical and Engineering Sciences*, 361(1806), 2003, 1001–1019.
99. F. Chen, H. Tan, J. Chen, and Z. Jiao, Influences of earthquake characteristics on seismic performance of anchored sheet pile quay with barrette piles, *Journal of Coastal Research, Special Issue No. 85: Proceedings of 15th International Coastal Symposium*, Haeundae, Busan, 13–18 May 2018, 701–705.

100. D. S. Battisti, A. C. Hirst, and E. S. Sarachik, Instability and predictability in coupled atmosphere-ocean models, *Philosophical Transactions of the Royal Society of London. Series A, Mathematical and Physical Sciences*, 329(1604), 1989, 237–247.
101. D. Cox, J. Kämpf, and M. Fernandes, Dispersion and connectivity of land-based discharges near the mouth of a coastal inlet, *Journal of Coastal Research*, 29(6A), 2013, 100–109.
102. F. Tisseur and K. Meerbergen, The quadratic eigenvalue problem, *SIAM Review*, 43(2), 2001, 235–286.
103. E. K.-W. Chu, T.-M. Huang, W.-W. Lin, and C.-T. Wu, Palindromic eigenvalue problems: A brief survey, *Taiwanese Journal of Mathematics*, 14(3A), 2010, 743–779.
104. J. P. Fay, S. Puria, and C. R. Steele, The discordant eardrum, *Proceedings of the National Academy of Sciences of the United States of America*, 103(52), 2006, 19743–19748.
105. H. Naderian, M. M. S. Cheung, M. Mohammadian, and E. Dragomirescu, Integrated finite strip flutter analysis of bridges, *Computers & Structures*, vol. 212, 2019, 145–161.
106. Z. Tian, L. Huo, W. Gao, H. Li, and G. Song, Modeling of the attenuation of stress waves in concrete based on the Rayleigh damping model using time-reversal and PZT transducers, *Smart Materials and Structures*, 26(10), 2017, 105030.
107. G. Iovane and A. V. Nasedkin, Finite element dynamic analysis of anisotropic elastic solids with voids, *Computers & Structures*, 87(15–16), 2009, 981–989.
108. L. A. S. Kouris, A. Penna, and G. Magenes, Seismic damage diagnosis of a masonry building using short-term damping measurements, *Journal of Sound and Vibration*, 394, 2017, 366–391.
109. X. D. Jin and L. J. Xia, *Ship Hull Vibration*, Shanghai Jiaotong University Press, Shanghai, China, 2011. In Chinese.

III

Fractional Euler-Bernoulli Beams

CHAPTER 17

Free Response to Longitudinal Vibrations of Uniform Circular Beam with Fractional Coordinates

THE CONTRIBUTIONS IN THIS chapter are in three folds. First, we propose an equivalent motion equation of the beam with the fractional coordinates of order α $(1<\alpha<3)$ regarding longitudinal vibrations. Second, we present the analytic representation, in closed form, of the general solution with respect to the free response to the longitudinal vibration of a beam with the fractional coordinates. Such a beam appears damping free in form but it is damped in nature, relying on the value of α. Third, we show that the beam with $2<\alpha<3$ may suffer from self-vibrations.

17.1 BACKGROUND

Let $u(x, t)$ be the displacement at position x and time t on a uniform circular beam (beam for short) with l length. Denote the material density of the beam by ρ. Let E be its elastic modulus. Consider longitudinal vibrations of a beam under the excitation of axial force. Regarding free response to longitudinal vibrations of a beam with the fractional coordinates of order α $(1<\alpha<3)$, the motion equation is expressed by

$$\frac{\partial^{\alpha} u(x,t)}{\partial t^{\alpha}} = a^2 \frac{\partial^2 u(x,t)}{\partial x^2}, \quad 1<\alpha<3, \tag{17.1}$$

where $a = \sqrt{\dfrac{E}{\rho}}$ is a propagation speed of stress wave in the beam.

DOI: 10.1201/9781003460947-20

By variable separation method, one has

$$u(x,t) = \varphi(x)p(t), \tag{17.2}$$

where $\varphi(x)$ is a vibration mode function while $p(t)$ is a coordinate function. Substituting the above into Eq. (17.1) results in

$$\frac{\varphi''(x)}{\varphi(x)} = \frac{p^{(\alpha)}(t)}{a^2 p(t)} = -k, \tag{17.3}$$

where k is a positive constant. From the above, we have a fractional differential equation for $p(t)$ in the form

$$p^{(\alpha)}(t) + ka^2 p(t) = 0, \tag{17.4}$$

and an ordinary differential equation for $\varphi(x)$ in the form

$$\varphi''(x) + k\varphi(x) = 0. \tag{17.5}$$

The quantity k is given by

$$k = \frac{j^2\pi^2}{l^2}, \quad j = 1, 2, \ldots \tag{17.6}$$

Denote by

$$\mu_j = \frac{j\pi}{l}. \tag{17.7}$$

Then, we have the fractional differential equation for the jth coordinate $p_j(t)$ in the form

$$p_j^{(\alpha)}(t) + \mu_j^2 a^2 p_j(t) = 0, \tag{17.8}$$

and the ordinary differential equation for the jth vibration mode $\varphi_j(x)$ in the form

$$\varphi_j''(x) + \mu_j^2 \varphi_j(x) = 0. \tag{17.9}$$

The abstract form of the general solution of total free response is given by

$$u(x,t) = \sum_{j=1}^{\infty} u_j(x,t) = \sum_{j=1}^{\infty} \varphi_j(x) p_j(t). \tag{17.10}$$

The literature for studying solution $p_j(t)$ of Eq. (17.8) is rich, see e.g., [1–31]. However, the closed-form solution of Eq. (17.8) is rarely seen. Consequently, the closed-form representation of $u(x, t)$ is seldom reported. In this chapter, we present an equivalent equation of (17.1) and the closed-form representation of the free response to Eq. (17.1). In addition, we shall reveal the self-vibrating phenomenon of such a beam.

The rest of the chapter is organized as follows. In Section 17.2, we present an equivalent equation of (17.1). We propose the closed-form expression of the free response to Eq. (17.1) in Section 17.3. Case study is given in Section 17.4. Self-vibrations are revealed in Section 17.5, which is followed by a summary.

17.2 EQUIVALENT EQUATION OF (17.1)

17.2.1 Preliminaries

Let $X(\omega)$ be the Fourier transform of $x(t)$. Using the Weyl fractional derivative, the Fourier transform of $x^{(v)}(t)$ for $v \geq 0$ is given by

$$\mathrm{F}\left[x^{(v)}(t)\right] = (i\omega)^v X(\omega), \tag{17.11}$$

where F the Fourier transform operator and $i = \sqrt{-1}$, see Uchaikin [1], Li [20], Miller and Ross [32], Lavoie et al. [33], Li and Cai [34].

Let $\mathrm{F}[f_1(t)] = F_1(\omega)$ and $\mathrm{F}[f_2(t)] = F_2(\omega)$. When $\mathrm{F}[f_1(t) - f_2(t)] = 0$, one says that $f_1(t) - f_2(t)$ is a null function. If $f_1(t) - f_2(t)$ is a null function, $f_1(t) = f_2(t)$ in the sense of $F_1(\omega) = F_2(\omega)$.

17.2.2 Equivalent Equation

We present the following theorem to describe an equivalent equation of (17.1).

Theorem 17.1. Equation (17.1) is equivalently expressed by

$$-\omega^{\alpha-2}\cos\frac{\alpha\pi}{2}\frac{\partial^2 u(x,t)}{\partial t^2} + \omega^{\alpha-1}\sin\frac{\alpha\pi}{2}\frac{\partial u(x,t)}{\partial t} = a^2\frac{\partial^2 u(x,t)}{\partial x^2}, \quad 1 < \alpha < 3. \tag{17.12}$$

Proof. Let $P(\omega)$ be the Fourier transform of $p(t)$. Let

$$f_1(t) = p^{(\alpha)}(t). \tag{17.13}$$

Doing the Fourier transform of $f_1(t)$ produces

$$\mathrm{F}\left[f_1(t)\right] = (i\omega)^\alpha P(\omega). \tag{17.14}$$

Substituting $u(x, t) = \varphi(x)p(t)$ into Eq. (17.12) yields

$$-\omega^{\alpha-2}\cos\frac{\alpha\pi}{2}\frac{d^2 p(t)}{dt^2} + \omega^{\alpha-1}\sin\frac{\alpha\pi}{2}\frac{dp(t)}{dt} = \frac{a^2}{\varphi(x)}\frac{\partial^2 u(x,t)}{\partial x^2}. \tag{17.15}$$

Denote by

$$f_2(t) = -\omega^{\alpha-2}\cos\frac{\alpha\pi}{2}\frac{d^2 p(t)}{dt^2} + \omega^{\alpha-1}\sin\frac{\alpha\pi}{2}\frac{dp(t)}{dt}. \tag{17.16}$$

Then,

$$\mathrm{F}\left[f_2(t)\right]=\omega^{\alpha}\cos\frac{\alpha\pi}{2}P(\omega)+i\omega^{\alpha}\sin\frac{\alpha\pi}{2}P(\omega). \tag{17.17}$$

With the principal value of i^{α}, we rewrite Eq. (17.17) by

$$\mathrm{F}\left[f_2(t)\right]=(i\omega)^{\alpha}P(\omega). \tag{17.18}$$

From Equations (17.14) and (17.18), we see that $\mathrm{F}[f_1(t)]=\mathrm{F}[f_2(t)]$. Thus, $f_1(t)=f_2(t)$ in the sense of $\mathrm{F}[f_1(t)-f_2(t)]=0$. Multiplying both sides of Eq. (17.15) by $\varphi(x)$ results in Eq. (17.12). Since Eq. (17.1) implies

$$p^{(\alpha)}(t)\varphi(x)=a^2p(t)\varphi''(x), \quad 1<\alpha<3. \tag{17.19}$$

Considering $f_1(t)=f_2(t)$ in the above, we see that Theorem 17.1 holds. This finishes the proof. □

Theorem 17.2. An equivalent equation of (17.4) is given by

$$-\omega^{\alpha-2}\cos\frac{\alpha\pi}{2}\frac{d^2p(t)}{dt^2}+\omega^{\alpha-1}\sin\frac{\alpha\pi}{2}\frac{dp(t)}{dt}+ka^2p(t)=0. \tag{17.20}$$

Proof. According to $f_1(t)=f_2(t)$, we have

$$\frac{d^{\alpha}p(t)}{dt^{\alpha}}=-\omega^{\alpha-2}\cos\frac{\alpha\pi}{2}\frac{d^2p(t)}{dt^2}+\omega^{\alpha-1}\sin\frac{\alpha\pi}{2}\frac{dp(t)}{dt}. \tag{17.21}$$

Thus, Theorem 17.2 holds. The proof is completed. □

Corollary 17.1. An equivalent equation of (17.8) is in the form

$$-\omega^{\alpha-2}\cos\frac{\alpha\pi}{2}\frac{d^2p_j(t)}{dt^2}+\omega^{\alpha-1}\sin\frac{\alpha\pi}{2}\frac{dp_j(t)}{dt}+\mu_j^2a^2p_j(t)=0. \tag{17.22}$$

Proof. According to Eq. (17.21), we write Eq. (17.8) by the above. The proof is finished. □

17.3 EXPRESSION OF FREE RESPONSE

Theorem 17.3. The jth coordinate function $p_j(t)$ is expressed by

$$p_j(t)=e^{-\zeta_{ej}\omega_{enj}t}\left(A_j\cos\omega_{endj}t+B_j\sin\omega_{endj}t\right),\quad t\geq 0, \tag{17.23}$$

where A_j and B_j are determined by initial conditions. The quantities ω_{enj}, ζ_{ej}, and ω_{endj} are given in Eqs. (17.27), (17.28), and (17.32), respectively.

Proof. Denote the natural angular frequency of the jth order vibration mode by

$$\omega_{nj}=a\mu_j=\frac{j\pi}{l}\sqrt{\frac{E}{\rho}}. \tag{17.24}$$

With it, we write Eq. (17.8) by

$$p_j^{(\alpha)}(t)+\omega_{nj}^2a^2p_j(t)=0. \tag{17.25}$$

Taking into account Corollary 17.1, the above is expressed by

$$\frac{d^2p_j(t)}{dt^2}-\omega\tan\frac{\alpha\pi}{2}\frac{dp_j(t)}{dt}-\omega^{2-\alpha}\sec\frac{\alpha\pi}{2}\mu_j^2a^2p_j(t)=0. \tag{17.26}$$

Let ω_{enj} the effective angular natural frequency of the jth order vibration mode. It is given by

$$\omega_{enj}=\sqrt{-\omega^{2-\alpha}\sec\frac{\alpha\pi}{2}}\mu_ja. \tag{17.27}$$

Denote by ζ_{ej} the effective damping ratio of the jth coordinate. Define it by

$$\varsigma_{ej}=-\frac{\omega\tan\frac{\alpha\pi}{2}}{2\omega_{enj}}. \tag{17.28}$$

Then, Eq. (17.26) can be rewritten in the form

$$\frac{d^2p_j(t)}{dt^2}+2\omega_{enj}\varsigma_{ej}\frac{dp_j(t)}{dt}+\omega_{enj}^2p_j(t)=0. \tag{17.29}$$

Let $s_{1,2}$ be the characteristic roots of Eq. (17.28). Then,

$$s_{1,2}=-\varsigma_{ej}\omega_{enj}\pm\omega_{enj}\sqrt{\varsigma_{ej}^2-1}. \tag{17.30}$$

In engineering, people are interested in small damping (Li [20], Harris [35], Palley et al. [36], Nakagawa and Ringo [37]). Thus, we consider $|\zeta_{ej}| < 1$ in what follows. In that case, we have

$$s_{1,2} = -\varsigma_{ej}\omega_{enj} \pm i\omega_{enj}\sqrt{1-\varsigma_{ej}^2}. \tag{17.31}$$

Denote by

$$\omega_{endj} = \omega_{enj}\sqrt{1-\varsigma_{ej}^2}. \tag{17.32}$$

We call ω_{endj} the effective damped natural angular frequency of the jth order vibration mode. Then,

$$s_{1,2} = -\varsigma_{ej}\omega_{enj} \pm i\omega_{endj}. \tag{17.33}$$

Therefore, we obtain the general solution $p_j(t)$ to Eq. (17.29) in the form

$$p_j(t) = e^{-\zeta_{ej}\omega_{enj}t}\left(A_j \cos\omega_{endj}t + B_j \sin\omega_{endj}t\right), \quad t \geq 0,$$

where A_j and B_j are determined by initial conditions. This completes the proof. □

Note that ω_{enj} is a function in terms of ω, α, and j. When $\alpha=2$, it reduces to the conventional natural angular frequency of the jth order vibration mode in the form $\omega_{nj}=\mu_j a$. On the other side, ζ_{ej} reduces to 0 if $\alpha=2$.

Theorem 17.4. The general solution of free response to a beam with the fractional coordinates is given by

$$u(x,t) = \sum_{j=1}^{\infty} \varphi_j(x) e^{-\zeta_{ej}\omega_{enj}t}\left(A_j \cos\omega_{endj}t + B_j \sin\omega_{endj}t\right), \quad t \geq 0, \tag{17.34}$$

where $\varphi_j(x)$ is the jth vibration mode function.

Proof. According to Theorem 17.3, we have

$$u_j(x,t) = \varphi_j(x) p_j(t) = \varphi_j(x) e^{-\zeta_{ej}\omega_{enj}t}\left(A_j \cos\omega_{endj}t + B_j \sin\omega_{endj}t\right). \tag{17.35}$$

As $u(x,t) = \sum_{j=1}^{\infty} u_j(x,t)$, we have Eq. (17.34). The proof is finished. □

17.4 CASE STUDY

We now consider the free response to a fixed-fixed beam in this section.

Theorem 17.5: Free Response to a Fixed-Fixed Beam. For a fixed-fixed beam with the fractional coordinates below

$$\begin{cases} \dfrac{\partial^\alpha u}{\partial t^\alpha} = a^2 \dfrac{\partial^2 u}{\partial x^2}, \\ u(0,t)=0, u(l,t)=0, \\ u(x,0)=f(x), \left.\dfrac{\partial u}{\partial t}\right|_{t=0} = g(x), \end{cases} \tag{17.36}$$

its solution of free response is given by

$$u(x,t)=\sum_{j=1}^{\infty} \sin\frac{j\pi x}{l} e^{-\zeta_{ej}\omega_{enj}t} \left[\begin{array}{l} \left(\dfrac{2}{l}\displaystyle\int_0^l f(x)\sin\frac{j\pi x}{l}dx \right)\cos\omega_{endj}t \\ +\left(\begin{array}{l} \dfrac{2}{l\omega_{endj}}\displaystyle\int_0^l g(x)\sin\frac{j\pi x}{l}dx \\ +\zeta_{ej}\dfrac{\omega_{enj}}{\omega_{endj}}\dfrac{2}{l}\displaystyle\int_0^l f(x)\sin\frac{j\pi x}{l}dx \end{array} \right)\sin\omega_{endj}t \end{array} \right], \quad t \geq 0. \tag{17.37}$$

Proof. From Eq. (17.36), one has

$$\begin{cases} \varphi''(x)+k\varphi(x)=0, \\ \varphi(0)=\varphi(l)=0. \end{cases} \tag{17.38}$$

Since $k=\dfrac{j^2\pi^2}{l^2}$ $(j=1, 2, \ldots)$, one has

$$\begin{cases} \varphi_j''(x)+\dfrac{j^2\pi^2}{l^2}\varphi_j(x)=0, \\ \varphi_j(0)=\varphi_j(l)=0. \end{cases} \tag{17.39}$$

Thus, $\varphi_j(x) \sim \sin\dfrac{j\pi x}{l}$. Therefore, we have the jth order free response in the form

$$\begin{aligned} u_j(x,t) &= \varphi_j(x)p_j(t) \\ &= \varphi_j(x)e^{-\zeta_{ej}\omega_{enj}t}\left(A_j\cos\omega_{endj}t + B_j\sin\omega_{endj}t\right), \quad t \geq 0. \end{aligned} \tag{17.40}$$

The general solution of total free response is given by

$$u(x,t)=\sum_{j=1}^{\infty}u_j(x,t)=\sum_{j=1}^{\infty}\varphi_j(x)e^{-\zeta_{ej}\omega_{enj}t}\left(A_j\cos\omega_{endj}t+B_j\sin\omega_{endj}t\right),\quad t\geq 0. \tag{17.41}$$

As the initial conditions are $u(x, 0)=f(x)$ and $u_t(x, 0)=g(x)$, from the above, we have

$$\begin{cases} f(x)=\sum_{j=1}^{\infty}A_j\varphi_j(x), \\ g(x)=\sum_{j=1}^{\infty}\varphi_j(x)\left(-\varsigma_{ej}\omega_{enj}A_j+B_j\omega_{endj}\right), \quad 0<x<l. \end{cases} \tag{17.42}$$

Considering the orthogonality of trigonometric functions, we have

$$A_j=\frac{2}{l}\int_0^l f(x)\varphi_j(x)dx=\frac{2}{l}\int_0^l f(x)\sin\frac{j\pi x}{l}dx, \tag{17.43}$$

and

$$\begin{aligned} B_j&=\frac{2}{l\omega_{endj}}\int_0^l g(x)\varphi_j(x)dx+\varsigma_{ej}\frac{\omega_{enj}}{\omega_{endj}}\frac{2}{l}\int_0^l f(x)\varphi_j(x)dx \\ &=\frac{2}{l\omega_{endj}}\int_0^l g(x)\sin\frac{j\pi x}{l}dx+\varsigma_{ej}\frac{\omega_{enj}}{\omega_{endj}}\frac{2}{l}\int_0^l f(x)\sin\frac{j\pi x}{l}dx. \end{aligned} \tag{17.44}$$

Thus, the general solution of total free response is given by

$$\begin{aligned} u(x,t)&=\sum_{j=1}^{\infty}u_j(x,t) \\ &=\sum_{j=1}^{\infty}\sin\frac{j\pi x}{l}e^{-\zeta_{ej}\omega_{enj}t}\left[\begin{array}{l}\left(\frac{2}{l}\int_0^l f(x)\sin\frac{j\pi x}{l}dx\right)\cos\omega_{endj}t \\ +\left(\begin{array}{l}\frac{2}{l\omega_{endj}}\int_0^l g(x)\sin\frac{j\pi x}{l}dx \\ +\varsigma_{ej}\frac{\omega_{enj}}{\omega_{endj}}\frac{2}{l}\int_0^l f(x)\sin\frac{j\pi x}{l}dx\end{array}\right)\sin\omega_{endj}t\end{array}\right],\quad t\geq 0. \end{aligned}$$

The above is Eq. (17.37). The proof is finished. □

With Theorem 17.5, we can find the solution $u(x, t)$ in the equation below

$$\begin{cases} \dfrac{\partial^2 u}{\partial t^2}=a^2\dfrac{\partial^2 u}{\partial x^2}, \\ u(0,t)=0, u(l,t)=0, \\ u(x,0)=f(x)=x(l-x), \left.\dfrac{\partial u}{\partial t}\right|_{t=0}=g(x)=0. \end{cases} \tag{17.45}$$

In fact,

$$A_j=\frac{2}{l}\int_0^l f(x)\sin\frac{j\pi x}{l}dx=\frac{l^3}{j\pi}(-1)^{j+1}+\left(\frac{l}{j\pi}\right)^2\left[(-1)^j-1\right]. \tag{17.46}$$

Besides,

$$B_j=\frac{\varsigma_{ej}\omega_{enj}}{\omega_{endj}}\frac{l^3}{j\pi}(-1)^{j+1}+\left(\frac{l}{j\pi}\right)^2\left[(-1)^j-1\right]. \tag{17.47}$$

Thus, we have

$$p_j(t)=e^{-\varsigma_{ej}\omega_{enj}t}\left\{\frac{l^3}{j\pi}(-1)^{j+1}+\left(\frac{l}{j\pi}\right)^2\left[(-1)^j-1\right]\right\}\begin{pmatrix}\cos\omega_{endj}t \\ +\dfrac{\varsigma_{ej}\omega_{enj}}{\omega_{endj}}\sin\omega_{endj}t\end{pmatrix}, \quad t\geq 0. \tag{17.48}$$

Finally, we have

$$\begin{aligned} u(x,t)&=\sum_{j=1}^{\infty}u_j(x,t) \\ &=\sum_{j=1}^{\infty}\sin\frac{j\pi x}{l}e^{-\varsigma_{ej}\omega_{enj}t}\left\{\frac{l^3}{j\pi}(-1)^{j+1}+\left(\frac{l}{j\pi}\right)^2\left[(-1)^j-1\right]\right\}\begin{pmatrix}\cos\omega_{endj}t \\ +\dfrac{\varsigma_{ej}\omega_{enj}}{\omega_{endj}}\sin\omega_{endj}t\end{pmatrix}, \quad t\geq 0. \end{aligned} \tag{17.49}$$

Figure 17.1 shows the plots of $p_j(t)$ for different values of α and j.

An interesting phenomenon of the beam with the fractional coordinates is that the beam is damping free in form but it is in nature damped as indicated by Figure 17.1. The function $p_j(t)$ reduces to the conventional one when $\alpha=2$ as shown in Figure 17.2.

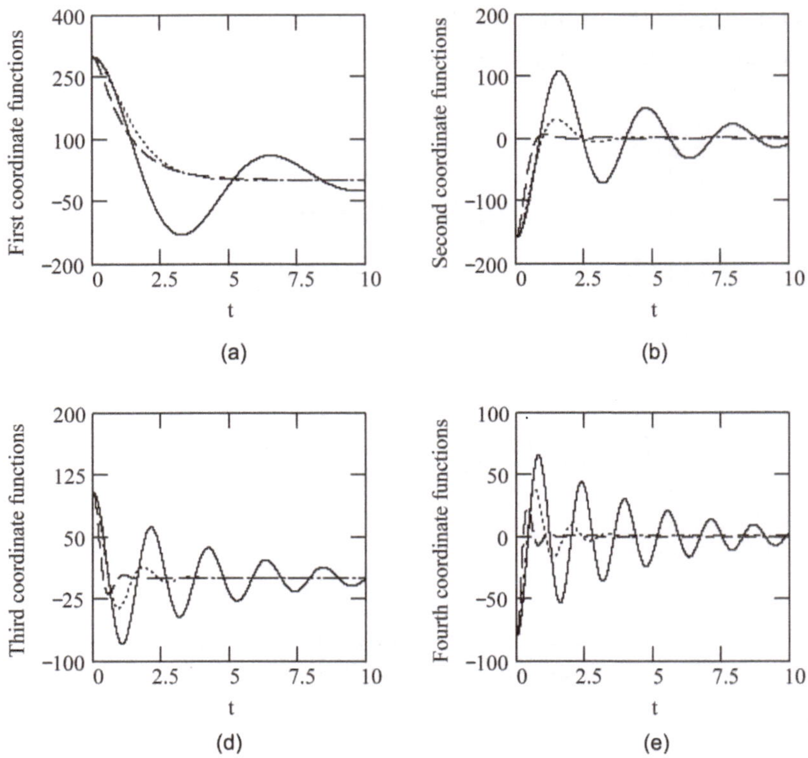

FIGURE 17.1 Plots of $p_j(t)$ of the fixed-fixed beam (17.45) with $\omega=\pi$ for $\alpha=1.9$ (solid), 1.6 (dot), 1.3 (dash). (a) $p_1(t)$. (b) $p_2(t)$. (c) $p_3(t)$. (d) $p_4(t)$.

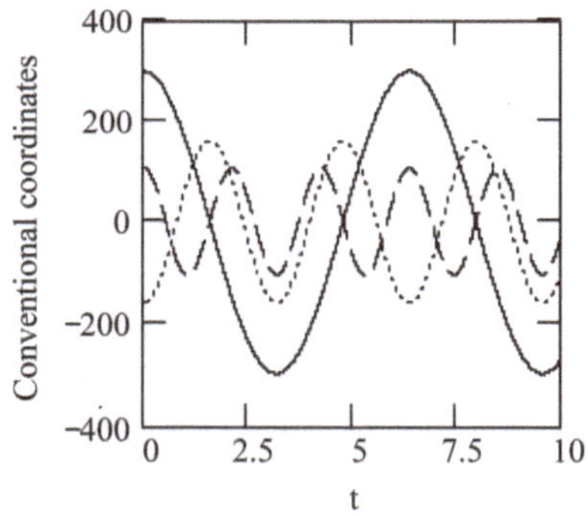

FIGURE 17.2 Plots of $p_j(t)$ of the fixed-fixed beam (17.45) with $\omega=\pi$ when $\alpha=2$ for $j=1$ (solid), 2 (dot), 3 (dash).

Now, we demonstrate the illustrations of the jth response $u_j(x, t)$. For facilitating the demonstrations, we denote $u_j(x, t)$ by uj(x, t). Figures 17.3–17.5 show the plots of u1(x, t), u3(x, t), and u5(x, t) of the fixed-fixed beam (17.45) for $\alpha=1.2$, 1.5, 1.8, and 2. Figure 17.6 shows the plots of total response $u(x,t)\approx\sum_{j=1}^{8}u_j(x,t)=u$ of the fixed-fixed beam (17.45).

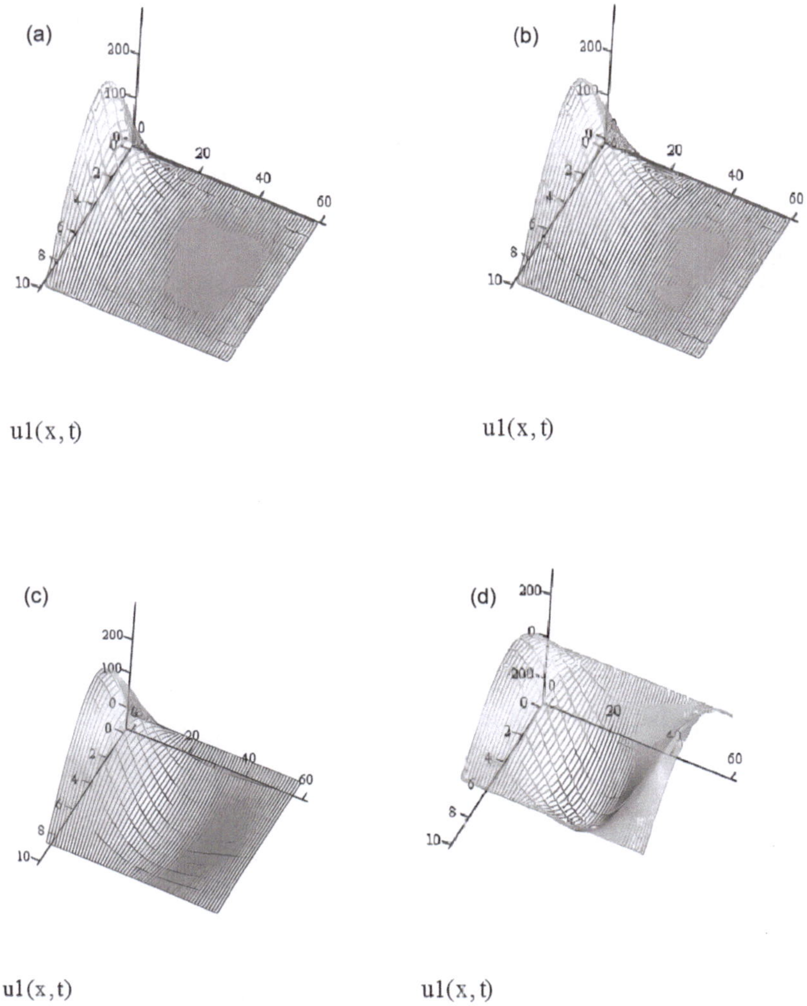

FIGURE 17.3 Plots of the first response ul(x, t) of the fixed-fixed beam (17.45) with $\omega=\pi$, $x=0, \ldots, 10$, and $t=0, \ldots, 60$. (a) ul(x, t) for $\alpha=1.2$. (b) ul(x, t) for $\alpha=1.5$. (c) ul(x, t) for $\alpha=1.8$. (d) ul(x, t) for $\alpha=2$.

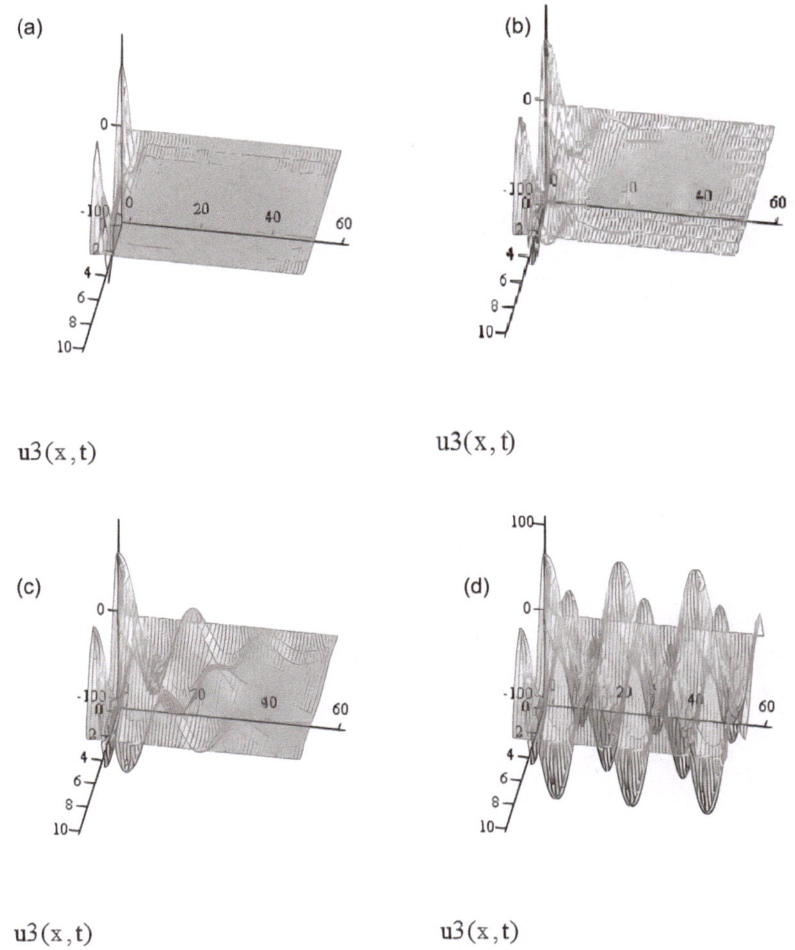

FIGURE 17.4 Plots of the third response u3(x, t) of the fixed-fixed beam (17.45) with $\omega=\pi$, $x=0, \ldots, 10$, and $t=0, \ldots, 60$. (a) u3(x, t) for $\alpha=1.2$. (b) u3(x, t) for $\alpha=1.5$. (c) u3(x, t) for $\alpha=1.8$. (d) u3(x, t) for $\alpha=2$.

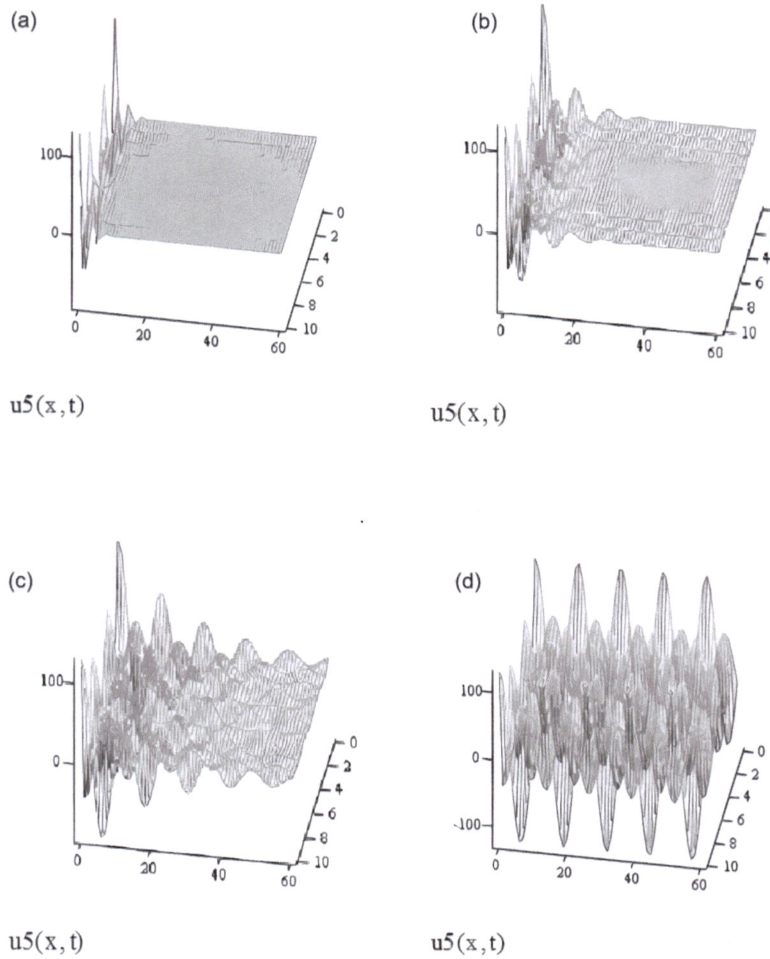

FIGURE 17.5 Plots of the fifth response u5(x, t) of the fixed-fixed beam (17.45) with $\omega=\pi$, $x=0, \ldots, 10$, and $t=0, \ldots, 60$. (a) u5(x, t) for $\alpha=1.2$. (b) u5(x, t) for $\alpha=1.5$. (c) u5(x, t) for $\alpha=1.8$. (d) u5(x, t) for $\alpha=2$.

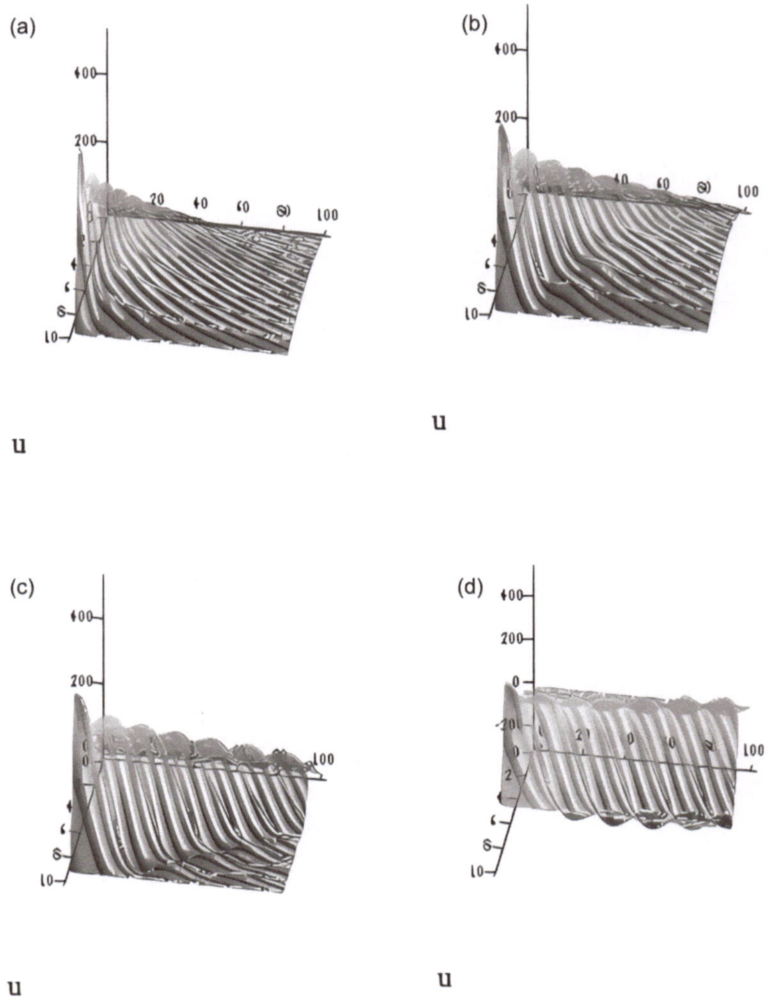

FIGURE 17.6 Plots of the response $u(x,t) \approx \sum_{j=1}^{8} u_j(x,t) = \mathrm{u}$ of the fixed-fixed beam (17.45) with $\omega=\pi$, $x=0, \ldots, 10$, and $t=0, \ldots, 60$. (a) u for $\alpha=1.2$. (b) u for $\alpha=1.5$. (c) u for $\alpha=1.8$. (d) u for $\alpha=2$.

17.5 SELF-VIBRATIONS

Previous discussions consider $1<\alpha\leq 2$. As a matter of fact, we can extend the range of α such that $1<\alpha<3$. When $2<\alpha<3$, $\zeta_{ej}<0$, see Figure 17.7. Consequently, the beam suffers from self-vibrations if $2<\alpha<3$. In Figure 17.8, we illustrate that the coordinates in Eq. (17.48) are increasing oscillations when $\alpha=2.1$. In Figure 17.9, we use $u(x, t)$ to show the self-vibration phenomenon when $\alpha=2.1$.

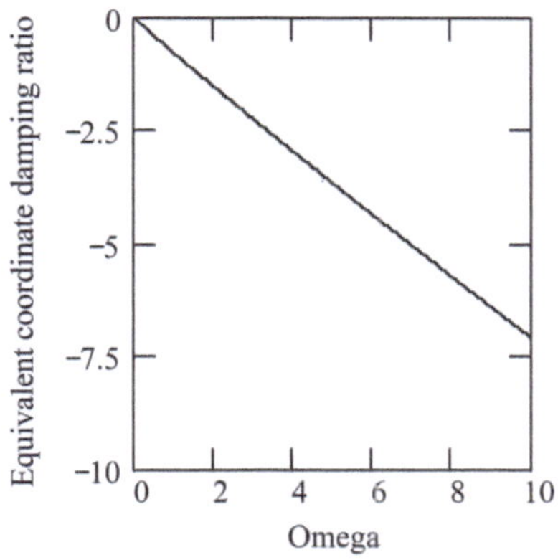

FIGURE 17.7 Illustration of negative damping ratio for $\alpha=2.1$ for the beam (17.45).

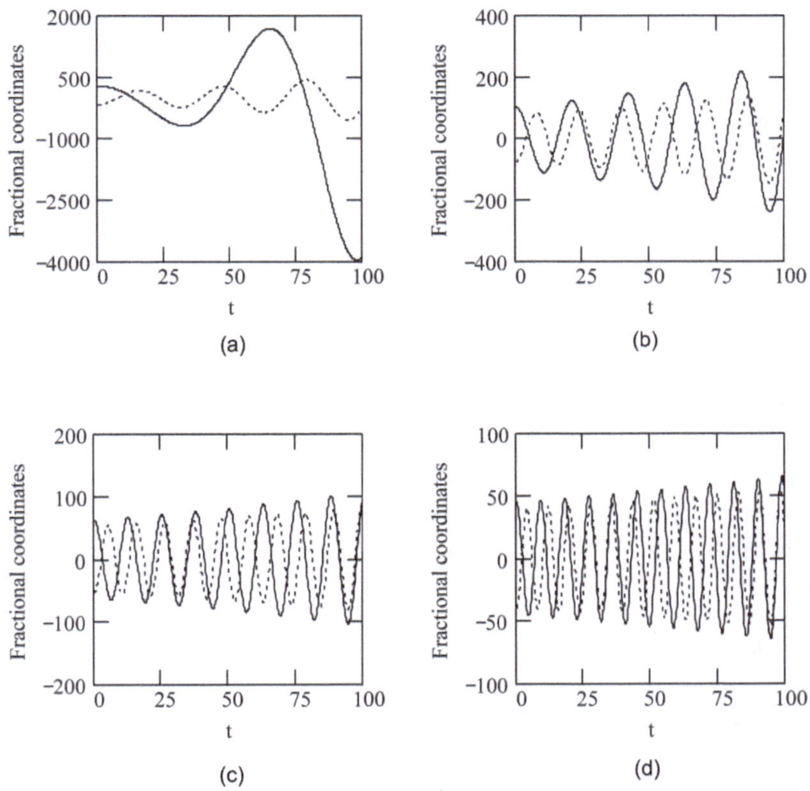

FIGURE 17.8 Self-vibration illustrations of the fixed-fixed beam (17.45) with the fractional coordinates when $\alpha=2.1$ and $\omega=0.314$. (a) $p_j(t)$ for $j=1$ (solid), 2 (dot). (b) $p_j(t)$ for $j=3$ (solid), 4 (dot). (c) $p_j(t)$ for $j=5$ (solid), 6 (dot). (d) $p_j(t)$ for $j=7$ (solid), 8 (dot).

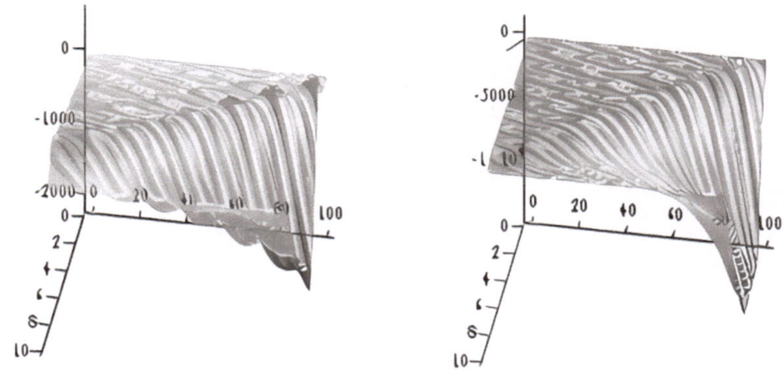

(a) (b)

u u

FIGURE 17.9 Using $u(x,t)\approx\sum_{i-1}^{8}u_j(x,t)=$ to show self-vibrations of the fixed-fixed beam (17.45) with the fractional coordinates when $2<\alpha<3$, $\omega=0.628$, $x=0$, …, 10, and $t=0$, …, 80. (a) u for $\alpha=2.3$. (b) u for $\alpha=2.6$.

17.6 SUMMARY

We have given the equivalent motion equation of the longitudinal vibration of the beam with the fractional coordinates in Theorem 17.1. We have presented the closed-form representation of the general solution with respect to the free response to the longitudinal vibration of the beam with the fractional coordinates in Theorem 17.4. Case study for the fixed-fixed beam with the fractional coordinates is described in Theorem 17.5 and Figures 17.1–17.6. The present results exhibit that the free response is exponentially decayed if $1<\alpha<2$, though the beam with the fractional coordinates appears damping free in form. The free response to the conventional longitudinal vibration of the beam is simply a special case of the present free response when $\alpha=2$. Self-vibrations of such a beam may occur when $2<\alpha<3$.

17.7 EXERCISES

17.1. Find the condition for $0<\zeta_{ej}<1$.

17.2. Find the condition for ω_{endj} being real.

17.3. Express the jth fractional coordinate function $p_j(t)$ of order α using the Mittag-Leffler functions.

17.4. Express the general solution of the jth free response to a beam the fractional coordinates of order α using the Mittag-Leffler functions.

REFERENCES

1. V. V. Uchaikin, *Fractional Derivatives for Physicists and Engineers*, vol. II, Springer, Berlin, 2013.
2. J.-S. Duan, The periodic solution of fractional oscillation equation with periodic input, *Advances in Mathematical Physics*, 2013, 2013, 6pp.
3. J.-S. Duan, Z. Wang, Y.-L. Liu, and X. Qiu, Eigenvalue problems for fractional ordinary differential equations, *Chaos, Solitons & Fractals*, 46, 2013, 46–53.
4. J.-S. Duan and Y.-Y. Xu, Vibration equation of fractional order describing viscoelasticity and viscous inertia, *Open Physics*, 17(1), 2019, 850–856.
5. S.-M. Yang and J.-S. Duan, Response analysis of six-parameter fractional constitutive model, *Physica Scripta*, 96(2), 2021, 025220.
6. Y. Li and J.-S. Duan, The periodic response of a fractional oscillator with a spring-pot and an inerter-pot, *Journal of Mechanics*, 37(37), 2021, 108–117.
7. J.-S. Duan, Vibration systems with fractional order and distributed order derivatives characterizing viscoinertia, *Fractal and Fractional*, 5(3), 2021, 67.
8. V. E. Tarasov, Exact solutions of Bernoulli and logistic fractional differential equations with power law coefficients, *Mathematics*, 8(12), 2020, 2231.
9. M. Zurigat, Solving fractional oscillators using Laplace homotopy analysis method, Annals of the University of Craiova, Mathematics and Computer Science Series, 38(4), 2011, 1–11.
10. T. Blaszczyk and M. Ciesielski, Fractional oscillator equation: Transformation into integral equation and numerical solution, *Applied Mathematics and Computation*, 257, 2015, 428–435.
11. T. Blaszczyk, M. Ciesielski, M. Klimek, and J. Leszczynski, Numerical solution of fractional oscillator equation, *Applied Mathematics and Computation*, 218(6), 2011, 2480–2488.
12. A. A. Stanislavsky, Fractional oscillator, *Physical Review E*, 70(5), 2004, 051103 (6 pp).
13. B. N. N. Achar, J. W. Hanneken, and T. Clarke, Damping characteristics of a fractional oscillator, *Physica A*, 339(3–4), 2004, 311–319.
14. B. N. N. Achar, J. W. Hanneken, and T. Clarke, Response characteristics of a fractional oscillator, *Physica A*, 309(3–4), 2002, 275–288.
15. B. N. N. Achar, J. W. Hanneken, T. Enck, and T. Clarke, Dynamics of the fractional oscillator, *Physica A*, 297(3–4), 2001, 361–367.
16. T. Blaszczyk, A numerical solution of a fractional oscillator equation in a non-resisting medium with natural boundary conditions, *Romanian Reports in Physics*, 67(2), 2015, 350–358.
17. T. Sandev and Z. Tomovski, The general time fractional wave equation for a vibrating string, *Journal of Physics A: Mathematical and Theoretical*, 43(5), 2010, 055204 (12pp).
18. H. Singh, H. M. Srivastava, and D. Kumar, A reliable numerical algorithm for the fractional vibration equation, *Chaos, Solitons & Fractals*, 103, 2017, 131–138.
19. M. Li, Three classes of fractional oscillators, *Symmetry-Basel*, 10(2), 2018 (91 pp).
20. M. Li, *Theory of Fractional Engineering Vibrations*, De Gruyter, Berlin/Boston, MA, 2021.
21. M. Li, Theory of vibrators with variable-order fractional forces, 7 July 2021, https://arxiv.org/abs/2107.02340.
22. A. Al-Rabtah, V. S. Ertürk, and S. Momani, Solutions of a fractional oscillator by using differential transform method, *Computers & Mathematics with Applications*, 59(3), 2010, 1356–1362.
23. M. H. Heydari, Z. Avazzadeh, and M. Razzaghi, Vieta-Lucas polynomials for the coupled nonlinear variable-order fractional Ginzburg-Landau equations, *Applied Numerical Mathematics*, 165, 2021, 442–458.
24. M. H. Heydari, A computational method for a class of systems of nonlinear variable-order fractional quadratic integral equations, *Applied Numerical Mathematics*, 153, 2020, 164–178.

25. M. H. Heydari, Numerical solution of nonlinear 2D optimal control problems generated by Atangana-Riemann-Liouville fractal-fractional derivative, *Applied Numerical Mathematics*, 150, 2020, 507–518.
26. M. H. Heydari and Z. Avazzadeh, New formulation of the orthonormal Bernoulli polynomials for solving the variable-order time fractional coupled Boussinesq-Burger's equations, *Engineering with Computers*, 37, 2021, 3509–3517.
27. M. H. Heydari, M. R. Hooshmandasl, A. Shakiba, and C. Cattani, Legendre wavelets Galerkin method for solving nonlinear stochastic integral equations, *Nonlinear Dynamics*, 86(2), 2016, 1185–1202.
28. M. H. Heydari, M. R. Hooshmandasl, C. Cattani, F. M. Maalek Ghaini, An efficient computational method for solving nonlinear stochastic Itô integral equations: Application for stochastic problems in physics, *Journal of Computational Physics*, 283, 2015, 148–168.
29. N. H. Tuan, A. Debbouche, and T. B. Ngoc, Existence and regularity of final value problems for time fractional wave equations, *Computers & Mathematics with Applications*, 78(5), 2019, 1396–1414.
30. D. Baleanu and H. K. Jassim, Approximate solutions of the damped wave equation and dissipative wave equation in fractal strings, *Fractal and Fractional*, 3(2), 2019, 26.
31. E. A. Abdel-Rehim, A. M. A. El-Sayed, and A. S. Hashem, Simulation of the approximate solutions of the time-fractional multi-term wave equations, *Computers & Mathematics with Applications*, 73(6), 2017, 1134–1154.
32. K. S. Miller and B. Ross, *An Introduction to the Fractional Calculus and Fractional Differential Equations*, John Wiley & Sons, Hoboken, NJ, 1993.
33. J. L. Lavoie, T. J. Osler, and R. Tremblay, Fractional derivatives and special functions, *SIAM Review*, 18(2), 1976, 240–268.
34. C. P. Li and M. Cai, *Theory and Numerical Approximations of Fractional Integrals and Derivatives*, SIAM, Philadelphia, 2019.
35. C. M. Harris, *Shock and Vibration Handbook*, 5th Edition, McGraw-Hill, New York, 2002.
36. O. M. Palley, G. B. Bahizov, and E. Я. Voroneysk, *Handbook of Ship Structural Mechanics*, National Defense Industry Publishing House, Beijing, China, 2002. In Chinese. Translated from Russian by B. H. Xu, X. Xu, and M. Q. Xu.
37. K. Nakagawa and M. Ringo, *Engineering Vibrations*, Shanghai Science and Technology Publishing House, 1981, Shanghai, China. In Chinese. Translated by S. R. Xia from Japanese.

CHAPTER 18

Free Response to Euler-Bernoulli Beam with Fractional Coordinates

THE CONTRIBUTIONS IN THIS chapter are in two folds. First, we propose the closed-form representation of the general solution of the free response to the Euler-Bernoulli beam with the fractional coordinates. Then, we present the analytical representations of the free responses to the simply supported Euler-Bernoulli beam and the free-free one with the fractional coordinates.

18.1 INTRODUCTION

The simplified motion equation of the conventional Euler-Bernoulli beam is in the form

$$EI\frac{\partial^4 w(x,t)}{\partial x^4}+\rho A\frac{\partial^2 w(x,t)}{\partial t^2}=q(x,t). \tag{18.1}$$

Its analytical expressions of responses are well studied as can be seen from Palley et al. [1], Faltinsen [2], Harris [3], Jin and Xia [4], Nakagawa and Ringo [5], Gao and Su [6], Li [7], Han et al. [8], Zhou et al. [9], Ruge and Birk [10], simply mentioning a few.

As far as the fractional Euler-Bernoulli beams were concerned, Alkhaldi et al. [11] studied the beam with fractional damping and Pirrotta et al [12] stated their research of the Euler-Bernoulli beam with fractional stress and strain. Note that Lewandowski and Baum [13], and Paola et al. [14] also explained their research regarding the Euler-Bernoulli beam with fractional stress and strain. However, reports regarding closed-form representation of the response to the fractional Euler-Bernoulli beam in the form

$$EI\frac{\partial^4 w(x,t)}{\partial x^4}+\rho A\frac{\partial^\alpha w(x,t)}{\partial t^\alpha}=q(x,t), \quad 1<\alpha<3, \tag{18.2}$$

DOI: 10.1201/9781003460947-21

is rarely seen. We present the closed-form representation of the free response to the above in this chapter.

The rest of this chapter is organized as follows. The closed-form representation of the free response to the fractional Euler-Bernoulli beam is presented in Section 18.2. The representations of the free responses to the simply supported and free-free fractional Euler-Bernoulli beams are proposed in Section 18.3. Self-vibrations are discussed in Section 18.4. Summary is given in Section 18.5.

18.2 GENERAL SOLUTION OF FREE RESPONSE

18.2.1 Equations of Vibration Mode and Fractional Coordinates

By using $w(x, t)=\varphi(x)p(t)$ and when $q(x, t)=0$, from Eq. (18.2), we have

$$\frac{EI}{\rho A\varphi(x)}\frac{d^4\varphi(x)}{dx^4}=-\frac{1}{p(t)}\frac{d^\alpha p(t)}{dt^\alpha}=\omega_n^2. \tag{18.3}$$

The quantity ω_n is the natural angular frequency (natural frequency for short). The above implies two differential equations. One is for vibration mode $\varphi(x)$ in the form

$$EI\frac{d^4\varphi(x)}{dx^4}-\rho A\omega_n^2\varphi(x)=0. \tag{18.4}$$

The other is in terms of coordinate $p(t)$ in the form

$$\frac{d^\alpha p(t)}{dt^\alpha}+\omega_n^2 p(t)=0. \tag{18.5}$$

The above is the equation with the fractional coordinates of order $1<\alpha<3$.

Rewrite the vibration mode equation by

$$\frac{d^4\varphi(x)}{dx^4}-\mu^4\varphi(x)=0, \tag{18.6}$$

where

$$\mu^4=\frac{\rho A}{EI}\omega_n^2, \tag{18.7}$$

and

$$\omega_n=\mu^2\sqrt{\frac{EI}{\rho A}}. \tag{18.8}$$

The parameter μ or ω_n is determined by system itself irrelevant of external environments.

18.2.2 General Solution of Fractional Coordinate Equation

Theorem 18.1. The general solution of the fractional coordinate equation (18.5) is given by

$$p(t)=e^{-\varsigma_e\omega_{en}t}\left(p_0\cos\omega_{end}t+\frac{v_0+\varsigma_e\omega_{en}p_0}{\omega_{end}}\sin\omega_{end}t\right),\quad t\geq 0, \tag{18.9}$$

where $p_0=p(0)$ and $v_0=p'(0)$ are the initial conditions. The parameters ω_{en}, ζ_e, and ω_{end} are given in Eqs. (18.15), (18.17), and (18.19), respectively.

Proof. Denote by $\delta(t)$ the Delta function. Following Chapter 7, we rewrite Eq. (18.5) by

$$p''(t)*\delta^{(\alpha-2)}(t)+\omega_n^2 p(t)=0, \tag{18.10}$$

where $*$ implies the convolution operation. Let $P(\omega)$ be the Fourier transform of $p(t)$. Performing the Fourier transform on Eq. (18.10) yields

$$(i\omega)^{\alpha-2}(i\omega)^2 P(\omega)+\omega_n^2 P(\omega)=0. \tag{18.11}$$

Taking into account $i^\alpha=\cos\frac{\alpha\pi}{2}+i\sin\frac{\alpha\pi}{2}$, the above can be written by

$$-\omega^{\alpha-2}\cos\frac{\alpha\pi}{2}(i\omega)^2 P(\omega)+\omega^{\alpha-1}\sin\frac{\alpha\pi}{2}(i\omega)P(\omega)+\omega_n^2 P(\omega)=0. \tag{18.12}$$

Rewrite the above by

$$(i\omega)^2 P(\omega)+\frac{\omega^{\alpha-1}\sin\frac{\alpha\pi}{2}}{-\omega^{\alpha-2}\cos\frac{\alpha\pi}{2}}(i\omega)P(\omega)+\frac{\omega_n^2}{-\omega^{\alpha-2}\cos\frac{\alpha\pi}{2}}P(\omega)=0. \tag{18.13}$$

The above implies

$$(i\omega)^2 P(\omega)+\left(-\omega\tan\frac{\alpha\pi}{2}\right)(i\omega)P(\omega)+\left(-\omega^{2-\alpha}\omega_n^2\sec\frac{\alpha\pi}{2}\right)P(\omega)=0. \tag{18.14}$$

Denote by ω_{en} the equivalent natural frequency regarding the fractional coordinate equation (18.5). It is in the form

$$\omega_{en}^2=-\omega^{2-\alpha}\omega_n^2\sec\frac{\alpha\pi}{2}. \tag{18.15}$$

Let ζ_e be the equivalent damping ratio in terms of Eq. (18.5). Denote by

$$2\varsigma_e\omega_{en}=-\omega\tan\frac{\alpha\pi}{2}. \tag{18.16}$$

Then,

$$\varsigma_e = -\frac{\omega}{2\omega_{en}}\tan\frac{\alpha\pi}{2}. \tag{18.17}$$

Therefore, Eq. (18.14) can be expressed by

$$(i\omega)^2 P(\omega) + 2\varsigma_e\omega_{en}(i\omega)P(\omega) + \omega_{en}^2 P(\omega) = 0. \tag{18.18}$$

From a practical view, we consider $|\zeta_e| \leq 1$. Denote by

$$\omega_{end} = \omega_{en}\sqrt{1-\varsigma_e^2}. \tag{18.19}$$

Thus, we obtain the general solution in terms of $p(t)$ in the form

$$p(t) = e^{-\varsigma_e\omega_{en}t}\left(p_0\cos\omega_{end}t + \frac{v_0 + \varsigma_e\omega_{en}p_0}{\omega_{end}}\sin\omega_{end}t\right), \quad t \geq 0.$$

The proof is completed. □

The above can be expressed by

$$p(t) = Ae^{-\varsigma_e\omega_{en}t}\cos(\omega_{end}t - \theta), \quad t \geq 0, \tag{18.20}$$

where

$$\begin{aligned} A &= \sqrt{p_0^2 + \left(\frac{v_0 + \varsigma_e\omega_{en}p_0}{\omega_{end}}\right)^2}, \\ \theta &= \arctan\left(\frac{v_0 + \varsigma_e\omega_{en}p_0}{p_0\omega_{end}}\right). \end{aligned} \tag{18.21}$$

Note 18.1: If $\alpha = 2$, $\omega_{end} = \omega_{en} = \omega_n$.

Note 18.2: When $\alpha = 2$, $\zeta_e = 0$.

Note 18.3: When $\alpha = 2$, $p(t)$ reduces to the conventional coordinate in the form $p(t) = p_0\cos\omega_n t + \frac{v_0}{\omega_n}\sin\omega_n t$.

18.2.3 General Solution of Free Response to Euler-Bernoulli Beam with Fractional Coordinates

Theorem 18.2. The general solution of the free response to the Euler-Bernoulli beam with fractional coordinates is expressed by

$$w(x,t) = \varphi(x)p(t) = \varphi(x)e^{-\varsigma_e\omega_{en}t}\left(p_0\cos\omega_{end}t + \frac{v_0 + \varsigma_e\omega_{en}p_0}{\omega_{end}}\sin\omega_{end}t\right). \tag{18.22}$$

Proof. Since $w(x, t) = \varphi(x)p(t)$ and due to Theorem 18.1, the above is true. This completes the proof. □

An alternative expression of Eq. (18.22) can be expressed by

$$w(x,t) = \varphi(x)Ae^{-\varsigma_e \omega_{en} t} \cos(\omega_{end} t - \theta). \tag{18.23}$$

18.3 CASE STUDY

18.3.1 Free Response to Simply Supported Euler-Bernoulli Beam with Fractional Coordinates

The characteristic equation of (18.6) is $s^4 - \mu^4 = 0$. Thus, we have $s_{1,2} = \pm i\mu$ and $s_{3,4} = \pm\mu$. Therefore, the vibration mode function $\varphi(x)$ is given by

$$\varphi(x) = B_1 \sin \mu x + B_2 \cos \mu x + B_3 \text{sh}\mu x + B_4 \text{ch}\mu x, \tag{18.24}$$

where four coefficients B_1, B_2, B_3, and B_4 are determined by boundary conditions.

For a simply supported beam, we have

$$\varphi(0) = B_2 + B_4 = 0. \tag{18.25}$$

In addition,

$$\varphi''(0) = \mu^2(-B_2 + B_4) = 0. \tag{18.26}$$

Equations (18.25) and (18.26) imply

$$B_2 = B_4 = 0. \tag{18.27}$$

Thus,

$$\varphi(x) = B_1 \sin \mu x + B_3 \text{sh}\mu x. \tag{18.28}$$

Substituting $\varphi(l) = 0$ and $\varphi''(l) = 0$ into the above produces

$$\begin{aligned} B_1 \sin \mu l + B_3 \text{sh}\mu l &= 0, \\ -B_1 \sin \mu l + B_3 \text{sh}\mu l &= 0. \end{aligned} \tag{18.29}$$

For non-zero B_1 and B_3, we have

$$\begin{vmatrix} \sin \mu l & \text{sh}\mu l \\ -\sin \mu l & \text{sh}\mu l \end{vmatrix} = 0. \tag{18.30}$$

The above is the frequency equation of the simply supported Euler-Bernoulli beam. Since $\mathrm{sh}\mu l \neq 0$, we must have

$$\sin \mu l = 0. \tag{18.31}$$

Denote by

$$\mu l \triangleq \mu_j l = j\pi, \quad j = 1, 2, \ldots \tag{18.32}$$

Then,

$$\mu_j = \frac{j\pi}{l}, \quad j = 1, 2, \ldots \tag{18.33}$$

The jth order natural angular frequency is given by

$$\omega_{nj} = \mu_j^2 \sqrt{\frac{EI}{\rho A}} = \left(\frac{j\pi}{l}\right)^2 \sqrt{\frac{EI}{\rho A}}, \quad j = 1, 2, \ldots \tag{18.34}$$

Therefore, considering $B_3 = 0$, the jth order vibration mode is expressed by

$$\varphi_j(x) = B_{1j} \sin \frac{j\pi x}{l}, \quad j = 1, 2, \ldots \tag{18.35}$$

For simplicity, we use the symbol $C_j = B_{1j}$. Thus, the above is rewritten by

$$\varphi_j(x) = C_j \sin \frac{j\pi x}{l}, \quad j = 1, 2, \ldots \tag{18.36}$$

Considering the jth coordinate, we have

$$p_j''(t) * \delta^{(\alpha-2)}(t) + \omega_{nj}^2 p_j(t) = 0. \tag{18.37}$$

In the frequency domain, the jth coordinate follows

$$(i\omega)^2 P_j(\omega) + 2\varsigma_{\mathrm{e}j}\omega_{\mathrm{en}j}(i\omega)P_j(\omega) + \omega_{\mathrm{en}j}^2 P_j(\omega) = 0, \tag{18.38}$$

where

$$\omega_{\mathrm{en}j} = \sqrt{-\omega^{2-\alpha}\omega_{nj}^2 \sec\frac{\alpha\pi}{2}}, \tag{18.39}$$

and

$$\varsigma_{\mathrm{e}j} = -\frac{\omega}{2\omega_{\mathrm{en}j}} \tan\frac{\alpha\pi}{2}. \tag{18.40}$$

Considering $|\zeta_{\mathrm{e}j}| \leq 1$, we denote

$$\omega_{\mathrm{end}j} = \omega_{\mathrm{en}j}\sqrt{1 - \varsigma_{\mathrm{e}j}^2}. \tag{18.41}$$

Then, we attain the jth coordinate $p_j(t)$ in the form

$$p_j(t)=A_j e^{-\varsigma_{ej}\omega_{enj}t}\cos\left(\omega_{endj}t-\theta_j\right),\quad t\geq 0, \tag{18.42}$$

where

$$A_j=\sqrt{p_{0j}^2+\left(\frac{v_{0,j}+\varsigma_{ej}\omega_{enj}p_{0j}}{\omega_{endj}}\right)^2},$$

$$\theta_j=\arctan\left(\frac{v_{0j}+\varsigma_{ej}\ \omega_{enj}p_0}{p_{0j}\omega_{endj}}\right), \tag{18.43}$$

where, $p_{0j}=p_j(0)$, $v_{0j}=p_j'(0)$, $\omega_{enj}=\sqrt{-\omega^{2-\alpha}\omega_{nj}^2\sec\frac{\alpha\pi}{2}}$.

Plots of vibration modes of a simply supported Euler-Bernoulli beam with $C_j=1$ are shown in Figure 18.1. Figure 18.2 indicates some plots of fractional coordinates.

The jth order normal vibration of a simply supported Euler-Bernoulli beam with the fractional coordinates is expressed by

$$w_j(x,t)=\varphi_j(x)p_j(t)=\varphi_j(x)A_j e^{-\varsigma_{ej}\omega_{enj}t}\cos\left(\omega_{endj}t-\theta_j\right). \tag{18.44}$$

The total free response to a simply supported Euler-Bernoulli beam with the fractional coordinates is in the form

$$w(x,t)=\sum_{j=1}^{\infty}w_j(x,t)=\sum_{j=1}^{\infty}C_j\sin\frac{j\pi x}{l}A_j e^{-\varsigma_{ej}\omega_{enj}t}\cos\left(\omega_{endj}t-\theta_j\right). \tag{18.45}$$

Some plots of $w_j(x,t)=w_j$ are shown in Figure 18.3 and the plots of $w(x,t)$ are illustrated in Figure 18.4.

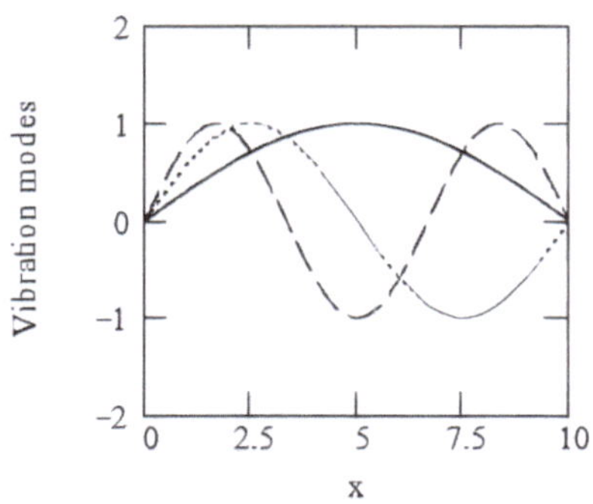

FIGURE 18.1 Plots of vibration modes of a simply supported Euler-Bernoulli beam of $l=10$. $\varphi_1(x)$ (solid), $\varphi_2(x)$ (dot), $\varphi_3(x)$ (dash).

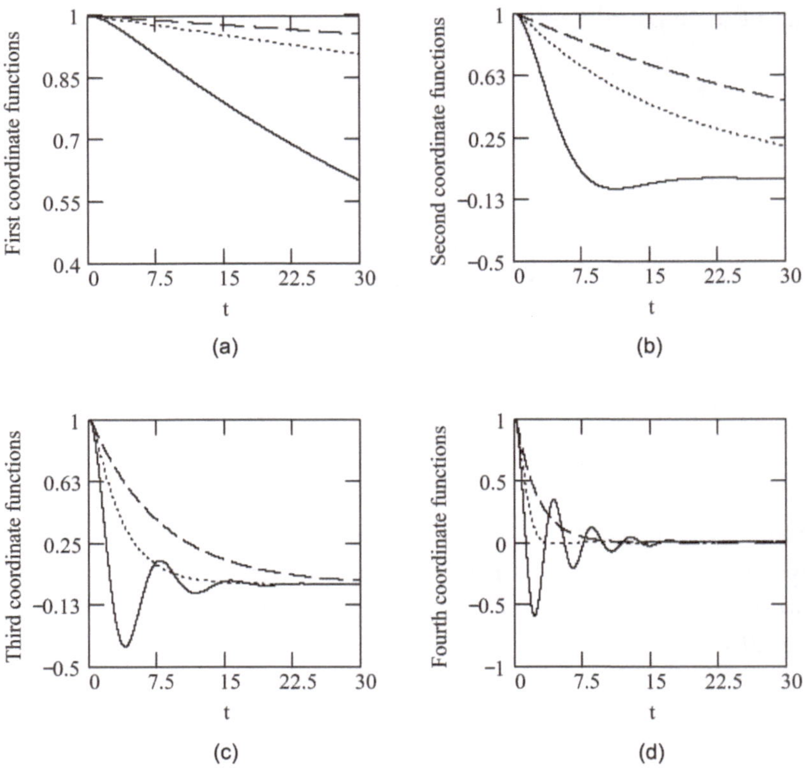

FIGURE 18.2 Plots of fractional coordinates of a simply supported Euler-Bernoulli beam of $l=10$ for $\alpha=1.9$ (solid), 1.6 (dot), and 1.3 (dash), when $p_0=1$ and $v_0=0$. (a) $p_1(t)$. (b) $p_2(t)$. (c) $p_3(t)$. (d) $p_4(t)$.

18.3.2 Free Response to Free-Free Euler-Bernoulli Beam with Fractional Coordinates

For a free-free Euler-Bernoulli beam, we first write

$$\varphi''(x)=\mu^2\left(-B_1\sin\mu x-B_2\cos\mu x+B_3\mathrm{sh}\mu x+B_4\mathrm{ch}\mu x\right), \tag{18.46}$$

and

$$\varphi'''(x)=\mu^3\left(-B_1\cos\mu x+B_2\sin\mu x+B_3\mathrm{ch}\mu x+B_4\mathrm{sh}\mu x\right). \tag{18.47}$$

Note that the boundary conditions of a free-free Euler-Bernoulli beam are with $\varphi''(0)=0$ and $\varphi'''(0)=0$. Thus, substituting $\varphi''(0)=0$ and $\varphi'''(0)=0$ into the above yields

$$\varphi''(0)=-B_2+B_4=0, \tag{18.48}$$

and

$$\varphi'''(0)=-B_1+B_3=0. \tag{18.49}$$

Therefore, we have

$$B_2=B_4, B_1=B_3. \tag{18.50}$$

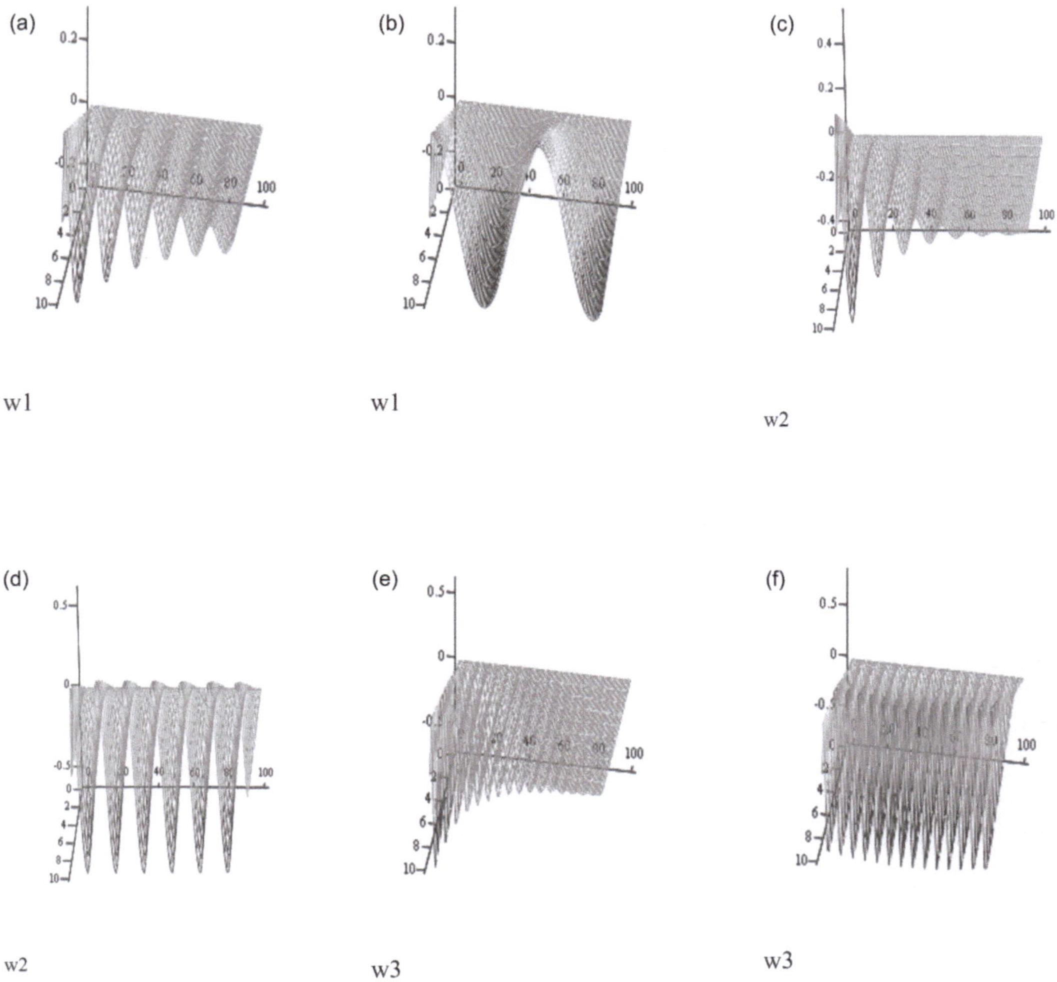

FIGURE 18.3 Plots of $w_j(x, t)$ of a simply supported Euler-Bernoulli beam of $l=10$ for $\omega=0.314$, $x=0, \ldots, 10$, and $t=0, \ldots, 100$, when $p_0=1$ and $v_0=0$. (a) $w_1(x, t)=\text{w1}$ for $\alpha=1.4$. (b) $w_1(x, t)=\text{w1}$ for $\alpha=2$. (c) $w_2(x, t)=\text{w2}$ for $\alpha=1.8$. (d) $w_2(x, t)=\text{w2}$ for $\alpha=2$. (e) $w_3(x, t)=\text{w3}$ for $\alpha=1.8$. (f) $w_3(x, t)=\text{w3}$ for $\alpha=2$.

Consequently, $\varphi(x)$ can be expressed by

$$\varphi(x)=B_1(\sin\mu x+\text{sh}\mu x)+B_2(\cos\mu x+\text{ch}\mu x). \tag{18.51}$$

With the above, one has

$$\varphi''(x)=\mu^2\left[B_1(\text{sh}\mu x-\sin\mu x)+B_2(\text{ch}\mu x-\cos\mu x)\right], \tag{18.52}$$

and

$$\varphi'''(x)=\mu^3\left[B_1(\text{ch}\mu x-\cos\mu x)+B_2(\text{sh}\mu x+\sin\mu x)\right]. \tag{18.53}$$

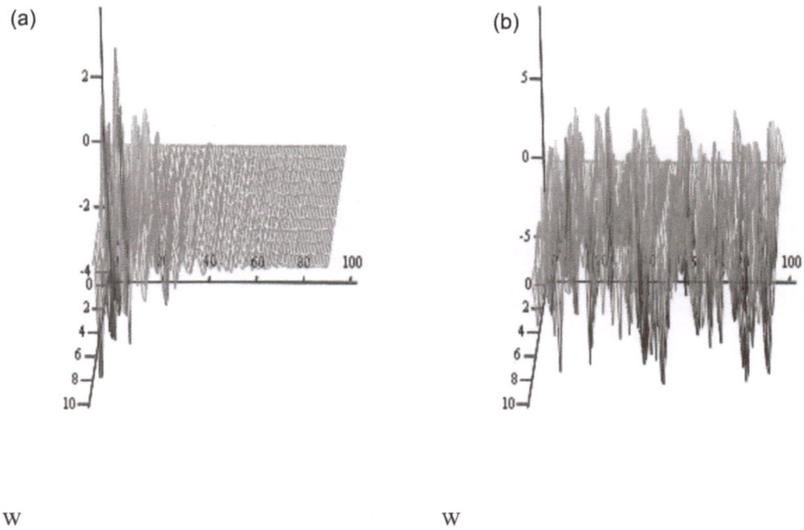

FIGURE 18.4 Free response $w(x,t) \approx \sum_{j=1}^{32} w_j(x,t)$ to a simply supported Euler-Bernoulli beam with $l=10$, $\omega=0.314$, $x=0, \ldots, 10$, and $t=0, \ldots, 100$, when $p_0=1$ and $v_0=0$. (a) $\sum_{j=1}^{32} w_j(x,t)=\mathrm{w}$ for $\alpha=1.8$. (b) $\sum_{j=1}^{32} w_j(x,t)=\mathrm{w}$ for $\alpha=2$.

Substituting $\varphi''(l)=0$ and $\varphi'''(l)=0$ into the above produces

$$B_1(\mathrm{sh}\mu l-\sin\mu l)+B_2(\mathrm{ch}\mu l-\cos\mu l)=0, \tag{18.54}$$

and

$$B_1(\mathrm{ch}\mu l-\cos\mu l)+B_2(\mathrm{sh}\mu l+\sin\mu l)=0. \tag{18.55}$$

Finding non-zero B_1 and B_2 requires

$$\begin{vmatrix} \mathrm{sh}\mu l-\sin\mu l & \mathrm{ch}\mu l-\cos\mu l \\ \mathrm{ch}\mu l-\cos\mu l & \mathrm{sh}\mu l+\sin\mu l \end{vmatrix}=0. \tag{18.56}$$

Expanding the above determinant results in the frequency equation in the form

$$\cos\mu l=\frac{1}{\mathrm{ch}\mu l}. \tag{18.57}$$

The numeric results about $\cos(\mu l)=1/\mathrm{ch}(\mu l)$ are given by

$$\frac{\mu_j l}{(2j+1)\dfrac{\pi}{2}}, \quad j=1,2,\ldots \tag{18.58}$$

Thus, we have natural frequencies given by

$$\omega_{nj} = \mu_j^2 \sqrt{\frac{EI}{\rho A}} = \left(\frac{(2j+1)\pi}{2l}\right)^2 \sqrt{\frac{EI}{\rho A}}, \quad j = 1, 2, \ldots \tag{18.59}$$

Substituting $\mu_j l$ into Eq. (18.54) yields the ratio expressed by

$$\frac{B_1}{B_2} = -\frac{\mathrm{ch}\mu_j l - \cos\mu_j l}{\mathrm{sh}\mu_j l - \sin\mu_j l}. \tag{18.60}$$

Note that a vibration mode $\varphi_j(x)$ multiplied by a constant is still a vibration mode. Thus, $\frac{\varphi(x)}{B_2}$ is a vibration mode. Taking into account $\varphi(x)$ in Eq. (18.60), we have

$$\frac{\varphi(x)}{B_2} = \frac{B_1}{B_2}(\sin\mu x + \mathrm{sh}\mu x) + (\cos\mu x + \mathrm{ch}\mu x). \tag{18.61}$$

Denoting it by $\varphi_j(x)$ yields

$$\varphi_j(x) = \frac{\mathrm{ch}\mu_j l - \cos\mu_j l}{\sin\mu_j l - \mathrm{sh}\mu_j l}(\sin\mu_j x + \mathrm{sh}\mu_j x) + (\cos\mu_j x + \mathrm{ch}\mu_j x). \tag{18.62}$$

Figure 18.5 indicates the plots of $\varphi_j(x)$.

The free response to a free-free Euler-Bernoulli beam with the fractional coordinates is therefore given by

$$\begin{aligned} w(x,t) &= \sum_{j=1}^{\infty} w_j(x,t) = \sum_{j=1}^{\infty} \varphi_j(x) p_j(t) \\ &= \sum_{j=1}^{\infty} \left[\frac{\mathrm{ch}\mu_j l - \cos\mu_j l}{\sin\mu_j l - \mathrm{sh}\mu_j l}(\sin\mu_j x + \mathrm{sh}\mu_j x) + (\cos\mu_j x + \mathrm{ch}\mu_j x) \right] A_j e^{-\varsigma_{ej}\omega_{enj}t} \cos(\omega_{endj}t - \theta_j), \end{aligned} \tag{18.63}$$

where ω_{enj}, ω_{endj}, and ζ_{ej}, are associated with ω_{nj} expressed by Eq. (18.34). Figure 18.6 illustrates of $w(x, t)$ when $j = 1, \ldots, 32$.

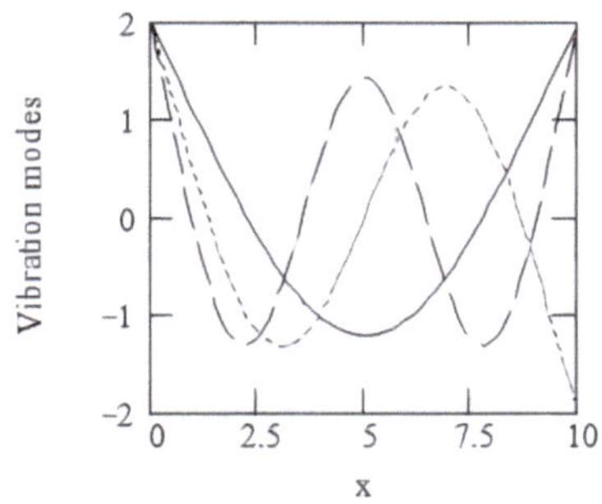

FIGURE 18.5 Plots of vibration modes for a free-free Euler-Bernoulli beam with $l = 10$. $\varphi_1(x)$ (solid), $\varphi_2(x)$ (dot), and $\varphi_3(x)$ (dash).

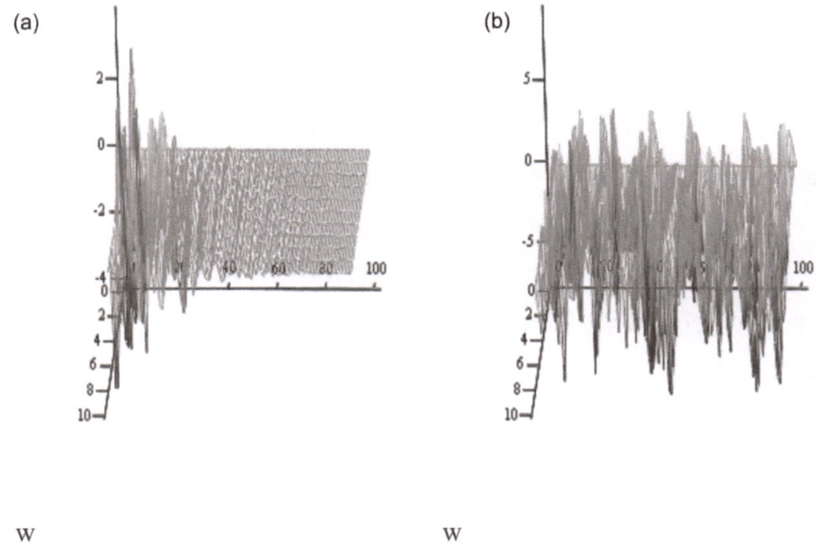

FIGURE 18.6 Free response $w(x,t) \approx \sum_{j=1}^{32} w_j(x,t)$ to a free-free Euler-Bernoulli beam with $l=10$, $\omega=0.0314$, $x=0, \ldots, 10$, and $t=0, \ldots, 100$, when $p_0=1$ and $v_0=0$. (a) $\sum_{j=1}^{32} w_j(x,t)=\mathrm{w}$ for $\alpha=1.8$. (b) $\sum_{j=1}^{32} w_j(x,t)=\mathrm{w}$ for $\alpha=2$.

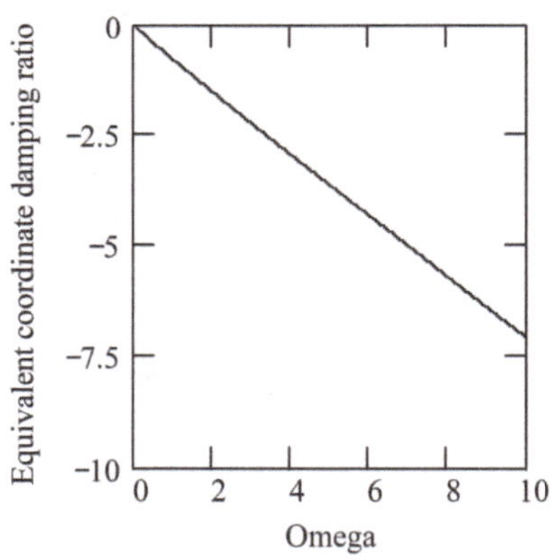

FIGURE 18.7 Illustration of negative damping ratio for $\alpha=2.1$ and $j=1$.

18.4 SELF-VIBRATIONS

Note that $\zeta_{ej}<0$ when $2<\alpha<3$. Thus, the Euler-Bernoulli beam with the fractional coordinates suffers from self-vibrations if $2<\alpha<3$. Figure 18.7 indicates an illustration of $\zeta_{e1}<0$ when $\alpha=2.1$. Figure 18.8 indicates the self-vibration phenomena using the fractional coordinates for $\alpha=2.1$. In Figure 18.9, we use $w(x, t)=\mathrm{w}$ to indicate the self-vibrating phenomenon when $\alpha=2.1$.

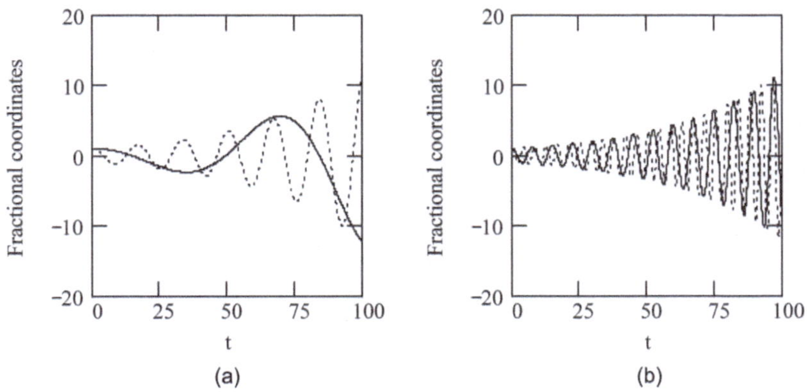

FIGURE 18.8 Self-vibration illustrations of a free-free Euler-Bernoulli beam with the fractional coordinates when $\alpha = 2.1$ and $\omega = 0.314$, when $p_0 = 1$ and $v_0 = 0$. (a). $p_j(t)$ for $j = 1$ (solid), 2 (dot). (b). $p_j(t)$ for j = 3 (solid), 4 (dot).

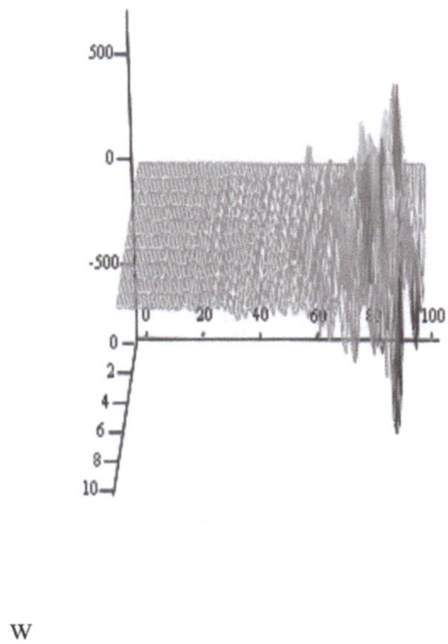

FIGURE 18.9 Illustrating self-vibration with $w(x, t) = \mathrm{w}$ for $\alpha = 2.2$ and $\omega = 0.314$, $x = 0, \ldots, 10$, and $t = 0, \ldots, 100$, when $\mathrm{p}_0 = 1$ and $v_0 = 0$.

18.5 SUMMARY

We have presented the analytical representation in closed form of the general solution of free response to an Euler-Bernoulli beam with the fractional coordinates. In case study, we have given the analytic representations of the free responses to the simply supported and free-free Euler-Bernoulli beams. The studied beam is damping free in form but, in fact, suffers from damping due to fractional coordinates. If $\alpha = 2$, the present results reduce to those of the conventional Euler-Bernoulli beam. Self-vibration case of the fractional Euler-Bernoulli beam has been mentioned when $2 < \alpha < 3$.

18.6 EXERCISES

18.1. For a simply supported Euler-Bernoulli beam with the fractional coordinates of order α when $p_0=1$ and $v_0=1$, draw the plots of the free response $w(x, t)$ for $\alpha=1.8$ and $\alpha=2.2$.

18.2. For a simply supported Euler-Bernoulli beam with the fractional coordinates of order α if $p_0=1$ and $v_0=1$, draw the plots of the free response $w_1(x, t)$ for $\alpha=1.8$ and $\alpha=2.2$.

18.3. For a free-free Euler-Bernoulli beam with the fractional coordinates of order α when $p_0=1$ and $v_0=1$, draw the plots of the free response $w(x, t)$ for $\alpha=1.8$ and $\alpha=2.2$.

18.4. For a free-free Euler-Bernoulli beam with the fractional coordinates of order α if $p_0=1$ and $v_0=1$, draw the plots of the free response $w_1(x, t)$ for $\alpha=1.8$ and $\alpha=2.2$.

18.5. For a simply supported Euler-Bernoulli beam with the fractional coordinates of order α when $p_0=0$ and $v_0=1$, draw the plots of the free response $w(x, t)$ for $\alpha=1.8$ and $\alpha=2.2$.

18.6. For a simply supported Euler-Bernoulli beam with the fractional coordinates of order α if $p_0=0$ and $v_0=1$, draw the plots of the free response $w_1(x, t)$ for $\alpha=1.8$ and $\alpha=2.2$.

REFERENCES

1. O. M. Palley, Г. B. Bahizov, and E. Я. Voroneysk, *Handbook of Ship Structural Mechanics*, National Defense Industry Publishing House, Beijing, China, 2002. In Chinese. Translated from Russian by B. H. Xu, X. Xu, and M. Q. Xu.
2. O. M. Faltinsen, *Hydrodynamics of High-Speed Marine Vehicles*, Cambridge University Press, Cambridge, 2005.
3. C. M. Harris, *Shock and Vibration Handbook*, 5th Ed., McGraw-Hill, New York, 2002.
4. X. D. Jin and L. J. Xia, *Ship Hull Vibration*, The Press of Shanghai Jiaotong University, Shanghai, China, 2011. In Chinese.
5. K. Nakagawa and M. Ringo, *Engineering Vibrations*, Shanghai Science and Technology Publishing House, Shanghai, China, 1981. In Chinese. Translated from Japanese by S. R. Xia.
6. X.-L. Gao and Y.-Y. Su, An analytical study on peeling of an adhesively bonded joint based on a viscoelastic Bernoulli-Euler beam model, *Acta Mechanica*, 226(9), 2015, 3059–3067.
7. M. Li, *Theory of Fractional Engineering Vibrations*, Walter de Gruyter, Berlin/Boston, MA, 2021.
8. S. M. Han, H. B. Benaroya, and T. Wei, Dynamics of transversely vibrating beams using four engineering theories, *Journal of Sound and Vibration*, 225(5), 1999, 935–988.
9. J. Zhou, K. Wang, D. Xu, and H. Ouyang, Multi-low-frequency flexural wave attenuation in Euler-Bernoulli beams using local resonators containing negative-stiffness mechanisms, *Physics Letters A*, 381(37), 2017, 3141–3148.
10. P. Ruge and C. Birk, A comparison of infinite Timoshenko and Euler-Bernoulli beam models on Winkler foundation in the frequency- and time-domain, *Journal of Sound and Vibration*, 304(3–5), 2007, 932–947.
11. H. S. Alkhaldi, I. M. Abu-Alshaikh, and A. N. Al-Rabadi, Vibration control of fractionally-damped beam subjected to a moving vehicle and attached to fractionally-damped multi-absorbers, *Advances in Mathematical Physics*, 2013, 2013, 12pp.
12. A. Pirrotta, S. Cutrona, and S. D. Lorenzo, Fractional visco-elastic Timoshenko beam from elastic Euler-Bernoulli beam, *Acta Mechanica*, 226(1), 2015, 179–189.
13. R. Lewandowski and M. Baum, Dynamic characteristics of multilayered beams with visco-elastic layers described by the fractional Zener model, *Archive of Applied Mechanics*, 82(15), 2015, 1793–1824.
14. M. D. Paola, R. Heuer, and A. Pirrotta, Fractional visco-elastic Euler-Bernoulli beam, *International Journal of Solids and Structures*, 50(22–23), 2013, 3505–3510.

CHAPTER 19

Forced Response to Euler-Bernoulli Beam with Fractional Coordinates

IN THIS CHAPTER, WE address the forced response to an Euler-Bernoulli beam with fractional coordinates. The contributed results are in three aspects. At first, we present the closed-form representation of the general solution of the forced response to an Euler-Bernoulli beam with fractional coordinates. After that, we propose the closed-form representations of the forced responses to simply supported and free-free Euler-Bernoulli beams with fractional coordinates. Finally, the phenomenon of increasing-vibrations of such a beam is revealed.

19.1 INTRODUCTION

For a beam at position x, we denote $\rho(x)$ as the material density, $A(x)$ as the cross-sectional area, $I(x)$ as the cross-sectional moment of inertia, and $EI(x)$ the bending rigidity, where E is elastic modulus. The motion equation of the conventional Euler-Bernoulli beam is given by

$$\rho A(x)\frac{\partial^2 w(x,t)}{\partial t^2}+\frac{\partial^2}{\partial x^2}\left[EI(x)\frac{\partial^2 w(x,t)}{\partial x^2}\right]=q(x,t), \tag{19.1}$$

where $q(x, t)$ is a driven force and $w(x, t)$ deflection at position x and time t (Palley et al. [1], Harris [2], Li [3], Nakagawa and Ringo [4]). When $A(x)$ and $I(x)$ are constants, the above is simplified in the form

$$EI\frac{\partial^4 w(x,t)}{\partial x^4}+\rho A\frac{\partial^2 w(x,t)}{\partial t^2}=q(x,t). \tag{19.2}$$

DOI: 10.1201/9781003460947-22

Taking into account fractional stress and strain (Shitikova [5], Rossikhin and Shitikova [6]), we write the fractional Euler-Bernoulli beam in the form

$$EI\frac{\partial^4 w(x,t)}{\partial x^4}+\rho A\frac{\partial^\alpha w(x,t)}{\partial t^\alpha}=q(x,t),\quad 1<\alpha<3. \tag{19.3}$$

The reports regarding the responses to the above are rare. In the last chapter, we present the free response to such a beam. In this chapter, we shall propose the closed-form representation of the forced response to the above.

The rest of this chapter is organized as follows. In Section 19.2, we propose the closed-form representation of the forced response to the Euler-Bernoulli beam with the fractional coordinates. Case study regarding the simply supported and free-free Euler-Bernoulli beams with the fractional coordinates is discussed in Section 19.3. Increasing-vibrations of the fractional Euler-Bernoulli beam are addressed in Section 19.4. Summary is given in Section 19.5.

19.2 FORCED RESPONSE

Theorem 19.1. The zero-state solution to Eq. (19.3) is expressed by

$$w(x,t)=\sum_{j=1}^{\infty}\varphi_j(x)\left[q_j(t)*\mathrm{F}^{-1}\left(-\omega^{2-\alpha}\sec\frac{\alpha\pi}{2}\right)*\frac{1}{M_j\omega_{\mathrm{end}j}}e^{-\varsigma_{ej}\omega_{\mathrm{en}j}t}\sin\omega_{\mathrm{end}j}t\right], \tag{19.4}$$

where F^{-1} is the operator of the inverse Fourier transform and M_j is the jth order generalized mass (see Eq. (6.144)) in the form

$$M_j=\rho A\int_0^l \varphi_j^2(x)dx. \tag{19.5}$$

In Eq. (19.4), $q_j(t)$ is

$$q_j(t)=\frac{Q_j(t)}{M_j}, \tag{19.6}$$

where $Q_j(t)$ is given by

$$Q_j(t)=\int_0^l q(x,t)\varphi_j(x)dx. \tag{19.7}$$

In Eq. (19.4), $\omega_{\mathrm{en}j}$ is the equivalent damping free natural frequency of the jth order vibration mode in the form

$$\omega_{\mathrm{en}j}=\sqrt{-\omega^{2-\alpha}\omega_{nj}^2\sec\frac{\alpha\pi}{2}}, \tag{19.8}$$

the quantity ζ_{ej} is the jth order equivalent damping ratio in the form

$$\varsigma_{ej} = -\frac{\omega}{2\omega_{enj}} \tan\frac{\alpha\pi}{2}, \tag{19.9}$$

ω_{endj} is the equivalent damped natural frequency of the jth order vibration mode, for $|\zeta_{ej}| \le 1$, in the form

$$\omega_{endj} = \omega_{enj}\sqrt{1-\varsigma_{ej}^2}. \tag{19.10}$$

Proof. Let $w(x, t) = \varphi(x)p(t)$. Then,

$$w(x,t) = \sum_{j=1}^{\infty} w_j(x,t) = \sum_{j=1}^{\infty} \varphi_j(x)p_j(t), \tag{19.11}$$

where $\varphi_j(x)$ denotes the jth vibration mode and $p_j(t)$ the jth coordinate. Substituting the above to Eq. (19.3) produces

$$EI\sum_{j=1}^{\infty} \frac{d^4\varphi_j(x)}{dx^4} p_j(t) + \rho A \sum_{j=1}^{\infty} \varphi_j(x) \frac{d^\alpha p_j(t)}{dt^\alpha} = q(x,t). \tag{19.12}$$

Since vibration modes are normal, we use $EI\frac{d^4\varphi(x)}{dx^4} = \rho A\omega_n^2\varphi(x)$ to express the above by

$$\rho A\sum_{j=1}^{\infty} \varphi_j(x) \frac{d^\alpha p_j(t)}{dt^\alpha} + \rho A \sum_{j=1}^{\infty} \omega_{nj}^2 \varphi_j(x) p_j(t) = q(x,t). \tag{19.13}$$

Multiplying $\varphi_m(x)$ on both sides of the above yields

$$\rho A\sum_{j=1}^{\infty} \varphi_j(x)\varphi_m(x) \frac{d^\alpha p_j(t)}{dt^\alpha} + \rho A \sum_{j=1}^{\infty} \omega_{nj}^2 \varphi_j(x)\varphi_m(x) p_j(t) = q(x,t)\varphi_m(x). \tag{19.14}$$

Doing the integral from 0 to l in terms of x on both sides of the above equation produces

$$\rho A\sum_{j=1}^{\infty} \int_0^l \varphi_j(x)\varphi_m(x)dx \frac{d^\alpha p_j(t)}{dt^\alpha} + \rho A \sum_{j=1}^{\infty} \omega_{nj}^2 \int_0^l \varphi_j(x)\varphi_m(x)dx p_j(t) = \int_0^l q(x,t)\varphi_m(x)dx. \tag{19.15}$$

Considering the orthogonality of vibration modes and the jth generalized mass M_j, we have

$$M_j \frac{d^\alpha p_j(t)}{dt^\alpha} + M_j\omega_{nj}^2 p_j(t) = \int_0^l q(x,t)\varphi_j(x)dx. \tag{19.16}$$

In the above, $M_j\omega_{nj}^2$ is the jth generalized stiffness in the form

$$K_j = M_j\omega_{nj}^2. \tag{19.17}$$

Due to $Q_j(t)$ in Eq. (19.7), (19.16) can be expressed by

$$\frac{d^\alpha p_j(t)}{dt^\alpha} + \omega_{nj}^2 p_j(t) = \frac{Q_j(t)}{M_j}, \tag{19.18}$$

where ω_{nj} is the damping free natural frequency of the jth coordinate in the form

$$\omega_{nj} = \sqrt{\frac{K_j}{M_j}}. \tag{19.19}$$

Using the Delta function $\delta(t)$, we rewrite Eq. (19.18) by

$$p_j''(t) * \delta^{(\alpha-2)}(t) + \omega_{nj}^2 p_j(t) = q_j(t), \tag{19.20}$$

where * implies the convolution operation and

$$q_j(t) = \frac{Q_j(t)}{M_j}. \tag{19.21}$$

Let $P_j(\omega) = \mathrm{F}[p_j(t)]$ and $Q_j(\omega) = \mathrm{F}[q_j(t)]$, where F is the operator of Fourier transform. Performing the Fourier transform on Eq. (19.20) yields

$$(i\omega)^{\alpha-2}(i\omega)^2 P_j(\omega) + \omega_{nj}^2 P_j(\omega) = Q_j(\omega). \tag{19.22}$$

As $i^\alpha = \cos\frac{\alpha\pi}{2} + i\sin\frac{\alpha\pi}{2}$, the above can be written by

$$-\omega^{\alpha-2}\cos\frac{\alpha\pi}{2}(i\omega)^2 P_j(\omega) + \omega^{\alpha-1}\sin\frac{\alpha\pi}{2}(i\omega)P_j(\omega) + \omega_{nj}^2 P_j(\omega) = Q_j(\omega). \tag{19.23}$$

Rewrite the above by

$$\left[(i\omega)^2 + \frac{\omega^{\alpha-1}\sin\frac{\alpha\pi}{2}}{-\omega^{\alpha-2}\cos\frac{\alpha\pi}{2}}(i\omega) + \frac{\omega_{nj}^2}{-\omega^{\alpha-2}\cos\frac{\alpha\pi}{2}}\right]P_j(\omega) = \frac{Q_j(\omega)}{-\omega^{\alpha-2}\cos\frac{\alpha\pi}{2}}. \tag{19.24}$$

Then, we have

$$\left[(i\omega)^2 + \left(-\omega\tan\frac{\alpha\pi}{2}\right)(i\omega) + \left(-\omega^{2-\alpha}\omega_{nj}^2\sec\frac{\alpha\pi}{2}\right)\right]P_j(\omega) = -\omega^{2-\alpha}\sec\frac{\alpha\pi}{2}Q_j(\omega). \tag{19.25}$$

Using ω_{enj} and Eq. (19.9), we have

$$2\varsigma_{ej}\omega_{enj} = -\omega\tan\frac{\alpha\pi}{2}, \tag{19.26}$$

where ζ_{ej} is the equivalent damping ratio of the jth coordinate in the form (19.9). Then, Eq. (19.25) can be expressed by

$$\left[(i\omega)^2+2\varsigma_{ej}\omega_{enj}(i\omega)+\omega_{enj}^2\right]P_j(\omega)=-\omega^{2-\alpha}\sec\frac{\alpha\pi}{2}Q_j(\omega). \tag{19.27}$$

From a practical view, we have $|\zeta_{ej}|\le 1$. Denote by ω_{endj} the jth equivalent damped natural frequency in the form (19.10). Considering the inverse Fourier transform of the above yields

$$p_j''(t)+2\varsigma_{ej}\omega_{enj}p_j'(t)+\omega_{nj}^2p_j(t)=q_j(t)*\mathrm{F}^{-1}\left(-\omega^{2-\alpha}\sec\frac{\alpha\pi}{2}\right). \tag{19.28}$$

The general solution in terms of $p_j(t)$ with zero initial conditions is attained in the form

$$p_j(t)=q_j(t)*\mathrm{F}^{-1}\left(-\omega^{2-\alpha}\sec\frac{\alpha\pi}{2}\right)*\frac{1}{M_j\omega_{endj}}e^{-\varsigma_{ej}\omega_{enj}t}\sin\omega_{endj}t,\quad t\ge 0. \tag{19.29}$$

Therefore, the general solution of the forced response to the Euler-Bernoulli beam with the coordinates of fractional order α is expressed by Eq. (19.4). The proof is completed. □

In what follows, zero initial conditions are defaulted, including exercises in Section 19.6, unless otherwise stated.

Note 19.1: If $\alpha=2$, $\omega_{endj}=\omega_{enj}=\omega_{nj}$.

Note 19.2: When $\alpha=2$, $\zeta_{ej}=0$.

Note 19.3: When $\alpha=2$, $p_j(t)$ reduces to the conventional coordinate in the form $p_j(t)=q_j(t)*\dfrac{1}{M_j\omega_{nj}}\sin\omega_{nj}t$.

19.3 CASE STUDY

19.3.1 Forced Response to Simply Supported Euler-Bernoulli Beam with Fractional Coordinates

For a simply supported Euler-Bernoulli beam, one has

$$\varphi_j(x)\sim\sin\frac{j\pi x}{l},\quad j=1,2,\ldots \tag{19.30}$$

Thus, the forced response in the zero state is expressed by

$$w(x,t)=\sum_{j=1}^{\infty}\sin\frac{j\pi x}{l}\left[q_j(t)*\mathrm{F}^{-1}\left(-\omega^{2-\alpha}\sec\frac{\alpha\pi}{2}\right)*\frac{1}{M_j\omega_{endj}}e^{-\varsigma_{ej}\omega_{enj}t}\sin\omega_{endj}t\right]. \tag{19.31}$$

Let

$$\mu_j=\frac{j\pi}{l},\quad j=1,2,\ldots \tag{19.32}$$

Then, the jth natural frequency is given by

$$\omega_{nj}=\mu_j^2\sqrt{\frac{EI}{\rho A}}=\left(\frac{j\pi}{l}\right)^2\sqrt{\frac{EI}{\rho A}},\quad j=1,2,\ldots \tag{19.33}$$

The quantities ω_{enj}, ζ_{ej}, and ω_{endj} are given by Eqs. (19.8)–(19.10), respectively.

Suppose $l=10$. When $q_j(t)=\delta(t)$, we illustrate the plots of $p_j(t)$ $(j=1, 2, 3, 4)$ in Figure 19.1. If $\alpha=2$, $p_j(t)$ reduces to the conventional coordinate functions as shown in Figure 19.2. The plots of $w(x, t)$ with the fractional coordinates are indicated in Figure 19.3. That with $\alpha=2$ is shown in Figure 19.4.

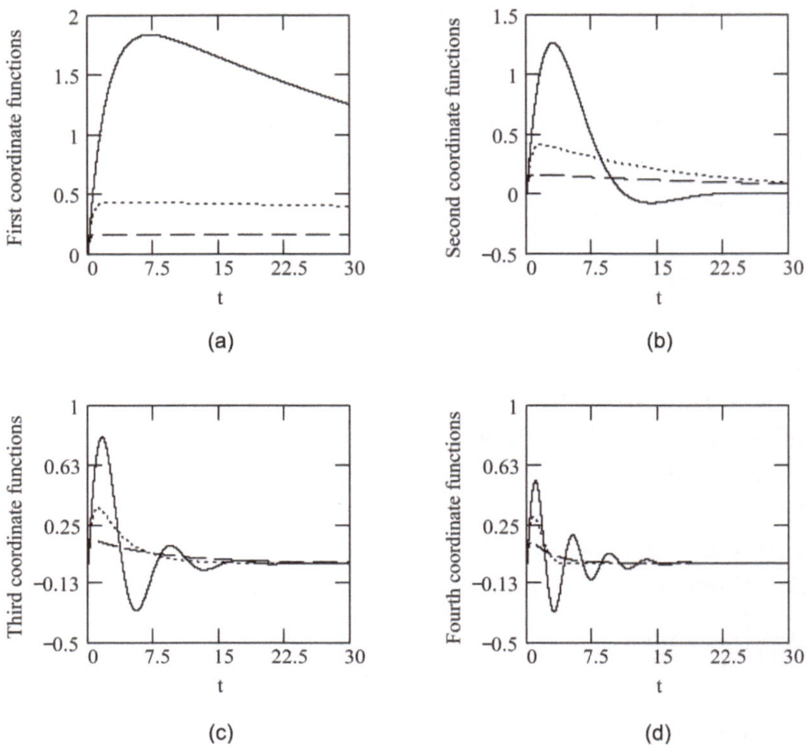

FIGURE 19.1 Plots of $p_j(t)$ $(j=1, 2, 3, 4)$ with $\omega=3.14$ for $\alpha=1.9$ (solid line), $\alpha=1.6$ (dot line), and $\alpha=1.3$ (dash line) when $q_j(t)=\delta(t)$ and $l=10$. (a) $p_1(t)$. (b) $p_2(t)$. (c) $p_3(t)$. (d) $p_4(t)$.

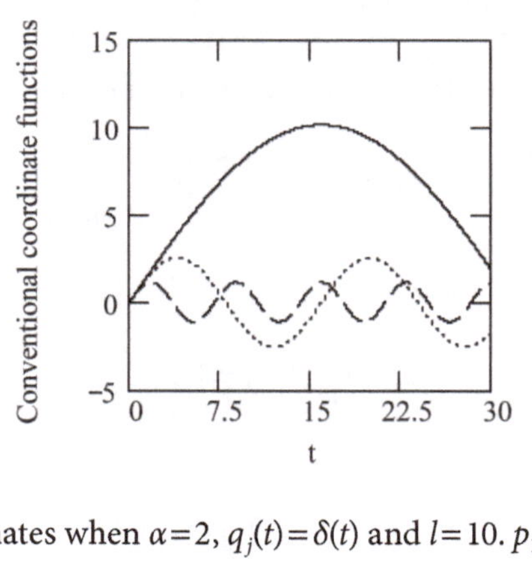

FIGURE 19.2 Plots of coordinates when $\alpha=2$, $q_j(t)=\delta(t)$ and $l=10$. $p_1(t)$ (solid), $p_2(t)$ (dot), and $p_3(t)$ (dash).

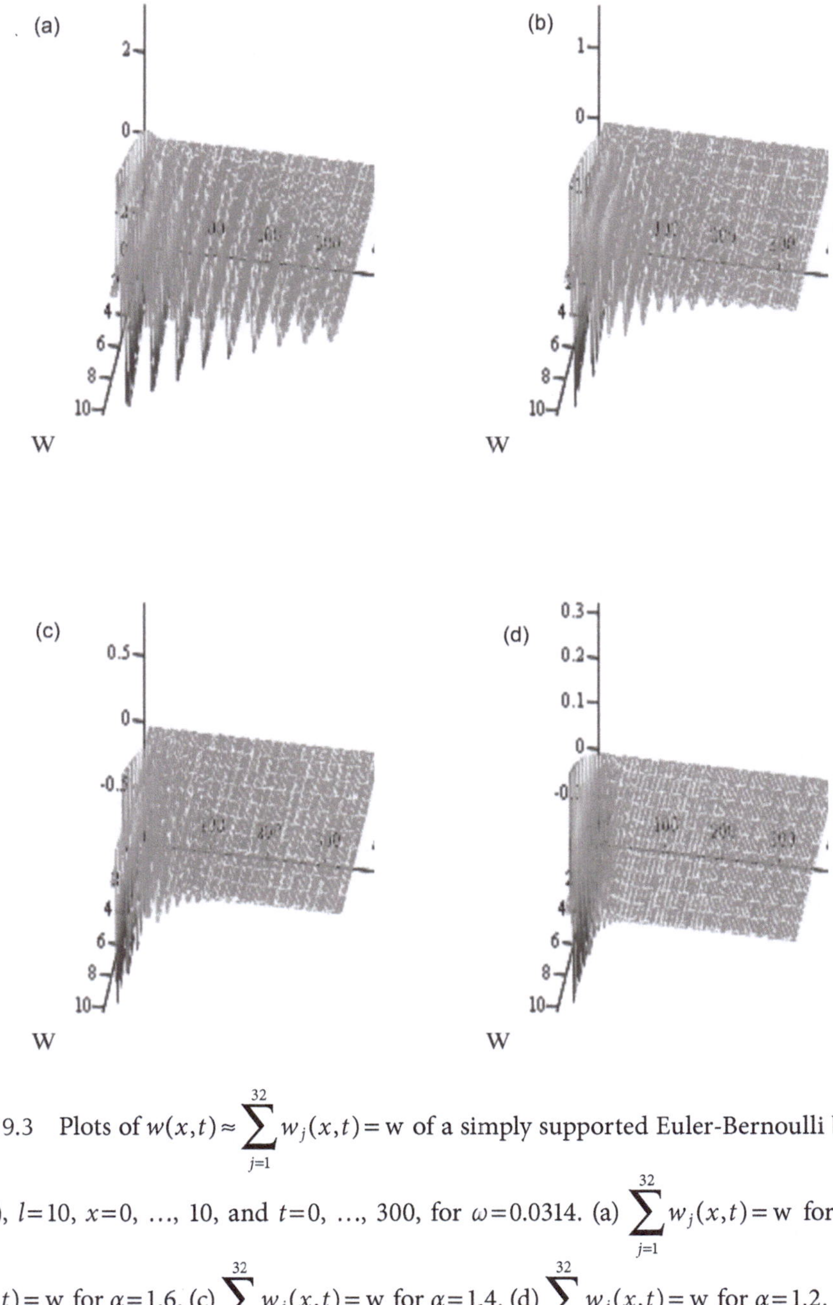

FIGURE 19.3 Plots of $w(x,t) \approx \sum_{j=1}^{32} w_j(x,t) = \mathrm{w}$ of a simply supported Euler-Bernoulli beam when $q_j(t) = \delta(t)$, $l = 10$, $x = 0, \ldots, 10$, and $t = 0, \ldots, 300$, for $\omega = 0.0314$. (a) $\sum_{j=1}^{32} w_j(x,t) = \mathrm{w}$ for $\alpha = 1.8$. (b) $\sum_{j=1}^{32} w_j(x,t) = \mathrm{w}$ for $\alpha = 1.6$. (c) $\sum_{j=1}^{32} w_j(x,t) = \mathrm{w}$ for $\alpha = 1.4$. (d) $\sum_{j=1}^{32} w_j(x,t) = \mathrm{w}$ for $\alpha = 1.2$.

19.3.2 Forced Response to Free-Free Euler-Bernoulli Beam with Fractional Coordinates

The vibration mode of a free-free Euler-Bernoulli beam is given by

$$\varphi_j(x) \sim \frac{\mathrm{ch}\mu_j l - \cos \mu_j l}{\sin \mu_j l - \mathrm{sh}\mu_j l}\left(\sin \mu_j x + \mathrm{sh}\mu_j x\right) + \left(\cos \mu_j x + \mathrm{ch}\mu_j x\right).$$

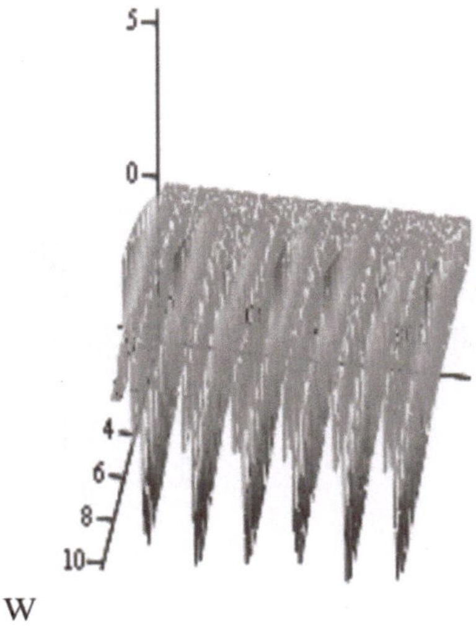

FIGURE 19.4 Plot of $w(x,t)\approx\sum_{j=1}^{32} w_j(x,t)=$ w of a simply supported Euler-Bernoulli beam when $q_j(t)=\delta(t)$, $l=10$, $\alpha=2$, $x=0, \ldots, 10$, and $t=0, \ldots, 300$, for $\omega=0.0314$.

Therefore, its forced response is expressed by

$$w(x,t)=\sum_{j=1}^{\infty}\frac{\left[\dfrac{\mathrm{ch}\mu_j l-\cos\mu_j l}{\sin\mu_j l-\mathrm{sh}\mu_j l}\left(\sin\mu_j x+sh\mu_j x\right)+\left(\cos\mu_j x+\mathrm{ch}\mu_j x\right)\right]}{\left[q_j(t)*\mathrm{F}^{-1}\left(-\omega^{2-\alpha}\sec\dfrac{\alpha\pi}{2}\right)*\dfrac{1}{M_j\omega_{\mathrm{end}j}}e^{-\varsigma_{ej}\omega_{\mathrm{en}j}t}\sin\omega_{\mathrm{end}j}t\right]}. \tag{19.34}$$

In this case, we use the numerical result of $\cos(\mu_j)=1/\mathrm{ch}(\mu_j)$ in the form

$$\frac{\mu_j l}{(2j+1)\dfrac{\pi}{2}},\quad j=1,2,\ldots \tag{19.35}$$

The jth natural frequency is given by

$$\omega_{nj}=\mu_j^2\sqrt{\frac{EI}{\rho A}}=\left(\frac{(2j+1)\pi}{2l}\right)^2\sqrt{\frac{EI}{\rho A}}. \tag{19.36}$$

The plots of $w_1(x, t)=$ w1 are illustrated in Figures 19.5 and 19.6.

19.4 INCREASING-VIBRATIONS

In the previous discussions, we use $1<\alpha\leq 2$. We now consider $2<\alpha<3$. One thing worth noting is that the beam will be increasing-vibrating if $2<\alpha<3$.

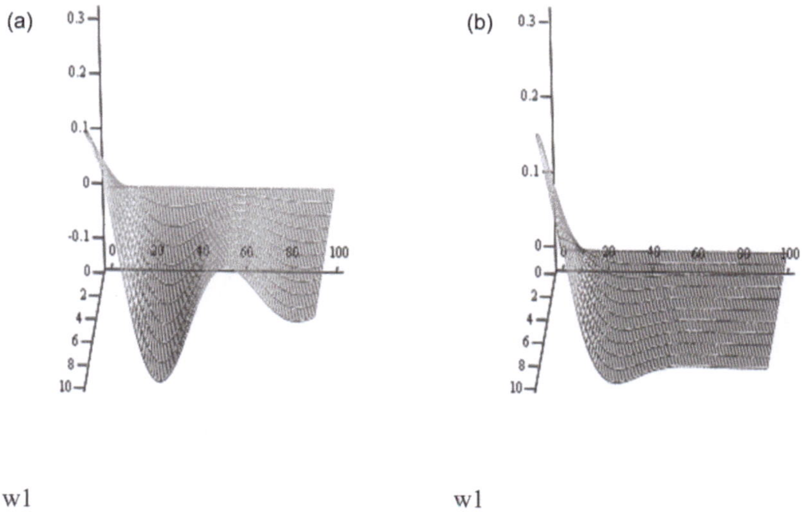

FIGURE 19.5 Plots of $w_1(x, t)$=w1 of a free-free Euler-Bernoulli beam with the fractional coordinates when $q_j(t)=\delta(t)$, $l=10$, $x=0, \ldots, 10$, and $t=0, \ldots, 100$, for $\omega=0.314$. (a) w1 for $\alpha=1.9$. (b) w1 for $\alpha=1.7$.

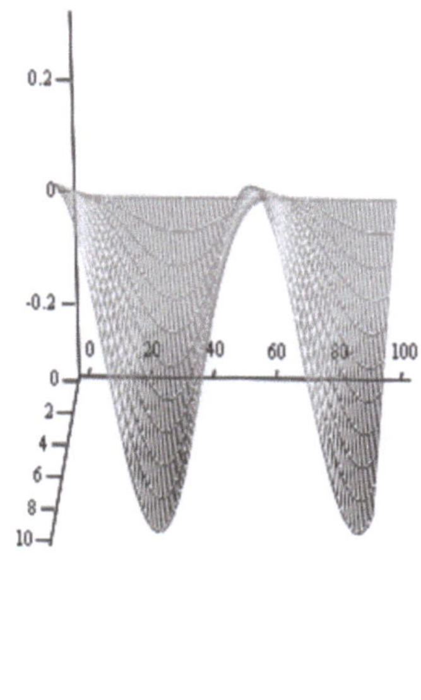

FIGURE 19.6 Plot of $w_1(x, t)$=w1 of a free-free Euler-Bernoulli beam with the fractional coordinates when $q_j(t)=\delta(t)$, $l=10$, $\alpha=2$, $x=0, \ldots, 10$, and $t=0, \ldots, 100$, for $\omega=0.314$.

Theorem 19.2. For an Euler-Bernoulli beam with the fractional coordinates, its equivalent coordinate damping ratio $\zeta_{ej}<0$ if $2<\alpha<3$.

Proof. From Eq. (19.9), we see that $\zeta_{ej}<0$ if $2<\alpha<3$. The proof is finished. □

The above theorem implies that an Euler-Bernoulli beam with the fractional coordinates of order $2<\alpha<3$ suffers from increasing-vibrations. Figure 19.7 shows the negative ζ_{ej}. In Figure 19.8, we indicate the plots of coordinates for a simply supported Euler-Bernoulli beam for $2<\alpha<3$. Plots of $w(x, t)$ in the case of increasing-vibrations when $2<\alpha<3$ are illustrated in Figure 19.9.

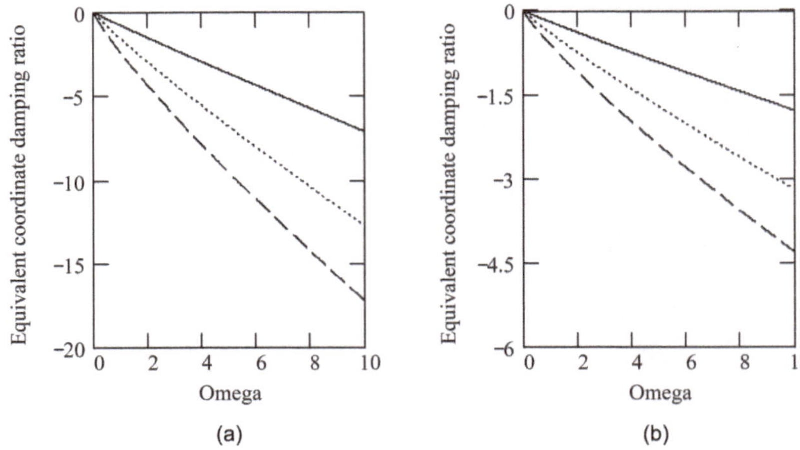

FIGURE 19.7 Illustrations of negative ζ_{ej} for $\alpha=2.1$ (solid), 2.2 (dot), 2.3 (dash). (a) ζ_{e1}. (b) ζ_{e2}.

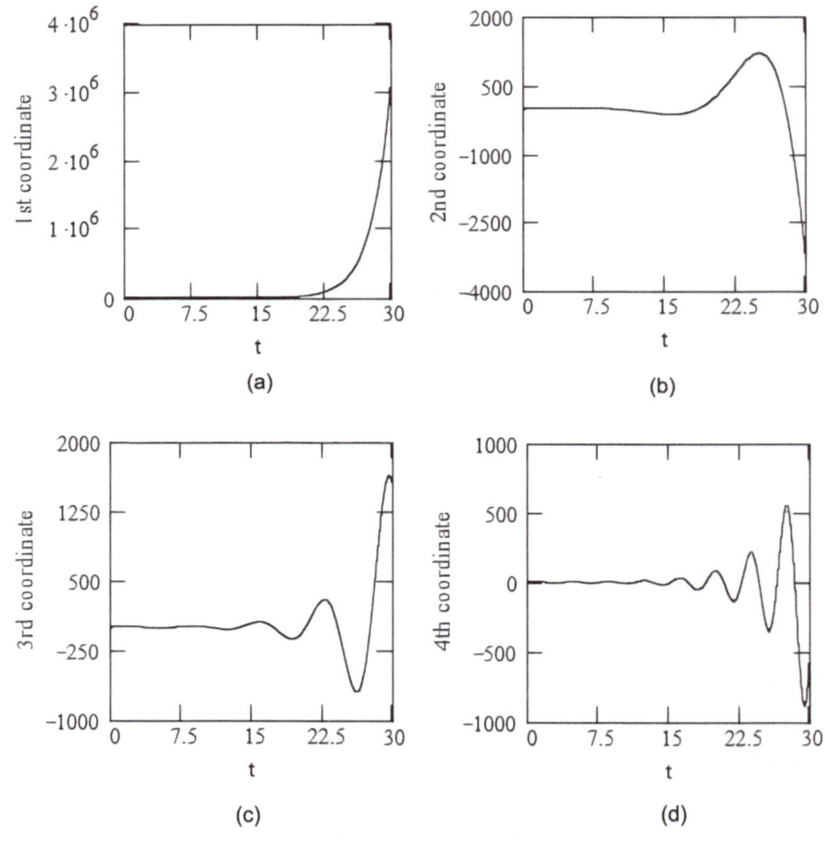

FIGURE 19.8 Plots of a simply supported Euler-Bernoulli beam with the fractional coordinates $p_j(t)$ for $\omega=3.14$ and $\alpha=2.1$ (increasing-vibration). (a) $p_1(t)$. (b) $p_2(t)$. (c) $p_3(t)$. (d) $p_4(t)$.

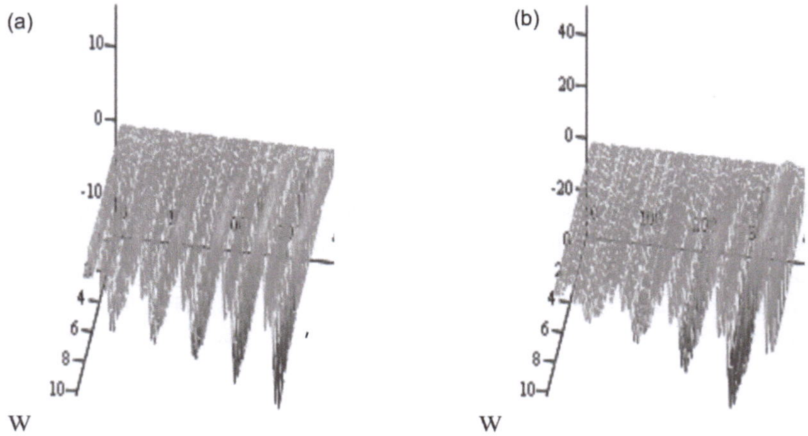

FIGURE 19.9 Plots of $w(x,t) \approx \sum_{j=1}^{32} w_j(x,t) = \text{w}$ of a simply supported Euler-Bernoulli beam with the fractional coordinates when $q_j(t) = \delta(t)$, $l = 10$, and $\omega = 0.0314$. (a) w for $\alpha = 2.1$ (increasing-vibration). (b) w for $\alpha = 2.2$ (increasing-vibration).

19.5 SUMMARY

We have presented the analytical expression of the general solution of forced response to a damping free Euler-Bernoulli beam with the fractional coordinates of order $1<\alpha<3$. When $1<\alpha<2$, the beam is damping free in form but positively damped in nature. If $2<\alpha<3$, the beam suffers from increasing-vibrations. It reduces to the conventional Euler-Bernoulli beam when $\alpha=2$.

19.6 EXERCISES

19.1. Consider a simply supported Euler-Bernoulli beam with the fractional coordinates of order α. Find the forced response $w(x, t)$ when $q_j(t)=u(t)$ $(j=1,\ldots)$, where $u(t)$ is the unit step function.

19.2. Consider a simply supported Euler-Bernoulli beam with the fractional coordinates of order α. Find the forced response $w(x, t)$ for $q_j(t)=\sin(\omega_0 t)$ $(j=1, \ldots)$, where ω_0 is a constant.

19.3. Consider a free-free Euler-Bernoulli beam with the fractional coordinates of order α. Find the forced response $w(x, t)$ if $q_j(t)=\sin(\omega_0 t)$ $(j=1, \ldots)$, where ω_0 is a constant.

19.4. Consider a free-free Euler-Bernoulli beam with the fractional coordinates of order α. Find the forced response $w(x, t)$ when $q_j(t)=u(t)$ $(j=1, \ldots)$, where $u(t)$ is the unit step function.

19.5. Consider a simply supported Euler-Bernoulli beam with the fractional coordinates of order α. Find the forced response $w(x, t)$ if $q_j(t)=u(t-\tau/2)-u(t+\tau/2)$ $(j=1, \ldots)$, where τ is a constant and $u(t)$ is the unit step function.

19.6. Consider a free-free Euler-Bernoulli beam with the fractional coordinates of order α. Find the forced response $w(x, t)$ for $q_j(t) = u(t - \tau/2) - u(t + \tau/2)$ $(j = 1, \ldots)$, where τ is a constant and $u(t)$ is the unit step function.

REFERENCES

1. O. M. Palley, Г. B. Bahizov, and E. Я. Voroneysk, *Handbook of Ship Structural Mechanics*, National Defense Industry Publishing House, Beijing, China, 2002. In Chinese. Translated from Russian by B. H. Xu, X. Xu, and M. Q. Xu.
2. C. M. Harris, *Shock and Vibration Handbook*, 5th Ed., McGraw-Hill, New York, 2002.
3. M. Li, *Theory of Fractional Engineering Vibrations*, Walter de Gruyter, Berlin/Boston, MA, 2021.
4. K. Nakagawa and M. Ringo, *Engineering Vibrations*, Shanghai Science and Technology Publishing House, Shanghai, China, 1981. In Chinese. Translated from Japanese by S. R. Xia.
5. M. V. Shitikova, Fractional operator viscoelastic models in dynamic problems of mechanics of solids: A review, *Mechanics of Solids*, 57(1), 2022, 1–33.
6. Y. A. Rossikhin and M. V. Shitikova, Classical beams and plates in a fractional derivative medium, Impact response, in: H. Altenbach and A. Öchsner (eds.), *Encyclopedia of Continuum Mechanics*, vol. 1, Springer, Berlin, 2020, 294–305.

CHAPTER 20

Seven Classes of Fractionally Damped Euler-Bernoulli Beams

THIS CHAPTER DESCRIBES MY classification of fractionally damped Euler-Bernoulli beams for Chapters 21–27.

20.1 INTRODUCTION

I shall address the forced responses to seven classes of fractionally damped Euler-Bernoulli beams from Chapters 21 to 27, respectively. For facilitating the discussions there, I write the classification of fractionally damped Euler-Bernoulli beams in this short chapter.

Following Palley et al. [1], Harris [2], Li [3], the motion equation of the conventionally damped Euler-Bernoulli is given by

$$\frac{\partial^2}{\partial x^2}\left(EI(x)\frac{\partial^2 w}{\partial x^2}+c_s I(x)\frac{\partial^3 w}{\partial x^2 \partial t}\right)+\rho(x)A\frac{\partial^2 w}{\partial t^2}+c\frac{\partial w}{\partial t}=q(x,t), \tag{20.1}$$

where $\rho(x)$ is material density at position x, $A(x)$ is cross-sectional area at x, $I(x)$ is cross-sectional moment of inertia at x, $EI(x)$ is bending rigidity at x, E is elastic modulus, c is external damping, c_s is internal damping, $q(x, t)$ is driven force at x and t, and $w(x, t)$ is deflection at x and t. In what follows, suppose that $\rho(x)$, $A(x)$, and $I(x)$ are constants. Hence, Eq. (20.1) is simplified in the form

$$\frac{\partial^2}{\partial x^2}\left(EI\frac{\partial^2 w}{\partial x^2}+c_s I\frac{\partial^3 w}{\partial x^2 \partial t}\right)+\rho A\frac{\partial^2 w}{\partial t^2}+c\frac{\partial w}{\partial t}=q(x,t). \tag{20.2}$$

DOI: 10.1201/9781003460947-23

20.2 SEVEN CLASSES

The following motion equation is for the class 1 type of fractionally damped Euler-Bernoulli beams

$$\frac{\partial^2}{\partial x^2}\left(EI\frac{\partial^2 w}{\partial x^2}+c_sI\frac{\partial^3 w}{\partial x^2\partial t}\right)+\rho A\frac{\partial^\alpha w}{\partial t^\alpha}+c\frac{\partial w}{\partial t}=q(x,t), \quad 1<\alpha<3. \tag{20.3}$$

The second class designates the motion equation below for fractionally damped Euler-Bernoulli beams in the form

$$\frac{\partial^2}{\partial x^2}\left(EI\frac{\partial^2 w}{\partial x^2}+c_sI\frac{\partial^3 w}{\partial x^2\partial t}\right)+\rho A\frac{\partial^2 w}{\partial t^2}+c\frac{\partial^\beta w}{\partial t^\beta}=q(x,t), \quad 0<\beta<2. \tag{20.4}$$

The third class means that the motion equation of fractionally damped Euler-Bernoulli beams is in the form

$$\frac{\partial^2}{\partial x^2}\left(EI\frac{\partial^2 w}{\partial x^2}+c_sI\frac{\partial^{2+\lambda} w}{\partial x^2\partial t^\lambda}\right)+\rho A\frac{\partial^2 w}{\partial t^2}+c\frac{\partial w}{\partial t}=q(x,t), \quad 0<\lambda<2. \tag{20.5}$$

Class 4 implies

$$\frac{\partial^2}{\partial x^2}\left(EI\frac{\partial^2 w}{\partial x^2}+c_sI\frac{\partial^{2+\lambda} w}{\partial x^2\partial t^\lambda}\right)+\rho A\frac{\partial^2 w}{\partial t^2}+c\frac{\partial^\beta w}{\partial t^\beta}=q(x,t), \quad 0<\beta<2,\ 0<\lambda<2. \tag{20.6}$$

The motion equation of the class 5 type fractionally damped Euler-Bernoulli beams is

$$\frac{\partial^2}{\partial x^2}\left(EI\frac{\partial^2 w}{\partial x^2}+c_sI\frac{\partial^3 w}{\partial x^2\partial t}\right)+\rho A\frac{\partial^\alpha w}{\partial t^\alpha}+c\frac{\partial^\beta w}{\partial t^\beta}=q(x,t), \quad 1<\alpha<3,\ 0<\beta<2. \tag{20.7}$$

The class 6 of fractionally damped Euler-Bernoulli beams means the motion equation given by

$$\frac{\partial^2}{\partial x^2}\left(EI\frac{\partial^2 w}{\partial x^2}+c_sI\frac{\partial^{2+\lambda} w}{\partial x^2\partial t^\lambda}\right)+\rho A\frac{\partial^\alpha w}{\partial t^\alpha}+c\frac{\partial w}{\partial t}=q(x,t), \quad 1<\alpha<3, 0<\lambda<2. \tag{20.8}$$

Class 7 is made up of fractionally damped Euler-Bernoulli beams with the motion equation in the form

$$\frac{\partial^2}{\partial x^2}\left(EI\frac{\partial^2 w}{\partial x^2}+c_sI\frac{\partial^{2+\lambda} w}{\partial x^2\partial t^\lambda}\right)+\rho A\frac{\partial^\alpha w}{\partial t^\alpha}+c\frac{\partial^\beta w}{\partial t^\beta}=q(x,t), \quad 1<\alpha<3,\ 0<\beta<2,\ 0<\lambda<2. \tag{20.9}$$

Each class may correspond to specific phenomena of vibrations. The analytic solution of the forced response of each class of fractionally damped Euler-Bernoulli beams in closed form remains unknown. I shall address the analytic solution for each class one by one from Chapters 21 to 27, respectively.

REFERENCES

1. O. M. Palley, Г. B. Bahizov, and E. Я. Voroneysk, *Handbook of Ship Structural Mechanics*, National Defense Industry Publishing House, Beijing, China, 2002. In Chinese. Translated from Russian by B. H. Xu, X. Xu, and M. Q. Xu.
2. C. M. Harris, *Shock and Vibration Handbook*, 5th Ed., McGraw-Hill, New York, 2002.
3. M. Li, *Theory of Fractional Engineering Vibrations*, Walter de Gruyter, Berlin/Boston, MA, 2021.

CHAPTER 21

Forced Response to Damped Euler-Bernoulli Beam with Fractional Inertia Force (Class 1)

THE CONTRIBUTIONS PRESENTED IN this chapter are in three aspects for the forced response to a damped Euler-Bernoulli beam with fractional inertia force. First, we propose an equivalent motion equation of fractional coordinates of such a beam in Theorem 21.1. Second, we present the analytic expression of the forced response to it in Theorem 21.6. Third, we reveal a new variation of the Rayleigh damping in Theorem 21.3.

21.1 INTRODUCTION

According to the Voigt assumption on materials by taking into account internal damping, the motion equation of the conventional damped Euler-Bernoulli beam is in the form

$$\frac{\partial^2}{\partial x^2}\left(EI\frac{\partial^2 w}{\partial x^2}+c_s I\frac{\partial^3 w}{\partial x^2 \partial t}\right)+\rho A\frac{\partial^2 w}{\partial t^2}+c\frac{\partial w}{\partial t}=q(x,t), \tag{21.1}$$

where c_s is internal damping and c is external damping (Palley et al. [1]). In Eq. (21.1), $c\frac{\partial w}{\partial t}$ is external damping force, $c_s I\frac{\partial^3 w}{\partial x^2 \partial t}$ is internal damping force, and $\rho A\frac{\partial^2 w}{\partial t^2}$ is inertia force. The forced response to the aforementioned equation under the Rayleigh damping assumption is known (Palley et al. [1], Jin and Xia [2]).

In this chapter, we use the theory of fractional vibrations in Chapter 8 and Li [3,4] to study a damped Euler-Bernoulli beam with the fractional inertia force. Such a beam is

DOI: 10.1201/9781003460947-24

called the class 1 fractionally damped Euler-Bernoulli beam in the last chapter. Its motion equation is in the form

$$\frac{\partial^2}{\partial x^2}\left(EI\frac{\partial^2 w}{\partial x^2}+c_sI\frac{\partial^3 w}{\partial x^2 \partial t}\right)+\rho A\frac{\partial^\alpha w}{\partial t^\alpha}+c\frac{\partial w}{\partial t}=q(x,t), \quad 1<\alpha<3. \tag{21.2}$$

The analytic solution in the closed form regarding forced response to the above equation remains unknown. This chapter presents its analytic solution in the closed form.

The rest of this chapter is organized as follows. The analytic solution to Eq. (21.2) is addressed in Section 21.2. Case study is in Section 21.3, which is followed by summary.

21.2 ANALYTIC EXPRESSION OF FORCED RESPONSE

Using $w(x, t)=\varphi(x)p(t)$ to rewrite Eq. (21.2) produces

$$\begin{aligned}&\sum_{j=1}^{\infty}\rho A\varphi_j(x)\frac{d^\alpha p_j(t)}{dt^\alpha}+\sum_{j=1}^{\infty}c\varphi_j(x)\frac{dp_j(t)}{dt}+\sum_{j=1}^{\infty}\frac{d^2}{dx^2}\left[c_sI\frac{d^2\varphi_j(x)}{dx^2}\right]\frac{dp_j(t)}{dt}\\&+\sum_{j=1}^{\infty}\frac{d^2}{dx^2}\left[EI\frac{d^2\varphi_j(x)}{dx^2}\right]p_j(t)=q(x,t).\end{aligned} \tag{21.3}$$

Multiplying $\varphi_m(x)$ on both sides of the above yields

$$\begin{aligned}&\sum_{j=1}^{\infty}\rho A\varphi_j(x)\varphi_m(x)\frac{d^\alpha p_j(t)}{dt^\alpha}+\sum_{j=1}^{\infty}c\varphi_j(x)\varphi_m(x)\frac{dp_j(t)}{dt}\\&+\sum_{j=1}^{\infty}\varphi_m(x)\frac{d^2}{dx^2}\left[c_sI\frac{d^2\varphi_j(x)}{dx^2}\right]\frac{dp_j(t)}{dt}\\&+\sum_{j=1}^{\infty}\varphi_m(x)\frac{d^2}{dx^2}\left[EI\frac{d^2\varphi_j(x)}{dx^2}\right]p_j(t)=\varphi_m(x)q(x,t).\end{aligned} \tag{21.4}$$

Performing the integral from 0 to l in terms of x on both sides of the aforementioned equation and considering the orthogonality of vibration modes results in

$$\begin{aligned}&M_j\frac{d^\alpha p_j(t)}{dt^\alpha}+\frac{M_jc}{\rho A}\frac{dp_j(t)}{dt}+\sum_{j=1}^{\infty}\int_0^l\varphi_m(x)\frac{d^2}{dx^2}\left[c_sI\frac{d^2\varphi_j(x)}{dx^2}\right]dx\frac{dp_j(t)}{dt}\\&+\sum_{j=1}^{\infty}\int_0^l\varphi_m(x)\frac{d^2}{dx^2}\left[EI\frac{d^2\varphi_j(x)}{dx^2}\right]dxp_j(t)=Q_j(t),\end{aligned} \tag{21.5}$$

where $M_j=\rho A\int_0^l\varphi_j^2(x)dx$.

Since (see Eqs. (6.136) or (6.174))

$$\frac{d^2}{dx^2}\left[EI\frac{d^2\varphi_j(x)}{dx^2}\right]=\rho A\omega_{nj}^2\varphi_j(x), \tag{21.6}$$

equation (21.5) can be written by

$$M_j\frac{d^\alpha p_j(t)}{dt^\alpha}+\frac{M_jc}{\rho A}\frac{dp_j(t)}{dt}+\sum_{j=1}^{\infty}\int_0^l\varphi_m(x)\frac{d^2}{dx^2}\left[c_sI\frac{d^2\varphi_j(x)}{dx^2}\right]dx\frac{dp_j(t)}{dt}$$
$$+M_j\omega_{nj}^2p_j(t)=Q_j(t). \tag{21.7}$$

Adopting the Rayleigh damping assumption, see Chapter 14, Palley et al. [1], Jin and Xia [2], we have

$$c=rAa, \tag{21.8}$$

where a is a coefficient with the unit of time since ρA is with the unit of mass and

$$c_s=Eb, \tag{21.9}$$

where b is a coefficient with the unit of frequency as E is with the unit [N/m].

Substituting Eqs. (21.8) and (21.9) into (21.7) and taking into account the orthogonality of vibration modes results in

$$M_j\frac{d^\alpha p_j(t)}{dt^\alpha}+M_ja\frac{dp_j(t)}{dt}+M_j\omega_{nj}^2b\frac{dp_j(t)}{dt}+M_j\omega_{nj}^2p_j(t)=Q_j(t). \tag{21.10}$$

Therefore, for the jth order forced vibration of the damped Euler-Bernoulli beam with fractional inertia, we have its coordinate equation in the form

$$\frac{d^\alpha p_j(t)}{dt^\alpha}+\left(a+\omega_{nj}^2b\right)\frac{dp_j(t)}{dt}+\omega_{nj}^2p_j(t)=f_j(t), \tag{21.11}$$

where

$$f_j(t)=\frac{Q_j(t)}{M_j}. \tag{21.12}$$

Theorem 21.1. An equivalent equation of (21.11) is given by

$$-\cos\frac{\alpha\pi}{2}\omega^{\alpha-2}\frac{d^2p_j(t)}{dt^2}+\left(\sin\frac{\alpha\pi}{2}\omega^{\alpha-1}+a+\omega_{nj}^2b\right)\frac{dp_j(t)}{dt}+\omega_{nj}^2p_j(t)=f_j(t). \tag{21.13}$$

Proof. Let $P_j(\omega)=\mathrm{F}[p_j(t)]$ and $F_j(\omega)=\mathrm{F}[f_j(t)]$. Doing the Fourier transform on both sides of Eq. (21.11) yields

$$\left[(i\omega)^{\alpha}+a(i\omega)+\omega_{nj}^{2}b(i\omega)+\omega_{nj}^{2}\right]P_{j}(\omega)=F_{j}(\omega). \tag{21.14}$$

Taking into account the principal value of i^{α}, we have

$$\begin{aligned}&(i\omega)^{\alpha}+a(i\omega)+\omega_{nj}^{2}b(i\omega)+\omega_{nj}^{2}\\&=-\cos\frac{\alpha\pi}{2}\omega^{\alpha-2}(i\omega)^{2}+\left(\sin\frac{\alpha\pi}{2}\omega^{\alpha-1}+a+\omega_{nj}^{2}b\right)(i\omega)+\omega_{nj}^{2}.\end{aligned} \tag{21.15}$$

Therefore, Eq. (21.14) can be expressed by

$$\begin{aligned}&\left[(i\omega)^{\alpha}+a(i\omega)+\omega_{nj}^{2}b(i\omega)+\omega_{nj}^{2}\right]P_{j}(\omega)\\&=\begin{bmatrix}-\cos\frac{\alpha\pi}{2}\omega^{\alpha-2}(i\omega)^{2}\\+\left(\sin\frac{\alpha\pi}{2}\omega^{\alpha-1}+a+\omega_{nj}^{2}b\right)(i\omega)+\omega_{nj}^{2}\end{bmatrix}P_{j}(\omega)=F_{j}(\omega).\end{aligned} \tag{21.16}$$

The above is the Fourier transform of Eq. (21.13). The proof is finished. □

Theorem 21.2. Denote by m_{e1} the equivalent mass of the system (21.11). Then,

$$m_{e1}=-\cos\frac{\alpha\pi}{2}\omega^{\alpha-2}. \tag{21.17}$$

Proof. The right side of the above is the coefficient of $\frac{d^{2}p_{j}(t)}{dt^{2}}$ in Eq. (21.13). Thus, Eq. (21.17) is true. The proof is finished. □

The quantity m_{e1} is positive for $1<\alpha<3$. It does not relate to j. If $\alpha=2$, $m_{e1}=1$.

Theorem 21.3. Denote by c_{e1j} the jth equivalent damping in the system (21.11). Then,

$$c_{e1j}=\sin\frac{\alpha\pi}{2}\omega^{\alpha-1}+a+\omega_{nj}^{2}b. \tag{21.18}$$

Proof. The right side of the above is the coefficient of $dp_j(t)/dt$ in Eq. (21.13). Thus, Eq. (21.18) is true. The proof is completed. □

The quantity c_{e1j} consists of two parts. One is $a+\omega_{nj}^{2}b$. The other is the one resulted from the fractional inertia, that is, $\sin\frac{\alpha\pi}{2}\omega^{\alpha-1}$. Thus, c_{e1j} is a variation of the Rayleigh damping. If $\alpha=2$, $c_{e1j}=a+\omega_{nj}^{2}b$.

Using Eqs. (21.17) and (21.18), (21.13) is rewritten by

$$m_{e1}\frac{d^{2}p_{j}(t)}{dt^{2}}+c_{e1j}\frac{dp_{j}(t)}{dt}+\omega_{nj}^{2}p_{j}(t)=f_{j}(t). \tag{21.19}$$

Besides, (21.16) can be rewritten by

$$\left[(i\omega)^2+\frac{c_{e1j}}{m_{e1}}(i\omega)+\frac{\omega_{nj}^2}{m_{e1}}\right]P_j(\omega)=\frac{F_j(\omega)}{m_{e1}}. \tag{21.20}$$

Denote by ζ_{e1j} the jth equivalent damping ratio in the system (21.11). It is defined by

$$\varsigma_{e1j}=\frac{c_{e1j}}{2\sqrt{m_{e1}}}=\frac{\sin\frac{\alpha\pi}{2}\omega^{\alpha-1}+a+\omega_{nj}^2 b}{2\sqrt{-\cos\frac{\alpha\pi}{2}\omega^{\alpha-2}}}. \tag{21.21}$$

Let ω_{en1j} be the jth equivalent damping free natural frequency of the system (21.11). Then, we have

$$\omega_{en1j}^2=\frac{\omega_{nj}^2}{m_{e1}}. \tag{21.22}$$

Thus, we can write Eq. (21.20) by

$$\left[(i\omega)^2+2\varsigma_{e1j}\omega_{en1j}(i\omega)+\omega_{en1j}^2\right]P_j(\omega)=\frac{F_j(\omega)}{m_{e1}}. \tag{21.23}$$

Practically, we adopt

$$\left|\varsigma_{e1j}\right|\leq 1. \tag{21.24}$$

Thus, the equivalent damped natural frequency regarding the system (21.11) is given by

$$\omega_{end1j}=\omega_{en1j}\sqrt{1-\varsigma_{e1j}^2}. \tag{21.25}$$

Theorem 21.4. Denote by $h_{1j}(t)$ the jth impulse response function in the system (21.19). Then,

$$h_{1j}(t)=\frac{1}{m_{e1}\omega_{end1j}}e^{-\varsigma_{e1j}\omega_{en1j}t}\sin\omega_{end1j}t,\quad t>0. \tag{21.26}$$

Proof. Since Eq. (21.19) is in the form of standard motion equation, $h_{1j}(t)$ takes the above form. This finishes the proof. □

When using the unit step function $u(t)$, the above can be expressed by

$$h_{1j}(t)=\frac{1}{m_{e1}\omega_{end1j}}e^{-\varsigma_{e1j}\omega_{en1j}t}\left(\sin\omega_{end1j}t\right)u(t). \tag{21.27}$$

Theorem 21.5. The jth coordinate with respect to Eq. (21.11) with zero initial conditions is represented by

$$p_j(t)=f_j(t)*h_{1j}(t)=\frac{1}{m_{e1}\omega_{end1j}}\int_{-\infty}^{\infty}e^{-\varsigma_{e1j}\omega_{en1j}\tau}\sin\omega_{end1j}\tau f_j(t-\tau)d\tau. \tag{21.28}$$

Proof. From the convolution theory, we see that the above is true. The proof is finished. □

Theorem 21.6. The zero-state forced response to a damped Euler-Bernoulli beam with the fractional inertia force is expressed by

$$w(x,t)=\sum_{j=1}^{\infty}\frac{\varphi_j(x)}{m_{e1}\omega_{end1j}}\int_{-\infty}^{\infty}e^{-\varsigma_{e1j}\omega_{en1j}\tau}\sin\omega_{end1j}\tau f_j(t-\tau)d\tau. \tag{21.29}$$

Proof. Because $w(x,t)=\sum_{j=1}^{\infty}\varphi_j(x)p_j(t)=\sum_{j=1}^{\infty}\varphi_j(x)\left[f_j(t)*h_{1j}(t)\right]$, using Theorem 21.5 yields the above. The proof is finished. □

If $\alpha=2$, $\omega_{en1j}=\omega_{nj}$ and $\varsigma_{e1j}=\dfrac{a+\omega_{nj}^2 b}{2}$. In what follows, zero initial conditions are defaulted, including exercises in Section 21.5.

21.3 CASE STUDY

For a simply supported Euler-Bernoulli beam, we have

$$\varphi_j(x)\sim\sin\frac{j\pi x}{l},\quad j=1,2,\ldots \tag{21.30}$$

Thus,

$$w(x,t)=\sum_{j=1}^{\infty}\sin\frac{j\pi x}{l}\left[f_j(t)*\frac{1}{m_{e1}\omega_{end1j}}e^{-\varsigma_{e1j}\omega_{en1j}t}\sin\omega_{end1j}t\right]. \tag{21.31}$$

Since

$$\mu_j=\frac{j\pi}{l},\quad j=1,2,\ldots \tag{21.32}$$

the jth order natural angular frequency is in the form

$$\omega_{n1j}=\mu_j^2\sqrt{\frac{EI}{\rho A}}=\left(\frac{j\pi}{l}\right)^2\sqrt{\frac{EI}{\rho A}},\quad j=1,2,\ldots \tag{21.33}$$

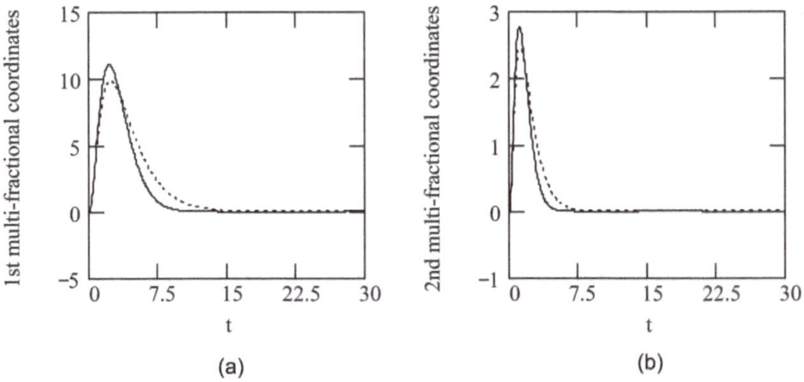

FIGURE 21.1 Plots of $p_j(t)$ ($j=1, 2$) of a simply supported damped Euler-Bernoulli beam with the fractional inertia force for $\alpha=1.8$ (solid line), 1.4 (dot line) when $f_j(t)=10\sin(0.314t)$, $l=10$, $\omega=0.314$, $a=1$, $b=1$, and $E=1$. (a) $p_1(t)$. (b) $p_2(t)$.

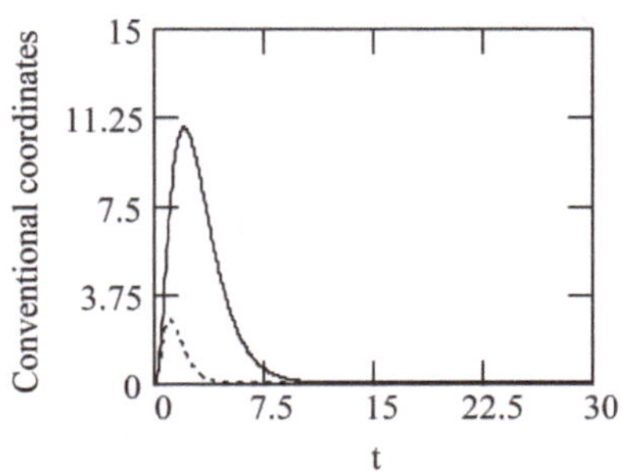

FIGURE 21.2 Plots of $p_1(t)$ (solid line) and $p_2(t)$ (dot line) when $\alpha=2$, $f_j(t)=10\sin(0.314t)$, $l=10$, $\omega=0.314$, $a=1$, $b=1$, and $E=1$.

The quantities ω_{en1j}, ζ_{e1j}, and ω_{end1j} are given by Eqs. (21.22), (21.21), and (21.25), respectively.

Let $l=10$. When $f_j(t)=10\sin(0.314t)u(t)$, we indicate the plots of $p_j(t)$ ($j=1, 2$) in Figure 21.1. If $\alpha=2$, $p_j(t)$ reduces to the conventional coordinate functions as shown in Figure 21.2. The plots of $w_1(x, t)=\mathrm{w1}$ are shown in Figure 21.3.

21.4 SUMMARY

We have given an equivalent motion equation of a damped Euler-Bernoulli beam with the fractional inertia force in Theorem 21.1. We have proposed the analytic expression of the forced response to a damped Euler-Bernoulli beam with the fractional inertia force under zero initial conditions in Theorem 21.6. The case study has been described. We have revealed a novel variation of the Rayleigh damping in Theorem 21.3 with the quantity c_{e1j}.

21.5 EXERCISES

21.1. Suppose the jth driven function $f_j(t)=u(t)$ (unit step function) ($j=1, 2, \ldots$). For a simply supported damped Euler-Bernoulli beam with the fractional inertia force of order $\alpha \in (1, 3)$, when $a=1$, $b=1$, $E=1$, and beam length $l=10$, find the forced response $w(x, t)$.

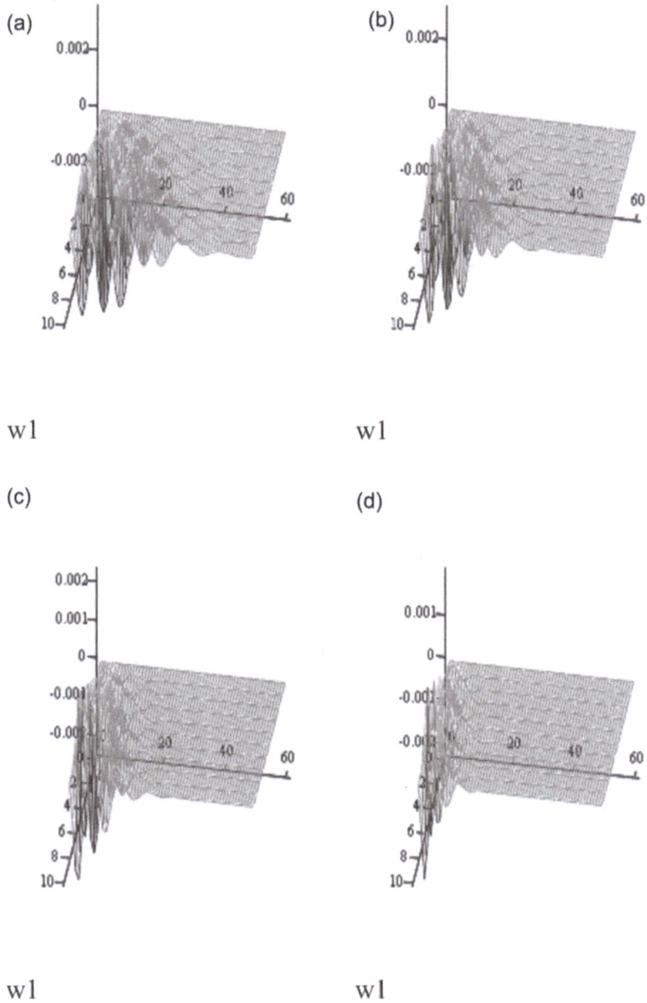

FIGURE 21.3 Plots of $w_1(x, t)$=w1 of a simply supported damped Euler-Bernoulli beam with the fractional inertia force when $f_j(t)=10\sin(0.314t)$, $l=10$, $\omega=0.0314$, $a=1$, $b=1$, $E=1$, $x=0, \ldots, 10$, and $t=0, \ldots, 60$. (a) w1 for $\alpha=1.6$. (b) w1 for $\alpha=1.8$. (c) w1 for $\alpha=2$. (d) w1 for $\alpha=2.4$.

21.2. Suppose the jth driven function $f_j(t)=u(t-\tau/2)-u(t+\tau/2)$ ($j=1, 2, \ldots$), where τ is a constant and $u(t)$ is the unit step function. For a simply supported damped Euler-Bernoulli beam with the fractional inertia force of order $\alpha \in (1, 3)$, when $a=1$, $b=1$, $E=1$, and beam length $l=10$, find the forced response $w(x, t)$.

21.3. Suppose the jth driven function $f_j(t)=\mathrm{A}\sin(\omega_0 t)$ ($j=1, 2, \ldots$), where A and ω_0 are constants. For a free-free damped Euler-Bernoulli beam with the fractional inertia force of order $\alpha \in (1, 3)$, when $a=1$, $b=1$, $E=1$, and beam length $l=10$, find the forced response $w(x, t)$.

21.4. Suppose the jth driven function $f_j(t)=u(t)$ ($j=1, 2, \ldots$). For a free-free damped Euler-Bernoulli beam with the fractional inertia force of order $\alpha \in (1, 3)$, when $a=1$, $b=1$, $E=1$, and beam length $l=10$, find the forced response $w(x, t)$.

21.5. Suppose the jth driven function $f_j(t) = u(t - \tau/2) - u(t + \tau/2)$ $(j = 1, 2, \ldots)$, where τ is a constant and $u(t)$ is the unit step function. For a free-free damped Euler-Bernoulli beam with the fractional inertia force of order $\alpha \in (1, 3)$, when $a = 1$, $b = 1$, $E = 1$, and beam length $l = 10$, find the forced response $w(x, t)$.

REFERENCES

1. O. M. Palley, Г. B. Bahizov, and E. Я. Voroneysk, *Handbook of Ship Structural Mechanics*, National Defense Industry Publishing House, Beijing, China, 2002. In Chinese. Translated from Russian by B. H. Xu, X. Xu, and M. Q. Xu.
2. X. D. Jin and L. J. Xia, *Ship Hull Vibration*, The Press of Shanghai Jiaotong University, Shanghai, China, 2011. In Chinese.
3. M. Li, *Theory of Fractional Engineering Vibrations*, Walter de Gruyter, Berlin/Boston, MA, 2021.
4. M. Li, Three classes of fractional oscillators, *Symmetry-Basel*, 10(2), 2018, 91 pp.

CHAPTER 22

Forced Response to Damped Euler-Bernoulli Beam with Fractional External Damping Force (Class 2)

WE CONTRIBUTE THREE THINGS in this chapter with respect to the forced response to a damped Euler-Bernoulli beam with fractional external damping force. (1) Equivalent equation of fractional coordinates of such a beam. (2) The closed-form expression of the forced response to such a beam. (3) A new variation of the Rayleigh damping based on such a beam.

22.1 BACKGROUND

The motion equation of the conventional damped Euler-Bernoulli beam is given by

$$\frac{\partial^2}{\partial x^2}\left(EI\frac{\partial^2 w}{\partial x^2}+c_s I\frac{\partial^3 w}{\partial x^2 \partial t}\right)+\rho A\frac{\partial^2 w}{\partial t^2}+c\frac{\partial w}{\partial t}=q(x,t), \tag{22.1}$$

where c_s is internal damping and c is external damping (Palley et al. [1]). In the above, $c\frac{\partial w}{\partial t}$ is external damping force, $c_s I\frac{\partial^3 w}{\partial x^2 \partial t}$ is internal damping force, and $\rho A\frac{\partial^2 w}{\partial t^2}$ is inertia one. The literature regarding the forced response to Eq. (22.1) under the Rayleigh damping assumption is rich, see e.g., Palley et al. [1], Jin and Xia [2]. However, the reports regarding the closed-form solution to the following equation are rarely seen.

$$\frac{\partial^2}{\partial x^2}\left(EI\frac{\partial^2 w}{\partial x^2}+c_s I\frac{\partial^3 w}{\partial x^2 \partial t}\right)+\rho A\frac{\partial^2 w}{\partial t^2}+c\frac{\partial^\beta w}{\partial t^\beta}=q(x,t), \quad 0<\beta<2. \tag{22.2}$$

DOI: 10.1201/9781003460947-25

To be precise, the aforementioned equation stands for a damped Euler-Bernoulli beam with the fractional external damping force $c\frac{\partial^\beta w}{\partial t^\beta}$. The above beam is called the class 2 type fractionally damped Euler-Bernoulli beam in Chapter 20. What its closed form forced response is remains a problem open. In this chapter, we aim at giving a solution to that problem.

The rest of this chapter is organized as follows. The analytic solution to Eq. (22.2) is proposed in Section 22.2. Case study is in Section 22.3. Discussions are given in Section 22.4, which is followed by summary.

22.2 ANALYTIC SOLUTION

By using $w(x, t)=\varphi(x)p(t)$, we rewrite Eq. (22.2) by

$$\sum_{j=1}^{\infty}\rho A\varphi_j(x)\frac{d^2p_j(t)}{dt^2}+\sum_{j=1}^{\infty}c\varphi_j(x)\frac{d^\beta p_j(t)}{dt^\beta}+\sum_{j=1}^{\infty}\frac{d^2}{dx^2}\left[c_sI\frac{d^2\varphi_j(x)}{dx^2}\right]\frac{dp_j(t)}{dt}$$
$$+\sum_{j=1}^{\infty}\frac{d^2}{dx^2}\left[EI\frac{d^2\varphi_j(x)}{dx^2}\right]p_j(t)=q(x,t). \tag{22.3}$$

Multiplying $\varphi_m(x)$ on both sides of the above produces

$$\sum_{j=1}^{\infty}\rho A\varphi_j(x)\varphi_m(x)\frac{d^2p_j(t)}{dt^2}+\sum_{j=1}^{\infty}c\varphi_j(x)\varphi_m(x)\frac{d^\beta p_j(t)}{dt^\beta}$$
$$+\sum_{j=1}^{\infty}\varphi_m(x)\frac{d^2}{dx^2}\left[c_sI\frac{d^2\varphi_j(x)}{dx^2}\right]\frac{dp_j(t)}{dt}$$
$$+\sum_{j=1}^{\infty}\varphi_m(x)\frac{d^2}{dx^2}\left[EI\frac{d^2\varphi_j(x)}{dx^2}\right]p_j(t)=\varphi_m(x)q(x,t). \tag{22.4}$$

Performing the integral from 0 to l in terms of x on both sides of the aforementioned equation and using the orthogonality of vibration modes yields

$$M_j\frac{d^2p_j(t)}{dt^2}+\frac{M_jc}{\rho A}\frac{d^\beta p_j(t)}{dt^\beta}+\sum_{j=1}^{\infty}\int_0^l\varphi_m(x)\frac{d^2}{dx^2}\left[c_sI\frac{d^2\varphi_j(x)}{dx^2}\right]dx\frac{dp_j(t)}{dt}$$
$$+\sum_{j=1}^{\infty}\int_0^l\varphi_m(x)\frac{d^2}{dx^2}\left[EI\frac{d^2\varphi_j(x)}{dx^2}\right]dxp_j(t)=Q_j(t), \tag{22.5}$$

where $M_j=\rho A\int_0^l\varphi_j^2(x)dx$.

Following Eq. (6.136) or (6.174),

$$\frac{d^2}{dx^2}\left[EI\frac{d^2\varphi_j(x)}{dx^2}\right]=\rho A\omega_{nj}^2\varphi_j(x). \tag{22.6}$$

Therefore, Eq. (22.5) can be written by

$$M_j\frac{d^2p_j(t)}{dt^2}+\frac{M_jc}{\rho A}\frac{d^\beta p_j(t)}{dt^\beta}+\sum_{j=1}^{\infty}\int_0^l\varphi_m(x)\frac{d^2}{dx^2}\left[c_sI\frac{d^2\varphi_j(x)}{dx^2}\right]dx\frac{dp_j(t)}{dt} \\ +M_j\omega_{nj}^2p_j(t)=Q_j(t). \tag{22.7}$$

Adopting the Rayleigh damping assumption, we have

$$c=\rho Aa, \tag{22.8}$$

where a is a coefficient with the unit of time since ρA is with the unit of mass and

$$c_s=Eb, \tag{22.9}$$

where b is a coefficient with the unit of frequency as E is with the unit [N/m].

Substituting Eqs. (22.8) and (22.9) into (22.7) and taking into account the orthogonality of vibration modes produces

$$M_j\frac{d^2p_j(t)}{dt^2}+M_ja\frac{d^\beta p_j(t)}{dt^\beta}+M_j\omega_{nj}^2b\frac{dp_j(t)}{dt}+M_j\omega_{nj}^2p_j(t)=Q_j(t). \tag{22.10}$$

Therefore, for the jth order forced vibration of a damped Euler-Bernoulli beam with the fractional external damping, we have

$$\frac{d^2p_j(t)}{dt^2}+\left(a\frac{d^\beta p_j(t)}{dt^\beta}+\omega_{nj}^2b\frac{dp_j(t)}{dt}\right)+\omega_{nj}^2p_j(t)=f_j(t), \tag{22.11}$$

where

$$f_j(t)=\frac{Q_j(t)}{M_j}. \tag{22.12}$$

Theorem 22.1. An equivalent equation of (22.11) is given by

$$\left(1-a\cos\frac{\beta\pi}{2}\omega^{\beta-2}\right)\frac{d^2p_j(t)}{dt^2}+\left(a\sin\frac{\beta\pi}{2}\omega^{\beta-1}+\omega_{nj}^2b\right)\frac{dp_j(t)}{dt}+\omega_{nj}^2p_j(t)=f_j(t). \tag{22.13}$$

Proof. Let $P_j(\omega)=\mathrm{F}[p_j(t)]$ and $F_j(\omega)=\mathrm{F}[f_j(t)]$. Doing the Fourier transform on the both sides of Eq. (22.11) yields

$$\left[(i\omega)^2+a(i\omega)^\beta+\omega_{nj}^2b(i\omega)+\omega_{nj}^2\right]P_j(\omega)=F_j(\omega). \tag{22.14}$$

Taking into account the principal value of i^β, we have

$$(i\omega)^2 + a(i\omega)^\beta + \omega_{nj}^2 b(i\omega) + \omega_{nj}^2 = \left(1 - a\cos\frac{\beta\pi}{2}\omega^{\beta-2}\right)(i\omega)^2 + \left(a\sin\frac{\beta\pi}{2}\omega^{\beta-1} + \omega_{nj}^2 b\right)(i\omega) + \omega_{nj}^2. \tag{22.15}$$

Thus, Eq. (22.14) can be expressed by

$$\left[(i\omega)^\alpha + a(i\omega)^\beta + \omega_{nj}^2 b(i\omega)^\lambda + \omega_{nj}^2\right]P_j(\omega) = \left[\left(1 - a\cos\frac{\beta\pi}{2}\omega^{\beta-2}\right)(i\omega)^2 + \left(a\sin\frac{\beta\pi}{2}\omega^{\beta-1} + \omega_{nj}^2 b\right)(i\omega) + \omega_{nj}^2\right]P_j(\omega) = F_j(\omega). \tag{22.16}$$

The above is the Fourier transform of Eq. (22.13). This completes the proof. □

Theorem 22.2. Denote by m_{e2} the equivalent mass of the system (22.11). Then,

$$m_{e2} = 1 - a\cos\frac{\beta\pi}{2}\omega^{\beta-2}. \tag{22.17}$$

Proof. From the first item on the left side of Eq. (22.13), we see that the above is true. This finishes the proof. □

Theorem 22.3. Denote by c_{e2j} the jth equivalent damping in the system (22.11). Then,

$$c_{e2j} = a\sin\frac{\beta\pi}{2}\omega^{\beta-1} + \omega_{nj}^2 b. \tag{22.18}$$

Proof. From the second item on the left side of Eq. (22.13), we see that the above is valid. This completes the proof. □

Note that c_{e2j} is actually a variation of the Rayleigh damping. Using Eqs. (22.17) and (22.18), (22.13) is rewritten by

$$m_{e2}\frac{d^2 p_j(t)}{dt^2} + c_{e2j}\frac{dp_j(t)}{dt} + \omega_{nj}^2 p_j(t) = f_j(t). \tag{22.19}$$

Besides, Eq. (22.16) can be rewritten by

$$\left[(i\omega)^2 + \frac{c_{e2j}}{m_{e2}}(i\omega) + \frac{\omega_{nj}^2}{m_{e2}}\right]P_j(\omega) = \frac{F_j(\omega)}{m_{e2}}. \tag{22.20}$$

Denote by ζ_{e2j} the jth equivalent damping ratio in the system (22.11). It is defined by

$$\varsigma_{e2j} = \frac{c_{e2j}}{2\sqrt{m_{e2}}} = \frac{a\sin\frac{\beta\pi}{2}\omega^{\beta-1} + \omega_{nj}^2 b}{2\sqrt{1 - a\cos\frac{\beta\pi}{2}\omega^{\beta-2}}}. \tag{22.21}$$

Let ω_{en2j} be the jth equivalent damped natural frequency in the system (22.11). Then,

$$\omega_{en2j}^2 = \frac{\omega_{nj}^2}{m_{e2}}. \tag{22.22}$$

Following the above, therefore, we can write Eq. (22.20) by

$$\left[(i\omega)^2 + 2\varsigma_{e2j}\omega_{en2j}(i\omega) + \omega_{en2j}^2\right]P_j(\omega) = \frac{F_j(\omega)}{m_{e2}}. \tag{22.23}$$

Practically, we adopt

$$\left|\varsigma_{e2j}\right| \le 1. \tag{22.24}$$

Thus, the equivalent damped natural frequency for the system (22.11), equivalently, Eq. (22.19), is given by

$$\omega_{end2j} = \omega_{en2j}\sqrt{1-\varsigma_{e2j}^2}. \tag{22.25}$$

Theorem 22.4. Denote by $h_{2j}(t)$ the jth impulse response function in the system (22.19). Then,

$$h_{2j}(t) = \frac{1}{m_{e2}\omega_{end2j}} e^{-\varsigma_{e2j}\omega_{en2j}t} \sin\omega_{end2j}t, \quad t > 0. \tag{22.26}$$

Proof. Because (22.19) takes the form of standard motion equation, the above holds. The proof is finished. □

Using the unit step function $u(t)$, we write (22.26) by

$$h_{2j}(t) = \frac{1}{m_{e2}\omega_{end2j}} e^{-\varsigma_{e2j}\omega_{en2j}t}\left(\sin\omega_{end2j}t\right)u(t). \tag{22.27}$$

Theorem 22.5. The jth coordinate with respect to Eq. (22.11) with zero initial conditions is represented by

$$p_j(t) = f_j(t) * h_{2j}(t) = \frac{1}{m_{e2}\omega_{end2j}} \int_{-\infty}^{\infty} e^{-\varsigma_{e2j}\omega_{en2j}\tau} \sin\omega_{end2j}\tau f_j(t-\tau)d\tau. \tag{22.28}$$

Proof. From the convolution theory, we see that the above is true. The proof is finished. □

Theorem 22.6. The zero-state forced response to a damped Euler-Bernoulli beam with the fractional external damping is expressed by

$$w(x,t)=\sum_{j=1}^{\infty}\frac{\varphi_j(x)}{m_{e2}\omega_{end2j}}\int_{-\infty}^{\infty}e^{-\varsigma_{e2j}\omega_{en2j}\tau}\sin\omega_{end2j}\tau f_j(t-\tau)d\tau. \tag{22.29}$$

Proof. Because $w(x,t)=\sum_{j=1}^{\infty}w_j(x,t)=\sum_{j=1}^{\infty}\varphi_j(x)p_j(t)=\sum_{j=1}^{\infty}\varphi_j(x)\left[f_j(t)*h_{2j}(t)\right]$, Theorem 22.6 is true. This finishes the proof. □

In what follows, we always suppose that the initial conditions are zero, including exercises in Section 22.6.

22.3 CASE STUDY

The jth vibration mode of a simply supported Euler-Bernoulli beam is in the form

$$\varphi_j(x)\sim\sin\frac{j\pi x}{l},\quad j=1,2,\ldots \tag{22.30}$$

Thus, the zero-state forced response is expressed by

$$w(x,t)=\sum_{j=1}^{\infty}\sin\frac{j\pi x}{l}\left[f_j(t)*\frac{1}{m_{e2}\omega_{end2j}}e^{-\varsigma_{e2j}\omega_{en2j}t}\sin\omega_{end2j}t\right]. \tag{22.31}$$

Since

$$\mu_j=\frac{j\pi}{l}, \tag{22.32}$$

the jth order natural angular frequency is in the form

$$\omega_{n2j}=\mu_j^2\sqrt{\frac{EI}{\rho A}}=\left(\frac{j\pi}{l}\right)^2\sqrt{\frac{EI}{\rho A}},\quad j=1,2,\ldots \tag{22.33}$$

The quantities ω_{en2j}, ζ_{e2j}, and ω_{end2j} are given by Eqs. (22.22), (22.21), and (22.25), respectively.

Let $l=10$. When $f_j(t)=10\sin(0.314t)$, we indicate the plots of $p_j(t)$ ($j=1, 2$) in Figure 22.1. The plots of $w_1(x, t)=$ w1 are shown in Figure 22.2.

22.4 DISCUSSIONS

For a damped Euler-Bernoulli beam with the fractional external damping force regarding forced response, the coordinate function $p_j(t)$ is associated with β, ω_{en2j}, ζ_{e2j}, ω_{end2j}, ω, and driven function $f_j(t)$. In this chapter, we revealed a novel form of the Rayleigh damping by c_{e2j}. Other variation forms of the Rayleigh damping may appear if using other types of fractional stress and strain, e.g., those stated by Shitikova [3], Rossikhin and Shitikova [4].

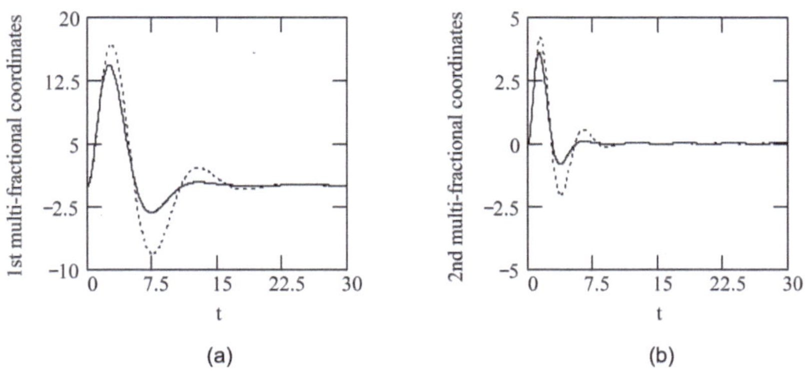

FIGURE 22.1 Plots of $p_j(t)$ ($j=1, 2$) of a simply supported damped Euler-Bernoulli beam with the fractional external damping force for $\beta=1.2$ (solid line), 1.4 (dot line) when $f_j(t)=10\sin(0.314t)$, $l=10$, $\omega=0.314$, $a=1$, $b=1$, and $E=1$. (a) $p_1(t)$. (b) $p_2(t)$.

22.5 SUMMARY

We have put forward an equivalent equation of fractional coordinates of a damped Euler-Bernoulli beam with the fractional external damping force in Theorem 22.1. We have proposed the analytic expression of the zero-state forced response to such a beam in Theorem 22.6. We have revealed a new variation form of the Rayleigh damping in Theorem 22.3.

22.6 EXERCISES

22.1. Suppose the jth driven function $f_j(t)=u(t)$ (unit step function) ($j=1, 2, \ldots$). For a simply supported damped Euler-Bernoulli beam with the fractional external damping force of order β, when $a=1$, $b=1$, $E=1$, and beam length $l=10$, find the forced response $w(x, t)$.

22.2. Suppose the jth driven function $f_j(t)=u(t-\tau/2)-u(t+\tau/2)$ ($j=1, 2,\ldots$), where τ is a constant and $u(t)$ is the unit step function. For a simply supported damped Euler-Bernoulli beam with the fractional external damping force of order β, when $a=1$, $b=1$, $E=1$, and beam length $l=10$, find the forced response $w(x, t)$.

22.3. Suppose the jth driven function $f_j(t)=\text{A}\sin(\omega_0 t)$ ($j=1, 2, \ldots$), where A and ω_0 are constants. For a free-free damped Euler-Bernoulli beam with the fractional external damping force of order β, when $a=1$, $b=1$, $E=1$, and beam length $l=10$, find the forced response $w(x, t)$.

22.4. Suppose the jth driven function $f_j(t)=u(t)$ ($j=1, 2, \ldots$). For a free-free damped Euler-Bernoulli beam with the fractional external damping force of order β, when $a=1$, $b=1$, $E=1$, and beam length $l=10$, find the forced response $w(x, t)$.

22.5. Suppose the jth driven function $f_j(t)=u(t-\tau/2)-u(t+\tau/2)$ ($j=1, 2, \ldots$), where τ is a constant and $u(t)$ is the unit step function. For a free-free damped Euler-Bernoulli beam with the fractional external damping force of order β, when $a=1$, $b=1$, $E=1$, and beam length $l=10$, find the forced response $w(x, t)$.

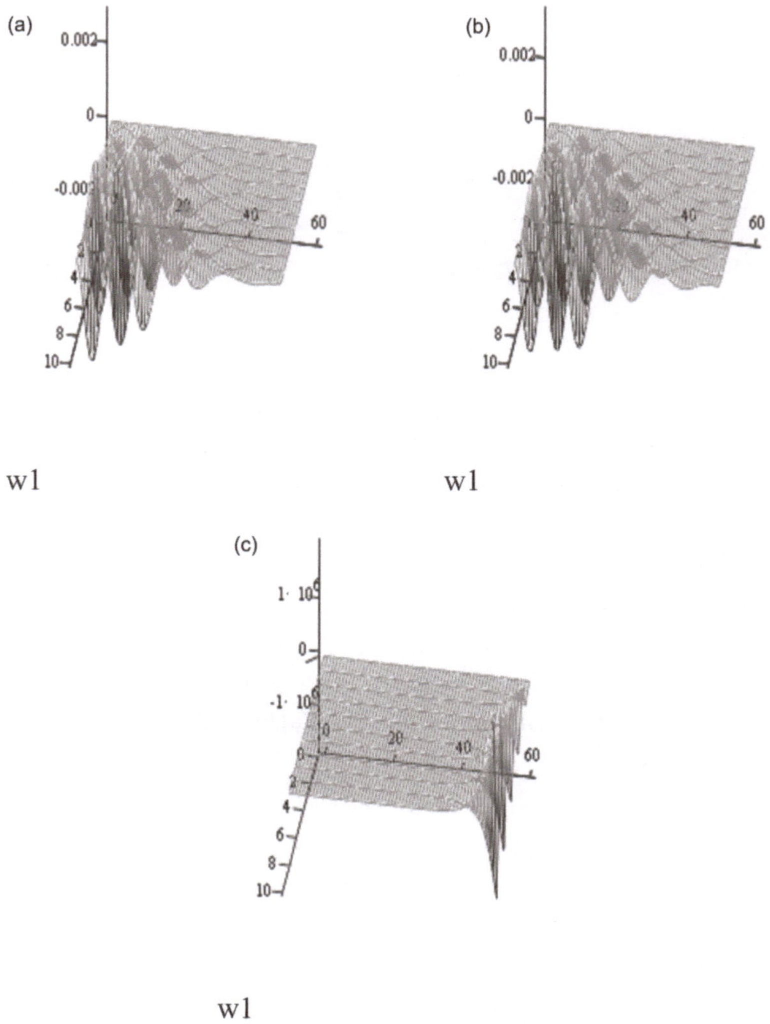

FIGURE 22.2 Plots of $w_1(x, t)$=w1 of a simply supported damped Euler-Bernoulli beam with the fractional external damping force when $f_j(t) = 10\sin(0.314t)$, $l=10$, $\omega=0.0314$, $a=1$, $b=1$, $E=1$, $x=0, \ldots, 10$, and $t=0, \ldots, 60$. (a) w1 for $\beta=1.2$. (b) w1 for $\beta=1.4$. (c) w1 for $\beta=0.8$.

REFERENCES

1. O. M. Palley, Г. B. Bahizov, and E. Я. Voroneysk, *Handbook of Ship Structural Mechanics*, National Defense Industry Publishing House, Beijing, China, 2002. In Chinese. Translated from Russian by B. H. Xu, X. Xu, and M. Q. Xu.
2. X. D. Jin and L. J. Xia, *Ship Hull Vibration*, The Press of Shanghai Jiaotong University, Shanghai, China, 2011. In Chinese.
3. M. V. Shitikova, Fractional operator viscoelastic models in dynamic problems of mechanics of solids: A review, *Mechanics of Solids*, 57(1), 2022, 1–33.
4. Y. A. Rossikhin and M. V. Shitikova, Applications of fractional calculus to dynamic problems of linear and nonlinear hereditary mechanics of solids, *Applied Mechanics Reviews*, 50(1), 1997, 15–67.

CHAPTER 23

Forced Response to Damped Euler-Bernoulli Beam with Fractional Internal Damping Force (Class 3)

THE CONTRIBUTIONS PRESENTED IN this chapter are in three aspects with respect to the forced response to a damped Euler-Bernoulli beam with fractional internal damping force as follows. (1) An equivalent equation of fractional coordinates of such a beam in Theorem 23.1. (2) The closed-form expression of the forced response to such a beam in Theorem 23.6. (3) A new variation form of the Rayleigh damping based on such a beam in Theorem 23.3.

23.1 BACKGROUND

In the conventional damped Euler-Bernoulli beam described below

$$\frac{\partial^2}{\partial x^2}\left(EI\frac{\partial^2 w}{\partial x^2}+c_sI\frac{\partial^3 w}{\partial x^2\partial t}\right)+\rho A\frac{\partial^2 w}{\partial t^2}+c\frac{\partial w}{\partial t}=q(x,t), \tag{23.1}$$

c_s is internal damping and c is external one. Therefore, $c\frac{\partial w}{\partial t}$ is external damping force, $c_sI\frac{\partial^3 w}{\partial x^2\partial t}$ is internal damping force, and $\rho A\frac{\partial^2 w}{\partial t^2}$ is inertia one. The forced response to the aforementioned equation under the Rayleigh damping assumption is known, see e.g., Chapter 6, Palley et al. [1], Jin and Xia [2]. Nevertheless, the closed-form solution to the following equation remains unknown

$$\frac{\partial^2}{\partial x^2}\left(EI\frac{\partial^2 w}{\partial x^2}+c_sI\frac{\partial^{2+\lambda} w}{\partial x^2\partial t^\lambda}\right)+\rho A\frac{\partial^2 w}{\partial t^2}+c\frac{\partial w}{\partial t}=q(x,t), \quad 0<\lambda<2. \tag{23.2}$$

DOI: 10.1201/9781003460947-26

Note that the above stands for the motion equation of a damped Euler-Bernoulli beam with the fractional internal damping force $c_s I \frac{\partial^{2+\lambda} w}{\partial x^2 \partial t^\lambda}$. We call the above beam in class 3 of fractional damped Euler-Bernoulli beams in Chapter 20. For Eq. (23.2), how to analytically represent its forced response in closed form is an open problem. I aim at presenting its analytic solution in this chapter.

The rest of this chapter is organized as follows. The analytic solution to Eq. (23.2) is put forward in Section 23.2. Case study is in Section 23.3. Discussions are given in Section 23.4, which is followed by summary.

23.2 ANALYTIC EXPRESSION OF FORCED RESPONSE

Due to $w(x, t)=\varphi(x)p(t)$, we may write Eq. (23.2) by

$$\sum_{j=1}^{\infty} \rho A \varphi_j(x) \frac{d^2 p_j(t)}{dt^2} + \sum_{j=1}^{\infty} c\varphi_j(x) \frac{dp_j(t)}{dt} + \sum_{j=1}^{\infty} \frac{d^2}{dx^2}\left[c_s I \frac{d^2\varphi_j(x)}{dx^2}\right] \frac{d^\lambda p_j(t)}{dt^\lambda}$$
$$+\sum_{j=1}^{\infty} \frac{d^2}{dx^2}\left[EI \frac{d^2\varphi_j(x)}{dx^2}\right] p_j(t) = q(x,t). \tag{23.3}$$

Multiplying $\varphi_m(x)$ on both sides of the aforementioned equation produces

$$\sum_{j=1}^{\infty} \rho A \varphi_j(x)\varphi_m(x) \frac{d^2 p_j(t)}{dt^2} + \sum_{j=1}^{\infty} c\varphi_j(x)\varphi_m(x) \frac{dp_j(t)}{dt}$$
$$+\sum_{j=1}^{\infty} \varphi_m(x) \frac{d^2}{dx^2}\left[c_s I \frac{d^2\varphi_j(x)}{dx^2}\right] \frac{d^\lambda p_j(t)}{dt^\lambda} \tag{23.4}$$
$$+\sum_{j=1}^{\infty} \varphi_m(x) \frac{d^2}{dx^2}\left[EI \frac{d^2\varphi_j(x)}{dx^2}\right] p_j(t) = \varphi_m(x) q(x,t).$$

Performing the integral from 0 to l in terms of x on both sides of the above and taking into account the orthogonality of vibration modes yields

$$M_j \frac{d^2 p_j(t)}{dt^2} + \frac{M_j c}{\rho A} \frac{dp_j(t)}{dt} + \sum_{j=1}^{\infty} \int_0^l \varphi_m(x) \frac{d^2}{dx^2}\left[c_s I \frac{d^2\varphi_j(x)}{dx^2}\right] dx \frac{d^\lambda p_j(t)}{dt^\lambda}$$
$$+\sum_{j=1}^{\infty} \int_0^l \varphi_m(x) \frac{d^2}{dx^2}\left[EI \frac{d^2\varphi_j(x)}{dx^2}\right] dx p_j(t) = Q_j(t), \tag{23.5}$$

where $M_j = \rho A \int_0^l \varphi_j^2(x) dx$.

According to Eq. (6.136) or (6.174),

$$\frac{d^2}{dx^2}\left[EI\frac{d^2\varphi_j(x)}{dx^2}\right]=\rho A\omega_{nj}^2\varphi_j(x). \tag{23.6}$$

Thus, Eq. (23.5) can be written by

$$M_j\frac{d^2p_j(t)}{dt^2}+\frac{M_jc}{\rho A}\frac{dp_j(t)}{dt}+\sum_{j=1}^{\infty}\int_0^l\varphi_m(x)\frac{d^2}{dx^2}\left[c_sI\frac{d^2\varphi_j(x)}{dx^2}\right]dx\frac{d^\lambda p_j(t)}{dt^\lambda} \\ +M_j\omega_{nj}^2p_j(t)=Q_j(t). \tag{23.7}$$

By using the Rayleigh damping assumption, c is in the form

$$c=\rho Aa, \tag{23.8}$$

where a is a coefficient with the unit of time since ρA is with the unit of mass and

$$c_s=Eb, \tag{23.9}$$

where b is a coefficient with the unit of frequency because E is with the unit [N/m].

Substituting Eqs. (23.8) and (23.9) into (23.7) and taking into account the orthogonality of vibration modes yields

$$M_j\frac{d^2p_j(t)}{dt^2}+M_ja\frac{dp_j(t)}{dt}+M_j\omega_{nj}^2b\frac{d^\lambda p_j(t)}{dt^\lambda}+M_j\omega_{nj}^2p_j(t)=Q_j(t). \tag{23.10}$$

Therefore, for the jth order forced vibration of a damped Euler-Bernoulli beam with the fractional external damping, we have

$$\frac{d^2p_j(t)}{dt^2}+\left(a\frac{dp_j(t)}{dt}+\omega_{nj}^2b\frac{d^\lambda p_j(t)}{dt^\lambda}\right)+\omega_{nj}^2p_j(t)=f_j(t), \tag{23.11}$$

where $f_j(t)$ is the jth generalized driven force. It is in the form

$$f_j(t)=\frac{Q_j(t)}{M_j}. \tag{23.12}$$

Theorem 23.1. An equivalent equation of (23.11) is given by

$$\left(1-\omega_{nj}^2b\cos\frac{\lambda\pi}{2}\omega^{\lambda-2}\right)\frac{d^2p_j(t)}{dt^2}+\left(a+\omega_{nj}^2b\sin\frac{\lambda\pi}{2}\omega^{\lambda-1}\right)\frac{dp_j(t)}{dt}+\omega_{nj}^2p_j(t)=f_j(t). \tag{23.13}$$

Proof. Let $P_j(\omega)=\mathrm{F}[p_j(t)]$ and $F_j(\omega)=\mathrm{F}[f_j(t)]$. Doing the Fourier transform on both sides of Eq. (23.11) yields

$$\left[(i\omega)^2+a(i\omega)+\omega_{nj}^2 b(i\omega)^\lambda+\omega_{nj}^2\right]P_j(\omega)=F_j(\omega). \tag{23.14}$$

Taking into account the principal value of i^λ, we have

$$\begin{aligned}&(i\omega)^2+a(i\omega)+\omega_{nj}^2 b(i\omega)^\lambda+\omega_{nj}^2\\ &=\left(1-\omega_{nj}^2 b\cos\frac{\lambda\pi}{2}\omega^{\lambda-2}\right)(i\omega)^2+\left(a+\omega_{nj}^2 b\sin\frac{\lambda\pi}{2}\omega^{\lambda-1}\right)(i\omega)+\omega_{nj}^2.\end{aligned} \tag{23.15}$$

Thus, we write Eq. (23.14) by

$$\begin{aligned}&\left[(i\omega)^2+a(i\omega)+\omega_{nj}^2 b(i\omega)^\lambda+\omega_{nj}^2\right]P_j(\omega)\\ &=\begin{bmatrix}\left(1-\omega_{nj}^2 b\cos\frac{\lambda\pi}{2}\omega^{\lambda-2}\right)(i\omega)^2\\ +\left(a+\omega_{nj}^2 b\sin\frac{\lambda\pi}{2}\omega^{\lambda-1}\right)(i\omega)+\omega_{nj}^2\end{bmatrix}P_j(\omega)=F_j(\omega).\end{aligned} \tag{23.16}$$

The above is the Fourier transform of Eq. (23.13). This completes the proof. □

Theorem 23.2. Denote by m_{e3j} the jth equivalent mass of the system (23.11). Then,

$$m_{e3j}=1-\omega_{nj}^2 b\cos\frac{\lambda\pi}{2}\omega^{\lambda-2}. \tag{23.17}$$

Proof. From the first item on the left side of Eq. (23.13), we see that the above is true. The proof is finished. □

Theorem 23.3. Denote by c_{e3j} the jth equivalent damping of the system (23.11). Then,

$$c_{e3j}=a+\omega_{nj}^2 b\sin\frac{\lambda\pi}{2}\omega^{\lambda-1}. \tag{23.18}$$

Proof. From the second item on the left side of Eq. (23.13), we see that the above is valid. This completes the proof. □

Using Eqs. (23.17) and (23.18), (23.13) is rewritten by

$$m_{e3j}\frac{d^2p_j(t)}{dt^2}+c_{e3j}\frac{dp_j(t)}{dt}+\omega_{nj}^2 p_j(t)=f_j(t). \tag{23.19}$$

Besides, Eq. (23.16) can be rewritten by

$$\left[(i\omega)^2+\frac{c_{e3j}}{m_{e3j}}(i\omega)+\frac{\omega_{nj}^2}{m_{e3j}}\right]P_j(\omega)=\frac{F_j(\omega)}{m_{e3j}}. \tag{23.20}$$

Let ζ_{e3j} be the jth equivalent damping ratio in the system (23.11). Define it by

$$\varsigma_{e3j}=\frac{c_{e3j}}{2\sqrt{m_{e3j}}}=\frac{a+\omega_{nj}^2 b\sin\frac{\lambda\pi}{2}\omega^{\lambda-1}}{2\sqrt{1-\omega_{nj}^2 b\cos\frac{\lambda\pi}{2}\omega^{\lambda-2}}}. \tag{23.21}$$

Let ω_{en3j} be the jth equivalent damping free natural frequency of the system (23.11). Define it by

$$\omega_{en3j}^2=\frac{\omega_{nj}^2}{m_{e3j}}. \tag{23.22}$$

From the above, we write Eq. (23.20) by

$$\left[(i\omega)^2+2\varsigma_{e3j}\omega_{en3j}(i\omega)+\omega_{en3j}^2\right]P_j(\omega)=\frac{F_j(\omega)}{m_{e3j}}. \tag{23.23}$$

In practice, we are interested in

$$\left|\varsigma_{e3j}\right|\le 1. \tag{23.24}$$

Thus, the equivalent damped natural frequency of the system (23.11) is

$$\omega_{end3j}=\omega_{en3j}\sqrt{1-\varsigma_{e3j}^2}. \tag{23.25}$$

Theorem 23.4. Denote by $h_{3j}(t)$ the jth impulse response function of the system (23.19) or equivalently, Eq. (23.11). Then,

$$h_{3j}(t)=\frac{1}{m_{e3j}\omega_{end3j}}e^{-\varsigma_{e3j}\omega_{en3j}t}\sin\omega_{end3j}t,\quad t>0. \tag{23.26}$$

Proof. Let $f_j(t)=\delta(t)$ in Eq. (23.19). Then, the above holds. The proof is finished. □

Using the unit step function $u(t)$, $h_{3j}(t)$ can be expressed by

$$h_{3j}(t)=\frac{1}{m_{e3}\omega_{end3j}}e^{-\varsigma_{e3j}\omega_{en3j}t}\left(\sin\omega_{end3j}t\right)u(t). \tag{23.27}$$

Theorem 23.5. The jth coordinate with respect to Eq. (23.11) with zero initial conditions is represented by

$$p_j(t)=f_j(t)*h_{3j}(t)=\frac{1}{m_{e3j}\omega_{end3j}}\int_{-\infty}^{\infty}e^{-\varsigma_{e3j}\omega_{en3j}\tau}\sin\omega_{end3j}\tau f_j(t-\tau)d\tau. \tag{23.28}$$

Proof. From the convolution theory, the above is true. This finishes the proof. □

Theorem 23.6. The zero-state forced response to a damped Euler-Bernoulli beam with the fractional internal damping is expressed by

$$w(x,t)=\sum_{j=1}^{\infty}\varphi_j(x)\left[f_j(t)*h_{3j}(t)\right]=\sum_{j=1}^{\infty}\frac{\varphi_j(x)}{m_{e3j}\omega_{end3j}}\int_{-\infty}^{\infty}e^{-\varsigma_{e3j}\omega_{en3j}\tau}\sin\omega_{end3j}\tau f_j(t-\tau)d\tau. \tag{23.29}$$

Proof. Because $w_j(x, t)=\varphi_j(x)p_j(t)$, the above is true. The proof is finished. □

We suppose that the initial conditions are zero in what follows, including exercises in Section 23.6.

23.3 CASE STUDY

For a free-free Euler-Bernoulli beam, we have

$$\varphi_j(x)\sim\frac{\mathrm{ch}\mu_jl-\cos\mu_jl}{\sin\mu_jl-\mathrm{sh}\mu_jl}\left(\sin\mu_jx+sh\mu_jx\right)+\left(\cos\mu_jx+\mathrm{ch}\mu_jx\right),\quad j=1,\dots \tag{23.30}$$

Thus,

$$w(x,t)=\sum_{j=1}^{\infty}\begin{bmatrix}\frac{\mathrm{ch}\mu_jl-\cos\mu_jl}{\sin\mu_jl-\mathrm{sh}\mu_jl}\left(\sin\mu_jx+sh\mu_jx\right)+\left(\cos\mu_jx+ch\mu_jx\right)\\ f_j(t)*\frac{1}{m_{e3j}\omega_{end3j}}e^{-\varsigma_{e3j}\omega_{en3j}t}\sin\omega_{end3j}t\end{bmatrix}. \tag{23.31}$$

Since

$$\frac{\mu_jl}{(2j+1)\frac{\pi}{2}},\quad j=1,2,\dots \tag{23.32}$$

the jth order natural angular frequency is given by

$$\omega_{n3j}=\mu_j^2\sqrt{\frac{EI}{\rho A}}=\left(\frac{(2j+1)\pi}{2l}\right)^2\sqrt{\frac{EI}{\rho A}},\quad j=1,2,\dots \tag{23.33}$$

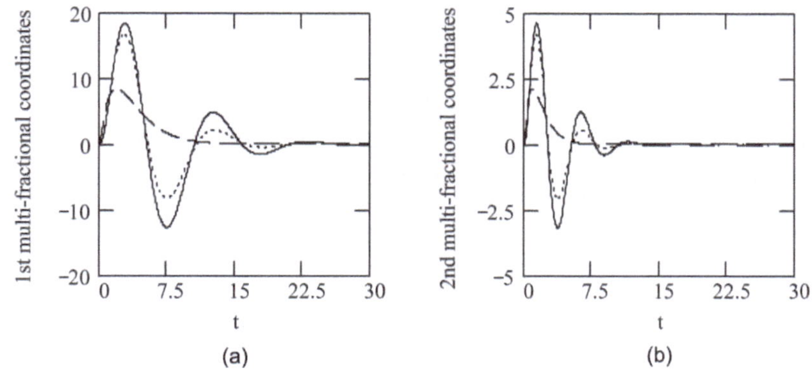

FIGURE 23.1 Plots of $p_j(t)$ ($j=1, 2$) of a free-free damped Euler-Bernoulli beam with the fractional internal damping for $\lambda=1.6$ (solid line), 1.4 (dot line), $\lambda=0.9$ (dash line) when $f_j(t)=10\sin(0.314t)$, $l=10$, $\omega=0.314$, $a=1$, $b=1$, $E=1$. (a) $p_1(t)$. (b) $p_2(t)$.

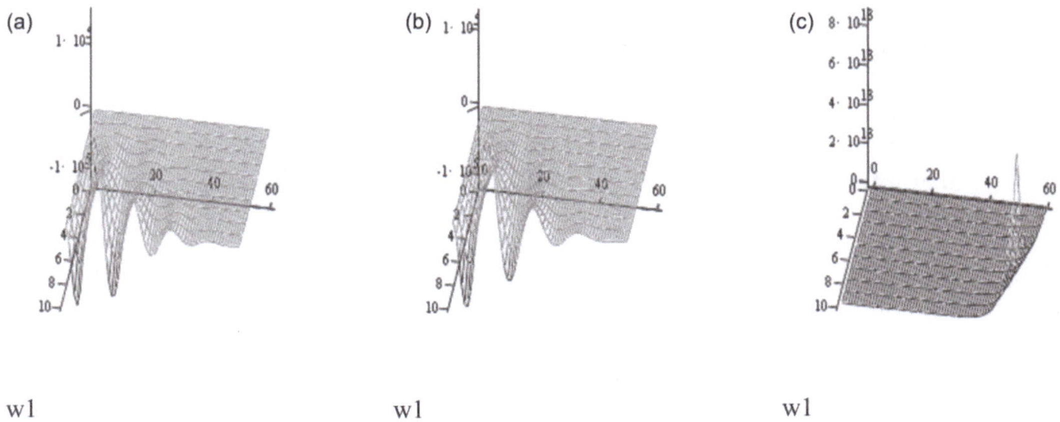

FIGURE 23.2 Plots of $w1(x, t)=$ w1 of a free-free damped Euler-Bernoulli beam with the fractional internal damping when $f_j(t)=10\sin(0.314t)$ and $l=10$ for $\omega=0.0314$, $a=1$, $b=1$, $E=1$, $x=0, \ldots, 10$, and $t=0, \ldots, 60$. (a) $w1$ for $\lambda=1.6$. (b) w1 for $\lambda=1.2$. (c) w1 for $\lambda=0.8$.

The quantities ω_{en3j}, ζ_{e3j}, and ω_{end3j} are given by Eqs. (23.22), (23.21), and (23.25), respectively.

When $l=10$ and $f_j(t)=10\sin(0.314t)$, we indicate the plots of $p_j(t)$ ($j=1, 2$) in Figure 23.1. The plots of $w_1(x, t)=$ w1 for a free-free damped Euler-Bernoulli beam with the fractional internal damping are shown in Figure 23.2.

23.4 DISCUSSIONS

For a damped Euler-Bernoulli beam with the fractional internal damping regarding forced response, the coordinate function $p_j(t)$ is associated with λ, ω_{en3j}, ζ_{e3j}, ω_{end3j}, ω, as well as driven function $f_j(t)$. In addition to the present expression of the forced response to a damped Euler-Bernoulli beam with the fractional internal damping, we reveal another novel form of the Rayleigh damping with c_{e3j}. If considering other models regarding fractional stress and strain, see e.g., Shitikova [3,4], other variation forms of the Rayleigh damping may be possible.

23.5 SUMMARY

In Theorem 23.1, we have given an equivalent equation of fractional coordinates for a damped Euler-Bernoulli beam with the fractional internal damping. In Theorem 23.6, we have proposed the analytic expression of the zero-state forced response to that beam. In Theorem 23.3, we have revealed a new variation form of the Rayleigh damping, that is, c_{e3j}.

23.6 EXERCISES

23.1. Suppose the jth driven function $f_j(t)=u(t)$ (unit step function) ($j=1, 2, \ldots$). For a simply supported damped Euler-Bernoulli beam with the fractional internal damping force, when $a=1$, $b=1$, $E=1$, and beam length $l=10$, find the forced response $w(x, t)$.

23.2. Suppose the jth driven function $f_j(t)=u(t-\tau/2)-u(t+\tau/2)$ ($j=1, 2, \ldots$), where τ is a constant and $u(t)$ is the unit step function. For a simply-supported damped Euler-Bernoulli beam with the fractional internal damping force, when $a=1$, $b=1$, $E=1$, and beam length $l=10$, find the forced response $w(x, t)$.

23.3. Suppose the jth driven function $f_j(t)=\text{A}\sin(\omega_0 t)$ ($j=1, 2, \ldots$), where A and ω_0 are constants. For a simply supported damped Euler-Bernoulli beam with the fractional internal damping force, when $a=1$, $b=1$, $E=1$, and beam length $l=10$, find the forced response $w(x, t)$.

23.4. Suppose the jth driven function $f_j(t)=u(t)$ ($j=1, 2, \ldots$). For a free-free damped Euler-Bernoulli beam with the fractional internal damping force, when $a=1$, $b=1$, $E=1$, and beam length $l=10$, find the forced response $w(x, t)$.

23.5. Suppose the jth driven function $f_j(t)=u(t-\tau/2)-u(t+\tau/2)$ ($j=1, 2, \ldots$), where τ is a constant and $u(t)$ is the unit step function. For a free-free damped Euler-Bernoulli beam with the fractional internal damping force, when $a=1$, $b=1$, $E=1$, and beam length $l=10$, find the forced response $w(x, t)$.

REFERENCES

1. O. M. Palley, Г. B. Bahizov, and E. Я. Voroneysk, *Handbook of Ship Structural Mechanics*, National Defense Industry Publishing House, Beijing, China, 2002. In Chinese. Translated from Russian by B. H. Xu, X. Xu, and M. Q. Xu.
2. X. D. Jin and L. J. Xia, *Ship Hull Vibration*, The Press of Shanghai Jiaotong University, Shanghai, China, 2011. In Chinese.
3. M. V. Shitikova, Fractional operator viscoelastic models in dynamic problems of mechanics of solids: A review, *Mechanics of Solids*, 57(1), 2022, 1–33.
4. M. V Shitikova, Impact response of a thin shallow doubly curved linear viscoelastic shell rectangular in plan, *Mathematics and Mechanics of Solids*, 27(9), 2022, 1721–1739.

CHAPTER 24

Forced Response to Damped Euler-Bernoulli Beam with Fractional External and Internal Damping Forces (Class 4)

WE CONTRIBUTE THREE THINGS in this chapter. First, we put forward an equivalent equation of fractional coordinates of a damped Euler-Bernoulli beam with fractional external and internal damping forces. Second, we present the closed-form expression of the forced response to that beam. The third is to reveal a new variation form of the Rayleigh damping based on such a beam.

24.1 BACKGROUND

Following Chapter 20, the motion equation of a damped Euler-Bernoulli beam with fractional external and internal damping forces is in the form

$$\frac{\partial^2}{\partial x^2}\left(EI\frac{\partial^2 w}{\partial x^2}+c_s I\frac{\partial^{2+\lambda} w}{\partial x^2 \partial t^\lambda}\right)+\rho A\frac{\partial^2 w}{\partial t^2}+c\frac{\partial^\beta w}{\partial t^\beta}=q(x,t), \quad 0<\beta<2, 0<\lambda<2, \quad (24.1)$$

where $c_s I\frac{\partial^{2+\lambda} w}{\partial x^2 \partial t^\lambda}$ is fractional internal damping force and $c\frac{\partial^\beta w}{\partial t^\beta}$ is fractional external damping one. Such a beam is called the class 4 fractional damped Euler-Bernoulli beam as described in Chapter 20. The closed-form analytic solution to the above equation in terms of forced response remains unknown. We aim at presenting its analytic solution in this chapter.

DOI: 10.1201/9781003460947-27

The rest of this chapter is organized as follows. The analytic solution in closed form to Eq. (24.1) is addressed in Section 24.2. Case study is in Section 24.3. Discussions are given in Section 24.4, which is followed by summary.

24.2 ANALYTIC SOLUTION

Using $w(x, t)=\varphi(x)p(t)$ to rewrite Eq. (24.1) produces

$$\sum_{j=1}^{\infty}\rho A\varphi_j(x)\frac{d^2p_j(t)}{dt^2}+\sum_{j=1}^{\infty}c\varphi_j(x)\frac{d^{\beta}p_j(t)}{dt^{\beta}}+\sum_{j=1}^{\infty}\frac{d^2}{dx^2}\left[c_sI\frac{d^2\varphi_j(x)}{dx^2}\right]\frac{d^{\lambda}p_j(t)}{dt^{\lambda}}$$
$$+\sum_{j=1}^{\infty}\frac{d^2}{dx^2}\left[EI\frac{d^2\varphi_j(x)}{dx^2}\right]p_j(t)=q(x,t). \tag{24.2}$$

Multiplying $\varphi_m(x)$ on both sides of the above results in

$$\sum_{j=1}^{\infty}\rho A\varphi_j(x)\varphi_m(x)\frac{d^2p_j(t)}{dt^2}+\sum_{j=1}^{\infty}c\varphi_j(x)\varphi_m(x)\frac{d^{\beta}p_j(t)}{dt^{\beta}}$$
$$+\sum_{j=1}^{\infty}\varphi_m(x)\frac{d^2}{dx^2}\left[c_sI\frac{d^2\varphi_j(x)}{dx^2}\right]\frac{d^{\lambda}p_j(t)}{dt^{\lambda}}$$
$$+\sum_{j=1}^{\infty}\varphi_m(x)\frac{d^2}{dx^2}\left[EI\frac{d^2\varphi_j(x)}{dx^2}\right]p_j(t)=\varphi_m(x)q(x,t). \tag{24.3}$$

Performing the integral from 0 to l in terms of x on both sides of the aforementioned equation and utilizing the orthogonality of vibration modes yields

$$M_j\frac{d^2p_j(t)}{dt^2}+\frac{M_jc}{\rho A}\frac{d^{\beta}p_j(t)}{dt^{\beta}}+\sum_{j=1}^{\infty}\int_0^l\varphi_m(x)\frac{d^2}{dx^2}\left[c_sI\frac{d^2\varphi_j(x)}{dx^2}\right]dx\frac{d^{\lambda}p_j(t)}{dt^{\lambda}}$$
$$+\sum_{j=1}^{\infty}\int_0^l\varphi_m(x)\frac{d^2}{dx^2}\left[EI\frac{d^2\varphi_j(x)}{dx^2}\right]dxp_j(t)=Q_j(t), \tag{24.4}$$

where M_j is the jth generalized mass in the form

$$M_j=\rho A\int_0^l\varphi_j^2(x)dx. \tag{24.5}$$

Following Eq. (6.136) or (6.174), we have

$$\frac{d^2}{dx^2}\left[EI\frac{d^2\varphi_j(x)}{dx^2}\right]=\rho A\omega_{nj}^2\varphi_j(x). \tag{24.6}$$

With the above, we write Eq. (24.4) by

$$M_j \frac{d^2 p_j(t)}{dt^2} + \frac{M_j c}{\rho A}\frac{d^\beta p_j(t)}{dt^\beta} + \sum_{j=1}^{\infty}\int_0^l \varphi_m(x)\frac{d^2}{dx^2}\left[c_s I \frac{d^2\varphi_j(x)}{dx^2}\right]dx \frac{d^\lambda p_j(t)}{dt^\lambda} + M_j\omega_{nj}^2 p_j(t) = Q_j(t). \tag{24.7}$$

Following the Rayleigh damping assumption, we write

$$c = \rho A a, \tag{24.8}$$

where a is a coefficient with the unit of time and

$$c_s = Eb, \tag{24.9}$$

where b is a coefficient with the unit of frequency.

Substituting Eqs. (24.8) and (24.9) into (24.7) and taking into account the orthogonality of vibration modes, Σ on the third item on the left side of Eq. (24.7) can be removed. When doing so, we have

$$M_j \frac{d^2 p_j(t)}{dt^2} + M_j a \frac{d^\beta p_j(t)}{dt^\beta} + M_j\omega_{nj}^2 b \frac{d^\lambda p_j(t)}{dt^\lambda} + M_j\omega_{nj}^2 p_j(t) = Q_j(t). \tag{24.10}$$

Therefore, for the jth order forced vibration of a damped Euler-Bernoulli beam with the fractional external damping and internal damping forces, we have its coordinate equation in the form

$$\frac{d^2 p_j(t)}{dt^2} + \left(a\frac{d^\beta p_j(t)}{dt^\beta} + \omega_{nj}^2 b \frac{d^\lambda p_j(t)}{dt^\lambda}\right) + \omega_{nj}^2 p_j(t) = f_j(t), \tag{24.11}$$

where

$$f_j(t) = \frac{Q_j(t)}{M_j}. \tag{24.12}$$

Theorem 24.1. An equivalent equation of (24.11) is given by

$$\begin{aligned}&\left(1 - a\cos\frac{\beta\pi}{2}\omega^{\beta-2} - \omega_{nj}^2 b\cos\frac{\lambda\pi}{2}\omega^{\lambda-2}\right)\frac{d^2 p_j(t)}{dt^2}\\ &+\left(a\sin\frac{\beta\pi}{2}\omega^{\beta-1} + \omega_{nj}^2 b\sin\frac{\lambda\pi}{2}\omega^{\lambda-1}\right)\frac{dp_j(t)}{dt} + \omega_{nj}^2 p_j(t) = f_j(t).\end{aligned} \tag{24.13}$$

Proof. Let $P_j(\omega) = \mathrm{F}[p_j(t)]$ and $F_j(\omega) = \mathrm{F}[f_j(t)]$. Doing the Fourier transform on both sides of Eq. (24.11) yields

$$\left[(i\omega)^2 + a(i\omega)^\beta + \omega_{nj}^2 b(i\omega)^\lambda + \omega_{nj}^2\right]P_j(\omega) = F_j(\omega). \tag{24.14}$$

Taking into account the principal values of i^β and i^λ, we have

$$\begin{aligned}&(i\omega)^2 + a(i\omega)^\beta + \omega_{nj}^2 b(i\omega)^\lambda + \omega_{nj}^2\\&=\left(1 - a\cos\frac{\beta\pi}{2}\omega^{\beta-2} - \omega_{nj}^2 b\cos\frac{\lambda\pi}{2}\omega^{\lambda-2}\right)(i\omega)^2\\&+\left(a\sin\frac{\beta\pi}{2}\omega^{\beta-1} + \omega_{nj}^2 b\sin\frac{\lambda\pi}{2}\omega^{\lambda-1}\right)(i\omega) + \omega_{nj}^2.\end{aligned} \tag{24.15}$$

Thus, Eq. (24.14) can be expressed by

$$\begin{aligned}&\left[(i\omega)^2 + a(i\omega)^\beta + \omega_{nj}^2 b(i\omega)^\lambda + \omega_{nj}^2\right]P_j(\omega)\\&=\left[\begin{array}{l}\left(1 - a\cos\frac{\beta\pi}{2}\omega^{\beta-2} - \omega_{nj}^2 b\cos\frac{\lambda\pi}{2}\omega^{\lambda-2}\right)(i\omega)^2\\+\left(a\sin\frac{\beta\pi}{2}\omega^{\beta-1} + \omega_{nj}^2 b\sin\frac{\lambda\pi}{2}\omega^{\lambda-1}\right)(i\omega) + \omega_{nj}^2\end{array}\right]P_j(\omega) = F_j(\omega).\end{aligned} \tag{24.16}$$

The above is the Fourier transform of Eq. (24.13). This completes the proof. □

Theorem 24.2. Denote by m_{e4j} the jth equivalent mass of the system (24.11). Then,

$$m_{e4j} = 1 - a\cos\frac{\beta\pi}{2}\omega^{\beta-2} - \omega_{nj}^2 b\cos\frac{\lambda\pi}{2}\omega^{\lambda-2}. \tag{24.17}$$

Proof. The right side of the above equation is the coefficient of $\frac{d^2 p_j(t)}{dt^2}$ in Eq. (24.13). Thus, the above is true. This finishes the proof. □

Theorem 24.3. Denote by c_{e4j} the jth equivalent damping of the system (24.11). Then,

$$c_{e4j} = a\sin\frac{\beta\pi}{2}\omega^{\beta-1} + \omega_{nj}^2 b\sin\frac{\lambda\pi}{2}\omega^{\lambda-1}. \tag{24.18}$$

Proof. The right side of the above is the coefficient of $\frac{dp_j(t)}{dt}$ in Eq. (24.13). Therefore, the above is valid. The proof is completed. □

Using Eqs. (24.17) and (24.18), we rewrite Eq. (24.13) by

$$m_{e4j}\frac{d^2 p_j(t)}{dt^2} + c_{e4j}\frac{dp_j(t)}{dt} + \omega_{nj}^2 p_j(t) = f_j(t). \tag{24.19}$$

Besides, Eq. (24.16) can be rewritten by

$$\left[(i\omega)^2 + \frac{c_{e4j}}{m_{e4j}}(i\omega) + \frac{\omega_{nj}^2}{m_{e4j}}\right]P_j(\omega) = \frac{F_j(\omega)}{m_{e4j}}. \tag{24.20}$$

Denote by ζ_{e4j} the jth equivalent damping ratio in the system (24.11). It is defined by

$$\varsigma_{e4j} = \frac{c_{e4j}}{2\sqrt{m_{e4j}}} = \frac{a\sin\frac{\beta\pi}{2}\omega^{\beta-1} + \omega_{nj}^2 b\sin\frac{\lambda\pi}{2}\omega^{\lambda-1}}{2\sqrt{1 - a\cos\frac{\beta\pi}{2}\omega^{\beta-2} - \omega_{nj}^2 b\cos\frac{\lambda\pi}{2}\omega^{\lambda-2}}}. \tag{24.21}$$

Let ω_{en4j} be the jth equivalent damping free natural frequency regarding the system (24.11). It is given by

$$\omega_{en4j}^2 = \frac{\omega_{nj}^2}{m_{e4j}}. \tag{24.22}$$

Hence, we can write Eq. (24.20) by

$$\left[(i\omega)^2 + 2\varsigma_{e4j}\omega_{en4j}(i\omega) + \omega_{en4j}^2\right]P_j(\omega) = \frac{F_j(\omega)}{m_{e4j}}. \tag{24.23}$$

From the point of view of practice,

$$\left|\varsigma_{e4j}\right| \le 1. \tag{24.24}$$

Therefore, the jth damped natural frequency in terms of the system (24.11) is given by

$$\omega_{end4j} = \omega_{en4j}\sqrt{1 - \varsigma_{e4j}^2}. \tag{24.25}$$

Theorem 24.4. Denote by $h_{4j}(t)$ the jth impulse response function of the system (24.19). Then,

$$h_{4j}(t) = \frac{1}{m_{e4j}\omega_{end4j}} e^{-\varsigma_{e4j}\omega_{en4j}t} \sin\omega_{end4j}t, \quad t > 0. \tag{24.26}$$

Proof. Let $f_j(t) = \delta(t)$. Then, from Eq. (24.19), we have the above. The proof is finished. □

The above can be expressed by

$$h_{4j}(t) = \frac{1}{m_{e4}\omega_{end4j}} e^{-\varsigma_{e4j}\omega_{en4j}t}\left(\sin\omega_{end4j}t\right)u(t), \tag{24.27}$$

where $u(t)$ is the unit step function.

Theorem 24.5. The jth coordinate with respect to Eq. (24.11) with initial conditions is represented by

$$p_j(t)=f_j(t)*h_{4j}(t)=\frac{1}{m_{e4j}\omega_{end4j}}\int_{-\infty}^{\infty}e^{-\varsigma_{e4j}\omega_{en4j}\tau}\sin\omega_{end4j}\tau f_j(t-\tau)d\tau. \qquad (24.28)$$

Proof. From the convolution theory, we see that the above is true. The proof is finished. □

Theorem 24.6. The zero-state forced response to a damped Euler-Bernoulli beam with the fractional external and internal damping forces is expressed by

$$w(x,t)=\sum_{j=1}^{\infty}\frac{\varphi_j(x)}{m_{e4j}\omega_{end4j}}\int_{-\infty}^{\infty}e^{-\varsigma_{e4j}\omega_{en4j}\tau}\sin\omega_{end4j}\tau f_j(t-\tau)d\tau. \qquad (24.29)$$

Proof. Because $w(x,t)=\sum_{j=1}^{\infty}w_j(x,t)=\sum_{j=1}^{\infty}\varphi_j(x)p_j(t)=\sum_{j=1}^{\infty}\varphi_j(x)\left[f_j(t)*h_{4j}(t)\right]$ and $p_j(t)$ is expressed by Theorem 24.5, the above is true. The proof is finished. □

We default that the initial conditions in what follows are zero, including exercises in Section 24.6.

24.3 CASE STUDY

For a free-free Euler-Bernoulli beam, the vibration mode function is

$$\varphi_j(x)\sim\frac{\mathrm{ch}\mu_j l-\cos\mu_j l}{\sin\mu_j l-\mathrm{sh}\mu_j l}\left(\sin\mu_j x+sh\mu_j x\right)+\left(\cos\mu_j x+\mathrm{ch}\mu_j x\right),\qquad j=1,2,\ldots \qquad (24.30)$$

Thus, the zero-state forced response is given by

$$w(x,t)=\sum_{j=1}^{\infty}\varphi_j(x)\left[f_j(t)*\frac{1}{m_{e4j}\omega_{end4j}}e^{-\varsigma_{e4j}\omega_{en4j}t}\sin\omega_{end4j}t\right]. \qquad (24.31)$$

Since

$$\frac{\mu_j l}{(2j+1)\frac{\pi}{2}},\qquad j=1,2,\ldots \qquad (24.32)$$

the jth order natural angular frequency is in the form

$$\omega_{n4j}=\mu_j^2\sqrt{\frac{EI}{\rho A}}=\left(\frac{(2j+1)\pi}{2l}\right)^2\sqrt{\frac{EI}{\rho A}},\qquad j=1,2,\ldots \qquad (24.33)$$

The quantities ω_{en4j}, ζ_{e4j}, and ω_{end4j} are given by Eqs. (24.22), (24.21), and (24.25), respectively.

We indicate the plots of $w_1(x, t) = \text{w1}$ in Figure 24.1 when $l = 10$ and $f_j(t) = 10\sin(0.314t)$.

24.4 DISCUSSIONS

For a damped Euler-Bernoulli beam with the fractional external and internal damping forces regarding forced response, the coordinate function $p_j(t)$ is associated with β, λ, ω_{en4j}, ζ_{e4j}, ω_{end4j}, ω, as well as driven function $f_j(t)$. In addition to the present analytic solution of such a beam, we also found one more novel form of the Rayleigh damping expressed by c_{e4j}. A number of fractional stress models are described in Shitikova [1,2]. Thus, applying those to investigate possible variations of the Rayleigh damping may be interesting.

24.5 SUMMARY

An equivalent equation of fractional coordinates of a damped Euler-Bernoulli beam with the fractional external and internal damping forces is described in Theorem 24.1. The closed-form expression of the zero-state forced response to such a beam is given in Theorem 24.6. A new variation form of the Rayleigh damping is revealed in Theorem 24.3.

24.6 EXERCISES

24.1. Suppose the jth driven function $f_j(t) = u(t)$ (unit step function) ($j = 1, 2, \ldots$). For a simply-supported damped Euler-Bernoulli beam with the fractional external and internal damping forces, when $a = 1$, $b = 1$, $E = 1$, and beam length $l = 10$, find the forced response $w(x, t)$.

24.2. Suppose the jth driven function $f_j(t) = u(t - \tau/2) - u(t + \tau/2)$ ($j = 1, 2, \ldots$), where τ is a constant and $u(t)$ is the unit step function. For a simply-supported damped Euler-Bernoulli beam with the fractional external and internal damping forces, when $a = 1$, $b = 1$, $E = 1$, and beam length $l = 10$, find the forced response $w(x, t)$.

24.3. Suppose the jth driven function $f_j(t) = \text{A}\sin(\omega_0 t)$ ($j = 1, 2, \ldots$), where A and ω_0 are constants. For a free-free damped Euler-Bernoulli beam with the fractional external and internal damping forces, when $a = 1$, $b = 1$, $E = 1$, and beam length $l = 10$, find the forced response $w(x, t)$.

24.4. Suppose the jth driven function $f_j(t) = u(t)$ ($j = 1, 2, \ldots$). For a free-free damped Euler-Bernoulli beam with the fractional external and internal damping forces, when $a = 1$, $b = 1$, $E = 1$, and beam length $l = 10$, find the forced response $w(x, t)$.

24.5. Suppose the jth driven function $f_j(t) = u(t - \tau/2) - u(t + \tau/2)$ ($j = 1, 2, \ldots$), where τ is a constant and $u(t)$ is the unit step function. For a free-free damped Euler-Bernoulli beam with the fractional external and internal damping forces, when $a = 1$, $b = 1$, $E = 1$, and beam length $l = 10$, find the forced response $w(x, t)$.

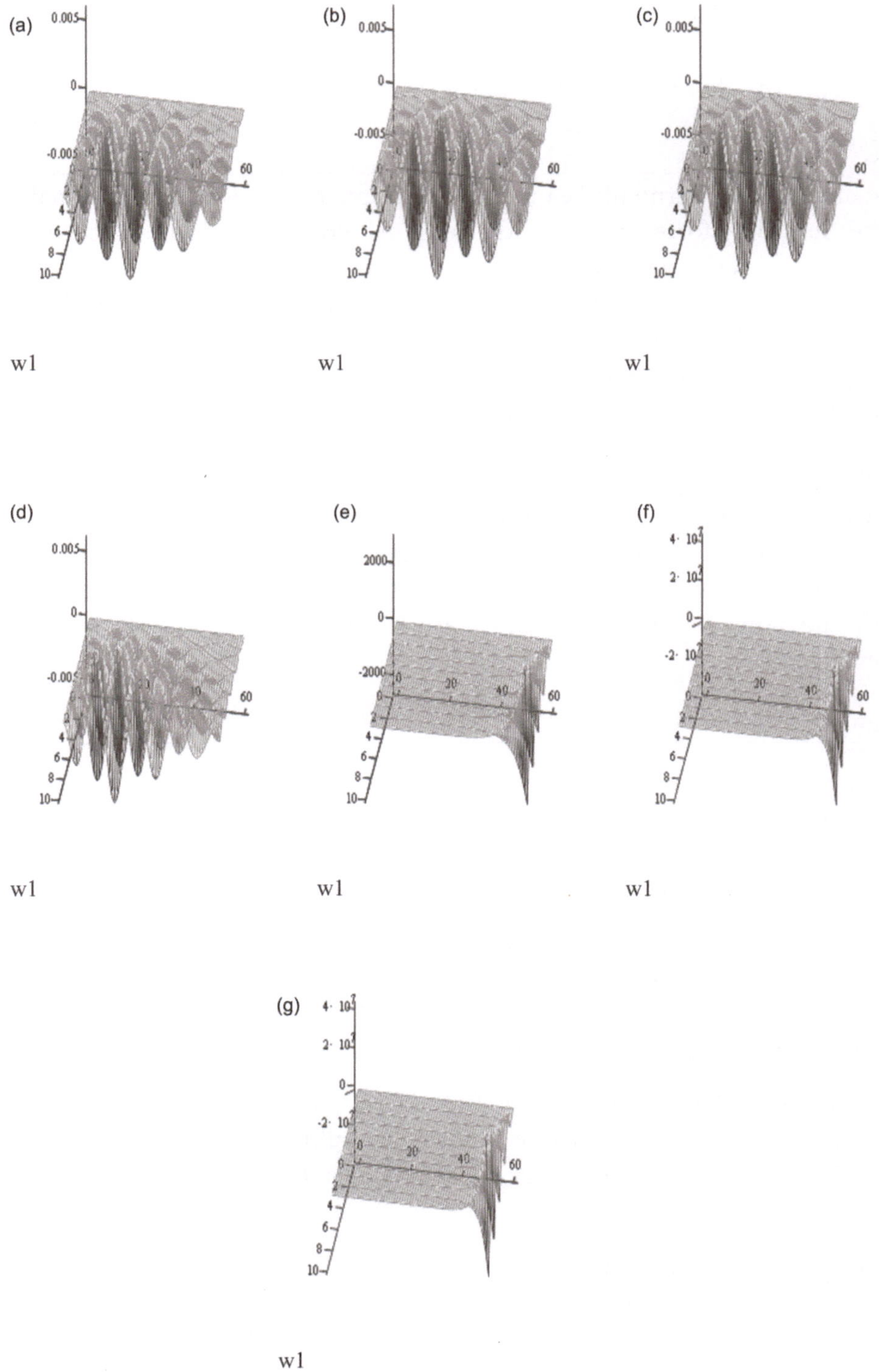

FIGURE 24.1 Plots of $w_1(x, t)$ = w1 of a free-free damped Euler-Bernoulli beam with the fractional external and internal damping forces when $f_j(t) = 10\sin(0.314t)$ and $l = 10$ for $\omega = 0.0314$, $a = 1$, $b = 1$, $E = 1$, $x = 0, \ldots, 10$, and $t = 0, \ldots, 60$. (a) w1 for $\beta = \lambda = 1.2$. (b) w1 for $\beta = 1.4$ and $\lambda = 1.2$. (c) w1 for $\beta = 1.2$ and $\lambda = 1.4$. (d) w1 for $\beta = 1.8$ and $\lambda = 1.1$. (e) w1 for $\beta = 0.8$ and $\lambda = 0.8$. (f) w1 for $\beta = 0.8$ and $\lambda = 1.2$. (g) w1 for $\beta = 1.2$ and $\lambda = 0.8$.

REFERENCES

1. M. V. Shitikova, Fractional operator viscoelastic models in dynamic problems of mechanics of solids: A review, *Mechanics of Solids*, 57(1), 2022, 1–33.
2. M. V. Shitikova, Wave theory of impact and professor Yury Rossikhin Contribution in the field (a memorial survey), *Journal of Materials Engineering and Performance*, 28, 2019, 3161–3173.

CHAPTER 25

Forced Response to Damped Euler-Bernoulli Beam with Fractional Inertia and External Damping Forces (Class 5)

THE CONTRIBUTIONS GIVEN IN this chapter are in three aspects with respect to the forced response to a damped Euler-Bernoulli beam with fractional inertia and external damping forces as follows. (1) An equivalent equation of fractional coordinates for that beam. (2) The closed form expression of the forced response to such a beam. (3) A new variation form of the Rayleigh damping resulting from such a beam.

25.1 INTRODUCTION

I now study the forced response to a damped Euler-Bernoulli beam with the fractional inertia force and fractional external damping force. The motion equation of such a beam is given by

$$\frac{\partial^2}{\partial x^2}\left(EI\frac{\partial^2 w}{\partial x^2}+c_s I\frac{\partial^3 w}{\partial x^2 \partial t^3}\right)+\rho A\frac{\partial^\alpha w}{\partial t^\alpha}+c\frac{\partial^\beta w}{\partial t^\beta}=q(x,t), \qquad 1<\alpha<3,\ 0<\beta<2. \tag{25.1}$$

As a matter of fact, the above stands for a beam with the fractional inertia force $\rho A\frac{\partial^\alpha w}{\partial t^\alpha}$ and the fractional external damping one $c\frac{\partial^\beta w}{\partial t^\beta}$. The above is called the class 5 fractionally damped Euler-Bernoulli beams in Chapter 20. Its analytic expression of forced response in closed form is unknown. I shall present its closed-form solution in this chapter.

DOI: 10.1201/9781003460947-28

The rest of this chapter is organized as follows. The solution to Eq. (25.1) is proposed in Section 25.2. Case study is in Section 25.3. Discussions are given is in Section 25.4, which is followed by summary.

25.2 CLOSED-FORM SOLUTION

By taking into account $w(x, t)=\varphi(x)p(t)$, we write Eq. (25.1) by

$$\sum_{j=1}^{\infty}\rho A\varphi_j(x)\frac{d^\alpha p_j(t)}{dt^\alpha}+\sum_{j=1}^{\infty}c\varphi_j(x)\frac{d^\beta p_j(t)}{dt^\beta}+\sum_{j=1}^{\infty}\frac{d^2}{dx^2}\left[c_sI\frac{d^2\varphi_j(x)}{dx^2}\right]\frac{dp_j(t)}{dt}$$
$$+\sum_{j=1}^{\infty}\frac{d^2}{dx^2}\left[EI\frac{d^2\varphi_j(x)}{dx^2}\right]p_j(t)=q(x,t). \tag{25.2}$$

Multiplying $\varphi_m(x)$ on both sides of the above produces

$$\sum_{j=1}^{\infty}\rho A\varphi_j(x)\varphi_m(x)\frac{d^\alpha p_j(t)}{dt^\alpha}+\sum_{j=1}^{\infty}c\varphi_j(x)\varphi_m(x)\frac{d^\beta p_j(t)}{dt^\beta}$$
$$+\sum_{j=1}^{\infty}\varphi_m(x)\frac{d^2}{dx^2}\left[c_sI\frac{d^2\varphi_j(x)}{dx^2}\right]\frac{dp_j(t)}{dt}+\sum_{j=1}^{\infty}\varphi_m(x)\frac{d^2}{dx^2}\left[EI\frac{d^2\varphi_j(x)}{dx^2}\right]p_j(t) \tag{25.3}$$
$$=\varphi_m(x)q(x,t).$$

Performing the integral from 0 to l in terms of x on both sides of the above equation and using the orthogonality of vibration modes yields

$$M_j\frac{d^\alpha p_j(t)}{dt^\alpha}+\frac{M_jc}{\rho A}\frac{d^\beta p_j(t)}{dt^\beta}+\sum_{j=1}^{\infty}\int_0^l\varphi_m(x)\frac{d^2}{dx^2}\left[c_sI\frac{d^2\varphi_j(x)}{dx^2}\right]dx\frac{dp_j(t)}{dt}$$
$$+\sum_{j=1}^{\infty}\int_0^l\varphi_m(x)\frac{d^2}{dx^2}\left[EI\frac{d^2\varphi_j(x)}{dx^2}\right]dxp_j(t)=Q_j(t), \tag{25.4}$$

where M_j is the jth generalized mass expressed by

$$M_j=\rho A\int_0^l\varphi_j^2(x)dx. \tag{25.5}$$

Following Eqs. (6.136) or (6.174),

$$\frac{d^2}{dx^2}\left[EI\frac{d^2\varphi_j(x)}{dx^2}\right]=\rho A\omega_{nj}^2\varphi_j(x). \tag{25.6}$$

Thus, Eq. (25.4) can be written by

$$M_j \frac{d^\alpha p_j(t)}{dt^\alpha} + \frac{M_j c}{\rho A}\frac{d^\beta p_j(t)}{dt^\beta} + \sum_{j=1}^{\infty}\int_0^l \varphi_m(x)\frac{d^2}{dx^2}\left[c_s I \frac{d^2\varphi_j(x)}{dx^2}\right]dx \frac{dp_j(t)}{dt}$$
$$+M_j\omega_{nj}^2 p_j(t) = Q_j(t). \tag{25.7}$$

According to the Rayleigh damping assumption, one has

$$c = \rho A a, \tag{25.8}$$

where a is a coefficient with the unit of time and

$$c_s = Eb, \tag{25.9}$$

where b is a coefficient with the unit of frequency.

After substituting Eqs. (25.8) and (25.9) into (25.7) and taking into account the orthogonality of vibration modes, Σ on the third item on the left side of Eq. (25.7) can be removed. In doing so, we have

$$M_j \frac{d^\alpha p_j(t)}{dt^\alpha} + M_j a \frac{d^\beta p_j(t)}{dt^\beta} + M_j\omega_{nj}^2 b \frac{dp_j(t)}{dt} + M_j\omega_{nj}^2 p_j(t) = Q_j(t). \tag{25.10}$$

Write the aforementioned equation by

$$\frac{d^\alpha p_j(t)}{dt^\alpha} + \left(a\frac{d^\beta p_j(t)}{dt^\beta} + \omega_{nj}^2 b\frac{dp_j(t)}{dt}\right) + \omega_{nj}^2 p_j(t) = f_j(t), \tag{25.11}$$

where

$$f_j(t) = \frac{Q_j(t)}{M_j}. \tag{25.12}$$

Theorem 25.1. An equivalent equation of (25.11) is given by

$$-\left(\cos\frac{\alpha\pi}{2}\omega^{\alpha-2} + a\cos\frac{\beta\pi}{2}\omega^{\beta-2}\right)\frac{d^2 p_j(t)}{dt^2}$$
$$+\left(\sin\frac{\alpha\pi}{2}\omega^{\alpha-1} + a\sin\frac{\beta\pi}{2}\omega^{\beta-1} + \omega_{nj}^2 b\right)\frac{dp_j(t)}{dt} + \omega_{nj}^2 p_j(t) = f_j(t). \tag{25.13}$$

Proof. Let $P_j(\omega) = \mathrm{F}[p_j(t)]$ and $F_j(\omega) = \mathrm{F}[f_j(t)]$. Doing the Fourier transform on both sides of Eq. (25.11) yields

$$\left[(i\omega)^\alpha + a(i\omega)^\beta + \omega_{nj}^2 b(i\omega) + \omega_{nj}^2\right]P_j(\omega) = F_j(\omega). \tag{25.14}$$

Since

$$(i\omega)^{\alpha}+a(i\omega)^{\beta}+\omega_{nj}^{2}b(i\omega)+\omega_{nj}^{2}=-\left(\cos\frac{\alpha\pi}{2}\omega^{\alpha-2}+a\cos\frac{\beta\pi}{2}\omega^{\beta-2}\right)(i\omega)^{2}$$
$$+\left(\sin\frac{\alpha\pi}{2}\omega^{\alpha-1}+a\sin\frac{\beta\pi}{2}\omega^{\beta-1}+\omega_{nj}^{2}b\right)(i\omega)+\omega_{nj}^{2}, \tag{25.15}$$

equation (25.14) can be expressed by

$$\left[(i\omega)^{\alpha}+a(i\omega)^{\beta}+\omega_{nj}^{2}b(i\omega)+\omega_{nj}^{2}\right]P_{j}(\omega)$$
$$=\left[\begin{array}{l}-\left(\cos\frac{\alpha\pi}{2}\omega^{\alpha-2}+a\cos\frac{\beta\pi}{2}\omega^{\beta-2}\right)(i\omega)^{2}\\+\left(\sin\frac{\alpha\pi}{2}\omega^{\alpha-1}+a\sin\frac{\beta\pi}{2}\omega^{\beta-1}+\omega_{nj}^{2}b\right)(i\omega)+\omega_{nj}^{2}\end{array}\right]P_{j}(\omega)=F_{j}(\omega). \tag{25.16}$$

The above is the Fourier transform of Eq. (25.13). This completes the proof. □

Theorem 25.2. Denote by m_{e5} the equivalent mass of the system (25.11). Then,

$$m_{e5}=-\left(\cos\frac{\alpha\pi}{2}\omega^{\alpha-2}+a\cos\frac{\beta\pi}{2}\omega^{\beta-2}\right). \tag{25.17}$$

Proof. From the first item on the left side of Eq. (25.13), we see that the above is true. This finishes the proof. □

Theorem 25.3. Denote by c_{e5j} the jth equivalent damping in the system (25.11). Then,

$$c_{e5j}=\sin\frac{\alpha\pi}{2}\omega^{\alpha-1}+a\sin\frac{\beta\pi}{2}\omega^{\beta-1}+\omega_{nj}^{2}b. \tag{25.18}$$

Proof. From the second item on the left side of Eq. (25.13), we see that the above is valid. The proof is completed. □

Using Eqs. (25.17) and (25.18), we write Eq. (25.13) by

$$m_{e5}\frac{d^{2}p_{j}(t)}{dt^{2}}+c_{e5j}\frac{dp_{j}(t)}{dt}+\omega_{nj}^{2}p_{j}(t)=f_{j}(t). \tag{25.19}$$

Besides, Eq. (25.16) can be written by

$$\left[(i\omega)^{2}+\frac{c_{e5j}}{m_{e5}}(i\omega)+\frac{\omega_{nj}^{2}}{m_{e5}}\right]P_{j}(\omega)=\frac{F_{j}(\omega)}{m_{e5}}. \tag{25.20}$$

Denote by ζ_{e5j} the jth equivalent damping ratio of the system (25.11). It is defined by

$$\varsigma_{e5j} = \frac{c_{e5j}}{2\sqrt{m_{e5}}} = \frac{\sin\frac{\alpha\pi}{2}\omega^{\alpha-1} + a\sin\frac{\beta\pi}{2}\omega^{\beta-1} + \omega_{nj}^2 b}{2\sqrt{-\left(\cos\frac{\alpha\pi}{2}\omega^{\alpha-2} + a\cos\frac{\beta\pi}{2}\omega^{\beta-2}\right)}}. \tag{25.21}$$

Let ω_{en5j} be the jth equivalent damping free natural frequency regarding the system (25.11). It is given by

$$\omega_{en5j}^2 = \frac{\omega_{nj}^2}{m_{e5}}. \tag{25.22}$$

Then, we write Eq. (25.20) by

$$\left[(i\omega)^2 + 2\varsigma_{e5j}\omega_{en5j}(i\omega) + \omega_{en5j}^2\right]P_j(\omega) = \frac{F_j(\omega)}{m_{e5}}. \tag{25.23}$$

Practically, we adopt

$$\left|\varsigma_{e5j}\right| \le 1. \tag{25.24}$$

Let ω_{end5j} be the jth equivalent damped natural frequency for the system (25.11). Then,

$$\omega_{end5j} = \omega_{en5j}\sqrt{1-\varsigma_{e5j}^2}. \tag{25.25}$$

Theorem 25.4. Denote by $h_{5j}(t)$ the jth impulse response function of the system (25.19). Then,

$$h_{5j}(t) = \frac{1}{m_{e5}\omega_{end5j}} e^{-\varsigma_{e5j}\omega_{en5j}t}\sin\omega_{end5j}t, \quad t>0. \tag{25.26}$$

Proof. Let $f_j(t) = \delta(t)$. Then, we have the above from Eq. (25.19). The proof is finished. □
The above can be expressed by

$$h_{5j}(t) = \frac{1}{m_{e5}\omega_{end5j}} e^{-\varsigma_{e5j}\omega_{en5j}t}\left(\sin\omega_{end5j}t\right)u(t), \tag{25.27}$$

where $u(t)$ is the unit step function.

Theorem 25.5. The jth coordinate with respect to Eq. (25.11) under zero initial conditions is represented by

$$p_j(t) = f_j(t) * h_{5j}(t) = \frac{1}{m_{e5}\omega_{end5j}} \int_{-\infty}^{\infty} e^{-\varsigma_{e5j}\omega_{en5j}\tau}\sin\omega_{end5j}\tau f_j(t-\tau)d\tau. \tag{25.28}$$

Proof. From the convolution theory, we see that the above is true. The proof is finished. □

Theorem 25.6. The zero-state forced response to a damped Euler-Bernoulli beam with the fractional inertia force and fractional external damping one is expressed by

$$w(x,t)=\sum_{j=1}^{\infty}\frac{\varphi_j(x)}{m_{e5}\omega_{end5j}}\int_{-\infty}^{\infty}e^{-\varsigma_{e5j}\omega_{en5j}\tau}\sin\omega_{end5j}\tau f_j(t-\tau)d\tau. \tag{25.29}$$

Proof. Because $w(x,t)=\sum_{j=1}^{\infty}w_j(x,t)=\sum_{j=1}^{\infty}\varphi_j(x)p_j(t)$ and $p_j(t)$ is expressed by Theorem 25.5, the above is true. The proof is finished. □

In what follows, zero initial conditions are defaulted, including exercises in Section 25.6.

25.3 CASE STUDY

Following Palley et al. [1], the jth vibration mode function of a simply-supported Euler-Bernoulli beam is given by

$$\varphi_j(x)\sim\sin\frac{j\pi x}{l},\quad j=1,2,\ldots \tag{25.30}$$

Thus, the forced response is

$$w(x,t)=\sum_{j=1}^{\infty}\sin\frac{j\pi x}{l}\left[f_j(t)*\frac{1}{m_{e5}\omega_{end5j}}e^{-\varsigma_{e5j}\omega_{en5j}t}\sin\omega_{end5j}t\right]. \tag{25.31}$$

Since

$$\mu_j=\frac{j\pi}{l},\quad j=1,2,\ldots \tag{25.32}$$

the jth order natural angular frequency is in the form

$$\omega_{n5j}=\mu_j^2\sqrt{\frac{EI}{\rho A}}=\left(\frac{j\pi}{l}\right)^2\sqrt{\frac{EI}{\rho A}},\quad j=1,2,\ldots \tag{25.33}$$

The quantities ω_{en5j}, ζ_{e5j}, and ω_{end5j} are given by Eqs. (25.22), (25.21), and (25.25), respectively.

In Figure 25.1, we indicate the plots of $w_1(x, t)=\text{w1}$ when $l=10$ and $f_j(t)=10\sin(0.314t)$.

25.4 DISCUSSIONS

The coordinate function $p_j(t)$ of a damped Euler-Bernoulli beam with the fractional inertia and fractional external damping forces regarding forced response is associated with α, β, ω_{en5j}, ζ_{e5j}, ω_{end5j}, ω, and driven function $f_j(t)$. In addition to the present solution of the forced response to such a beam, we found a novel form of the Rayleigh damping expressed by c_{e5j}, see Theorem 25.3. When using other models of fractional stress (Shitikova [2], Rossikhin [3]), it might be possible to obtain interesting results with respect to the Rayleigh damping.

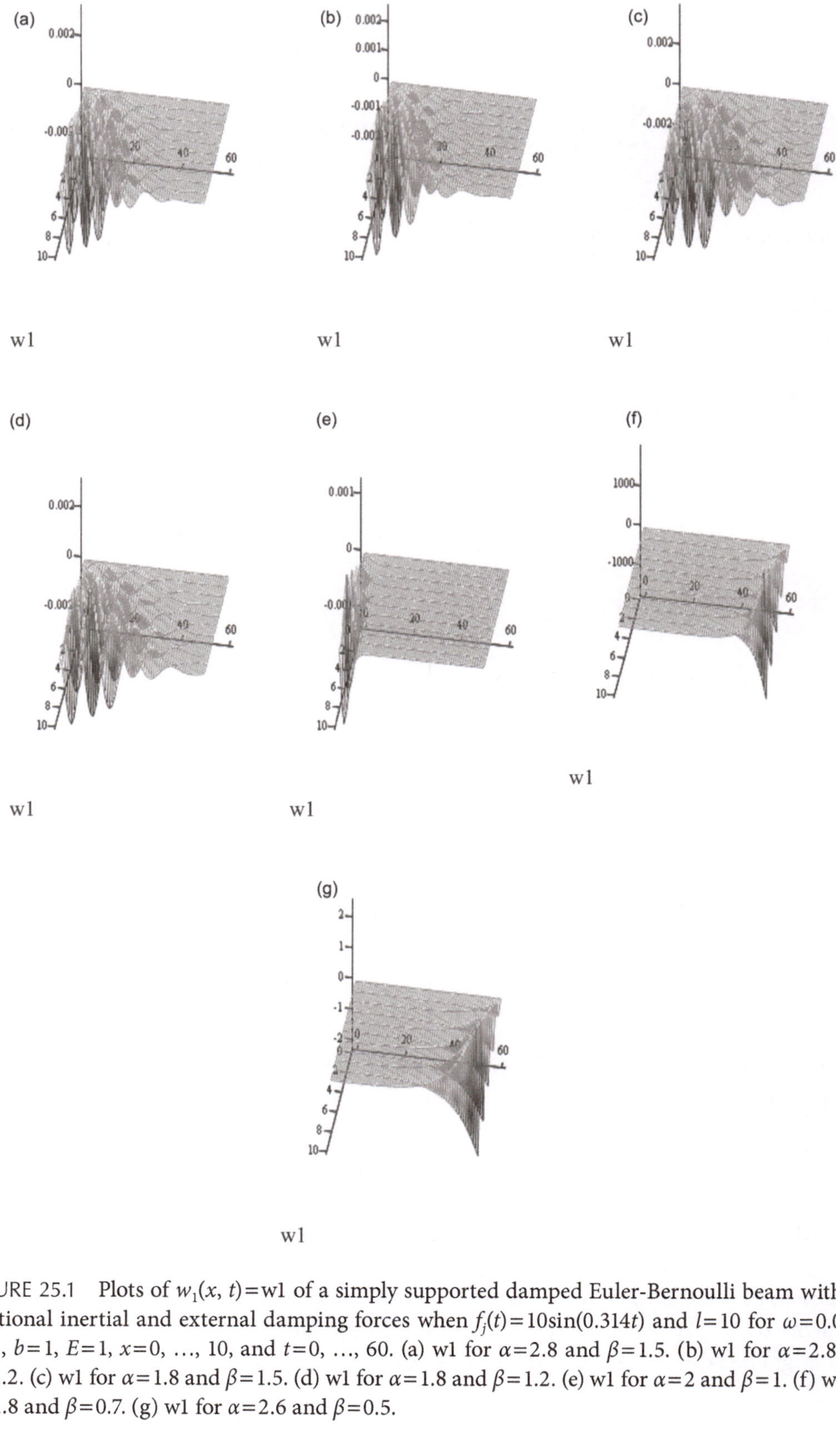

FIGURE 25.1 Plots of $w_1(x, t)$=w1 of a simply supported damped Euler-Bernoulli beam with the fractional inertial and external damping forces when $f_j(t) = 10\sin(0.314t)$ and $l = 10$ for $\omega = 0.0314$, $a = 1$, $b = 1$, $E = 1$, $x = 0, \ldots, 10$, and $t = 0, \ldots, 60$. (a) w1 for $\alpha = 2.8$ and $\beta = 1.5$. (b) w1 for $\alpha = 2.8$ and $\beta = 1.2$. (c) w1 for $\alpha = 1.8$ and $\beta = 1.5$. (d) w1 for $\alpha = 1.8$ and $\beta = 1.2$. (e) w1 for $\alpha = 2$ and $\beta = 1$. (f) w1 for $\alpha = 1.8$ and $\beta = 0.7$. (g) w1 for $\alpha = 2.6$ and $\beta = 0.5$.

25.5 SUMMARY

An equivalent equation of fractional coordinates of a damped Euler-Bernoulli beam with the fractional inertia and fractional external damping forces is addressed in Theorem 25.1. The analytic expression of the zero-state forced response to that beam is given in Theorem 25.6. A new variation form of the Rayleigh damping is given in Theorem 25.3.

25.6 EXERCISES

25.1. Denote by $f_j(t)$ the jth driven function. Let $f_j(t)=u(t)$ (unit step function) ($j=1, 2, \ldots$). For a simply-supported damped Euler-Bernoulli beam with the fractional inertia force and fractional external damping one, when $a=1$, $b=1$, $E=1$, and beam length $l=10$, find the forced response $w(x, t)$.

25.2. Denote by $f_j(t)$ the jth driven function. Let $f_j(t)=u(t-\tau/2)-u(t+\tau/2)$ ($j=1, 2, \ldots$), where τ is a constant and $u(t)$ is the unit step function. For a simply supported damped Euler-Bernoulli beam with the fractional inertia force and fractional external damping one, when $a=1$, $b=1$, $E=1$, and beam length $l=10$, find the forced response $w(x, t)$.

25.3. Let $f_j(t)=\mathrm{A}\sin(\omega_0 t)$ ($j=1, 2, \ldots$), where A and ω_0 are constants. For a free-free damped Euler-Bernoulli beam with the fractional inertia force and fractional external damping one, when $a=1$, $b=1$, $E=1$, and beam length $l=10$, find the forced response $w(x, t)$.

25.4. Let $f_j(t)=u(t)$ ($j=1, 2, \ldots$). For a free-free damped Euler-Bernoulli beam with the fractional inertia force and fractional external damping one, when $a=1$, $b=1$, $E=1$, and beam length $l=10$, find the forced response $w(x, t)$.

25.5. Denote by $f_j(t)$ the jth driven function. Let $f_j(t)=u(t-\tau/2)-u(t+\tau/2)$ ($j=1, 2, \ldots$), where τ is a constant and $u(t)$ is the unit step function. For a free-free damped Euler-Bernoulli beam with the fractional inertia force and fractional external damping one, when $a=1$, $b=1$, $E=1$, and beam length $l=10$, find the forced response $w(x, t)$.

REFERENCES

1. O. M. Palley, Г. B. Bahizov, and E. Я. Voroneysk, *Handbook of Ship Structural Mechanics*, National Defense Industry Publishing House, Beijing, China, 2002. In Chinese. Translated from Russian by B. H. Xu, X. Xu, and M. Q. Xu.
2. M. V. Shitikova, Fractional operator viscoelastic models in dynamic problems of mechanics of solids: A review, *Mechanics of Solids*, 57(1), 2022, 1–33.
3. Y. A. Rossikhin, Reflections on two parallel ways in progress of fractional calculus in mechanics of solids, *Applied Mechanics Reviews*, 63(1), 2010, 010701.

CHAPTER 26

Forced Response to Damped Euler-Bernoulli Beam with Fractional Inertia and Internal Damping Forces (Class 6)

THE CONTRIBUTIONS GIVEN IN this chapter are in three folds with respect to the forced response to a damped Euler-Bernoulli beam with fractional inertia and fractional internal damping forces. (1) An equivalent equation of fractional coordinates of that beam in Theorem 26.1. (2) The closed-form expression of the forced response to such a beam in Theorem 26.6. (3) A new variation of the Rayleigh damping based on such a beam in Theorem 26.3.

26.1 INTRODUCTION

In this chapter, we study the forced response to a damped Euler-Bernoulli beam with the fractional inertia and fractional internal damping forces. The motion equation of such a beam is given by

$$\frac{\partial^2}{\partial x^2}\left(EI\frac{\partial^2 w}{\partial x^2}+c_s I\frac{\partial^{2+\lambda} w}{\partial x^2\,\partial t^\lambda}\right)+\rho A\frac{\partial^\alpha w}{\partial t^\alpha}+c\frac{\partial w}{\partial t}=q(x,t),\ \ 1<\alpha<3,\ 0<\lambda<2, \tag{26.1}$$

where $\rho A\frac{\partial^\alpha w}{\partial t^\alpha}$ is fractional inertia force and $c_s I\frac{\partial^{2+\lambda} w}{\partial x^2\,\partial t^\lambda}$ is fractional internal damping force. Following Chapter 20, Eq. (26.1) stands for the class 6 fractionally damped Euler-Bernoulli

DOI: 10.1201/9781003460947-29

beams. Note that the closed-form expression of such a forced response remains a problem open. In this chapter, we aim at giving the solution to that problem.

The rest of this chapter is organized as follows. The analytic solution to Eq. (26.1) is addressed in Section 26.2. Case study is in Section 26.3. Discussions are given in Section 26.4, which is followed by summary.

26.2 ANALYTIC EXPRESSION OF FORCED RESPONSE

By using $w(x, t) = \varphi(x)p(t)$, Eq. (26.1) can be written by

$$\sum_{j=1}^{\infty} \rho A \varphi_j(x) \frac{d^{\alpha} p_j(t)}{dt^{\alpha}} + \sum_{j=1}^{\infty} c \varphi_j(x) \frac{dp_j(t)}{dt} + \sum_{j=1}^{\infty} \frac{d^2}{dx^2}\left[c_s I \frac{d^2 \varphi_j(x)}{dx^2}\right] \frac{d^{\lambda} p_j(t)}{dt^{\lambda}}$$
$$+ \sum_{j=1}^{\infty} \frac{d^2}{dx^2}\left[EI \frac{d^2 \varphi_j(x)}{dx^2}\right] p_j(t) = q(x,t). \tag{26.2}$$

Multiplying $\varphi_m(x)$ on both sides of the above equation produces

$$\sum_{j=1}^{\infty} \rho A \varphi_j(x) \varphi_m(x) \frac{d^{\alpha} p_j(t)}{dt^{\alpha}} + \sum_{j=1}^{\infty} c \varphi_j(x) \varphi_m(x) \frac{dp_j(t)}{dt}$$
$$+ \sum_{j=1}^{\infty} \varphi_m(x) \frac{d^2}{dx^2}\left[c_s I \frac{d^2 \varphi_j(x)}{dx^2}\right] \frac{d^{\lambda} p_j(t)}{dt^{\lambda}} + \sum_{j=1}^{\infty} \varphi_m(x) \frac{d^2}{dx^2}\left[EI \frac{d^2 \varphi_j(x)}{dx^2}\right] p_j(t) \tag{26.3}$$
$$= \varphi_m(x) q(x,t).$$

Performing the integral from 0 to l in terms of x on both sides of the above and taking into account the orthogonality of vibration modes yields

$$M_j \frac{d^{\alpha} p_j(t)}{dt^{\alpha}} + \frac{M_j c}{\rho A} \frac{dp_j(t)}{dt} + \sum_{j=1}^{\infty} \int_0^l \varphi_m(x) \frac{d^2}{dx^2}\left[c_s I \frac{d^2 \varphi_j(x)}{dx^2}\right] dx \frac{d^{\lambda} p_j(t)}{dt^{\lambda}}$$
$$+ \sum_{j=1}^{\infty} \int_0^l \varphi_m(x) \frac{d^2}{dx^2}\left[EI \frac{d^2 \varphi_j(x)}{dx^2}\right] dx p_j(t) = Q_j(t), \tag{26.4}$$

where

$$M_j = \rho A \int_0^l \varphi_j^2(x) dx \tag{26.5}$$

is the jth generalized mass.

From Chapter 6, we have

$$\frac{d^2}{dx^2}\left[EI \frac{d^2 \varphi_j(x)}{dx^2}\right] = \rho A \omega_{nj}^2 \varphi_j(x). \tag{26.6}$$

Thus, Eq. (26.4) can be written by

$$M_j \frac{d^\alpha p_j(t)}{dt^\alpha} + \frac{M_j c}{\rho A}\frac{dp_j(t)}{dt} + \sum_{j=1}^{\infty} \int_0^l \varphi_m(x) \frac{d^2}{dx^2}\left[c_s I \frac{d^2 \varphi_j(x)}{dx^2}\right] dx \frac{d^\lambda p_j(t)}{dt^\lambda}$$
$$+ M_j \omega_{nj}^2 p_j(t) = Q_j(t). \tag{26.7}$$

Following the Rayleigh damping assumption, the external damping c is in the form

$$c = \rho A a, \tag{26.8}$$

where a is a coefficient with the unit of time. The internal damping c_s is given by

$$c_s = Eb, \tag{26.9}$$

where b is a coefficient with the unit of frequency.

Substituting Eqs. (26.8) and (26.9) into (26.7) and considering the orthogonality of vibration modes yields

$$M_j \frac{d^\alpha p_j(t)}{dt^\alpha} + M_j a \frac{dp_j(t)}{dt} + M_j \omega_{nj}^2 b \frac{d^\lambda p_j(t)}{dt^\lambda} + M_j \omega_{nj}^2 p_j(t) = Q_j(t). \tag{26.10}$$

For the jth order forced vibration of a damped Euler-Bernoulli beam with the fractional inertia and fractional internal damping forces, we have

$$\frac{d^\alpha p_j(t)}{dt^\alpha} + \left(a \frac{dp_j(t)}{dt} + \omega_{nj}^2 b \frac{d^\lambda p_j(t)}{dt^\lambda}\right) + \omega_{nj}^2 p_j(t) = f_j(t), \tag{26.11}$$

where

$$f_j(t) = \frac{Q_j(t)}{M_j}. \tag{26.12}$$

Theorem 26.1. An equivalent equation of (26.11) is given by

$$-\left(\cos\frac{\alpha\pi}{2}\omega^{\alpha-2} + \omega_{nj}^2 b \cos\frac{\lambda\pi}{2}\omega^{\lambda-2}\right)\frac{d^2 p_j(t)}{dt^2}$$
$$+\left(\sin\frac{\alpha\pi}{2}\omega^{\alpha-1} + a + \omega_{nj}^2 b \sin\frac{\lambda\pi}{2}\omega^{\lambda-1}\right)\frac{dp_j(t)}{dt} + \omega_{nj}^2 p_j(t) = f_j(t). \tag{26.13}$$

Proof. Let $P_j(\omega) = \mathrm{F}[p_j(t)]$ and $F_j(\omega) = \mathrm{F}[f_j(t)]$. Doing the Fourier transform on both sides of Eq. (26.11) yields

$$\left[(i\omega)^\alpha + a(i\omega) + \omega_{nj}^2 b (i\omega)^\lambda + \omega_{nj}^2\right] P_j(\omega) = F_j(\omega). \tag{26.14}$$

Since

$$(i\omega)^{\alpha}+a(i\omega)+\omega_{nj}^{2}b(i\omega)^{\lambda}+\omega_{nj}^{2}=-\left(\cos\frac{\alpha\pi}{2}\omega^{\alpha-2}+\omega_{nj}^{2}b\cos\frac{\lambda\pi}{2}\omega^{\lambda-2}\right)(i\omega)^{2}$$
$$+\left(\sin\frac{\alpha\pi}{2}\omega^{\alpha-1}+a+\omega_{nj}^{2}b\sin\frac{\lambda\pi}{2}\omega^{\lambda-1}\right)(i\omega)+\omega_{nj}^{2}, \tag{26.15}$$

equation (26.14) can be expressed by

$$\left[(i\omega)^{\alpha}+a(i\omega)+\omega_{nj}^{2}b(i\omega)^{\lambda}+\omega_{nj}^{2}\right]P_j(\omega)$$
$$=\begin{bmatrix}-\left(\cos\frac{\alpha\pi}{2}\omega^{\alpha-2}+\omega_{nj}^{2}b\cos\frac{\lambda\pi}{2}\omega^{\lambda-2}\right)(i\omega)^{2}\\+\left(\sin\frac{\alpha\pi}{2}\omega^{\alpha-1}+a+\omega_{nj}^{2}b\sin\frac{\lambda\pi}{2}\omega^{\lambda-1}\right)(i\omega)+\omega_{nj}^{2}\end{bmatrix}P_j(\omega)=F_j(\omega). \tag{26.16}$$

The above is the Fourier transform of Eq. (26.13). This completes the proof. □

Theorem 26.2. Denote by m_{e6j} the jth equivalent mass in the system (26.11). Then,

$$m_{e6j}=-\left(\cos\frac{\alpha\pi}{2}\omega^{\alpha-2}+\omega_{nj}^{2}b\cos\frac{\lambda\pi}{2}\omega^{\lambda-2}\right). \tag{26.17}$$

Proof. The right side of the above equation is the coefficient of $\frac{d^2p_j(t)}{dt^2}$ in Eq. (26.13). Thus, the above is true. This finishes the proof. □

Theorem 26.3. Denote by c_{e6j} the jth equivalent damping of the system (26.11). Then,

$$c_{e6j}=\sin\frac{\alpha\pi}{2}\omega^{\alpha-1}+a+\omega_{nj}^{2}b\sin\frac{\lambda\pi}{2}\omega^{\lambda-1}. \tag{26.18}$$

Proof. The right side of the aforementioned equation is the coefficient of $\frac{dp_j(t)}{dt}$ in Eq. (26.13). Hence, the above is valid. The proof is completed. □

Using Eqs. (26.17) and (26.18), we rewrite Eq. (26.13) by

$$m_{e6j}\frac{d^2p_j(t)}{dt^2}+c_{e6j}\frac{dp_j(t)}{dt}+\omega_{nj}^{2}p_j(t)=f_j(t). \tag{26.19}$$

Additionally, Eq. (26.16) can be rewritten by

$$\left[(i\omega)^2+\frac{c_{e6j}}{m_{e6j}}(i\omega)+\frac{\omega_{nj}^{2}}{m_{e6j}}\right]P_j(\omega)=\frac{F_j(\omega)}{m_{e6j}}. \tag{26.20}$$

Denote by ζ_{e6j} the jth equivalent damping ratio of the system (26.11). Define it by

$$\varsigma_{e6j} = \frac{c_{e6j}}{2\sqrt{m_{e6j}}} = \frac{\sin\frac{\alpha\pi}{2}\omega^{\alpha-1} + a\omega_{nj}^2 b\sin\frac{\lambda\pi}{2}\omega^{\lambda-1}}{2\sqrt{-\left(\cos\frac{\alpha\pi}{2}\omega^{\alpha-2} + \omega_{nj}^2 b\cos\frac{\lambda\pi}{2}\omega^{\lambda-2}\right)}}. \tag{26.21}$$

Let ω_{en6j} be the jth equivalent damping free natural frequency regarding the system (26.11). It is given by

$$\omega_{en6j}^2 = \frac{\omega_{nj}^2}{m_{e6j}}. \tag{26.22}$$

Then, we write Eq. (26.20) by

$$\left[(i\omega)^2 + 2\varsigma_{e6j}\omega_{en6j}(i\omega) + \omega_{en6j}^2\right]P_j(\omega) = \frac{F_j(\omega)}{m_{e6j}}. \tag{26.23}$$

Practically, we adopt

$$\left|\varsigma_{e6j}\right| \le 1. \tag{26.24}$$

Let ω_{en6j} be the jth equivalent damped natural frequency regarding the system (26.11). Then,

$$\omega_{end6j} = \omega_{en6j}\sqrt{1-\varsigma_{e6j}^2}. \tag{26.25}$$

Theorem 26.4. Denote by $h_{6j}(t)$ the jth impulse response function of the system (26.19). Then,

$$h_{6j}(t) = \frac{1}{m_{e6j}\omega_{end6j}} e^{-\varsigma_{e6j}\omega_{en6j}t}\sin\omega_{end6j}t, \quad t>0. \tag{26.26}$$

Proof. Let $f_j(t)=\delta(t)$ in Eq. (26.19). Then, we have the above. The proof is finished. □

When using the unit step function $u(t)$, we express $h_{6j}(t)$ by

$$h_{6j}(t) = \frac{1}{m_{e6}\omega_{end6j}} e^{-\varsigma_{e6j}\omega_{en6j}t}\left(\sin\omega_{end6j}t\right)u(t). \tag{26.27}$$

Theorem 26.5. The jth coordinate with respect to Eq. (26.11) under zero initial conditions is represented by

$$p_j(t) = f_j(t) * h_{6j}(t) = \frac{1}{m_{e6j}\omega_{end6j}}\int_{-\infty}^{\infty} e^{-\varsigma_{e6j}\omega_{en6j}\tau}\sin\omega_{end6j}\tau f_j(t-\tau)d\tau. \tag{26.28}$$

Proof. From the convolution theory (Harris [1]), we see that the above is true. The proof is finished. □

Theorem 26.6. The zero-state forced response to a damped Euler-Bernoulli beam with the fractional inertia and fractional internal damping forces is expressed by

$$w(x,t)=\sum_{j=1}^{\infty}\frac{\varphi_j(x)}{m_{e6j}\omega_{end6j}}\int_{-\infty}^{\infty}e^{-\varsigma_{e6j}\omega_{en6j}\tau}\sin\omega_{end6j}\tau f_j(t-\tau)d\tau. \tag{26.29}$$

Proof. Because $w(x,t)=\sum_{j=1}^{\infty}\varphi_j(x)p_j(t)=\sum_{j=1}^{\infty}\varphi_j(x)\left[f_j(t)*h_{6j}(t)\right]$, the above is true. This finishes the proof. □

We suppose that initial conditions are zero in what follows, containing exercises in Section 26.6.

26.3 CASE STUDY

For a simply supported Euler-Bernoulli beam [2], its vibration mode function is given by

$$\varphi_j(x)\sim\sin\frac{j\pi x}{l},\quad j=1,2,\ldots \tag{26.30}$$

Thus,

$$w(x,t)=\sum_{j=1}^{\infty}\sin\frac{j\pi x}{l}\left[f_j(t)*\frac{1}{m_{e6j}\omega_{end6j}}e^{-\varsigma_{e6j}\omega_{en6j}t}\sin\omega_{end6j}t\right]. \tag{26.31}$$

Since

$$\mu_j=\frac{j\pi}{l},\quad j=1,2,\ldots \tag{26.32}$$

the jth order natural angular frequency is in the form

$$\omega_{n6j}=\mu_j^2\sqrt{\frac{EI}{\rho A}}=\left(\frac{j\pi}{l}\right)^2\sqrt{\frac{EI}{\rho A}},\quad j=1,2,\ldots \tag{26.33}$$

The quantities ω_{en6j}, ζ_{e6j}, and ω_{end6j} are given by Eqs. (26.22), (26.21), and (26.25), respectively.

Let $l=10$. When $f_j(t)=10\sin(0.314t)$, we illustrate the plots of $w_1(x,t)=\text{w1}$ in Figure 26.1.

26.4 DISCUSSIONS

With respect to the forced response to a damped Euler-Bernoulli beam with the fractional inertia force and fractional internal damping one, the coordinate function $p_j(t)$ is associated with α, λ, ω_{en6j}, ζ_{e6j}, ω_{end6j}, ω, and driven function $f_j(t)$. Our finding of a novel form of the Rayleigh damping is expressed by c_{e6j}. One thing worth noting is that the premise of the variation of the Rayleigh damping c_{e6j} is the Voigt assumption on materials. If using other models of fractional stress and strain (Shitikova [3], Ogorodnikov et al. [4]), interesting results of forced response to such a beam may come out.

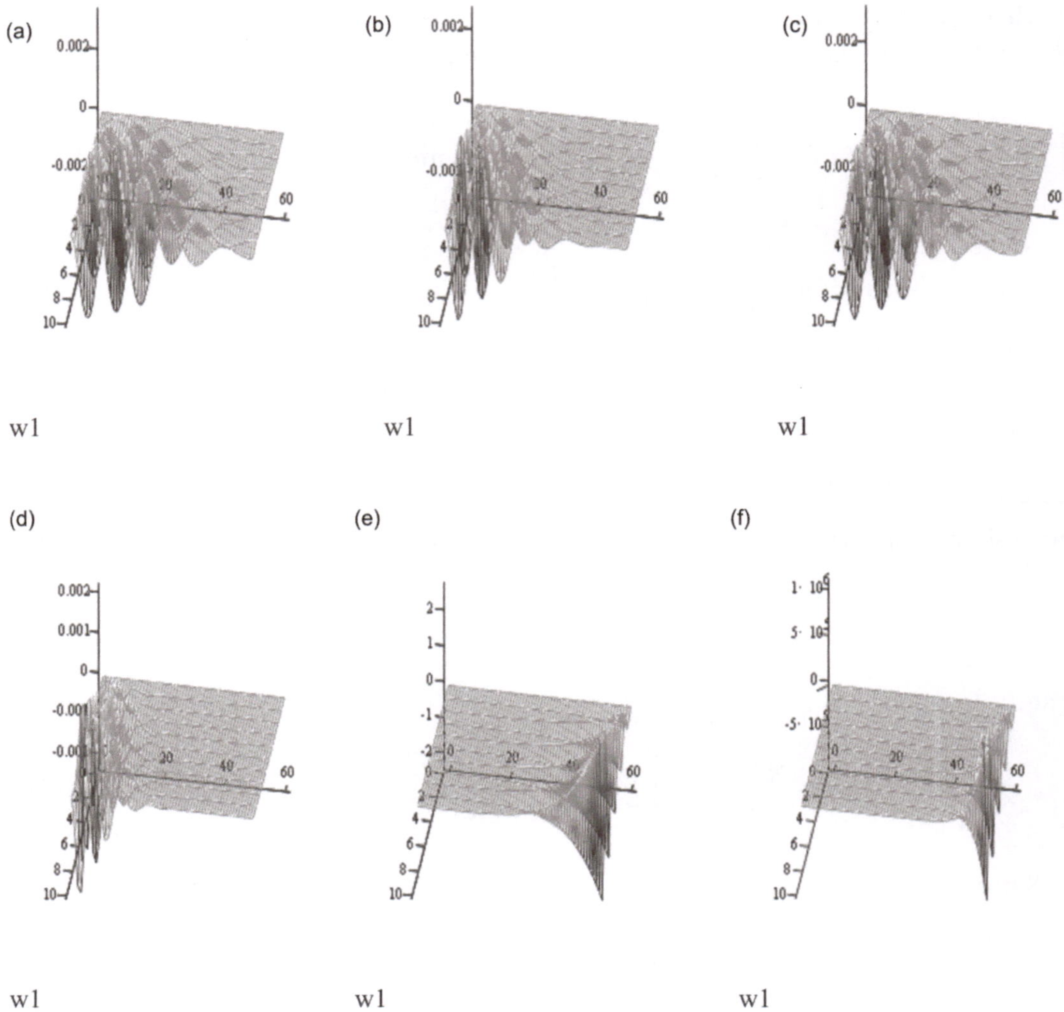

FIGURE 26.1 Plots of $w_1(x, t)$=w1 of the simply supported damped Euler-Bernoulli beam with the fractional inertia force and fractional internal damping one when $f_j(t) = 10\sin(0.314t)$ and $l = 10$ for $\omega = 0.0314$, $a = 1$, $b = 1$, $E = 1$, $x = 0, \ldots, 10$, and $t = 0, \ldots, 60$. (a) w1 for $\alpha = 1.8$ and $\lambda = 1.2$. (b) w1 for $\alpha = 2.8$ and $\lambda = 1.2$. (c) w1 for $\alpha = 1.2$ and $\lambda = 1.8$. (d) w1 for $\alpha = 2.8$ and $\lambda = 1.8$. (e) w1 for $\alpha = 1.8$ and $\lambda = 0.5$. (f) w1 for $\alpha = 2.8$ and $\lambda = 0.8$.

26.5 SUMMARY

We have given an equivalent equation of fractional coordinates of a damped Euler-Bernoulli beam with fractional inertia and fractional internal damping forces. We have proposed the analytic expression of the zero-state forced response to that beam. We have revealed a new variation of the Rayleigh damping based on such a beam.

26.6 EXERCISES

26.1. Denote by $f_j(t)$ the jth driven function. Let $f_j(t)=u(t)$ (unit step function) ($j=1, 2, \ldots$). For a simply supported damped Euler-Bernoulli beam with the fractional inertia and fractional internal damping forces, when $a=1$, $b=1$, $E=1$, and beam length $l=10$, find the forced response $w(x, t)$.

26.2. Denote by $f_j(t)$ the jth driven function. Let $f_j(t)=u(t-\tau/2)-u(t+\tau/2)$ ($j=1, 2, \ldots$), where τ is a constant and $u(t)$ is the unit step function. For a simply supported damped Euler-Bernoulli beam with the fractional inertia and fractional internal damping forces, when $a=1$, $b=1$, $E=1$, and beam length $l=10$, find the forced response $w(x, t)$.

26.3. Let $f_j(t)=A\sin(\omega_0 t)$ ($j=1, 2, \ldots$), where A and ω_0 are constants. For a free-free damped Euler-Bernoulli beam with the fractional inertia and fractional internal damping forces, when $a=1$, $b=1$, $E=1$, and beam length $l=10$, find the forced response $w(x, t)$.

26.4. Let $f_j(t)=u(t)$ ($j=1, 2, \ldots$). For a free-free damped Euler-Bernoulli beam with the fractional inertia and fractional internal damping forces, when $a=1$, $b=1$, $E=1$, and beam length $l=10$, find the forced response $w(x, t)$.

26.5. Denote by $f_j(t)$ the jth driven function. Let $f_j(t)=u(t-\tau/2)-u(t+\tau/2)$ ($j=1, 2, \ldots$), where τ is a constant and $u(t)$ is the unit step function. For a free-free damped Euler-Bernoulli beam with the fractional inertia and fractional internal damping forces, when $a=1$, $b=1$, $E=1$, and beam length $l=10$, find the forced response $w(x, t)$.

REFERENCES

1. C. M. Harris, *Shock and Vibration Handbook*, 5th Ed., McGraw-Hill, New York, 2002.
2. O. M. Palley, Г. B. Bahizov, and E. Я. Voroneysk, *Handbook of Ship Structural Mechanics*, National Defense Industry Publishing House, Beijing, China, 2002. In Chinese. Translated from Russian by B. H. Xu, X. Xu, and M. Q. Xu.
3. M. V. Shitikova, Fractional operator viscoelastic models in dynamic problems of mechanics of solids: A review, *Mechanics of Solids*, 57(1), 2022, 1–33.
4. E. N. Ogorodnikov, V. P. Radchenko, and L. G. Ungarova, Mathematical models of nonlinear viscoelasticity with operators of integro-differentiation, *PNRPU Mechanics Bulletin (Vestnik PNIPU. Mekh.)*, (2), 2018, 147–161. In Russian.

CHAPTER 27

Forced Response to Multi-Fractional Damped Euler-Bernoulli Beam (Class 7)

IN THIS CHAPTER, WE study the forced response to a multi-fractional damped Euler-Bernoulli beam. By multi-fractional, we mean that inertia force, internal and external damping forces are of fractional orders. The contributions presented in this chapter are in three aspects as follows. (1) An equivalent equation of fractional coordinates of a multi-fractional damped Euler-Bernoulli beam in Theorem 27.1. (2) The analytic representation of the forced response to such a beam in Theorem 27.6. (3) A new variation of the Rayleigh damping in Theorem 27.3 based on such a beam.

27.1 INTRODUCTION

The motion equation of the conventional damped Euler-Bernoulli beam is given by

$$\frac{\partial^2}{\partial x^2}\left(EI\frac{\partial^2 w}{\partial x^2}+c_s I\frac{\partial^3 w}{\partial x^2\partial t}\right)+\rho A\frac{\partial^2 w}{\partial t^2}+c\frac{\partial w}{\partial t}=q(x,t), \tag{27.1}$$

where c_s is internal damping and c is external one (Palley et al. [1]). In the above, $c\frac{\partial w}{\partial t}$ is external damping force, $c_s I\frac{\partial^3 w}{\partial x^2\partial t}$ is internal damping force, and $\rho A\frac{\partial^2 w}{\partial t^2}$ is inertia force. The forced response to the aforementioned equation under the Rayleigh damping assumption is known, see e.g., Palley et al. [1], Jin and Xia [2].

The aforementioned equation takes the Voigt assumption on materials about internal damping into account. In this chapter, I study a multi-fractional damped Euler-Bernoulli beam in the form

 DOI: 10.1201/9781003460947-30

$$\frac{\partial^2}{\partial x^2}\left(EI\frac{\partial^2 w}{\partial x^2}+c_sI\frac{\partial^{2+\lambda}w}{\partial x^2\partial t^\lambda}\right)+\rho A\frac{\partial^\alpha w}{\partial t^\alpha}+c\frac{\partial^\beta w}{\partial t^\beta}=q(x,t),\quad 1<\alpha<3,\ 0<\beta<2,\ 0<\lambda<2. \tag{27.2}$$

Precisely, the above stands for a beam with the fractional inertia force $\rho A\frac{\partial^\alpha w}{\partial t^\alpha}$, fractional internal damping force $c_sI\frac{\partial^{2+\lambda}w}{\partial x^2\partial t^\lambda}$, and fractional external damping one $c\frac{\partial^\beta w}{\partial t^\beta}$. The beam expressed by Eq. (27.2) is called the class 7 fractional damped Euler-Bernoulli beam in Chapter 20. Its analytic expression with respect to the forced response is unknown. In this chapter, we will give its analytic solution in closed form.

The rest of this chapter is organized as follows. The analytic solution to Eq. (27.2) is addressed in Section 27.2. Case study is in Section 27.3. Discussions are given in Section 27.4, which is followed by summary.

27.2 ANALYTIC EXPRESSION OF FORCED RESPONSE

Substituting $w(x, t)=\varphi(x)p(t)$ into Eq. (27.2) yields

$$\begin{aligned}&\sum_{j=1}^{\infty}\rho A\varphi_j(x)\frac{d^\alpha p_j(t)}{dt^\alpha}+\sum_{j=1}^{\infty}c\varphi_j(x)\frac{d^\beta p_j(t)}{dt^\beta}+\sum_{j=1}^{\infty}\frac{d^2}{dx^2}\left[c_sI\frac{d^2\varphi_j(x)}{dx^2}\right]\frac{d^\lambda p_j(t)}{dt^\lambda}\\&+\sum_{j=1}^{\infty}\frac{d^2}{dx^2}\left[EI\frac{d^2\varphi_j(x)}{dx^2}\right]p_j(t)=q(x,t).\end{aligned} \tag{27.3}$$

Multiplying $\varphi_m(x)$ on both sides of the aforementioned equation produces

$$\begin{aligned}&\sum_{j=1}^{\infty}\rho A\varphi_j(x)\varphi_m(x)\frac{d^\alpha p_j(t)}{dt^\alpha}+\sum_{j=1}^{\infty}c\varphi_j(x)\varphi_m(x)\frac{d^\beta p_j(t)}{dt^\beta}\\&+\sum_{j=1}^{\infty}\varphi_m(x)\frac{d^2}{dx^2}\left[c_sI\frac{d^2\varphi_j(x)}{dx^2}\right]\frac{d^\lambda p_j(t)}{dt^\lambda}+\sum_{j=1}^{\infty}\varphi_m(x)\frac{d^2}{dx^2}\left[EI\frac{d^2\varphi_j(x)}{dx^2}\right]p_j(t)\\&=\varphi_m(x)q(x,t).\end{aligned} \tag{27.4}$$

Doing the integral from 0 to l in terms of x on both sides of the aforementioned equation and using the orthogonality of vibration modes results in

$$\begin{aligned}&M_j\frac{d^\alpha p_j(t)}{dt^\alpha}+\frac{M_jc}{\rho A}\frac{d^\beta p_j(t)}{dt^\beta}+\sum_{j=1}^{\infty}\int_0^l\varphi_m(x)\frac{d^2}{dx^2}\left[c_sI\frac{d^2\varphi_j(x)}{dx^2}\right]dx\frac{d^\lambda p_j(t)}{dt^\lambda}\\&+\sum_{j=1}^{\infty}\int_0^l\varphi_m(x)\frac{d^2}{dx^2}\left[EI\frac{d^2\varphi_j(x)}{dx^2}\right]dxp_j(t)=Q_j(t),\end{aligned} \tag{27.5}$$

where $M_j = \rho A \int_0^l \varphi_j^2(x)dx.$

According to Eq. (6.136) or (6.174),

$$\frac{d^2}{dx^2}\left[EI\frac{d^2\varphi_j(x)}{dx^2}\right] = \rho A\omega_{nj}^2\varphi_j(x). \tag{27.6}$$

Thus, Eq. (27.5) can be written by

$$M_j\frac{d^\alpha p_j(t)}{dt^\alpha}+\frac{M_j c}{\rho A}\frac{d^\beta p_j(t)}{dt^\beta}+\sum_{j=1}^{\infty}\int_0^l \varphi_m(x)\frac{d^2}{dx^2}\left[c_s I\frac{d^2\varphi_j(x)}{dx^2}\right]dx\frac{d^\lambda p_j(t)}{dt^\lambda}$$
$$+M_j\omega_{nj}^2 p_j(t)=Q_j(t). \tag{27.7}$$

Following the Rayleigh damping assumption, one has

$$c=\rho Aa, \tag{27.8}$$

where a is a coefficient with the unit of time since ρA is with the unit of mass and

$$c_s = Eb, \tag{27.9}$$

where b is a coefficient with the unit of frequency as E is with the unit [N/m].

After substituting Eqs. (27.8) and (27.9) into (27.7) and taking into account the orthogonality of vibration modes, Σ on the third item on the left side of Eq. (27.7) can be removed. When doing so, we have

$$M_j\frac{d^\alpha p_j(t)}{dt^\alpha}+M_j a\frac{d^\beta p_j(t)}{dt^\beta}+M_j\omega_{nj}^2 b\frac{d^\lambda p_j(t)}{dt^\lambda}+M_j\omega_{nj}^2 p_j(t)=Q_j(t). \tag{27.10}$$

Therefore, for the jth order coordinate function, we have

$$\frac{d^\alpha p_j(t)}{dt^\alpha}+\left(a\frac{d^\beta p_j(t)}{dt^\beta}+\omega_{nj}^2 b\frac{d^\lambda p_j(t)}{dt^\lambda}\right)+\omega_{nj}^2 p_j(t)=f_j(t), \tag{27.11}$$

where

$$f_j(t)=\frac{Q_j(t)}{M_j}. \tag{27.12}$$

Theorem 27.1. An equivalent equation of (27.11) is given by

$$\begin{aligned}&-\left(\cos\frac{\alpha\pi}{2}\omega^{\alpha-2}+a\cos\frac{\beta\pi}{2}\omega^{\beta-2}+\omega_{nj}^2 b\cos\frac{\lambda\pi}{2}\omega^{\lambda-2}\right)\frac{d^2p_j(t)}{dt^2}\\&+\left(\sin\frac{\alpha\pi}{2}\omega^{\alpha-1}+a\sin\frac{\beta\pi}{2}\omega^{\beta-1}+\omega_{nj}^2 b\sin\frac{\lambda\pi}{2}\omega^{\lambda-1}\right)\frac{dp_j(t)}{dt}+\omega_{nj}^2 p_j(t)=f_j(t).\end{aligned} \tag{27.13}$$

Proof. Let $P_j(\omega)=\mathrm{F}[p_j(t)]$ and $F_j(\omega)=\mathrm{F}[f_j(t)]$. Doing the Fourier transform on both sides of Eq. (27.11) yields

$$\left[(i\omega)^\alpha+a(i\omega)^\beta+\omega_{nj}^2 b(i\omega)^\lambda+\omega_{nj}^2\right]P_j(\omega)=F_j(\omega). \tag{27.14}$$

Because

$$\begin{aligned}&(i\omega)^\alpha+a(i\omega)^\beta+\omega_{nj}^2 b(i\omega)^\lambda+\omega_{nj}^2\\&=-\left(\cos\frac{\alpha\pi}{2}\omega^{\alpha-2}+a\cos\frac{\beta\pi}{2}\omega^{\beta-2}+\omega_{nj}^2 b\cos\frac{\lambda\pi}{2}\omega^{\lambda-2}\right)(i\omega)^2\\&+\left(\sin\frac{\alpha\pi}{2}\omega^{\alpha-1}+a\sin\frac{\beta\pi}{2}\omega^{\beta-1}+\omega_{nj}^2 b\sin\frac{\lambda\pi}{2}\omega^{\lambda-1}\right)(i\omega)+\omega_{nj}^2,\end{aligned} \tag{27.15}$$

equation (27.14) can be expressed by

$$\begin{aligned}&\left[(i\omega)^\alpha+a(i\omega)^\beta+\omega_{nj}^2 b(i\omega)^\lambda+\omega_{nj}^2\right]P_j(\omega)\\&=\begin{bmatrix}-\left(\cos\frac{\alpha\pi}{2}\omega^{\alpha-2}+a\cos\frac{\beta\pi}{2}\omega^{\beta-2}+\omega_{nj}^2 b\cos\frac{\lambda\pi}{2}\omega^{\lambda-2}\right)(i\omega)^2\\+\left(\sin\frac{\alpha\pi}{2}\omega^{\alpha-1}+a\sin\frac{\beta\pi}{2}\omega^{\beta-1}+\omega_{nj}^2 b\sin\frac{\lambda\pi}{2}\omega^{\lambda-1}\right)(i\omega)+\omega_{nj}^2\end{bmatrix}P_j(\omega)=F_j(\omega).\end{aligned} \tag{27.16}$$

The above is the Fourier transform of Eq. (27.13). The proof is completed. □

Theorem 27.2. Denote by m_{e7j} the jth equivalent mass in the system (27.11). Then,

$$m_{e7j}=-\left(\cos\frac{\alpha\pi}{2}\omega^{\alpha-2}+a\cos\frac{\beta\pi}{2}\omega^{\beta-2}+\omega_{nj}^2 b\cos\frac{\lambda\pi}{2}\omega^{\lambda-2}\right). \tag{27.17}$$

Proof. From the first item on the left side of Eq. (27.13), we see that the above is true. The proof is finished. □

Theorem 27.3. Denote by c_{e7j} the jth equivalent damping in the system (27.11). Then,

$$c_{e7j} = \sin\frac{\alpha\pi}{2}\omega^{\alpha-1} + a\sin\frac{\beta\pi}{2}\omega^{\beta-1} + \omega_{nj}^2 b\sin\frac{\lambda\pi}{2}\omega^{\lambda-1}. \tag{27.18}$$

Proof. From the second item on the left side of Eq. (27.13), we see that the above is valid. The proof is completed. □

Using Eqs. (27.17) and (27.18), (27.13) is rewritten by

$$m_{e7j}\frac{d^2p_j(t)}{dt^2} + c_{e7j}\frac{dp_j(t)}{dt} + \omega_{nj}^2 p_j(t) = f_j(t). \tag{27.19}$$

In addition, Eq. (27.16) can be rewritten by

$$\left[(i\omega)^2 + \frac{c_{e7j}}{m_{e7j}}(i\omega) + \frac{\omega_{nj}^2}{m_{e7j}}\right]P_j(\omega) = \frac{F_j(\omega)}{m_{e7j}}. \tag{27.20}$$

Denote by ζ_{e7j} the jth equivalent damping ratio in the system (27.11). Define it by

$$\varsigma_{e7j} = \frac{c_{e7j}}{2\sqrt{m_{e7j}}} = \frac{\sin\frac{\alpha\pi}{2}\omega^{\alpha-1} + a\sin\frac{\beta\pi}{2}\omega^{\beta-1} + \omega_{nj}^2 b\sin\frac{\lambda\pi}{2}\omega^{\lambda-1}}{2\sqrt{-\left(\cos\frac{\alpha\pi}{2}\omega^{\alpha-2} + a\cos\frac{\beta\pi}{2}\omega^{\beta-2} + \omega_{nj}^2 b\cos\frac{\lambda\pi}{2}\omega^{\lambda-2}\right)}}. \tag{27.21}$$

Let ω_{en7j} be the jth equivalent damping free natural frequency regarding the system (27.11). It is given by

$$\omega_{en7j}^2 = \frac{\omega_{nj}^2}{m_{e7j}}. \tag{27.22}$$

Then, we write Eq. (27.20) by

$$\left[(i\omega)^2 + 2\varsigma_{e7j}\omega_{en7j}(i\omega) + \omega_{en7j}^2\right]P_j(\omega) = \frac{F_j(\omega)}{m_{e7j}}. \tag{27.23}$$

Practically, we are interested in

$$\left|\varsigma_{e7j}\right| \le 1. \tag{27.24}$$

Let ω_{end7j} be the jth equivalent damped natural frequency regarding the system (27.11). Then,

$$\omega_{end7j} = \omega_{en7j}\sqrt{1-\varsigma_{e7j}^2}. \tag{27.25}$$

Theorem 27.4. Denote by $h_{7j}(t)$ the jth impulse response function of the system (27.19). Then,

$$h_{7j}(t)=\frac{1}{m_{e7j}\omega_{end7j}}e^{-\varsigma_{e7j}\omega_{en7j}t}\sin\omega_{end7j}t,\quad t>0. \tag{27.26}$$

Proof. Let $f_j(t)=\delta(t)$ in Eq. (27.19). Then, we have the above. This finishes the proof. □

The above can be expressed by

$$h_{7j}(t)=\frac{1}{m_{e7}\omega_{end7j}}e^{-\varsigma_{e7j}\omega_{en7j}t}\left(\sin\omega_{end7j}t\right)u(t), \tag{27.27}$$

where $u(t)$ is the unit step function.

Theorem 27.5. The jth coordinate with respect to Eq. (27.11) under zero initial conditions is represented by

$$p_j(t)=f_j(t)*h_{7j}(t)=\frac{1}{m_{e7j}\omega_{end7j}}\int_{-\infty}^{\infty}e^{-\varsigma_{e7j}\omega_{en7j}\tau}\sin\omega_{end7j}\tau f_j(t-\tau)d\tau. \tag{27.28}$$

Proof. From the convolution theory, we see that the above is true. This completes the proof. □

Theorem 27.6. The zero-state forced response to a multi-fractional damped Euler-Bernoulli beam is expressed by

$$w(x,t)=\sum_{j=1}^{\infty}\frac{\varphi_j(x)}{m_{e7j}\omega_{end7j}}\int_{-\infty}^{\infty}e^{-\varsigma_{e7j}\omega_{en7j}\tau}\sin\omega_{end7j}\tau f_j(t-\tau)d\tau. \tag{27.29}$$

Proof. Because $w(x,t)=\sum_{j=1}^{\infty}w_j(x,t)=\sum_{j=1}^{\infty}\varphi_j(x)p_j(t)$ and $p_j(t)$ is expressed by Theorem 27.5, the above is true. The proof is finished. □

If $\alpha=2$, $\beta=1$, and $\lambda=1$, $\omega_{en7j}=\omega_{nj}$. In that case, a multi-fractional damped Euler-Bernoulli beam reduces to the conventional one. In what follows, zero initial conditions are considered unless otherwise stated, including exercises in Section 27.6.

27.3 CASE STUDY

For a simply supported Euler-Bernoulli beam, we have

$$\varphi_j(x)\sim\sin\frac{j\pi x}{l},\qquad j=1,2,\ldots \tag{27.30}$$

Thus,

$$w(x,t)=\sum_{j=1}^{\infty}\sin\frac{j\pi x}{l}\left[f_j(t)*\frac{1}{m_{e7j}\omega_{end7j}}e^{-\varsigma_{e7j}\omega_{en7j}t}\sin\omega_{end7j}t\right]. \quad (27.31)$$

Since

$$\mu_j=\frac{j\pi}{l},\quad j=1,2,... \quad (27.32)$$

the jth order natural angular frequency is in the form

$$\omega_{n7j}=\mu_j^2\sqrt{\frac{EI}{\rho A}}=\left(\frac{j\pi}{l}\right)^2\sqrt{\frac{EI}{\rho A}},\quad j=1,2,... \quad (27.33)$$

The quantities ω_{en7j}, ζ_{e7j}, and ω_{end7j} are given by Eqs. (27.22), (27.21), and (27.25), respectively.

Let $l=10$. When $f_j(t)=10\sin(0.314t)$, we indicate the plots of $p_j(t)$ $(j=1,\ 2)$ in Figure 27.1. If $\alpha=2$, and $\beta=\lambda=1$, $p_j(t)$ reduces to the conventional coordinate functions as shown in Figure 27.2. The plots of $w(x,\ t)$ with the multi-fractional coordinates are shown in Figure 27.3.

27.4 DISCUSSIONS

For a multi-fractional damped Euler-Bernoulli beam regarding forced response, $p_j(t)$ is associated with α, β, λ, ω_{en7j}, ζ_{e7j}, ω_{end7j}, ω, as well as driven function $f_j(t)$, see Eq. (27.28). Therefore, things may be complicated with respect to increasing-vibrations. For instance, when $f_j(t)=10\sin(0.314t)$, $\omega=0.314$, $\alpha=\beta=2$, $\lambda=1.6$, increasing-vibrations occur, see Figure 27.4. A particular thing is that increasing-vibrations are related to a driven force. For example, if $f_j(t)=u(t)$ (unit step function) and when $\omega=0.314$, $\alpha=\beta=2$, $\lambda=1.8$, increasing-vibrations do not take place, see Figure 27.5.

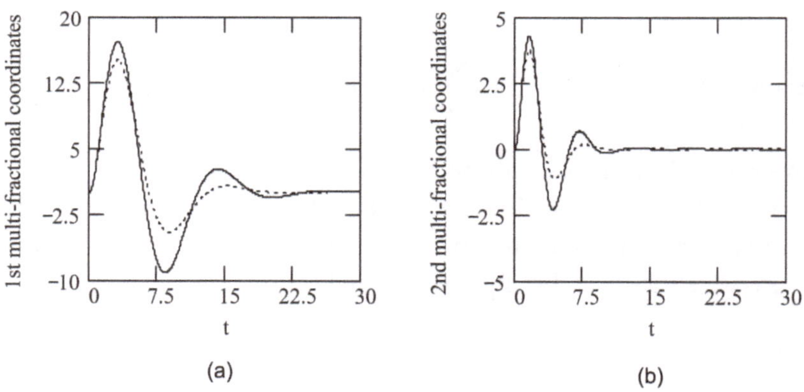

FIGURE 27.1 Plots of $p_j(t)$ $(j=1,\ 2)$ of a simply supported multi-fractional damped Euler-Bernoulli beam for $\alpha=1.8$ (solid line), $\alpha=1.4$ (dot line) when $f_j(t)=10\sin(0.314t)$, $l=10$, $\beta=\lambda=1.2$, $\omega=0.314$, $a=1$, $b=1$, and $E=1$. (a) $p_1(t)$. (b) $p_2(t)$.

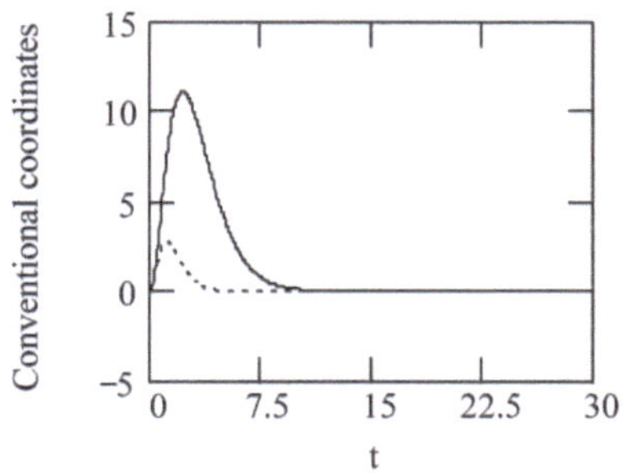

FIGURE 27.2 Plots of $p_1(t)$ (solid) and $p_2(t)$ (dot) of a simply supported multi-fractional damped Euler-Bernoulli beam for when $\alpha=2$, $\beta=\lambda=1$, $f_j(t)=10\sin(0.314t)$, $l=10$ for $\omega=0.314$, $a=1$, $b=1$, and $E=1$.

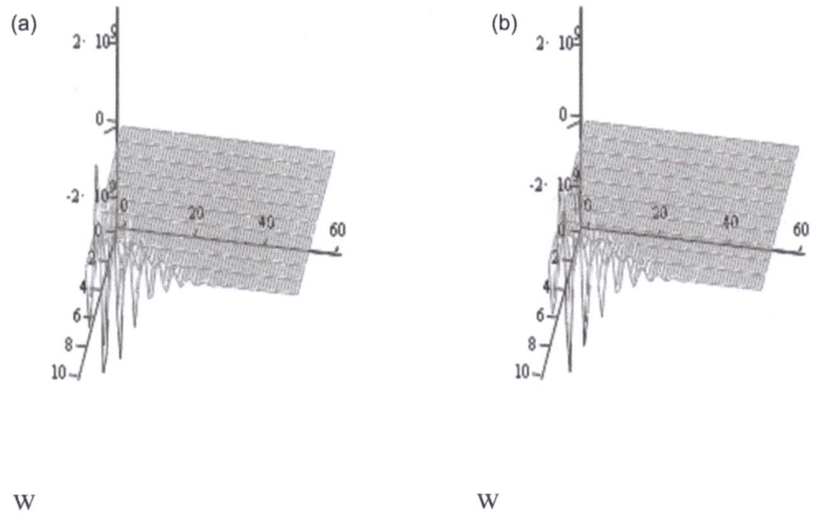

FIGURE 27.3 Plots of $w(x,t) \approx \sum_{j=1}^{6} w_j(x,t) = \mathrm{w}$ of a simply supported multi-fractional damped Euler-Bernoulli beam when $f_j(t)=10\sin(0.314t)$ and $l=10$ for $\omega=0.0314$, $a=1$, $b=1$, $E=1$, $\lambda=1.2$, $x=0, \ldots, 10$, and $t=0, \ldots, 60$. (a) w for $\alpha=1.8$ and $\beta=1.2$. (b) w for $\alpha=1.6$ and $\beta=1.4$.

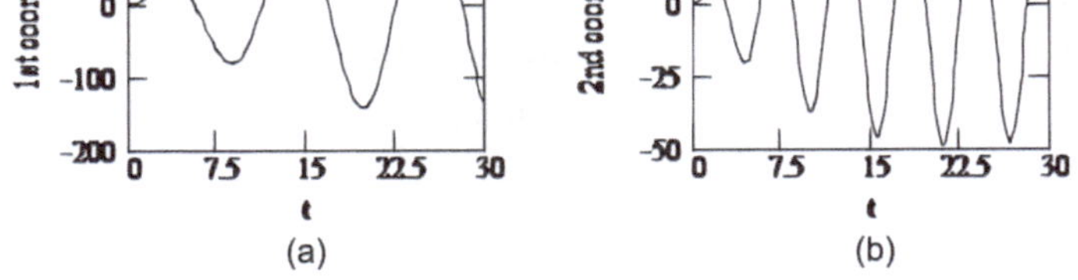

FIGURE 27.4 Illustrating plots of coordinates for increasing-vibrations for a simply supported multi-fractional damped Euler-Bernoulli beam when $f_j(t)=10\sin(0.314t)$, $\omega=0.314$, $\alpha=\beta=2$, $\lambda=1.8$. (a) $p_1(t)$. (b) $p_2(t)$.

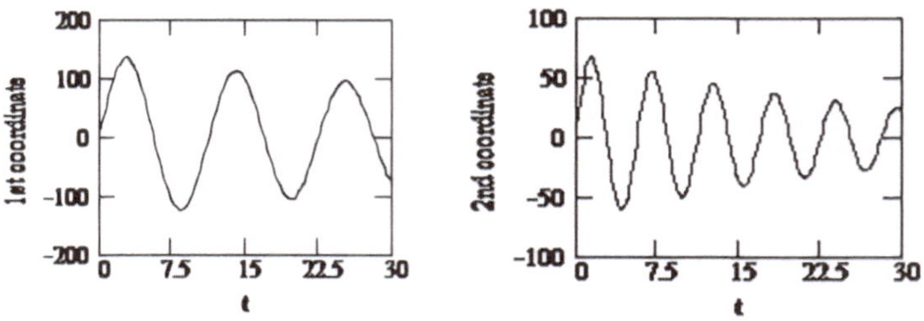

(a) (b)

FIGURE 27.5 Plots of fractional coordinates for a simply supported multi-fractional damped Euler-Bernoulli beam when $f_j(t) = u(t)$, $\omega = 0.314$, $\alpha = \beta = 2$, $\lambda = 1.8$. (a) $p_1(t)$. (b) $p_2(t)$.

If using other fractional models of stress-strain (Shitikova [3], Rossikhin and Shitikova [4]), more interesting vibration phenomena may be observed.

27.5 SUMMARY

We have given an equivalent equation of fractional coordinates of a multi-fractional damped Euler-Bernoulli beam. We have proposed the analytic expression of the zero-state forced response to such a beam. In addition, we have revealed a new variation of the Rayleigh damping (Theorem 27.3). Increasing-vibrations of such a beam are associated with a driven force function.

27.6 EXERCISES

27.1. Denote by $f_j(t)$ the jth driven function. Let $f_j(t)=u(t)$ (unit step function) ($j=1, 2, \ldots$). For a simply supported multi-fractional damped Euler-Bernoulli beam, when $a=1$, $b=1$, $E=1$, and beam length $l=10$, find the forced response $w(x, t)$.

27.2. Denote by $f_j(t)$ the jth driven function. Let $f_j(t)=u(t-\tau/2)-u(t+\tau/2)$ ($j=1, 2, \ldots$), where τ is a constant and $u(t)$ is the unit step function. For a simply supported multi-fractional damped Euler-Bernoulli beam, when $a=1$, $b=1$, $E=1$, and beam length $l=10$, find the forced response $w(x, t)$.

27.3. Let $f_j(t)=\text{A}\sin(\omega_0 t)$ ($j=1, 2, \ldots$), where A and ω_0 are constants. For a free-free multi-fractional damped Euler-Bernoulli beam, when $a=1$, $b=1$, $E=1$, and beam length $l=10$, find the forced response $w(x, t)$.

27.4. Let $f_j(t)=u(t)$ ($j=1, 2, \ldots$). For a free-free multi-fractional damped Euler-Bernoulli beam, when $a=1$, $b=1$, $E=1$, and beam length $l=10$, find the forced response $w(x, t)$.

27.5. Denote by $f_j(t)$ the jth driven function. Let $f_j(t)=u(t-\tau/2)-u(t+\tau/2)$ ($j=1, 2, \ldots$), where τ is a constant and $u(t)$ is the unit step function. For a free-free multi-fractional damped Euler-Bernoulli beam, when $a=1$, $b=1$, $E=1$, and beam length $l=10$, find the forced response $w(x, t)$.

REFERENCES

1. O. M. Palley, Г. B. Bahizov, and E. Я. Voroneysk, *Handbook of Ship Structural Mechanics*, National Defense Industry Publishing House, Beijing, China, 2002. In Chinese. Translated from Russian by B. H. Xu, X. Xu, and M. Q. Xu.
2. X. D. Jin and L. J. Xia, *Ship Hull Vibration*, The Press of Shanghai Jiaotong University, Shanghai, China, 2011. In Chinese.
3. M. V. Shitikova, Fractional operator viscoelastic models in dynamic problems of mechanics of solids: A review, *Mechanics of Solids*, 57(1), 2022, 1–33.
4. Y. A. Rossikhin and M. V. Shitikova, Application of fractional operators to the analysis of damped vibrations of viscoelastic single-mass systems, *Journal of Sound and Vibration*, 199(4), 1997, 567–586.

CHAPTER 28

Notes on Fractional Vibrations

BEFORE DISCUSSING TEST TECHNIQUES in the next chapters, I need to talk about something associated with vibrations in both engineering and academic research.

28.1 ENGINEERING CONSIDERATIONS

28.1.1 Classroom Teaching vs. Practice

As is known, actual vibration events exhibit that damping may be frequency dependent for ship hull as can be seen from an actual response to a real ship hull described by Jin and Xia [1, Chapter 4]. Although we mentioned the Rayleigh damping, which is frequency dependent, in Chapter 6, in most cases in Chapters 1–6, we assume that damping is viscous and constant. Since $\varsigma = \frac{c}{2\sqrt{mk}}$, when observing frequency varying ζ, we might say that m or k is also frequency varying in addition to c. As a matter of fact, frequency-dependent elements in mass, and or damping, and or stiffness attract the interest of researchers, see e.g., Banerjee [2], White et al. [3], Hamdaoui et al. [4], Dumont and de Oliveira [5], Wu and Hsieh [6], Gross and Weltin [7], just mentioning a few. In short, there is a noticeable gap between the basic study in classroom and practice. Classroom teaching simply gives an introduction to some subjects. Many phenomena in practice are not unsophisticated in understanding, which may not be explained by the conventional theory of linear vibrations with m, c, and k being constants.

28.1.2 Importance of Measurement and Testing

When measuring external load, such as ocean wave or wind speed, the quality of measured or recorded data, loading spectrum on a test machine, and spectrum estimation are crucial to the vibration analysis of a structure of interest.

While recording data, it is essential to assure the record length such that spectrum estimation may achieve the required quality (Bendat and Piersol [8], Laning and Battin [9]).

DOI: 10.1201/9781003460947-31

Although my paper [10] is for the record length requirement of ocean waves and [11,12] for network traffic time series, they may yet be a reference for Gaussian processes in general in measurement.

In laboratory vibration tests, one needs setting load spectrum on a control computer. However, a load spectrum exerting on a specimen structure may suffer from considerable distortion and reduce the test quality accordingly. That may be particularly true since ship structures are on a large scale. Therefore, loading control is desired for laboratory vibration tests. It is required not only for vibration tests but also for laboratory fatigue tests [13–15], sea keeping tests [16], and others as well, such as wind tunnel tests.

An essential technique used in laboratory tests is about random data simulation. Simulated random data should be consistent with required probability density function [17], or required power spectrum [13–16,18–20], or required autocorrelation function [12,21].

As known, while doing spectrum estimation in a vibration test, selecting a window function is demanded [22]. The references with respect to windowing are immense, citing [23] just for vibrations. Tukey considered that windowing may not improve the quality of spectrum estimation [24,25]. I agree with Tukey's opinion. However, it is noted that, in vibration data processing, selecting a window is a routine [22,23]. I would like to stress that classroom teaching only paves the way towards engineering vibrations and vibration theory. A lot of techniques, which are not covered in classroom teaching, are necessary in engineering as well as academic research.

28.1.3 Challenges Regarding Fractional Vibrations in Engineering

When a vibration system is of fractional order, a critical thing in engineering is to determine what a value of a fractional order is (Rossikhin [26,27], Shitikova [28,29]). However, techniques and instrumentations in measurement of acceleration, velocity, and displacement only give measurement readings with integer order [23]. Consequently, a bottleneck in applying the theory of fractional vibrations to engineering is how to determine a value of a fractional order of a fractional vibration system. As a matter of fact, that remains a challenging problem [26–29].

28.2 NONLINEARITY AND NON-NEWTONIAN FORCES IN FRACTIONAL VIBRATIONS

We use a vibrator of class VI to discuss non-Newtonian forces encountered in fractional vibrations. The motion equation of a class VI vibrator is given by Eq. (13.1), where $m\frac{d^{\alpha}x_6(t)}{dt^{\alpha}}$ is fractional inertia force, $c\frac{d^{\beta}x_6(t)}{dt^{\beta}}$ is fractional damping force, and $k\frac{d^{\lambda}x_6(t)}{dt^{\lambda}}$ is fractional restoration one. The unit of $m\frac{d^{\alpha}x_6(t)}{dt^{\alpha}}$ is non-Newtonian unless $\alpha=2$. The unit of $c\frac{d^{\beta}x_6(t)}{dt^{\beta}}$ is non-Newtonian either if $\beta\neq1$. Similarly, the unit of fractional restoration force $k\frac{d^{\lambda}x_6(t)}{dt^{\lambda}}$ is non-Newtonian for $\lambda\neq0$. Thus, we say that non-Newtonian forces may be an attribute of fractional forces in fractional vibrations in general. That in fact reflects nonlinearity of fractional vibration systems.

28.3 NONLINEARITY AND FREQUENCY VARYING VIBRATION PARAMETERS

Fractional vibration is in the domain of nonlinear vibrations. In this book, we use the skill of linearization. By linearization, nonlinearity of a fractional vibrator is reflected in the aspect of frequency varying mass or frequency varying damping or frequency varying stiffness. It is worth noting that a frequency varying mass might be a bridge to connect the time varying mass described by $m_{\text{Einstein eq}}$ from the point of view of fractional vibrations. That might be an interesting issue.

28.4 FRACTIONALLY VIBRATIONAL MOTIONS

The motion $x''(t)$ is acceleration while $x'(t)$ is velocity. However, for $0<\alpha<3$, $x^{(\alpha)}(t)$ stands for three kinds of motions. When $\alpha=2$, $x^{(\alpha)}(t)$ is the conventional acceleration. We introduce two terms for $2<\alpha<3$ and $1<\alpha<2$ with respect to $x^{(\alpha)}(t)$. When $2<\alpha<3$, we say that $x^{(\alpha)}(t)$ designates a motion called super-acceleration. If $1<\alpha<2$, the motion of $x^{(\alpha)}(t)$ is called sub-acceleration or super-velocity. Super-acceleration may produce self-vibrations, see Chapter 8, but super-velocity does not, see Chapter 9.

For $x^{(\beta)}(t)$ $(0<\beta<2)$, we call $x^{(\beta)}(t)$ super-velocity when $1<\beta<2$ and sub-velocity if $0<\beta<1$ or super-displacement. It becomes the conventional velocity when $\beta=1$.

For $x^{(\lambda)}(t)$ $(0<\lambda<1)$, we call $x^{(\lambda)}(t)$ super-displacement. It reduces to the conventional displacement when $\lambda=0$.

The fractionally vibrational motions introduced in this chapter are for vibration systems rather than those from the point of view of random functions as Eliazar and Shlesinger addressed in [30]. The fractional motions discussed in this section imply that motions may be with super-/sub-acceleration or super-/sub-velocity or super-displacement while the ones Eliazar and Shlesinger discussed are fractional and or fractal random processes (Li [12]).

REFERENCES

1. X. D. Jin and L. J. Xia, *Ship Hull Vibration*, The Press of Shanghai Jiaotong University, Shanghai, China, 2011. In Chinese.
2. J. R. Banerjee, Frequency dependent mass and stiffness matrices of bar and beam elements and their equivalency with the dynamic stiffness matrix, *Computers & Structures*, 254, 2021, 106616.
3. R. E. White, J. H. G. Macdonald, and N. A. Alexander, A nonlinear frequency-dependent spring-mass model for estimating loading caused by rhythmic human jumping, *Engineering Structures*, 240, 2021, 112229.
4. M. Hamdaoui, G. Robin, M. Jrad, and E. M. Daya, Optimal design of frequency dependent three-layered rectangular composite beams for low mass and high damping, *Composite Structures*, 120, 2015, 174–182.
5. N. A. Dumont and R. de Oliveira, From frequency-dependent mass and stiffness matrices to the dynamic response of elastic systems, *International Journal of Solids and Structures*, 38(10–13), 2001, 1813–1830.
6. J.-S. Wu and M. Hsieh, An experimental method for determining the frequency-dependent added mass and added mass moment of inertia for a floating body in heave and pitch motions, *Ocean Engineering*, 28(4), 2001, 417–438.

7. J. R. Gross and U. Weltin, A computational approach to calculating frequency-dependent structural operators using the spectral finite element method and a wave number-based gridding technique, *Journal of Sound and Vibration*, 332(25), 2013, 6659–6664.
8. J. S. Bendat and A. G. Piersol, *Random Data: Analysis and Measurement Procedure*, 4th Ed., John Wiley & Sons, Hoboken, NJ, 2010.
9. J. H. Laning and R. H. Battin, *Random Processes in Automatic Control*, McGraw Hill, New York, 1956.
10. M. Li, A method for requiring block size for spectrum measurement of ocean surface waves, *IEEE Transactions on Instrumentation and Measurement*, 55(6), 2006, 2207–2215.
11. M. Li, Record length requirement of long-range dependent teletraffic, *Physica A*, 472, 2017, 164–187.
12. M. Li, *Multi-Fractal Traffic and Anomaly Detection in Computer Communications*, CRC Press, Boca Raton, FL, 2022.
13. M. Li, An iteration method to adjusting random loading for a laboratory fatigue test, *International Journal of Fatigue*, 27(7), 2005, 783–789.
14. M. Li, Experimental stability analysis of test system for doing fatigue test under random loading, *Journal of Testing and Evaluation*, 34(4), 2006, 364–367.
15. M. Li, B.-H. Xu, and Y.-S. Wu, An H_2-optimal control of random loading for a laboratory fatigue test, *Journal of Testing and Evaluation*, 26(6), 1998, 619–625.
16. M. Li, An optimal controller of an irregular wave maker, *Applied Mathematical Modelling*, 29(1), 2005, 55–63.
17. R. Saucier, Computer generation of statistical distributions, *Technical Report*, ARL-TR-2168, Army Research Laboratory, USA, 2000.
18. J. N. Yang, Simulation of random envelope processes, *Journal of Sound and Vibration*, 21(1), 1972, 73–85.
19. S. K. Chakrabarti, *Offshore Structure Modeling*, World Scientific, Singapore, 1994.
20. R. M. Stanislaw, *Ocean Surface Waves: Their Physics and Prediction*, World Scientific, Singapore, 1996.
21. M. Li, Generation of teletraffic of generalized Cauchy type, *Physica Scripta*, 81(2), 2010, 025007 (10pp).
22. Schlumberger Electronics Ltd., *1200 Real Time Signal Processor Operating Manual*, UK, 1983.
23. C. M. Harris, *Shock and Vibration Handbook*, 5th Ed., McGraw-Hill, New York, 2002.
24. J. W. Tukey, Modern techniques of power spectrum estimation, *IEEE Transactions on Information Theory*, 15(2), 1967, 56–66.
25. J. W. Tukey, Data analysis and the frontiers of geophysics, *Science*, 148(3675), 1965, 1283–1289.
26. Y. A. Rossikhin and M. V. Shitikova, Application of fractional operators to the analysis of damped vibrations of viscoelastic single-mass systems, *Journal of Sound and Vibration*, 199(4), 1997, 567–586.
27. Y. A. Rossikhin, Reflections on two parallel ways in progress of fractional calculus in mechanics of solids, *Applied Mechanics Reviews*, 63(1), 2010, 010701.
28. M. V. Shitikova, Fractional operator viscoelastic models in dynamic problems of mechanics of solids: A review, *Mechanics of Solids*, 57(1), 2022, 1–33.
29. M. V Shitikova, The fractional derivative expansion method in nonlinear dynamic analysis of structures, *Nonlinear Dynamics*, 99(1), 2020, 109–122.
30. I. I. Eliazar and M. F. Shlesinger, Fractional motions, *Physics Reports*, 527(2), 2013, 101–129.

IV

Some Techniques in Vibrations

CHAPTER 29

Sampling, Aliasing, Anti-Aliasing Filtering and Time Signal Leakage

In this chapter, we address four issues in vibration data processing. The first is sampling. The second is aliasing. The third is anti-aliasing filtering. The last is the concept of time signal leakage.

29.1 SAMPLING AND ALIASING

An actual signal, such as ocean wave series, is often a continuous function $f(t)$ for $t \in (0, \infty)$. However, practically used computers and some instruments, such as Fourier Analyzers for vibrations, are of digital. To be precise, the data we really process are the output of Figure 29.1 instead of actually measured continuous signal $f(t)$ at the input of Figure 29.1.

A binary signal is a binary sequence consisting of two numbers, 0 and 1. When a binary code is lengthy enough, errors caused by binary coding can be ignored. The main error source results from sampling sector.

The functionality of a sampling system is just to multiply $f(t)$ by sampling pulse $p(t)$, see Figure 29.2. Intuitively, when the pulse width of $p(t)$ is sufficiently narrow and the sampling period T_s of $p(t)$ is sufficiently small, $f_s(t) = f(t)p(t)$ may contain enough information of $f(t)$.

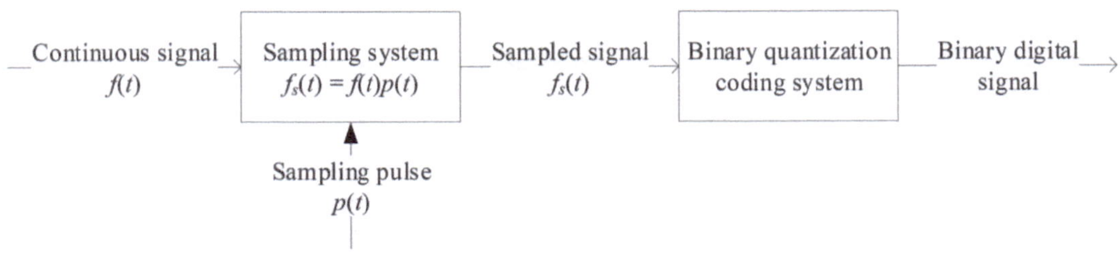

FIGURE 29.1 Sampling diagram.

DOI: 10.1201/9781003460947-33

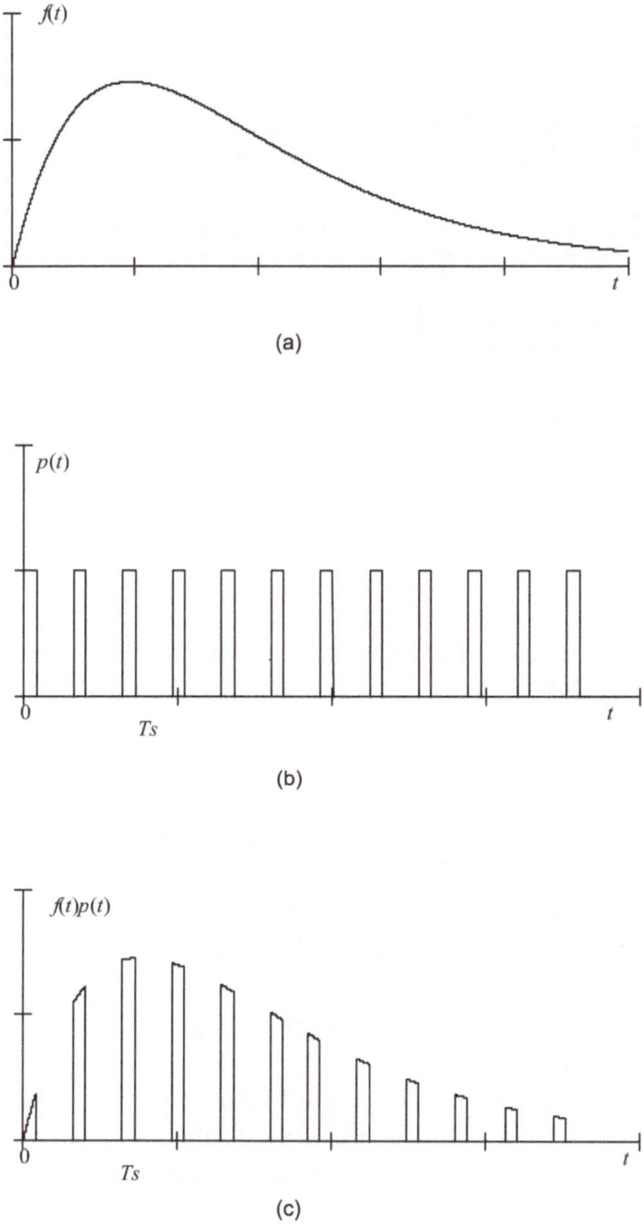

FIGURE 29.2 Sampling procedure. (a) Original signal $f(t)$. (b) Sampling pulse $p(t)$. (c) Sampled signal $f(t)p(t)$.

In what follows, we suppose that the pulse width of $p(t)$ is sufficiently narrow. Therefore, the question is how to quantitatively require the sampling period T_s such that sampled signal $f_s(t)$ can contain the complete information of $f(t)$. Harry Nyquist answered that question [1] (also see Gabel and Roberts [2]).

Proposition 29.1: Nyquist's Sampling Rule. Let $f(t)$ be a band-limited signal. By band-limited signal, one means that its spectrum $F(\omega)=0$ for $\omega>\omega_m$, where ω_m is the maximum frequency of $f(t)$. Then, if

$$f_s \geq 2 f_m, \tag{29.1}$$

$f_s(t)$ contains all information of $f(t)$, where $f_s = 1/T_s$, $f_m = \omega_m/2\pi$.

Proof. Without the generality losing, $p(t)$ is assumed to be a delta function train in the form

$$p(t)=\delta_{T_s}=\sum_{n=-\infty}^{\infty}\delta(t-nT_s). \tag{29.2}$$

Thus, the sampled signal $f_s(t)$ is given by

$$f_s(t)=f(t)p(t)=f(t)\delta_{T_s}=\sum_{n=-\infty}^{\infty}f(t)\delta(t-nT_s). \tag{29.3}$$

Let $f_s(t) \leftrightarrow F_s(\omega)$ and $f(t) \leftrightarrow F(\omega)$. According to the convolution theorem in the frequency domain, one has

$$\begin{aligned} F_s(\omega) &= \frac{1}{2\pi}F(\omega)*P(\omega)=\sum_{n=-\infty}^{\infty}F(\omega)*\int_{-\infty}^{\infty}\delta(t-nT_s)e^{-i\omega t}\,dt \\ &= \frac{1}{2\pi T_s}\sum_{n=-\infty}^{\infty}F(\omega)*\delta(\omega-n\omega_s)=\frac{1}{2\pi T_s}\sum_{n=-\infty}^{\infty}F(\omega-n\omega_s). \end{aligned} \tag{29.4}$$

Note that $f(t)$ is band limited, see $F(\omega)$ in Figure 29.3. Thus, if $f_s \geq 2f_m$ or $\omega_s \geq 2\omega_m$, $\sum_{n=-\infty}^{\infty}F(\omega-n\omega_s)$ stands for a series of spectra $F(\omega-n\omega_s)$, which are never overlapped, see Figure 29.4. With a low-pass filter or band-pass one, we can completely recover $f(t)$ from $F_s(\omega)$. This finishes the proof. □

If $\omega_s < 2\omega_m$, a series of spectra $F(\omega-n\omega_s)$ are overlapped. Accordingly, aliasing takes place, see Figure 29.5. In the case of aliasing, it is impossible to recover the original signal $f(t)$ from the aliased spectrum $F_s(\omega)$.

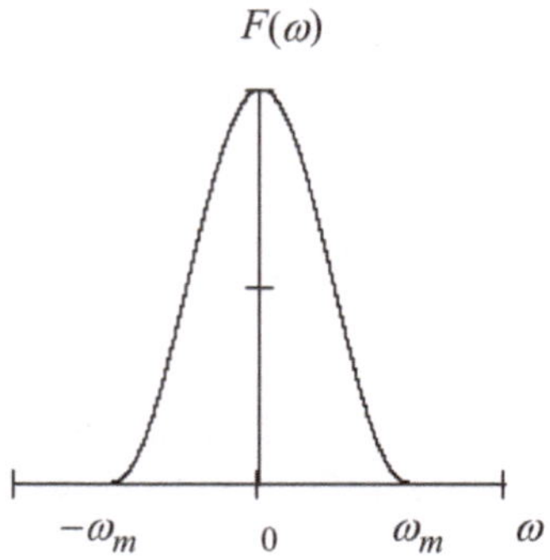

FIGURE 29.3 Illustration of spectrum of a band-limited signal.

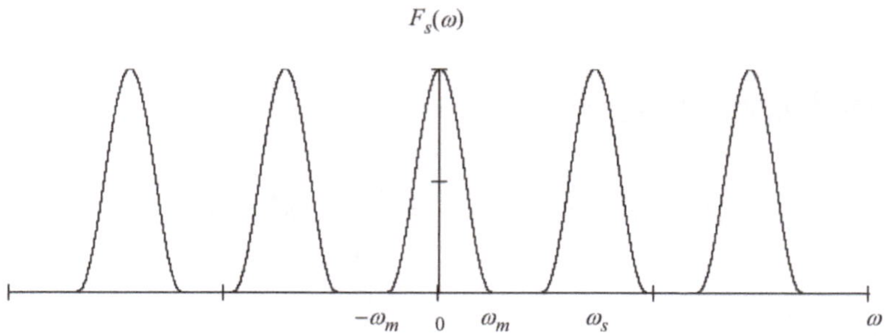

FIGURE 29.4 Illustration of spectrum of sampled signal $F_s(\omega)$ with $f_s \geq 2f_m$.

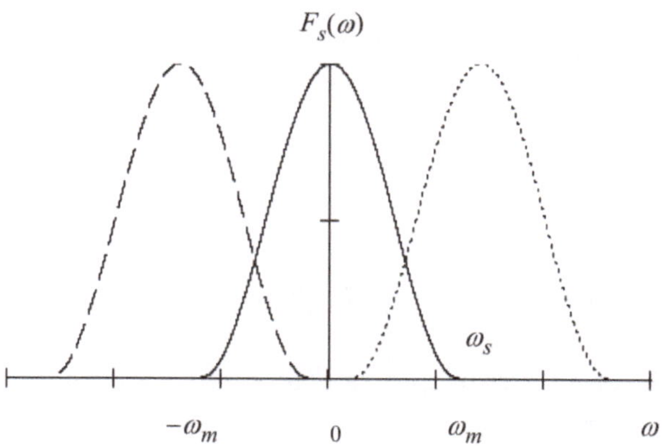

FIGURE 29.5 Spectra aliasing when $f_s < 2f_m$.

29.2 ANTI-ALIASING FILTERING

The Nyquist's sampling rule requires that a signal $f(t)$ is band limited. That requirement may be too strict in practice. There are many spectra of practical signals that are not band limited. Examples include ocean surface waves (Li [3]). Thus, if $f(t)$ is not band limited, its sampling must result in aliasing. Consequently, to recover the original signal $f(t)$ from the sampled signal $f_s(t)$ becomes impossible.

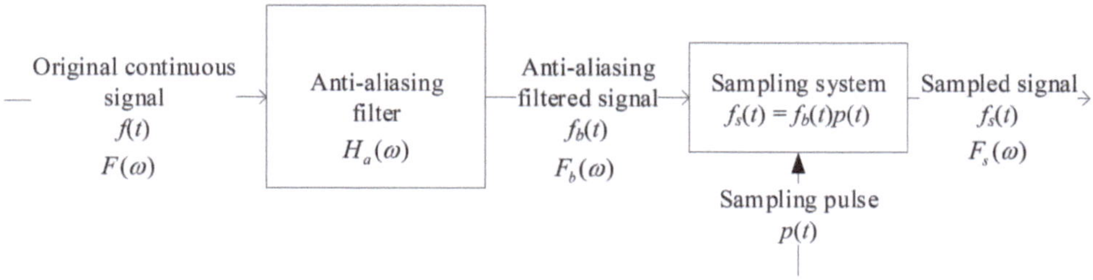

FIGURE 29.6 Diagram of anti-aliasing filter used in sampling.

To smooth away aliasing due to sampling, one may process the signal $f(t)$ to be sampled using anti-aliasing filter. Let $H_a(\omega)$ be the frequency transfer function of an anti-aliasing filter. The ideal $H_a(\omega)$ may be in the form

$$H_a(\omega)=\begin{cases} 1, & -\omega_m<\omega<\omega_m, \\ 0, & \text{otherwise}, \end{cases} \tag{29.5}$$

where ω_m is the maximum effective frequency of $f(t)$. Denote by $f_b(t)$ the output of the anti-aliasing filter, see Figure 29.6. Its Fourier transform is denoted by $F_b(\omega)$. It is

$$F_b(\omega)=H_a(\omega)F(\omega). \tag{29.6}$$

The above clearly indicates that $f_b(t)$ is a band limited signal with the maximum effective frequency ω_m. Thus, once $\omega_s \geq 2\omega_m$, aliasing never occurs in sampling. Since $f_b(t)$ contains all frequency components within $(-\omega_m, \omega_m)$, it may be regarded as a satisfactory approximation of $f(t)$ in engineering. Anti-aliasing filter is usually a routine technique in vibrations (Harris [4]).

Example 29.1. Consider the rectangular pulse in the form

$$\text{rect}(t)=\begin{cases} E, \ -\dfrac{\tau}{2}<t<\dfrac{\tau}{2}, & E>0. \\ 0, \ \text{elsewhere}, \end{cases} \tag{29.7}$$

Its Fourier transform is

$$F(\omega)=E\tau\text{Sa}\left(\frac{\omega\tau}{2}\right). \tag{29.8}$$

Figure 29.7 indicates the plots of rect(t) and $F(\omega)$. Its maximum effective frequency $\omega_m=2\pi/\tau$.

The signal rect(t) is not band limited. Thus, if not anti-aliasing, its sampling results in aliasing of $F_s(\omega)$ no matter how large of sampling frequency ω_s is, even if $\omega_s \geq 2\omega_m$ as indicated in Figure 29.8. Let $F_{\text{snon}}(\omega)$ be the spectrum of the sampled rect(t) that is not anti-aliasing filtered. Then, Figure 29.8 shows $F_{\text{snon}}(\omega)$ for $\omega_s \geq 2\omega_m$.

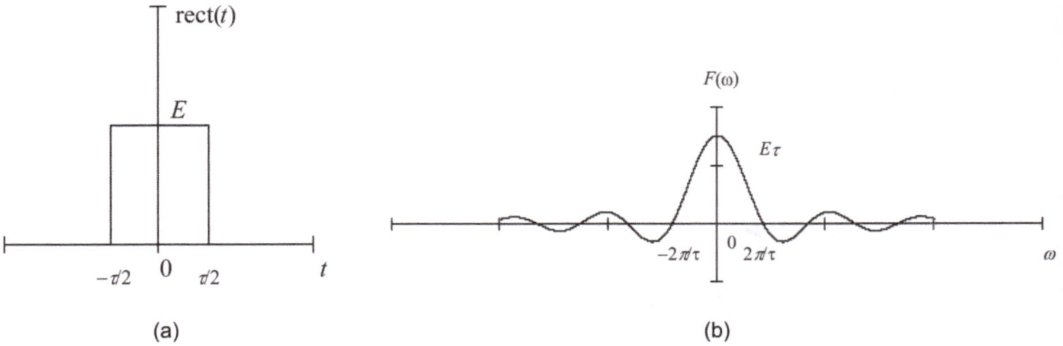

FIGURE 29.7 Illustrations of rectangular pulse and its spectrum. (a) Plot of rect(t). (b) Plot of $F(\omega)$.

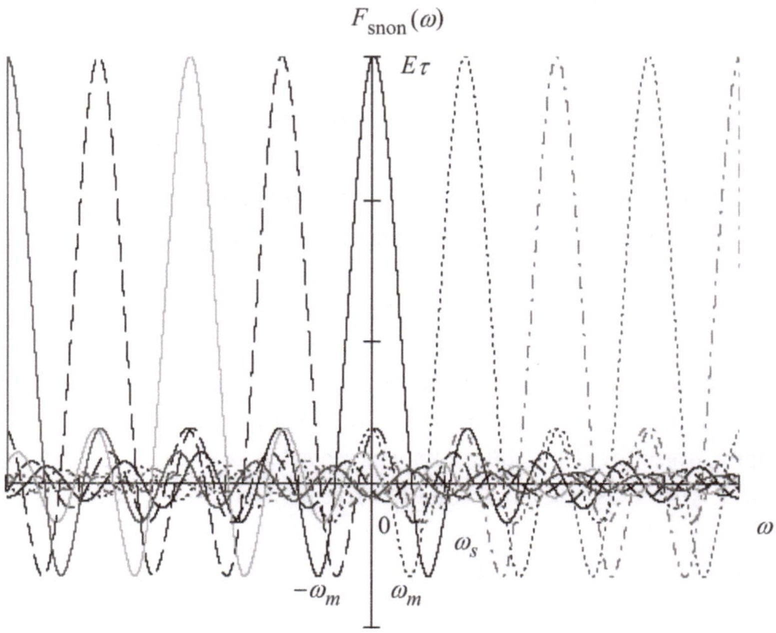

FIGURE 29.8 Spectrum of sampled signal without anti-aliasing filtering when $\omega_s \geq 2\omega_m$.

When an anti-aliasing filter is used, there is no aliasing in $F_s(\omega)$ so long as $\omega_s \geq 2\omega_m$, see Figure 29.9.

Denote by $F_{sf}(\omega)$ the spectrum that results from $F_s(\omega)$ in Figure 29.9 passes through a low-pass filter with the cutoff frequency $\omega_m = 2\pi/\tau$. Then, $F_{sf}(\omega)$ is the spectrum that rect(t) is anti-aliasing filtered, see Figure 29.10. In comparison with the original signal $F(\omega)$ in Figure 29.7b, $F_{sf}(\omega)$ is certainly a distortion of $F(\omega)$. Let $F_{sna}(\omega)$ be the spectrum that is the output of a low-pass filter with the cutoff frequency $\omega_m = 2\pi/\tau$ with the input $F_{snon}(\omega)$ in Figure 29.8. Then, by eye, we see that a sampled signal that is not anti-aliasing filtered contains more distortion than the one that suffers from the anti-aliasing filtered as can be seen from Figures 29.10 and 29.11. I use this example to enhance that anti-aliasing filtering does make effect on sampling.

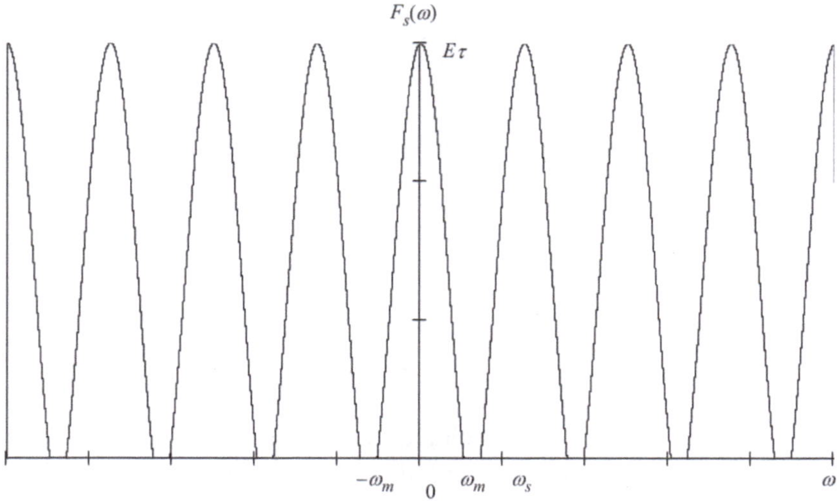

FIGURE 29.9 Spectrum of sampled signal with anti-aliasing filtering when $\omega_s \geq 2\omega_m$.

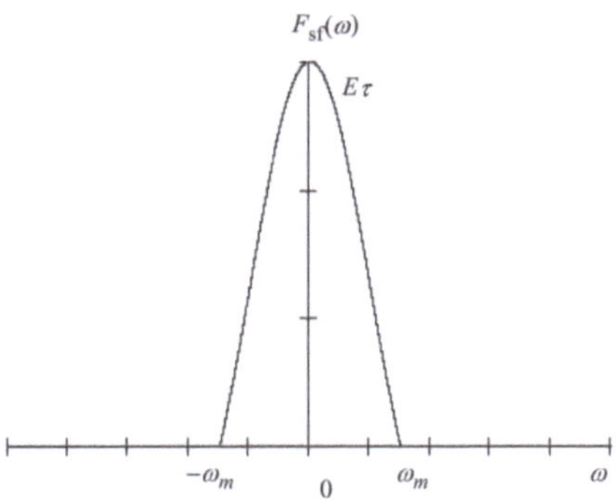

FIGURE 29.10 Spectrum of sampled anti-aliasing filtered signal passing through low-pass filter.

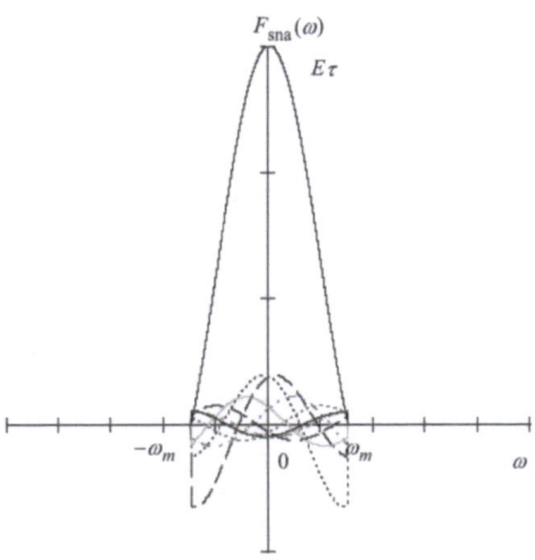

FIGURE 29.11 Spectrum of sampled signal without anti-aliasing filtering passing through low-pass filter.

29.3 TIME SIGNAL LEAKAGE

Denote by $h_a(t)$ the impulse response function of an anti-aliasing filter in Eq. (29.5). Then,

$$h_a(t) = \mathrm{F}^{-1}[H_a(\omega)] = \frac{\omega_m}{2\pi}\mathrm{Sa}\left(\frac{\omega_m t}{2}\right). \tag{29.9}$$

Denote by $f_b(t)$ the actual signal to be sampled in Eq. (29.6). Then, according to the convolution rule in the time domain,

$$f_b(t) = h_a(t) * f(t) = \frac{\omega_m}{2\pi}\mathrm{Sa}\left(\frac{\omega_m t}{2}\right) * f(t). \tag{29.10}$$

Thus, $f_b(t)$ is the original signal $f(t)$ convoluted by $h_a(t)$. We call $f(t) - f_b(t)$ the time signal leakage. Consequently, we say that the technique of anti-aliasing filtering used in sampling is at the cost of sacrificing the time resolution of $f(t)$ due to leakage.

Assume that the maximum effective frequency of $f(t)$ approaches infinity. Then, we have

$$h_a(t) = \lim_{\omega_m \to \infty} \frac{\omega_m}{2\pi}\mathrm{Sa}\left(\frac{\omega_m t}{2}\right) = \delta(t). \tag{29.11}$$

The above implies that leakage disappears if $\omega_m \to \infty$. However, the Nyquist's sampling rule fails for signals with $\omega_m \to \infty$ because we cannot manage the sampling rate ω_s for assuring $\omega_s > 2\omega_m$.

Theoretically, there are signals with the property of $\omega_m \to \infty$. An example is $\delta(t)$. Another example is fractional Gaussian noise with short-range dependence (Li [5–7]). In engineering, we consider signals with finite ω_m. For a signal with sufficiently large ω_m, its leakage in time may be ignored because

$$h_a(t) \approx \delta(t) \quad \text{if } \omega_m \text{ is large enough.} \tag{29.12}$$

However, when ω_m is large enough, the performance requirement of a sampling circuit is high. On the contrary, if ω_m is small enough, for any $f(t)$, we may have

$$f_b(t) \approx h_a(t) = \frac{\omega_m}{2\pi}\mathrm{Sa}\left(\frac{\omega_m t}{2}\right). \tag{29.13}$$

The above says that $h_a(t)$ may dominate $f_b(t)$ if ω_m of $f(t)$ is sufficiently small. Therefore, even in the case of $\omega_s > 2\omega_m$, selecting large ω_s is beneficial.

29.4 SUMMARY

We have discussed four things regarding data processing in vibration engineering. The first one is sampling. The second is about aliasing. The third is about anti-aliasing filtering. The last one is about time signal leakage. The importance of anti-aliasing filtering has been emphasized. The usefulness of why selecting large ω_s even if $\omega_s > 2\omega_m$ has been explained.

REFERENCES

1. H. Nyquist, Certain topics in telegraph transmission theory, *Transactions of the American Institute of Electrical Engineers,* 47(2), 1928, 617–644.
2. R. A. Gabel and R. A. Roberts, *Signals and Linear Systems,* John Wiley & Sons, Hoboken, NJ, 1973.
3. M. Li, A method for requiring block size for spectrum measurement of ocean surface waves, *IEEE Transactions on Instrumentation and Measurement*, 55(6), 2006, 2207–2215.
4. C. M. Harris, *Shock and Vibration Handbook*, 5th Ed., McGraw-Hill, New York, 2002.
5. M. Li, *Multi-Fractal Traffic and Anomaly Detection in Computer Communications,* CRC Press, Boca Raton, FL, 2022.
6. M. Li, *Fractal Teletraffic Modeling and Delay Bounds in Computer Communications*, CRC Press, Boca Raton, FL, 2022.
7. M. Li and W. Zhao, On bandlimitedness and lag-limitedness of fractional Gaussian noise, *Physica A*, 392(9), 2013, 1955–1961.

CHAPTER 30

A Method for Requiring Block Size for Spectrum Measurement of Ocean Surface Waves

SPECTRUM MEASUREMENT OF FIELD data of ocean surface waves is essential for the performance research, analysis and design of ship structure, and laboratory testing. There are two problems in measuring a power spectrum of wave data. One is how to estimate the upper bound of the mean square error of a spectrum measurement of a block of wave data at a given wind speed (sea state). The other is how to quantitatively require the lower bound of the block size according to a given wind speed and a given degree of the mean square error of spectrum measurement of a block of wave data. This chapter presents four theorems and two corollaries with respect to the solutions to the above two problems. Theorems 30.1–30.4 describe the general solutions to the above problems based on two general spectral models of wave data. Corollaries 30.1 and 30.2 show the specific solutions to the problems based on the Pierson–Moskowitz spectrum. With Corollaries 30.1 and 30.2, the effect of large wind speed on spectrum measurement of wave data is analyzed. The significance of the present results is examined and demonstrated by spectrum measurement of real field data of ocean waves.

30.1 INTRODUCTION

Wind-generated ocean surface waves (surface waves or waves for short) are random processes. Actual wave loads acting on a ship are random. Though there have been many advances in numerical computation, testing is yet an essential way to investigate many critical phenomena [1].

DOI: 10.1201/9781003460947-34

There are two types of testing. One is real ship trial and the other is laboratory testing [2]. Let us take fatigue test of ship structure as an example. Fatigue test on real ship trial can provide reliable data of the fatigue performance of a ship at a given sea state but the testing cost is too high in addition to the risk that the trial ship may be fractured during testing. Therefore, laboratory fatigue test under random loading has received much attention [1,3].

For laboratory fatigue test under random loading, an essential thing is to simulate random loading (real environment) in laboratory according to a measured power spectrum of field data as accurately as possible [2,3]. Hence, how to assure a high-quality spectrum measurement of wave data is vital.

Results of the laboratory fatigue test are used to evaluate the fatigue life of a structure in the standard sense. Therefore, spectrum measurement of wave data should be reliable and traceable in the standard sense too. For that reason, we estimate the spectrum with recognized instruments (e.g., the Solartron 1200 Real Time Signal Processor). From a perspective of measurement, the estimated spectrum of interest is actually measured by measurement instruments. Measurement errors of a spectrum are the errors of that spectrum estimation. Hence, the term spectrum measurement used in the chapter is similar in meaning to that of spectrum estimation in signal processing.

A raw field data series is usually collected as an analog signal, but it is sampled to be a discrete series when its spectrum is measured by digital instruments or processed with digital computers. Practically, a spectrum is measured on a block-by-block basis [4]. Therefore, there are errors (e.g., truncation error) in spectrum measurement. To reduce errors, the spectrum is usually measured by averaging spectra estimates of blocks of data with a certain window weight function [4,5]. Let L be the block size and M be the number of average count, respectively. Suppose sectioned data are non-overlapped. Then, the total data length for measurement is estimated by

$$\text{Total data length} = L \times M. \tag{30.1}$$

As far as the parameters in Eq. (30.1) are concerned, approaches for requiring M can be found in many references, e.g., [5]. Nevertheless, how to quantitatively require L at a given wind speed with a given accuracy for spectrum measurement is a problem (P2).

The solution to P2 is critical because a measured spectrum may not be accurate enough if the block size is too small while it may cost too much if the block size is overlong (this is especially true in field measurement on real ship trial in open sea). The inverse problem of P2 is denoted as P1, which concerns how to quantitatively evaluate the accuracy of the spectrum measurement of a wave data series with L length at a given wind speed. This chapter focuses on the solutions to P1 and P2. Here, the statements of two problems emphasize wind speed in spectrum measurement because it is a primary factor reflecting the roughness of waves and an equivalent parameter describing sea state [6,7].

Let std^2 be the upper bound of the mean square error of the spectrum measurement of a wave series with L length. Let V be the wind speed at an elevation of a certain height above the sea surface. Then, to solve the problems P1 and P2 is to find the expressions std(L, V) and L(V, std). The expression std(L, V) is for the solution to P1, while L(V, std) to P2. Mathematically, L(V, std) appears simply the inverse of std(L, V) in terms of L. From a view of measurement, however, they are two related aspects in spectrum measurement. In fact, std(L, V) implies what an upper bound of the accuracy of a spectrum measurement of a wave series with L length should be at a given wind speed V. On the other hand, L(V, std) explains what a lower bound of block size should be at a given V so as to achieve a given accuracy.

There is diversity for various spectral models of wave data reported, see for examples [6,8]. Therefore, studying spectrum measurement with the general spectral models of wave data is meaningful. By general model, we mean that the function form of a model abstractly covers most of wave spectra reported in the literature. Obviously, the general solutions to P1 and P2 with respect to the general spectral models of wave data can provide a basis to derive specific solutions to P1 and P2 for most specific wave spectra. Although there are many spectral models of wave data, the specific model called the Pierson–Moskowitz (P-M) spectrum is yet the most popular one [6]. For these reasons, the present formulas regarding std(L, V) and L(V, std) are in two catalogues. One catalogue is for two general spectral models of wave data and the other is for the P-M spectrum.

This chapter is organized as follows. Section 30.2 describes the research background. Section 30.3 discusses std(L, V). Section 30.4 explains L(V, std). Discussions are given in Section 30.5. Section 30.6 is the summary. The proofs of Theorems 30.1 and 30.2 and Corollary 30.1 are given in Appendices A–C.

30.2 RESEARCH BACKGROUND

30.2.1 Wave Spectra

There are about 20 kinds of wave spectra reported in Refs. [6,8] and the references therein. They are measurement-based findings. Most of them can be abstractly expressed by

$$S(\omega)=A_1\omega^{-p_1}\exp\left(-B_1\omega^{-q}\right)+A_2\omega^{-p_2}\exp\left(-B_2\omega^{-q}\right), \tag{30.2}$$

where A_j, B_j, p_j ($j=1$, 2) and q are free parameters.

Different parameters in Eq. (30.2) correspond to different forms of wave spectra. Letting $A_2=0$ yields

$$S(\omega)=A\omega^{-p}\exp\left(-B\omega^{-q}\right). \tag{30.3}$$

The above expression is true for all spectra reported in Refs. [6,8] except the Ochi–Hubble spectrum [9]. Hence, it is still a kind of general spectral form of wave data in the sense that it does not contain the Ochi–Hubble spectrum.

Eleven kinds of spectra are recommended in the final report of the *23rd International Towing Tank Conference (ITTC)*, 2002, in the field of ship science and technology [8].

All spectra reported in Ref. [8] have the case of $A_2=0$ and $q=4$ except the Ochi–Hubble spectrum and Scott spectrum [10]. Thus, the 23rd ITTC expresses the general spectral model of wave data as

$$S(\omega)=A\omega^{-p}\exp\left(-B\omega^{-4}\right). \tag{30.4}$$

In the rest of this chapter, however, we consider Eqs. (30.2) and (30.3) as the general forms of wave spectra as they cover the form of Eq. (30.4).

Three commonly reported spectra are briefed below.

30.2.1.1 The Pierson–Moskowitz Spectrum

Among the wave spectra recommended by the 23rd ITTC, the first item is the P-M spectrum [11]. Pierson and Moskowitz proposed a spectrum, based on the measured field data and the theoretical results of Phillips [12], for statistically describing fully developed wind seas when phase speed equals the wind speed. The P-M spectrum is given by

$$S(\omega;V)=\alpha g^2\omega^{-5}e^{-\beta\left(\frac{g}{V}\right)^4\frac{1}{\omega^4}}, \tag{30.5}$$

where $\alpha=8.1\times10^{-3}$, $\beta=0.74$, g is the acceleration of gravity (m/s^2) and V wind speed (m/s) at an elevation of 19.5 m above the sea surface [6, p. 79]. The unit of S is m^{2}s.

The experimental analysis of wave data obtained by Pierson and Moskowitz produces [6, p. 79]

$$\frac{V\omega_p}{g}=0.879, \tag{30.6}$$

where ω_p is angular frequency of spectral peak. Therefore, the P-M spectrum can be expressed by

$$S(\omega;\omega_p)=\alpha g^2\omega^{-5}e^{-1.25\omega_p^4\frac{1}{\omega^4}}. \tag{30.7}$$

Denote H_s as significant wave height. Then, $H_s=0.0214V^2$. Thus, the P-M spectrum is equivalently given by

$$S(\omega;\omega_p)=\alpha g^2\omega^{-5}e^{-\left(\frac{3.11}{H_s^2}\right)\frac{1}{\omega^4}}. \tag{30.8}$$

Equation (30.8) is also termed the ITTC Single Parameter Spectrum (1966). Figure 30.1 indicates the P-M spectrum at different wind speeds.

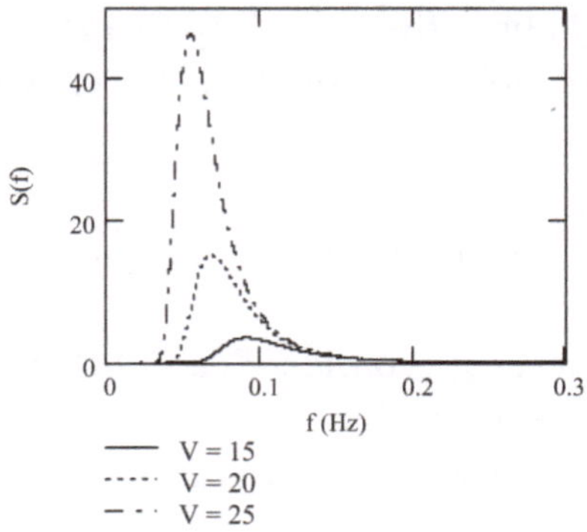

FIGURE 30.1 Plots of the P-M spectrum at $V=15, 20, 25\,\text{m/s}$.

30.2.1.2 Joint North Sea Wave Project (JONSWAP) Spectrum

The JONSWAP spectrum was established from the results obtained in a wave measurement program called Joint North Sea Wave Project [13]. It is given by

$$S(\omega;\gamma,\omega_p)=\alpha g^2\omega^{-5}e^{-1.25\left(\frac{\omega_p}{\omega}\right)^4}\gamma^{\exp\left[\frac{(\omega-\omega_p)^2}{2\tau^2\omega^2}\right]}. \tag{30.9}$$

In Eq. (30.9), $1<\gamma<7$, $\tau=0.07$ for $\omega\le\omega_p$, $\tau=0.09$ for $\omega>\omega_p$, and $\alpha=0.0081$ or $0.076\left(\frac{gF}{V^2}\right)^{-0.22}$, where F is fetch and V the wind speed at an elevation of 10-m above the sea surface [8, p. 539]. The relationship between ω_p, V, and F is given by

$$\omega_p=2\pi\frac{g}{V}\left(\frac{V^2}{gF}\right)^{1/3}. \tag{30.10}$$

Thus,

$$S(\omega;V)=\alpha g^2\omega^{-5}e^{-1.25\left(\frac{2\pi\frac{g}{V}\left(\frac{V^2}{gF}\right)^{1/3}}{\omega}\right)^4}\gamma^{\exp\left[\frac{\left[\omega-2\pi\frac{g}{V}\left(\frac{V^2}{gF}\right)^{1/3}\right]^2}{2\tau^2\omega^2}\right]}. \tag{30.11}$$

For the JONSWAP spectrum, γ describes the degree of peakedness and its exponent describes the width of the peak region. It is easily seen that the JONSWAP spectrum is an extension of the P-M spectrum.

30.2.1.3 The Ochi–Hubble Spectrum

The Ochi–Hubble spectrum takes a relatively complicated form given by

$$S(\omega;H_{s1},H_{s2},\lambda_1,\lambda_2,\omega_{p1},\omega_{p2})=\frac{1}{4}\sum_{j=1,2}\frac{\left(\frac{4\lambda_j+1}{4}\omega_{pj}^4\right)^{\lambda_j}}{\Gamma(\lambda_j)}\frac{H_{sj}}{\omega^{4\lambda_j+1}}\exp\left[-\frac{4\lambda_j+1}{4}\left(\frac{\omega_{pj}}{\omega}\right)^4\right]. \tag{30.12}$$

The shape of the spectrum is relevant to six parameters $(H_{s1},H_{s2},\lambda_1,\lambda_2,\omega_{p1},\omega_{p2})$ but it follows the form of Eq. (30.2) in general. All parameters except experiential parameter λ_1 are characterized by H_s, see [8, pp. 541], while H_s can be equivalently expressed by wind speed. The Ochi–Hubble spectrum is also an extension of the P-M spectrum.

For all spectra describing wind-generated waves, one thing in common is that each takes wind speed as its radical parameter [6,8]. Some (e.g., the P-M spectrum) take wind speed (or its equivalent) as the unique parameter. Others (e.g., the JONSWAP spectrum) regard it as the primary parameter though it is not unique. A noticeable thing in wave modeling is that most reported wave spectra have a root in the P-M spectrum.

30.2.2 Problem Statement and Research Thoughts

Let $x(t)$ be a wave series. Let $r(\tau)$ be the "true" autocorrelation function of $x(t)$. Practically, $x(t)$ is a series of single history. In this case [14], $r(\tau)$ is computed by

$$r(\tau)=\lim_{L\to\infty}\frac{1}{L}\int_0^L x(t)x(t+\tau)dt. \tag{30.13}$$

According to the Wiener–Khintchine theorem, the power spectrum of $x(t)$ is given by

$$S(\omega)=\frac{1}{2\pi}\int_{-\infty}^{\infty} r(\tau)e^{-j\omega\tau}d\tau. \tag{30.14}$$

Mathematically, $r(\tau)$ holds the same position as $S(\omega)$ does for statistically describing $x(t)$. For that reason and facilitating this chapter, we consider the mean square error of autocorrelation estimation of $x(t)$.

A physically measured $x(t)$ is of not only single history but finite length. Hence, the true autocorrelation or spectrum of $x(t)$ may never be achieved exactly. Let $R(\tau)$ be the estimate of $r(\tau)$ with L length. Then,

$$R(\tau)=\frac{1}{L}\int_{t_0+mL}^{t_0+(m+1)L} x(t)x(t+\tau)dt,\quad m=0,1,\ldots \tag{30.15}$$

where t_0 is the starting time. The function R for each m in Eq. (30.15) is an estimate of r but there is a certain error between two. Generally, for $m\neq n$,

$$\frac{1}{L}\int_{t_0+mL}^{t_0+(m+1)L} x(t)x(t+\tau)dt\neq\frac{1}{L}\int_{t_0+nL}^{t_0+(n+1)L} x(t)x(t+\tau)dt. \tag{30.16}$$

In fact, R is a random variable [14]. Let $\mathrm{M}^2(R)$ be the mean square error of R. Then,

$$\mathrm{M}^2(R)=\mathrm{E}\left[(R-r)^2\right]=\mathrm{Var}(R)+\mathrm{b}^2(R), \tag{30.17}$$

where E is the expectation operator and $\mathrm{b}(R)$ the bias of R. Since

$$\mathrm{E}(R)=\frac{1}{L}\int_{t_0+mL}^{t_0+(m+1)L}\mathrm{E}[x(t)x(t+\tau)]dt=\frac{1}{L}\int_{t_0+mL}^{t_0+(m+1)L}r(\tau)dt=r, \tag{30.18}$$

R is an unbiased estimate of r. Therefore, $\mathrm{M}^2(R)=\mathrm{Var}(R)$. That is, the variance of R equals to the mean square error of R. For that reason, $\mathrm{Var}(R)$ equals to $\mathrm{M}^2(R)$ in the rest of this chapter unless otherwise stated.

By comparing Eq. (30.13) with (30.15), it is seen that the larger the L the smaller the $\mathrm{Var}(R)$. Hence, $\mathrm{Var}(R)$ is a function of L. On the other hand, $\mathrm{Var}(R)$ is also a function of V because it is a radical parameter of $S(\omega)$. Therefore, std can be written as std(L, V), which will be a solution to P1 if its representation is found. The inverse of std in terms of L yields L(V, std), implying a lower bound of L with the constraints of accuracy and wind speed. Obviously, L(V, std) will be a solution to P2 if its expression is discovered.

30.3 UPPER BOUND OF $[\mathrm{VAR}(R)]^{0.5}$

For simplicity, continuous signals are considered for the derivations in the chapter (see Appendices of this chapter), although signals are sampled in spectrum measurement. This does not lose the generality of the results discussed in the chapter.

Theorem 30.1. Let x be a wave series with Gaussian distribution. Let r be the "true" autocorrelation function of x. Let R be an estimate of r with L length. Let $\mathrm{Var}(R)$ be the variance of R. Suppose that the power spectrum of x takes the general spectral form of Eq. (30.3). Then,

$$\mathrm{Var}(R)\le\frac{2A^2}{\pi qL}(2B)^{\frac{1-2p}{q}}\Gamma\left(\frac{2p-1}{q}\right). \tag{30.19}$$

The proof is given in Appendix A. □

Theorem 30.2. Let x be a wave series with Gaussian distribution. Let r be the "true" autocorrelation function of x. Let R be an estimate of r with L length. Let $\mathrm{Var}(R)$ be the variance of R. Suppose that the power spectrum of x takes the general spectral form of Eq. (30.2). Then,

$$\begin{aligned}\mathrm{Var}(R)\le&\frac{2A_1^2}{\pi qL}(2B_1)^{\frac{1-2p_1}{q}}\Gamma\left(\frac{2p_1-1}{q}\right)+\frac{2A_2^2}{\pi qL}(2B_2)^{\frac{1-2p_2}{q}}\Gamma\left(\frac{2p_2-1}{q}\right)\\&+\frac{4A_1A_2}{\pi qL}(B_1+B_2)^{\frac{1-(p_1+p_2)}{q}}\Gamma\left(\frac{p_1+p_2-1}{q}\right).\end{aligned} \tag{30.20}$$

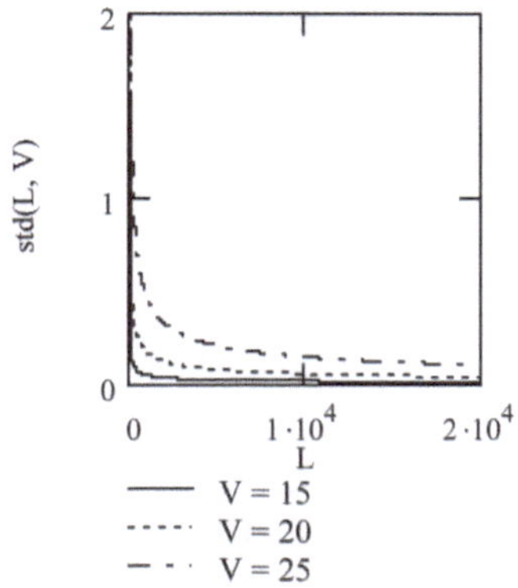

FIGURE 30.2 The locus of std(L, V) at V=15, 20 and 25 m/s, respectively.

The proof is given in Appendix B. □

Corollary 30.1. Let x be a wave series with the P-M spectrum. Let r be the "true" autocorrelation function of x. Let R be an estimate of r with L length. Let Var(R) be the variance of R. Then,

$$\mathrm{Var}(R) \leq \frac{\alpha^2 V^9}{4\pi 2^{5/4}(0.74)^{9/4} g^5 L}\Gamma(9/4). \tag{30.21}$$

The proof is given in Appendix C. □

From Eq. (30.21), the upper bound of $\sqrt{\mathrm{Var}(R)}$ based on the P-M spectrum is given by

$$\mathrm{std}(L,V) = \sqrt{\frac{\alpha^2 V^9}{4\pi 2^{5/4}(0.74)^{9/4} g^5 L}\Gamma(9/4)}. \tag{30.22}$$

Figure 30.2. The locus of upper bound of $\sqrt{\mathrm{Var}(R)}$ in terms of L at V=15, 20 and 25 m/s (P-M spectrum).

30.4 LOWER BOUND OF *L*

Theorem 30.3. Let x be a wave series with Gaussian distribution. Let r be the "true" autocorrelation function of x. Let R be an estimate of r with L length. Let Var(R) be the variance of R. Let std be the upper bound of $\sqrt{\mathrm{Var}(R)}$. Let $L(\cdot, \mathrm{std})$ be the lower bound of L. Suppose that the power spectrum of x takes the form of Eq. (30.3). Then,

$$L(\cdot,\mathrm{std}) = \frac{2A^2}{\pi q \,\mathrm{std}^2}(2B)^{\frac{1-2p}{q}}\Gamma\left(\frac{2p-1}{q}\right). \tag{30.23}$$

According to Theorem 30.1, the proof of Theorem 30.3 is straightforward. □

Theorem 30.4. Let x be a wave series with Gaussian distribution. Let r be the "true" autocorrelation function of x. Let R be an estimate of r with L length. Let Var(R) be the variance of R. Let std be the upper bound of $\sqrt{\mathrm{Var}(R)}$. Let $L(\cdot, \text{std})$ be the lower bound of L. Suppose that the power spectrum of x takes the form of Eq. (30.2). Then,

$$L(\cdot,\text{std})=\frac{2A_1^2}{\pi q\,\text{std}^2}(2B)^{\frac{1-2p_1}{q}}\Gamma\left(\frac{2p_1-1}{q}\right)+\frac{2A_2^2}{\pi q\,\text{std}^2}(2B)^{\frac{1-2p_2}{q}}\Gamma\left(\frac{2p_2-1}{q}\right)$$
$$+\frac{4A_1A_2}{\pi q\,\text{std}^2}(B_1+B_2)^{\frac{1-(p_1+p_2)}{q}}\Gamma\left(\frac{2(p_1+p_2)-1}{q}\right). \tag{30.24}$$

Considering Theorem 30.2, the proof of Theorem 30.4 is straightforward. □

In Theorems 30.3 and 30.4, "·" implies wind speed.

Corollary 30.2. Let x be a wave series with the P-M spectrum. Let r be the "true" autocorrelation function of x. Let R be an estimate of r with L length. Let Var(R) be the variance of R. Let std be the upper bound of $\sqrt{\mathrm{Var}(R)}$. Let $L(\cdot, \text{std})$ be the lower bound of L. Then,

$$L(V,\text{std})=\frac{\alpha^2V^9}{4\pi 2^{5/4}(0.74)^{9/4}g^5\text{std}^2}\Gamma(9/4). \tag{30.25}$$

The proof of Eq. (30.25) is straightforward from Corollary 30.1. □

With the help of Eq. (30.25), it is seen that the larger the V the larger the L is required for a given accuracy. In fact, L is an increasing function of V for a given std. This is reasonable because a large V corresponds to a rough sea state. For the P-M spectrum, the effect of V on requiring L can be described by

$$\frac{\partial L(V,\text{std})}{\partial V}=\frac{9\alpha^2V^8}{4\pi 2^{5/4}(0.74)^{9/4}g^5\text{std}^2}\Gamma(9/4). \tag{30.26}$$

Equation (30.26) indicates a meaningful phenomenon: the larger the wind speed, the larger the effect of V on requiring L because $\frac{\partial L(V,\text{std})}{\partial V}$ is an increasing function of V for a given std.

30.5 DISCUSSIONS

The following discussions are based on the P-M spectrum.

30.5.1 Accuracy Selection

As indicated in Eq. (30.25), L is also constrained by std. We plot $L(V, \text{std})$ in the high-accuracy case, see Figure 30.3. In practice, however, the selection of the concrete value of std is related to the other factors, such as the dynamics of sensors, the accuracy of measurement instruments, and the total requirement of accuracy for measurement and test concerned in experiments.

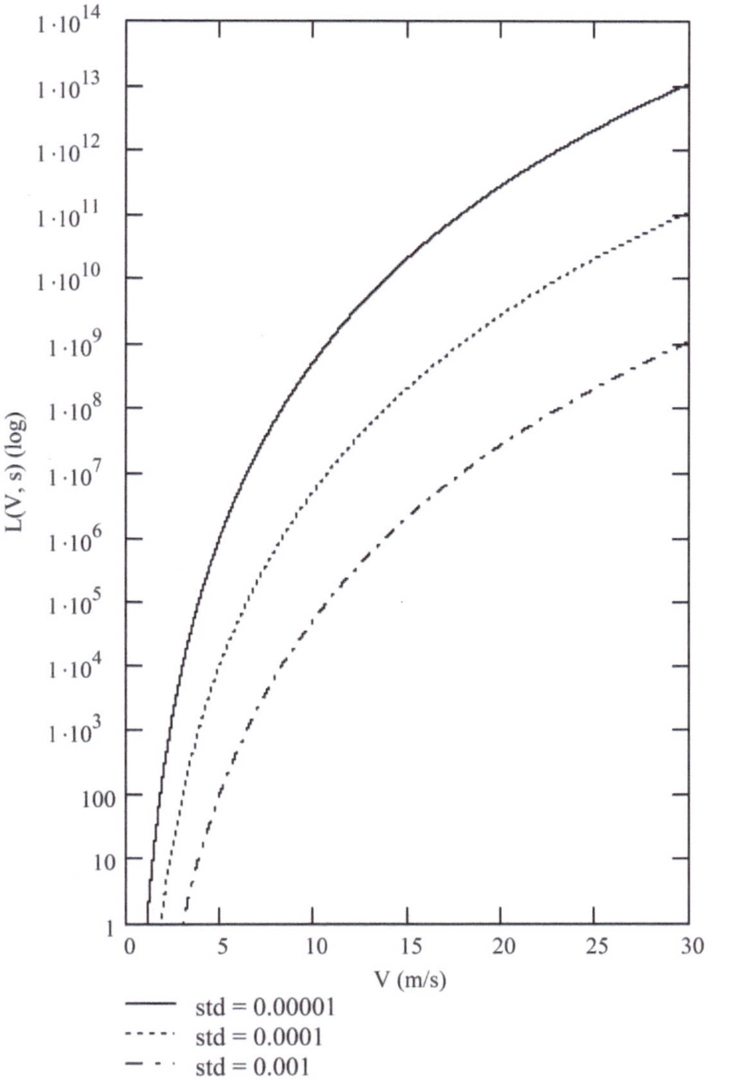

FIGURE 30.3 The locus of $L(V, \text{std})$ (log) versus V in high accuracy case (P-M spectrum).

30.5.2 Small Wind Speed

The effect of small wind speed on requiring L may be negligible if a selected value of std is not too small as can be seen from Eqs. (30.25), (30.26) and Figure 30.3. That is quite natural, as waves at small wind speed are calm. The case of small wind speed is less interesting from a research view. Let it be outside the scope of this research.

30.5.3 Effect of Large Wind Speed on Requiring L

Figure 30.4 illustrates three curves of L versus V for std=0.01, 0.10, and 1.00, respectively. An example is used to explain the effect of large wind speed on requiring L.

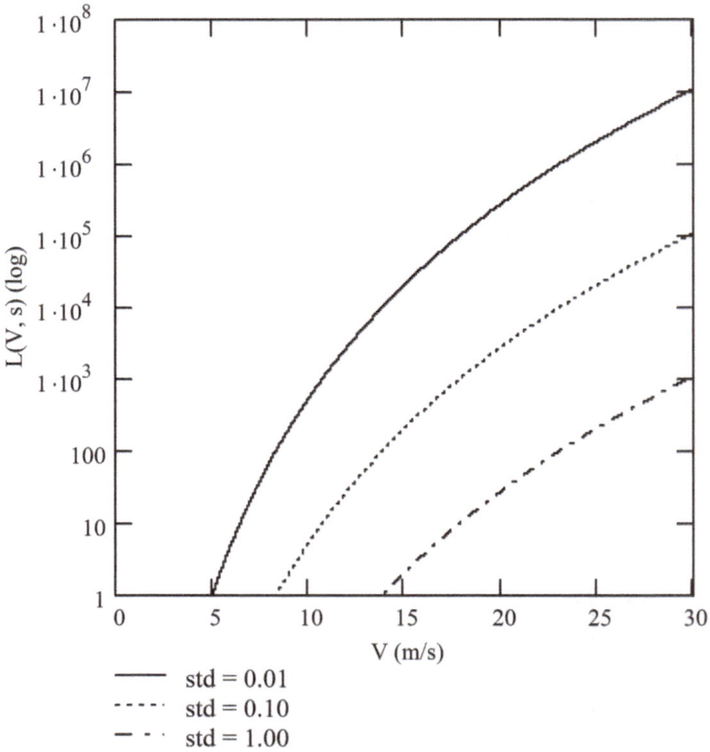

FIGURE 30.4 The locus of $L(V, \text{std})$ (log) versus V for std = 0.01, 0.10, and 1.00 (P-M spectrum).

Example 30.1. Suppose std = 0.01. Then, for V = 10 and 20 m/s, the corresponding bounds are $L(10, 0.01) \approx 542$ and $L(20, 0.01) \approx 2.8 \times 10^5$, respectively. That $L(20, 0.01)$ is much greater than $L(10, 0.01)$ implies the noteworthy effect of large wind speed on requiring L.

30.5.4 Applicability

The Eqs. (30.19), (30.20), (30.23), and (30.24) in Theorems 30.1–30.4 are based on general wave spectra (30.2) and (30.3), respectively. Replacing A_j, B_j, p_j, and q in Eq. (30.2) with specific parameters of a specific spectral model immediately yields the formulas corresponding to the specific spectral model concerned for error analysis and requiring L. Therefore, the significance of Eqs. (30.19), (30.20), (30.23), and (30.24) is that they serve as a fundamental to derive the upper bound of $\sqrt{\mathrm{Var}(R)}$ and the lower bound of L for most practical wave spectra. In this chapter, we omit to list the derivations from Eqs. (30.19), (30.20), (30.23), and (30.24) for other types of wave spectra but the P-M spectrum based on which Corollaries 30.1 and 30.2 are derived.

It is often the case that the spectrum measurement of field data that actually acts on a ship in an area where the normal cruise of a ship is expected at a specific sea state may differ from any reported spectral models in the best fitting sense. That is a reason why field measurement of wave is still crucial today to ship science and technology. That also implies that there is usually *no a priori* information of which wave spectrum best fits really measured field data at a specific sea state. Without such *a priori* information, from the point of view of engineering, it is reasonable to suggest that Corollaries 30.1 and 30.2 can yet serve as a guideline in spectrum measurement due to the simplicity of the P-M spectrum.

30.5.5 Case Study

Figure 30.5 indicates a real wave series $x(n)$ measured in the South China Sea at $H_s=2.44\,\text{m}$, which corresponds to $V=10.79\,\text{m/s}$ when using the P-M spectrum. The wave data were measured with wave buoys, see [15,16] for facilities used in measurements and the stress analysis of a ship structure.

Suppose std is 0.005 ± 0.001. Then, $L\geq 4{,}300$ is suggested according to Eq. (30.25). On Solartron 1200 Real Time Signal Processor, a Hanning widow was set and $M=200$ was selected. Figure 30.6 illustrates the measured spectrum $S_1(f)$ of $x(n)$ with $L_1=4{,}300$ while Figure 30.7 indicates $S_2(f)$ with $L_2=512$ that is much smaller than L_1. Both are in the normalized case.

Since std(L_1, 10.79)$\approx 5\times10^{-3}$ while std(L_2, 10.79)≈ 0.014, $S_2(f)$ is a reference standard to $S_2(f)$. Judging from Figure 30.8, however, $S_2(f)$ considerably deviates from $S_1(f)$ The spectrum $S_1(f)$ is of single peak (Figure 30.6) but $S_2(f)$ appears multipeak (Figure 30.7). Moreover, the frequencies of spectral peaks of $S_2(f)$ greatly differ from those of $S_1(f)$ (Figure 30.8).

It was reported that more than ten Chinese vessels were wrecked in the North Pacific Ocean within several years around 1990 due to inaccurate computations of wave load as can be seen from [17, Section 2]. In laboratory test, on the one hand, the simulated data

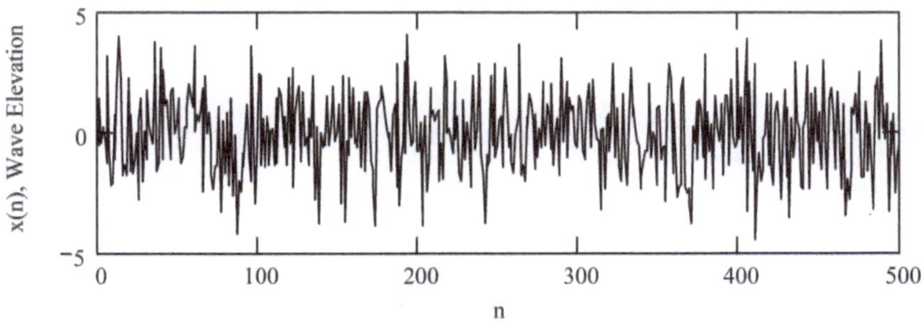

FIGURE 30.5 A real wave series $x(n)$.

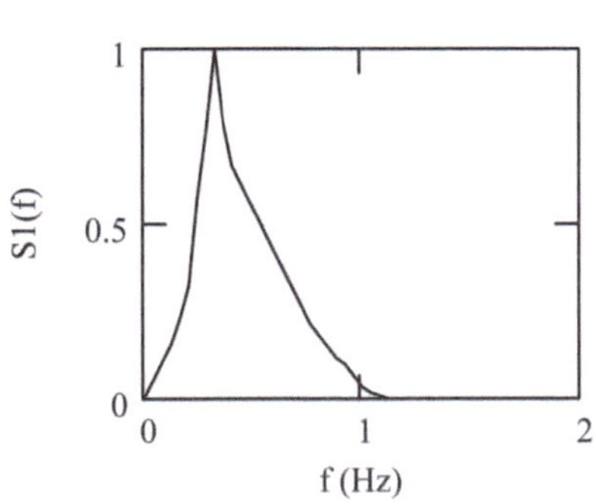

FIGURE 30.6 Spectrum measurement $S_1(f)$ with L_1.

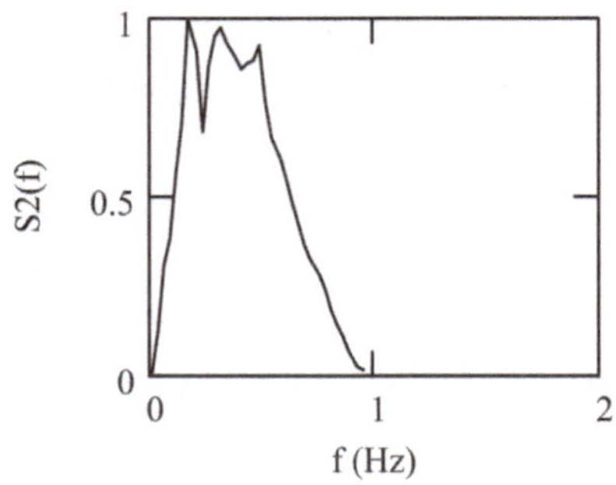

FIGURE 30.7 Spectrum measurement $S_2(f)$ with L_2.

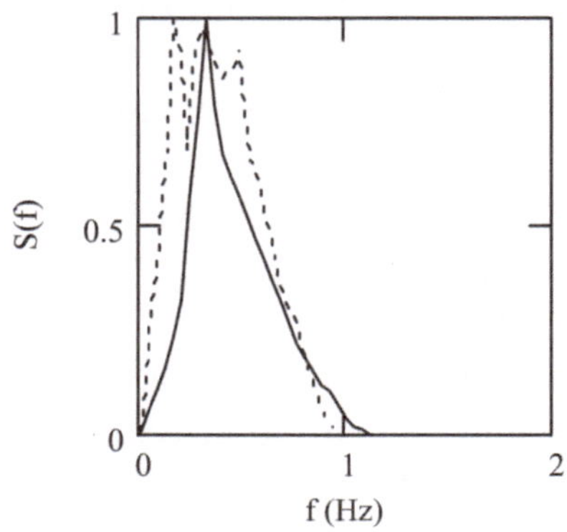

FIGURE 30.8 Comparison between $S_1(f)$ (solid line) and $S_2(f)$ (dot line).

series is given by $z(t)=\sum\sqrt{2S_{zz}(\omega_k)\Delta\omega_k}\cos(\omega_k t+\psi_k)$, where ψ_k is a random variable in uniform distribution and S_{zz} is the prerequired power spectrum of z, according to the random data generation in the software package [3,18]. Therefore, using considerably distorted load spectrum for synthesizing random series implies inappropriate load series for laboratory tests, likely noticeably reducing the reliability of test results. On the other hand, from a view of computational mechanics, considerably under- or overestimating the frequency of a spectral peak may introduce apparent errors of computation results relating to measured spectrum, such as structure response and wave bending moment as can be seen from, e.g., [19–21]. Hence, the above case study demonstrates the practicality and significance in engineering of the present results in the aspect of assuring the quality of a spectrum measurement at a given wind speed with a given accuracy.

30.5.6 Future Work

Following the results represented by Theorems 30.1–30.4 and Corollaries 30.1 and 30.2, it may be quite natural to ask for how tight the present bound is. Let it be the topic of future work.

30.6 SUMMARY

The formulas to evaluate the upper bound of the mean square error of a spectrum measurement of a block of wave data according to a given wind speed and a given block size for both the general forms of wave spectra and the P-M spectrum have been derived. The formulas to require the lower bound of the block size at a given wind speed and with a given mean square error for both the general forms of wave spectra and the P-M spectrum have been proposed. The effect of large wind speed on spectrum measurement of ocean surface waves has been analyzed and demonstrated.

APPENDIX A: PROOF OF THEOREM 30.1

The variance of $R(\tau)$ is computed by $\mathrm{Var}(R)=\mathrm{E}\{[R-\mathrm{E}(R)]^2\}=\mathrm{E}(R^2)-\mathrm{E}^2(R)$. According to Eq. (30.18), [14,22],

$$\mathrm{Var}(R)=\mathrm{E}(R^2)-r^2(\tau). \tag{A.1}$$

Expanding $\mathrm{E}(R^2)$ yields

$$\begin{aligned}
\mathrm{E}\left(R^2\right)&=\mathrm{E}\left\{\left[\frac{1}{L}\int_{t_0}^{t_0+L} x(t)x(t+\tau)dt\right]^2\right\}\\
&=\mathrm{E}\left[\frac{1}{L^2}\int_{t_0}^{t_0+L} x(t_1)x(t_1+\tau)dt_1\int_{t_0}^{t_0+L} x(t_2)x(t_2+\tau)dt_2\right]\\
&=\mathrm{E}\left[\frac{1}{L^2}\int_{t_0}^{t_0+L}\int_{t_0}^{t_0+L} x(t_1)x(t_2)x(t_1+\tau)x(t_2+\tau)dt_1dt_2\right]\\
&=\frac{1}{L^2}\int_{t_0}^{t_0+L}\int_{t_0}^{t_0+L} \mathrm{E}\left[x(t_1)x(t_2)x(t_1+\tau)x(t_2+\tau)\right]dt_1dt_2.
\end{aligned}$$

Therefore, Eq. (A.1) becomes

$$\mathrm{Var}(R)=\frac{1}{L^2}\int_{t_0}^{t_0+L}\int_{t_0}^{t_0+L} \mathrm{E}\left[x(t_1)x(t_2)x(t_1+\tau)x(t_2+\tau)\right]dt_1dt_2-r^2(\tau). \tag{A.2}$$

Let

$$\begin{cases}
X_1=x(t_1),\\
X_2=x(t_2),\\
X_3=x(t_1+\tau),\\
X_4=x(t_2+\tau).
\end{cases} \tag{A.3}$$

Then, $\mathrm{E}[x(t_1)x(t_2)x(t_1+\tau)x(t_2+\tau)] = \mathrm{E}(X_1X_2X_3X_4)$. Since x is Gaussian, X_1, X_2, X_3, and X_4 have a joint-normal distribution and $\mathrm{E}(X_1X_2X_3X_4) = m_{12}m_{34} + m_{13}m_{24} + m_{14}m_{23}$, where

$$\begin{cases} m_{12} = \mathrm{E}\left[x(t_1)x(t_2)\right] = r(t_2 - t_1), \\ m_{13} = \mathrm{E}\left[x(t_1)x(t_1+\tau)\right] = r(\tau), \\ m_{14} = \mathrm{E}\left[x(t_1)x(t_2+\tau)\right] = r(t_2 - t_1 + \tau), \\ m_{23} = \mathrm{E}\left[x(t_2)x(t_1+\tau)\right] = r(t_1 - t_2 + \tau), \\ m_{24} = \mathrm{E}\left[x(t_2)x(t_2+\tau)\right] = r(\tau), \\ m_{34} = \mathrm{E}\left[x(t_1)x(t_2+\tau)\right] = r(t_2 - t_1). \end{cases} \tag{A.4}$$

Therefore,

$$\begin{aligned} \frac{1}{L^2}\int_{t_0}^{t_0+L}\int_{t_0}^{t_0+L} \mathrm{E}\left[x(t_1)x(t_2)x(t_1+\tau)x(t_2+\tau)\right]dt_1dt_2 &= \frac{1}{L^2}\int_{t_0}^{t_0+L}\int_{t_0}^{t_0+L} \mathrm{E}(X_1X_2X_3X_4)dt_1dt_2 \\ &= \frac{1}{L^2}\int_{t_0}^{t_0+L}\int_{t_0}^{t_0+L} (m_{12}m_{34} + m_{13}m_{24} + m_{14}m_{23})dt_1dt_2 \\ &= \frac{1}{L^2}\int_{t_0}^{t_0+L}\int_{t_0}^{t_0+L} \begin{bmatrix} r^2(t_2-t_1) + r^2(\tau) \\ +r(t_2-t_1+\tau)r(t_1-t_2+\tau) \end{bmatrix} dt_1dt_2 \\ &= \frac{1}{L^2}\int_{t_0}^{t_0+L}\int_{t_0}^{t_0+L} \begin{bmatrix} r^2(t_2-t_1) + r(t_2-t_1+\tau) \\ r(t_1-t_2+\tau) \end{bmatrix} \\ &\quad dt_1dt_2 + r^2(\tau). \end{aligned}$$

Hence, Eq. (A.2) is expressed as

$$\mathrm{Var}(R) = \frac{1}{L^2}\int_{t_0}^{t_0+L}\int_{t_0}^{t_0+L} \left[r^2(t_2-t_1) + r(t_2-t_1+\tau)r(t_1-t_2+\tau)\right] dt_1dt_2.$$

Replacing $(t_2 - t_1)$ with t in the above expression yields

$$\mathrm{Var}(R) = \frac{1}{L^2}\int_{t_0}^{t_0+L} dt_1 \int_{t_0-t_1}^{t_0-t_1+L} \left[r^2(t) + r(t+\tau)r(-t+\tau)\right]dt.$$

Let $f(t)=r^2(t)+r(t+\tau)r(-t+\tau)$. Then,

$$\mathrm{Var}(R)=\frac{1}{L^2}\int_{t_0}^{t_0+L}dt_1\int_{t_0-t_1}^{t_0-t_1+L}f(t)dt. \tag{A.5}$$

Without losing the generality, let $t_0=0$. Then,

$$\mathrm{Var}(R)=\frac{1}{L^2}\int_0^L dt_1\int_{-t_1}^{L-t_1}f(t)dt=\frac{1}{L^2}\int_0^L f(t)dt\int_0^{L-t}dt_1+\frac{1}{L^2}\int_{-L}^0 f(t)dt\int_{-t}^L dt_1$$

$$=\frac{1}{L^2}\int_0^L(L-t)f(t)dt+\frac{1}{L^2}\int_{-L}^0(L+t)f(t)dt.$$

Since $f(t)$ is an even function, the above expression is written as

$$\begin{aligned}\mathrm{Var}(R)&=\frac{2}{L^2}\int_0^L(L-t)f(t)dt=\frac{2}{L^2}\int_0^L(L-t)\left[r^2(t)+r(t+\tau)r(-t+\tau)\right]dt\\&=\frac{2}{L}\int_0^L(1-t/L)\left[r^2(t)+r(t+\tau)r(-t+\tau)\right]dt.\end{aligned} \tag{A.6}$$

By using inequality

$$\begin{aligned}\mathrm{Var}(R)&\leq\frac{2}{L}\int_0^L\left|1-\frac{t}{L}\right|\left|r^2(t)+r(t+\tau)r(-t+\tau)\right|dt\leq\frac{2}{L}\int_0^L\left|r^2(t)+r(t+\tau)r(-t+\tau)\right|dt\\&\leq\frac{2}{L}\int_0^L r^2(t)dt+\frac{2}{L}\int_0^L\left|r(t+\tau)r(-t+\tau)\right|dt,\end{aligned} \tag{A.7}$$

the function $r(t+\tau)r(-t+\tau)$ is even in terms of t. Since

$$\begin{aligned}\frac{2}{L}\int_0^L\left|r(t+\tau)r(-t+\tau)\right|dt&=\frac{1}{L}\int_{-L}^L\left|r(t+\tau)r(-t+\tau)\right|dt\\&\leq\frac{1}{L}\sqrt{\int_{-L}^L r^2(t+\tau)dt\int_{-L}^L r^2(-t+\tau)dt},\end{aligned} \tag{A.8}$$

$$\int_{-L}^{L} r^2(t+\tau)dt \le \int_{-\infty}^{\infty} r^2(t)dt, \tag{A.9}$$

$$\int_{-L}^{L} r^2(-t+\tau)dt \le \int_{-\infty}^{\infty} r^2(t)dt, \tag{A.10}$$

the following holds:

$$\mathrm{Var}(R) \le \frac{4}{L}\int_{0}^{\infty} r^2(t)dt. \tag{A.11}$$

According to the Parsval theorem, we attain

$$\mathrm{Var}(R) \le \frac{4}{L}\int_{0}^{\infty} r^2(t)dt = \frac{2}{\pi L}\int_{0}^{\infty} S^2(\omega)d\omega. \tag{A.12}$$

Substituting $S(\omega)$ in Eq. (A.12) for $A\omega^{-p}\exp\left(-B\omega^{-q}\right)$ yields

$$\mathrm{Var}(R) \le \frac{2A^2}{\pi q L}(2B)^{\frac{1-2p}{q}}\Gamma\left(\frac{2p-1}{q}\right). \tag{A.13}$$

Thus, Theorem 30.1 holds.

APPENDIX B: PROOF OF THEOREM 30.2

Denote $S_1(\omega) = A_1\omega^{-p_1}\exp\left(-B_1\omega^{-q}\right)$ and $S_2(\omega) = A_2\omega^{-p_2}\exp\left(-B_2\omega^{-q}\right)$. Then,

$$\begin{aligned} S^2(\omega) &= S_1^2(\omega) + S_2^2(\omega) + 2S_1(\omega)S_2(\omega) \\ &= \left[A_1\omega^{-p_1}\exp\left(-B_1\omega^{-q}\right) + A_2\omega^{-p_2}\exp\left(-B_2\omega^{-q}\right)\right]^2 \\ &= A_1^2\omega^{-2p_1}\exp\left(-2B_1\omega^{-q}\right) + A_2^2\omega^{-2p_2}\exp\left(-2B_2\omega^{-q}\right) \\ &+ 2A_1A_2\omega^{-(p_1+p_2)}\exp\left[-2(B_1+B_2)\omega^{-q}\right]. \end{aligned}$$

According to Eq. (A.13),

$$\frac{2}{\pi L}\int_{0}^{\infty} S_1^2(\omega)d\omega = \frac{2A_1^2}{\pi q L}(2B_1)^{\frac{1-2p_1}{q}}\Gamma\left(\frac{2p_1-1}{q}\right),$$

$$\frac{2}{\pi L}\int_0^{\infty} S_2^2(\omega)d\omega = \frac{2A_2^2}{\pi qL}(2B_2)^{\frac{1-2p_2}{q}}\Gamma\left(\frac{2p_2-1}{q}\right).$$

On the other hand,

$$\frac{2}{\pi L}\int_0^{\infty} 2S_1(\omega)S_2(\omega)d\omega = \frac{4A_1A_2}{\pi qL}(B_1+B_2)^{\frac{1-(p_1+p_2)}{q}}\Gamma\left(\frac{p_1+p_2-1}{q}\right).$$

Thus,

$$\begin{aligned}\mathrm{Var}(R) \le &\frac{2A_1^2}{\pi qL}(2B_1)^{\frac{1-2P_1}{q}}\Gamma\left(\frac{2p_1-1}{q}\right)+\frac{2A_2^2}{\pi qL}(2B_2)^{\frac{1-2p_2}{q}}\Gamma\left(\frac{2p_2-1}{q}\right)\\ &+\frac{4A_1A_2}{\pi qL}(B_1+B_2)^{\frac{1-(p_1+p_2)}{q}}\Gamma\left(\frac{p_1+p_2-1}{q}\right).\end{aligned} \tag{B.1}$$

Therefore, Theorem 30.2 follows.

APPENDIX C: PROOF OF COROLLARY 30.1

The P-M spectrum is based on the assumption of Gaussian distribution [11]. Hence, Eq. (A.12) is valid when data follows the P-M spectrum. Replacing A, B, p, and q in Eq. (A.13) with αg^2, $0.74\left(\frac{g}{V}\right)^4$, 5 and 4, respectively, yields

$$\mathrm{Var}(R) \le \frac{2}{\pi L}\int_0^{\infty} S^2(\omega)d\omega = \frac{\alpha^2 V^9}{4\pi 2^{5/4}(0.74)^{9/4} g^5 L}\Gamma(9/4). \tag{C.1}$$

Corollary 30.1 results.

REFERENCES

1. S. R. Swanson, Ed., *Handbook of Fatigue Testing*, STP 566, American Society for Testing and Materials, Philadelphia, PA, 1974.
2. M. Li, An optimal controller of an irregular wave maker, *Applied Mathematical Modelling*, 29(1), 2005, 55–63.
3. M. Li, An iteration method to adjusting random loading for a laboratory fatigue test, *International Journal of Fatigue*, 27(7), 2005, 783–789.
4. Schlumberger Electronics Ltd., *1200 Real Time Signal Processor Operating Manual*, U.K., 1983.
5. S. K. Mitra and J. F. Kaiser, *Handbook for Digital Signal Processing*, John Wiley & Sons, Hoboken, NJ, 1993.
6. S. R. Massel, *Ocean Surface Waves: Their Physics and Prediction*, World Scientific, Singapore, 1997, Chapter 3.

7. S. K. Chairabarti, *Offshore Structure Modeling*, World Scientific, Singapore, 1994.
8. The Specialist Committee on Waves, Final Report and Recommendations to the 23rd ITTC, *Proceedings of the 23rd ITTC,* vol. II, Venice, Italy, 2002, pp. 497–543.
9. M. K. Ochi and E. N. Hubble, Six parameter wave spectra, *Proceedings of the 15th Coastal Engineering Conference*, Honolulu, HI, 1976, pp. 301–328.
10. J. R. Scott, A sea spectrum for model tests and long-term ship prediction, *Journal of Ship Research*, 9(2), 1965, 145–152.
11. W. J. Pierson and L. Moskowitz, A proposed spectral form for fully developed wind seas based on the similarity theory of S. A. Kitaigorodskii, *Journal of Geophysical Research*, 69(24), 1964, 5181–5190.
12. O. M. Phillips, The equilibrium range in the spectrum of wind-generated waves, *Journal of Fluid Mechanics*, 4(4), 1958, 426–434.
13. K. Hasselmann et al., Measurements of wind-wave growth and swell decay during the Joint North Sea Wave Project (JONSWAP), *Deutsche Hydrographische Zeitschrift*, 8(12), 1973, 95.
14. J. S. Bendat and A. G. Piersol, *Random Data: Analysis and Measurement Procedure*, 3rd Ed., John Wiley & Sons, Hoboken, NJ, 2000.
15. J.-W. Shen, Structure navigability test for 051 ship in the South China Sea, Technical Report 76015, China Ship Scientific Research Center, Wuxi, China, 1976. In Chinese.
16. J.-J. Hu, Report on field measurement analysis for navigability test of 051B(167) ship-stress measurement of hull, Technical Report 2000038, China Ship Scientific Research Center, Wuxi, China, 2000. In Chinese.
17. Z.-M. Zhen, H. Zhou, H.-X. Zhang, K.-Z. Huang, and Y.-L. Bai, Trends of development in mechanics in the early 21st century, *Advances in Mechanics*, 25(4), 1995. In Chinese.
18. Technical Software Consultants Limited, FLAPS: Fatigue Laboratory Applications Package, Version 01:03, Instron 2165035, 1985, U.K.
19. J. J. Jensen and M. Dogliani, Wave-induced ship hull vibrations in stochastic seaways, *Marine Structures*, 9(3–4), 1996, 353–387.
20. X.-K. Gu, J.-J. Hu, and J.-W. Shen, The frequency characteristics of ship vertical wave bending moments, *Shipbuilding of China*, 41(1), 2000, 31–38. In Chinese.
21. D. Benasciutti and R. Tovo, Spectral methods for lifetime prediction under wide-band stationary random processes, *International Journal of Fatigue*, 27(8), 2005, 867–877.
22. J. H. Laning and R. H. Battin, *Random Processes in Automatic Control*, McGraw-Hill, New York, 1956.

CHAPTER 31

Time-Frequency Distributions of Encountered Waves Using Hilbert-Huang Transform

TIME-FREQUENCY DISTRIBUTION (TFD) OF signals gains increasing applications in vibration engineering. This chapter presents our work of time-frequency (TF) analysis of encountered wave signal for studying vibration responses of a marine structure. The results in this chapter exhibit that the Hilbert-Huang transform (HHT) based TFD of encountered wave signal has satisfactory resolution in comparison with those obtained from the traditional methods, such as short-time Fourier transform (STFT), wavelet transform (WT) TFD, and Choi-Williams TFD.

31.1 BACKGROUND

Time-frequency (TF) description of nonstationary and/or nonlinear signals is important in vibration engineering (Bendat and Piersol [1], Elishakoff and Lyon [2], Preumont [3]). Spectral response to encountered waves exerting on a ship hull plays a role in the design of ship structures, see e.g., Troesch [4,5], Jensen [6], Gu et al. [7]. Hence, TF distribution (TFD) of encountered wave signal in good quality is greatly desired in ship mechanics (Gu et al. [7], Xu and Gu [8]).

The power spectrum density (PSD) of a nonstationary signal is time varying. That is, it is also a function of time in addition to frequency. More precisely, the PSD of a nonstationary process $\{x(t)\}$ should be expressed by $S_{xx}(t, \omega)$, which is called evolutionary spectrum in vibrations [2]. In the field of signal processing, it is usually termed TFD. By analogy with stationary processes, the PSD of a nonstationary process is defined as

$$S_{xx}(t,\omega)=\mathrm{F}\left[r_{xx}(t,\tau)\right]=\int_{-\infty}^{\infty} r_{xx}(t,\tau)e^{-j\omega\tau}d\tau, \quad (31.1)$$

DOI: 10.1201/9781003460947-35

where $r_{xx}(t,\tau)$ is the autocorrelation function of $x(t)$ and F the operator of the Fourier transform. Equation (31.1) implies that the PSD of a nonstationary process should be analyzed in a TF plane.

Some early work regarding the mathematical properties of TFD was reported by Priestley [9], Loynes [10], Mark [11], Shinozuka [12]. Several types of methods are commonly used in engineering to describe $S_{xx}(t, \omega)$, such as Priestley's evolutionary spectrum [9], Mark's physical spectrum [11], short-time Fourier transform (STFT) [13,14], TFDs in the Cohen class [15], e.g., Chio-Williams distribution (CWD), and wavelet-based TFD [16]. Note that Huang et al. introduced a new type of TFD based on the Hilbert-Huang transform (HHT) [17,18]. Its applications in practice are noticeable, see e.g., Du and Yang [19], Yang and Gao [20]. This chapter aims at presenting our TF analysis of encountered wave signal using HHT in ship mechanics and demonstrating the resolution comparison with the TFDs based on STFT, WT, and CWD. The results imply that the HHT-based TFD for encountered wave signal has satisfactory resolution in comparison with other three.

The rest of this chapter is organized as follows. In Section 31.2, the environment of signal measurement is briefed. We demonstrate the TFD comparisons in Section 31.3 together with the brief description of each computational method. Section 31.4 gives the summary.

31.2 ENVIRONMENT OF SIGNAL MEASUREMENT

The signal was measured during a model test of a large-scale oil tanker in the seakeeping test basin in the China Ship Scientific Research Center. It is in the size of 69 m (length) × 46 m (width) × 4 m (depth). The wave-height gauge was mounted at the point 2 m away in front of the prow. The recorded data history of the encountered waves is indicated in Figure 31.1. We shall discuss its TFD in the next section.

31.3 TFDs OF ENCOUNTERED WAVES WITH COMPARISON DEMONSTRATIONS

In this section, the TFDs of the encountered waves in Figure 31.1 are discussed based on STFT, WT, CWD, and HHT in turn.

31.3.1 STFT of Encountered Wave Signal

The basic idea behind Gabor's work [21] regarding TF description of a signal is to obtain a TFD of the signal by performing Fourier analysis on the signal as it is when observed

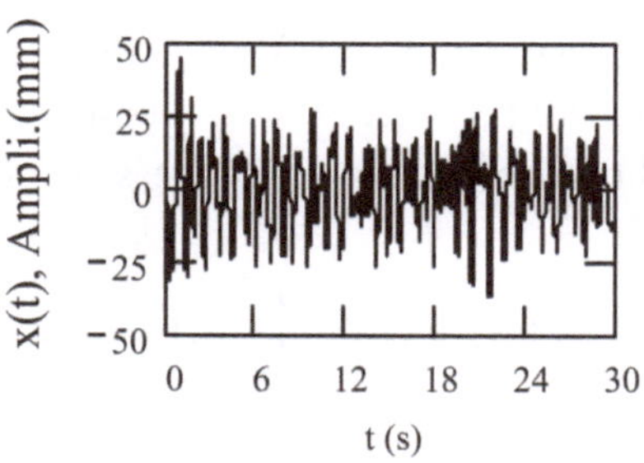

FIGURE 31.1 Encountered waves: signal to be processed.

through a set of identical windows that are translated with respect to each other in time. The functionality of windows is to localize the signal in the TF plane. The window function Gabor suggested is Gaussian.

STFT is a kind of linear transforms. It is an extension of Gabor's transform [22]. Denote $S_{xx}(t, \omega)$ the STFT of $x(t)$. Then,

$$S_{xx}(t,\omega)=\int_{-\infty}^{\infty} x(\tau)h(t-\tau)e^{-j\omega\tau}d\tau, \tag{31.2}$$

where $h(t)$ is a window function. Using the Hamming window of the window size 127, we obtain the STFT-based TFD of the encountered wave signal $x(t)$ in Figure 31.1 as indicated in Figure 31.2.

Note that STFT stands for a set of methods. Different methods use different window functions. To select a window function so that it is optimally suitable for the signal to be processed may be an uneasy task in practice. In addition, the bandwidth of the analyzing functions is a constant that is independent of center frequency. Similarly, the time duration or window size of the analyzing functions is also a constant. To choose concrete values of the bandwidth and the time duration so that they are optimally suitable for the signal to be processed when a window function is given and appears to be a hard problem.

31.3.2 CWD of Encountered Wave Signal

Different from STFT, a class of TFDs is discussed by Cohen [15,23]. TFDs in the Cohen class are generalizations of the Wigner-Ville distribution that was first introduced by Wigner [24] in 1932 in quantum mechanics and Ville [25] in 1948 for TF analysis.

Denote by $\mathrm{WD}_x(t, \omega)$ the Wigner distribution (WD) of a real-valued signal $x(t)$. Then,

$$\mathrm{WD}_x(t,\omega)=\int_{-\infty}^{\infty} x\left(t+\frac{\tau}{2}\right)x\left(t-\frac{\tau}{2}\right)e^{-j\omega\tau}d\tau. \tag{31.3}$$

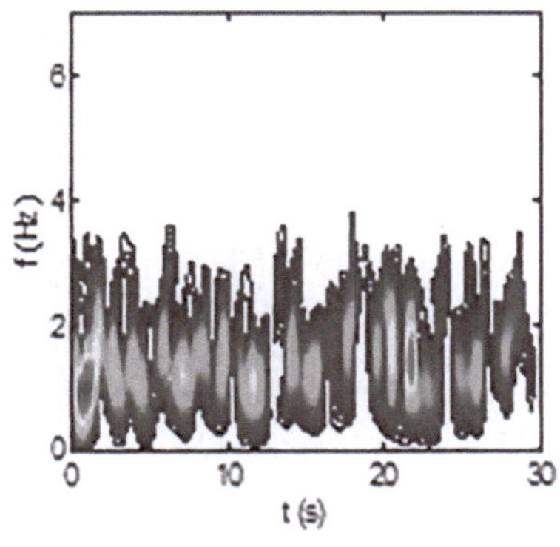

FIGURE 31.2 STFT-based TFD of the encountered waves.

The main unsatisfactory point of WD in practice is that it produces cross-terms. In fact, using the Fourier series, one has $x(t)=\sum A_n \cos(n\omega_1 t+\varphi_n)$, where A_n and φ_n are the amplitude and the initial phase of the nth harmonic of $x(t)$, respectively. Thus,

$$x(t+\tau/2)x(t-\tau/2)=\sum A_n \cos\left[n\omega_1(t+\tau/2)+\varphi_n\right]\sum A_n \cos\left[n\omega_1(t-\tau/2)+\varphi_n\right]. \quad (31.4)$$

The above clearly interprets why there exist cross-terms in a WD.

To suppress cross-terms and to obtain auto-terms as many as possible, a kernel is utilized in the Cohen class. Denote by $\Phi(u, \tau)$ a kernel function. In the general sense, we denote a TFD of x in the Cohen class by $\mathrm{GTFD}_x(t, \omega)$, which is written by

$$\mathrm{GTFD}_x(t,\omega)=\frac{1}{2\pi}\int_{-\infty}^{\infty}\int_{-\infty}^{\infty}\int_{-\infty}^{\infty} x(t+\tau/2)x(t-\tau/2)\Phi(u,\tau)e^{j(ut-\tau\omega-ut)}dt\,d\tau\,du, \quad (31.5)$$

where $\Phi(u, \tau)$ satisfies the conditions mentioned in Ref. [15].

The literature regarding kernel design is rich. Several commonly used kernels are Gaussian function discussed by Choi and Williams [26], Bessel function studied by Guo et al. [27], corn-shaped kernel explained by Oh and Marks [28], and others [15]. The Gaussian function used in CWD is given by

$$\Phi_{\mathrm{CWD}}(u,\tau;\sigma)=e^{-\frac{u^2\tau^2}{\sigma}}, \quad (31.6)$$

where σ is a scaling factor to control its attenuation rate. Figure 31.3 shows the CWD of the encountered waves for $\sigma=1$.

Note that to select the optimal kernel for a signal being processed may be difficult since kernel selection is signal-dependent (Baraniuk and Jones [29,30]).

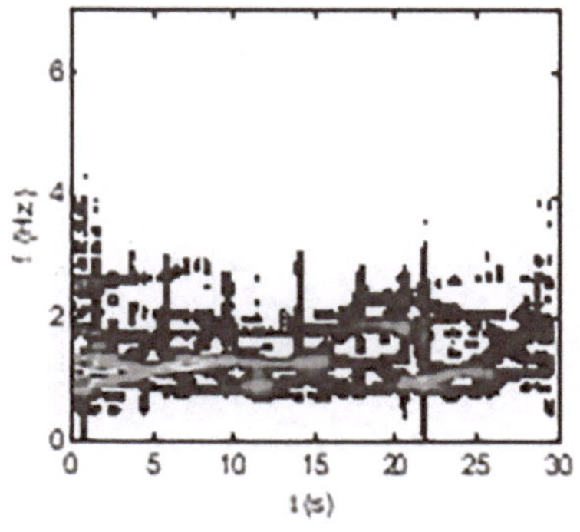

FIGURE 31.3 CWD of the encountered waves for $\sigma=1$.

31.3.3 WT-Based TFD of Encountered Waves

WT of a signal $x(t)$ is defined by

$$\mathrm{WT}_x(t,a)=\frac{1}{\sqrt{a}}\int_{-\infty}^{\infty}x(\tau)g^*\left(\frac{\tau-t}{a}\right)d\tau, \tag{31.7}$$

where $g(t)$ is the basic wavelet or the mother function, * represents the complex conjugate, $a>0$ is a scalar. The constant $1/\sqrt{a}$ is used for energy normalization. Different values of a and t cause the property of multiresolution.

Note that $g(t)$ in Eq. (31.7) can be taken as an impulse function of a linear system. Let f_0 be the central frequency of the analysis system. Then, the scalar a can be expressed by

$$a=\frac{f_0}{f}. \tag{31.8}$$

Hence, WT is a tool for TF analysis of signals with the local frequency $f=af_0$, see e.g., Rioul and Vetterli [31], Strang and Nguyen [32], and Daubechies [33].

There are a number of mother functions in the field, such as Daubechies's 4-coefficient wavelet, Haar wavelet, Morlet wavelet, and so on. We use the Morlet wavelet to obtain WT-based TFD of the encountered wave signal $x(t)$ as indicated in Figure 31.4. Note that the selection of the optimal mother function for a signal being processed may be a hard issue.

31.3.4 HHT-Based TFD of Encountered Waves

By using HHT, a signal being analyzed can be represented in the TF domain by combining the empirical mode decomposition (EMD) with the Hilbert transform (Huang et al. [17,18]). Different from the three approaches mentioned previously, EMD is adaptively data-driven, see e.g., Flandrin and Gonçalvès [34,35].

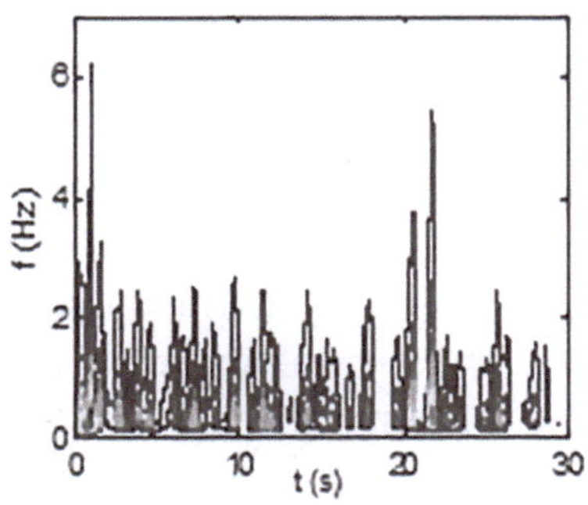

FIGURE 31.4 WT-based TFD of the encountered wave signal.

Denote H the operator of the Hilbert transform (HT). Then, H[$x(t)$] is given by

$$\mathrm{H}[x(t)] = y(t) = \frac{1}{\pi}\int_{-\infty}^{\infty}\frac{x(\tau)}{t-\tau}d\tau. \tag{31.9}$$

As an analytic signal $z(t)$ can be expressed by the sum of its real part $x(t)$ and the imaginary part $y(t)$ that is the Hilbert transform of the real part (Papoulis [36, Chapter 7]), one has

$$z(t) = x(t) + jy(t). \tag{31.10}$$

In the polar coordinate system, the above can be written by

$$z(t) = a(t)\exp\left[j\vartheta(t)\right], \tag{31.11}$$

where $a(t) = [x^2(t) + y^2(t)]^{0.5}$ and $\vartheta(t) = \tan^{-1}[y(t)/x(t)]$ are the instantaneous amplitude and the instantaneous phase of $z(t)$, respectively. Thus, $x(t)$ can be recovered from $z(t)$ by

$$x(t) = \mathrm{Re}[z(t)] = \mathrm{Re}\left\{a(t)\exp[j\vartheta(t)]\right\}. \tag{31.12}$$

Based on HT, therefore, the instantaneous frequency $\omega(t)$ is expressed by

$$\omega(t) = \frac{d\vartheta(t)}{dt} = \frac{\dot{y}(t)x(t) - y(t)\dot{x}(t)}{x^2(t) + y^2(t)}. \tag{31.13}$$

Note that to assure of the physical meaning of $\omega(t)$ requires that $\vartheta(t)$ must be a single-valued function over time, i.e., a mono-component function. A signal being processed in general, however, may not be mono-component but multi-component. Thus, a method to decompose $x(t)$ into a series of mono-functions is desired. In this aspect, Huang et al. developed a method called EMD that decomposes $x(t)$ into a series of mono-functions termed intrinsic mode functions (IMFs).

It is noted that, physically speaking, the necessary conditions to define a meaningful instantaneous frequency are that the signal being processed must be symmetric regarding the local zero mean, and have the same number of zero crossings and extrema. This implies that, in an IMF, the number of extrema and the number of zero crossings must be either equal or different at most by one in the whole data set, and the mean value of the envelope defined by the local maxima and the envelope defined by the local minima is zero at every point. Those conditions are so strict that the resulting IMF may not satisfy them exactly in general. Hence, the result is generally nearly a mono-component function instead of a perfect one. However, this does not matter in practice when one considers that signals to be processed usually include a certain amount of noise or measurement errors in addition to computation errors in signal processing.

The procedure to decompose $x(t)$ into a series of IMFs is as follows. First, identify all local maxima from $x(t)$ and then connect them with the cubic spline line to form the upper envelope of $x(t)$. Denote the upper envelope of $x(t)$ by $x_{up}(t)$. Second, identify all local minima from $x(t)$ and then connect them with the cubic spline line to form the lower envelope of $x(t)$. Denote the lower envelope of $x(t)$ by $x_{low}(t)$. Third, compute the mean by

$$m_{11}(t)=\left[x_{up}(t)+x_{low}(t)\right]/2 \tag{31.14}$$

and construct a new signal $h_{11}(t)$ by

$$h_{11}(t)=x(t)-m_{11}(t). \tag{31.15}$$

In the ideal case, $h_{11}(t)$ is an IMF since it satisfies all the conditions of IMF. In practice, however, there may exist overshoots and undershoots during processing. This is particularly true for processing encountered wave signals. Those overshoots or undershoots may distort the mean values, accordingly making the envelope mean differ from the true local mean, and, as a result, make $h_{11}(t)$ asymmetric. To deal with this practical issue, Huang et al. suggested the fourth step below.

Fourth, repeat the shifting process (31.15) by taking $h_{11}(t)$ as a new signal. After the kth iteration, we have

$$h_{1k}(t)=h_{1(k-1)}(t)-m_{1k}(t), \tag{31.16}$$

where $m_{1k}(t)$ is the mean envelope after the kth iteration, and $h_{1(k-1)}(t)$ is the difference between the signal and the mean envelope at the $(k-1)$th iteration. Define $h_{1k}(t)$ as the first IMF component. Express it by

$$c_1(t)=h_{1k}(t). \tag{31.17}$$

Fifth, having separated $c_1(t)$ from $x(t)$, one has the residue given by

$$r_1(t)=x(t)-c_1(t). \tag{31.18}$$

The criterion for stopping the iteration suggested by Huang et al. is like this. Given the standard deviation expressed by

$$\text{SD}=\sum_{t=0}^{N}\frac{\left|h_{1(k-1)}(t)-h_{1k}(t)\right|^2}{h_{1(k-1)}^2(t)}. \tag{31.19}$$

Then, the iteration stops when SD is equal to or less than a predetermined value. Huang et al. suggested that, typically, $\text{SD}\approx 0.2\sim 0.3$, which is very rigorous limit for the difference between two consecutive iterations.

Sixth, treating $r_1(t)$ as a new signal and the above iteration procedure is repeated to extract the rest of IMFs to the signal $x(t)$ as

$$\begin{cases} r_1(t) - c_2(t) = r_2(t), \\ \vdots \\ r_{n-1}(t) - c_n(t) = r_n(t). \end{cases} \tag{31.20}$$

Seventh, the signal decomposition procedure ends when $r_n(t)$ becomes a monotonic function or a constant, which implies that no further IMFs can be extracted from $x(t)$.

Replacing (31.20) with (31.18), a series of IMFs of $x(t)$ are obtained. Therefore, $x(t)$ can be expressed as the combination of $c_i(t)$ plus the residue $r_n(t)$. That is,

$$x(t) = \sum_{i=1}^{n} c_i(t) + r_n(t). \tag{31.21}$$

Now, performing HT on $c_i(t)$ yields

$$a_i(t) = \sqrt{[c_i(t)]^2 + \mathrm{H}[c_i(t)]^2}, \quad \vartheta_i(t) = \tan^{-1}\left\{\frac{\mathrm{H}[c_i(t)]}{c_i(t)}\right\}. \tag{31.22}$$

The instantaneous frequency is given by

$$\omega_i(t) = \frac{d\vartheta_i(t)}{dt}. \tag{31.23}$$

In the polar coordinate system, $x(t)$ is expressed by

$$x(t) = \mathrm{Re}\left(\sum_{i=1}^{n} a_i(t) \exp\left[j \int \omega_i(t) dt\right]\right) + r_n(t). \tag{31.24}$$

Ignoring the residue is practically allowed since it is either a monotonic function or a constant. Doing so yields

$$x(t) \approx \mathrm{Re}\left(\sum_{i=1}^{n} a_i(t) \exp\left[j \int \omega_i(t) dt\right]\right). \tag{31.25}$$

Let $a_i(t, \omega)$ be the combination of the amplitude $a_i(t)$ and the instantaneous frequency $\omega_i(t)$ of the ith IMF. Denote the HHT of $x(t)$ by HHT(t, ω). Then,

$$\mathrm{HHT}(t,\omega) = \sum_{i=1}^{n} a_i(t,\omega). \tag{31.26}$$

Figure 31.5 indicates the IMFs and the residue of the encountered wave signal $x(t)$. Figure 31.6 is its HHT(t, ω).

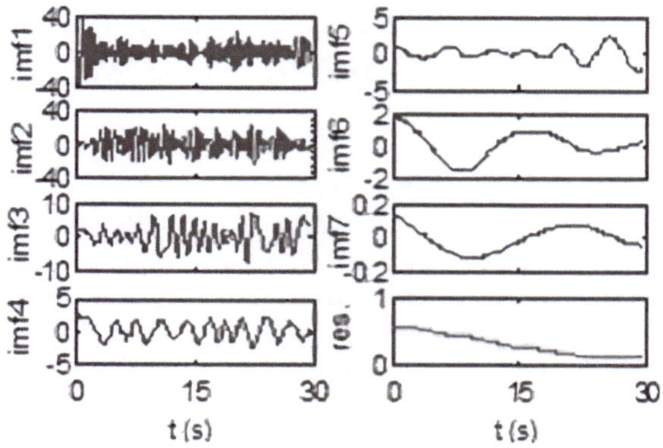

FIGURE 31.5 IMFs of the encountered wave signal.

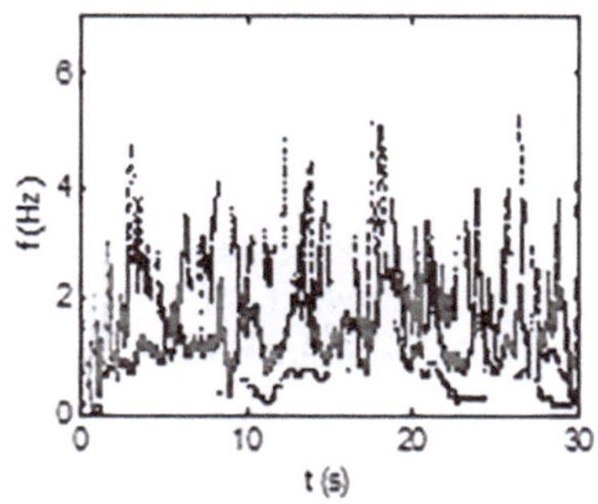

FIGURE 31.6 HHT-based TFD of the encountered wave signal.

Judging from Figures 31.3–31.6, one sees that HHT(t, ω) clearly has satisfactory resolution in comparison with those using STFT, CWD, and WT. As a matter of fact, the basis of HHT is adaptively constructed according to the signal being analyzed while others are not. This advantage makes HHT superior to other three without the limitations caused by those like various window functions in STFT or different mother function in WT or different transform kernels in GTFD. In addition, HHT has no requirement in record length of signal being analyzed since its decomposition accuracy is adaptively satisfied with a given SD expressed in Eq. (31.19). For the Fourier type PSD, however, in order to obtain the desired resolution of PSD of ocean waves, the record length of wave signal is strictly required as can be seen from Chapter 30 or Li [37].

31.4 SUMMARY

We have discussed our TF analysis of the encountered wave signal. The results indicate that the HHT-based TFD has satisfactory resolution.

REFERENCES

1. J. S. Bendat and A. G. Piersol, *Random Data: Analysis and Measurement Procedure*, 3rd Ed., John Wiley & Sons, Hoboken, NJ, 2000.
2. I. Elishakoff and R. H. Lyon, *Random Vibration Status and Recent Developments*, Studies in Applied Mechanics, vol. 14, Elsevier, New York, 1986.
3. A. Preumont, *Random Vibration and Spectral Analysis*, Studies in Applied Mechanics, vol. 33, Kluwer, London, 1993.
4. A. W. Troesch, Wave-induced hull vibrations: An experimental and theoretical study, *Journal of Ship Research*, 28(2), 1984, 141–150.
5. A. W. Troesch, Effects of nonlinearities on hull springing, *Marine Technology*, 21(4), 1984, 256–363.
6. J. J. Jensen, Wave-induced ship hull vibrations in stochastic seaways, *Marine Structure*, 9(3–4), 1996, 353–387.
7. X. K. Gu, J. W. Shen, and T. Moan, Experimental and theoretical investigation of higher order harmonic components of nonlinear bending moments of ships, *Ship Technology Research*, 47(4), 2000, 143–154.
8. C. Xu and X. K. Gu, Application of continuous wavelet transform in the investigation of ship hull springing, *Journal of Ship Mechanics*, 10(4), 2006, 72–79. In Chinese.
9. M. B. Priestley, Evolutionary spectra and non-stationary processes, *Journal of the Royal Statistical Society: Series B*, 27(2), 1965, 204–237.
10. R. M. Loynes, On the concept of the spectrum for non-stationary processes, *Journal of the Royal Statistical Society: Series B*, 30(1), 1968, 1–30.
11. W. D. Mark, Spectral analysis of the convolution and filtering of non-stationary stochastic processes, *Journal of Sound and Vibration*, 11(1), 1970, 19–63.
12. M. Shinozuka, Locally stationary random processes, *IRE Transactions on Information Theory/IEEE Transactions on Information Theory*, 3(3), 1970, 182–187.
13. Y. Yorozu, M. Hirano, K. Oka, and Y. Tagawa, Electron spectroscopy, time-frequency representation of digital signals and systems based on short-time Fourier analysis, *IEEE Transactions on Acoustics, Speech, and Signal Processing*, 28(1), 1980, 55–69.
14. C. R. Pinnegar and L. Mansinha, Time-local spectral analysis for non-stationary time series: The S-transform for noisy signals, *Fluctuation and Noise Letters*, 3(3), 2003, L357–64.
15. L. Cohen, *Time-Frequency Analysis: Theory and Applications*, Prentice Hall, Hoboken, NJ, 1995.
16. S. E. Qian, *Introduction to Time-Frequency and Wavelet Transforms*, 1st ed., Prentice Hall, Hoboken, NJ, 2001.
17. N. E. Huang, Z. Shen, S. R. Long, M. L. Wu, H. H. Shih, Q. Zheng, N. C. Yen, C. C. Tung, and H. H. Liu, The empirical mode decomposition method and the Hilbert spectrum for non-stationary time series analysis, *Proceedings of the Royal Society of London: Series A*, 454(1971), 1998, 903–995.
18. N. E. Huang and S. S. P. Shen, (eds.), *Hilbert-Huang Transform and Its Applications*, World Scientific, Singapore, 2005.
19. Q. H. Du and S. N. Yang, Application of the EMD method in the vibration analysis of ball bearings, *Mechanical Systems and Signal Processing*, 21(6), 2007, 2634–2644.
20. R. Q. Yang and X. Gao, Hilbert-Huang transform-based vibration signal analysis for machine health monitoring, *IEEE Transactions on Instrumentation and Measurement*, 55(6), 2006, 2320–2329.

21. D. Gabor, Theory of communication, *Journal of the Institution of Electrical Engineers*, 93, 1945, 429–457.
22. J. B. Allen and L. R. Rabiner, A unified approach to short-time Fourier analysis and synthesis, *Proceedings of the IEEE*, 65(11), 1977, 1558–1564.
23. L. Cohen, Time-frequency distributions: A review, *Proceedings of the IEEE*, 77(7), 1989, 941–981.
24. E. Wigner, On the quantum correction for thermodynamic equilibrium, *Physical Review*, 40(5), 1932, 749–759.
25. J. Ville, Theorie at applications de la notion de signal analytique, *Cables et Transmission*, 2A(1), 1948, 61–74.
26. H. I. Choi and J. W. Williams, Improved time-frequency representation of multicomponent signals using exponential kernels, *IEEE Transactions on Signal Processing*, 37(6), 1989, 862–871.
27. Z.-Y. Guo, L.-G. Durand, and H. C. Lee, The time-frequency distributions of non-stationary signals based on a Bessel kernel, *IEEE Transactions on Signal Processing*, 42(7), 1994, 1700–1745.
28. S. Oh and R. J. Marks, Some properties of the generalized time frequency representation with corn-shaped kernel, *IEEE Transactions on Signal Processing*, 40(7), 1992, 1735–1745.
29. R. G. Baraniuk and D. L. Jones, Signal-dependent time-frequency analysis using a radially Gaussian kernel, *Signal Processing*, 32(3), 1993, 263–284.
30. R. G. Baraniuk and D. L. Jones, A signal dependent time-frequency representation: Optimal kernel design, *IEEE Transactions on Signal Processing*, 41(4), 1993, 1589–1602.
31. O. Rioul and M. Vetterli, Wavelet and signal processing, *IEEE Signal Processing Magazine*, 8(4), 1991, 14–38.
32. G. Strang and T. Nguyen, *Wavelets and Filter Banks*, Wellesley-Cambridge Press, Cambridge, MA, 1996.
33. I. Daubechies, The wavelet transform, time-frequency localization, and signal analysis, *IEEE Transactions on Signal Processing*, 36(5), 1990, 961–1005.
34. P. Flandrin and P. Gonçalvès, Empirical mode decompositions as data-driven wavelet-like expansions, *International Journal of Wavelets, Multiresolution and Information Processing*, 2(4), 2004, 477–496.
35. P. Flandrin and P. Gonçalvès, Empirical mode decomposition as a filterbank, *IEEE Signal Processing Letters*, 11(2), 2004, 112–114.
36. A. Papoulis, *Signal Analysis*, McGraw-Hill, New York, 1977.
37. M. Li, A method for requiring block size for spectrum measurement of ocean surface waves, *IEEE Transactions on Instrumentation and Measurement*, 55(6), 2006, 2207–2215.

CHAPTER 32

An Optimal Controller of an Irregular Wave Maker

LABORATORY SEAKEEPING TEST UNDER irregular wave load is essential for studying seakeeping performances of marine and offshore structures, which is associated with vibrations. One of key techniques in such a sort of test is how to assure the quality of wave load. By assuring the quality of wave load, we mean that man-made irregular wave series exerting on a test model should have a given power spectrum. From a view of optimal control, the goal to assure the quality of wave load is to manage wave maker such that produced irregular wave series exerting on a test model has a given power spectrum in the optimization sense. This chapter gives a control model to optimize wave maker to achieve the mentioned goal. Its application to wave maker control in a seakeeping model test of an offshore structure is demonstrated.

32.1 BACKGROUND

Wind-generated ocean surface waves are usually random processes [1–3]. Therefore, actual wave loads acting on a marine structure are usually random. Though there have been many advances in numerical computation, the laboratory seakeeping test is yet an essential way to investigate seakeeping performances of a marine structure. In fact, seakeeping performances of a marine structure are associated with its vibration performances.

There are two types of seakeeping tests. One is based on real marine structure trial and the other laboratory testing. The seakeeping test on real marine structure trial can provide reliable data of seakeeping performances of that structure at a given sea state but the testing cost is too high. Therefore, generating irregular surface wave series in basin is vital for studying seakeeping performances of a marine structure [1,3]. Figure 32.1 indicates the diagram of a wave maker.

Man-made random wave series is produced by a mechatronics system called wave maker [1,4]. Without losing the generality in control, this chapter only discusses an optimal control model without considering the directions of waves.

DOI: 10.1201/9781003460947-36

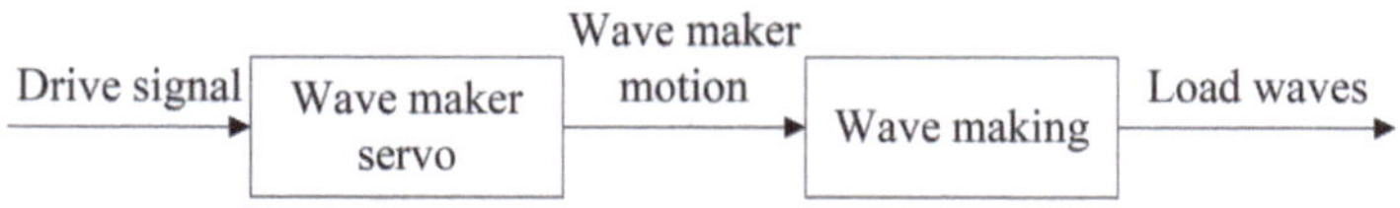

FIGURE 32.1 Block diagram of wave maker.

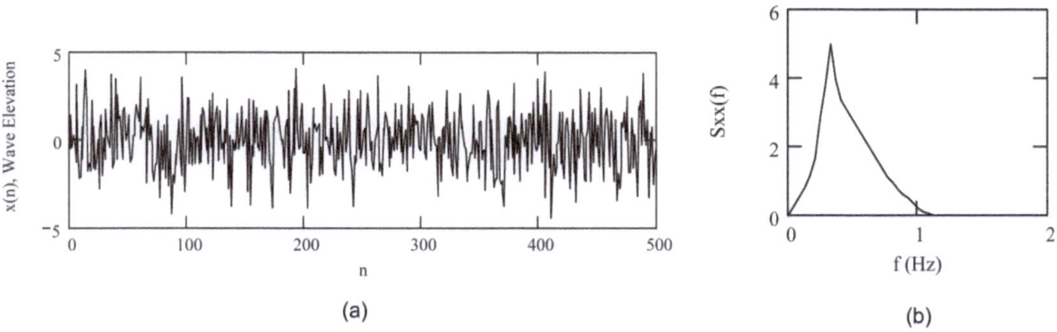

FIGURE 32.2 Real wave data. (a) Time series $x(n)$. (b) Target power spectrum.

Surface wave series can be characterized by a power spectrum [1–3]. As for wave generation in basin, one essential thing in laboratory seakeeping test is to generate random series in computer such that wave load series, which acts on a model, has a given power spectrum without considerable spectrum distortion [1,3,5,6].

A method for computer generation of random data with a given power spectrum is explained as follows. For a given a power spectrum $S_{xx}(f)$, its corresponding time series is given by

$$x(t) = \sum_{k} \sqrt{2S_{xx}(k\Delta f)\Delta f} \cos\left[2\pi k\Delta f t + \psi(k\Delta f)\right], \tag{32.1}$$

where $\psi(k\Delta f)$ is a uniformly distributed random variable and $S_{xx}(f)$ the power spectrum of x [1,2].

In practice, when a computer-generated series that follows a target power spectrum passes through the wave maker, the power spectrum of the wave load series acting on a model may often differ significantly from the target one due to the transfer characteristics and uncertainties of test devices. Figure 32.2a indicates a real wave series x measured in the South China Sea at a significant wave height of 2.44 m and Figure 32.2b is its power spectrum $S_{xx}(f)$ that is taken as the target spectrum in our test. However, when taking $S_{xx}(f)$ as the input spectrum, the load spectrum, denoted as $S_{yy}(f)$, is indicated in Figure 32.3a. Figure 32.3b shows its corresponding load time series. By eye, a noteworthy spectrum distortion can be observed.

The pattern similarity between $S_{xx}(f)$ and $S_{yy}(f)$ can be measured by correlation coefficient according to the theory of pattern recognition [7]. Denote corr$[S_{xx}, S_{yy}]$ as the

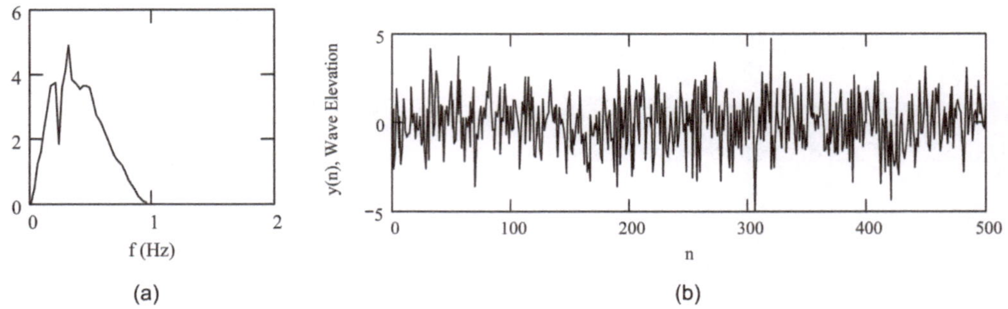

FIGURE 32.3 Load on test model of an offshore structure: corr[S_{xx}, S_{yy}]=0.776. (a) Load spectrum. (b) Load time series.

correlation coefficient between S_{xx} and S_{yy}. Then, for the spectra shown in Figure 32.2b and Figure 32.3a, we obtained corr[S_{xx}, S_{yy}]=0.776.

The above discussions imply that an original wave maker may transform an excitation series into load series with a certain amount distortion in spectrum. Thus, in order to make load spectrum match the target one as closely as possible, the input spectrum should be adjusted according to a certain control strategy. In this aspect [5,6], explained an iteration method to adjust wave without taking optimal control into account. The author's previous work [8] gives a theoretical explanation of the existence of optimal control of wave maker. This chapter aims at presenting an optimal controller of wave maker. To the best of this author's knowledge, the present control model is rarely seen in control of wave makers.

The rest of chapter is organized as follows. Section 32.2 discusses the model of wave maker from a view of control. Section 32.3 explicates the representation of the optimal controller in the frequency domain. Section 32.4 demonstrates the application of the present controller to a model test. Section 32.5 is the summary.

32.2 MODEL OF WAVE MAKER

A wave maker consists of computer, electric console, data acquisition, hydraulic system, and so on. Abstractly, the basic function of the wave maker is to transfer a drive signal into a wave load series exerting on a test model. Errors of measurement and any other uncertainties would affect the performances of wave maker. According to control theory [9,10], all errors and uncertainties can be abstracted as a model uncertainty of a linear system. Thus, a wave maker can be expressed in Figure 32.4.

Hence, one may regard a wave maker as a linear system contaminated by uncertainty $\varepsilon(t)$. The relationship between the input and output in Figure 32.4 in the time domain is given by

$$y = h * (x + \varepsilon), \tag{32.2}$$

where * stands for the operation of convolution. The above expression implies two factors that result in spectrum distortion. One is h and the other ε.

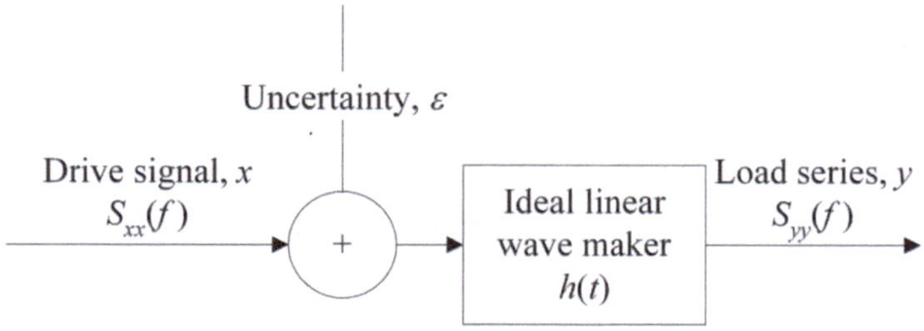

FIGURE 32.4 Model of original wave maker.

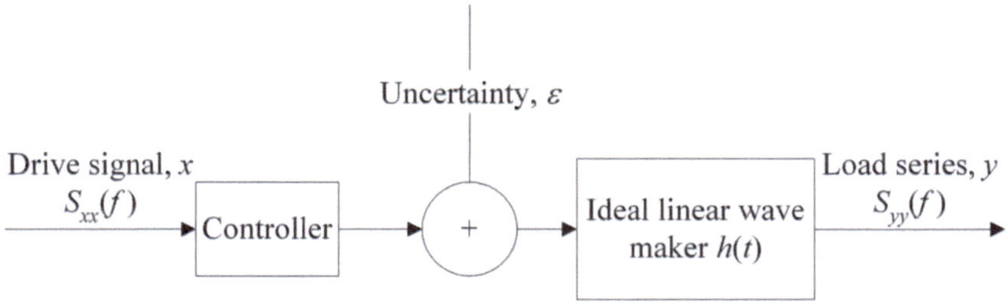

FIGURE 32.5 Wave maker model with additional controller.

32.3 AN OPTIMAL CONTROLLER IN FREQUENCY DOMAIN

Suppose we insert a controller between drive signal and original wave maker as indicated by Figure 32.5. Let $\hat{h}$ be the impulse function of controller. Then, one has

$$y = \hat{h} * h * (x + \varepsilon). \tag{32.3}$$

In practice, x and ε are energy-limited. Assume that ε and x are uncorrelated. Then,

$$r_{x\varepsilon}(t) = x(t) * \varepsilon(-t) = 0, \tag{32.4}$$

where $r_{x\varepsilon}(t)$ is the cross correlation between x and ε [11, p. 121]. Here, the correlation function is expressed by the convolution according to Refs. [12,13]. Doing the autocorrelation operation on both sides of Eq. (32.3) produces

$$\begin{aligned} r_{yy}(t) &= y(t) * y(-t) \\ &= \left[h(t) * \hat{h}(t) * x(t) + h(t) * \varepsilon(t)\right] * \left[h(-t) * \hat{h}(-t) * x(-t) + h(-t) * \varepsilon(-t)\right] \\ &= r_{hh}(t) * \hat{r}_{hh}(t) * r_{xx}(t) + r_{hh}(t) * r_{\varepsilon\varepsilon}(t). \end{aligned} \tag{32.5}$$

Let y in Eq. (32.3) be x. Then, Eq. (32.5) becomes

$$r_{xx}(t) = x(t) * x(-t) = r_{hh}(t) * \hat{r}_{hh}(t) * r_{xx}(t) + r_{hh}(t) * r_{\varepsilon\varepsilon}(t). \tag{32.6}$$

Applying the convolution theorem to Eq. (32.6) yields

$$S_{xx}(f) = |H(f)|^2 |\hat{H}(f)|^2 S_{xx}(f) + |H(f)|^2 S_{\varepsilon\varepsilon}(f). \tag{32.7}$$

From Eq. (32.7), we attain the following optimal controller

$$|\hat{H}(f)|^2 = \left| \frac{1}{|H(f)|^2} - \frac{S_{\varepsilon\varepsilon}(f)}{S_{xx}(f)} \right|. \tag{32.8}$$

Therefore, instead of feeding a target spectrum directly into the computer, we feed a pre-processed spectrum, say In(f),

$$\text{In}(f) = |\hat{H}(f)|^2 S_{xx}(f) \tag{32.9}$$

for the purpose of optimizing wave maker. Theoretically, in this case, the output of a wave maker under the excitation of x should have the power spectrum S_{xx}. In practice, however, there is still a certain error because $S_{\varepsilon\varepsilon}$ may be difficult (if not impossible) to be obtained exactly. In our practice, it is estimated by

$$S_{\varepsilon\varepsilon}(f) = \sqrt{|S_{xx}^2(f) - S_{yy}^2(f)|}, \tag{32.10}$$

where $S_{yy}(f)$ is the load spectrum under the excitation of $S_{xx}(f)$ without taking $|\hat{H}(f)|^2$ into account.

32.4 APPLICATION TO A MODEL TEST

32.4.1 Experiment Conditions

The experiment was done in a seakeeping test basin at China Ship Scientific Research Center. The test model stands for an offshore structure. The irregular wave has a significant wave height of 0.713 m and current velocity of 0.13 m/s.

32.4.2 Transfer Function

To perform the optimization of wave maker, we need the transfer function $H(f) = |H(f)|^{-j\varphi(f)}$, where $j = \sqrt{-1}$. To identify $H(f)$, a set of sinusoidal waves with constant amplitudes was selected to scan the frequency response. In order to excite the transfer mode of the wave maker sufficiently and at the same time assure the accuracy of signal processing, excitation signals were arranged to last for a relatively long time at each frequency point. We arranged for 30 minutes at each point and $\Delta f = 0.1$ Hz. The scanning band was from 0 to 2.5 Hz. A Hanning window was used, block size was 4,096 and average count 150. A Real Time Signal Processor ([14]) recorded the data and the transfer function $H(f)$ was obtained as indicated in Figure 32.6.

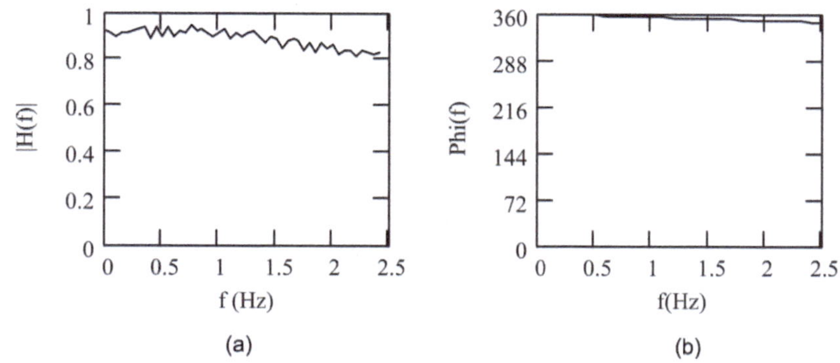

FIGURE 32.6 Transfer function of wave maker. (a) Amplitude of transfer function. (b) Phase of transfer function.

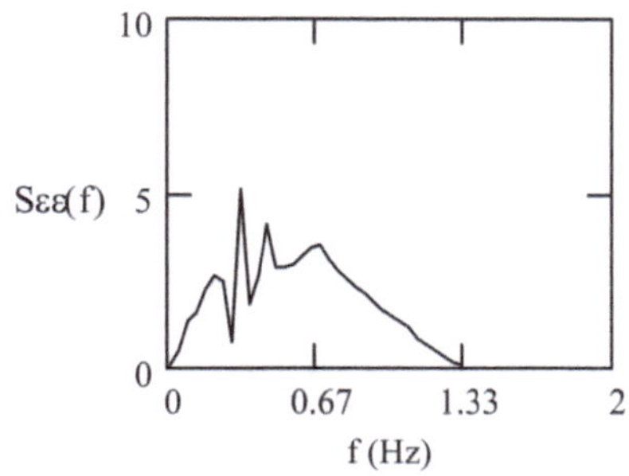

FIGURE 32.7 Error spectrum.

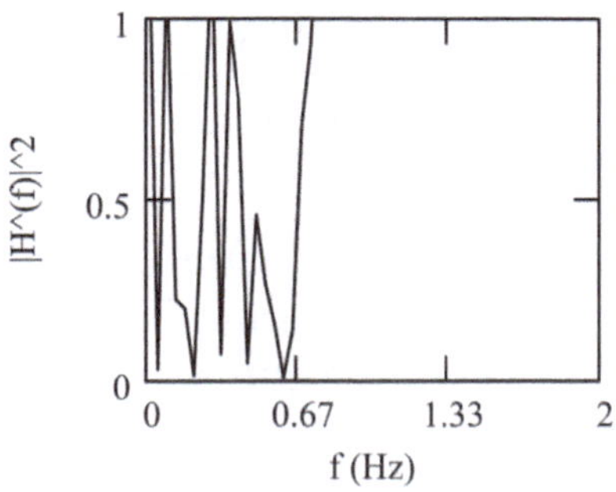

FIGURE 32.8 Controller $\left|\hat{H}(f)\right|^2$.

32.4.3 Control Procedure

Let $\hat{x}$ be a load series actually measured on the test model when the controller (32.8) is used. Let $S_{\hat{x}\hat{x}}$ be the power spectrum of $\hat{x}$. The concrete value of the pattern similarity between S_{xx} and $S_{\hat{x}\hat{x}}$ depends on the requirements desired by marine researchers. In the test, $\mathrm{corr}[S_{xx}, S_{\hat{x}\hat{x}}] \geq 0.90$ is required. Following Eq. (32.10), the error spectrum was obtained as shown in Figure 32.7. According to Eq. (32.8), we attained the controller illustrated in Figure 32.8. Based on Eq. (32.9), we got an adjusted input spectrum shown in Figure 32.9.

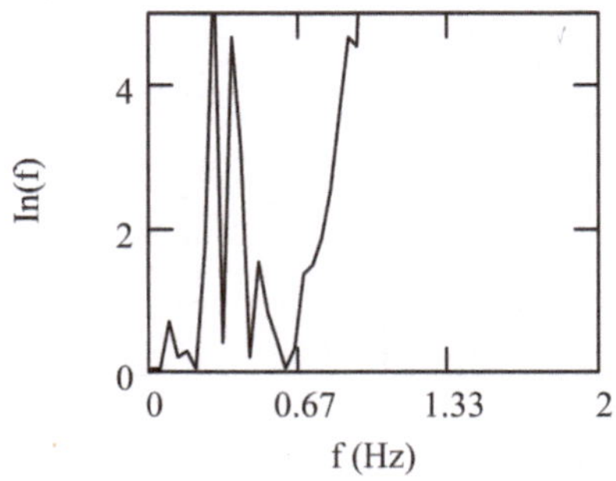

FIGURE 32.9 Adjusted input spectrum In(*f*).

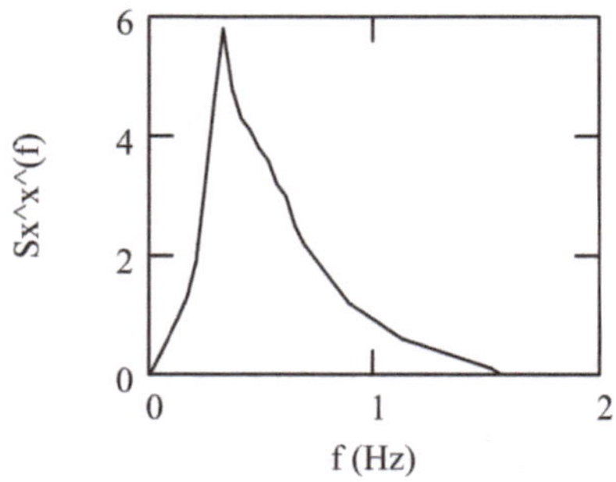

FIGURE 32.10 Optimized load spectrum: corr$[S_{xx}, S_{\hat{x}\hat{x}}] = 0.989$.

Under the excitation of In(*f*), we obtained $S_{\hat{x}\hat{x}}$ (Figure 32.10). The correlation coefficient between S_{xx} and $S_{\hat{x}\hat{x}}$ is 0.989, implying a satisfactory optimization.

32.5 SUMMARY

An optimal controller of wave maker in the frequency domain for laboratory seakeeping test under irregular wave load has been derived and discussed. The control procedure in a model test has been shown. Results in this chapter suggest that the present model may yet be an applicable control method of wave maker in laboratory seakeeping tests.

REFERENCES

1. S. K. Chakrabarti, *Offshore Structure Modeling*, World Scientific, Singapore, 1994.
2. S. R. Massel, *Ocean Surface Waves: Their Physics and Prediction*, World Scientific, Singapore, 1996.
3. The Specialist Committee on Waves, Final report and recommendations to the 23rd ITTC, *Proceedings of the 23rd ITTC, Volume II*, 2002.
4. S. A. Hughes, *Physical Models and Laboratory Techniques in Coastal Engineering*, World Scientific, Singapore, 1993.
5. Y. S. Tao, (Ed.), *Ship Seakeeping*, Shanghai Jiaotong University Press, Shanghai, China, 1985. (In Chinese).
6. S. X. Yu, et al., *Experiment Techniques of Ship Performances*, Shanghai Jiaotong University Press, Shanghai, China, 1991. (In Chinese).

7. K. S. Fu, *Digital Pattern Recognition*, Springer-Verlag, Berlin, 1976.
8. M. Li, Control analysis of the wave generator in ocean engineering, In *Proceedings of IEEE Conference on Mechatronics and Machine Vision in Practice*, August 2001, Hong Kong, pp. 173–175.
9. H. S. Tsien and J. Song, *Engineering Cybernetics*, Book 2, Science Press, China, 1980. (In Chinese).
10. S. Thompson, Robust control: An introduction, *Measurement + Control*, 26, 1993, 235–241.
11. J. S. Bendat and A. G. Piersol, *Random Data Analysis and Measurement Procedures*, 3rd Ed., John Wiely & Sons, Hoboken, NJ, 2000.
12. A. Papoulis, *Signal Analysis*, McGraw-Hill, New York, 1977.
13. A. Papoulis, *The Fourier Integral and Its Applications*, McGraw-Hill, New York, 1962.
14. Schlumberger Electronics Ltd., *1200 Real Time Signal Processor Operating Manual*, U.K., 1983.

CHAPTER 33

On von Kármán Spectrum from a View of Fractal

Von Kármán originally deduced his spectrum of wind speed fluctuation based on the Stokes–Navier equation. That derivation, however, is lacking in exhibiting the fractal information of wind velocity fluctuation. This chapter gives a novel derivation of the von Kármán spectrum based on fractional Langevin equation. The present results exhibit that a time series that follows the von Kármán spectrum can be taken as a specifically fractional Ornstein–Uhlenbeck process with the fractal dimension 5/3, providing a new view of the famous spectrum of von Kármán's from the point of view of fractals. Besides, we propose a relationship between two famous spectra in fluid mechanics, namely, the Kolmogorov's spectrum and the von Kármán's. Consequently, this chapter may yet be useful in practice, such as ocean engineering and ship mechanics.

33.1 INTRODUCTION

The power spectral density (PSD) functions introduced by von Kármán [1], known as the von Kármán spectra (VKS), have been widely used in diverse fields, ranging from marine science to structure vibrations, see e.g., Goedecke et al. [2] and the references therein. Among the von Kármán spectra, the spectrum (VKSW for short) expressed in Eq. (33.1) is particularly useful in the field of wind engineering for the modeling of wind speed fluctuation [3–11],

$$S_{\text{von}}(f)=\frac{4u_f^2 b_v w}{f\left(1+70.8w^2\right)^{5/6}},\quad w=\frac{fL_u^x}{U},\tag{33.1}$$

where f is frequency (Hz), L_u^x is turbulence integral scale, U is mean speed, u_f is friction velocity (m/s), and b_v is friction velocity coefficient such that the variance of wind speed is $\sigma_u^2=b_v u_f^2$.

DOI: 10.1201/9781003460947-37

Note that von Kármán derived (33.1) based on the Stokes–Navier equation [1]. That spectrum does not originally relate to the concept of fractal dimension, which is crucial to characterizing fluctuations from the point of view of fractals. Reports regarding turbulence's fractal dimension derived directly based on the Stokes–Navier equation are rarely seen as Gaoan stated in Ref. [12, p. 55]. In this chapter, we propose a rigorous but concise derivation of Eq. (33.1) based on the fractional Langevin equation. The present result is not only an alternative derivation of Eq. (33.1) but also provides a new outlook of Eq. (33.1) from the point of view of fractal time series. We will show that the process that follows (33.1) is actually in the class of fractional Ornstein–Uhlenbeck (OU) processes, and it has the fractal dimension 5/3.

The rest of this chapter is organized as follows. In Section 33.2, the preliminaries are briefed. In Section 33.3, we give a new derivation of Eq. (33.1). Discussions are in Section 33.4, which is followed by conclusions.

33.2 PRELIMINARIES

Let $v>0$ and $f(t)$ be a piecewise continuous function on $(0, \infty)$. It is integrable on any finite subinterval of $[0, \infty)$. For $t>0$, we call ${}_0D_t^{-v}$ the Riemann–Liouville integral operator of order v [13, p. 45]. It is given by

$$ {}_0D_t^{-v} f(t) = \frac{1}{\Gamma(v)} \int_0^t (t-u)^{v-1} f(u) du, \tag{33.2} $$

where Γ is the Gamma function. For simplicity, we write ${}_0D_t^{-v}$ by D^{-v} below.

Let $v_p, v_{p-1}, \ldots, v_0$ and $u_q, u_{q-1}, \ldots, u_0$ be two strictly decreasing sequences of nonnegative numbers. Then, for the constants a_i and b_i, we have

$$ \sum_{i=0}^{p} a_{p-i} D^{v_i} y(t) = \sum_{i=0}^{q} b_{q-i} D^{u_i} w(t), \tag{33.3} $$

where $w(t)$ is the white noise. The above is a stochastically fractional differential equation with constant coefficients of order v_p if $v_p > u_q$. It corresponds to a stochastically fractional linear filter of order v_p for $v_p > u_q$. The transfer function of this filter expressed by the Laplace transform is (Ortigueira [14])

$$ G(s) = \frac{1 + \sum_{i=1}^{q} b_{q-i} s^{-u_i}}{1 + \sum_{i=1}^{p} a_{p-i} s^{-v_i}}. \tag{33.4} $$

In the discrete case, the transfer function is expressed in the z domain by Ortigueira [15]

$$G(z)=\frac{1+\sum_{i=1}^{q} b_{q-i}z^{-u_i}}{1+\sum_{i=1}^{p} a_{p-i}z^{-v_i}}. \tag{33.5}$$

Substitute the inverse Laplace transform and the inverse z transform by L^{-1} and Z^{-1}, respectively. Then, the impulse response of the filter expressed by Eq. (33.3) in the continuous and discrete cases are given by

$$\begin{aligned} g(t)&=L^{-1}[G(s)],\\ g(n)&=Z^{-1}[G(z)]. \end{aligned} \tag{33.6}$$

Without loss of generality, we reduce (33.3) by letting $b_i=0$ except $b_i=1$ to the following

$$\sum_{i=0}^{p} a_{p-i}D^{v_i}y(t)=w(t). \tag{33.7}$$

Consequently, Eqs. (33.4) and (33.5) are reduced to

$$G(s)=1+\sum_{i=1}^{q} b_{q-i}s^{-u_i} \tag{33.8}$$

and

$$G(z)=1+\sum_{i=1}^{q} b_{q-i}z^{-u_i}. \tag{33.9}$$

The response to the system in the zero initial state is expressed by

$$y(t)=w(t)*g(t), \tag{33.10}$$

where * implies the operation of convolution [16–20]. In the discrete case, one has

$$y(n)=w(n)*g(n). \tag{33.11}$$

Substitute $Y(f)=F[y(t)]$, $G(f)=F[g(t)]$, and $W(f)=F[w(t)]$, respectively, where F is the operator of the Fourier transform. Then, according to the convolution theorem, we have

$$Y(f)=W(f)G(f). \tag{33.12}$$

Substitute the PSDs of $y(t)$ and $w(t)$ respectively by $S_{yy}(f)$ and $S_{ww}(f)$. Let $w(t)$ be the normalized white noise. Then, $S_{ww}(f)=1$. Therefore, we have

$$S_{yy}(f)=G(f)\left[G(f)\right]^{*}=\left|G(f)\right|^{2}, \tag{33.13}$$

where $[G(f)]^{*}$ is the complex conjugate of $G(f)$. Denote the autocorrelation function (ACF) of $y(t)$ by $r_{yy}(\tau)$. Then,

$$r_{yy}(\tau)=\mathrm{F}^{-1}\left[S_{yy}(f)\right]=\mathrm{F}^{-1}\left[\left|G(f)\right|^{2}\right]. \tag{33.14}$$

33.3 DERIVATION

For simplicity in discussions and without losing the generality, we replace w in Eq. (33.1) by $\frac{fL_u^x}{U}$ and $b_v u_f^2$ by σ_u^2 so that (33.1) becomes

$$S_{\mathrm{vk}}(f)=\frac{A_{\mathrm{vk}}}{\left[(B_{\mathrm{vk}})^{2}+f^{2}\right]^{5/6}}, \tag{33.15}$$

where

$$A_{\mathrm{vk}}=\frac{4\sigma_u^2}{70.8^{5/6}\left(\frac{L_u^x}{U}\right)^{2/3}} \tag{33.16}$$

and

$$B_{\mathrm{vk}}=\frac{U}{70.8^{1/2}L_u^x}. \tag{33.17}$$

The Langevin equation is expressed by

$$\left(\frac{d}{dt}+\lambda\right)y_{\mathrm{OU}}(t)=w(t), \tag{33.18}$$

where λ is a positive constant (Coffey et al. [21]). Its solution is the Ornstein–Uhlenbeck (OU) process. The fractional Langevin equation with the coefficient A, which was addressed by Lim and Muniandy [22], is given by

$$A\left(\frac{d}{dt}+\lambda\right)^{\beta}y_{\mathrm{fOU}}(t)=w(t), \quad \beta>0. \tag{33.19}$$

Consequently, its solution is the fractional OU process stated by Lim and Muniandy [22].

Denote $g_{\mathrm{fOU}}(t)$ the impulse response function of the system expressed by Eq. (33.19). Then, it is the solution to the following equation with zero initial conditions

$$A\left(\frac{d}{dt}+\lambda\right)^{\beta} g_{\mathrm{fOU}}(t)=\delta(t), \tag{33.20}$$

where $\delta(t)$ is the delta function. Doing the Fourier transforms on both sides of the above equation yields

$$G_{\mathrm{fOU}}(f)=\frac{A}{(\lambda-j2\pi f)^{\beta}}, \tag{33.21}$$

where $G_{\mathrm{fOU}}(f)$ is the Fourier transform of $g_{\mathrm{fOU}}(t)$.

Let $S_{y\mathrm{fOU}}(f)$ be the PSD of $y_{\mathrm{fOU}}(t)$. According to Eq. (33.13), therefore, $S_{y\mathrm{fOU}}(f)$ is given by

$$S_{y\mathrm{fOU}}(f)=\left|G_{\mathrm{fOU}}(f)\right|^{2}=\frac{A^{2}}{\left[\lambda^{2}+(2\pi f)^{2}\right]^{\beta}}. \tag{33.22}$$

Thus, we have the theorem below.

Theorem 33.1. Let $X_{\mathrm{vk}}(t)$ be the random function that obeys VKSW (Eq. (33.1) or (33.15)). Then, $X_{\mathrm{vk}}(t)$ is governed by the fractional Langevin equation of order 5/6. It is given by

$$\sqrt{A_{\mathrm{vk}}}\left(\frac{d}{dt}+B_{\mathrm{vk}}\right)^{5/6} X_{\mathrm{vk}}(t)=w(t). \tag{33.23}$$

Its solution in the frequency domain is given by

$$S_{\mathrm{vk}}(f)=\frac{A_{\mathrm{vk}}}{\left[(B_{\mathrm{vk}})^{2}+(2\pi f)^{2}\right]^{5/6}}. \tag{33.24}$$

Proof. Replacing A, λ and β in Eq. (33.19) with $\sqrt{A_{\mathrm{vk}}}$, B_{vk}, and 5/6, respectively, yields (33.23). Substituting A, λ and β in Eq. (33.22) with $\sqrt{A_{\mathrm{vk}}}$, B_{vk}, and 5/6, respectively, produces (33.24). Thus, Theorem 33.1 results. □

For the purpose of exhibiting the solution in the time domain, a lemma is given first.

Lemma 33.1. The inverse Fourier transform of $\frac{1}{\left[1+(2\pi f)^{2}\right]^{\beta}}$ for $\beta>1/2$ is given by

$$\mathrm{F}^{-1}\left\{\frac{1}{\left[1+(2\pi f)^{2}\right]^{\beta}}\right\}=\frac{2\sqrt{\pi}}{\Gamma(\beta)}\left(\frac{|\tau|}{2}\right)^{\beta-1/2} K_{\beta-1/2}(|\tau|), \tag{33.25}$$

where $K_{\nu}(z)$ is the modified Bessel function of the second kind or the MacDonald function. The proof is referred to Gelfand and Vilenkin [23, p. 188, in Section 2, Chapter 2]. □

Corollary 33.1. Denote $r_{vk}(\tau)$ the ACF of $X_{vk}(t)$. Then, the solution to Eq. (33.23) in the time domain is given by

$$r_{vk}(\tau)=\frac{2\sqrt{\pi}A_{vk}}{\Gamma(5/6)(B_{vk})^{1/3}}\left(\frac{|\tau|}{2}\right)^{1/3}K_{1/3}\left(B_{vk}|\tau|\right), \tag{33.26}$$

where $K_{1/3}(\cdot)$ is the modified Bessel function of the second kind $K_v(z)$ for $v=1/3$.

Proof. Note that

$$S_{vk}(f)=\frac{A_{vk}}{\left[(B_{vk})^2+(2\pi f)^2\right]^{5/6}}=\frac{A_{vk}}{(B_{vk})^{5/3}\left[1+\left(\frac{2\pi f}{B_{vk}}\right)^2\right]^{5/6}}.$$

Thus, by letting $\beta=5/6$ in the above lemma and taking into account the property of the time scaling property of the Fourier integral (Papoulis [24, p. 14]), we have $F^{-1}\left[S_{vk}(f)\right]$ given in Corollary 33.1. This completes the proof. □

33.4 DISCUSSIONS

Note that the Fourier transform of $|t|^{\alpha}$ is expressed by Ref. [23, Chap. 2], Li and Lim [25]

$$F\left(|t|^{\alpha}\right)=-2\sin\left(\alpha\pi/2\right)\Gamma(\alpha+1)|\omega|^{-\alpha-1}, \tag{33.27}$$

where $\alpha\neq-1, -3, \ldots$

Since

$$S_{vk}(f)\sim\frac{1}{f^{5/3}} \quad \text{for} \quad f\to\infty, \tag{33.28}$$

we have

$$r_{vk}(\tau)\sim|\tau|^{2/3} \quad \text{for} \quad \tau\to 0. \tag{33.29}$$

The fractal dimension of a process can be determined by its ACF for $\tau\to 0$ (Hall and Roy [26]). Thus,

$$r_{vk}(0)-r_{vk}(\tau)\sim\left(|\tau|^{a}\right)\Big|_{a=2/3} \quad \text{as} \quad \tau\to 0. \tag{33.30}$$

With the probability one [26], the fractal dimension of the von Kármán process is given by

$$D_{vk}=\left(2-\frac{a}{2}\right)\Big|_{a=2/3}=\frac{5}{3}. \tag{33.31}$$

Note that $S_{vk}(f)$ is convergent at $f=0$. Thus, the von Kármán process is short-range dependent (SRD). Nevertheless, it is a type of fractal time series (Li [27]). Denote the fractal dimension of a sample path by D. Then,

$$1 \leq D < 2, \tag{33.32}$$

see e.g., Mandelbrot [28], Beran [29]. Since a high value of fractal dimension of a sample path implies a high irregularity of that path [30–32], the value of $D_{vk}=5/3$ means that the von Kármán process has considerable local irregularity although it is SRD.

In the end of this section, we mention that VKSW follows the Kolmogorov's −5/3 law described in Kolmogorov [33] when $f \to \infty$, as can be seen from Eq. (33.28). As a matter of fact, it approximately obeys the Kolmogorov's −5/3 law if $B_{vk} \ll f$ from a view of engineering. Therefore, this research provides a relationship between the Kolmogorov's spectrum and the von Kármán's from a view of fractal analysis.

33.5 CONCLUSIONS

We have given the derivation of the von Kármán spectrum of wind speed solely based on the fractional Langevin equation. The results suggest that the process obeying VKSW is in the class of fractional OU processes. The fractal dimension 5/3 of the series following VKSW has been obtained, which implies a considerable local irregularity of the series obeying VKSW. Besides, we have established the relationship between the Kolmogorov's spectrum and the von Kármán's.

REFERENCES

1. T. von Kármán, Progress in the statistical theory of turbulence, *Proceedings of the National Academy of Sciences of the United States of America*, 34(11), 1948, 530–539.
2. G. H. Goedecke, V. E. Ostashev, D. K. Wilson, and H. J. Auvermann, Quasi-wavelet model of von Kármán spectrum of turbulent velocity fluctuations, *Boundary-Layer Meteorology*, 112(1), 2004, 33–56.
3. E. E. Morfiadakis, G. L. Glinou, and M. J. Koulouvari, The suitability of the von Karman spectrum for the structure of turbulence in a complex terrain wind farm, *Journal of Wind Engineering and Industrial Aerodynamics*, 62(2–3), 1996, 237–257.
4. M. C. H. Hui, A. Larsen, and H. F. Xiang, Wind turbulence characteristics study at the stonecutters bridge site: part II wind power spectra, integral length scales and coherences, *Journal of Wind Engineering and Industrial Aerodynamics*, 97(1), 2009, 48–59.
5. G. Huang and X. Chen, Wavelets-based estimation of multivariate evolutionary spectra and its application to nonstationary downburst winds, *Engineering Structures*, 31(4), 2009, 976–989.
6. D. K. Wilson, V. E. Ostashev, and G. H. Goedecke, Quasi-wavelet formulations of turbulence and other random fields with correlated properties, *Probabilistic Engineering Mechanics*, 24(3), 2009, 343–357.
7. G. Li and Q. Li, *Theory of Time-Varying Reliability for Engineering Structures and Its Applications*, Science Press, Beijing, China, 2001. In Chinese.
8. J. Pang, Z. Lin, and Y. Lu, Discussion on the simulation of atmospheric boundary layer with spires and roughness elements in wind tunnels, *Experiments and Measurements in Fluid Mechanics*, 18(2), 2004, 32–37. In Chinese.

9. Y.-Q. Xiao, J.-C. Sun, and Q. Li, Turbulence integral scale and fluctuation wind speed spectrum of typhoon: An analysis based on field measurements, *Journal of Natural Disasters*, 15(5), 2006, 45–53. In Chinese.
10. J. C. Kaimal, J. C. Wyngaard, Y. Yzumi, and O. R. Cote, Spectral characteristics of surface layer turbulence, *Quarterly Journal of the Royal Meteorological Society*, 98(147), 1972, 563–589.
11. H. Panofsky, D. Larko, R. Lipschutz, and G. Stone, Spectra of velocity components over complex terrain, *Quarterly Journal of the Royal Meteorological Society*, 108(455), 1982, 215–230.
12. Gaoan Xiushu (Japan), *Fractal*, Seism Press, China, 1989. Translated by by B. M. Shen and Z. W. Chang. In Chinese.
13. K. S. Miller and B. Ross, *An Introduction to the Fractional Calculus and Fractional Differential Equations*, John Wiley & Sons, Hoboken, NJ, 1993.
14. M. D. Ortigueira, Introduction to fractional linear systems. Part I: Continuous-time systems, *IEE Proceedings: Vision, Image and Signal Processing*, 147(1), 2000, 62–70.
15. M. D. Ortigueira, Introduction to fractional linear systems. Part II: Discrete-time systems, *IEE Proceedings: Vision, Image and Signal Processing*, 147(1), 2000, 71–78.
16. I. Podlubny, Fractional-order systems and $PI^{\lambda}D^{\mu}$-controllers, *IEEE Transactions on Automatic Control*, 44(1), 1999, 208–213.
17. J. A. T. Machado and A. M. S. Galhano, Statistical fractional dynamics, *ASME Journal of Computational and Nonlinear Dynamics*, 3(2), 2008, 021201-1–021201-5.
18. Y. Q. Chen and K. L. Moore, Discretization schemes for fractional order differentiators and integrators, *IEEE Transactions on Circuits and Systems I Fundamental Theory and Applications*, 49(3), 2002, 363–367.
19. B. M. Vinagre, Y. Q. Chen, and I. Petras, Two direct Tustin discretization methods for fractional-order differentiator/integrator, *Journal of the Franklin Institute*, 340(5), 2003, 349–362.
20. J. Van de Vegte, *Fundamentals of Digital Signal Processing*, Prentice Hall, Hoboken, NJ, 2003.
21. W. T. Coffey, Y. P. Kalmykov, and J. T. Waldron, *The Langevin Equation*, 2nd ed., World Scientific, Singapore, 2004.
22. S. C. Lim and S. V. Muniandy, Generalized Ornstein-Uhlenbeck processes and associated self-similar processes, *Journal of Physics A: Mathematical and General*, 36(14), 2003, 3961–3982.
23. I. M. Gelfand and K. Vilenkin, *Generalized Functions*, vol. 1, Academic Press, San Diego, 1964. Translated by J. B. Lin. In Chinese.
24. A. Papoulis, *The Fourier Integral and Its Applications*, McGraw-Hill, New York, 1962.
25. M. Li and S. C. Lim, A rigorous derivation of power spectrum of fractional Gaussian noise, *Fluctuation and Noise Letters*, 6(4), 2006, C33–C36.
26. P. Hall and R. Roy, On the relationship between fractal dimension and fractal index for stationary stochastic processes, *The Annals of Applied Probability*, 4(1), 1994, 241–253.
27. M. Li, *Multi-Fractal Traffic and Anomaly Detection in Computer Communications*, CRC Press, Boca Raton, FL, 2022.
28. B. B. Mandelbrot, *Gaussian Self-Affinity and Fractals*, Springer, New York, 2001.
29. J. Beran, *Statistics for Long-Memory Processes*, Chapman & Hall, New York, 1994.
30. S. C. Lim, M. Li, and L. P. Teo, Locally self-similar fractional oscillator processes, *Fluctuation and Noise Letters*, 7(2), 2007, L169–L179.
31. M. Li and S. C. Lim, Modeling network traffic using generalized Cauchy process, *Physica A*, 387(11), 2008, 2584–2594.
32. M. Li, *Fractal Teletraffic Modeling and Delay Bounds in Computer Communications*, CRC Press, Boca Raton, FL, 2022.
33. A. Kolmogorov, The local structure of turbulence in incompressible fluid for very large Reynolds numbers, *Proceedings of the Royal Society of London: Series A*, 434(1890), 1991, 9–13.

Index